The Hippocampus from Cells to Systems

Deborah E. Hannula • Melissa C. Duff

Editors

The Hippocampus from Cells to Systems

Structure, Connectivity, and Functional Contributions to Memory and Flexible Cognition

 Springer

Editors
Deborah E. Hannula
Department of Psychology
University of Wisconsin - Milwaukee
Milwaukee, WI, USA

Melissa C. Duff
Department of Hearing and Speech Sciences
Vanderbilt University Medical Center
Nashville, Tennessee
USA

ISBN 978-3-319-50405-6   ISBN 978-3-319-50406-3   (eBook)
DOI 10.1007/978-3-319-50406-3

Library of Congress Control Number: 2017932221

Printed on acid-free paper

This Springer imprint is published by Springer Nature
The registered company is Springer International Publishing AG
The registered company address is: Gewerbestrasse 11, 6330 Cham, Switzerland

Preface

The hippocampus has long been considered a critical substrate in the neurobiology, neuropsychology, and cognitive neuroscience of memory. Over the past few decades, a number of groundbreaking theoretical and methodological advancements have radically enhanced our understanding of the structure and function of the hippocampus and revolutionized the neuroscientific study of memory. Cutting across disciplines and approaches, these advances offer novel insights into the molecular and cellular structure and physiology of the hippocampus and the role of the hippocampus in the formation, (re)consolidation, enhancement, and retrieval of memory across time and development and permit investigators to address questions about how the hippocampus interacts, functionally and anatomically, with other neural systems in service of memory. In addition, recent investigations also suggest that the mechanistic properties and functional processing features of the hippocampus permit broader contributions to cognition, beyond memory, in the domains of decision-making, language, social cognition, and a variety of other capacities that are critical for flexible cognition and behavior. These advances have considerable implications for the neurobiology and cognitive neuroscience of hippocampus-dependent cognition and for numerous psychiatric and neurological diseases and disorders for which hippocampal pathology is a hallmark, such as schizophrenia and Alzheimer's disease.

This book combines, in a single source, an integrated review of these advances providing state-of-the-art treatment on the structure and function of the hippocampus. Contributors examine the hippocampus from a variety of levels (from cells to systems) using a wide range of methods (from neurobiological approaches in nonhuman animals to neuroimaging and neuropsychological work in humans). We hope that this timely collection of chapters from leading experts in the field will offer novices and scholars alike a framework for more complete appreciation of the various advances in the scientific study of the hippocampus and will permit readers to bridge the rich history of these endeavors with current perspectives and theories.

The impetus for this book was a symposium that we organized for the Annual Meeting of the Cognitive Neuroscience Society in 2014. The focus of that session was the relational memory theory, originally proposed by Neal Cohen and Howard Eichenbaum in 1993. The relational memory theory has had considerable impact in the domains of learning, memory, and hippocampal function and continues to inspire novel predictions more than two decades post-inception. Indeed, much of the research described in this book was influenced somehow by the theoretical tenets of hippocampal function articulated by Cohen and Eichenbaum. We would like to extend our deepest gratitude to Neal Cohen, who served as our graduate mentor and instilled in us great interest and passion for the empirical study of the human hippocampus. We are tremendously thankful for his guidance in our professional careers and for his friendship over the years. Our work is also a direct product of interactions with a vast network of additional mentors and collaborators, particularly our postdoctoral advisors—Charan Ranganath and Dan Tranel. To all of these individuals, we would like to say thank you for your time, effort, and interest in our work. We hope that readers find this book engaging and informative; it has been our great privilege to work with the contributors. Many thanks to the anonymous reviewers who provided helpful suggestions along the way.

Milwaukee, WI   Deborah E. Hannula
Nashville, TN   Melissa C. Duff

Acknowledgements

To David and to my parents, your enthusiasm for my work and patience when demands of academic life wear thin is so very much appreciated—couldn't do it without you. To Jack, you amaze me every day, what a great number of adventures we will have.

Deborah E. Hannula

To Robert, everything that matters most in my life you gave me. Thank you.

To Eleanor, thank you for being interested in what I do and for quizzing your pediatricians on what the hippocampus does. You're right; they didn't all know!

Melissa C. Duff

Contents

Part I
Cellular Structure and Function

The Nonhuman Primate Hippocampus: Neuroanatomy and Patterns of Cortical Connectivity

R. Insausti, M.P. Marcos, A. Mohedano-Moriano, M.M. Arroyo-Jiménez, M. Córcoles-Parada, E. Artacho-Pérula, M.M. Ubero-Martínez, and M. Muñoz-López

Abstract The purpose of this chapter is to review the neuroanatomical basis of nonhuman primate Hippocampal Formation. The term "Hippocampal Formation" is defined to include the dentate gyrus, hippocampal CA fields and subiculum, on one hand, and on the other hand, presubiculum and parasubiculum and entorhinal cortex. The circuitry supporting hippocampal function starts at the upper layers (II-III) of the entorhinal cortex to the dentate gyrus, and after a series of steps, the flow of information reaches the deep layers (V-VI) of the entorhinal cortex. The characterization of the commissural and cortical inputs and outputs of the entorhinal cortex is emphasized (no subcortical connections have been included). The main cortical input and output is with the parahippocampal region (temporal pole cortex, perirhinal and posterior parahippocampal cortices), which along with the retrosplenial cortex account for the vast majority of the direct cortical relationship of the Hippocampal Formation. Further cortical areas are the insular, medial frontal and orbitofrontal cortices, as well as the polysensory cortex at the dorsal bank of the superior temporal sulcus. This set of connections shows that only polysensory (polymodal) association cortices connect directly with the Hippocampal Formation. Secondary cortico-cortical connections increase the possibility of more extensive (i.e. unimodal association cortices) cortical connectivity. This network is closely related to hippocampal function in nonhuman primates, and possibly in human as well.

List of Abbreviations

7 area 7 of Brodmann
9 area 9 of Brodmann
12 area 12 of Brodmann
13 area 13 of Brodmann

R. Insausti (✉) • M.P. Marcos • A. Mohedano-Moriano • M.M. Arroyo-Jiménez •
M. Córcoles-Parada • E. Artacho-Pérula • M.M. Ubero-Martínez • M. Muñoz-López
Human Neuroanatomy Laboratory, Department of Health Sciences and CRIB, School of
Medicine, University of Castilla-la Mancha, Almansa 14, 02006 Albacete, Spain
e-mail: ricardo.insausti@uclm.es

© Springer International Publishing AG 2017
D.E. Hannula, M.C. Duff (eds.), *The Hippocampus from Cells to Systems*,
DOI 10.1007/978-3-319-50406-3_1

14 area 14 of Brodmann
23 area 23 of Brodmann
24 area 24 of Brodmann
25 area 25 of Brodmann
29 area 29 of Brodmann
30 area 30 of Brodmann
30v ventral portion of the area 30 of Brodmann
32 area 32 of Brodmann
35 area 35 of Brodmann, perirhinal cortex
36 area 36 of Brodmann, perirhinal cortex
36r rostral portion of the area 36 of Brodmann, perirhinal cortex
38_{DL} dorsal lateral division of the temporal pole
45 area 45 of Brodmann
46 area 46 of Brodmann
A Amygdala
B body of the hippocampus
CA cornu ammonis
CA1 CA1 field of the hippocampus
CA2 CA2 field of the hippocampus
CA3 CA3 field of the hippocampus
CA4 CA4 field of the hippocampus
DG dentate gyrus
EC entorhinal cortex
E_{LR} entorhinal cortex, lateral rostral subfield
E_O entorhinal cortex, olfactory subfield
E_R entorhinal cortex, rostral subfield
E_I entorhinal cortex, intermediate subfield
E_C entorhinal cortex, caudal subfield
E_{CL} entorhinal cortex, caudal limiting subfield
HF hippocampal formation
hf hippocampal fissure
GIL *Gyrus intralimbicus*
LAS lateral sulcus
LPFC lateral prefrontal cortex
lv lateral ventricle
MFC medial frontal cortex
MTL medial temporal lobe
OFC caudal orbitofrontal cortex
ots temporo occipital sulcus
PaS parasubiculum
PHR parahippocampal region
PFC prefrontal cortex
PHC parahippocampal cortex
PM polymodal association cortex
pmts posterior middle temporal sulcus
PRC areas 35 and 36 of the perirhinal cortex

PrS presubiculum
rs rhinal sulcus
RSC retrosplenial cortex
S subiculum
sr sulcus rhinalis
sts superior temporal sulcus
TE area TE of Von Bonin and Bailey (1947)
TEO area TEO of Von Bonin and Bailey (1947)
TF area TF of von Bonin and Bailey (1947)
TH area TH of von Bonin and Bailey (1947)
TPC temporal polar cortex
UMa unimodal auditory association cortex
UMv unimodal visual association cortex
U uncus of hippocampus
TPO area TPO of Seltzer and Pandya (1978, 1989)
V1 striate cortex
V4 area V4 of Von Bonin and Bailey (1947)

Introduction

Purpose and Justification

The Hippocampal Formation (HF) is a neural system more extended than what is commonly known as the hippocampus. The hippocampus itself (or hippocampus proper) as described by Ramón y Cajal (1893) and his disciple Lorente de Nó (1934) can be considered in the nonhuman primate as the final destination of a complex network that maintains reciprocal connections with polymodal cortical association areas. Only olfactory sensory input is the exception to this rule (Amaral et al. 1983, 1987; Insausti et al. 1987a). In contrast to nonhuman and human primates, rodents present direct, reciprocal connections, not only with association cortices, but with primary sensory and motor areas as well.

In this chapter we begin with an overview of the basic anatomical organization of the HF, and some of the nomenclature in use. This will be followed by a description of the neurochemical phenotype characterization of the cell population in the hippocampus. Finally, the core section will be devoted to gross anatomy and architectural organization, including substantial consideration of HF connectivity—both the intrinsic connectivity of this structure and extrinsic connections with other systems in the brain related to memory function. The subcortical connections of the HF are not addressed here. Interested readers can see information existing in the literature on this topic elsewhere (Insausti et al. 1987b; Witter et al. 1989a; Amaral and Lavenex 2007; Insausti and Amaral 2012; Christiansen et al. 2016).

Part 1: Gross Anatomy of the HF

Overview of the Organization of the HF. Anatomical Organization and Nomenclature

The hippocampus is a phylogenetically old structure and is preserved along the vertebrate scale. It reaches maximal size in human and nonhuman primates. According to allometric determinations, the hippocampus is 4.2 times larger in humans than insectivores of equal weight (Stephan and Andy 1970). In contrast to other six-layered cortical areas, the hippocampus presents only three layers. For this reason, and in a very classical but informative and effective nomenclature, the hippocampus is considered a primitive type of cortex called allocortex (allo-meaning odd, different or strange cortex) or archicortex (archi-meaning old cortex) (Filimonoff 1947). The hippocampus proper is allocortex, and it includes the dentate gyrus (DG), CA fields (CA3, CA2, CA1) and the Subiculum (S). In our nomenclature we have dropped the term CA4 (see Insausti and Amaral 2012 for details). The periallocortex lies in continuation with the allocortex towards the isocortex (neocortex), and is made up of the Presubiculum (PrS), Parasubiculum (PaS) and entorhinal cortex (EC). We will not use the term "subicular complex", because, in contrast to S, the PrS and PaS present more than the three layers typical of archicortex (Braak 1980). For this reason we propose the term "**Juxtasubicular**" cortex (JsC) to encompass PrS and PaS together.

Note that all three of the above structures (PrS, PaS, and EC) are also classically named "schizocortex" because of the presence of a fiber layer (*lamina dissecans*) that splits the thickness of the cortex into two parts (Stephan and Andy 1970; Stephan 1975). The PrS presents outer and inner principal layers separated by this cell-free space (i.e., the *lamina dissecans*). The PaS is much the same, but the inner principal layer is more laminated, thus resembling the deep layers of the EC.

The number of layers in the EC increases up to six (not to be mistaken with the layers of the isocortex). The EC has the same number of layers throughout in macaques as in humans (Amaral et al. 1987; Insausti et al. 1995), and remains constant. The EC is surrounded by proisocortex, which includes structures in the parahippocampal region described below. Proisocortex is a type of cortex closer to the typical isocortex, but that still retains a somewhat different appearance in lamination (Suzuki and Amaral 2003). Finally, the proisocortex is continued by the isocortex (including, for example, inferotemporal cortex), which presents with a fully developed architectural organization in six layers, although it may have different architectonic organization in specialized areas of the cortex (i.e. primary sensory cortex is different from multimodal association cortex).

The term "Hippocampal Formation" has been used to refer only to the hippocampus proper. However, in the present chapter, as in our previous work, we use the term HF to denote the combination of the hippocampus proper and the periallocortex (PrS, PaS, EC). This convention follows the studies of Ramón y Cajal (1893) and Lorente de Nó (1934) and rests on the characteristic unity of hippocampal connectivity provided by the unidirectional set of intrinsic connections along with the connections that link the hippocampus to PrS, PaS, and EC (Amaral and Lavenex 2007; Insausti and Amaral 2012). We therefore include under the term HF, the DG, CA fields, S, PrS, PaS, and EC (Insausti and Amaral 2012). The HF fields are characterized by their stepwise, unidirectional closed loop connections beginning in the superficial layers of the EC, with an intermediate point in the DG where most of the EC input terminates, and an end point in the deep layers of the EC again. The EC is considered the gateway of cortical input to the hippocampus because it is the recipient of most of the cortical input.[1]

The HF and adjacent areas such as the perirhinal and parahippocampal cortices receive altogether the name medial temporal lobe (MTL, Amaral et al. 1987; Insausti and Amaral 2012). As suggested by this convention, the MTL is made up of both the HF and a strip of cortex that surrounds the HF as far as the caudal end of the hippocampus. This strip of cortex is called the parahippocampal region (PHR, see Witter and Wouterlood 2002 for details). The most anterior part of the PHR is the tip of the temporal lobe or temporopolar cortex (TPC); the TPC is continued by the perirhinal cortex (PRC), which closely surrounds the entorhinal cortex, but is separated from this structure by the rhinal sulcus in nonhuman primates or by the collateral sulcus in humans.

The PRC is made up of areas 35 and 36 (perirhinal cortex and ectorhinal cortex respectively, after Brodmann 1909). The Parahippocampal Cortex (PHC) is made up two areas, TH and TF, according to the nomenclature of Von Bonin and Bailey (1947), who in turn, followed the cytoarchitectonic parcellation of Von Economo (1929). The PHC boundaries are marked by its start in continuation with the caudal limit of the EC and PRC, and its end at the transition with the retrosplenial cortex. The PHR is anatomically linked to the HF, and in particular to the EC, and it contributes more than two-thirds of the cortical polysensory afferent input to the EC (Insausti et al. 1987a). Additional cortical input is directed to the S (Van Hoesen et al. 1979).

The PHR belongs to the "proisocortex", as it lacks the more defined laminations of other isocortical areas, which present six layers well defined morphologically. As indicated above, the PHR is continued by the retrosplenial cortex (RSC), which also provides a heavy projection to the EC (see below). Therefore, the PHR and RSC, both providing input to the EC, form a sizeable portion of the "limbic" cortex system of Broca (1878). The cingulate cortex, which is situated above and around the *corpus callosum*, represents the anterior continuation of the limbic cortex. The subdivisions of the PHR are depicted in Fig. 1b.

---

[1]Temporal cortex input also reaches the S at the oblique interface with CA1 (van Hoesen et al. 1979).

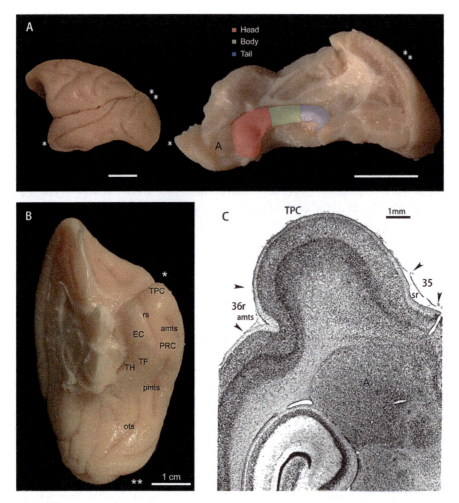

Fig. 1 (a) Photograph of the lateral surface of a *Macaca fascicularis* monkey brain. In panels **a** and **b**, the stars point to the anterior tip of the temporal lobe (*single star*) and the occipital pole (*double star*). The right hand-side in panel **a** is a gross anatomical dissection of the temporal lobe in which the hippocampus has been exposed by removing the brain on top. The hippocampus has been colored for easier visualization in *red* (head), body (*green*) and tail (*blue*). Note that the medial bend of the hippocampal head in the picture corresponds to the uncus. Panel **b** shows a ventral view of a *Macaca fascicularis* brain in which the EC, as well as components of the PHR and related sulci are indicated. Panel **c** is a horizontal section of a *Macaca fascicularis* monkey brain in which the temporopolar cortex, the amygdala and the hippocampus can be appreciated sequentially from top to bottom. Scale bars in **a** and **b** are 1 cm. Scale bar in **c** is 1 mm

Gross Anatomy of the HF

From a gross anatomical perspective, the hippocampus is located at the floor of the lateral ventricle, where it resembles a horn, and based on this appearance, it received the classical name "Ammon's horn" or *"Cornu Ammonis"*.[2] The hippocampus can be seen in the gross anatomical dissection of a monkey brain in Fig. 1a. When the nonhuman primate brain is seen from the lateral side, the hippocampus is hidden by the temporal cortex. One can imagine it, however, as an elongated structure with a straight, oblique long axis, dorsally oriented in the temporal lobe. In comparison, the EC occupies a rostral location relative to the hippocampus in such a way that approximately the rostral one-half of the EC is anterior to the hippocampus, where it surrounds the amygdala. In the nonhuman primate, the caudal part of the EC is situated under the head of the hippocampus (Fig. 2a) and is related anatomically with the beginning of the hippocampal fissure (hf), which runs in between these structures (see below). The identification of this gross morphology of the hippocampus in humans and nonhuman primates is of importance in neuroimaging studies, as they are the only anatomical references visible in magnetic resonance images of the brain.

The subdivision of the hippocampus along its longitudinal axis into head, body and tail is widely accepted. At the level of the anterior tip of the lateral ventricle and underneath the caudal one-half of the amygdala the hippocampal head forms the uncus (sometimes this part of the hippocampus is known as "uncal hippocampus"). An important anatomical landmark located at the caudal end of the hippocampal head is marked by a round prominence called the *Gyrus intralimbicus*, visible at the depths of the MTL. A cross section depiction of the *Gyrus intralimbicus* is presented in Fig. 2D. While the hippocampal head has a smooth appearance in the nonhuman primate, in humans it shows a more complicated structure (for details, see Amaral et al. 1987; Insausti and Amaral 2012).

The EC is situated external (lateral) to the hippocampus at the surface of the MTL, and it can be identified more easily than other HF regions, which do not have obvious anatomical landmarks (i.e. CA1). The EC extent can even be recognized macroscopically, as it shows neat medial and lateral boundaries: rostrally, the *sulcus semiannularis* makes the medial boundary with the peryamigdaloid cortex; caudally, the medial boundary is the hf. The rhinal sulcus (rs, in non-human primates; collateral sulcus in humans) forms the lateral boundary between EC and the PRC as far as the posterior limit of these structures. From there, the hippocampus continues caudally, medial to the PHC. The transition between the body and the tail of the hippocampus is inconspicuous, and takes place approximately at the level of the caudal end of the lateral geniculate nucleus of the thalamus. The hippocampal tail ends at the level of the posterior columns of the fornix (Fig. 1a).

The hippocampal fields are present all along the longitudinal extent of the macroscopically defined hippocampus. The DG, CA fields and S extend from the hippocampal head as far back as the *splenium* of the *corpus callosum*. In contrast, the EC, PaS and PrS are not present at every level of the HF. The PHR surrounds the

---

[2]The abbreviation of *Cornu Ammonis* (CA) gives the name to the CA fields commonly used (Lorente de Nó, personal communication).

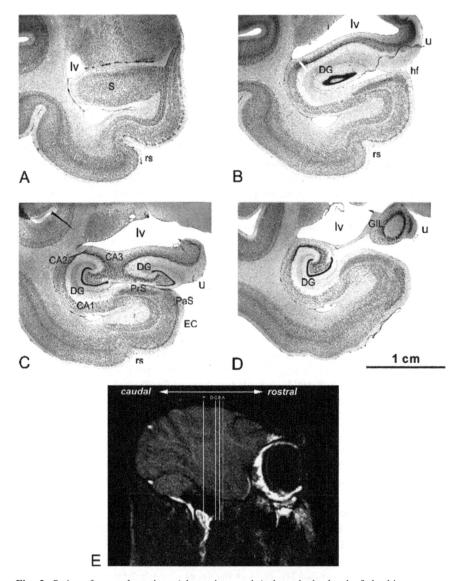

Fig. 2 Series of coronal sections (photomicrographs) through the head of the hippocampus arranged from rostral (**a**) to caudal (**d**). Panel **a** is a section tangential to the head of the hippocampus, and only S is apparent. Under S, lies the intermediate EC, and on top of S is the amygdala. In panel **b**, the level of the section passes through the anterior part of the hippocampal head, and a tangential section through the DG stands out as a dark band at the center. Panel **c** is at the level of the uncus a bit more caudally than panel **b**. and shows the EC at a caudal level, as well as the DG, hippocampal CA fields, S, juxtasubicular fields and EC. Panel **d** is a section at the end of the uncus, in which the *Gyrus intralimbicus* is clearly evident as small rounded structure. Note also the caudal pole of the EC underneath the *Gyrus intralimbicus*. Panel **e** is a lateral view of a 1.5 T MRI of a monkey brain, in which the longitudinal extent and orientation of the hippocampus can be appreciated. *Vertical lines* point to the approximate different levels of the sections shown in panels **a–d**. The *asterisk* corresponds to the section shown in Fig. 3. Scale bar is 1 cm

HF from the temporal pole and along the lateral aspect of the EC and hippocampus as far as the transition with the RSC. Figure 1c shows a horizontal section of the rostral (anterior) part of the monkey temporal lobe. It begins with the TPC, and continues to PRC. Structures that are located caudal to the PRC (e.g., the PHC, tail of the hippocampus) are not visible in this image.

The nonhuman primate hippocampal head results from a rostral and medial bend of the hippocampal fields, although these are not as pronounced and convoluted as they are in the human brain. All of the HF fields (DG, CA3, CA2, CA1 and S) are present in the hippocampal head. The rostral-most field is the S and the end of the medial bend, which is marked by the appearance of *Gyrus intralimbicus* and corresponds to the caudal end of the hippocampal head, is made up of hippocampal field CA3. The relative positions of the HF fields can be seen in a series of coronal sections, arranged rostral to caudal, in Fig. 2. Although the hippocampal head in the monkey does not reach the development and complexity of the human, it is sometimes difficult to establish boundaries among fields at this level in both species.

Caudal to the head, in the hippocampal body, the EC is no longer present. At the level of the hippocampal body and anterior part of the tail, hippocampus shows the more usual appearance as an interlocking structure of bands of neurons, which start with the blades of the DG and is followed by CA3 (between the DG blades) along with the remaining hippocampal fields (CA2, CA1, S), and the juxtasubicular fields (PrS and PaS). As indicated above, the transition between the body and tail of the hippocampus is indistinct and only approximate limits can be established. One is the *Gyrus intralimbicus* at the end of the lateral geniculate nucleus, where the beginning of the tail of the hippocampus starts. The hippocampus extends for about 2 cm in a rostrocaudal direction as far as the splenium of the *corpus callosum*, and keeps the typical appearance almost as far as the caudal end in the hippocampal tail (Fig. 2).

Anatomical Organization of Hippocampal Fields

Under the microscope, the components of the HF are easily distinguishable at mid rostro-caudal level, i.e. in the body of the hippocampus. However, the flexures of the hippocampal head (uncal part) and the curvature of hippocampal tail towards the splenium of the *corpus callosum* make it challenging to precisely delineate cytoarchitectonic boundaries in those subdivisions.

A section through the mid-level of the hippocampus in Fig. 3 displays the more common C-shaped view of this structure. Note that this section corresponds to just one part of the hippocampal body, and the EC is not present at this level.[3] One of the

---

[3]The depiction of the hippocampal body section very often includes the EC as the cortex under the hippocampus. However, the EC ends approximately at the *Gyrus intralimbicus*, and it is likely that the cortex at this level corresponds the PHC.

Fig. 3 Cross-section through the body of the hippocampus (B). The hippocampus fields take the typical C-shape. The opened hf is on top of the PrS, and it continues to the left already fused. Note the S/CA1 interface at the boundary between CA1 and the S. The EC is not present at this level. The PHC can be appreciated under the PaS. Scale bar is 1 mm

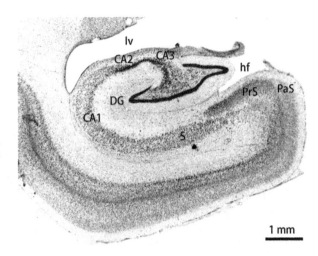

main features of the hippocampus, visible in the figure, is its "folded" appearance. The embryological growth of the hippocampal primordium determines the interlocking of the DG and the closely related CA3. This interlocking takes place all along the longitudinal axis of the hippocampus.

Dentate Gyrus, CA Fields and Subiculum

The DG is easily identifiable in the mammalian hippocampus as a dense and narrow, open band of darkly stained neurons that surround a less dense central region (i.e., the polymorphic cell layer or *hilus*) where the proximal end of CA3 "interlocks" with the DG (Fig. 3). It is worth mentioning that Lorente de Nó (1934) separated this area as a distinct field called CA4 that, as stated above, is not recognized here (for details of the justification to drop the field CA4 from our nomenclature see Insausti and Amaral 2012).

The DG shows three layers and is therefore categorized as allocortex. The principal cell layer is the granule cell layer, very distinctive because of the packing and staining density of the granule cells. The outer part of the DG is the molecular layer, a largely acellular band made up by the dendrites of the DG granule cells. The inner layer is the polymorphic cell layer, which borders CA3 neurons. The molecular layer is separated from the most distal part of the apical dendrites of the pyramidal cells of the CA fields (see below) by the hf. The hf is more noticeable medially, where it broadens and can be subdivided into a dorsal part (mostly S and PrS/PaS) and a ventral part (mostly PHC). The hf is a result of the fusion of two different sheets of neuroblasts during embryological development of the hippocampus, and is therefore covered by piamater. This means that the hf is largely reduced to a line of pial cells that separates the hippocampal CA fields from the molecular layer of the DG (Gloor 1997).[4] More specifically, the fused hf separates the molecular layer of the DG from the *stratum lacunosum-moleculare* primarily of CA1.

The medial boundary of the CA3 field is the polymorphic cell layer of the DG. This field takes a curved shape away from the DG at the boundary with CA2. The CA2 field is made up of dark and tightly packed neurons, although its boundary with the neighboring fields CA3 and CA1 is not always distinct. Finally, field CA1 is adjacent to field CA2 in the transversal plane and extends as far as the border with the S. The boundary between CA1 and S is oblique (S/CA1 interface), and is important in relation to the cortical input and output of the HF (see below and Insausti and Muñoz 2001). The overlap between the S and CA1 has been named "Prosubiculum", a term that we do not use (for details, see Amaral and Lavenex 2007; Insausti and Amaral 2012).

Like the DG, fields CA3, CA2 and CA1 also are allocortex and have three layers. These basic layers are further subdivided to constitute the well-known layers of the hippocampus. From the ventricular surface towards the hf we find *alveus, stratum oriens, stratum pyramidale, stratum radiatum*, and *stratum lacunosum-moleculare*. The axons of the pyramidal cell layer neurons or *stratum pyramidale* form the *stratum oriens*, in which the basal dendrites and many interneurons lie (see below). These axons collect in the *alveus*, a fiber layer that abuts the ependymal covering of the temporal horn of the lateral ventricle, and runs backwards to form the fimbria. Part of the alveus is the prominence seen in the floor of the lateral ventricle. The cell bodies form the *stratum pyramidale*, and their dendrites form the *stratum radiatum*. The apical portion of the dendrites forms the *stratum lacunosum-moleculare*. Collectively, these layers make a prominence in the floor of the lateral ventricle (Fig. 3), and while this layered organization applies to all of the CA fields, CA3 shows an extra layer the *stratum lucidum* in which the axons of the DG (i.e., the mossy fibers) terminate.

The S is the final part of the allocortex and continues the CA1 field towards the already opened hf. The molecular layer is the outer layer of the S and usually faces the hf. The S consists of a pyramidal cell layer, subdivided into principal and polymorphic layers (Bakst and Amaral 1984; Kobayashi and Amaral 1999). The

---

[4]Medially, it opens to the subarachnoid space, where the anterior choroidal artery runs. The space in which the anterior choroidal artery travels receives the name of choroidal fissure, and it extends along the longitudinal extent of the hippocampus facing the PrS, PaS, and extending as far as the caudalmost part (tail of the hippocampus). The anterior choroidal artery is a principal branch of the arterial circle of Willis although it may arise directly from the internal carotid artery, and provides most of the vascularization to the hippocampus. A small complement of the posterior choroidal artery, a branch of the posterior cerebral artery also exists. The branches of the anterior choroidal artery are seen as vascular holes in the "fused" hippocampal fissure among which run the perforant pathway fibers.

polymorphic layer of the S abuts the white matter, largely made up by the angular bundle that carries, among others, the perforant path fibers. The perforant path owes its name to the fact that it "perforates" (i.e., traverses through) the S cell layers.

Presubiculum and Parasubiculum

The PrS is a continuation of the S toward the midline and presents an outer principal layer of small neurons that partially overlaps the S, a cell free layer, or *lamina dissecans*, and an inner principal layer. In contrast to the human PrS in which the outer principal layer is broken into discontinuous clumps, the nonhuman primate PrS is continuous and begins at the boundary with the PaS. The PrS extends as far caudally as the junction with the RSC.

The PaS also presents an outer cell layer made up of neurons of various sizes, which take a more laminar distribution, a cell free *lamina dissecans*, and an inner cell layer that resembles the deep layers of the EC. Topographically, the PaS starts at the level of the hippocampal head and extends approximately the whole length of the hippocampus body.

Entorhinal Cortex (EC)

The EC can be subdivided into two parts: a rostral part anterior to the hippocampus and lateral ventricle, and a caudal part, which lies under the hippocampus, approximately as far as the *Gyrus intralimbicus* (i.e. the end of the uncus). Like the PrS and PaS the nonhuman primate EC is periallocortex. However, the EC presents a more complex and better definition of the lamination than the PrS and the PaS. The laminar structure of the EC consists of six layers, named I to VI. It should be stressed though, that these layers are more rudimentary than, and therefore not equivalent to, the layers I-VI of the isocortex.

The outermost layer is the molecular layer or layer I. While in the nonhuman primate the external surface is smooth, in humans the outer layer presents characteristic "bumps" or "verrucae" visible macroscopically (for details, see Insausti and Amaral 2012). Layer I is the recipient of dense cortical input (Insausti and Amaral 2008). Layer II is characterized by the presence of big and darkly stained neurons, which originate laminar projections to the molecular layer of the DG (Witter and Amaral 1991). The appearance of layer II is not homogeneous throughout the extent of the EC, and presents variations (for details see Amaral et al. 1987 for nonhuman primate, and Insausti et al. 1995 for humans). Layer III is made up of medium or small pyramids, more irregularly distributed rostrally, but columnar caudally. Deep to layer III stands *lamina dissecans* (layer IV),[5] one of the more distinctive features

---

[5]Layer IV is used as equivalent to layer V in publications that do not recognize *lamina dissecans* as another layer of the EC. For details see Amaral et al. 1987.

of the EC. This cell free layer of the EC is made up of basal dendrites of the pyramidal neurons of layer III (which is not in continuation with the homologous layers in the PrS and PaS). *Lamina dissecans* is not present throughout the EC, but rather, is better observed at an intermediate level of the EC (Amaral et al. 1987; Insausti et al. 1995). Layer V presents large and darkly stained pyramidal cells and originates much of the output from the HF to the cortex (Muñoz and Insausti 2005). Finally, layer VI is a multiform layer with neurons of various sizes and orientations, although at the caudal level it takes a "coiled" appearance (Amaral et al. 1987).

It is important to note that the cytoarchitectonic structure of the EC is not homogeneous; mediolateral as well as rostrocaudal differences are present. These differences justify a subdivision of up to seven subfields, namely the olfactory (E_O), Lateral Rostral (E_{LR}), Rostral (E_R), Intermediate (E_I), Lateral Caudal (E_{LC}), Caudal (E_C), and Caudal limiting (E_{CL}) subfields (for details, see Amaral et al. 1987). Notably, EC organization is homologous in the non-human primate and the human brain (Insausti and Amaral 2012) and stands in contrast to the rodent where the topological relationship among the HF structures is less comparable (Stephan 1975 Insausti 1993; Insausti et al. 1997).

Neurochemical Phenotype of the HF Neuronal Population

The main purpose of this section is to provide an overview of the neurochemical phenotype of cell populations in the HF. This information is complementary to, and provides additional insight about, the architectural and cellular organization of the HF, taking into account the functional aspects of cellular interaction, a phenomenon fundamental for memory function.

Neurons of the cellular layers of the HF can be inhibitory (interneurons) or excitatory (principal cells). These cells have been extensively studied since Ramón y Cajal (1893) and Lorente de Nó (1934). Principal cells are located primarily in the granular layer of the DG and the pyramidal layer of the CA fields (Storm-Mathisen 1981; Ottersen and Storm-Mathisen 1985). In contrast, interneurons are found throughout the HF, although they are especially concentrated in the polymorphic layer of the DG and the *stratum oriens* of the CA fields.

In rats, a detailed account of the great variety of interneurons of the HF according to their locations has been described. Additional information about these interneurons—i.e., the pattern of discharge, dendritic arbors, neurochemical content, and a variety of other characteristics—can be found in Freund and Buzsaki (1996). Homologous neurons in the nonhuman primate HF are likely similar; however, studies performed on humans are rather scarce (for review, see Kobayashi and Amaral 1999; Graterón et al. 2003; Cebada-Sánchez et al. 2014).

The classification of different neuronal populations in the HF according to their neurochemical phenotype has been one of the main criteria in past years for understanding how neuronal populations can elicit a response. The knowledge of neurochemistry of the HF is thus important given that it provides information about possible neuronal function: glutamatergic (principal) neurons are excitatory; GABAergic neurons (interneurons) are inhibitory. By using immunohistochemical methods the neurochemical phenotype of a cell can be detected, although the immunohistochemical detection of GABA is particularly difficult. The discovery that GABA coexists in the same neuron with calcium-binding proteins and/or several neuropeptides has provided an easier way to identify interneurons and increases the possibility of pinning down the interplay among principal cells and interneurons (Leranth and Ribak 1991; Seress et al. 1991, 1992, 1993a, b, 1994; Braak et al. 1991; Sloviter et al. 1991; Pitkanen and Amaral 1993; Hornung and Celio 1992; Tuñon et al. 1992; Nitsch an Ohm 1995; Solodkin and Van Hoesen 1996; Brady and Mufson 1997).

Neurochemical studies have also shown that in the HF some neuroactive substances are useful as markers for specific fiber tracts or neuronal populations. For example, the opiate neuropeptide dynorphin A is considered a marker of the mossy fibers (Chavkin et al. 1985) and neurotensin has been detected only in fibers in the CA fields, PrS, PaS and EC (Sakamoto et al. 1986; Mai et al. 1987; Gaspar et al. 1990; Berger and Alvarez 1994; Kobayashi and Amaral 1999). The density of immunoreactivity for these substances varies depending on the region and/or layer where they are located. For example, the PrS is the region of HF that displays the highest amount of GABAergic cell bodies and fibers (Kobayashi and Amaral 1999).

Not every substance considered to be a unique marker for interneurons is present in each inhibitory cell; instead, interneuron markers can be different depending on the HF region. In the DG and CA fields, GABA may co-localize with calcium-binding proteins (Freund and Buzsaki 1996), and with neuropeptides such as substance P (Del Fiacco et al. 1987; Sakamoto et al. 1987; Iritani et al. 1989; Nitsch and Leranth 1994), with somatostatin (Bakst et al. 1985; Chan-Palay et al. 1986; Chan-Palay 1987; Bouras et al. 1987; Amaral et al. 1988; Cebada-Sánchez et al. 2014) and with neuropeptide Y (Lotstra et al. 1989; Nitsch and Leranth 1991; Cebada-Sánchez et al. 2014). In contrast, in the S, PrS, PaS and EC calcium-binding proteins co-localize most commonly with GABAergic cells. Somatostatin and neuropeptide Y also co-localize with GABAergic cells (Carboni et al. 1990; Friederich-Ecsy et al. 1988; Dournaud et al. 1994; Solodkin and Van Hoesen 1996; Kobayashi and Amaral 1999).

Identification of the cellular neurochemical phenotype has also provided new insights into the morphological basis for possible interactions between different neurotransmitters and neuromodulators, or between neuropeptides and calcium-binding proteins. Some of these interactions are very specific, such as the serotonergic input to the DG, which is specifically directed toward calbindin cells, but not to other neurons in the same layer even if they contain calcium-binding proteins (Kobayashi and Amaral 1999). Notably, the distribution of neurochemical populations varies along the rostrocaudal axis of the HF. For example, this is the case for somatostatin and neuropeptide Y in the DG (Cebada-Sánchez et al. 2014), which suggests that the regulatory role of interneurons on the activity of principal cells may be different at different levels of the HF.

Whereas the majority of neuropeptides are considered to be interneuron markers, the neuropeptide neurotensin has been detected in layers containing cells that are typically excitatory, such as the granule cell layer of DG and the pyramidal layer of the S (Sakamoto et al. 1986) and PrS (Mai et al. 1987; Berger and Alvarez 1994. Inhibitory neurons are not evident in these regions (Kobayashi and Amaral 1999). Consequently, this neuropeptide may be an important intracellular regulator of the activity of excitatory cells, an example that stands in contrast to the external influence and control of principal cells by interneurons. It is important to note that the presence of neuroactive substances in HF neurons is not always an indicator of cell function since they have been found in layers containing interneurons as well as in layers with principal cells in the same HF region. This is the case for calbindin, which is present in both principal cells and interneurons of the HF (Baimbridge and Miller 1982; Baimbridge et al. 1982; Sloviter 1989; Toth and Freund 1992). In summary, the determination of the cellular neurochemical phenotype, as well as the study of the distribution of neurotransmitters and neuromodulators in different regions of the HF, can contribute to the knowledge of the morphological basis underlying HF functions.

Part 2: Connectivity of the HF

One of the most characteristic and functionally relevant features of the HF is the organization of its connections, which form a unidirectional stepwise closed loop (Fig. 4). This loop is connected with both cortical (Insausti et al. 1987a) and subcortical structures (Insausti et al. 1987b), but here the focus is limited to intrinsic and cortical connectivity (descriptions of subcortical connections are provided elsewhere—see Amaral and Lavenex 2007 for review). Specifically, this section focuses on intrinsic HF connections (EC, DG, CA fields and S) along with HF output from the EC and the CA1/S interface to the cerebral cortex. Subsequently, cortical inputs to the nonhuman primate HF and the pathways that give rise to those inputs are considered.

Intrinsic HF Connectivity

EC Projections to the Hippocampus Proper (DG, CA Fields, and S)

The nonhuman primate HF can be seen as a structure that consists of a closed loop of intrinsic connections. It has been known for a long time that the EC directly targets the molecular layer of the DG, and subsequent work has indicated that there are also direct projections from EC to the CA fields and the S (these connections are outlined in detail below). The progression of projections in the hippocampus proper is such that the DG projects to CA3 via the mossy fibers. This projection is

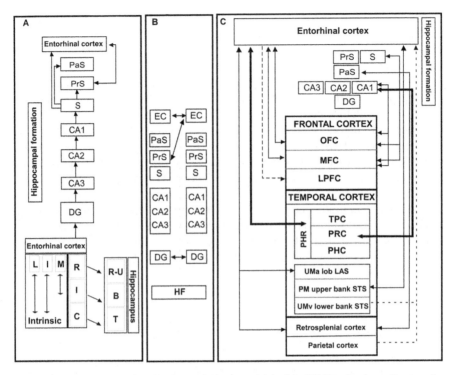

Fig. 4 Summary diagram showing the cortical connectivity of the HF. Panel **a** shows the stepwise connectivity organization of the HF; the lower part of the panel is the schematic representation of the intrinsic EC connections, divided into Rostral, Intermediate and Caudal portions of the EC, and Lateral, Intermediate and Medial bands in the EC. Also, on the same panel, the projections to the Uncus (hippocampal head), Body and Tail of the hippocampus is summarized. Panel **b** represents the commissural projections of the HF. Panel **c** summarizes the cortical connectivity of the HF. The size of the *arrows* orients to the density of the connections. Notice the heavy connections between the PHR and the EC. *Dashed lines* represent weaker projections to or from the EC

exclusively limited to CA3 and does not reach CA2. In turn, CA3 and CA2 project to CA1 via the Schaffer collaterals and CA1 projects to the S. Finally, the S projects to HF structures—namely, the PrS, PaS and deep layers (V and VI) of the EC— effectively closing the loop of stepwise intrinsic connectivity. In turn, deep layers of EC originate the cortical and subcortical output.

One of the first descriptions of intrinsic HF connectivity was provided by Ramon y Cajal (1893). He described the perforant path as a fiber bundle that links the EC with the DG in the hippocampus proper. Subsequent work has indicated that EC projects not only to DG, but also to the CA fields and the S. Fibers that make up the perforant pathway originate in EC layers II and III. Schematically, layer II neurons project to the molecular layer of the DG, CA3 and CA2, while layer III projects to CA1 and S. Projections from layers VI (DG and CA2–CA3) and layer V (CA1) have been also described (Witter and Amaral 1991).

Additional work has indicated that there is also a *topographic* mapping of EC outputs to the hippocampus proper. Specifically, it has been reported that EC neurons that are part of a rostro-medial band project to DG neurons that are localized to the head (or uncal part) of the hippocampus. In contrast, an intermediate EC band projects to the body of the hippocampus and a lateral EC band, which is adjacent to the rhinal sulcus, projects to caudal hippocampus. A notable difference between rodents and primates can be found at the termination site of EC projections in the molecular layer of the DG. Specifically, while rodent EC projections to DG show clear separation between the middle and outer one-third of the DG molecular layer, the same separation is not present in the nonhuman primate.

As indicated above, the EC also projects to the CA fields and the S of the hippocampus proper and topographic mapping of EC termination points have also been observed in these regions. For example, it has been reported that the rostral subfields of the EC project to the S/CA1 interface region, whereas more caudal subfields of the EC project to two separate locations—namely, distal S (nearest the PrS) and proximal CA1 (adjacent to CA2). In contrast to this topographical convention, the whole of the EC projects to the *stratum lacunosum-moleculare* of CA2 and CA3 without apparent topography. In sum, connectivity is evident between all cellular layers and subfields of the EC to the DG, CA fields and the S. Whether or not the EC projects directly to PrS and PaS remains uncertain (Witter and Amaral 1991).

Intrinsic Connectivity in the Hippocampus Proper (DG, CA Fields, and S)

The DG projects to the field CA3 through the mossy fiber system. This projection can be revealed by different histochemical and immunohistochemical techniques, most characteristically by the Timm method for the demonstration of heavy metals, which permits visualization of this projection. The polymorphic (*hilus*) cell layer of the DG also projects to the inner third of the molecular layer of the DG, a projection that contains the peptide somatostatin (Bakst et al. 1985). The mossy fibers travel within the pyramidal cell layer of CA3 in the transverse plane of the hippocampus—i.e., orthogonal to its long axis. At the distal end of CA3 (closer to CA2), the mossy fibers course longitudinally in a rostral direction, for about 3–5 mm (Kondo et al. 2008). On their way, they send collaterals to the polymorphic cell layer of the DG and the inner one-third of the molecular layer. In this way, an association system is formed, running in both rostral and caudal directions for approximately up to 80% of the hippocampal length (Kondo et al. 2008). Finally, there are also local projections from the mossy fibers to the polymorphic layer of the DG, which also present a longitudinal orientation.

The set of connections continues in a stepwise fashion as CA3 projects to fields CA2 and CA1. This projection forms the system of Schaffer's collaterals. The CA3 projections innervate *strata oriens, pyramidale* and *radiatum* of CA1 in the transverse axis. Along the longitudinal axis, the CA3 to CA1 projection covers about three-fourths of the whole extent of CA1 (Kondo et al. 2009). In addition, both CA2 and CA3 give rise to intrinsic association projections that innervate *strata oriens, pyramidale* and *radiatum,* although they are shorter in the longitudinal axis and less dense than the projections to CA1. The *stratum lacunosum-moleculare* does not receive innervation from the Schaffer's collateral system of CA2, or CA3 (Kondo et al. 2009). CA1 projections are directed to the S and EC, although the exact topography of these connections is poorly understood (Rosene and Van Hoesen 1987; Saunders and Rosene 1988). The final output from the hippocampus is then sent to deep layers, V and VI, of the EC. Interestingly, these are the same layers that originate the return projections to cortical and subcortical targets. In contrast to the rodent, CA2 and CA3 fields do not project to the EC in nonhuman primates.

Intrinsic Connectivity in the Entorhinal Cortex

The EC itself maintains a heavy intrinsic set of connections, particularly well developed in the caudal EC (Chrobak and Amaral 2007). Anterograde tracer studies (Chrobak and Amaral 2007) have indicated that the intrinsic set of EC connections is organized into three rostrocaudal bands oriented longitudinally. The lateral band of the EC provides the most extensive intrinsic connectivity, extending for up to one-half of the length of the EC. Intermediate EC gives rise to a mid-mediolateral longitudinal band that, interestingly, does not project to the lateral band of EC. The medial-most part of the EC, largely subfield E_O, has within band connections, but does not receive, or send projections to either of the remaining two bands. The densest intrinsic projection terminates in layer III, although layer I is also recipient of a dense projection. Layer II receives a much lighter projection, but given that layer III contains dendrites of the layer II neurons in the EC, it is possible that a synaptic contact with layer II neurons may exist. However, tissue in the vicinity of the tracer is dense, so this cannot be stated with certainty. It is worth noting that the intrinsic connections of the EC are largely coincident with the topographical arrangement of EC projections to the DG and hippocampal CA fields (Witter et al. 1989a, b; Witter and Amaral 1991).

HF Commissural Connections

The nonhuman primate HF has a rather meager commissural fiber system relative to the rodent. Retrograde and anterograde tracing studies indicate that only the head of the hippocampus (CA3 and DG) shows a true commissural pathway, and it is restricted to the contralateral hippocampal head. The cells of origin for this commissural projection are the polymorphic cell layer of the DG and the *stratum pyramidale* of CA3. The PrS is the HF region that contributes most substantially to commissural connectivity with the contralateral EC. The PrS cells that project to the contralateral EC are located mostly in the deep part of *lamina principalis externa*, although no specific topography seems to be present. Data from neuronal tracing studies (Amaral et al. 1984; Demeter et al. 1985; Saunders and Rosene 1988) show that the PrS sends a robust projection to the caudal part of the contralateral EC, but not to the rostral part. The S has a less dense, reciprocal, connection with the caudal part of the EC, where it terminates more heavily in layers III and V. The PaS does not show commissural connections. The EC itself has a weak projection to layer III of the contralateral caudal EC, as well as to CA1 and the outer part of the molecular layer of caudal DG (i.e., the crossed temporo-ammonic pathway). In contrast, the rostral part of the EC does not show contralateral projections.

Cortical Connections of the HF

Overview

As indicated in section "Overview of the Organization of the HF. Anatomical Organization and Nomenclature" (above) basic descriptions of EC inputs to DG were articulated long ago (Ramón y Cajal 1893). However, questions about where inputs to the EC (and the PHR) had come from were left unresolved—in particular, points of origin in the cerebral cortex had not been identified. Indeed, it was not until 1975 that Van Hoesen and colleagues (Van Hoesen and Pandya 1975a, b; Van Hoesen et al. 1975) clarified some of the cortical inputs to the EC. Subsequently, and as described in more detail below, anterograde and retrograde tracing studies were performed that have provided considerably more information about patterns of cortical connectivity with the HF.

In general, cortical connectivity of the HF is characterized primarily by afferent input from high-order polysensory (polymodal) association cortical areas where inputs from different sensory modalities converge and can then be relayed to the HF. In contrast to rodents, cortical connectivity of the HF in nonhuman primates is unique in the sense that *only* polysensory, high-order association cortical areas send afferents to the HF. The one exception to this rule is a direct input from the olfactory bulb that terminates in subfield E_O without any intervening synapse.

Significantly, most of the cortical connectivity with the HF is directed to the PHR and RSC, rather than the HF itself. As indicated earlier, input from the PHR accounts for about two-thirds of the total cortical input to EC (Insausti et al. 1987a), and it is likely that this percentage is maintained in the cortical output of the HF (Suzuki and Amaral 1994b; Muñoz and Insausti 2005; Legidos-García 2014). Afferents that do project directly to the HF terminate in EC and the S/CA1 interface (Rosene and Van Hoesen 1977; Iwai and Yukie 1988; Suzuki and Amaral 1990; Barbas and Blatt 1995; Blatt and Rosene 1998; Insausti and Muñoz 2001). Likewise, S/CA1 output reciprocates many of these inputs, in particular to the temporal lobe (Rosene and van Hoesen 1977). Figure 4 shows a summary of the extrinsic cortical connectivity of the HF.

Finally, before discussing specific connectivity patterns, it is worth noting that in nonhuman primates no other hippocampal field (DG, CA3, CA2, CA1) receives direct input from, or projects directly to, the cerebral cortex. That said, cortical connections with the HF are distributed throughout all lobules of the brain—i.e., frontal, temporal, parietal, and occipital association areas are all connected with the HF. Therefore, direct cortical connectivity of the HF is broad. Still, when just one more synapse via the PHR is added to the connectivity profile, the number of cortical connections increases dramatically. If we put aside primary sensory and motor cortices, the remainder of the cortical mantle is connected with the HF (either directly or via the PHR), except the dorsolateral prefrontal cortex. A selective overview of cortical-HF connectivity patterns is provided in the subsections that follow.

Olfactory

Olfaction is the only primary sensory modality that has direct connectivity with subfield E_O of the EC, which accounts for approximately 10% of the cortical surface of this structure (Amaral et al. 1987). Interestingly, the extent of E_O in the nonhuman primate is very similar to humans, where one architectonic subfield (E_O) closely resembles that described in nonhuman primates (Insausti et al. 1995; Insausti and Amaral 2012). Olfactory fibers terminate in layer I of E_O.

Temporal Lobe

The temporal lobe shows a great deal of the cortical connectivity of the HF. The combination of TPC, PRC and PHC cortices and that of the upper bank of the superior temporal sulcus account for the majority of the direct cortical afferents to the HF.

Visually Related Connections with the HF

The cortical pathways for visual processing originate in the striate cortex and subdivide into ventral and dorsal processing streams. While the dorsal stream, mainly located in the parietal lobe shows only a meager projection to the S/CA1 interface, PrS and EC (see below section on parietal cortex projections), the ventral stream contributes extensively to the HF. Processing supported by the ventral stream contributes to object identification and recognition (Mishkin and Ungerleider 1982; Kravitz et al. 2013). The ventral stream is organized anatomically in a hierarchical series of connections that are characterized functionally by the processing of increasingly complex visual information at progressively more rostral areas (Mishkin and Ungerleider 1982; Desimone 1996; Nakamura and Kubota 1996; Tanaka 1996; Kravitz et al. 2013). This processing stream originates in the striate cortex (V1) and courses through the occipitotemporal cortex (V4, TEO) to its anterior temporal target (area TE, Kravitz et al. 2013). Area TE then projects into the HF mostly through the PRC (although PHC is also connected with V4 and TEO areas, Suzuki and Amaral 1994a, b), and a small contribution to E_{LR}[6] (Saleem and Tanaka 1996). Projections from the PRC and PHC then reach the EC, which, in turn, projects to the HF (Insausti et al. 1987a; Suzuki and Amaral 1994a). The importance of the entry of this cortical input to the HF is highlighted by impairment on high-level visual perception and recognition memory tests as a consequence of damage to the PHR (Meunier et al. 1993; Malkova et al. 2001; Baxter 2009); a much larger medial temporal lobe lesion does not result in comparably greater impairment.

The precise topography of HF output to visual processing areas along the ventral pathway is still poorly understood. However, anterograde and retrograde tracing studies suggest that the EC and the S/CA1 interface reciprocate temporal lobe input to the PHR (Rosene and Van Hoesen 1977; Iwai and Yukie 1988; Suzuki and Amaral 1994b; Barbas and Blatt 1995; Blatt and Rosene 1998; Insausti and Muñoz 2001; Muñoz and Insausti 2005).

Auditory Related Connections with the HF

It is known that cortical auditory processing in primates starts in primary auditory areas located in the supratemporal plane and, as in vision, soon diverts into ventral and dorsal processing streams (Romanski et al. 1999). However, details about the anatomy and function of the auditory processing pathways have been relatively underexplored in comparison to vision, and much work remains to be done. In the text that follows a basic overview of auditory processing pathways, and their projections to the HF, is provided.

The auditory ventral stream originates in primary auditory areas (core A1/R/RT fields of Kaas and Hackett 2000) and courses in a rostral direction through the supratemporal plane. Information is passed in parallel to secondary and tertiary association areas (Kaas and Hackett 2000; Galaburda and Pandya 1983; Yeterian and Pandya 1985). These areas, in contrast to auditory dorsal stream, project to the dorsolateral surface of the superior temporal gyrus and make their way as far rostral as the dorsolateral temporal polar cortex (TPC; Suzuki and Amaral 1994b; Kondo et al. 2003; Muñoz et al. 2003; Mohedano-Moriano et al. 2008; Muñoz-López et al. 2010, 2015). In turn, TPC projects heavily into the rostral EC (Insausti et al. 1987a; Insausti and Amaral 2008; Muñoz-López et al. 2015) as well as the S/CA1 interface

---

[6]E_{LR} subfield is recipient of converging cortical input from polysensory cortical areas (i.e. related to visual and auditory information) and from other memory related areas of the cortex (Mohedano-Moriano et al. 2008).

(Rosene and Van Hoesen 1977). Compared with visual information, auditory input to the PHR is less dense (Amaral et al. 1983; Insausti et al. 1987a; Mohedano-Moriano et al. 2008) although there are meager projections to PRC and another minor entry from the caudal part of superior temporal gyrus to area TH of the PHC (Tranel et al. 1988; Suzuki and Amaral 1994b). From these PHR structures, information is routed to the EC and S/CA1 fields of the HF.

Available data suggest that the rostrally directed stimulus processing dorsal stream in the superior temporal gyrus (area 38_{DL} in Muñoz-López et al. 2015) supports high-level auditory processing, and codes complex auditory information including, for example, species-specific calls. Consistent with this proposal neuro-imaging studies conducted with nonhuman primates show activity differences in the rostral supratemporal plane and dorsolateral temporal pole when species-specific calls are presented (Gil da Costa et al. 2004; Poremba et al. 2004; Petkov et al. 2008). This region of the temporal pole sends direct projections to the HF.

In nonhuman primates and humans, there is a remarkable capability to form long-term visual and tactile memories (Murray and Mishkin 1984; Goulet and Murray 2001), but surprisingly poor ability to store auditory sensory information (primates in Fritz et al. 2005; Scott et al. 2012, humans in Bigelow and Poremba 2014). Our own work had suggested an anatomical pathway that could add a critical piece to the anatomical puzzle of auditory memory and may help explain memory differences across perceptual modalities. The areas that form the rostral superior temporal gyrus make up to 70% of the total cortical input to the dorsolateral temporal pole area 38_{DL}, which in turn projects to the S/CA1 interface, EC, rostral area 35, and areas TH and TF (Muñoz-López et al. 2015). However, and in striking contrast to the pathway important for visual memory, this projection bypasses most of area 36 of the PRC. One possibility then is that poor recognition memory of rhesus monkeys for auditory events is a consequence of these connectivity differences. This possibility might also explain relatively poor auditory memory performance in humans (Bigelow and Poremba 2014). While this possibility is worth further investigation, we note that auditory anatomy and function remain (relatively) poorly understood.

Superior Temporal Sulcus Polysensory Association Cortex

The superior temporal gyrus has a polysensory area in the dorsal bank of the superior temporal sulcus (sts, area TPO of Seltzer and Pandya 1978, or STGi and STGm of Insausti et al. 1987a, b). The sts integrates polysensory information, from primarily visual and auditory sensory modalities (Perrett et al. 1982; Baylis et al. 1987). The cortex on the dorsal bank of the superior temporal gyrus shows direct connections to the intermediate and caudal parts of the EC (Amaral et al. 1983; Mohedano-Moriano et al. 2008; Insausti and Amaral 2008). Interestingly, this afferent projection is not present at the S/CA1 interface region of the hippocampus (Insausti and Muñoz 2001; Insausti and Amaral 2008). Likewise, the EC is the only HF structure that sends a weak projection to the sts. Little is known about the HF output connections to this polymodal area, although anterograde tracing experiments seem to indicate that these connections may exist (Legidos-García 2014).

Somatosensory Related Connections

There is not a direct projection from somatosensory areas to the HF. Instead, projections are routed through the PRC. Consistent with this circuitry, research studies suggest that the ability of monkeys and humans to retain tactile information across long delays is impaired when lesions involve the PRC (Goulet and Murray 2001; Bigelow and Poremba 2014). The main pathway described in the literature so far is a projection from higher order somatosensory processing area SII of the insular cortex, which in turn projects to area 35 of the PRC (Murray and Mishkin 1984; Schneider et al. 1993; Friedman et al. 1986). This pathway, which ultimately terminates in the EC (ELr field) (Suzuki and Amaral 1994a) and the S/CA1 interface region (Rosene and Van Hoesen 1977), is likely important for tactile memory.

Frontal Lobe Cortical Connections with the HF

The prefrontal cortex (PFC) is an extensive region, not only in humans, but in nonhuman primates as well. HF connectivity with this region ranges from extensive to modest. For example, while there is considerable connectivity with medial frontal cortex (MFC) and caudal orbitofrontal cortex (OFC), connectivity with the lateral PFC is quite sparse (i.e. weak with ventrolateral PFC, and virtually non-existent with dorsolateral PFC). These projections are described in more detail below.

Medial Frontal Cortex[7]

Anterograde and retrograde tracing studies have provided much information about the rich connectivity patterns of MFC with the HF and broader MTL. For example, retrograde tracing studies in nonhuman primates have indicated that most projections to the EC come from MFC areas 24 (anterior cingulate cortex), 32 (prelimbic cortex) and 25 (infralimbic cortex), as well as area 14, which is located in the *Gyrus rectus*. Studies have also indicated that the density of retrograde labeling in the MFC varies within and across subregions (e.g., rostral labeling is different from caudal labeling). Specifically, MFC areas 24 and 25 show the greatest number of labeled neurons, while area 32 has far fewer.

---

[7]We depart from previous reports (Insausti et al. 1987a) and include area 24 of Brodmann among MFC areas for the sake of complementarity with more rostrally situated areas 14, 32 and 25.

The termination points of MFC afferents in the EC are not homogeneous either. While rostral EC receives a moderate to heavy projection, the caudal part of the EC receives a far less dense projection (see Table 1 in Insausti et al. 1987a, b; Insausti and Amaral 2008). Specifically, rostral and mid-EC (subfields E_O, E_R, E_I) receive projections from areas 25, 32, and 24, which also provides input to caudal EC (subfields E_C and E_{CL}). In short, MFC output is directed primarily to anterior EC, with fewer projections to posterior EC (Insausti and Amaral 2008; Morecraft et al. 2012). No information exists for the non-human primate about where exactly these MFC projections might terminate in the hippocampus proper.

Research studies have suggested that there is a crude topographical gradient that characterizes projections from the temporal lobe to the PFC. Several studies, for example, have proposed a rostro-caudal organization of temporal lobe projections to this region. Specifically, data suggest that more caudal temporal areas project preferentially to dorsolateral prefrontal regions while increasingly rostral temporal areas project to progressively more ventral and medial PFC regions (Yeterian and Pandya 1985; Petrides and Pandya 1988; Seltzer and Pandya 1989; Romanski et al. 1999; Kondo et al. 2003). Indeed, research suggests that the primary frontal lobe target of EC outputs is the MFC (Legidos-García 2014).

Retrograde tracing studies (Muñoz and Insausti 2005) show that these projections to MFC originate primarily in the deep layers of the rostral two-thirds of the EC, and reciprocate afferents from the same MFC areas that send inputs to EC. Together with the rostral part of the PHR (Carmichael and Price 1995; Barbas et al. 1999; Lavenex et al. 2002; Kondo et al. 2003; Mohedano-Moriano et al. 2015), the EC is likely to be the part of the MTL that projects most heavily to caudal PFC regions, especially areas 24 and 25, but also area 32 (Insausti et al. 1987a). The number of neurons projecting to frontal areas 10 and 14, and to the rostral part of area 32 from EC and the rostral PHR is lower than to those projecting to more caudal MFC areas (Barbas et al. 1999; Legidos-García 2014). Similarly, the medial region of area 9 in the frontal cortex receives few, if any, projections from any medial temporal cortical area (Barbas et al. 1999; Legidos-García 2014).

Efferents that originate in hippocampus proper and terminate in MFC have been identified as well. The highest density of direct projections from hippocampus proper to MFC (areas 25 and 32) originate in the head of the hippocampus, specifically in CA1, mainly from pyramidal cells in its superficial half. Projections also originate from the S, especially from the body of the hippocampus and to a lesser extent from the hippocampal head (Barbas and Blatt 1995; Insausti and Muñoz 2001). A sizeable projection to MFC from the caudal part of the PaS is evident as well, while the PrS projection is relatively restricted. In contrast to areas 32 and 25, area 24 of the MFC does not receive efferent inputs from the hippocampal head or from the PaS. Instead, projections to area 24 have been identified that originate in the S at the level of the body of the hippocampus.

In summary, the MFC is one of the most important input and output of the HF. The preferential site of termination in the EC is the anterior one-half of the EC, which connects with the anterior part of the hippocampus proper. Similarly, the EC output to MFC is also concentrated in the anterior part of the EC.

Orbitofrontal Cortex

The OFC cortex is a large, heterogeneous region that covers the ventral surface of the frontal lobe (Walker 1940; Carmichael and Price 1994). The OFC has dense reciprocal connections with the HF (EC, S/CA1 interface) and with the PHR (van Hoesen and Pandya 1975a, b; Amaral et al. 1984; Insausti et al. 1987a; Morecraft et al. 1992; Barbas 1993; Suzuki and Amaral 1994b; Barbas and Blatt 1995; Carmichael and Price 1995; Ongur and Price 2000; Cavada et al. 2000; Insausti and Amaral 2008).

The OFC projects primarily to the rostral and intermediate portions of the EC (subfields E_O, E_R and E_I), and represents a significant proportion (about 5%) of all the cortical input to the HF (van Hoesen and Pandya 1975a, b; Insausti et al. 1987a; Insausti and Amaral 2008). Area 13 provides the bulk of the afferents to the EC, while areas 12 and 14 provide far fewer inputs (Van Hoesen et al. 1975; Insausti et al. 1987a).

The OFC projection to the rostral part of the EC is dense (subfields E_O, E_R and E_{LR}, (Insausti et al. 1987a; Insausti and Amaral 2008). OFC afferents to the S/CA1 interface and PrS have been described, but many details are still unknown (Rosene and Van Hoesen 1977; Morecraft et al. 1992; Barbas and Blatt 1995; Carmichael and Price 1995; Cavada et al. 2000).

The HF projection to OFC is ipsilateral and arises principally from CA1 and PrS (Cavada et al. 2000; Insausti and Muñoz 2001). While it is not very dense, this projection originates from up to the rostral one-half of the hippocampus. The rostral PrS and PaS also project to the OFC, although the latter to a lesser extent (Insausti and Muñoz 2001).

The OFC network receives sensory inputs from several modalities, including olfaction, taste, touch, and vision, which appear to be especially related to food or eating and reward (Carmichael and Price 1994, 1995). The variety of inputs to OFC indicates that it processes polysensory information that maintains heavy connections with the HF.

In summary, the OFC sends and receives heavy projections with the anterior part of the EC, CA1, S, and juxtasubicular complex. The functional implications of this organization suggest that the anterior part of the HF is one main component of the emotional aspect of memories.

Lateral Frontal Cortex

In sharp contrast to the HF connectivity with the MFC and OFC, the lateral PFC (i.e., Brodmann's areas 45, 46, 8 and 9) provides just a minor input to the EC (Insausti et al. 1987a). Specifically, while a very light projection to the EC seems to originate in the vLPFC (area 12 and 45), it is negligible in area 46, and areas 8 and 9 provide very light input to the rostrolateral EC (E_{LR} subfield, Insausti et al. 1987a; Insausti and Amaral 2008). There are no projections, in either direction, with the S or PrS (Insausti and Muñoz 2001). Similarly, entorhinal and non-entorhinal (i.e. S/CA1 interface, PrS, and PaS) HF output to LPFC is also very meager, and it also originates in the lateral subfields of the EC (E_{LR} and E_{LC}, Insausti and Muñoz 2001). The paucity of connections between LPFC and the HF is remarkable, and these observations extend to the rostral part of the PHR (Suzuki and Amaral 1994a, b; Mohedano-Moriano et al. 2015). The functional significance of this connectivity profile (or rather, the lack thereof) remains to be determined.

Retrosplenial and Parietal Cortices

Much like MFC and OFC above, the retrosplenial cortex (RSC) serves as yet another source of substantial, direct input to the EC. Indeed, projections from RSC account for nearly 20% of the total cortical input to the EC (Insausti et al. 1987a), which suggests that this structure may have a substantial influence on HF function.

The RSC is situated caudal to PHC and curves dorsally behind the splenium of the corpus callosum (from where it takes its name). It is continued by the posterior cingulate cortex, area 23 of Brodmann (Kobayashi and Amaral 2003). Briefly, the dorsal bank of the callosal sulcus and its ventral extension correspond to RSC cortical areas 29 and 30; most of the medial surface of the posterior cingulate gyrus and the ventral bank of the posterior cingulate sulcus consist of area 23. On the ventral surface there is a transitional zone, area 30v, which separates RSC from prestriate visual cortex (Kobayashi and Amaral 2003).

The RSC receives afferents from the HF, the PHR, other cingulate cortical areas, LPFC, and parietal cortex (Goldman-Rakic et al. 1984; Kobayashi and Amaral 2003). Therefore, the RSC has potential to exert a substantial influence on HF function. The cytoarchitectonic organization of the macaque monkey RSC and caudal portion of the cingulate cortices has been previously described (Kobayashi and Amaral 2000).

Neuroanatomical tracing data in the nonhuman primate show that the RSC has strong connections with the EC and PrS. Afferents to the EC originate in both areas 29 and 30, although these projections are restricted to the caudal one-half of the EC (namely E_I, and particularly so E_C and E_{CL} subfields). The rostral one-half of the EC receives little or no projections from the RSC. RSC afferents to the EC terminate mostly in layer I in a restricted topographical manner (Insausti and Amaral 2008).

The EC reciprocates the projection to the RSC. The termination pattern of EC efferent projections to the RSC are typically concentrated in layer I of area 29; lighter projections are directed to area 30, and layer III of the posterior cingulate cortex. The caudal part of the EC (subfield E_{CL}) provides more widespread RSC projections (Morris et al. 1999a; Kobayashi and Amaral 2003; Aggleton et al. 2012). In addition, retrograde tracing experiments demonstrate that other fields of the HF such as the S, PrS and PaS send projections to the RSC (Vogt and Pandya 1987; Morris et al. 1999b; Kobayashi and Amaral 2003; Aggleton et al. 2012). Projections from the S and PrS terminate principally in layers I and III of area 29.

While the uncal part of the S projects primarily to the ventral RSC, mid and caudal levels of the S have denser projections to both the caudal and dorsal RSC. Both, caudal S and PrS project to dorsal area 30 (layer III) (Rosene and Van Hoesen 1977; Aggleton et al. 2012). Retrograde transport studies have also shown that both PRC (area 35; Morris et al. 1999a; Kobayashi and Amaral 2003) and PHC (areas TH and TF) project to RSC as well (Kobayashi and Amaral 2003; Lavenex et al. 2002).

The RSC plays a key role in spatial memory and navigation, which means that this area represents a very important node in a broader spatial processing network that includes the HF. The RSC projections to the EC terminate near those from the parietal cortex (area 7 of Brodmann), where their overlap might be important for spatial information. Several investigators (Morris et al. 1999a, b; Kobayashi and Amaral 2003, 2007) have pointed out the importance of reciprocal connectivity between the HF and RSC for spatial information processing and spatial memory.

The parietal cortex (area 7) is also reciprocally connected with the HF, primarily with EC (Insausti et al. 1987a; Ding et al. 2000). Area 7 projections to the EC terminate in the caudal part of the EC (E_C and E_{CL} divisions and therefore overlap with the afferents from RSC); in addition the parietal cortex shows heavy projections to the PrS (Ding et al. 2000). Non-entorhinal cortex HF output to the parietal cortex originates in the PrS almost exclusively (Insausti and Munoz 2001). The EC itself shows reciprocal connections with area 7 coming from subfields E_C and E_{CL}, and, interestingly, lateral and rostral EC (subfield E_{LR}), and these EC fields originate projections to the caudal part of the hippocampus.

Part 3. Functional Implications and Concluding Remarks

Most likely, HF cortical connections with the HF are implicated in different aspects of memory function, both at the spatial and object recognition levels. Much of the evidence supporting this assumption comes from lesion studies and neuropsychological testing after lesions in patients.

One of the central features of declarative episodic memory is that our memories of events are formed by information received via all different sensory modalities and, consequently, episodic memory is often said to be multimodal, providing our episodic memories with rich, complex contextual information. The limbic system is at the core of this complex memory system and includes a wide variety of structures in the HF, as well as outside the HF. However, to fully understand the role of the cortical connectivity of the HF in memory, it is important to determine how information reaches directly the HF. The anatomical studies in primates described above reveal opportunities and constraints on the manner by which information reaches brain regions involved in memory processing.

It is reasonable to suppose that the interaction with other cortical areas is crucial for the cortical-hippocampal axis (Insausti 1993), which supports the interaction between both systems for encoding, consolidation and retrieval of declarative memory. The role of subcortical connections still needs to be put in context with the basic general outline here exposed to gain in the understanding of memory processes.

Acknowledgments Supported by grants BFI2003-09581, BFU 2006-12964 and BFU 2009-14705, Ministry of Science and Education (Spain).

References

Aggleton JP, Wright NF, Vann SD, Saunders RC (2012) Medial temporal lobe projections to the retrosplenial cortex of the macaque monkey. Hippocampus 22(9):1883–1900

Amaral DG, Lavenex P (2007) Hippocampal neuroanatomy. The Hippocampus Book. Oxford University Press, Oxford, pp. 37–114

Amaral DG, Insausti R, Cowan WM (1983) Evidence for a direct projection from the superior temporal gyrus to the entorhinal cortex in the monkey. Brain Res 275(2):263–277

Amaral DG, Insausti R, Cowan WM (1984) The commissural connections of the monkey hippocampal formation. J Comp Neurol 224(3):307–336

Amaral DG, Insausti R, Cowan WM (1987) The entorhinal cortex of the monkey: I. Cytoarchitectonic organization. J Comp Neurol 264(3):326–355

Amaral DG, Insausti R, Campbell MJ (1988) Distribution of somatostatin immunoreactivity in the human dentate gyrus. J Neurosci Off J Soc Neurosci 8(9):3306–3316

Baimbridge KG, Miller JJ (1982) Immunohistochemical localization of calcium-binding protein in the cerebellum, hippocampal formation and olfactory bulb of the rat. Brain Res 245 (2):223–229

Baimbridge KG, Miller JJ, Parkes CO (1982) Calcium-binding protein distribution in the rat brain. Brain Res 239(2):519–525

Bakst I, Amaral DG (1984) The distribution of acetylcholinesterase in the hippocampal formation of the monkey. J Comp Neurol 225(3):344–371

Bakst I, Morrison JH, Amaral DG (1985) The distribution of somatostatin-like immunoreactivity in the monkey hippocampal formation. J Comp Neurol 236(4):423–442

Barbas H (1993) Organization of cortical afferent input to orbitofrontal areas in the rhesus monkey. Neuroscience 56(4):841–864

Barbas H, Blatt GJ (1995) Topographically specific hippocampal projections target functionally distinct prefrontal areas in the rhesus monkey. Hippocampus 5(6):511–533

Barbas H, Ghashghaei H, Dombrowski SM, Rempel-Clower NL (1999) Medial prefrontal cortices are unified by common connections with superior temporal cortices and distinguished by input from memory-related areas in the rhesus monkey. J Comp Neurol 410(3):343–367

Baxter MG (2009) Involvement of medial temporal lobe structures in memory and perception. Neuron 61(5):667–677

Baylis GC, Rolls ET, Leonard CM (1987) Functional subdivisions of the temporal lobe neocortex. J Neurosci Off J Soc Neurosci 7(2):330–342

Berger B, Alvarez C (1994) Neurochemical development of the hippocampal region in the fetal rhesus monkey. II. Immunocytochemistry of peptides, calcium-binding proteins, DARPP-32, and monoamine innervation in the entorhinal cortex by the end of gestation. Hippocampus 4 (1):85–114

Bigelow J, Poremba A (2014) Achilles' ear? Inferior human short-term and recognition memory in the auditory modality. PLoS One 9(2):e89914

Blatt GJ, Rosene DL (1998) Organization of direct hippocampal efferent projections to the cerebral cortex of the rhesus monkey: projections from CA1, prosubiculum, and subiculum to the temporal lobe. J Comp Neurol 392(1):92–114

Bouras C, Magistretti PJ, Morrison JH, Constantinidis J (1987) An immunohistochemical study of pro-somatostatin-derived peptides in the human brain. Neuroscience 22(3):781–800

Braak H (1980) Architectonics of the human telencephalic cortex. Springer, Berlin

Braak E, Strotkamp B, Braak H (1991) Parvalbumin-immunoreactive structures in the hippocampus of the human adult. Cell Tissue Res 264(1):33–48

Brady DR, Mufson EJ (1997) Parvalbumin-immunoreactive neurons in the hippocampal formation of Alzheimer's diseased brain. Neuroscience 80(4):1113–1125

Broca P (1878) Anatomie comparée des circonvolutions cérébrales. Le grand lobe limbique et la scissure limbique dans la série des mammifères. Revue d'Anthropologie Ser 2(1):385–498

Brodmann K (1909) Vergleichende Lokalisationslehre der Grosshimrinde in ihren Principien dargestellt auf des Grund des Zellenbayes. Barth, Leipzig

Carboni AA, Lavelle WG, Barnes CL, Cipolloni PB (1990) Neurons of the lateral entorhinal cortex of the rhesus monkey: a Golgi, histochemical, and immunocytochemical characterization. J Comp Neurol 291(4):583–608

Carmichael ST, Price JL (1994) Architectonic subdivision of the orbital and medial prefrontal cortex in the macaque monkey. J Comp Neurol 346(3):366–402

Carmichael ST, Price JL (1995) Limbic connections of the orbital and medial prefrontal cortex in macaque monkeys. J Comp Neurol 363(4):615–641

Cavada C, Company T, Tejedor J, Cruz-Rizzolo RJ, Reinoso-Suarez F (2000) The anatomical connections of the macaque monkey orbitofrontal cortex. A review. Cereb Cortex 10 (3):220–242

Cebada-Sanchez S, Insausti R, Gonzalez-Fuentes J, Arroyo-Jimenez MM, Rivas-Infante E, Lagartos MJ, Martinez-Ruiz J, Lozano G, Marcos P (2014) Distribution of peptidergic populations in the human dentate gyrus (somatostatin [SOM-28, SOM-12] and neuropeptide Y [NPY]) during postnatal development. Cell Tissue Res 358(1):25–41

Chan-Palay V (1987) Somatostatin immunoreactive neurons in the human hippocampus and cortex shown by immunogold/silver intensification on vibratome sections: coexistence with neuropeptide Y neurons, and effects in Alzheimer-type dementia. J Comp Neurol 260 (2):201–223

Chan-Palay V, Kohler C, Haesler U, Lang W, Yasargil G (1986) Distribution of neurons and axons immunoreactive with antisera against neuropeptide Y in the normal human hippocampus. J Comp Neurol 248(3):360–375

Chavkin C, Shoemaker WJ, McGinty JF, Bayon A, Bloom FE (1985) Characterization of the prodynorphin and proenkephalin neuropeptide systems in rat hippocampus. J Neurosci Off J Soc Neurosci 5(3):808–816

Christiansen K, Dillingham CM, Wright NF, Saunders RC, Vann SD, Aggleton JP (2016) Complementary subicular pathways to the anterior thalamic nuclei and mammillary bodies in the rat and macaque monkey brain. Eur J Neurosci 43:1044–1061

Chrobak JJ, Amaral DG (2007) Entorhinal cortex of the monkey: VII. Intrinsic connections. J Comp Neurol 500(4):612–633

Del Fiacco M, Levanti MC, Dessi ML, Zucca G (1987) The human hippocampal formation and parahippocampal gyrus: localization of substance P-like immunoreactivity in newborn and adult post-mortem tissue. Neuroscience 21(1):141–150

Demeter S, Rosene DL, Van Hoesen GW (1985) Interhemispheric pathways of the hippocampal formation, presubiculum, and entorhinal and posterior parahippocampal cortices in the rhesus monkey: the structure and organization of the hippocampal commissures. J Comp Neurol 233 (1):30–47

Desimone R (1996) Neural mechanisms for visual memory and their role in attention. Proc Natl Acad Sci U S A 93(24):13494–13499

Ding SL, van Hoesen G, Rockland KS (2000) Inferior parietal lobule projections to the presubiculum and neighboring ventromedial temporal cortical areas. J Comp Neurol 425:510–530

Dournaud P, Cervera-Pierot P, Hirsch E, Javoy-Agid F, Kordon C, Agid Y, Epelbaum J (1994) Somatostatin messenger RNA-containing neurons in Alzheimer's disease: an in situ hybridization study in hippocampus, parahippocampal cortex and frontal cortex. Neuroscience 61 (4):755–764

Filimonoff IN (1947) A rational subdivision of the cerebral cortex. Arch Neurol Psychiatr 58:296–311

Freund TF, Buzsaki G (1996) Interneurons of the hippocampus. Hippocampus 6(4):347–470

Friederich-Ecsy B, Braak E, Braak H, Probst A (1988) Somatostatin-like immunoreactivity in non-pyramidal neurons of the human entorhinal region. Cell Tissue Res 254(2):361–367

Friedman DP, Murray EA, O'Neill JB, Mishkin M (1986) Cortical connections of the somatosensory fields of the lateral sulcus of macaques: evidence for a corticolimbic pathway for touch. J Comp Neurol 252(3):323–347

Fritz J, Mishkin M, Saunders RC (2005) In search of an auditory engram. Proc Natl Acad Sci U S A 102(26):9359–9364

Galaburda AM, Pandya DN (1983) The intrinsic architectonic and connectional organization of the superior temporal region of the rhesus monkey. J Comp Neurol 221(2):169–184

Gaspar P, Berger B, Febvret A (1990) Neurotensin innervation of the human cerebral cortex: lack of colocalization with catecholamines. Brain Res 530(2):181–195

Gil-da-Costa R, Braun A, Lopes M, Hauser MD, Carson RE, Herscovitch P, Martin A (2004) Toward an evolutionary perspective on conceptual representation: species-specific calls activate visual and affective processing systems in the macaque. Proc Natl Acad Sci U S A 101 (50):17516–17521

Gloor P (1997) The temporal lobe and limbic system. Oxford University Press, New York

Goldman-Rakic PS, Selemon LD, Schwartz ML (1984) Dual pathways connecting the dorsolateral prefrontal cortex with the hippocampal formation and parhippocampal cortex in the rhesus monkey. Neuroscience 12:719–743

Goulet S, Murray EA (2001) Neural substrates of crossmodal association memory in monkeys: the amygdala versus the anterior rhinal cortex. Behav Neurosci 115(2):271–284

Grateron L, Cebada-Sanchez S, Marcos P, Mohedano-Moriano A, Insausti AM, Munoz M, Arroyo-Jimenez MM, Martinez-Marcos A, Artacho-Perula E, Blaizot X, Insausti R (2003) Postnatal development of calcium-binding proteins immunoreactivity (parvalbumin, calbindin, calretinin) in the human entorhinal cortex. J Chem Neuroanat 26(4):311–316

Hornung JP, Celio MR (1992) The selective innervation by serotoninergic axons of calbindin-containing interneurons in the neocortex and hippocampus of the marmoset. J Comp Neurol 320(4):457–467

Insausti R (1993) Comparative anatomy of the entorhinal cortex and hippocampus in mammals. Hippocampus. 3 Spec No:19–26

Insausti R, Amaral DG (2008) Entorhinal cortex of the monkey: IV. Topographical and laminar organization of cortical afferents. J Comp Neurol 509(6):608–641

Insausti R, Amaral DG (2012) The hippocampal formation. In: JK M, Paxinos G (eds) The human nervous system, 3rd edn. Elsevier Academic Press, Amsterdam; Boston, pp 896–942

Insausti R, Munoz M (2001) Cortical projections of the non-entorhinal hippocampal formation in the cynomolgus monkey (Macaca fascicularis). Eur J Neurosci 14(3):435–451

Insausti R, Amaral DG, Cowan WM (1987a) The entorhinal cortex of the monkey: II. Cortical afferents. J Comp Neurol 264(3):356–395

Insausti R, Amaral DG, Cowan WM (1987b) The entorhinal cortex of the monkey: III. Subcortical afferents. J Comp Neurol 264(3):396–408

Insausti R, Tuñon T, Sobreviela T, Insausti AM, Gonzalo LM (1995) The human entorhinal cortex: a cytoarchitectonic analysis. J Comp Neurol 335(2):171–198

Insausti R, Herrero MT, Witter MP (1997) Entorhinal cortex of the rat: cytoarchitectonic subdivisions and the origin and distribution of cortical efferents. Hippocampus 7(2):146–183

Iritani S, Fujii M, Satoh K (1989) The distribution of substance P in the cerebral cortex and hippocampal formation: an immunohistochemical study in the monkey and rat. Brain Res Bull 22(2):295–303

Iwai E, Yukie M (1988) A direct projection from hippocampal field CA1 to ventral area TE of inferotemporal cortex in the monkey. Brain Res 444(2):397–401

Kaas JH, Hackett TA (2000) How the visual projection map instructs the auditory computational map. J Comp Neurol 421(2):143–145

Kobayashi Y, Amaral DG (1999) Chemical neuroanatomy of the hippocampal formation and the perirhinal and parahippocampal cortices. Handbook of chemical neuroanatomy (vol 15). Amsterdam: Elsevier Science Publishers, pp 285–337

Kobayashi Y, Amaral DG (2000) Macaque monkey retrosplenial cortex: I. three-dimensional and cytoarchitectonic organization. J Comp Neurol 426(3):339–365

Kobayashi Y, Amaral DG (2003) Macaque monkey retrosplenial cortex: II. Cortical afferents. J Comp Neurol 466(1):48–79

Kobayashi Y, Amaral DG (2007) Macaque monkey retrosplenial cortex: III. Cortical efferents. J Comp Neurol 502(5):810–833

Kondo H, Saleem KS, Price JL (2003) Differential connections of the temporal pole with the orbital and medial prefrontal networks in macaque monkeys. J Comp Neurol 465(4):499–523

Kondo H, Lavenex P, Amaral DG (2008) Intrinsic connections of the macaque monkey hippocampal formation: I. Dentate gyrus. J Comp Neurol 511(4):497–520

Kondo H, Lavenex P, Amaral DG (2009) Intrinsic connections of the macaque monkey hippocampal formation: II. CA3 connections. J Comp Neurol 515(3):349–377

Kravitz E, Gaisler-Salomon I, Biegon A (2013) Hippocampal glutamate NMDA receptor loss tracks progression in Alzheimer's disease: quantitative autoradiography in postmortem human brain. PLoS One 8(11):e81244

Lavenex P, Suzuki WA, Amaral DG (2002) Perirhinal and parahippocampal cortices of the macaque monkey: projections to the neocortex. J Comp Neurol 447(4):394–420

Legidos-Garcia ME (2014) Bases esteructurales de la memoria declarativa. Estudio de la interacción entre la formación del hipocampo y corteza cerebral en el primate *Macaca fascicularis*. PhD Tesis

Leranth C, Ribak CE (1991) Calcium-binding proteins are concentrated in the CA2 field of the monkey hippocampus: a possible key to this region's resistance to epileptic damage. Exp Brain Res 85(1):129–136

Lorente de Nó R (1934) Studies onn the structure of the cerebral cortex II. Continuation of the study of the ammonic system. J Psychol Neurol 46:113–177

Lotstra F, Schiffmann SN, Vanderhaeghen JJ (1989) Neuropeptide Y-containing neurons in the human infant hippocampus. Brain Res 478(2):211–226

Mai JK, Triepel J, Metz J (1987) Neurotensin in the human brain. Neuroscience 22(2):499–524

Malkova L, Bachevalier J, Mishkin M, Saunders RC (2001) Neurotoxic lesions of perirhinal cortex impair visual recognition memory in rhesus monkeys. Neuroreport 12(9):1913–1917

Meunier M, Bachevalier J, Mishkin M, Murray EA (1993) Effects on visual recognition of combined and separate ablations of the entorhinal and perirhinal cortex in rhesus monkeys. J Neurosci Off J Soc Neurosci 13(12):5418–5432

Mishkin M, Ungerleider LG (1982) Contribution of striate inputs to the visuospatial functions of parieto-preoccipital cortex in monkeys. Behav Brain Res 6(1):57–77

Mohedano-Moriano A, Martinez-Marcos A, Pro-Sistiaga P, Blaizot X, Arroyo-Jimenez MM, Marcos P, Artacho-Perula E, Insausti R (2008) Convergence of unimodal and polymodal sensory input to the entorhinal cortex in the fascicularis monkey. Neuroscience 151 (1):255–271

Mohedano-Moriano A, Munoz-Lopez M, Sanz-Arigita E, Pro-Sistiaga P, Martinez-Marcos A, Legidos-García ME, Insausti AM, Cebada Sánchez S, Arroyo-Jimenez-MM, Marcos P, Artacho-Pérula E, Insausti R (2015) Prefrontal cortex afferents to the anterior temporal lobe in the *Macaca fascicularis* monkey. J Comp Neurol 523(17):2570–2598

Morecraft RJ, Geula C, Mesulam MM (1992) Cytoarchitecture and neural afferents of orbitofrontal cortex in the brain of the monkey. J Comp Neurol 323(3):341–358

Morecraft RJ, Stilwell-Morecraft KS, Cipolloni PB, Ge J, McNeal DW, Pandya DN (2012) Cytoarchitecture and cortical connections of the anterior cingulate and adjacent somatomotor fields in the rhesus monkey. Brain Res Bull 87(4–5):457–497

Morris R, Petrides M, Pandya DN (1999a) Architecture and connections of retrosplenial area 30 in the rhesus monkey (Macaca mulatta). Eur J Neurosci 11(7):2506–2518

Morris R, Pandya DN, Petrides M (1999b) Fiber system linking the mid-dorsolateral frontal cortex with the retrosplenial/presubicular region in the rhesus monkey. J Comp Neurol 407 (2):183–192

Muñoz M, Insausti R (2005) Cortical efferents of the entorhinal cortex and the adjacent parahippocampal region in the monkey (Macaca fascicularis). Eur J Neurosci 22 (6):1368–1388

Muñoz M, Mishkin M, Saunders, RC (2003) Lateral temporal pole, input from the superior temporal gyrus and output to the medial temporal cortex in the rhesus monkey. Soc Neurosci. Abstract: 939.1. New Orleans, LA, USA

Muñoz-Lopez M, Mohedano-Moriano A, Insausti R (2010) Anatomical pathways for auditory memory in primates. Front Neuroanat 4:129

Muñoz-Lopez M, Insausti R, Mohedano-Moriano A, Mishkin M, Saunders RC (2015) Anatomical pathways for auditory memory II: information from rostral superior temporal gyrus to dorsolateral temporal pole and medial temporal cortex. Front Neurosci 9:158

Murray EA, Mishkin M (1984) Relative contributions of SII and area 5 to tactile discrimination in monkeys. Behav Brain Res 11(1):67–83

Nakamura K, Kubota K (1996) The primate temporal pole: its putative role in object recognition and memory. Behav Brain Res 77(1–2):53–77

Nitsch R, Leranth C (1991) Neuropeptide Y (NPY)-immunoreactive neurons in the primate fascia dentata; occasional coexistence with calcium-binding proteins: a light and electron microscopic study. J Comp Neurol 309(4):430–444

Nitsch R, Leranth C (1994) Substance P-containing hypothalamic afferents to the monkey hippocampus: an immunocytochemical, tracing, and coexistence study. Exp Brain Res 101 (2):231–240

Nitsch R, Ohm TG (1995) Calretinin immunoreactive structures in the human hippocampal formation. J Comp Neurol 360(3):475–487

Ongur D, Price JL (2000) The organization of networks within the orbital and medial prefrontal cortex of rats, monkeys and humans. Cereb Cortex 10(3):206–219

Ottersen OP, Storm-Mathisen J (1985) Different neuronal localization of aspartate-like and glutamate-like immunoreactivities in the hippocampus of rat, guinea-pig and Senegalese baboon (Papio papio), with a note on the distribution of gamma-aminobutyrate. Neuroscience 16(3):589–606

Perrett DI, Rolls ET, Caan W (1982) Visual neurones responsive to faces in the monkey temporal cortex. Exp Brain Res 47(3):329–342

Petkov CI, Kayser C, Steudel T, Whittingstall K, Augath M, Logothetis NK (2008) A voice region in the monkey brain. Nat Neurosci 11(3):367–374

Petrides M, Pandya DN (1988) Association fiber pathways to the frontal cortex from the superior temporal region in the rhesus monkey. J Comp Neurol 273(1):52–66

Pitkanen A, Amaral DG (1993) Distribution of parvalbumin-immunoreactive cells and fibers in the monkey temporal lobe: the hippocampal formation. J Comp Neurol 331(1):37–74

Poremba A, Malloy M, Saunders RC, Carson RE, Herscovitch P, Mishkin M (2004) Species-specific calls evoke asymmetric activity in the monkey's temporal poles. Nature 427 (6973):448–451

Ramón y Cajal S (1893) Estructura del asta de Ammon y fascia dentata. Ann Soc Esp His Nat. 22

Romanski LM, Bates JF, Goldman-Rakic PS (1999) Auditory belt and parabelt projections to the prefrontal cortex in the rhesus monkey. J Comp Neurol 403(2):141–157

Rosene DL, Van Hoesen GW (1977) Hippocampal efferents reach widespread areas of cerebral cortex and amygdala in the rhesus monkey. Science 198(4314):315–317

Rosene DL, Van Hoesen GW (1987) Hippocampal formation primate brain: some comparative aspects of architecture and connection. In: The cortex. Plenum Press, New York and London, pp 345–456

Sakamoto N, Michel JP, Kiyama H, Tohyama M, Kopp N, Pearson J (1986) Neurotensin immunoreactivity in the human cingulate gyrus, hippocampal subiculum and mammillary bodies. Its potential role in memory processing. Brain Res 375(2):351–356

Sakamoto N, Michel JP, Kopp N, Tohyama M, Pearson J (1987) Substance P- and enkephalin-immunoreactive neurons in the hippocampus and related areas of the human infant brain. Neuroscience 22(3):801–811

Saunders RC, Rosene DL (1988) A comparison of the efferents of the amygdala and the hippocampal formation in the rhesus monkey. I. Convergence in the entorhinal, prorhinal and perirhinal cortices. J Comp Neurol 271:153–182

Saleem KS, Tanaka K (1996) Divergent projections from the anterior inferotemporal area TE to the perirhinal and entorhinal cortices in the macaque monkey. J Neurosci Off J Soc Neurosci 16 (15):4757–4775

Schneider RJ, Friedman DP, Mishkin M (1993) A modality-specific somatosensory area within the insula of the rhesus monkey. Brain Res 621(1):116–120

Scott BH, Mishkin M, Yin P (2012) Monkeys have a limited form of short-term memory in audition. Proc Natl Acad Sci U S A 109(30):12237–12241

Seltzer B, Pandya DN (1978) Afferent cortical connections and architectonics of the superior temporal sulcus and surrounding cortex in the rhesus monkey. Brain Res 149(1):1–24

Seltzer B, Pandya DN (1989) Intrinsic connections and architectonics of the superior temporal sulcus in the rhesus monkey. J Comp Neurol 290(4):451–471

Seress L, Gulyas AI, Freund TF (1991) Parvalbumin- and calbindin D28k-immunoreactive neurons in the hippocampal formation of the macaque monkey. J Comp Neurol 313 (1):162–177

Seress L, Gulyas AI, Freund TF (1992) Pyramidal neurons are immunoreactive for calbindin D28k in the CA1 subfield of the human hippocampus. Neurosci Lett 138(2):257–260

Seress L, Gulyas AI, Ferrer I, Tunon T, Soriano E, Freund TF (1993a) Distribution, morphological features, and synaptic connections of parvalbumin- and calbindin D28k-immunoreactive neurons in the human hippocampal formation. J Comp Neurol 337(2):208–230

Seress L, Nitsch R, Leranth C (1993b) Calretinin immunoreactivity in the monkey hippocampal formation—I. Light and electron microscopic characteristics and co-localization with other calcium-binding proteins. Neuroscience 55(3):775–796

Seress L, Leranth C, Frotscher M (1994) Distribution of calbindin D28k immunoreactive cells and fibers in the monkey hippocampus, subicular complex and entorhinal cortex. A light and electron microscopic study. J Hirnforsch 35(4):473–486

Sloviter RS (1989) Calcium-binding protein (calbindin-D28k) and parvalbumin immunocyto-chemistry: localization in the rat hippocampus with specific reference to the selective vulner-ability of hippocampal neurons to seizure activity. J Comp Neurol 280(2):183–196

Sloviter RS, Sollas AL, Barbaro NM, Laxer KD (1991) Calcium-binding protein (calbindin-D28K) and parvalbumin immunocytochemistry in the normal and epileptic human hippocam-pus. J Comp Neurol 308(3):381–396

Solodkin A, Van Hoesen GW (1996) Entorhinal cortex modules of the human brain. J Comp Neurol 365(4):610–617

Stephan H (1975) Allocortex. Springer, Berlin, Heidelberg, New York

Stephan H, Andy OJ (1970) In: Noback CR, Montagna W (eds) The primate brain. Appleton-Century-Crofts, New York

Storm-Mathisen J (1981) Glutamate in hippocampal pathways. Adv Biochem Psychopharmacol 27:43–55

Suzuki WA, Amaral DG (1990) Cortical inputs to the CA1 field of the monkey hippocampus originate from the perirhinal and parahippocampal cortex but not from area TE. Neurosci Lett 115(1):43–48

Suzuki WA, Amaral DG (1994a) Perirhinal and parahippocampal cortices of the macaque monkey: cortical afferents. J Comp Neurol 350(4):497–533

Suzuki WA, Amaral DG (1994b) Topographic organization of the reciprocal connections between the monkey entorhinal cortex and the perirhinal and parahippocampal cortices. J Neurosci Off J Soc Neurosci 14(3 Pt 2):1856–1877

Suzuki WA, Amaral DG (2003) Perirhinal and parahippocampal cortices of the macaque monkey: cytoarchitectonic and chemoarquitectonic organization. J Comp Neurol 463(1):67–91

Tanaka K (1996) Inferotemporal cortex and object vision. Annu Rev Neurosci 19:109–139

Toth K, Freund TF (1992) Calbindin D28k-containing nonpyramidal cells in the rat hippocampus: their immunoreactivity for GABA and projection to the medial septum. Neuroscience 49 (4):793–805

Tranel D, Brady DR, Van Hoesen GW, Damasio AR (1988) Parahippocampal projections to posterior auditory association cortex (area Tpt) in Old-World monkeys. Exp Brain Res 70 (2):406–416

Tunon T, Insausti R, Ferrer I, Sobreviela T, Soriano E (1992) Parvalbumin and calbindin D-28 K in the human entorhinal cortex. An immunohistochemical study. Brain Res 589(1):24–32

Van Hoesen G, Pandya DN (1975a) Some connections of the entorhinal (area 28) and perirhinal (area 35) cortices of the rhesus monkey. I. Temporal lobe afferents. Brain Res 95(1):1–24

Van Hoesen GW, Pandya DN (1975b) Some connections of the entorhinal (area 28) and perirhinal (area 35) cortices of the rhesus monkey. III. Efferent connections. Brain Res 95(1):39–59

Van Hoesen G, Pandya DN, Butters N (1975) Some connections of the entorhinal (area 28) and perirhinal (area 35) cortices of the rhesus monkey. II. Frontal lobe afferents. Brain Res 95 (1):25–38

Van Hoesen GW, Rosene DL, Mesulam MM (1979) Subicular input from temporal cortex in the rhesus monkey. Science 205(4406):608–610

Vogt BA, Pandya DN (1987) Cingulate cortex of the rhesus monkey: II. Cortical afferents. J Comp Neurol 262(2):271–289

Von Bonin G, Bailey P (1947) The neocortex of *Macaca mulatta*. University of Illinois Press, Urbana

Von Economo C (1929) The cytoarchitectonics of human cerebral cortex. Oxford University Press, London

Walker AE (1940) A cytoarchitectural study of the prefrontal area of the macaque monkey. J Comp Neurol 73:59–86

Witter MP, Amaral DG (1991) Entorhinal cortex of the monkey: V. Projections to the dentate gyrus, hippocampus, and subicular complex. J Comp Neurol 307(3):437–459

Witter MP, Wouterlood FG (2002) The parahippocampal Region. Organization and role in cognitive fuction. University Press Oxford, Oxford, pp. 3–19

Witter MP, Groenewegen HJ, Lopes Da Silva FH, Lohman AHM (1989a) Functional organization of the extrinsic and intrinsic circuitry of the parahippocampal region. Prog Neurobiol 33 (3):161–253

Witter MP, Van Hoesen GW, Amaral DG (1989b) Topographical organization of the entorhinal projection to the dentate gyrus of the monkey. J Neurosci Off J Soc Neurosci 9(1):216–228

Yeterian EH, Pandya DN (1985) Corticothalamic connections of the posterior parietal cortex in the rhesus monkey. J Comp Neurol 237(3):408–426

Human Hippocampal Theta Oscillations: Distinctive Features and Interspecies Commonalities

Joshua Jacobs, Bradley Lega, and Andrew J. Watrous

Abstract The hippocampus, along with its characteristic theta oscillation, has been widely implicated in various aspects of animal memory and behavior. Given the important roles that hippocampal theta oscillations have in theoretical models of brain function, it might be considered surprising that these signals have not been reported as often in humans as in animals. In this chapter we review recent research on hippocampal theta oscillations in humans, focusing on brain recordings from neurosurgical patients, which provide a key opportunity for observing hippocampal oscillations during cognition. The emerging theme of this body of work is that humans do indeed have hippocampal oscillations that are similar overall compared to the theta oscillation that is commonly found in rodents. Most notably, the human theta oscillation exhibits correlations with sensorimotor, navigation, and memory processing in the same general fashion as expected from rodents. However, some of the details of theta's relationship with behavior differ significantly compared to such signals in rodents—such as having a lower amplitude, frequency, and duration—which can make this signal less readily observable. Thus, theta oscillations are a key component of hippocampal processing in humans, but the patterns it exhibits compared to rodents point out distinctive aspects of human brain processes.

Introduction

In recent years there is growing evidence throughout neuroscience research that neuronal oscillations have an important role in how the brain supports behavior and cognition. Brain oscillations with critical functional roles are present in many brain regions, in both complex and simple behaviors (Buzsáki and Draguhn 2004), and

J. Jacobs (✉) • A.J. Watrous
Department of Biomedical Engineering, Columbia University, 351 Engineering Terrace, Mail Code 8904, 1210 Amsterdam Avenue, New York, NY 10027, USA
e-mail: joshua.jacobs@columbia.edu

B. Lega
Department of Neurosurgery, UT Southwestern, Dallas, TX, USA

© Springer International Publishing AG 2017
D.E. Hannula, M.C. Duff (eds.), *The Hippocampus from Cells to Systems*,
DOI 10.1007/978-3-319-50406-3_2

have even been found to evolve independently through multiple paths during evolution (Shein-Idelson et al. 2016). Understanding the roles of neuronal oscillations in brain processing is important on multiple levels. In any one neuron, the presence of membrane-potential oscillations can reveal when that cell is active (Klausberger et al. 2003). Across a neuronal network, the properties of oscillations reveal that rhythmic activations are an important part of neuronal computation (Gray et al. 1989; Fries 2005). At the largest scale, brain-wide patterns of neuronal oscillations illustrate which brain regions are active at any particular moment and how information moves between brain areas (Jensen et al. 2007). In addition, neuronal oscillations are useful as a practical research tool in humans because they can be recorded easily and because they demonstrate patterns related to high-level cognitive behaviors (Kahana 2006; Jacobs and Kahana 2010).

The hippocampal theta oscillation, in particular, is one brain oscillation that has an especially notable role. The hippocampus is a key brain region for high-level cognitive processing, including memory and spatial cognition (Burgess et al. 2002; Buzsáki and Moser 2013). Research has linked theta oscillations to virtually every aspect of hippocampal processing throughout these behaviors. Notably, theta oscillations in rodents, which generally occur at 4–10 Hz, relate to hippocampal processing at multiple levels, including the timing of neuronal spiking, the modulation of synaptic plasticity, the properties of neuronal computation within the hippocampus, and the hippocampus's interactions with other brain structures (Buzsáki 2002, 2005). Theta oscillations are also associated with many behavioral processes, including memory, spatial navigation, and sensorimotor processing. Owing to the diverse array of neural processes in which theta is involved, it seems that understanding theta's functional role is important for showing how the brain as a whole operates, especially including any behavioral or neural process that is linked to the hippocampus.

Historically, our understanding of hippocampal theta oscillations has been primarily derived from studies in rodents (e.g., Vanderwolf 1969). Of course, the general tactic of using research findings from rodents to make insights regarding biological processes in other species has traditionally been very successful and impactful. However, several key features of rodent theta oscillations have not been observed in humans (Jacobs 2014). Some papers even suggested that hippocampal theta oscillations may not exist in various species, including humans (Halgren et al. 1978), monkeys (Skaggs et al. 2007) and bats (Yartsev et al. 2011)!

As a result, there currently seems to be a large conceptual disconnect between research in rodents, which suggests that theta oscillations underlie most every aspect of hippocampal function (Buzsáki 2005), versus studies of brain recordings in humans and other species that do not find evidence of theta. In light of this uncertainty, it seems important to evaluate the literature regarding human theta oscillations in a comparative cross-species manner. The goal of this chapter is to provide this evaluation by describing our current understanding of human theta amidst the larger literature on theta oscillations in rodents and other simpler animals.

We organize our review of human data around the main hypothesized functions of theta that were identified in rodents. We also discuss data from non-human primates when possible. Our overarching conclusion is that humans and monkeys do have hippocampal theta oscillations and that these signals are largely analogous in function to the theta oscillations commonly observed in rodents. However, as we explain below, human hippocampal oscillations generally are slower in frequency compared with theta oscillations in rodents. This difference caused some previous studies to refer to them as "delta" oscillations. Nonetheless, here we continue to use our preferred term "theta" to refer to human hippocampal oscillations across the wider band of 1–10 Hz because it emphasizes their important functional similarities with rodent 4–10 Hz theta (Jacobs 2014).

As we explain, most of the key findings from theta in rodents are relevant for understanding human hippocampal neuronal activity and the role of neuronal oscillations. However, when extending the rodent literature towards humans it is important to keep in mind that human behavior is more diverse in these tasks compared to the behavior of rodents. Throughout the course of a given experiment, humans must process language information and deal with nonspatial distractions such as verbal interactions with experimenters or reading task instructions. Therefore, as we explore below, there are multiple potential causes of differences in theta oscillations between humans and rodents, including those rooted in behavior as well as physiology. Nonetheless, despite these differences in the detailed properties of theta across species—which are notable especially in frequency and amplitude—we feel overall that the all-important functional similarities outweigh the differences.

Recording Hippocampal Theta Oscillations in Humans

Until the last couple of decades, the vast majority of research on the electrical activity in the human brain recorded electroencephalographic (EEG) signals from the scalp, which primarily measured activity from the outer layers of the cortex (Niedermeyer and da Silva 2005). This research was generally of limited use for directly understanding activity in deep brain structures such as the hippocampus, although there is evidence of links between scalp-recorded midline theta and hippocampal theta (Mitchell et al. 2008). Unlike laboratory studies in animals, where scientists place electrodes directly in a region of their choosing, for obvious ethical reasons it is not straightforward to characterize electrical activity from deep brain structures in humans.

Neurosurgery provides a unique opportunity to record neuronal activity from deep human brain structures because clinicians already place electrodes in these regions for clinical purposes (Fig. 1). A particular type of neurosurgery that lends itself well to recording human brain activity is the monitoring of patients who have severe epilepsy. These patients have electrodes implanted directly in various brain structures as part of a clinical mapping procedure to characterize the brain regions

A   B   C

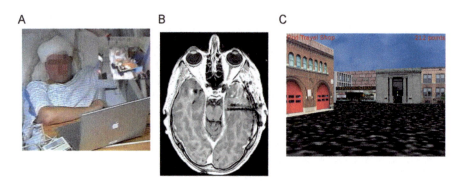

Fig. 1 Methods for examining human hippocampal brain signals. (**a**) Epilepsy patient with implanted electrodes performing a cognitive task on a bedside laptop computer. (**b**) Magnetic resonance image of a patient's brain showing depth electrodes targeting the hippocampus. (**c**) Screen image from a patient performing a virtual spatial navigation task. Images with permission from Jacobs and Kahana (2010); Jacob et al. (2012)

involved in seizures. The hippocampus is frequently targeted in this procedure because it plays an important role in seizure propagation (S.S. Spencer and Spencer 1994). The electrodes often remain in place for up to several weeks while patients remain in their hospital beds. Because patients are usually well functioning cognitively during these periods, it provides a unique opportunity to directly measure the human neural signals that underlie cognition (Jacobs and Kahana 2010). In time between clinical procedures, researchers can ask patients to perform cognitive tasks and the data recorded provides a direct view of the electrical activity underlying human behavior.

Several types of electrodes are commonly implanted in the brains of epilepsy patients who undergo this procedure: surface grid and strip electrodes, which record from the surface of the neocortex, and depth electrodes, which target deep structures including the hippocampus. Here our focus is on the depth electrodes that target the hippocampus. By incorporating the latest advances in high resolution MR scans, it is often possible to localize individual electrodes to small regions within the hippocampus, including small regions such as the subiculum and Cornu Ammonis areas (Ekstrom et al. 2009; Yushkevich et al. 2010).

Neurosurgical patients are generally limited to their hospital bed while electrodes are implanted. Therefore, any cognitive tasks must be performed from their bedside, generally on a laptop computer. Many behaviors can be examined using a laptop computer, with the use of computerized tasks that measure memory or perception of visually presented stimuli. However, some behaviors are more challenging to study in this setting, such as those that require movement or locomotion. Given this constraint, to study movement and navigation researchers developed video-game-like virtual-reality paradigms. In these paradigms patients are presented with a view of a virtual environment on a computer screen and use a keyboard or joystick to control movement. Although there are certainly important differences between the brain signals underlying real-world and virtual navigation (e.g., Taube et al. 2013; Minderer et al. 2016), this approach has proven useful because it has identified a number of key similarities between how the brain supports real and virtual navigation (Ekstrom et al. 2003; Jacobs et al. 2010a, 2013).

In addition to recordings from neurosurgical patients, magnetoencephalography (MEG) has emerged as an alternate approach for studying activity in deep human brain structures (Tesche 1997). By combining extracranial measurements of the brain's magnetic fields with advanced source localization methods based on beamforming, researchers have been able to characterize neuronal oscillations in deep brain structures (Quraan et al. 2011). In many cases this approach has replicated and extended findings of theta oscillations that appeared in direct brain recordings (Cornwell et al. 2008) where they were definitively localized to the hippocampus. This suggests that MEG is potentially viable for noninvasively probing hippocampal theta oscillations, although a key challenge is to establish definitively whether source-localized MEG signals are the same as directly measured hippocampal theta oscillations (Crespo-Garcí a et al. 2016).

Role of Theta in Sensorimotor Processing

Some of the first work identifying a functional role for hippocampal neuronal oscillations in any species focused on the fact that that these signals appeared prominently when animals processed sensory information. Green and Arduini (1954) provided seminal evidence in this area by showing that prominent oscillations at ~5–7 Hz appear in field potentials from the hippocampi of rabbits, cats, and monkeys when they perceive new sensory information. Notably, these signals were fairly "high level" such that they appeared in a similar form regardless of whether the animal received visual or auditory input. In this way, sensory-evoked hippocampal oscillations differed dramatically from sensory oscillations in neocortex, which are generally limited to a particular modality that is specific to the brain region where they are measured (Miller et al. 2007). A further feature that distinguished human hippocampal theta is that these signals were linked to arousal or attention, such that in drowsy animals sensory-evoked hippocampal theta was sometimes missing or diminished.

A largely separate line of research linked hippocampal oscillations to movement and motor planning. Vanderwolf (1969) found that oscillatory activity in the hippocampus often appeared before animals made movements such as walking, jumping, or rearing. In many cases these oscillations were sustained through the duration of these movements. These movement-related theta oscillations seem to be distinct from the sensory patterns mentioned above because they are driven by internal brain events rather than being responses to sensory cues.

It should be noted that even this early research on theta oscillations contained discrepancies regarding theta's sensorimotor role. Although both studies examined theta and sensory processing (Green and Arduini 1954; Vanderwolf 1969), only one of these studies actually found positive evidence for this pattern (Green and Arduini

1954). Later work confirmed that hippocampal theta has a role in perception, but nonetheless this type of discrepancy emphasizes the challenges of characterizing theta, foreshadowing issues that would be described subsequently (Buzsáki 2005).

The emerging theme from early animal research on hippocampal theta oscillations is that theta relates to sensory and motor processing in a complex manner that is linked to internal brain events and the relevance of external percepts. Owing to these distinctive high-level properties, researchers became interested in probing hippocampal theta in humans, in part because it seemed that this distinctive signal could be key for demonstrating the neural basis of more complex behaviors.

In humans, some of the earliest work on theta oscillations was performed by Halgren et al. (1978), who examined theta oscillations in a neurosurgical patient during verbal and motor tasks. In contrast to expectations of task-related theta elevations from studies in animals, this patient exhibited theta-band oscillations with *decreased* amplitude when processing new information in a memory task or when performing movements. This patient's hippocampus was indeed capable of exhibiting hippocampal theta oscillations because these signals were present in a baseline rest condition. Therefore, these results were interpreted as suggesting that theta's functional role may be fundamentally different in humans compared to animals (see also Brazier 1968).

Although this apparent difference in theta's properties between humans and animals was noteworthy, there was only limited follow-up work in this area for a number of years. In addition to the practical challenge of collecting human data from this deep brain structure, the follow-up experiments showed patterns of results that were harder to interpret than expected. One study examined hippocampal brain activity during different behaviors and found a divergent pattern of theta signals between behaviors (Meador et al. 1991). This study found differences in theta power between visual and auditory conditions, which was surprising because animal studies showed similar levels of theta activity irrespective of sensory modality. A different study examined MTL theta activity with MEG during a sensory task and also found lower than expected theta rhythmicity (Tesche 1997).

A surprisingly diverse pattern emerged from this early research, suggesting that human theta oscillations have a more complex behavioral role compared to analogous signals in rodents. It seemed that to understand human theta it would be important to use sophisticated analyses that measure behavioral correlates of theta precisely with regards to frequency and the timing of behavioral events, rather than coarser-grained approaches used earlier, which averaged signals across long intervals (Halgren et al. 1978; Tesche 1997). An example of this is an ahead-of-its-time study by Arnolds et al. (1980), who measured the properties of hippocampal theta oscillations in a temporally resolved fashion and found that the signal increased in amplitude and frequency during memory search as compared to reading and speaking.

In addition to behavioral variations in theta amplitude, later research showed that theta phase exhibits important relations with sensorimotor processing. One example of this is that in humans, the phase of hippocampal theta oscillations is linked to the onset of eye movements during visual search (Hoffman et al. 2013). Such a pattern is also evident in monkeys, as demonstrated by one study that examined links between hippocampal theta oscillations, sensory processing, and memory (Jutras et al. 2013). These two studies on hippocampal activity and eye movements therefore show an important new way that theta oscillations relate to sensory processing, by showing precise alterations (resets) in theta phase. These results are important because they suggest that hippocampal theta phase rather than amplitude is the critical functional variable for certain behaviors (see also Mormann et al. 2005; Lopour et al. 2013). In addition, establishing these behavior-related phase patterns could be important for understanding theta's role at the neuronal level, where oscillatory phase is a strong determinant of single-neuron activity (Jacobs et al. 2007). Across this work it is clear that theta has a rich link to behavior that includes rapid changes in amplitude and phase that are precisely coupled to behavioral events, which emphasizes the need for advanced analysis methods to characterize these patterns effectively.

Although studies of hippocampal theta oscillations in humans and animals usually show some type of significant correlation with sensorimotor behavior, there are key inconsistencies across this body of work even with the use of advanced analysis techniques. Therefore, going forward researchers took a richer approach to understanding theta's behavioral role by probing its relation to more complex behaviors. In particular, spatial navigation caught the attention of researchers because during navigation rodents exhibit hippocampal theta signals that are extremely robust and well studied (Buzsáki 2005). Given the strength of this signal and the broad literature on this topic, it seemed that spatial tasks could provide hope for characterizing theta's functional role more precisely in humans.

Role of Theta in Navigation and Path Integration

Spatial navigation is an important form of sensorimotor processing that is considered to be closely linked to hippocampal theta oscillations. This link between theta and spatial navigation builds off several findings. First, when a rodent runs through an environment it reliably elicits theta oscillations that have a large amplitude and that are reliably sustained for extended periods (Vanderwolf 1969; Winson 1978). Second, the hippocampus contains place cells that encode an animal's current spatial location (O'Keefe and Dostrovsky 1971) and whose firing is modulated by ongoing theta activity (O'Keefe and Recce 1993). Third, hippocampal theta activity is necessary for the activation of entorhinal "grid cells" in rodents (Bonnevie et al. 2013; but see Yartsev et al. 2011), which seem to play a role in allowing the animal to keep track of their location during movement (Steffenach et al. 2005; McNaughton et al. 2006).

Owing to the strength of these correlational links between theta and navigational behavior in rodents, researchers proposed various functional models to causally explain the role of theta oscillations in navigation. These models, which are in many cases not mutually exclusive, emphasize different aspects of theta's functional roles. Spatial navigation in complex environments inherently requires both sensory and motor processes—the planning of movements through an environment and the acquisition of sensory information to refine trajectories. To support the multimodal requirements of navigation, Bland and Oddie (2001) suggested that theta oscillations were important because they built dynamic functional links between the networks that support sensory and motor processing. According to this model, theta oscillations are important because they create a diverse sensorimotor neural network across a broad range of regions where neurons have distinct functional roles, including the brain stem, hippocampus, and cortical areas.

A different type of model focused on explaining theta's computational function, by suggesting that theta supports the detailed process of keeping track of the current spatial location by the entorhinal–hippocampal network. The oscillatory interference (OI) model explains that grid cells in the entorhinal cortex keep track of an animal's location during movement on the basis of differences in the phase of intracellular and extracellular theta oscillations (Burgess et al. 2007). This type of theory is innovative because it proposes for the first time that the theta oscillation is critically involved in neuronal computation for high-level behavioral information. Given the strong interest in this type of model (e.g., Zilli et al. 2009; Domnisoru et al. 2013; Schmidt-Hieber and Häusser 2013), it seems important to test the relevance of the OI model to humans. The OI model relies on several specific properties of theta oscillations that are robust in rodents, including a positive correlation between theta frequency and movement speed, and the stability of theta phase across time. Below we describe the features of human hippocampal theta oscillations in navigation, focusing on comparing these patterns with theta in rodents and the needs of the OI model.

Studies of human hippocampal oscillations during spatial navigation revealed an unexpectedly diverse pattern of results compared to findings from animals. Consistent with rodents, there is an overall tendency for theta-like oscillations in the human hippocampus during spatial navigation. This pattern was first suggested by Kahana et al. (1999), who showed the first evidence of human theta oscillations in a spatial navigation task by recording from cortex overlying the hippocampus. However, unlike navigation-related theta in rodents, these oscillations were transient and appeared in distinct episodes that generally lasted less than one second (Fig. 2). An additional finding from this work is that the amplitude and duration of theta episodes was greater in more challenging task conditions versus simpler ones (Caplan et al. 2001), which suggested a link between theta oscillations and task difficulty that was unexpected from rodent studies. Overall, this research was impactful because not only did it provide the first link between human theta oscillations and navigation, but it also suggested for the first time that in humans these signals relate to richer aspects of cognition and behavior.

Follow-up work in human navigation identified additional distinctive features of hippocampal theta. In rodents the link between movement and hippocampal theta is so robust that it can be observed easily by visual inspection of raw recordings, in which theta oscillations emerge strongly around the time of movement onset (Green and Arduini 1954). Instead, in humans the link between hippocampal theta and

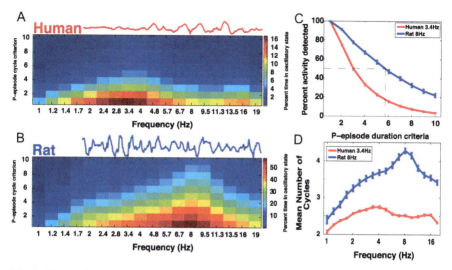

Fig. 2 Theta is less continuous and occurs at a lower frequency in humans compared to rodents during spatial navigation. (**a**) Example of human hippocampal raw LFP trace over 3 s (upper) and average proportion of time oscillatory activity was detected as a function of frequency and increasing P-episode duration criteria for 284 human hippocampal recordings (lower). (**b**) Similar to (**a**), but for 30 rodent recordings during a Barnes maze. Note the difference in color scale for humans and rodents. (**c**) Proportion of time oscillatory activity was detected at the human peak frequency of 3.4 Hz and the rodent peak frequency of 8 Hz. *Gray* dashed lines correspond to the half-maximum P-episode duration criteria at the human and rodent peak frequency. (**d**) Average number of cycles for each detected oscillatory event for humans and rodents. Error bars indicate the standard error of the mean across electrodes (Figure modified with permission from Watrous et al. 2013a)

movement is less consistent (although still very robust statistically). The first study to quantitatively characterize the prevalence of hippocampal oscillations to (virtual) movement found that although theta oscillations were more prominent during movement than stillness, they were only present in a fairly small proportion (~15 %) of movement trials (Ekstrom et al. 2005). Furthermore, this signal lacked specificity, as there were some moments when the patient was still but electrodes nonetheless exhibited robust theta oscillations. Follow-up work demonstrated that hippocampal theta power correlated with movement speed in humans, but that it also correlates with other more abstract task variables (Watrous et al. 2011). Overall, this work illustrated the challenge in comparing theta oscillations between humans and rodents. Clearly there is a correlation between human hippocampal theta and navigational behavior, but this relation is multifaceted and not as strong as in rodents. The fact that the link between movement and theta in humans is imperfect would seem to be an issue for OI models, which had proposed that a consistent theta phase pattern was vital for path integration. Therefore, one prediction of OI models is that people would become disoriented when theta power is low. This could be tested by comparing path integration performance with hippocampal theta power on a trial-by-trial basis.

Perhaps the most surprising difference between findings from human and rodent studies of theta oscillations in navigation concerns the frequency where these signals are present (Fig. 2). In humans hippocampal theta oscillations during navigation generally appear at the slower frequency of ~3 Hz. This frequency is significantly slower than ~8 Hz where these signals are present in rodents (Watrous et al. 2013a) as well as 4–8 Hz where oscillations often appear in human neocortex (Jacobs et al. 2007). Human hippocampal theta oscillations are also seemingly more variable in frequency compared to rodents, with some clear examples of theta-like activity at frequencies as slow as ~1 Hz during navigation (Jacobs et al. 2007; Rutishauser et al. 2010). One possibility is that this frequency difference is task-specific: Although most observations of human hippocampal theta oscillations occur during navigation at ~3 Hz (Watrous et al. 2013a; Jacobs 2014), there is evidence for theta in non-spatial tasks at ~7 Hz (Axmacher et al. 2010). It should be noted that a methodological issue when comparing human and rodents navigation-related brain signals is that humans are generally studied with virtual reality, in contrast to rodents who generally move through real-world environments. Research in rodents found that navigation in a virtual rather than real environment slows theta frequency by ~1 Hz (Ravassard et al. 2013). However, this difference is small in magnitude and thus does not explain fully the much slower theta frequency in humans. Nonetheless, the existence of this trend suggests that researchers should consider the possibility of changes in theta properties associated with the use of virtual reality.

An additional surprising set of findings emerging from this work is that human hippocampal theta oscillations related to other aspects of navigational behavior beyond movement. Ekstrom et al. (2005) found that hippocampal theta oscillations were more prevalent when a person searched for particular objects (see also Watrous et al. 2011). Other studies found similar findings regarding theta in neighboring brain regions (Caplan et al. 2003; Jacobs et al. 2010b). These findings parallel work in monkeys showing that hippocampal oscillations were specifically linked to visual perception (Jutras et al. 2013). This link between theta and visual search seems sensible in virtual-navigation paradigms where vision is the sole source of task input, in contrast to real-world navigation where humans receive proprioceptive input or where rodents use olfaction to inform navigation (Ekstrom 2015). Nonetheless, this pattern underscores the possibility that human theta is linked to other types of cognitive processing beyond the neural coding of location. This adds to the emerging view that the hippocampus, as well as theta, could have a broad, high-level role in cognition beyond spatial coding.

The two most prominent ideas concerning the functional role of theta oscillations during spatial navigation point to sensorimotor processing (Bland and Oddie 2001) and path integration (Burgess et al. 2007). It can be challenging to distinguish which of these processes are more closely coupled to navigation because vision and movement are usually active simultaneously in navigation (but see Markus et al. 1994). Using an innovative virtual-reality task, a recent study provided new data to elucidate this issue, by distinguishing whether theta was more closely tied to sensory processing or to path integration (Vass et al. 2016). Vass et al. (2016)

Fig. 3 Data from a hippocampal electrode that exhibited similar levels of theta activity during (virtual) navigation (*top*) and during teleportation (*bottom*). Figure from Vass et al. (2016)

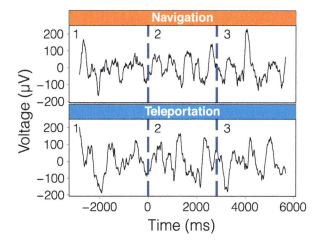

had patients perform a navigation task that included a condition where the screen went blank and the patient was "teleported" through an environment without receiving visual information. The teleportation condition provided key data to distinguish the role of theta by testing how this signal differed between spatial movements that were and were not accompanied by visual information. If theta oscillations were present during teleportation then it would indicate that these signals were coupled to the high-level behavioral function of path integration. Alternatively, if theta oscillations were absent during teleportation then it would indicate that these signals were linked to sensory processing.

The results from this study support the view that human hippocampal oscillations are involved in path integration. At most sites, the level of theta activity did not drop significantly during teleportation (Fig. 3). Thus, sensory input is not required to elicit hippocampal theta oscillations. Instead, internal cognitive events during navigation are indeed sufficient for activating theta. Together with the work described above, the view from this literature is that many types of behaviors are capable of activating human hippocampal theta (Watrous et al. 2011). In humans, the behaviors that elicit theta include all the signals identified in rodents, but also extend to a broader range that include some human-specific behaviors.

Roles of Theta in Memory Encoding

Building off seminal findings showing a link between hippocampal theta oscillations and synaptic plasticity (Huerta and Lisman 1993; Hölscher et al. 1997), a different line of research focused on the link between theta and memory. Following a series of groundbreaking experiments into human memory centered around the neurosurgical patient HM (who had undergone bilateral temporal lobe resections, including the hippocampi) the view emerged that humans utilize a mix of different memory *systems* (Tulving 1985). In particular, central figures in this work, including Endel Tulving, synthesized phenomenological properties of memory and experimental data to propose the existence of *episodic memory* ("mental time travel" that allows humans to re-experience past events within specific temporal units) and working memory (items immediately held in consciousness and used for decision making). This portion of our chapter will discuss the role of theta oscillations in these processes.

Some of the first and strongest behavioral evidence for the involvement of theta oscillations in memory encoding comes from classical conditioning experiments where some stimuli were deliberately presented during a "theta state," which was defined as a period of high-amplitude theta oscillations (Berry and Swain 1989; Griffin et al. 2004). This has proven to be a reproducible and robust effect in simpler animals. However, pre-stimulus hippocampal theta oscillations in humans have not yet been linked to this type of learning (Merkow et al. 2014). There is some evidence for pre-stimulus theta-frequency oscillations that predict successful encoding in neocortex (Addante et al. 2015) and oscillatory synchrony related to memory between hippocampal–neocortical connections (Haque et al. 2015; Guderian et al. 2009). Nonetheless, it remains to be demonstrated whether human behavior-related cortico-hippocampal synchrony is essentially the same phenomenon as the large-amplitude theta oscillations that are coupled to memory in rodents, or whether these behavioral patterns in humans instead involve more fine-grained changes in theta properties (Watrous et al. 2013b).

Episodic memory is thought to be the most hippocampally dependent memory process: it is most fragile in the face of degenerative diseases affecting the mesial temporal structures, it is specifically degraded in lesions such as those of HM, and it is cognitively demanding (Tulving 2002). Given this link, oscillatory processes during human episodic memory might be expected to most strongly relate to the theta signals seen in rodents, which appear whenever the hippocampus is active (Buzsáki 2002).

A common paradigm for testing episodic memory is the Free Recall task. Here participants are shown a list of stimuli (such as words) and, after a delay, are asked to speak aloud as many items as they can remember. If rodent behavior is analogous to episodic memory, successful encoding for items in this task should be associated with increased amplitude of theta oscillations. An early study examining human hippocampal activity during free recall observed robust *gamma*-band power increases during memory encoding without a memory-related power increase in the theta band (Sederberg et al. 2007). In fact, a large fraction of electrodes in the hippocampus exhibit a theta power *decrease* during successful item encoding. This was not an isolated finding. Across many studies encompassing hundreds of human subjects, a substantial fraction of hippocampal sites exhibited rodent-type oscillatory power *increases* in the theta band, but this is generally outweighed by a more widespread oscillatory power decrease for all frequencies below 24 Hz (Sederberg et al. 2003, 2007; Lega et al. 2012). Further, the electrodes that do exhibit oscillatory power increases do so at frequencies lower than that observed in rodents, centered at 3 Hz or even slower, as compared to the 4–10-Hz frequency of rodent

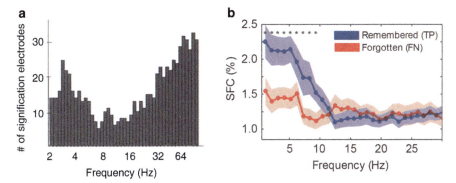

Fig. 4 Theta oscillations and episodic memory encoding. (**a**) Data from Lega et al. (2012) showing the count of electrodes where increased oscillatory power correlates with successful memory encoding. Note the peak in the theta band. (**a**) Analysis of coupling between hippocampal spiking and neuronal oscillations from Rutishauser et al. (2010). Spike-field coherence (SFC) reveals the temporal coordination between spike timing and oscillatory phase, separately computed for trials of good or bad memory. Figures obtained with permission

theta (Lega et al. 2012). Overall, a number of hippocampal sites do exhibit memory-related theta signals (Fig. 4a), but they are lower in number than expected.

A strong link between memory and navigation vis-a-vis theta oscillations was articulated by Buzsáki (2005) and later elaborated by Buzsáki and Moser (2013). This theory suggests that there is a direct analogy between item–place associations formed in traversing a two dimensional path in space and the resultant item–*time* and item–place associations that are the basic units of episodic memory. The prediction of this theory is that human oscillatory activity during episodic memory should demonstrate the same neurophysiological patterns as rodent spatial memory. The view that the hippocampal theta oscillations are important for remembering spatial locations was first supported by a study in rodents that compared memory performance on a task where the animals were taught to run to a particular spatial goal location (Winson 1978). One set of animals in this task had their theta oscillations disrupted with electrolytic lesions in their medial septum. Compared to controls, animals with disrupted theta were impaired in navigation accuracy. This was the first evidence for theta in remembering spatial locations.

In recent years similar evidence for hippocampal theta in spatial memory has emerged in humans. Cornwell et al. (2008) examined theta activity in patients performing a spatial learning task while they underwent MEG scanning. Here the amplitude of theta activity correlated positively with spatial memory performance (Fig. 5). After processing with a spatial filtering method, the region that exhibited this pattern appeared to include the posterior hippocampus. A subsequent study by Kaplan et al. (2012) supported this finding, and also showed that theta activity in human spatial memory perhaps spanned a broader set of cortical brain areas rather than being limited to the hippocampus. Given the potential challenge of using noninvasive MEG data to measure activity in deep brain structures, a key question from this work is how closely the identified theta signal was specifically localized to

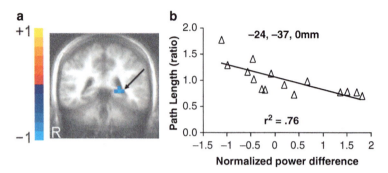

Fig. 5 Theta oscillations and spatial memory. Data from Cornwell et al. (2008) showing theta signals related to spatial memory accuracy. (**a**) Spatial localization of memory-related theta signals. (**b**) Relation between excess path length in a spatial memory task and theta power. Figures obtained with permission

the hippocampus. Because spatial localization is challenging with MEG data, one possibility is that the memory-related theta in these studies was not specifically localized to the hippocampus and was instead present in surrounding cortex. To bolster the link between hippocampal theta and memory performance it will be helpful for a study with direct hippocampal recordings to examine the link between theta and spatial memory. Nonetheless, the pattern of these results bolsters the view that theta oscillations are involved in spatial cognition in a general manner that is linked to memory, extending beyond pure path integration or sensory processing.

Aside from power changes, there are several other memory-related patterns of theta oscillations in the human hippocampus. One example is phase–amplitude coupling (PAC). PAC is a phenomenon where high-frequency oscillations increase in amplitude at particular low-frequency (theta) phases (Bragin et al. 1995; Canolty et al. 2006). PAC related to memory and theta oscillations appears in numerous human cortical and hippocampal structures (Canolty and Knight 2010; Kendrick et al. 2011; Lega et al. 2015). Phase resetting is another property of theta oscillations that has been linked to memory encoding in rodents (McCartney et al. 2004). Here, memory encoding is enhanced when theta oscillations are organized to reset to a specific phase after a stimulus is observed. Memory-related theta phase resetting has been observed in human hippocampus and neocortex (Mormann et al. 2005; Rizzuto et al. 2003), as well as in monkeys (Jutras et al. 2013).

Further evidence for the contribution of theta oscillations to memory processing in humans comes from hippocampal recordings that include measures of single-neuron spiking. Rutishauser et al. (2010) found that hippocampal neurons that fire at preferred phases of the 3–8 Hz theta oscillation exhibited increased firing when participants encoded memory items that were subsequently remembered (Fig. 4b). This suggests that theta oscillations are important for memory and efficient hippo-campal processing because they help to temporally coordinate neuronal activity (Jacobs et al. 2007).

Data linking human hippocampal oscillations in the 4–8-Hz frequency range to memory is probably most robust in "working memory" behaviors. In rodents, theta oscillations have been observed linking hippocampal and frontal lobe activity (Jones and Wilson 2005; Hyman et al. 2011). Increases in theta amplitude and theta–gamma coupling in the human hippocampus have been observed in the 4–8-Hz range during working memory tasks (Mormann et al. 2005; Axmacher et al. 2010). This work on hippocampal theta oscillations related to working memory is intriguing because, in fact, some researchers suggested that working memory occurs entirely independently of the hippocampus (Jeneson and Squire 2012). Thus, going forward, hippocampal theta oscillations related to working memory must be examined further to test whether these signals are causally important or if they are epiphenomena.

Human hippocampal EEG have been recorded for decades now in patients performing cognitive tasks (Jacobs and Kahana 2010). In spite of the numerous experiments looking for evidence of robust theta activation in humans, the properties of human hippocampal theta for episodic, spatial, and working memory are not fully understood because they are sometimes weaker than expected and not fully consistent with animal models. Although many hippocampal sites exhibit oscillations that increase in power for successful memory encoding, this pattern is not present at all sites. Furthermore, it remains unclear if memory-related hippocampal oscillations are most closely coupled to oscillations at ~3 Hz (Lega et al. 2012) or 7 Hz (Rutishauser et al. 2010; Axmacher et al. 2010). A further question is which aspects of theta oscillations are most important for memory coding: oscillatory power at a single site, interregion phase synchrony, or PAC. It will be important to examine these issues going forward with multiple types of memory tasks, as well as advanced statistical methods that can dissociate the contributions of these signals towards behavior.

Theta Dynamically Coordinates Within and Across Brain Regions

Although research in rodents primarily focused on characterizing the amplitude of theta in relation to behavior, there is key evidence that theta phase rather than power has a strong role in modulating local neuronal activity. In both humans and rodents the activity level of individual neurons is modulated by the phase of ongoing theta oscillations (Jacobs et al. 2007; Patel et al. 2012; Sirota et al. 2008; Rutishauser et al. 2010). To the extent that oscillations such as theta are simultaneously present in multiple brain regions, examining the timing of these signals could reveal how neurons in these regions interact on a rapid timescale to support cognition.

Rather than being a unitary signal that is always coherent across a fixed spatial topography, research has shown multiple theta sources in the hippocampus and MTL that dynamically coordinate with aspects of behaviors (Montgomery et al.

2009; Watrous et al. 2011). Based on empirical findings in rodents, we suggest that the phase of hippocampal theta is a rich signal that coordinates both local and distributed cell assemblies, dynamically linking hippocampus and neocortical regions to support wide-ranging types of cognition. These findings suggest that the hippocampus is part of a network that consists of both large- and small-scale theta signals that becomes dynamically synchronized to support rich behaviors. Consistent with this, rodent slice data provided evidence of theta generators that can operate independently of other brain regions (Goutagny et al. 2009), as well as similar data from the human medial temporal lobe (Mormann et al. 2008).

In line with these ideas, in rodents the phase of hippocampal theta modulates neuronal activity in both parietal and prefrontal cortices (Sirota et al. 2008; Jones and Wilson 2005; Siapas et al. 2005; Hyman et al. 2005; Benchenane et al. 2010). These observations are consistent with the theoretical role that low-frequency oscillations have in coordinating neural activity across larger spatial scales (von Stein and Sarnthein 2000; Fries 2005; Womelsdorf et al. 2007; Fell and Axmacher 2011). Consistent with this animal electrophysiological work, a rich body of human cognitive studies hypothesized that the hippocampus and neocortex are coordinated to support memory and navigation (Teyler and DiScenna 1986; Miller 1991; McClelland et al. 1995; Buzsáki 1996; Nadel and Moscovitch 1997; Eichenbaum 2000; Norman and O'Reilly 2003). These cognitive studies present a diverse landscape of information processing across the hippocampus and surrounding regions that may be coordinated by theta.

Extending these ideas to humans, a few studies in epilepsy patients show phase synchronization between the hippocampus and extra-hippocampal locations. Recording from surface electrodes overlying the MTL, Foster et al. (2013) showed that memory retrieval was associated with MTL–retrosplenial theta-band connectivity, followed by increased gamma-band activity in the retrosplenial cortex, putatively reflecting increases in firing rate. Another study by Anderson et al. (2009) also showed enhanced theta synchronization between parahippocampal and frontal electrodes during free recall. Watrous et al. (2013b) observed increased theta phase synchronization between frontal, parietal, and parahippocampal cortices during correct memory retrieval of both spatial and temporal information. Interestingly, in this latter study, whereas retrieval of spatial information led to synchronization at ~2 Hz, retrieval of temporal information led to synchronization at ~8 Hz. The PHG showed the strongest pattern of connectivity in both cases. Taken together, these three studies demonstrate theta oscillations in the human medial temporal lobe that coordinate with other cortical sites in a task-dependent manner.

The observation of frequency-specific information retrieval (Watrous et al. 2013b) is broadly consistent with the notion of frequency multiplexing. This emerging idea, which has gained increasing empirical support in animal experiments (Belitski et al. 2010; Cabral et al. 2014; Colgin 2016; Panzeri et al. 2010; Bieri et al. 2014; Zheng et al. 2016) suggests that distinct, frequency-specific oscillatory brain states subserve distinct behaviors (Donner and Siegel 2011; Siegel et al. 2012; Colgin 2016). Such a multiplexed representational scheme could allow multiple processing streams (Giraud and Poeppel 2012; Watrous et al. 2013b;

Benchenane et al. 2011) to process information with enhanced capacity (Akam and Kullmann 2014; Panzeri et al. 2010). Moreover, during rest the direction of information flow between brain areas in humans also shows frequency specificity (Hillebrand et al. 2016). Taken together, these studies highlight the possibility that hippocampal coupling at varying slow frequencies may subserve distinct behaviors (A.D. Ekstrom and Watrous 2014), including memory but also other domains (Schyns et al. 2011; Gross et al. 2013; Daitch et al. 2013; Freudenburg et al. 2014).

Role of Theta Oscillations in Neuronal Coding

In addition to having properties correlated with high-level behaviors, theta oscillations in rodents have also received substantial attention because they are involved in the representation of information by single neurons. The strongest example of this is the phenomonenon of phase precession in rodents (O'Keefe and Recce 1993). Here, hippocampal neurons systematically vary the theta phase of their spiking such that when an animal runs through a place field, the cell spikes at earlier phases over successive theta cycles. This phenomenon is viewed as an example of phase coding because individual neurons represent behavioral information by varying the timing of their spiking relative to the peaks and troughs of ongoing oscillations. In many cases the information that is signaled by spike phase is significantly more informative than information conveyed by the cell's overall firing rate (Jensen and Lisman 2000; Huxter et al. 2003). From rodent models, theta phase coding has widely been considered to be a key feature of hippocampal coding, not only because phase-based coding is an efficient manner for representing information but also because theta-phase coding causes neurons to activate on a timescale consistent with spike-timing-dependent plasticity (Skaggs et al. 1996; Mehta et al. 2002). An additional feature of this phenomenon is that spike phase is informative about multiple aspects of the animal's behavior, including the location of the animal relative to the center of a neuron's place field, as well as whether a cell's place field was traversed in the past or will be encountered in the future (Lubenov and Siapas 2009).

Although theta phase coding has not been identified directly in humans, some studies in humans provide evidence for related phenomena. Jacobs et al. (2007) examined simultaneous recording of spiking and oscillations from the hippocampus and entorhinal cortex of epilepsy patients with implanted microelectrodes. Across the brain the firing of many neurons was modulated by the ongoing phase of neuronal oscillations. For oscillations at most frequencies individual neurons reliably spiked at the trough of the oscillation. However, in the hippocampus at 1–4 Hz, individual neurons fired at a range of phases. The phase diversity of this theta pattern suggested that individual human hippocampal neurons robustly use theta phase as a mechanism for information coding. Going forward it will be useful to more precisely compare this pattern with phase precession in rodents by testing whether individual cells vary the theta phase of their spiking on a cycle-by-cycle basis as in place cells that use phase precession to code location.

The importance of theta oscillations for neuronal coding is bolstered by the new finding that human hippocampal oscillations are travelling waves. In rodents theta oscillations are not a static signal that exists identically across the whole hippocampus, but instead they move gradually along the long axis of the hippocampus (Lubenov and Siapas 2009; Patel et al. 2012). Because individual hippocampal neurons use theta phase as a mechanism for encoding the timing of behavioral events (Lubenov and Siapas 2009), these travelling waves have been interpreted as allowing sections of the hippocampus to simultaneously represent different times. In this way, travelling theta waves can be thought of as encoding time across the hippocampus, in much the same way that separate parts of the world exist in different time zones. A recent study showed that hippocampal theta oscillations in humans are also travelling waves (Zhang and Jacobs 2015). This study found that theta oscillations move across the hippocampus in a posterior-to-anterior direction, which is analogous to the dorsal-to-ventral movement in rodents (Fig. 6). However, this work went further and compared the properties of hippocampal travelling

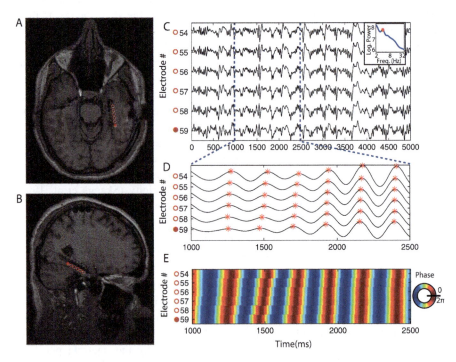

Fig. 6 Theta oscillations in the human hippocampus are travelling waves. Example data from one patient. Figure from Zhang and Jacobs (2015), with permission. (**a** and **b**) Images depicting electrode locations. (**c**) Raw recordings from several hippocampal contacts. Inset shows mean power spectrum with peak in the theta band. *Circles* on left side of plot denote electrode correspondence with panels (**a** and **b**). (**d**) Theta-filtered signal from these contacts. *Red asterisks* denote oscillatory peaks. Note the progressively positive phase shift as one moves from posterior to anterior sites, which is the key indication of a travelling wave. (**e**) Phase analysis of the theta signals at these electrodes. Color denotes cosine of theta phase

waves between theta oscillations at different frequencies, providing information on how these waves propagate.

Individual theta waves move through the hippocampus at a rate that correlates with the oscillation's temporal frequency. This link indicates that theta oscillations travel along the hippocampus by following the principles of coupled oscillator models (Ermentrout and Kleinfeld 2001). Following these models, phase differences are continually preserved as the oscillation moves across space in the hippocampus, regardless of the oscillation's temporal frequency. This distinctive phase pattern suggests that it may be possible for hippocampal neurons in humans to participate in theta phase-based coding despite the different properties of our theta oscillations.

Theta Oscillations and Internal Cognitive Events

The research reviewed above focused on the relation between theta and clearly defined external events such as sensorimotor, memory, and navigational processing. As one probes further into the literature on human hippocampal theta, it seems there is also evidence that theta oscillations relate to rich aspects of cognition that have only a weak or even no outwardly visible behavioral correlate. This suggests the intriguing possibility that theta oscillations could be a way to study internal brain events.

Some of the first evidence that theta indexed internal cognitive events was from early work on human hippocampal theta oscillations, which showed that theta amplitude not only increased when patients listened to auditory stimuli, but also in a section of a task when patients closed their eyes and rested (Meador et al. 1991). This finding was interpreted to suggest that theta oscillations in humans could be important for internal thought dynamics. One possibility is that subjects were attending to aspects of their own thoughts during this period (e.g., "day dreaming"). Additional evidence that theta oscillations related to internal thought processes was provided by Caplan et al. (2001), who studied brain signals underlying maze learning. Here the amplitude and duration of medial-temporal theta oscillations increased on task trials that were generally more difficult. Interestingly, these signals seemed to be a high-level neural correlate of cognitive processing because they only correlated to the trial's overall high-level difficulty and could not be estimated from the patient's instantaneous reaction time to each stimulus.

Other evidence for the involvement of theta oscillations in internal brain dynamics comes from patients performing self-initiated memory recall tasks. Foster et al. (2013) showed that prior to the moment of peak neural activity there was a transient burst of 1–3-Hz oscillations that synchronized the medial-temporal lobe and retrosplenial cortex. Because this theta burst occurred *before* the primary neuronal activation, it indicates that hippocampal theta oscillations mark the success of the memory search process by demonstrating what seems to be the first indication that a new memory is retrieved. It should be noted that hippocampal theta oscillations have been observed not only during wakefulness but also during sleep, in both rodents (Montgomery et al. 2008) and humans (Bódizs et al. 2001). Thus one speculative possibility is that hippocampal theta oscillations underlie a general set of cognitive processes related to bringing new thoughts into mind, including both memory retrieval during wakefulness as well as dreaming during sleep.

A final piece of evidence for the potential role of theta oscillations in internal cognitive events comes from work in rodents that found high-amplitude hippocampal theta oscillations in tasks where animals attentively wait for fixed periods of time (Pastalkova et al. 2008; MacDonald et al. 2013). Rodents in timed waiting tasks initiate their own movements after delays rather than by responding to external events. This suggests that hippocampal theta oscillations could play a role in representing sequences of internal cognitive states that sequentially represent different time periods in the absence of external stimuli. In the same way that theta oscillations support the hippocampal representation of separate spatial locations during navigation, this same signal could represent different moments in time (Eichenbaum 2014) to support both fixed-interval timing tasks as well as the temporal associations between events that occur at different time periods (Howard et al. 2005).

Conclusion

The literature on human hippocampal theta oscillations is complex, containing a combination of similarities and differences compared to similar signals in rodents. Whereas there is evidence that the strongest features of theta from rodents are preserved in humans—increases in theta amplitude during sensorimotor and memory tasks—some of these effects are smaller in magnitude (although still robust statistically). These types of differences make it challenging to distinguish which features of theta from rodent studies are relevant in humans.

Whereas we have a good grasp of human behavior in many of the memory and cognitive tasks where hippocampal theta appears, we currently have only a fairly coarse understanding of the core electrophysiological properties of human theta, including its spatial, temporal, and spectral properties. Perhaps the biggest mystery concerning theta in the human literature is that whereas the amplitude of these oscillations positively correlates with navigational behavior (Ekstrom et al. 2005), this link is weaker compared to the tight coupling between theta and movement found in rodents (Vanderwolf 1969). One possibility is that human hippocampal theta is spatially diverse, with signals in a subset of the hippocampus that behave as expected from animal studies and with oscillations in other subregions that behave differently, perhaps relating to non-spatial behaviors. Supporting this idea, there is evidence from rodent studies that hippocampal theta oscillations consist of many separable components, including both Type 1 and Type 2 theta at different frequencies (Kramis et al. 1975), as well as spatially distinct theta generators across hippocampal subfields (Sirota et al. 2008). To identify multiple theta sources it will be useful to measure the spatial extent of theta in detail using high-resolution methods for mapping hippocampal substructures (Ekstrom et al. 2009; Yushkevich et al. 2010). Research in rodents suggested that dorsal and ventral theta oscillations behave very differently (Sabolek et al. 2009; Royer et al. 2010). This result could be relevant to understanding human theta given that the primate posterior and anterior hippocampal areas are analogous structurally to the dorsal and ventral hippocampi of rodents (Strange et al. 2014). Thus, to test this pattern it would be useful to compare the properties of theta more fully along the anterior–posterior axis of the human hippocampus, expanding on an approach used recently (Mormann et al. 2008; Zhang and Jacobs 2015).

A distinctive feature of human hippocampal theta oscillations is that these signals exist in short episodes (Kahana et al. 1999; Watrous et al. 2013a). In this way human theta seems to be qualitatively different compared to the long-lasting and consistent theta oscillations in rodents (Vanderwolf 1969). A possibility is that the shorter, transient nature of human theta oscillations is the result of human multitasking, such that humans periodically switch the target of their attention, in contrast to rodents that stay focused on a single task (usually spatial processing). Theta oscillations exhibit phase resets when a person processes new attended events (Rizzuto et al. 2003; Mormann et al. 2005). These disruptions could cause human theta oscillations to have an apparent shorter duration because each phase reset caused the initiation of a new theta episode. It seems likely that epilepsy patients performing a rich behavioral task in a virtual setting on a laptop computer would have more distractions and interruptions compared to a rodent performing a simpler navigation task and who has weaker vision. If each distraction caused a theta phase reset, then it could cause short, disrupted theta oscillations. To characterize this potential explanation for human theta's short duration, it would be helpful for human behavioral studies to use video- and eye-tracking (Hoffman et al. 2013) to monitor shifts in attention and multitasking and test for correlations with theta pauses or resets.

One reason that the hippocampal theta oscillation is so strongly visible in rodents seems to be that this signal is expressed synchronously across a large number of neurons that have aligned dipoles (Buzsáki 2002). If the intracellular oscillations at each neuron were not perfectly synchronized, then the theta oscillation in the aggregate extracellular field potential would be smaller (Mitzdorf 1985). One recent study in rodents showed that hippocampal theta oscillations involve complex spatial patterns of phase shifts that encode cognitively relevant information (Agarwal et al. 2014). Thus, one possibility towards explaining theta's complex structure in humans is that human behavior, with its heavy information content, requires a more diverse set of electrophysiological processes to organize incoming information into neural assemblies. The human hippocampus receives projections from a broader range of diverse brain areas compared to rodents (Strange et al. 2014). As a result, individual hippocampal neurons could be involved in coding a very broad range of cognitive variables, and this may require their theta signals to express a larger range of spatial phase patterns. According to this idea, the smaller apparent amplitude of theta oscillations in extracellular recordings could be the result of these intracellular sources expressing various oscillatory phases at each moment. To test this idea it will be necessary to analyze the coding properties of human hippocampal theta with high resolution multi-contact recordings.

Future Directions

A key challenge going forward is to understand how the theta oscillation underlies such a broad range of behaviors in humans. One possibility is that the theta oscillation represents a single general computational and electrophysiological process. Alternatively, theta may not be a single phenomenon and, instead, could be the result from many distinct hippocampal electrophysiological and computational processes.

To address these issues as hippocampal theta research in humans moves forward, it seems research in this area will fall into two categories: (1) Interspecies comparisons of the properties of hippocampal theta oscillations between humans and rodents and (2) Characterizing theta in humans performing advanced behaviors where comparable animal data are rare. One way of performing work in the first category would be to measure the physiological characteristics of human theta oscillations in a relatively simple task with similar demands as rodent navigation. Although it is impossible to perfectly match human and rodent behavior, by comparing human and rodent brain activity in similar tasks it will provide useful data for distinguishing whether apparent interspecies differences in theta stem from behavioral or physiological differences. Then, it will be important to characterize the electrophysiology of these datasets in precise spatial, temporal, and spectral detail to reveal whether human and animal theta signals have similar properties. It will also be useful to use single-neuron recordings to compare across species the nature of theta oscillations' use for phase coding.

Equally important to comparing rodent and human theta is to understand features of theta oscillations that underlie complex behaviors that are seemingly specific to humans. One example of this is episodic memory retrieval, which exhibits some characteristics that are human-specific (Tulving 1985). A growing number of researchers suggested that the hippocampus performs a general computational process that underlies both memory and navigation (Eichenbaum 1999; Howard et al. 2005), so testing whether human theta's properties differ between these behaviors is an important step. Future work on episodic memory and other types of cognition should test the degree to which the properties of theta in these behaviors is consistent with one core set of principles of theta oscillations that underlie both navigation and memory (Buzsáki 2005). This can be done by comparing the phase, frequency, amplitude, and spatial topography of hippocampal theta oscillations in periods before and during the retrieval of episodic (autobiographical) memory episodes and comparing the results with analogous signals in navigation.

Studying theta oscillations in humans provides several new avenues for research tasks that could not be performed in rodents. The teleportation study by Vass et al. (2016) provides an example of a particularly powerful approach for assessing the fundamental functional nature of human theta oscillations. This study took advantage of humans' ability to understand a complex type of behavior—teleportation— that might not be intuitive to animals. Vass et al. study used this manipulation to dissociate whether theta was linked to high-level (path integration) or low-level (sensory) processing. Future experiments might build off this approach by using richer, human-specific experimental manipulations to identify the specific aspects of behavior that contribute most directly to theta, such as dissociating between theta contributions from motor, sensory, or internal brain processes. A different possibility for probing theta is to take advantage of humans' language and introspective abilities. One could envision an experiment whereby patients are asked to verbally describe their brain states during periods when different types of theta oscillations were evident. This approach, if applied in a careful and rigorous fashion, could provide unique data to distinguish unexpected variations in human theta signals.

An aspect of hippocampal theta oscillations that bears further investigation is the existence of inter-hemispheric differences. For some time now researchers have hypothesized that the right and left hippocampi support spatial and language processing, respectively (O'Keefe and Nadel 1978). While this view has garnered some empirical support (Spiers et al. 2001; Iglói et al. 2010), it is unclear how such specialization might have arisen evolutionary, given the conservation of these structures' other functions. A further challenge is that some work appears to show situations where this lateralization is not preserved entirely (Lacruz et al. 2010) as well as suggesting that there are situations where one hippocampus compensates for the other (Spencer and Huh 2008; Baxendale 2008). Nonetheless, this hemisphere-specific processing idea is important theoretically because it suggested that the two hippocampi could operate independently, at least in certain situations. The theta oscillation would seem to be a potential mechanism for testing this notion with a higher level of detail, if it was possible to compare data between the two hippocampi of one patient who performed both spatial and non-spatial tasks. An existing study in this area was inconclusive (Jacobs et al. 2010b) but it seems possible that an improved electrophysiological dataset that included both spatial and nonspatial tasks could shed key light on this laterality hypothesis.

Some of the strongest evidence for the importance of theta oscillations comes from animal studies showing that theta has a *causal* functional role (Seager et al. 2002). If hippocampal theta oscillations are shown to have this type of direct mechanistic role in cognition then it could lay the groundwork for key translational work to improve human behavior by modulating theta oscillations, perhaps with neural manipulation methods based on optogenetic (Sohal et al. 2009; Pastoll et al. 2013) or electrical stimulation (Wetzel et al. 1977; Kim et al. 2016). Researchers and engineers have tried many approaches to create devices to improve human cognition. It seems challenging to meaningfully alter human behavior by manipulating individual neurons, due to the large number of cells in the brain. However, owing to the large-scale nature of the human theta oscillation, this network signal may indeed have the potential for being modified with external methods to achieve meaningful behavioral outputs. Large-scale network manipulations, including theta alterations, may be one of the best short-term hopes we have for achieving human cognitive enhancement (Kim et al. 2016).

Acknowledgements This work was supported by NIH grants MH104606 & MH61975.

Competing Financial Interests The authors have no competing financial interests.

References

Addante RJ, de Chastelaine M, Rugg MD (2015) Pre-stimulus neural activity predicts successful encoding of inter-item associations. NeuroImage 105:21–31

Agarwal G, Stevenson IH, Berényi A, Mizuseki K, Buzsáki G, Sommer FT (2014) Spatially distributed local fields in the hippocampus encode rat position. Science 344(6184):626–630

Akam T, Kullmann DM (2014) Oscillatory multiplexing of population codes for selective communication in the mammalian brain. Nat Rev Neurosci 15(2):111–122

Anderson KL, Rajagovindan R, Ghacibeh GA, Meador KJ, Ding M (2009) Theta oscillations mediate interaction between prefrontal cortex and medial temporal lobe in human memory. Cereb Cortex 20(7):1604–1612

Arnolds DEAT, Lopes Da Silva FH, Aitink JW, Kamp A, Boeijinga P (1980) The spectral properties of hippocampal EEG related to behaviour in man. Electroencephalogr Clin Neurophysiol 50:324–328

Axmacher N, Henseler MM, Jensen O, Weinreich I, Elger CE, Fell J (2010, February) Cross-frequency coupling supports multi-item working memory in the human hippocampus. Proc Natl Acad Sci U S A 107(7):3228–3233

Baxendale S (2008) The impact of epilepsy surgery on cognition and behavior. Epilepsy Behav 12(4):592–599

Belitski A, Panzeri S, Magri C, Logothetis NK, Kayser C (2010) Sensory information in local field potentials and spikes from visual and auditory cortices: time scales and frequency bands. J Comput Neurosci 29(3):533–545

Benchenane K, Peyrache A, Khamassi M, Tierney PL, Gioanni Y, Battaglia FP, Wiener SI (2010) Coherent theta oscillations and reorganization of spike timing in the hippocampal-prefrontal network upon learning. Neuron 66(6):921–936

Benchenane K, Tiesinga PH, Battaglia FP (2011) Oscillations in the prefrontal cortex: a gateway to memory and attention. Curr Opin Neurobiol 21(3):475–485

Berry SD, Swain RA (1989) Water deprivation optimizes hippocampal activity and facilitates nictitating membrane conditioning. Behav Neurosci 103(1):71

Bieri KW, Bobbitt KN, Colgin LL (2014) Slow and fast gamma rhythms coordinate different spatial coding modes in hippocampal place cells. Neuron 82(3):670–681

Bland BH, Oddie SD (2001) Theta band oscillation and synchrony in the hippocampal formation and associated structures: the case for its role in sensorimotor integration. Behav Brain Res 127(1):119–136

Bódizs R, Kántor S, Szabó G, Szűcs A, Erőss L, Halász P (2001) Rhythmic hippocampal slow oscillation characterizes REM sleep in humans. Hippocampus 11(6):747–753

Bonnevie T, Dunn B, Fyhn M, Hafting T, Derdikman D, Kubie JL et al (2013) Grid cells require excitatory drive from the hippocampus. Nat Neurosci 16(3):309–317

Bragin A, Jando G, Nadasdy Z, Hetke J, Wise K, Buzsáki G (1995) Gamma (40–100 Hz) oscillation in the hippocampus of the behaving rat. J Neurosci 15:47–60

Brazier M (1968) Studies of the EEG activity of limbic structures in man. Electroencephalogr Clin Neurophysiol 25(4):309

Burgess N, Barry C, O'Keefe J, London U (2007) An oscillatory interference model of grid cell firing. Hippocampus 17(9):801–812

Burgess N, Maguire E, O'Keefe J (2002) The human hippocampus and spatial and episodic memory. Neuron 35:625–641

Buzsáki G (1996) The hippocampo-neocortical dialogue. Cereb Cortex 6:81–92

Buzsáki G (2002) Theta oscillations in the hippocampus. Neuron 33(3):325–340

Buzsáki G (2005) Theta rhythm of navigation: link between path integration and landmark navigation, episodic and semantic memory. Hippocampus 15:827–840

Buzsáki G, Draguhn A (2004) Neuronal oscillations in cortical networks. Science 304(5679): 1926–1929. doi:10.1126/science.1099745

Buzsáki G, Moser E (2013) Memory, navigation and theta rhythm in the hippocampal-entorhinal system. Nat Neurosci 16(2):130–138

Cabral HO, Vinck M, Fouquet C, Pennartz CM, Rondi-Reig L, Battaglia FP (2014) Oscillatory dynamics and place field maps reflect hippocampal ensemble processing of sequence and place memory under nmda receptor control. Neuron 81(2):402–415

Canolty RT, Edwards E, Dalal SS, Soltani M, Nagarajan SS, Kirsch HE et al (2006) High gamma power is phase-locked to theta oscillations in human neocortex. Science 313(5793):1626–1628

Canolty RT, Knight RT (2010) The functional role of cross-frequency coupling. Trends Cogn Sci 14(11):506–515

Caplan JB, Madsen JR, Raghavachari S, Kahana MJ (2001) Distinct patterns of brain oscillations underlie two basic parameters of human maze learning. J Neurophysiol 86:368–380

Caplan JB, Madsen JR, Schulze-Bonhage A, Aschenbrenner-Scheibe R, Newman EL, Kahana MJ (2003) Human theta oscillations related to sensorimotor integration and spatial learning. J Neurosci 23:4726–4736

Colgin LL (2016) Rhythms of the hippocampal network. Nat Rev Neurosci 17(4):239–249

Cornwell B, Johnson L, Holroyd T, Carver F, Grillon C (2008) Human hippocampal and parahippocampal theta during goal-directed spatial navigation predicts performance on a virtual Morris water maze. J Neurosci 28(23):5983–5990

Crespo-García M, Zeiller M, Leupold C, Kreiselmeyer G, Rampp S, Hamer HM, Dalal SS (2016) Slow-theta power decreases during item-place encoding predict spatial accuracy of subsequent context recall. NeuroImage. doi:10.1016/j.neuroimage.2016.08.021

Daitch AL, Sharma M, Roland JL, Astafiev SV, Bundy DT, Gaona CM et al (2013) Frequency-specific mechanism links human brain networks for spatial attention. Proc Natl Acad Sci U S A 110(48):19585–19590

Domnisoru C, Kinkhabwala AA, Tank DW (2013) Membrane potential dynamics of grid cells. Nature 495(7440):199–204

Donner TH, Siegel M (2011) A framework for local cortical oscillation patterns. Trends Cogn Sci 15(5):191–199

Eichenbaum H (1999) The hippocampus and mechanisms of declarative memory. Behav Brain Res 103:123–133

Eichenbaum H (2000, October) A cortical-hippocampal system for declarative memory. Nat Rev Neurosci 1(1):41–50

Eichenbaum H (2014) Time cells in the hippocampus: a new dimension for mapping memories. Nat Rev Neurosci 15(11):732–744

Ekstrom A, Bazih A, Suthana N, Al-Hakim R, Ogura K, Zeineh M et al (2009) Advances in high-resolution imaging and computational unfolding of the human hippocampus. NeuroImage 47(1):42–49

Ekstrom AD (2015) Why vision is important to how we navigate. Hippocampus 25(6):731–735

Ekstrom AD, Caplan J, Ho E, Shattuck K, Fried I, Kahana M (2005) Human hippocampal theta activity during virtual navigation. Hippocampus 15:881–889

Ekstrom AD, Kahana MJ, Caplan JB, Fields TA, Isham EA, Newman EL, Fried I (2003) Cellular networks underlying human spatial navigation. Nature 425:184–187

Ekstrom AD, Watrous AJ (2014) Multifacited roles for low-frequency oscillations in bottom-up and top-down processing during navigation and memory. NeuroImage 85(2):667–677

Ermentrout G, Kleinfeld D (2001) Traveling electrical waves in cortex insights from phase dynamics and speculation on a computational role. Neuron 29(1):33–44

Fell J, Axmacher N (2011, February) The role of phase synchronization in memory processes. Nat Rev Neurosci 12(2):105–118

Foster BL, Kaveh A, Dastjerdi M, Miller KJ, Parvizi J (2013) Human retrosplenial cortex displays transient theta phase locking with medial temporal cortex prior to activation during autobiographical memory retrieval. J Neurosci 33(25):10439–10446

Freudenburg ZV, Gaona CM, Sharma M, Bundy DT, Breshears JD, Pless RB, Leuthardt EC (2014) Fast-scale network dynamics in human cortex have specific spectral covariance patterns. Proc Natl Acad Sci U S A 111(12):4602–4607

Fries P (2005) A mechanism for cognitive dynamics: neuronal communication through neuronal coherence. Trends Cogn Sci 9(10):474–480. doi:10.1016/j.tics.2005.08.011

Giraud A-L, Poeppel D (2012) Cortical oscillations and speech processing: emerging computational principles and operations. Nat Neurosci 15(4):511–517

Goutagny R, Jackson J, Williams S (2009) Self-generated theta oscillations in the hippocampus. Nat Neurosci 12(12):1491–1493

Gray CM, König P, Engel AK, Singer W (1989) Oscillatory responses in cat visual cortex exhibit inter-columnar synchronization which reflects global stimulus properties. Nature 338:334–337

Green JD, Arduini AA (1954) Hippocampal electrical activity in arousal. J Neurophysiol 17(6): 533–557

Griffin AL, Yukiko A, Darling RD, Berry SD (2004) Theta-contingent trial presentation accelerates learning rate and enhances hippocampal plasticity during trace eyeblink conditioning. Behav Neurosci 118(2):403–411

Gross J, Hoogenboom N, Thut G, Schyns P, Panzeri S, Belin P, Garrod S (2013). Speech rhythms and multiplexed oscillatory sensory coding in the human brain. PLoS Biol. 11(12):e1001752

Guderian S, Schott B, Richardson-Klavehn A, Duzel E (2009) Medial temporal theta state before an event predicts episodic encoding success in humans. Proc Natl Acad Sci U S A 106(13): 5365

Halgren E, Babb TL, Crandall PH (1978) Human hippocampal formation EEG desynchronizes during attentiveness and movement. Electroencephalogr Clin Neurophysiol 44:778–781

Haque RU, Wittig JH, Damera SR, Inati SK, Zaghloul KA (2015) Cortical low-frequency power and progressive phase synchrony precede successful memory encoding. J Neurosci 35(40): 13577–13586

Hillebrand A, Tewarie P, van Dellen E, Yu M, Carbo EW, Douw L et al (2016) Direction of information flow in large-scale resting-state networks is frequency-dependent. Proc Natl Acad Sci U S A 113(14):3867–3872

Hoffman KL, Dragan MC, Leonard TK, Micheli C, Montefusco-Siegmund R, Valiante TA (2013) Saccades during visual exploration align hippocampal 3–8 Hz rhythms in human and non-human primates. Front Syst Neurosci 7

Hölscher C, Anwyl R, Rowan MJ (1997) Stimulation on the positive phase of hippocampal theta rhythm induces long-term potentiation that can be depotentiated by stimulation on the negative phase in area CA1 *in vivo*. J Neurosci 17:6470–6477

Howard MW, Fotedar MS, Datey AV, Hasselmo ME (2005) The temporal context model in spatial navigation and relational learning: toward a common explanation of medial temporal lobe function across domains. Psychol Rev 112(1):75–116

Huerta PT, Lisman JE (1993) Heightened synaptic plasticity of hippocampal CA1 neurons during a cholinergically induced rhythmic state. Nature 364(6439):723–725

Huxter J, Burgess N, O'Keefe J (2003) Independent rate and temporal coding in hippocampal pyramidal cells. Nature 425:828–832

Hyman J, Zilli E, Paley A, Hasselmo M (2005) Medial prefrontal cortex cells show dynamic modulation with the hippocampal theta rhythm dependent on behavior. Hippocampus 15(6): 739–749

Hyman JM, Hasselmo ME, Seamans JK (2011) What is the functional relevance of prefrontal cortex entrainment to hippocampal theta rhythms? Front Neurosci 5:24

Iglói K, Doeller CF, Berthoz A, Rondi-Reig L, Burgess N (2010) Lateralized human hippocampal activity predicts navigation based on sequence or place memory. Proc Natl Acad Sci U S A 107(32):14466–14471

Jacobs J (2014) Hippocampal theta oscillations are slower in humans than in rodents: implications for models of spatial navigation and memory. Philos Trans R Soc B Biol Sci 369(1635): 20130304

Jacobs J, Kahana MJ (2010) Direct brain recordings fuel advances in cognitive electrophysiology. Trends Cogn Sci 14(4):162–171

Jacobs J, Kahana MJ, Ekstrom AD, Fried I (2007) Brain oscillations control timing of single-neuron activity in humans. J Neurosci 27(14):3839–3844

Jacobs J, Kahana MJ, Ekstrom AD, Mollison MV, Fried I (2010a) A sense of direction in human entorhinal cortex. Proc Natl Acad Sci U S A 107(14):6487–6482

Jacobs J, Korolev I, Caplan J, Ekstrom A, Litt B, Baltuch G et al (2010b) Right-lateralized brain oscillations in human spatial navigation. J Cogn Neurosci 22(5):824–836

Jacobs J, Lega B, Anderson C (2012) Explaining how brain stimulation can evoke memories. J Cogn Neurosci 24(3):553–563

Jacobs J, Weidemann CT, Miller JF, Solway A, Burke JF, Wei X et al (2013) Direct recordings of grid-like neuronal activity in human spatial navigation. Nat Neurosci 16:1188–1190

Jeneson A, Squire LR (2012) Working memory, long-term memory, and medial temporal lobe function. Learn Mem 19(1):15–25

Jensen O, Kaiser J, Lachaux J (2007) Human gamma-frequency oscillations associated with attention and memory. Trends Neurosci 30(7):317–324

Jensen O, Lisman JE (2000) Position reconstruction from an ensemble of hippocampal place cells: contribution of theta phase coding. J Neurophysiol 83:2602–2609

Jones MW, Wilson MA (2005) Phase precession of medial prefrontal cortical activity relative to the hippocampal theta rhythm. Hippocampus 15:867–873

Jutras MJ, Fries P, Buffalo EA (2013) Oscillatory activity in the monkey hippocampus during visual exploration and memory formation. Proc Natl Acad Sci U S A 10(32):13144–13149

Kahana MJ (2006) The cognitive correlates of human brain oscillations. J Neurosci 26(6): 1669–1672. doi:10.1523/JNEUROSCI.3737-05c.2006

Kahana MJ, Sekuler R, Caplan JB, Kirschen M, Madsen JR (1999) Human theta oscillations exhibit task dependence during virtual maze navigation. Nature 399:781–784

Kaplan R, Doeller CF, Barnes GR, Litvak V, Düzel E, Bandettini PA, Burgess N (2012) Movement-related theta rhythm in humans: coordinating self-directed hippocampal learning. PLoS Biol 10(2):e1001267

Kendrick KM, Zhan Y, Fischer H, Nicol AU, Zhang X, Feng J (2011) Learning alters theta amplitude, theta-gamma coupling and neuronal synchronization in inferotemporal cortex. BMC Neurosci 12(1):1

Kim K, Ekstrom AD, Tandon N (2016) A network approach for modulating memory processes via direct and indirect brain stimulation: toward a causal approach for the neural basis of memory. Neurobiol Learn Mem 134 Pt A:162–177

Klausberger T, Magill PJ, Marton LF, Roberts JD, Cobden PM, Buzsáki G, Somogyi P (2003) Brain-state- and cell-type-specific firing of hippocampal interneurons in vivo. Nature 421: 844–848

Kramis R, Vanderwolf C, Bland B (1975) Two types of hippocampal rhythmical slow activity in both the rabbit and the rat: relations to behavior and effects of atropine, diethyl ether, urethane, and pentobarbital. Exp Neurol 49(1 Pt 1):58–85

Lacruz ME, Valentín A, Seoane JJG, Morris RG, Selway RP, Alarcón G (2010, October) Single pulse electrical stimulation of the hippocampus is sufficient to impair human episodic memory. Neuroscience 170(2):623–632

Lega B, Burke JF, Jacobs J, Kahana MJ (2015) Slow theta-to-gamma phase amplitude coupling in human hippocampus supports the formation of new episodic memories. Cereb Cortex 26(1): 268–278

Lega B, Jacobs J, Kahana M (2012) Human hippocampal theta oscillations and the formation of episodic memories. Hippocampus 22(4):748–761

Lopour BA, Tavassoli A, Fried I, Ringach DL (2013) Coding of information in the phase of local field potentials within human medial temporal lobe. Neuron 79(3):594–606

Lubenov EV, Siapas AG (2009) Hippocampal theta oscillations are travelling waves. Nature 459 (7246):534–539

MacDonald CJ, Carrow S, Place R, Eichenbaum H (2013) Distinct hippocampal time cell sequences represent odor memories in immobilized rats. J Neurosci 33(36):14607–14616

Markus EJ, Barnes CA, McNaughton BL, Gladden VL, Skaggs WE (1994) Spatial information content and reliability of hippocampal ca1 neurons: effects of visual input. Hippocampus 4(4): 410–421

McCartney H, Johnson AD, Weil ZM, Givens B (2004) Theta reset produces optimal conditions for long-term potentiation. Hippocampus 14:684–687

McClelland JL, McNaughton BL, O'Reilly RC (1995) Why there are complementary learning systems in the hippocampus and neocortex: insights from the successes and failures of connectionist models of learning and memory. Psychol Rev 102(3):419–457

McNaughton BL, Battaglia FP, Jensen O, Moser EI, Moser M-B (2006) Path integration and the neural basis of the 'cognitive map'. Nat Rev Neurosci 7:663–678

Meador KJ, Thompson JL, Loring DW, Murro AM, King DW, Gallagher BB et al (1991) Behavioral state-specific changes in human hippocampal theta activity. Neurology 41:869–872

Mehta MR, Lee AK, Wilson MA (2002, June) Role of experience and oscillations in transforming a rate code into a temporal code. Nature 417(6890):741–746. doi:10.1038/nature00807

Merkow MB, Burke JF, Stein JM, Kahana MJ (2014) Prestimulus theta in the human hippocampus predicts subsequent recognition but not recall. Hippocampus 24:1562–1569

Miller KJ, Leuthardt EC, Schalk G, Rao RPN, Anderson NR, Moran DW et al (2007) Spectral changes in cortical surface potentials during motor movement. J Neurosci 27:2424–2432

Miller R (1991) Cortico-hippocampal interplay and the representation of contexts in the brain. Springer, Berlin, NY

Minderer M, Harvey CD, Donato F, Moser EI (2016) Neuroscience: virtual reality explored. Nature 533(7603):324–325

Mitchell D, McNaughton N, Flanagan D, Kirk I (2008) Frontal-midline theta from the perspective of hippocampal theta. Prog Neurobiol 86(3):156

Mitzdorf U (1985) Current source-density method and application in cat cerebral cortex: investigation of evoked potentials and EEG phenomena. Physiol Rev 65(1):37–100

Montgomery S, Betancur M, Buzsaki G (2009) Behavior-dependent coordination of multiple theta dipoles in the hippocampus. J Neurosci 29(5):1381

Montgomery S, Sirota A, Buzsaki G (2008) Theta and gamma coordination of hippocampal networks during waking and rapid eye movement sleep. J Neurosci 28(26):6731

Mormann F, Fell J, Axmacher N, Weber B, Lehnertz K, Elger C, Ferna'ndez G (2005) Phase/amplitude reset and theta–gamma interaction in the human medial temporal lobe during a continuous word recognition memory task. Hippocampus 15(7):890–900

Mormann F, Kornblith S, Quiroga R, Kraskov A, Cerf M, Fried I, Koch C (2008) Latency and selectivity of single neurons indicate hierarchical processing in the human medial temporal lobe. J Neurosci 28(36):8865

Nadel L, Moscovitch M (1997) Memory consolidation, retrograde amnesia and the hippocampal complex. Curr Opin Neurobiol 7(2):217–227

Niedermeyer E, da Silva FL (2005) Electroencephalography: basic principles, clinical applications, and related fields. Lippincott Williams & Wilkins, Philadelphia; London

Norman KA, O'Reilly RC (2003) Modeling hippocampal and neocortical contributions to recognition memory: a complementary learning systems approach. Psychol Rev 110:611–646

O'Keefe J, Dostrovsky J (1971) The hippocampus as a spatial map: preliminary evidence from unit activity in the freely-moving rat. Brain Res 34:171–175

O'Keefe J, Nadel L (1978) The hippocampus as a cognitive map. Oxford University Press, New York

O'Keefe J, Recce ML (1993) Phase relationship between hippocampal place units and the EEG theta rhythm. Hippocampus 3:317–330

Panzeri S, Brunel N, Logothetis NK, Kayser C (2010) Sensory neural codes using multiplexed temporal scales. Trends Neurosci 33(3):111–120

Pastalkova E, Itskov V, Amarasingham A, Buzsáki G (2008) Internally generated cell assembly sequences in the rat hippocampus. Science 321:1322–1327

Pastoll H, Solanka L, van Rossum MC, Nolan MF (2013) Feedback inhibition enables theta-nested gamma oscillations and grid firing fields. Neuron 77(1):141–154

Patel J, Fujisawa S, Berényi A, Royer S, Buzsáki G (2012) Traveling theta waves along the entire septotemporal axis of the hippocampus. Neuron 75(3):410–417

Quraan MA, Moses SN, Hung Y, Mills T, Taylor MJ (2011) Detection and localization of hippocampal activity using beamformers with meg: a detailed investigation using simulations and empirical data. Hum Brain Mapp 32(5):812–827

Ravassard P, Kees A, Willers B, Ho D, Aharoni D, Cushman J et al (2013) Multisensory control of hippocampal spatiotemporal selectivity. Science 340(6138):1342–1346

Rizzuto D, Madsen JR, Bromfield EB, Schulze-Bonhage A, Seelig D, Aschenbrenner-Scheibe R, Kahana MJ (2003) Reset of human neocortical oscillations during a working memory task. Proc Natl Acad Sci U S A 100:7931–7936

Royer S, Sirota A, Patel J, Buzsáki G (2010) Distinct representations and theta dynamics in dorsal and ventral hippocampus. J Neurosci 30(5):1777–1787

Rutishauser U, Ross I, Mamelak A, Schuman E (2010) Human memory strength is predicted by theta-frequency phase-locking of single neurons. Nature 464(7290):903–907

Sabolek HR, Penley SC, Hinman JR, Bunce JG, Markus EJ, Escabi M, Chrobak JJ (2009) Theta and gamma coherence along the septotemporal axis of the hippocampus. J Neurophysiol 101(3):1192–1200

Schmidt-Hieber C, Häusser M (2013) Cellular mechanisms of spatial navigation in the medial entorhinal cortex. Nat Neurosci 16(3):325–331

Schyns PG, Thut G, Gross J (2011) Cracking the code of oscillatory activity. PLoS Biol 9(5): e1001064

Seager MA, Johnson LD, Chabot ES, Asaka Y, Berry SD (2002) Oscillatory brain states and learning: impact of hippocampal theta-contingent training. Proc Natl Acad Sci U S A 99: 1616–1620

Sederberg PB, Kahana MJ, Howard MW, Donner EJ, Madsen JR (2003) Theta and gamma oscillations during encoding predict subsequent recall. J Neurosci 23(34):10809–10814

Sederberg PB, Schulze-Bonhage A, Madsen JR, Bromfield EB, McCarthy DC, Brandt A et al (2007) Hippocampal and neocortical gamma oscillations predict memory formation in humans. Cereb Cortex 17(5):1190–1196

Shein-Idelson M, Ondracek JM, Liaw H-P, Reiter S, Laurent G (2016) Slow waves, sharp waves, ripples, and rem in sleeping dragons. Science 352(6285):590–595

Siapas A, Lubenov E, Wilson M (2005) Prefrontal phase locking to hippocampal theta oscillations. Neuron 46:141–151

Siegel M, Donner TH, Engel AK (2012) Spectral fingerprints of large-scale neuronal interactions. Nat Rev Neurosci 13(2):121–134

Sirota A, Montgomery S, Fujisawa S, Isomura Y, Zugaro M, Buzsáki G (2008) Entrainment of neocortical neurons and gamma oscillations by the hippocampal theta rhythm. Neuron 60(4): 683–697

Skaggs WE, McNaughton BL, Permenter M, Archibeque M, Vogt J, Amaral DG, Barnes CA (2007) Eeg sharp waves and sparse ensemble unit activity in the macaque hippocampus. J Neurophysiol 98(2):898–910

Skaggs WE, McNaughton BL, Wilson MA, Barnes CA (1996) Theta phase precession in hippocampal neuronal populations and the compression of temporal sequences. Hippocampus 6: 149–172

Sohal VS, Zhang F, Yizhar O, Deisseroth K (2009) Parvalbumin neurons and gamma rhythms enhance cortical circuit performance. Nature 459(7247):698–702

Spencer S, Huh L (2008) Outcomes of epilepsy surgery in adults and children. Lancet Neurol 7(6): 525–537

Spencer SS, Spencer DD (1994) Entorhinal-hippocampal interactions in medial temporal lobe epilepsy. Epilepsia 35(4):721–727

Spiers HJ, Burgess N, Maguire EA, Baxendale SA, Hartley T, Thompson PJ, O'Keefe J (2001) Unilateral temporal lobectomy patients show lateralized topographical and episodic memory deficits in a virtual town. Brain 124(Pt 12):2476–2489

Steffenach H, Witter M, Moser M, Moser E (2005) Spatial memory in the rat requires the dorsolateral band of the entorhinal cortex. Neuron 45(2):301–313

Strange BA, Witter MP, Lein ES, Moser EI (2014) Functional organization of the hippocampal longitudinal axis. Nat Rev Neurosci 15(10):655–669

Taube JS, Valerio S, Yoder RM (2013) Is navigation in virtual reality with fmri really navigation? J Cogn Neurosci 25(7):1008–1019

Tesche CD (1997) Non-invasive detection of ongoing neuronal population activity in normal human hippocampus. Brain Res 749(1):53–60

Teyler TJ, DiScenna P (1986) The hippocampal memory indexing theory. Behav Neurosci 100(2): 147

Tulving E (1985) How many memory systems are there? Am Psychol 40(4):385

Tulving E (2002) Episodic memory: from mind to brain. Annu Rev Psychol 53:1–25

Vanderwolf C (1969) Hippocampal electrical activity and voluntary movement of the rat. Electroencephalogr Clin Neurophysiol 26:407–418

Vass LK, Copara MS, Seyal M, Shahlaie K, Farias ST, Shen PY, Ekstrom AD (2016) Oscillations go the distance: low-frequency human hippocampal oscillations code spatial distance in the absence of sensory cues during teleportation. Neuron 89(6):1180–1186

von Stein A, Sarnthein J (2000) Different frequencies for different scales of cortical integration: from local gamma to long range alpha/theta synchronization. Int J Psychophysiol 38(3): 301–313

Watrous AJ, Fried I, Ekstrom AD (2011) Behavioral correlates of human hippocampal delta and theta oscillations during navigation. J Neurophysiol 105(4):1747–1755

Watrous AJ, Lee DJ, Izadi A, Gurkoff GG, Shahlaie K, Ekstrom AD (2013a) A comparative study of human and rat hippocampal low-frequency oscillations during spatial navigation. Hippocampus 23(8):656–661

Watrous AJ, Tandon N, Conner CR, Pieters T, Ekstrom AD (2013b) Frequency-specific network connectivity increases underlie accurate spatiotemporal memory retrieval. Nat Neurosci 16(3): 349–356

Wetzel W, Ott T, Matthies H (1977) Post-training hippocampal rhythmic slow activity ("theta") elicited by septal stimulation improves memory consolidation in rats. Behav Biol 21(1):32–40

Winson J (1978) Loss of hippocampal theta rhythms in spatial memory deficit in the rat. Science 201:160–163

Womelsdorf T, Schoffelen J, Oostenveld R, Singer W, Desimone R, Engel A, Fries P (2007) Modulation of neuronal interactions through neuronal synchronization. Science 316(5831): 1609

Yartsev M, Witter M, Ulanovsky N (2011) Grid cells without theta oscillations in the entorhinal cortex of bats. Nature 479(7371):103–107

Yushkevich PA, Wang H, Pluta J, Das SR, Craige C, Avants BB et al (2010) Nearly automatic segmentation of hippocampal subfields in *in vivo* focal T2-weighted MRI. NeuroImage 53(4): 1208–1224

Zhang H, Jacobs J (2015) Traveling theta waves in the human hippocampus. J Neurosci 35(36): 12477–12487

Zheng C, Bieri KW, Hwaun E, Colgin LL (2016) Fast gamma rhythms in the hippocampus promote encoding of novel object-place pairings. *eNeuro*, ENEURO-0001

Zilli EA, Yoshida M, Tahvildari B, Giocomo LM, Hasselmo ME (2009) Evaluation of the oscillatory interference model of grid cell firing through analysis and measured period variance of some biological oscillators. PLoS Comput Biol 5(11):e1000573

Elements of Information Processing in Hippocampal Neuronal Activity: Space, Time, and Memory

Howard Eichenbaum

Abstract The earliest studies on the firing properties of hippocampal neurons revealed coding of both spatial and non-spatial dimensions of experience. Since then, distinct lines of investigation have elaborated these findings to provide compelling evidence that the hippocampal neurons represent the events we remember within spatial as well as temporal frameworks. This characterization suggests that neural networks in the hippocampus underlie a "memory space" that organizes the features of memory dependent on hippocampal function.

A comprehensive understanding of the hippocampus requires identifying the nature of information encoded by its information processing elements combined with interpretation of the overall network representations that underlie cognitive and memory functions. Here I will attempt an overview of our knowledge about information processing by hippocampal neurons and networks. This will not be a comprehensive review—there have been several recent collections that survey the firing properties of hippocampal neurons in behaving animals and humans (Hartley et al. 2013; Mizumori 2007; Derdikman and Knierim 2014). Rather, here I will provide examples of the broad variety of hippocampal coding properties and attempt a synthesis of what these findings tell us about single neuron and network coding mechanisms that underlie memory representations.

Ancient History: The Early Studies on Firing Patterns of Hippocampal Neurons in Behaving Animals

In the early 1970s, several investigators adopted newly developed methods using single sharp electrodes or bundles of small-diameter flexible wires to record the activity of principal neurons in the hippocampus. Their studies pre-dated the advent of digitized recordings and computerized data analysis, and so depended on human

H. Eichenbaum (✉)
Center for Memory and Brain, Boston University, Boston, MA, USA
e-mail: hbe@bu.edu

© Springer International Publishing AG 2017
D.E. Hannula, M.C. Duff (eds.), *The Hippocampus from Cells to Systems*,
DOI 10.1007/978-3-319-50406-3_3

observation to correlate auditory artifacts of neuronal spiking with ongoing behavior or simple automated averaging of spiking over time to compute firing rates time-locked to specific stimuli. These papers identified both spatial and non-spatial correlates of hippocampal neural activity that we still struggle to reconcile today.

The first of these publications was a short communication by O'Keefe and Dostrovsky (1971) that described the firing properties of neurons recorded from the dorsal hippocampus in rats using sharp electrodes as they moved through or were positioned within an open field environment. They focused on the activity patterns of eight hippocampal neurons that fired solely or maximally when a rat was in a particular part of the open field. The activity of most of these cells was also dependent on specific sensory stimuli (e.g. a tactile or visual stimulus) and the direction of orientation within the environment. The more extensive follow-up study by O'Keefe (1976) described many more hippocampal cells whose activity was dependent on spatial location and emphasized a distinction between "place cells" that fired when the rat occupied or ran past a particular location and "misplace cells" whose spatially specific activity depended on exploratory sniffing, usually when the rat did not find an expected object at the location. So, while these firing patterns were immediately interpreted as supporting the idea that hippocampal neurons map space, the data were equally clear that hippocampal neuronal firing patterns also encoded specific stimuli, behaviors, and cognitive states.

Quite independently, and around the same time, James Olds and his colleagues recorded from single neurons using fine wire electrodes positioned in various brain areas. They established an approach to identifying "learning centers" in the rat brain defined as areas where neurons developed short-latency responses time locked to stimuli (tones) as animals were classically conditioned to expect food delivery following the tones (Olds et al. 1972). Using this paradigm they identified neuronal responses to the conditioned tone observed throughout the hippocampus (Segal and Olds 1972). In these studies no effort was employed to control or determine the location of the animal within the small conditioning chamber. However, typically the neurons did not respond to the tones or reward delivery during a preliminary pseudo-conditioning session, suggesting that the stimulus-driven responses depended specifically on the learned association and not solely other aspects of sensory experience, behavior, or location.

In 1973 James Ranck published an extensive analysis of hippocampal neuron firing patterns observed in rats performing a variety of behaviors in an open field, including eating, drinking, grooming, being held, bar pressing, and sleeping. He observed correlations between neural activity and ongoing behavior in almost all hippocampal neurons, and reported that no two principal cells had the identical behavioral correlate. Four main types emerged from his analysis: "approach-consummate cells" that fired during the approach to and consumption of food, "approach-consummate mismatch cells" that fired similarly during approach and also during exploration of a missing water bottle (like O'Keefe's misplace cells), "appetitive cells" that fired during orienting movements and approach but not consummatory behavior, and "motion-punctuate cells" that fired at the end of orienting movements or change in direction of movement. No effort was made to control for spatial location in this study, and Ranck acknowledged that, "perhaps spatial characteristics are the entire basis of firing in these cells" (Ranck 1973). However, the distinctions between the different behavioral correlates of these cell types seems unlikely explained purely by differences in where the behaviors occurred.

Finally, Theodore Berger, Richard Thompson, and their colleagues recorded multi-units and single neurons in the hippocampus of rabbits undergoing tone-cued classical eye-blink conditioning (Berger et al. 1976, 1983). They reported the emergence of tone-evoked conditioned responses of hippocampal neurons that paralleled both success in learning across trials and the time course of the conditioned eye-blink within trials. In these studies position within space was strictly controlled in that the animals were immobilized within a restraining device throughout learning. Thus the learning and behavioral correlates of conditioned eye-blinks cannot be attributed to spatial coding.

In many ways these early observations already provided insights into the broad scope of information that is encoded by hippocampal neural activity patterns that are evident in current studies. Place is a major determinant of the firing patterns of hippocampal neurons in animals that freely move through the environment. This property of hippocampal neurons was recognized in the awarding of the 2015 Nobel Prize to O'Keefe, who discovered the spatial firing patterns of hippocampal neurons. However, differences in spatial location do not account fully for firing patterns of many neurons, such as the misplace/mismatch neurons of O'Keefe and Ranck suggestive of additional correlates of cognitive and memory function. In addition, the coding of specific sensory stimuli was implicated in O'Keefe's original study and more systematically in Olds and colleagues' observations on conditioned neural responses. And, just as Ranck's observations are strongly suggestive that specific actions (e.g., approach behavior) seem to play some role, the findings of conditioned eye-blink related responses by Berger & Thompson strongly indicate that learned actions are encoded by hippocampal neurons in immobilized animals where location cannot explain the neural firing patterns.

Subsequent work on hippocampal neuron firing patterns in behaving animals and humans has expanded in four main directions. First, many studies have explored the spatial firing properties of hippocampal neurons, identifying cues that control, as well as other factors that modulate, spatial firing patterns. Second, many other studies have explored how learning of non-spatial information or actions is encoded by hippocampal neurons, along with or independent of spatial information. Third, recent evidence has indicated that hippocampal neurons encode time much like they encode space, suggesting a parallel dimension for mapping experiences. Fourth, another new direction involves explorations of how hippocampal neuronal ensembles integrate representations of multiple related experiences into networks of memories (also called "schemas"). These directions will be examined in turn. As you read this review, note that, while the coding of position in space has received the greatest attention in this literature, there is considerable evidence that position coding is often subordinate to other abstract features (the "context") of a behavioral task, and the finding of robust temporal coding indicates that space may be only one of the dimensions employed by hippocampal networks to organize memories.

Spatial Coding by Hippocampal Neurons

As can been deduced from the early observations, a mixture of spatial and non-spatial parameters influences hippocampal neural activity. In particular, it is clear that non-spatial events must be considered because of the findings on classical eyelid conditioning that show coding of learned behavior when space is held constant. Thus, when animals are freely moving in space, it might be that overt or subtle distinctions in ongoing perception, behavior, or cognition are confounded with, and drive the observation of position correlates of hippocampal neurons.

This issue was addressed by Olton et al. (1978) who identified clear place fields of hippocampal neurons in rats performing a task where they traversed the arms of an 8-arm radial maze and were required to remember visited arms. Despite the behavioral sequence being identical on all maze arms, many hippocampal neurons fired as the animal ran through particular locations on only one or a few of the arms, thus distinguishing the spatial correlate on some arms from the absence of activity during matched behavior on all arms. Another way the issue was addressed employed a clever behavioral paradigm created by Muller and colleagues (1987) that involved recording from hippocampal neurons as rats foraged for small bits of food dropped within an open field. The aim of this approach was to control for potential behavioral influences by testing whether a position correlate would emerge in a situation where foraging behavior is constant over all locations in the environment, thereby experimentally "subtracting" its influence. The results were striking: many hippocampal neurons had clear-cut place fields during random foraging in an open field. The observation of strong position coding when behavior is constant, involving either continuous foraging throughout a two-dimensional open field or identical movement sequences through linear tracks or mazes, have been replicated many times.

Variants of these linear maze and open field paradigms have been employed to characterize the sensory cues that determine position coding by hippocampal neurons. These findings can be summarized as follows. Nearly all of our information on hippocampal neuronal firing patterns comes from data on CA1 and CA3 pyramidal cells in the dorsal hippocampus of rats and to some degree in mice, monkeys, and humans (see Muller 1996; Eichenbaum et al. 1999, for more detailed reviews). As the animal explores or merely traverses a large environment, one can readily correlate dramatic increases in a cell's firing rate when the rat arrives at a particular location, called the "place field", and these cells are called "place cells". From a baseline of less than 1 spikes/s, the firing rate can exceed 100 Hz, although during some passes through the place field the cell may not fire at all. Typically a large fraction of cells, perhaps 40–75%, have place fields in any environment, although the low baseline firing rates may let many cells without place fields go undetected. Place fields vary in size from quite small to half the size of an environment and are dispersed throughout the environment, although they may be concentrated at areas of particular salience such as where rewards occur (e.g., Hollup et al. 2001; McKenzie et al. 2013). In most of the environments used to date, most hippocampal cells have only one or two place fields, although in large environments they can have many place fields (Rich et al. 2014).

Sensory Cues That Govern the Spatial Firing Patterns of Place Cells Many studies have focused on identifying the environmental cues that drive spatially specific activity. O'Keefe (1979) defined place cells as neurons whose activity is not dependent on any particular stimulus, but rather reflects the presence and topography of multiple environmental cues. Several studies have shown that a variety of visual and nonvisual cues can determine the location of place fields (e.g., Hill and Best 1981; Muller et al. 1987; Save et al. 2000; Gener et al. 2013; but see Cressant et al. 1997). O'Keefe and Conway (1978) performed the first study where multiple spatial cues were provided and then manipulated to determine which cues controlled spatial representations, and found that some cells were controlled by only one or two of the cues and others by any subset of the cues. More recent studies indicate that place cells are driven by relatively few relatively proximal cues. O'Keefe and Burgess (1996) showed that the shape and locus of most place fields within a simple rectangular chamber are determined by the dimensions of, and spatial relations between, only a few nearby walls of the environment (see also Hetherington and Shapiro 1997). Several other studies have shown that place cells can encode subsets of the spatial cues and that these representations are independent of the spatial representations of other cells in the same environment. Shapiro, Tanila, and colleagues (Shapiro et al. 1997; Tanila et al. 1997a, b, c) and Knierim (2002) examined the responses of hippocampal cells to systematic manipulations of a large set of spatial cues, including both distant cues outside a maze and proximal cues on the floors of maze arms. Different place cells encoded individual proximal and distant stimuli, combinations of proximal or distant stimuli, or relations between proximal and distant cues. The place fields of some cells were fully controlled by as little as a single cue within a very complex environment, and most cells were controlled by different subsets of the controlled cues. More recently Leutgeb et al. (2005) examined firing patterns of hippocampal neurons as rats explored multiple small environments (boxes) within multiple large environments (rooms) and reported that whether or not place cells fire and the locations of place fields depend on distant ("global") cues that lie outside of the small environment, whereas the firing rate, but not location of place fields depends on proximal cues (called "rate coding"). However, when distant cues are minimized, place fields can be entirely determined by local cues (Young et al. 1994; Hetherington and Shapiro 1997).

Not Necessarily Location Per Se: Length and Distance Place fields do not necessarily represent specific locations but rather can reflect continuous spatial dimensions of length and distance. O'Keefe and Burgess (1996) recorded from rats as they foraged in rectangular chambers whose walls varied in length. They found that place fields stretch along a wall of an environment that is elongated, indicating that when environmental cues are continuously variable, place cells represent spatial dimensions continuously. Gothard et al. (1996a, b) found that when a particularly salient cue or enclosure within an open field is moved repeatedly and randomly, the spatial firing patterns of some cells become tied to that cue. When rats were trained to shuttle between a mobile starting box and a goal location defined by landmarks in an open field, some cells fired relative to the static environmental cues, whereas others fired relative to a landmark-defined goal site, or in relation to the start box. When rats were trained to shuttle between a movable start-end box and goal site on a linear track, the anchor of the spatial representation of many cells switched between these two cues, depending on which was closer. Under these conditions the majority of the activated hippocampal cells did not exhibit location-specific activity that was associated with fixed environmental cues. Instead, their activity could be characterized as "spatial" only to the extent that they fired at specific distances from a particular stimulus or goal. Distance coding has also been observed in rats running on a treadmill where external spatial cues signaling motion are absent (Kraus et al. 2013) and in a task where spatial cues are variable and distance provides salient information about location (Ravassard et al. 2013; Aghajan et al. 2015).

Place Cells Encode Both the Similarities and Differences Between Environments That Share Spatial Features Several studies have shown that place cells are not linked together to form a cohesive map of the environment. Tanila et al. (1997b) found that ensembles of simultaneously recorded place cells changed their firing patterns independently associated with distinct subsets of the cues, indicating that the spatial representation was not cohesive but instead coded for spatial cues that were common to and distinct in multiple environments. In several cases where two cells had overlapping place fields associated with one configuration of the cues, each cell responded differently when the same cues were rearranged. This finding shows that each cell was controlled by a different subset of the cues at the same time, and that their differential encodings are not due to shifts between two different spatial "reference frames" used by all cells at different times (Gothard et al. 1996b). Skaggs and McNaughton (1998) confirmed this finding by recording from a large number of place cells simultaneously in rats foraging randomly in two identical enclosures, between which they could move freely. Each hippocampal ensemble contained cells that had similar place fields and others that had distinct spatial firing patterns between the two enclosures. In this situation, some cells encoded the physical cues, whereas the activity of others at the same time reflected the knowledge that the two environments were distinct.

Spatial Representations Are Context Dependent One view of place cells is that they compose a representation of the context in which specific events occur. What constitutes a "context", as opposed to a set of individual cues is not clear, and whether its domain includes spatial and temporal, as well as other aspects of the situation in which events occur is also not clear. The data suggests that all aspects of the background context in which specific events occur and when places are occupied can dramatically affect hippocampal neural activity. For example, the spatial firing patterns, and the extent to which firing is dependent on spatial orientation, are dramatically different when a rat forages randomly or produces repeated paths as it traverses the identical environment (Markus et al. 1995). Similarly, when different starting points in a radial maze determined the locations of goals, the firing patterns of place cells changed dramatically (Smith and Mizumori 2006). Notably, some places cells fire similarly in the two situational contexts whereas others change dramatically—showing that the hippocampus represents both the commonalities and differences in the two context-defined situations.

Seemingly subtle changes in environmental cues can also produce dramatic changes in the spatial firing patterns of hippocampal neurons. For example, changes in the background color or background odor of an environment can dramatically change the spatial firing patterns of individual hippocampal neurons (Anderson and Jeffrey 2003). Notably, again some cells do not change for each contextual shift, whereas others do. What cues and the extent of situational change that causes changes in firing patterns is not clear, but several studies have examined the dynamics of firing pattern changes when cues are gradually altered. When the shape of an environment is gradually altered (Wills et al. 2005), or critical cues are gradually changed (Rotenberg and Muller 1997), most place cells do not alter their firing patterns initially, but at some level of change, dramatically alter their firing patterns. This sudden switch of firing patterns when a threshold of cue alteration is passed suggests an attractor state dynamic (not unlike that of many other brain areas) in which the contextual representation switches from pattern completion to pattern separation. Area CA3 demonstrates a particularly sharp discrimination gradient in making this switch (Leutgeb et al. 2004; Lee et al. 2004). It appears that hippocampal cell assemblies can rapidly switch between spatial representations as animals perform different tasks within the same environment (Fenton et al. 1998; Jackson and Redish 2007).

Spatial firing patterns can also dramatically change when the affective association of a constant spatial environment is altered. Several studies have reported major alternations in hippocampal spatial representations of previously neutral environments when a rat is shocked in the environment, thus altering the meaning of the environment to evoke fear (Moita et al. 2004; Wang et al. 2012) or vice versa (Wang et al. 2015).

Several other recent studies have focused on changes in context defined by the behavioral demands of a task. In several of these studies, rats alternate routes that involve left and right turns through a T-maze where they traverse a part of the maze that is common to both routes. In this and similar tasks, many hippocampal neurons have distinct firing patterns, even when the rat traverses the common maze area depending on whether the rat is performing a left-turn or right-turn trial (Wood et al. 2000; Frank et al. 2000; Ferbinteanu and Shapiro 2003; Ainge et al. 2007; Bower et al. 2005; Lee et al. 2006; Griffin et al. 2007; reviewed in Shapiro et al. 2006). Importantly, some cells fire similarly as the rat performs both routes, indicating the hippocampus represents both the distinct paths and the common elements among them. Furthermore, the distinct firing patterns of place cells predict success in the alternation task (Robitsek et al. 2013). Also, the same pattern of findings occurs when the choice of different goals is guided by motivational context (hunger or thirst), indicating that the distinctions in firing patterns are not due to the accumulated movements (i.e., path integration) prior to the overlapping segment of the maze, but rather to the cognitive state associated with different routes through the maze (Kennedy and Shapiro 2009). A recent extension on these findings showed that, when the alternation task is separated into distinct sample and choice phases, most hippocampal neurons have different spatial firing patterns in the distinct trial phases, and within that, some cells also differentiate the two routes within each phase (Griffin et al. 2007). These data are consistent with other findings discussed above showing that different cognitive states within a single overall behavioral task are represented distinctly and linked by representations of their common features by hippocampal neurons.

Finally, new findings suggest that the ventral hippocampus, not examined in the studies described above, may represent large scale space that constitutes a meaningful spatial "context". Kjelstrup et al. (2008) compared the sizes of place fields in the dorsal and ventral hippocampus and found that place fields become larger as one records along the dorsal to ventral portions of the hippocampus. More recently, Komorowski et al. (2013) also recorded along this axis as rats performed a task where they were required to employ their current spatial context (one of two chambers) to remember which of two objects contained a reward, and found that ventral hippocampal neurons had large place fields, many of which filled most of all of one of the contexts. However, these fields never bridged between contexts in animals successfully performing the task, suggesting that ventral hippocampal networks code for representations of spatial and meaningful contexts.

Where the Rat "Thinks" It Is Notably, the spatial activity patterns of place cells may be more determined by where the rat may "think" it is rather than being explicitly driven by spatial cues. This possibility is consistent with the observation that the spatial firing patterns of place cells can persist even when all of the spatial cues are removed or the room is darkened (O'Keefe and Speakman 1987; Muller and Kubie 1987; Quirk et al. 1990), although the selectivity of spatial firing may be degraded in the dark (Markus et al. 1994). Also, the findings discussed above showing that place cells form categorical representations even in circumstances of ambiguous spatial cues (Skaggs and McNaughton 1998) or continuously changing spatial cues (Leutgeb et al. 2004), indicates that the animal's perspective on where it is can dominate over the actual spatial cues. Also, when a rat is first introduced into a new environment, place cells may continue firing associated with the cues of a former highly experienced environment, and then suddenly "re-map" after successive exposures (Bostock et al. 1991; see also Sharp et al. 1990). In a direct test of whether the animal's conception of its location can govern place cell activity, O'Keefe and Speakman (1987) tested rats in a task where they had to remember where removed spatial cues had been. They found that errors in their choice behavior predicted shifts of their hippocampal place fields, suggesting that these codings were determined by the orientation of the maze remembered by the rat, thus providing a compelling link between hippocampal spatial coding and spatial memory but also showing that place cells reflect an internal representation of space rather than a representation that depends on external cues.

Direction of Movement Influences Place Cells When Movements Through Space Are Meaningfully Directional According to O'Keefe (1979) true place cells fire whenever an animal is in the place field, regardless of its orientation or ongoing behavior. However, the only situation where large numbers of true place cells are observed is when animals forage by random walk through an environment, where behavior is held constant and the meaning of movement directions is homogeneous. However, in contrast to this open field foraging, in virtually any situation where movement directions are meaningfully different, distinct movement directions influence spatially specific activity. For example, in the radial maze task where animals regularly perform runs outward on each maze arm to obtain a reward, and then return to the central platform to initiate the next choice, outward and inward arm movements reflect meaningfully distinct behavioral episodes that occur repetitively. Correspondingly, hippocampal neurons reflect the relevant "directional structure" imposed by this protocol, and almost all place cells fire only during outward or inward journeys (McNaughton et al. 1983), and directionality is also observed when animals perform the same task in an open field, indicating that directionality is not due to the constraints of location by walls of the arms on a radial maze (Weiner et al. 1989). Similarly, place cells are activated selectively during distinct approach and return episodes and from variable goal and start locations in open fields and linear tracks. Furthermore, Muller et al. (1994) showed that the same place cells that are non-directional during random foraging are highly directional in a radial maze. Most impressively, Markus et al. (1995) directly compared the directionality of place cells under different task demands, and found that place cells that were non-directional when rats foraged randomly in an open field, were directional when they systematically visited a small number of reward locations. Taken together, these findings emphasize that place cells exhibit movement-related firing patterns whenever particular movements are associated with meaningfully different events. Also, directionality of place fields is obtained only following experience in directional movements (Navratilova et al. 2012).

Conclusions About Spatial Coding in Hippocampal Neurons The phenomenon of place cells in freely moving animals is highly robust and observed both in situations where the hippocampus is necessary for memory performance (e.g., the radial maze) and where it is not (foraging for food in an open field). A broad variety of individual spatial and non-spatial cues and cognitive states can drive or strongly influence place cells, so they do not provide a simple cohesive map of coordinate locations within a space defined by geometric relations among spatial cues as O'Keefe (1979) originally envisioned. On the other hand, perhaps the most straightforward explanation of place cells is that they reflect where an animal "thinks" it is in space as well as where it "thinks" it is going. This view is consistent with the notion that the hippocampal representation of space is "cognitive" as opposed to stimulus driven. A critical remaining question is whether the function of this cognitive map of space is dedicated to navigation, as some have suggested (McNaughton et al. 1996, 2006; Moser et al. 2008; Hartley et al. 2013) or whether the purpose of the map is to represent where events occur in spatial context, as has been suggested by recent studies on humans and animals (Eichenbaum et al. 2007; Davachi 2006; Diana et al. 2007). Much of the evidence that place cells are components of a dedicated spatial mapping system rest on the observation that hippocampal cells (and other cells in neighboring regions) can encode spatial parameters (location, head direction, borders, distance traveled; Hartley et al. 2013), but these findings may well just reflect the relevant dimensions of specific experiences that are dominated by spatial dimensions and lack non-spatial stimuli and behavioral demands. Deciding between these views rests instead on the extent to which hippocampal neurons encode specific stimuli, behavioral actions, and non-spatial cognitive events that fall outside the domain of spatial navigation and instead are consistent with a spatial framework for memories.

Representation of Stimuli, Behavioral Actions, and Cognitive States Independent of, or Along With Position

The Berger & Thompson studies described above indicate that hippocampal neurons can have clear learning and behavioral correlates in animals entirely restrained within a specific location. However, it may well be that space still plays a role even in this highly controlled task, because the same behavior related firing pattern may depend upon the location where conditioning occurs, as does the behavior in this kind of classical conditioning (Penick and Solomon 1991). To address this possibility, many studies employed learning and memory tasks where *explicitly distinct* sensory or behavioral events occur in multiple positions in an environment, with the aim of distinguishing the extent to which firing patterns are dependent on the nature of the event, on where it occurs, or both. These studies have revealed that hippocampal neuronal firing patterns distinguish both the different events and the positions and spatial contexts where they occur.

Sensory Driven Responses Many studies in rodents, monkeys, and humans have described hippocampal neuronal activity associated with a very broad range of non-spatial stimuli and behavioral events. In rodents, many studies have observed robust activation of hippocampal neurons associated with visual, tactile, olfactory, and auditory cues in several learning and memory paradigms (reviewed in Eichenbaum et al. 1999; Eichenbaum 2004). These findings join with many other reports of robust activation of hippocampal neurons associated with combinations of specific stimuli, match/non-match stimulus comparisons, and the locations of these events in animals performing discrimination and recognition memory tasks (Eichenbaum et al. 1987; Wood et al. 1999; Wiebe and Staubli 1999; Deadwyler et al. 1995; Otto and Eichenbaum 1992; Hampson et al. 1993; Wible et al. 1986). The extent to which non-spatial and spatial cues are represented depends on the context of behavioral demands. For example, in the same environment with the same olfactory cues, hippocampal neurons strongly encode location when rewards are associated with the location of the cue, but fire associated with the odors when the odor identity is associated with reward (Muzzio et al. 2009). Similarly, Lee and

Kim (2010) reported that hippocampal neuronal activity shifted from spatially determined to stimulus determined as learning about the stimuli developed. In addition, hippocampal neurons signal learned behavioral actions. Lenck-Santini et al. (2008) described hippocampal neurons that fire during learned "jump" avoidance responses, reminiscent of Ranck's (1973) pioneering descriptions of a variety of behavioral correlates of hippocampal neurons in rats and the findings on conditioned eye-blink related responses described by Berger et al. (1976), a finding extended in recent studies on classical eye-blink conditioning (Hattori et al. 2015; McEchron and Disterhoft 1997).

Consistent with these findings in rodents, a large fraction of hippocampal neurons in head-fixed monkeys fire robustly associated with learned associations between specific visual stimuli and eye-movement responses (Wirth et al. 2003). Similarly, a large fraction of hippocampal neurons in monkeys respond to visual stimuli modulated by their familiarity in the naturalistic recognition task described above (Jutras and Buffalo 2010). Furthermore, multiple studies have reported that hippocampal neurons in humans also respond to visual stimuli and their responses are modulated by familiarity in recognition tasks (Fried et al. 1997) and distinguish the stimuli that are recalled from those forgotten (Rutishauser et al. 2008). Hippocampal neuronal responses also predict memory for learned verbal paired associates (Cameron et al. 2007). Human hippocampal neurons exhibit sparse and distributed coding of individual remembered stimuli (Wixted et al. 2014) and rapidly develop as humans learn associations between objects and locations (Ison et al. 2015), and many hippocampal neurons generalize across closely related stimuli (Quiroga et al. 2005; Krieman et al. 2000a) and fire while the subject is imagining a cued stimulus (Krieman et al. 2000b). These studies provide strong evidence that many hippocampal neurons fire associated with specific stimuli and actions when space is held constant (e.g. eye-blink conditioning) and are driven by conditioned stimuli when the animal is immobile (Olds et al. 1972; the studies in monkeys and humans).

Conjoint Sensory-Behavioral and Spatial Responses Several other studies have shown that hippocampal neurons conjoin sensory-behavioral events and positions where they occur. The most striking of these studies also involve tracking learning about sensory stimuli and related conditioned behavioral responses. These studies show that hippocampal neuronal activation that occurs during the exploration of specific objects is embedded within the spatial firing patterns (place fields) of those neurons. For example, following tone-cued fear conditioning, hippocampal neurons come to be driven by the conditioned tone stimulus when the animal is within the place field of that neuron (Moita et al. 2003; Wang et al. 2012). Also, in rats performing a variant of the novel object exploration task, hippocampal neurons fired associated with specific objects and their familiarity embedded within the spatial firing patterns (place fields) of these neurons (Manns and Eichenbaum 2009). In rats performing a context-guided object-reward association task, hippocampal neurons fire when animals sample specific objects within particular locations and spatial contexts. In this experiment, the spatial specificity of responses occurred early and the object related activity paralleled learning to respond to different objects in only one context (Komorowski et al. 2009). Similarly, after training on somatosensory or auditory discrimination tasks, hippocampal neurons encode tactile and auditory cues along with the locations where they were experienced and rewarded (Itskov et al. 2011, 2012; Vinnik et al. 2012). This combination of studies clarifies that position-related firing precedes the adoption of stimulus or action specificity and suggests that the hippocampal network constitutes a spatial framework onto which memories of stimuli are incorporated. This conclusion is consistent with a large literature that positions the hippocampus as convergence site for streams of information processing about objects and space (reviewed in Davachi 2006; Eichenbaum et al. 2007), and suggests the mechanism for coding objects and events in space is conjunctive object and place coding by single hippocampal neurons.

Conclusions About Non-spatial Coding in Hippocampal Neurons There is considerable evidence that a broad range of specific significant stimuli can drive hippocampal neuronal activity and that hippocampal neurons fire associated with specific learned behaviors. At the same time, however, whenever these sensory and behavioral events occur in multiple locations, these activity patterns differ across locations. Thus, sensory-behavioral responses of hippocampal neurons are embedded within a spatial framework of hippocampal representation.

Time as an Additional Framework for Encoding Memories

There is considerable recent evidence that the hippocampus is involved in representing the flow of events in time, in parallel to its representation of the organization of events in space (Eichenbaum 2013, 2014), and indeed it has been suggested that bridging between successive events to link them in time may be a fundamental function of hippocampal circuitry (Rawlins 1985; Levy 1989; Wallenstein et al. 1998; Howard et al. 2014). Consistent with this idea, hippocampal lesions impair memory for the order of sequences of events (Fortin et al. 2002; Kesner et al. 2002) and ensemble activity patterns of CA1 neurons gradually change while rats sample sequences of odors, and this signal of continuously evolving temporal context predicted success in remembering the odor sequence (Manns et al. 2007). These findings, and more discussed below, suggest that temporal coding by the hippocampus is not merely representing the passage of time, but supports representation of the order of events in experiences, which can be used to guide subsequent behavior.

Several studies have now identified hippocampal principal neurons that fire at a particular moments in time of a temporally structured event, composing temporal maps of specific experiences. Across these studies, the location of the animal is held constant or firing patterns associated with elapsed time are distinguished from those associated with spatial and behavioral variables, and the firing patterns of these cells are dependent on the critical temporal parameters that characterize the task. Because these properties parallel those of place cells in coding locations in spatially structured experiences, we called these neurons "time cells" (MacDonald et al.

2011), even though these neurons are the same cells that exhibit spatial firing specificity in other circumstances.

Time cells have now been observed in several experiments. Pastalkova et al. (2008) recorded from single CA1 neurons as rats performed a spatial T-maze task where alternating left-turns and right-turns, and trials were separated by a fixed period of wheel running. They were the first to report that hippocampal neurons fire reliably at specific moments during wheel running and the entire period of each wheel run was filled by a sequence of brief neuronal activations. Importantly, the firing sequences differed between trials in which the rat subsequently turned left or right—even though the rat was largely in the same location (that is, in the running wheel) and performing the same behavior (that is, running)—but they were consistent between left-turn trials and consistent between right-turn trials, suggesting that a sequence was linked to the content of the trial. Subsequently, Kraus et al. (2013) also observed time cells in rats running in place on a treadmill in between trials on a T-maze, and showed that these cells are influenced independently and conjunctively by elapsed time and distance traveled on the treadmill (Fig. 1).

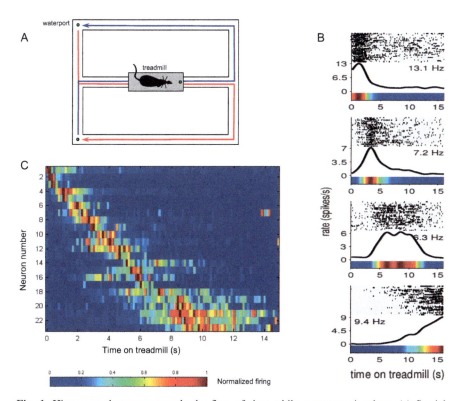

Fig. 1 Hippocampal neurons encode the flow of time while a rat runs in place. (**a**) Spatial alternation task in which on each trial the rat runs in place on a treadmill for 15 s. (**b**) Raster display, histogram, and heat plot of the time related firing pattern of four hippocampal neurons that fire at different moments during treadmill running. (**c**) Normalized firing rates of 23 hippocampal neurons over the course of treadmill running period. Adapted from Kraus et al. (2013)

Another study (Gill et al. 2011) examined activity patterns of CA1 neurons in rats performing a place-reversal task. In the first half of each daily session, trials began at any of three arms of a plus-maze and the rats had to go to the remaining arm to obtain a reward; in the second half of the session, another arm became the 'reward arm' and trials started from any of the other three arms. In between trials, rats were placed on a small platform outside the maze for several seconds. During the course of training, time-specific firing patterns emerged during the inter-trial periods, and the firing sequences differed between the two sessions. The rats could move freely during the delay, but cells that had reliable place fields were excluded from the analysis, indicating that the measured activity patterns encoded time rather than place.

In another study MacDonald et al. (2011) examined whether CA1 neurons also fired at specific moments in a non-spatial task where rats learned to associate each of two visually distinct objects with one of two cups of scented sand (Fig. 2a). On each trial, rats approached and sampled one of the two objects and, after a fixed delay, were exposed to one of the two odor cups. If the odor matched the object, the rat had to dig in the sand to retrieve a buried reward. During the delay period, individual neurons fired at successive moments that fill out the entire period, and firing patterns differed depending on which object the rats had to remember and were consistent between trials in which the same object had to be remembered. Extensive general linear model (GLM) analysis was used to distinguish activity patterns associated with the animal's location, speed and head direction during the delay period from the time elapsed. Although these spatial and behavioral parameters contributed to the activity patterns of many of the recorded cells, the analysis also revealed a contribution of time that was independent of these variables. Furthermore, the firing patterns of many of these neurons changed (i.e. they 're-timed') when the delay was increased. This happened even though the behavior and locations of the animal during the initial period did not change, indicating that the firing patterns of these cells reflected the passage of time rather than variations in behavior or place. Importantly, the cells firing later in the delay period were active for longer durations (i.e. had larger "time-fields"; also see Kraus et al. 2013, Fig. 1; MacDonald et al. 2013). This pattern suggests a scalar coding of time, which parallels a hallmark property of time judgments in humans and animals (Howard and Eichenbaum 2013). Each of these studies provided evidence for the existence of an evolving temporal signal that takes the form of a succession of briefly firing neurons.

Further evidence supporting the existence of temporal signals that are independent of place or distance has come from recent studies showing time cells in head-fixed animals in which the animal's location and behavior were kept constant and movement was eliminated. For example, in one study (MacDonald et al. 2013) rats performed an odor-cued delayed matching to sample task in which each trial began with the presentation of one of multiple sample odors for 1 s. Following a fixed delay, a test odor was presented. In order to receive a reward, the animal had to respond only to the test odor that matched the sample on that trial. We found that approximately 30 % of hippocampal cells encoded specific moments during the

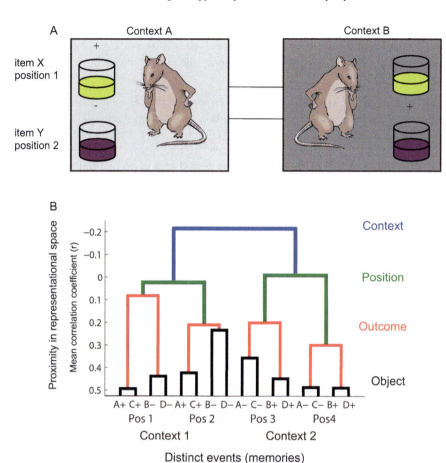

Fig. 2 Context-guided object memory task. (**a**) Rats enter either of two spatial contexts (A and B) in which they are presented with the same two object stimuli (X and Y) in either of the two positions shown. In Context A, object X contains a reward whereas in Context B, object Y contains a reward. (**b**) Dendrogram illustrating the relationships between representations of each type of event (x-axis) as linked (y-axis) by specific task dimensions (*right*). At the top of this schema, events that occur in different contexts are widely separated in representational space, indicated by anti-correlation between events that occur in different contexts, putting context as the highest superordinate dimension. Within each context-based network, events are then separated by positions within a context, i.e., positions are subordinate to contexts. Next, within positions, events are separated by different reward associations, i.e., reward association is subordinate to position. Finally, closest together in this schema are events that involve different objects that have the same reward association in the same location and context, i.e., object identity is subordinate to reward values. Adapted from McKenzie et al. (2014)

delay. Another study in head-fixed animals used two-photon calcium imaging to investigate the evolution of firing patterns among large ensembles of hippocampal neurons as mice underwent classical conditioning (Modi et al. 2014). On each trial, mice heard a brief tone that was followed, after a temporal gap, by an air-puff to the eye. During acquisition of the conditioned eye-blink response, CA1 cells developed time-locked firing sequences throughout the trial, including during the temporal gap.

Conclusions About Temporal Coding by Hippocampal Neurons Time cells have been observed in a range of behavioral conditions, including during delay periods in maze tasks in which rats alternate goals (Gill et al. 2011; Pastalkova et al. 2008; Kraus et al. 2013), bridging temporal gaps between associated non-spatial cues (MacDonald et al. 2011), during the delay period in a in non-spatial matching to sample task (MacDonald et al. 2013), and throughout trials in trace eyelid conditioning (Modi et al. 2014). Importantly, in some of these studies, the animal is immobilized and thus space plays no role in ongoing behavior or memory (MacDonald et al. 2013; Modi et al. 2014; Naya and Suzuki 2011). The findings of these studies establish a broad scope of temporally structured episodes in which the hippocampus encodes the temporal organization of specific experiences. Furthermore, some of the studies in animals have closely linked the emergence of time cell sequences to the encoding of specific memories and to subsequent memory accuracy (Gill et al. 2011; Modi et al. 2014; MacDonald et al. 2013), thus indicating a causal role of time cell firing patterns to memory performance. Also, the representation of temporally ordered sequences of events by the hippocampus extends to monkeys and humans. In monkeys, hippocampal neuronal activity signals elapsed time in a memory delay between associated objects (Naya and Suzuki 2011). In humans, hippocampal neurons fire in sequence associated with learning (Paz et al. 2010) and memory (Gelbard-Sagiv et al. 2008) of the flow of events experienced in movie clips.

Combined Spatial and Temporal Coding In addition, many studies have reported that ensembles of simultaneously recorded place cells that fire in sequential locations as animals traverse a path through a maze, subsequently also 'replay' the corresponding sequence of firings during 'off-line' periods, including sleep and quiet wakefulness when the animal is not moving through those locations (Carr et al. 2011; Karlsson & Frank 2009). Thus, spatial coding observed as rats actively run through a maze is recapitulated in temporal coded firing sequences when the rat is not moving. Disruption of these replay events impairs subsequent memory of the path (Jadhav et al. 2012). Moreover, field potentials associated with replays of sequences associated with alternative choice paths in a maze predict acquisition of learned performance (Singer et al. 2013). In addition, replay can be observed in sequential firing patterns associated with place-cell sequences that are about to occur as a rat takes a novel path in an open field (Pfeiffer and Foster 2013), and these replays converge on the target goal location (Pfeiffer and Foster 2015). The findings on replay strongly indicate a temporal coding of spatial representations relevant to memory.

The significance of prominent temporal representation as an aspect of non-spatial coding in the hippocampus is high in two ways. First, as introduced by Tulving (1983) episodic memories are defined by a temporal organization that embodies the temporal organization of events in personal experiences. We know that the hippocampus is critical to episodic memory and to memory for the temporal order of events, even when space is not relevant. Now the existence of time cells provides a mechanism by which the hippocampus organizes memories for events in time. Second, the existence of time cells offers a parallel temporal organizing mechanism to the spatial organizing mechanism offered by place cells. Therefore, the hippocampus could support representations of episodes by mapping objects and events within a framework of space and time, conferring upon those memories connections that reflect the spatial and temporal associations between distinct but related events embodied within a mapping by place and time cells (Eichenbaum 2013, 2014).

Linking Related Experiences into Memory Networks

McClelland et al. (1995) suggested that a key function of the hippocampus is to integrate new memories with the existing organization of related knowledge. Experimental evidence supporting this idea came from studies showing that rats integrate related memories and this capacity depends on the hippocampus (Dusek and Eichenbaum 1997, 1998; Bunsey and Eichenbaum 1996). More recently, Tse et al. (2007) showed that when rats learn to find specific food flavors in particular places in an open field, they develop an organized representation of the spatial relations among the objects in a particular environment and rely on the hippocampus for rapid assimilation of new flavor-place associations within the relational representation. Relating these findings to place cells, McKenzie et al. (2013) reported that hippocampal neurons encode multiple reward locations and rapidly assimilate and reorganize the overall network representation to incorporate new reward locations (see also Dupret et al. 2010).

In a more ambitious study, McKenzie et al. (2014) characterized hippocampal neural activity in a task where rats learned multiple context-dependent object-reward associations (Fig. 2a). Analyses of single neuron firing patterns revealed considerable variation in the types of non-spatial and spatial information encoded in hippocampal neural activity patterns, showing that hippocampal neuronal activity in complex tasks is "high-dimensional" in the sense that hippocampal neurons exhibit considerable mixed selectivity to multiple relevant non-spatial and spatial dimensions that are salient in a large range of memory tasks. In an effort to understand how these dimensions are organized in hippocampal networks, McKenzie et al. characterized the neural ensemble representations using a Representational Similarity Analysis (RSA) that compared population vectors accumulated during each type of event defined as a particular object in a specific position associated with reward or non-reward value within one of two spatial contexts. The RSA generated correlation coefficients that characterized the similarity of ensemble firing patterns among all pairs of event types. Then a hierarchical clustering analysis was used to determine the pairs of events that were most similar, then iteratively, the combined pairs of events that were most similar, and so on (Fig. 2b). This analysis revealed a hierarchy of relations among events: Events that involved the different objects of the same value were lowest in the hierarchy and embedded within specific positions. Next, events that involved different values were embedded within positions. Next, events at each position within a context were embedded within each context. Finally, representations of events across contexts were anticorrelated. Thus, hippocampal ensemble coding represented the identity of the objects, their reward assignments, the positions within a context in which they were experienced, and the context in which they occurred and networked these representations to form a systematic "map" of relations between the different types of memories.

Furthermore, after initial learning of one set of object associations, new object associations were rapidly assimilated into the relational structure that was established by initial learning. In addition, within the overall representation, items that had in common their reward associations in particular positions had strongly similar representations, even when they were never experienced together. These results suggest that, at the time of learning, new information is encoded within extant networks that stored related information, consistent with the view that new information is assimilated within networks of related memory traces to form hippocampal networks of related experiences (Eichenbaum 2004; McKenzie and Eichenbaum 2011). Similarities in hippocampal coding between familiar and novel conditions likely reflects the integration of related memories, arguably a primary purpose of memory systems in schema development and memory consolidation (McClelland et al. 1995; Tse et al. 2007). This overlapping code at the time of learning builds relational representations that can support transitive associations between separately learned experiences via of their common associations with a behaviorally relevant context (Dusek and Eichenbaum 1997; Bunsey and Eichenbaum 1996; Zeithamova et al. 2012).

The notion of relational representations that link memories in space can be readily extended to the linking of memories that are characterized by their flow in time. Thus, in studies described above where rats traverse different but overlapping routes through a T-maze, a typical finding is that some neurons represent the distinct memories that correspond to specific routes, even when rats traverse the overlapping segment of the maze, whereas other neurons fire similarly in the common segment thus providing a link between the distinct memories (Wood et al. 2000). Indeed, even in situations where animals traverse similarly structured routes in different mazes, whereas most neurons fire at distinct places in each maze, some fire similarly at positions that are functionally equivalent in the different mazes (Singer et al. 2010) or in different locations in the same maze (McKenzie et al. 2013). Thus, hippocampal networks create schemas that link spatial-temporal memories in situations where different routes have common features. Thus, the mechanism for interleaving of memories may be hippocampal neurons that encode overlapping features of multiple memories.

Conclusions: The Hippocampus as a Memory Space

The above review on hippocampal neuronal firing patterns allows me to address the following key questions: (1) What is the function of strong position coding by hippocampal neurons? And, (2) how are the various non-spatial and temporal coding properties of hippocampal neurons integrated with spatial coding?

It is remarkable that, after 40 years of research following the pioneering discoveries about hippocampal neurons in the 1970s, we have yet to reach a consensus on the nature of the hippocampal code. The early observations on hippocampal neurons in behaving animals revealed both behavioral and spatial firing properties. Each is quite apparent when the other is tightly controlled. Thus, in the studies following the early work, when behavior was held constant over locations, cells that exhibit spatial coding (place cells) are prevalent. Conversely, when space is held constant by immobilization, behavioral and temporal correlates of hippocampal activity are readily apparent in a variety of learning paradigms. Importantly, in a broad variety of testing paradigms when space, time, and sensory and behavioral events are salient, hippocampal neurons encode and integrate all of these dimensions of experience. The hundreds of studies on hippocampal neurons over these years has confirmed and extended these fundamental features of information coding by hippocampal neurons and networks. It is not too simplistic to conclude that the hippocampal network reflects all the salient events in attended experience, just as it should as indicated by its core function in memory. But how should we conceive the organization of information that supports this mirror of experience?

These properties support the notion that the hippocampus creates a "memory space" that binds in memory the elements of experiences and links memories via their common elements (Eichenbaum et al. 1999). By rapidly forming associations among any subset of its inputs, and between its inputs and reactivated relational memories, the hippocampus plays a critical role in the generation, recombination, and flexible use of information of all kinds. The representational schemes that underlie the memory space include representations of events as the relations among objects within the context in which they occur, representations of episodes as the flow of events across time, and representations that interleave events and episodes into relational networks, supporting the ability to draw novel inferences from memory (Eichenbaum 2004). This interpretation applies equally well to spatial and non-spatial domains of memory (Eichenbaum and Cohen 2014).

Considering the original definition of cognitive maps might provide progress towards a clarification of hippocampal function. According to Tolman (1948), a cognitive map is a form of mental organization of cognition, a tool for systematic organization of information across multiple domains of life. O'Keefe and Nadel (1978) interpreted the notion of a cognitive map narrowly to refer to a mental mapping of physical space and argued that the hippocampus performs spatial computations and represents geographical maps of the real world. The principals of cognitive mappings, however, can very well apply to episodic memories by viewing events as items organized in a spatial-temporal context (Butterly et al. 2012; Eichenbaum and Cohen 2014; Tavares et al. 2015). The memory space hypothesis takes the view that hippocampal networks map our location and movements within a broad range of life-spaces, supporting our ability to navigate spatial, temporal, and associational dimensions of personal experience (Eichenbaum et al. 1999; Eichenbaum 2004; see also Buzsaki and Moser 2013; Milivojevic and Doeller 2013).

Acknowledgements NIH MH094263, MH51570, MH052090.

References

Aghajan ZM, Acharya L, Moore JJ, Cushman JD, Vuong C, Mehta MR (2015) Impaired spatial selectivity and intact phase precession in two-dimensional virtual reality. Nat Neurosci 18:121–128

Ainge JA, Tamosiunaite M, Woergoetter F, Dudchencko PA (2007) Hippocampal CA1 place cells encode intended destination on a maze with multiple choice points. J Neurosci 27:9769–9779

Anderson MI, Jeffery KJ (2003) Heterogeneous modulation of place cell firing by changes in context. J Neurosci 23:8827–8835

Berger TW, Alger BE, Thompson RF (1976) Neuronal substrates of classical conditioning in the hippocampus. Science 192:483–485

Berger TW, Rinaldi PC, Weisz DJ, Thompson RF (1983) Single-unit analysis of different hippocampal cell types during classical conditioning of rabbit nictitating membrane response. J Neurophsiol 50:1197–1219

Bostock E, Muller RU, Kubie JL (1991) Experience-dependent modifications of hippocampal place cell firing. Hippocampus 1:193–206

Bower MR, Euston DR, McNaughton BL (2005) Sequential-context dependent hippocampal activity is not necessary to learn sequences with repeated elements. J Neurosci 15:1313–1323

Bunsey M, Eichenbaum H (1996) Conservation of hippocampal memory function in rats and humans. Nature 379:255–257

Butterly DA, Petroccione MA, Smith DM (2012) Hippocampal context processing is critical for interference free recall of odor memories in rats. Hippocampus 22:906–913

Buzsáki G, Moser EI (2013) Memory, navigation and theta rhythm in the hippocampal-entorhinal system. Nat Neurosci 16:130–138

Cameron KA, Yashar S, Wilson CL, Fried I (2007) Human hippocampal neurons predict how well word pairs will be remembered. Neuron 30:289–298

Carr MF, Jadhav SP, Frank LM (2011) Hippocampal replay in the awake state: a potential substrate for memory consolidation and retrieval. Nat Neurosci 14:147–153

Cressant A, Muller RU, Poucet B (1997) Failure of centrally placed objects to control the firing fields of hippocampal place cells. J Neurosci 17:2531–2542

Davachi L (2006) Item, context and relational episodic encoding in humans. Curr Opin Neurobiol 16:693–700

Deadwyler SA, Bunn T, Hampson RE (1995) Hippocampal ensemble activity during spatial delayed-nonmatch-to-sample performance in rats. J Neurosci 16:354–372

Derdikman D, Knierim JJ (eds) (2014) Space, time and memory in the hippocampal formation. Springer, Vienna

Diana RA, Yonelinas AP, Ranganath C (2007) Imaging recollection and familiarity in the medial temporal lobe: a three-component model. Trends Cogn Sci 11:379–386

Dupret D, O'Neill J, Pleydell-Bouverie B, Csicsvari J (2010) The reorganization and reactivation of hippocampal maps predict spatial memory performance. Nat Neurosci 13:995–1002

Dusek JA, Eichenbaum H (1997) The hippocampus and memory for orderly stimulus relations. Proc Natl Acad Sci U S A 94:7109–7114

Dusek JA, Eichenbaum H (1998) The hippocampus and transverse patterning guided by olfactory cues. Behav Neurosci 112:762–771

Eichenbaum H (2004) Hippocampus: cognitive processes and neural representations that underlie declarative memory. Neuron 44:109–120

Eichenbaum H (2013) Memory on time. Trends Cogn Sci 17:81–88

Eichenbaum H (2014) Time cells in the hippocampus: a new dimension for mapping memories. Nat Rev Neurosci 15:732–744

Eichenbaum H, Cohen NJ (2014) Can we reconcile the declarative memory and spatial navigation views of hippocampal function? Neuron 83:764–770

Eichenbaum H, Kuperstein M, Fagan A, Nagode J (1987) Cue-sampling and goal-approach correlates of hippocampal unit activity in rats performing an odor discrimination task. J Neurosci 7:716–732

Eichenbaum H, Dudchencko P, Wood E, Shapiro M, Tanila H (1999) The hippocampus, memory, and place cells: is it spatial memory or a memory space? Neuron 23:209–226

Eichenbaum H, Yonelinas AR, Ranganath C (2007) The medial temporal lobe and recognition memory. Ann Rev Neurosci 30:123–152

Fenton AA, Wsierska M, Kaminsky Y, Bures J (1998) Both here and there: simultaneous expression of autonomous spatial memories in rats. Proc Natl Acad Sci U S A 95:11493–11498

Ferbinteanu J, Shapiro ML (2003) Prospective and retrospective memory coding in the hippocampus. Neuron 40:1227–1239

Fortin NJ et al (2002) Critical role of the hippocampus in memory for sequences of events. Nat Neurosci 5:458–462

Frank LM, Brown EN, Wilson M (2000) Trajectory encoding in the hippocampus and entorhinal cortex. Neuron 27:169–178

Fried I, MacDonald KA, Wilson CL (1997) Single neurons activity in human hippocampus and amygdala during recognition of faces and objects. Neuron 18:753–765

Gelbard-Sagiv H, Mukamel R, Harel M, Malach R, Fried I (2008) Internally generated reactivation of single neurons in human hippocampus during free recall. Science 322:96–101

Gener T, Perez-Mendez L, Sanchez-Vives MV (2013) Tactile modulation of hippocampal place fields. Hippocampus 23:1453–1462

Gill PR, Mizumori SJ, Smith DM (2011) Hippocampal episode fields develop with learning. Hippocampus 21:1240–1249

Gothard KM, Skaggs WE, Moore KM, McNaughton BL (1996a) Binding of hippocampal CA1 neural activity to multiple reference frames in a landmark-based navigation task. J Neurosci 16:823–835

Gothard KM, Skaggs WE, McNaughton BL (1996b) Dynamics of mismatch correction in the hippocampal ensemble code for space: interaction between path integration and environmental cues. J Neurosci 16:8027–8040

Griffin AL, Eichenbaum H, Hasselmo ME (2007) Spatial representations of hippocampal CA1 neurons are modulated by behavioral context in a hippocampus-dependent memory task. J Neurosci 27:2416–2423

Hampson RE, Heyser CJ, Deadwyler SA (1993) Hippocampal cell firing correlates of delayed-match-to-sample performance in the rat. Behav Neurosci 107:715–739

Hartley T, Lever C, Burgess N, O'Keefe J (2013) Space in the brain: how the hippocampal formation supports spatial cognition. Philos Trans R Soc B Biol Sci 369(1635):20150510

Hattori S, Chen L, Weiss C, Disterhoft JF (2015) Robust hippocampal responsivity during retrieval of consolidated associative memory. Hippocampus 25:655–669

Hetherington PA, Shapiro ML (1997) Hippocampal place fields are altered by the removal of single visual cues in a distance-dependent manner. Behav Neurosci 111:20–34

Hill AJ, Best PJ (1981) Effects of deafness and blindness on the spatial correlates of hippocampal unit activity in the rat. Exp Neurol 74:204–217

Hollup SA, Molden S, Donnett JG, Moser M-B, Moser EI (2001) Accumulation of hippocampal place fields at the goal location in an annular watermaze task. J Neurosci 21:1635–1644

Howard MW, Eichenbaum H (2013) The hippocampus, time, and memory across scales. J Exp Psychol Gen 142:1211–1230

Howard MW, MacDonald CJ, Tiganj Z, Shankar KH, Du Q, Hasselmo ME, Eichenbaum H (2014) A unified mathematical framework for coding time, space, and sequences in the hippocampal region. J Neurosci 34:4692–4707

Itskov PM, Vinnik E, Diamond ME (2011) Hippocampal representation of touch-guided behavior in rats: persistent and independent traces of stimulus and reward location. PLoS One 6(1): e16462. doi:10.1371/journal.pone.0016462

Ison MJ, Quian Quiroga R, Fried I (2015) Rapid encoding of new memories by individual neurons in the human brain. Neuron 87:220–230

Itskov PM, Vinnik E, Honey C, Schnupp J, Diamond ME (2012) Sound sensitivity of neurons in rat hippocampus during performance of a sound-guided task. J Neurophysiol 107:1822–1834

Jackson J, Redish AD (2007) Network dynamics of hippocampal cell-assemblies resemble multiple spatial maps within single tasks. Hippocampus 17:1209–1229

Jadhav SP, Kemere C, German PW, Frank LM (2012) Awake hippocampal sharp-wave ripples support spatial memory. Science 336(6087):1454–1458

Jutras MJ, Buffalo EA (2010) Recognition memory signals in the macaque hippocampus. Proc Natl Acad Sci U S A 107:401–406

Karlsson MP, Frank LM (2009) Awake replay of remote experiences in the hippocampus. Nat Neurosci 12:913–918

Kennedy PJ, Shapiro ML (2009) Contextual memory retrieval: motivational states activate distinct hippocampal representations. Proc Natl Acad Sci U S A 106:10805–10810

Kesner RP et al (2002) The role of the hippocampus in memory for the temporal order of a sequence of odors. Behav Neurosci 116:286–290

Kjelstrup KB, Solstad T, Brun VH, Hafting T, Leutgeb S, Witter MP, Moser EI, Moser MB (2008) Finite scale of spatial representation in the hippocampus. Science 321:140–143

Komorowski RW, Manns JR, Eichenbaum H (2009) Robust conjunctive item-place coding by hippocampal neurons parallels learning what happens. J Neurosci 29:9918–9929

Komorowski RW, Garcia CG, Wilson A, Hattori S, Howard MW, Eichenbaum H (2013) Ventral hippocampal neurons are shaped by experience to represent behaviorally relevant contexts. J Neurosci 33:8079–8087

Kraus BJ, Robinson RJ II, White JA, Eichenbaum H, Hasselmo ME (2013) Hippocampal 'time cells': time versus path integration. Neuron 78:1090–1101

Kreiman G, Koch C, Fried I (2000a) Category-specific visual responses of single neurons in the human medial temporal lobe. Nat Neurosci 3:946–953

Kreiman G, Koch C, Fried I (2000b) Imagery neurons in the human brain. Nature 408:357–361

Lee I, Kim J (2010) The shift from a response strategy to object-in-place strategy during learning is accompanied by a matching shift in neural firing correlates in the hippocampus. Learn Mem 17:381–393

Lee I, Yoganarasimha D, Rao G, Knierim JJ (2004) Comparison of population coherence of place cells in hippocampal subfields CA1 and CA3. Nature 430:456–459

Lee I, Griffin AL, Zilli EA, Eichenbaum HM (2006) Gradual translocation of spatial correlates of neuronal firing in the hippocampus toward prospective reward locations. Neuron 51:539–650

Lenck-Santini PP, Fenton AA, Muller RU (2008) Discharge properties of hippocampal neurons during performance of a jump avoidance task. J Neurosci 28:6773–6786

Leutgeb S, Leutgeb JK, Treves A, Moser MB, Moser EI (2004) Distinct ensemble codes in hippocampal areas CA3 and CA1. Science 305:1295–1298

Leutgeb S, Leutgeb JK, Barnes CA, Moser EI, McNaughton BL, Moser MB (2005) Independent codes for spatial and episodic memory in hippocampal neuronal ensembles. Science 309:619–623

MacDonald CJ, Lepage KQ, Eden UT, Eichenbaum H (2011) Hippocampal "time cells" bridge the gap in memory for discontiguous events. Neuron 71:737–749

MacDonald CJ, Carrow S, Place R, Eichenbaum H (2013) Distinct hippocampal time cell sequences represent odor memories in immobilized rats. J Neurosci 33:14607–14616

Manns J, Eichenbaum H (2009) A cognitive map for object memory in the hippocampus. Learn Mem 16:616–624

Manns JR, Howard M, Eichenbaum H (2007) Gradual changes in hippocampal activity support remembering the order of events. Neuron 56:530–540

Markus EJ, Barnes CA, McNaughton BL, Gladden VL, Skaggs WE (1994) Spatial information content and reliability of hippocampal CA1 neurons: effects of visual input. Hippocampus 4:410–421

Markus EJ, Qin YL, Leonard B, Skaggs WE, McNaughton BL, Barnes CA (1995) Interactions between location and task affect the spatial and directional firing of hippocampal neurons. J Neurosci 15:7079–7094

McClelland JL, McNaughton BL, O'Reilly RC (1995) Why are there complementary learning systems in the hippocampus and neocortex: insights from the successes and failures of connectionist models of learning and memory. Psychol Rev 102:419–457

McEchron MD, Disterhoft JF (1997) Sequence of single neuron changes in CA1 hippocampus of rabbits during acquisition of trace eyeblink conditioned responses. J Neurophysiol 78:1030–1044

McKenzie S, Eichenbaum H (2011) Consolidation and reconsolidation: two lives of memories? Neuron 71:224–233

McKenzie S, Robinson NTM, Herrera L, Churchill JC, Eichenbaum H (2013) Learning causes reorganization of neuronal firing patterns to represent related experiences within a hippocampal schema. J Neurosci 33:10243–10256

McKenzie S, Frank AJ, Kinsky NR, Porter B, Rivière PD, Eichenbaum H (2014) Hippocampal representation of related and opposing memories develop within distinct, hierarchically-organized neural schemas. Neuron 83:202–215

McNaughton BL, Barnes CA, O'Keefe J (1983) The contributions of position, direction, and velocity to single unit activity in the hippocampus of freely-moving rats. Exp Brain Res 52:41–49

McNaughton BL, Barnes CA, Gerrard JL, Gothard M, Jung MW, Knierim JJ, Kudrimoti H, Qin Y, Skaggs WE, Suster M, Weaver KL (1996) Deciphering the hippocampal polyglot: the hippocampus as a path integration system. J Exp Biol 199:173–185

McNaughton BL, Battaglia FP, Jensen O, Moser EI, Moser MB (2006) Path-integration and the neural basis of the 'cognitive map'. Nat Rev Neurosci 7:663–678

Milivojevic B, Doeller CF (2013) Mnemonic networks in the hippocampal formation: from spatial maps to temporal and conceptual codes. J Exp Psychol Gen 142:1231–1241

Mizumori SJY (2007) Hippocampal place fields: relevance to learning and memory. Oxford University Press, Oxford

Modi MN, Dhawale AK, Bhalla US (2014) CA1 cell activity sequences emerge after reorganization of network correlation structure during associative learning. eLife 3:e01982

Moita MAP, Moisis S, Zhou Y, LeDoux JE, Blair HT (2003) Hippocampal place cells acquire location specific location specific responses to the conditioned stimulus during auditory fear conditioning. Neuron 37:485–497

Moita MA, Rosis S, Zhou Y, LeDoux JE, Blair HT (2004) Putting fear in its place: remapping of hippocampal place cells during fear conditioning. J Neurosci 24:7015–7023

Moser EI, Kropff K, Moser MB (2008) Place cells, grid cells, and the brain's spatial representation system. Ann Rev Neurosci 31:69–89

Muller RU (1996) A quarter of a century of place cells. Neuron 17:813–822

Muller RU, Kubie JL (1987) The effects of changes in the environment on the spatial firing of hippocampal complex-spike cells. J Neurosci 7:1951–1968

Muller RU, Kubie JL, Ranck JB Jr (1987) Spatial firing patterns of hippocampal complex spike cells in a fixed environment. J Neurosci 7:1935–1950

Muller RU, Bostock E, Taube JS, Kubie JL (1994) On the directional firing properties of hippocampal place cells. J Neurosci 14:7235–7251

Muzzio IA, Levita L, Kulkarni J, Monaco J, Kentros C, Stead M, Abbott LF, Kandel ER (2009) Attention enhances the retrieval and stability of visuospatial and olfactory representations in the dorsal hippocampus. PLoS Biol 7(6):e1000140. doi:10.1371/journal.pbio.1000140

Navratilova Z, Hoang LT, Schwindel CD, Tatsuno M, McNaughton BL (2012) Experience-dependent firing rate remapping generates directional selectivity in hippocampal place cells. Front Neural Circuits 6:6. doi:10.3389/fncir.2012.00006

Naya Y, Suzuki WA (2011) Integrating what and when across the primate medial temporal lobe. Science 333:773–776

O'Keefe J (1976) Place units in the hippocampus of the freely moving rat. Exp Neurol 51:78–109

O'Keefe J (1979) A review of hippocampal place cells. Prog Neurobiol 13:419–439

O'Keefe J, Burgess N (1996) Geometric determinants of the place fields of hippocampal neurons. Nature 381:425–428

O'Keefe J, Conway DH (1978) Hippocampal place units in the freely moving rat: why they fire when they fire. Exp Brain Res 31:573–590

O'Keefe J, Dostrovsky J (1971) The hippocampus as a spatial map. Preliminary evidence from unit activity in the freely-moving rat. Brain Res 34:171–175

O'Keefe J, Nadel L (1978) The Hippocampus as a Cognitive Map. Oxford University Press, New York

O'Keefe J, Speakman A (1987) Single unit activity in the rat hippocampus during a spatial memory task. Exp Brain Res 68:1–27

Olds J, Disterhoft JF, Segal M, Kornblith CL, Hirsh R (1972) Learning centers of rat brain mapped by latencies of conditioned unit responses. J Neurophysiol 35:202–219

Olton DS, Branch M, Best PJ (1978) Spatial correlates hippocampal unit activity. Exp Neurol 58:387–409

Otto T, Eichenbaum H (1992) Neuronal activity in the hippocampus during delayed non-match to sample performance in rats: evidence for hippocampal processing in recognition memory. Hippocampus 2:323–334

Pastalkova E, Itskov V, Amarasingham A, Buzsaki G (2008) Internally generated cell assembly sequences in the rat hippocampus. Science 321(5894):1322–1327

Paz R, Gelbard-Sagiv H, Mukamel R, Harel M, Malach R, Fried I (2010) A neural substrate in the human hippocampus for linking successive events. Proc Natl Acad Sci U S A 107:6046–6051

Penick S, Solomon PR (1991) Hippocampus, context, and conditioning. Behav Neurosci 105:611–617

Pfeiffer BE, Foster DJ (2013) Hippocampal place cell sequences depict future paths to remembered goals. Nature 497:74–79

Pfeiffer BE, Foster DJ (2015) Autoassociative dynamics in the generation of sequences of hippocampal place cells. Science 349:180–183

Quirk GJ, Muller RU, Kubie JL (1990) The firing of hippocampal place cells in the dark depends on the rat's recent experience. J Neurosci 10:2008–2017

Quiroga RQ, Reddy L, Kreiman G, Koch C, Fried I (2005) Invariant visual representation by single neurons in the human brain. Nature 435:1102–1107

Ranck JB Jr (1973) Studies on single neurons in dorsal hippocampal formation and septum in unrestrained rats. Part I. Behavioral correlates and firing repertoires. Exp Neurol 41:461–531

Ravassard P, Kees A, Willers B, Ho D, Aharoni D, Cushman J, Aghajan ZM, Mehta MR (2013) Multisensory control of hippocampal spatiotemporal selectivity. Science 340:1342–1346

Rawlins JNP (1985) Associations across time: the hippocampus as a temporary memory store. Behav Brain Sci 8:479–496

Rich PD, Liaw HP, Lee AK (2014) Place cells Large environments reveal the statistical structure governing hippocampal representations. Science 345:814–817

Levy WB (1989) A computational approach to hippocampal function. In: Hawkins RD, Bowers GH (eds) Computational models of learning in simple neural systems. Academic Press, Orlando, FL, pp 243–305

Robitsek JR, White J, Eichenbaum H (2013) Place cell activation predicts subsequent memory. Behav Brain Res 254:65–72

Rotenberg A, Muller RU (1997) Variable place-cell coupling to a continuously viewed stimulus: evidence that the hippocampus acts as a perceptual system. Philos Trans R Soc Lond B352:1505–1513

Rutishauser U, Schuman EM, Mamelak AN (2008) Activity of human hippocampal and amygdala neurons during retrieval of declarative memories. Proc Natl Acad Sci U S A 105:329–334

Save E, Nerad L, Poucet B (2000) Contribution of multiple sensory information to place field stability in hippocampal place cells. Hippocampus 10:64–76

Segal M, Olds J (1972) Behavior of units in hippocampal circuit of the rat during learning. J Neurophysiol 35:680–690

Shapiro ML, Tanila H, Eichenbaum H (1997) Cues that hippocampal place cells encode: dynamic and hierarchical representation of local and distal stimuli. Hippocampus 7:624–642

Shapiro ML, Kennedy P, Ferbinteanu J (2006) Representing episodes in the mammalian brain. Curr Opin Neurobiol 16:701–709

Sharp PE, Kubie JL, RU M (1990) Firing properties of hippocampal neurons in a visually symmetrical environment: contributions of multiple sensory cues and mnemonic processes. J Neurosci 10:3093–3105

Singer AC, Karlsson MP, Nathe AR, Carr MF, Frank LM (2010) Experience dependent development of coordinated hippocampal spatial activity representing the similarity of related locations. J Neurosci 30:11586–11604

Singer AC, Carr MF, Karlsson MP, Frank LM (2013) Hippocampal SWR activity predicts correct decisions during the initial learning of an alternation task. Neuron 77:1163–1173

Skaggs WE, McNaughton BL (1998) Spatial firing properties of hippocampal CA1 populations in an environment containing two visually identical regions. J Neurosci 18:8455–8466

Smith DM, Mizumori SJ (2006) Learning-related development of context-specific neuronal responses to places and events: the hippocampal role in context processing. J Neurosci 26:3154–3163

Tanila H, Shapiro M, Gallagher M, Eichenbaum H (1997a) Brain aging: impaired coding of novel environmental cues. J Neurosci 17:5167–5174

Tanila H, Shapiro ML, Eichenbaum HE (1997b) Discordance of spatial representation in ensembles of hippocampal place cells. Hippocampus 7:613–623

Tanila H, Sipila P, Shapiro M, Eichenbaum H (1997c) Brain aging: changes in the nature of information coding by the hippocampus. J Neurosci 17:5155–5166

Tavares RM, Mendelsohn A, Grossman Y, Williams CH, Shapiro M, Trope Y, Schiller D (2015) A map for social navigation in the human brain. Neuron 87:231–243

Tolman EC (1948) Cognitive maps in rats and men. Psychol Rev 55:189–208

Tse D, Langston RF, Kakeyama M, Bethus I, Spooner PA, Wood ER, Witter MP, Morris RGM (2007) Schemas and memory consolidation. Science 316:76–82

Tulving E (1983) Elements of Episodic Memory. Oxford University Press, New York

Vinnik E, Antopolskiy S, Itskov PM, Diamond ME (2012) Auditory stimuli elicit hippocampal neuronal responses during sleep. Front Syst Neurosci 6:49. doi:10.3389/fnsys.2012.00049

Wallenstein GV, Eichenbaum H, Hasselmo ME (1998) The hippocampus as an associator of discontiguous events. Trends Neurosci 21:315–365

Wang ME, Wann EG, Yuan RK, Ramos Álvarez MM, Stead SM, Muzzio IA (2012) Long-term stabilization of place cell remapping produced by a fearful experience. J Neurosci 32:15802–15814

Wang ME, Yuan RK, Keinath AT, Ramos Álvarez MM, Muzzio IA (2015) Extinction of learned fear induces hippocampal place cell remapping. J Neurosci 35:9122–9136

Wible CG, Findling RL, Shapiro M, Lang EJ, Crane S, Olton DS (1986) Mnemonic correlates of unit activity in the hippocampus. Brain Res 399:97–110

Wiebe SP, Stäubli UV (1999) Dynamic filtering of recognition codes in the hippocampus. J Neurosci 19:10562–10574

Wiener SI, Paul CA, Eichenbaum H (1989) Spatial and behavioral correlates of hippocampal neuronal activity. J Neurosci 9:2737–2763

Wills TJ, Lever C, Cacucci F, Burgess N, O'Keefe J (2005) Attractor dynamics in the hippocampal representation of the local environment. Science 308:873–876

Wirth S, Yanike M, Frank LM, Smith AC, Brown EN, Suzuki WA (2003) Single neurons in the monkey hippocampus and learning of new associations. Science 300:1578–1581

Wixted JT, Squire LR, Jang Y, Papesh MH, Goldinger SD, Kuhn JR, Smith KA, Treiman DM, Steinmetz PN (2014) Sparse and distributed coding of episodic memory ini neurons of the human hippocampus. Proc Natl Acad Sci U S A 111:9621–9626

Wood E, Dudchenko PA, Eichenbaum H (1999) The global record of memory in hippocampal neuronal activity. Nature 397:613–616

Wood E, Dudchenko PA, Robitsek JR, Eichenbaum H (2000) Hippocampal neurons encode information about different types of memory episodes occurring in the same location. Neuron 27:623–633

Young BJ, Fox GD, Eichenbaum H (1994) Correlates of hippocampal complex-spike cell activity in rats performing a nonspatial radial maze task. J Neurosci 14:6553–6563

Zeithamova D, Dominick AL, Preston AR (2012) Hippocampal and ventral medial prefrontal activation during retrieval-mediated learning supports novel inference. Neuron 75:168–179

Hippocampal Neurogenesis and Forgetting

Axel Guskjolen, Jonathan R. Epp, and Paul W. Frankland

Abstract Neurogenesis persists throughout life in the hippocampus, and there is a lot of interest in how the continuous addition of new neurons impacts hippocampal memory function. Behavioral studies have shown that artificially elevating hippocampal neurogenesis often facilitates new memory formation. However, since the integration of new neurons remodels existing hippocampal circuits, it has been hypothesized that hippocampal neurogenesis may also promote the degradation (or forgetting) of memories already stored in those circuits. Consistent with this idea, we have recently discovered that elevating rates of hippocampal neurogenesis after memory formation leads to forgetting. This finding changes the way we think about how hippocampal neurogenesis contributes to memory function, suggesting that it regulates a balance between encoding new memories and clearing out old memories.

A. Guskjolen
Program in Neurosciences and Mental Health, The Hospital for Sick Children, Toronto, Canada, M5G 1X8

Department of Physiology, University of Toronto, Toronto, Canada, M5S 1A8

J.R. Epp
Program in Neurosciences and Mental Health, The Hospital for Sick Children, Toronto, Canada, M5G 1X8

P.W. Frankland (✉)
Program in Neurosciences and Mental Health, The Hospital for Sick Children, Toronto, Canada, M5G 1X8

Department of Physiology, University of Toronto, Toronto, Canada, M5S 1A8

Department of Psychology, University of Toronto, Toronto, Canada, M5S 1A8

Institute of Medical Science, University of Toronto, Toronto, Canada, M5S 1A8
e-mail: paul.frankland@sickkids.ca

© Springer International Publishing AG 2017
D.E. Hannula, M.C. Duff (eds.), *The Hippocampus from Cells to Systems*,
DOI 10.1007/978-3-319-50406-3_4

Introduction

Neurogenesis, the production of new neurons, was believed to be completely absent in the mature nervous system. That is, all the neurons that an organism possessed were generated developmentally and there was no possibility for renewal or replacement. However, we now know that in at least two regions of the brain, new neurons continue to be produced throughout adult life. Research into the possibility of neurogenesis in the adult mammalian brain has a relatively brief history, but the field of functional neurogenesis has blossomed over the past two decades. In this chapter, we will outline the role neurogenesis plays in memory, with an emphasis on the newly discovered function of neurogenesis in regulating forgetting. First, we summarize the anterograde effects of neurogenic manipulation on subsequent learning and memory. These experiments have illuminated the role played by neurogenesis in the encoding, storage, and retrieval of hippocampal-dependent memories. We then outline recent experiments from our lab that have examined the retrograde effects of neurogenesis; that is, the effect that manipulating neurogenesis has on already established memories. Specifically, we detail the newly discovered role played by neurogenesis in regulating forgetting in mammalian species. We discuss three theories of how neurogenesis might cause forgetting, which memories are degraded, whether neurogenesis-induced forgetting is adaptive, and how this novel forgetting function of neurogenesis offers a new lens through which to understand and interpret the role neurogenesis plays in memory.

Historical Background

Early anatomists such as the Italian pathologist Giulio Bizzozero dismissed the possibility of neurogenesis (or cell genesis for that matter) in the adult mammalian brain. In 1894, Bizzozero proposed a classification scheme for different tissues based on their proliferative capacities (Bizzozero 1894). The three categories included tissues with constant proliferative capabilities such as red blood cells, tissues that could proliferate under certain conditions or for some time after birth (e.g. epithelium), and finally tissues that completely lacked proliferative activity. The tissues of the nervous system were placed as the key examples in this final category. Bizzozero's classification scheme was revised by others (Messier and Leblond 1960; Leblond 1964) during the following 70 years but in each classification system cells of the nervous system were placed into categories that described them as static or non-renewing.

In the early days of neurobiology, the functional units of the brain, namely neurons and glia, had not yet been resolved. In 1888, Santiago Ramon Y Cajal provided the first strong evidence that the nervous system was in fact made up of discrete neurons (Cajal 1888). Using the silver staining methodology developed by Golgi, he was able to stain and observe individual neurons in various parts of the nervous system. His observations, which became the foundation of what is known as the neuron doctrine, earned Cajal and Golgi the Nobel Prize in 1906. The neuron doctrine provided an extension of the existing cell theory; the idea that all living organisms are comprised of cells and that these cells arise from the division of other cells. This was a crucial step towards the discovery of adult neurogenesis although this was not to occur for many years to come. Cajal is often quoted as stating "Once the development was ended, the founts of growth and regeneration of the axons and dendrites dried up irrevocably. In the adult centers, the nerve paths are something fixed, ended, and immutable. Everything may die, nothing may be regenerated." However, the following sentence that shows a cautious degree of optimism that the preceding statement may not be true, is often omitted: "It is for the science of the future to change, if possible, this harsh decree"(Cajal 1914). Cajal's challenge has been met but not without many years of harsh skepticism.

In 1912, Ezra Allen described what he believed to be the existence of mitotic cells along the walls of the lateral ventricles of adult rats (Allen 1912). His identification was made purely based on the morphological appearance of the dividing cells in tissue sections stained with the basic nucleic acid stain thionin. Although this report provided no indication that the resulting daughter cells were neurons, it was the first evidence of cell division in the adult central nervous system and the first description of one of the two main germative zones in the adult brain. The first description of cell proliferation in other regions of the brain did not occur for another 50 years when Joseph Altman published a report using a relatively new tool called autoradiography to look for glial cell proliferation in response to cortical lesions (Altman 1962). In this technique, tritiated thymidine is administered to an animal. Any dividing cells will incorporate the radioactive thymidine into their DNA leaving a signature that can be detected later. Looking at the slides, Altman discovered cells labeled with thymidine, including what appeared to be neurons sparsely located throughout the cortex. This was the first identification of new neurons in the adult brain. In order to determine whether this neuronal labeling was simply a response to the injury, Altman conducted a follow up study using tritiated thymidine to investigate the brains of non-lesioned adult rats. Even without neural injury, Altman saw sparse labeling of neurons and glial cells in the neocortex (Altman 1963). However, this time an even greater discovery was made: the second major proliferative zone in the brain was finally discovered. Within the hippocampus, the dentate gyrus appeared to give rise to a large number of new neurons.

Although Altman had clearly identified adult neurogenesis, his findings were met with either criticism or indifference. Previous autoradiography studies had failed to identify adult neurogenesis, but perhaps more importantly, the belief that neurons were not produced in the adult brain was so engrained that even Altman himself remained cautiously skeptical of his findings. Altman conceded at first that perhaps some or most of the labeled neurons seen may actually represent "uptake by perineuronal satelites situated over the nuclei of neurons" or that "uptake of thymidine might reflect some process of DNA turnover that does not lead to cell multiplication" (Altman 1963).

Michael Kaplan and James Hinds expanded the findings of Altman by way of electron microscopy. Beginning with similar autoradiography techniques, they too located what appeared to be new neurons in the olfactory bulb, hippocampus (Kaplan and Hinds 1977), as well as the visual cortex (Kaplan 1981) of the adult rat. They then utilized electron microscopy to examine the fine structure of the proposed neurons. They were able to identify numerous neuronal characteristics such as axons, synapses, and synaptic vesicles. Kaplan not only helped prove the existence of neurogenesis in the adult brain, but also provided compelling evidence regarding the experience dependency of adult neurogenesis (Kaplan 1981), the population dynamics of the germative zones (Kaplan et al. 1985), as well as the long term viability of the new neurons (Kaplan 1985). Unfortunately, despite this growing evidence, there was still strong resistance to the idea of ongoing neurogenesis in the adult mammalian brain.

The most prominent and vocal skeptic of adult neurogenesis was Pasko Rakic. His continued opposition to the building evidence centered on four core arguments. Perhaps his most fundamental argument was that the cells being identified as neurons were in fact glia. Without the availability of neuron specific markers, the identity of the new cells was based on anatomical characteristics and was therefore subject to criteria that were not necessarily agreed upon by all researchers. For example, Michael Kaplan and Pasko Rakic had both independently performed similar autoradiography studies, but where Kaplan identified many new neurons, Rakic was unwilling to accept that the cells he saw met the criteria to be considered neurons (Rakic 1985a). A second (and related) argument was that although neurogenesis may occur in lower species such as rodents, it was highly unlikely to occur in the primate or human brain and was therefore largely irrelevant. To this end, Rakic, performed autoradiography studies on adult rhesus monkeys (Rakic 1985b). Although he identified a number of new glial cells, he failed to find any new cells that he considered to be neurons, even in the dentate gyrus. Some viewed neurogenesis (even in lower species) as a vestigial remnant or epiphenomenon; a phenomenon that lacked functional consequences, that was therefore not worth studying. A third argument focused on the relatively low numbers of new neurons on a given brain section. How could such small numbers of new neurons influence circuit function in meaningful ways? A final argument against the idea that neurogenesis persists in the adult brain centered on the notion that remembered experiences require stable neuron connectivity. Specifically, how could stable memories be maintained in a brain region where neurogenesis was inducing constant change? According to Pasko Rakic, a stable neuronal population was likely critical for the maintenance of long-term memory (Rakic 1985a, b). This statement was intended to be a nail in the coffin to the idea of continuous neurogenesis, but as we will soon discuss may actually have been an early clue to the function of adult-generated neurons.

What evidence was needed to put an end to this 'central dogma'? Three key items remained missing. First, specific neuronal markers were needed to convince the skeptics that the new cells were actually neurons and not glia. Second, it had to be demonstrated that new neurons survived long enough to have a functional impact. And finally, neurogenesis had to be demonstrated in primates, or more ideally, human brains. Elizabeth Gould provided the first of these missing items in 1992 (Gould et al. 1992). The authors of this study combined tritiated thymidine labeling with immunohistochemical labeling of a neuron specific protein called NSE (neuronal specific enolase). This finding confirmed that many of the new cells in the rat dentate gyrus were neurons.

Fernando Nottebohm's studies of neurogenesis in the songbird brain provided fodder for the second missing link in the 1980s. Work in his lab identified that new neurons were produced in the high vocal center (HVC) of the songbird brain. Interestingly, this region was also known to grow and shrink seasonally with song learning. In 1984, a study by Paton and Nottebohm (Paton and Nottebohm 1984) showed that the new neurons in the HVC were functional by combining electro-physiological labeling with autoradiography. After recording from random neurons, the cells were then filled with horseradish peroxidase to later identify the neurons. When comparing the electrophysiological recordings from the new neurons and the older neurons it was evident that both populations of neurons were functional (i.e., both fired action potentials in response to electrophysiological input). This was a particularly important finding because it was the first evidence that not only were adult generated neurons present but that they were actually connected to the mature circuitry in a meaningful way.

Towards the third missing link, several studies were performed in different labs to determine whether new neurons could be found in the adult primate brain. Although Pasko Rakic initially failed to identify adult neurogenesis in the primate brain (Rakic 1985b), he later published a study showing the presence of new neurons in the primate dentate gyrus (Kornack and Rakic 1999). Furthermore, Elizabeth Gould also demonstrated the presence of neurogenesis in the primate brain (Gould et al. 1999b). To date, adult neurogenesis has been identified in every mammalian species examined including humans. The groundbreaking work of Eriksson and colleagues put to rest the notion that neurogenesis only occurred in lower mammalian species (Eriksson et al. 1998). In their seminal paper, the authors obtained postmortem brain tissue from terminal cancer patients that had been administered Bromodeoxyuridine (BrdU; a synthetic analog of thymidine that incorporates into divided cells) to track the growth of their tumors. Evidence of labeled cells were found in both the subventricular zone as well as the subgranular zone of the dentate gyrus. The authors assessed the phenotype of the new cells using neuron specific proteins and confirmed that many of the adult generated cells were in fact neurons.

Maturation and Integration of Adult-Generated Neurons

Part of the reason why adult neurogenesis remained controversial for such a long time related to the ambiguity of how the new neurons were generated. In 1992, Reynolds and Weiss first demonstrated the existence of neural stem cells in the mouse striatum and thus identified a likely source of adult generated neurons in the dentate gyrus (Reynolds and Weiss 1992). In vivo, neural stem cells begin as a population of largely quiescent radial glial cells known as type 1 cells. Found in the subgranular zone (SGZ; a 2–3 cell-width layer in between the granule cell layer [GCL] and hilus), these cells can divide to produce highly proliferative type 2 progenitors that act as transit-amplifying cells to increase the pool of dividing cells. As early as 4 days after being generated, the type 2 cells migrate a short distance into the inner layers of the GCL where the vast majority become excitatory granule neurons (Cameron et al. 1993; Hastings and Gould 1999) (but see (Liu et al. 2003) for evidence that a small proportion become GABAergic inhibitory neurons) (see Fig. 1). The maturation and integration of newborn neurons follows a fairly stereotyped course (Espósito et al. 2005). Two weeks following their birth, the neurons have successfully extended dendrites into the molecular layer and axons into region CA3. Functional afferent and efferent synapses begin forming around 2.5 weeks, although full synaptic maturity and integration is not complete for several more weeks (Carlén et al. 2002; van Praag et al. 2002; Espósito et al. 2005; Toni et al. 2007, 2008; Faulkner et al. 2008). During this period, the new synapses exist preferentially alongside (Toni et al. 2007, 2008), and may even compete with and eliminate (Yasuda et al. 2011), pre-existing synaptic connections.

Modulation of Neurogenesis

Adult neurogenesis can be regulated by a variety of different internal and external factors that act to either increase or decrease the number of new neurons that are generated. The list of factors that affect the rate of neurogenesis is extensive and includes neurotransmitters, growth factors, hormones, and pharmacological agents. Environmental factors such as enrichment, exercise and stress are also potent modulators of neurogenesis and act through various transmitters and signaling cascades to modulate neurogenesis. Neurogenesis can be modulated at a number of different developmental milestones. The first step in the process of producing new neurons is cell proliferation. Cell proliferation may be increased or decreased, ultimately leading to a change in the number of neurons that are generated. However, not all cells that are born become neurons, and many of the cells that are born will ultimately not survive. Heather Cameron and colleagues first demonstrated that a large percentage of the new cells die between 1–2 weeks following proliferation (Cameron et al. 1993). Thus, three additional developmental stages that are modulated by internal and external factors are cell differentiation, survival, and integration into surrounding circuitry. However, regardless of the means through which adult neurogenesis is modulated, it is ultimately the survival and integration of the new neurons that is likely to have the most functional relevance.

One of the most potent negative modulators of adult neurogenesis is aging. Although neurogenesis continues throughout adult life, levels of cell proliferation and survival are not static, and instead decrease with age—a discovery first made by

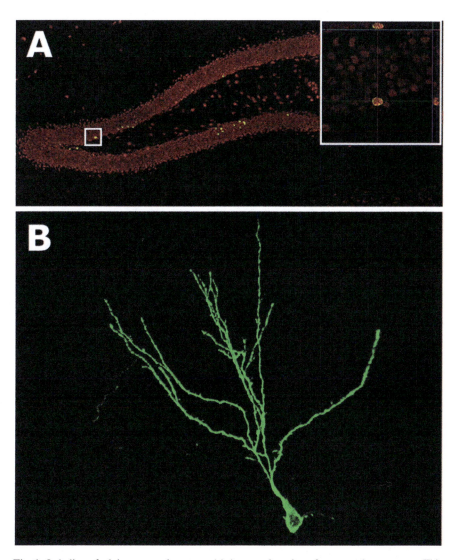

Fig. 1 Labeling of adult-generated neurons. (**a**) A coronal section of a mouse dentate gyrus. This animal was injected I.P. with BrdU 1 month earlier. The tissue has been stained immunohistochemically stained for BrdU (in *green*) and NeuN (a neuron-specific marker; in *red*). The BrdU positive cells appear in yellow because they are positive for both BrdU and NeuN, indicating that these cells are in fact adult-generated neurons. (**b**) A 28 day old adult-generated granule neuron. A retrovirus-CAG-GFP, which infects dividing cells, was stereotaxially injected into the dentate gyrus 1 month earlier. Unlike BrdU (which labels only the cell body), a retrovirus will infect the whole neuron. This allows for detailed morphological quantification of infected neurons in terms of dendritic complexity, spine characteristics, mossy fiber terminal characteristics, etc.

Georg Kuhn (Kuhn et al. 1996). The rate of adult neurogenesis is dependent on the particular species examined. In young adult rats the number of new neurons produced per day is approximately 9000. Between 5 weeks and 1 year of age the rate of both cell proliferation and new neuron production decrease by approximately 90 % (McDonald and Wojtowicz 2005). In humans it is much more difficult to determine the rates of neurogenesis at various ages. However, it has been shown that neurogenesis persists to some extent in subjects up to the age of 100 years (Knoth et al. 2010). Furthermore, using carbon-14 dating as a label for dividing cells, it was recently shown that human hippocampus sustains substantial levels of hippocampal neurogenesis throughout adulthood, generating approximately 1400 new hippocampal neurons per day (Spalding et al. 2013). This study found only a modest decline in rates of neurogenesis with aging. The Knoth and Spalding papers both suggest that rates of neurogenesis in humans and rodents are approximately equal to each other. This reinforces the belief that adult neurogenesis should be functionally important to hippocampal function in humans as it is in rodents.

Another powerful negative regulator of adult neurogenesis is chronic stress. Numerous studies have demonstrated that a variety of stress paradigms reduce the proliferation of new neurons in both the rodent and primate dentate gyrus (Tanapat et al. 1998; Coe et al. 2003; Willner 2005). Interestingly, the same stress paradigms can be used to produce rodent models of affective disorders such as depression. Naturally, these findings led to the hypothesis that depression may be caused by reduced neurogenesis and hippocampal atrophy. More evidence of the link between neurogenesis and depression came from Jessica Malberg who discovered that chronic administration of the antidepressant fluoxetine increased proliferation and new neuron survival in the dentate gyrus (Malberg et al. 2000; Malberg and Duman 2003). Subsequently, it has been shown that all major classes of antidepressant drugs also increase neurogenesis. In post mortem tissue from humans diagnosed with major depressive disorder, neurogenesis is also decreased. Although the link between neurogenesis and depression is strong, there have been studies that do not support a causal relationship, including one study that demonstrated the behavioural effects of antidepressents were independent of an increase in neurogenesis (Holick et al. 2008). As a result the status of the neurogenic hypothesis of depression remains uncertain, although both stress and antidepressants clearly have strong regulatory effects on cell proliferation.

Interestingly, the earliest observation that adult neurogenesis could be positively modulated by external factors may have come from Michael Kaplan in the early 1970s. Although his findings were ultimately never published, he claims to have found that exposure to an enriched environment increased the number of new neurons (Kaplan 2001). However, it was not until 1997 that data were published which demonstrated the potent pro-neurogenic effect of rearing animals in an enriched environment (Kempermann et al. 1997). This study built on a large amount of existing data showing that environmental enrichment (e.g., larger cages, tunnels, toys, more cagemates, and/or access to running wheel) produced a large number of structural changes in the brain. The authors housed mice in either an enriched environment versus standard laboratory housing and administered

BrdU. Several weeks later when they examined the brains they found that mice in the enriched environment had increased cell survival. Many additional studies have confirmed these effects (Brown et al. 2003; Olson et al. 2006) but see (Gregoire et al. 2014).

Exercise is another powerful modulator of adult neurogenesis. In rodents, voluntary running causes a fairly drastic (approximately twofold) increase in the number of new neurons produced in the dentate gyrus. Using a similar experimental protocol to Kempermann's enrichment study, this effect was first demonstrated by Henriette van Praag in 1999 (van Praag et al. 1999). Similar to environmental enrichment, voluntary exercise causes a variety of plasticity induced changes in the brain, but the increase in neurogenesis is an unambiguously strong effect.

Given that adult neurogenesis occurs only in restricted areas of the brain, it stands to reason that the process may be related to the specific functions performed by these neural regions. The hippocampus itself is crucial for a number of types of learning and memory, most notably spatial and contextual memory. Elizabeth Gould provided the first evidence of a functional link between learning and neurogenesis in 1999 (Gould et al. 1999a). In this study, Gould and colleagues took advantage of the fact that many of the new neurons that are produced in the dentate gyrus die during the first 1–2 weeks after proliferation. First, the researchers injected rats with BrdU to mark proliferating cells. One week later, they trained mice in a hippocampus dependent spatial learning task known as the Morris water maze. Following training, they examined the brains of the rats and discovered that the animals that were trained in the spatial task had an increase in the number of surviving new neurons. In addition, the same survival enhancing effect was found in another hippocampus dependent task known as trace eyeblink conditioning, thereby suggesting a generalized effect of learning on new neuron survival. A critical period spanning from day 6–10 after proliferation was later found during which learning enhances new neuron survival. Learning before or after this timeframe does not affect cell survival, likely due to the fact that the new neurons will either be too immature or mature to be impacted by the stimulation (Epp et al. 2007). The learning induced enhancement of survival appears to require a certain degree of hippocampal activation because exposure to the test environment itself does not enhance survival (Gould et al. 1999a). It has also been shown that the task must be made sufficiently difficult in order to promote survival (Epp et al. 2010).

Adult neurogenesis is now a widely accepted phenomenon. The general acceptance of adult neurogenesis has now led to another debate over the functional importance, if any, of producing new neurons in the adult brain. Although this debate continues today, much progress has been made in our understanding of the functional implications of adult neurogenesis. Thanks to many recent technological advances it is now possible to utilize numerous approaches, including the use of transgenic mice, to manipulate the levels of neurogenesis. Identification of new neurons has become remarkably easy due to an increase in the number of available proliferative and immature neuron markers. Viral labeling of dividing neurons has provided another powerful tool to label new neurons but also to directly activate or silence new neurons using chemogenetic or optogenetic constructs. In vivo imaging techniques now allow us to watch the maturation of neurons in real time to further understand how these new neurons make a functional contribution to the hippocampus. In the remainder of this chapter we will review the functional consequences and importance of adult generated neurons in the hippocampus, in particular as it relates to learning and memory.

Functional Consequences of Hippocampal Neurogenesis

Considerable support exists for a role of hippocampal neurogenesis in modulating learning and memory processing. This is perhaps not surprising given its evolutionary conservation across species and the specific location of neurogenesis within the adult mammalian brain. Not only does neurogenesis occur in the hippocampus—a structure often conceptualized as the "gateway to memory" (Kempermann 2002)—but also within the dentate gyrus, which is the main entry point into the tri-synaptic circuit of the hippocampus (i.e., entorhinal cortex to dentate gyrus, dentate gyrus to CA3, and CA3 to CA1). In this way, neurogenesis occurs in the region of the hippocampus through which a significant amount of information must pass before it can be encoded. This 'bottleneck' location makes strategic sense as it readily allows newly born dentate granule neurons to contribute to the processing of environmental input and to hippocampal plasticity more generally (see Fig. 2). As

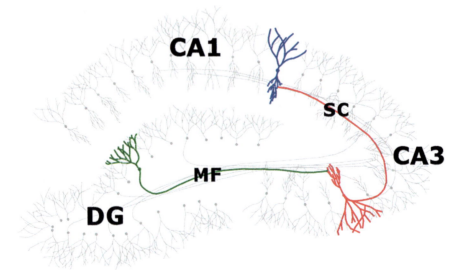

Fig. 2 Simplified hippocampal anatomy. Upon receiving input from the entorhinal cortex (not shown), granule cells of the dentate gyrus (DG) send axons (known as mossy fibers; MF) to neurons of CA3. Neurons in CA3 send axons (known as schaffer collaterals; SC) and form synapses with pyramidal neurons of CA1. Neurons in CA1 then synapse with the neurons in the subiculum (not shown), which subsequently send information back into the cortex. For ease of visualization, one dentate granule neuron is colored *green*, one CA3 neuron is colored *red*, and one CA1 neuron is colored *blue*

outlined above, the generation, maturation, and integration of new dentate granule neurons has a substantial impact on hippocampal circuitry. However, most theories attempting to explain the functional role played by hippocampal neurogenesis in memory have focused primarily on the unique physiological properties possessed by immature dentate granule neurons. In particular, these new neurons are highly excitable relative to their mature counterparts (Schmidt-Hieber et al. 2004; Ge et al. 2007; Mongiat et al. 2009; Marín-Burgin et al. 2012). As such, newly generated neurons are more likely to fire an action potential in response to mnemonically-relevant environmental (or internally generated) input. By having a subpopulation of hippocampal neurons that preferentially respond to inputs received while they develop, the dentate gyrus will have a high probability of always possessing a subpopulation of neurons that will be capable of responding to any environment the animal experiences (Aimone et al. 2009, 2011). The number of adult generated neurons eventually reaches ~1,000,000 in rodent brains. Given that each granule neuron contacts around 10–15 CA3 pyramidal neurons (Acsády et al. 1998), each of the approximately 500,000 CA3 pyramidal cells in a mature rat could potentially have a direct connection from an adult generated granule neuron (Snyder and Cameron 2012). Additionally, the connection between granule neurons and CA3 pyramidal neurons is so powerful that a single granule neuron is able to trigger activity in its CA3 targets (Henze et al. 2002). These facts, in conjunction with the high degree of intrinsic excitability present in immature granule neurons (Schmidt-Hieber et al. 2004; Ge et al. 2007; Mongiat et al. 2009; Marín-Burgin et al. 2012), suggest that even a small number of adult-generated granule neurons can have a disproportionately large impact on hippocampus-dependent memory function. In other words, the unique anatomical and electrophysiological properties (i.e. high excitability) of newborn neurons helps address the concern that too few new neurons are produced in the adult brain to functionally impact learning and memory processing.

A general consensus has emerged that adult generated dentate granule neurons participate in hippocampus-dependent memory processes. It is surprising that, almost without exception, the same general experimental design has been used to study the function of newborn neurons in relation to memory. Typically, researchers have first manipulated levels of neurogenesis in some way (e.g., experimentally increasing or decreasing rates of neurogenesis), and then evaluated whether animals learn better or worse. In other words, virtually all experimental designs up to this point have focused on the *anterograde effects* of manipulating hippocampal neurogenesis. These anterograde experiments have been successful in determining the role that hippocampal neurogenesis plays in the formation of memory. Intuitively, however, new neurons can have both anterograde and retrograde effects on memory, depending on when these neurons are generated relative to memory encoding. What are the *retrograde effects* of hippocampal neurogenesis? For example, how does the integration of new neurons into surrounding neural circuitry impact information already stored in those circuits?

Below, we outline evidence that neurogenesis impacts hippocampal-dependent memory in both the anterograde and retrograde directions, and focus, in particular, on the recently discovered role of neurogenesis in mediating forgetting (Akers et al. 2014).

Anterograde Effects: The Impact of Hippocampal Neurogenesis on Encoding, Storing and Retrieving New Memories

Memory Encoding As the main entry point of the hippocampus, the dentate gyrus is in an ideal position to contribute to the processing and integration of incoming information. Interestingly, the dentate gyrus has a sparse neural coding scheme (Chawla et al. 2005), with the majority of mature granule neurons being effectively silent in most of the environments the animal experiences (Alme et al. 2010). Given the silence of its mature neural population, possessing a subpopulation of highly excitable immature neurons (Schmidt-Hieber et al. 2004; Ge et al. 2007; Mongiat et al. 2009; Marín-Burgin et al. 2012) would be more amenable to the processing of incoming information and the subsequent formation of a memory. Consistent with this, it has been reported using a variety of methods that adult generated neurons functionally integrate into hippocampal-dependent memory circuitry (Carlén et al. 2002; van Praag et al. 2002; Ramirez-Amaya et al. 2006; Kee et al. 2007; Tashiro et al. 2007; Trouche et al. 2009; Stone et al. 2011). Thus, under normal conditions, these new neurons help form the physical substrate of memory. But are newly generated neurons *necessary* for memory encoding? If having a subpopulation of excitable immature neurons plays an essential role in memory encoding, then suppressing hippocampal neurogenesis prior to a learning experience would be predicted to inhibit subsequent learning. Conversely, increasing the generation or survival of new neurons prior to a learning experience would be predicted to enhance subsequent learning. Indeed, this is often found to be the case (Shors et al. 2001a), particularly in cases where spatial or contextual discrimination is required of the animal (i.e., 'behavioural pattern separation' or 'memory resolution' (Clelland et al. 2009; Creer et al. 2010; Sahay et al. 2011b; Kheirbek et al. 2012; Niibori et al. 2012; Tronel et al. 2012) see also (Aimone et al. 2011; Sahay et al. 2011a; Santoro 2013). However, the above pattern of results is not always found (Shors et al. 2001b; Snyder et al. 2005; Meshi et al. 2006; Saxe et al. 2006; Dupret et al. 2008; Zhang et al. 2008; Jaholkowski et al. 2009; Drew et al. 2010; Arruda-Carvalho et al. 2011; Denny et al. 2012; Jedynak et al. 2012; Groves et al. 2013; Martinez-Canabal et al. 2013; Urbach et al. 2013), indicating that developmentally generated and/or mature neurons are often capable of compensating for a lack of adult neurogenesis. That is, in the absence a of highly excitable population of immature neurons, the less excitable mature neurons are often sufficient in supporting hippocampal-dependent learning. The cause of these discrepant behavioural results are likely multifaceted, being dependent upon the age and sex of animal tested, the developmental stage of the neurons involved, the number of neurons targeted, and task difficulty. Regardless, it is now clear that adult-generated dentate granule neurons participate in the encoding of new hippocampal-dependent memories, and are sometimes (but not always) necessary for successful memory formation.

Memory Storage Logically, neurons which underlie memory encoding are also required (or at least a subset of them are required) in the subsequent storage and retrieval of that memory—an idea that has received support through "tag-and-ablate" studies in the amygdala (Han et al. 2007, 2009; Yiu et al. 2014) as well as "tag-and-activate" experiments in the hippocampus (Garner et al. 2012; Liu et al. 2012; Ramirez et al. 2013). Given that adult-generated neurons participate in memory encoding (Carlén et al. 2002; van Praag et al. 2002; Ramirez-Amaya et al. 2006; Kee et al. 2007; Tashiro et al. 2007; Trouche et al. 2009; Stone et al. 2011), it is reasonable to predict that the integrity of these neurons would be important in the maintenance and expression of these memories as well. One way to test this prediction is to ablate or inhibit these new neurons only after they have become a part of the memory trace (Frankland 2013). Using this experimental design, two papers have investigated this hypothesis. In this first paper (Arruda-Carvalho et al. 2011), a "tag-and-ablate" transgenic strategy was used that allowed the researchers to tag adult-generated neurons and then ablate them either before or after the neurons were incorporated into a memory trace. This tag-and-ablate strategy had the added benefit of allowing the authors to target a particular subgroup of adult-generated neurons rather than producing a global disruption in neurogenesis. Consistent with the idea that mature dentate granule neurons are often capable of supporting learning in the absence of an immature neural population, no evidence of encoding failure was found when the 'tagged' population of adult-generated neurons was ablated prior to memory encoding. However, Arruda-Carvalho and colleagues found that ablating the tagged population of neurons *after* learning produced a memory impairment in three distinct hippocampal-dependent memory tasks. Interestingly, ablating these neurons even 1 month after training impaired memory expression. These results indicate that upon committing to the memory trace, adult-generated hippocampal neurons form an essential and enduring component of the memory trace.

Memory Retrieval In Arruda-Carvalho et al. (2011), the ablation of the tagged population of neurons occurred over a period of 7 days. As such, this experiment design does not allow us to determine whether these neurons play a role only in the storage of the memory (i.e., no role in memory retrieval per se), or whether the integrity of these neurons are required for both the storage *and* retrieval of the memory trace. To investigate this issue, Gu and colleagues used a combination of retrovirus and inhibitory optogenetic techniques to 'birthdate' dividing cells and exert control over adult-generated neurons (Gu et al. 2012). First, the inhibitory opsin, ArchT, was retrovirally expressed in new dentate gyrus neurons. The mice were then trained on one of two hippocampal-dependent tasks either 2, 4, or

8 weeks later. One day after training, the ability of mice to retrieve the memory was tested. Silencing 4 week old adult-generated neurons (but not 2 or 8 week old neurons) suppressed retrieval of hippocampal dependent memories. Crucially, because the tagged population of adult-generated neurons was silenced only during the retrieval episode (i.e., the neurons are uninterrupted during the consolidation period), this experiment permits the conclusion that these immature neurons play a role in memory retrieval, independent of their role in memory storage. Gu and colleagues also found that silencing tagged adult-generated neurons during training had no adverse effect on memory encoding, a result consistent this the notion that mature dentate granule cells are capable of forming a memory trace when new neurons are absent. In summary, the work of Arruda-Carvalho et al. (2011) and Gu et al. (2012) indicates that (1) under normal circumstances, adult-generated neurons participate in memory formation, and (2) the integrity of immature neurons committed to a memory trace is essential in both the storage and retrieval of the memory.

Retrograde Effects: The Impact of Hippocampal Neurogenesis on Forgetting

While the anterograde method of inquiry has been successful in determining the role that neurogenesis plays in underlying memory encoding, it has potentially come at the expense of addressing *retrograde effects*, or the effects neurogenesis has on *already established memories*. By turning our attention in the retrograde direction, our lab has recently uncovered roles for neurogenesis in the forgetting of hippocampal memories.

The available data in the field of functional neurogenesis suggests that if neurogenesis has any effect on learning and memory, it does so in a way that promotes the formation and stability of memory. Exceptions are found in cases where ectopic or otherwise unhealthy neurons are generated, as in the case of epileptic seizures (Parent et al. 1997). We seem to be biased towards believing that these neurons have some positive effect on memory (if not, why has hippocampal neurogenesis been evolutionarily conserved?). However, as discussed above, a few early researchers in the field toyed with the idea that having no-to-low levels of neurogenesis is required for memory stability across time (Rakic 1985b; Eckenhoff and Rakic 1988). Furthermore, a variety of computational models have predicted that neurogenesis should cause degradation of previously stored information (i.e., retrograde effects) (Chambers et al. 2004; Deisseroth et al. 2004; Meltzer et al. 2005; Weisz and Argibay 2012). While some of these models argue that the continual addition of new neurons should cause forgetting of old information (additive models), others predict forgetting by removing and replacing the neurons after they have become a critical part of the memory trace (replacement models)—a result which is perhaps neither particularly surprising nor biologically relevant, given that neurons involved in a hippocampal-dependent memory trace see a survival advantage (Leuner et al. 2004; Epp et al. 2007). However, one can readily imagine a scenario in which forgetting occurs not by removing whole neurons, but rather by removing mnemonically relevant synapses from these neurons—a more biologically realistic possibility. Indeed, the main cause of forgetting is thought not to be passive trace decay due to time, but rather the onslaught of synaptic activity in response to new experience and learning (Wixted 2004; Fusi et al. 2005). Adult neurogenesis is a powerful means through which a constant onslaught of synaptic rearrangement occurs in the hippocampus, one that is upregulated in response to enriched experience and learning (Kempermann et al. 1997; Gould et al. 1999a; Leuner et al. 2004; Epp et al. 2007, 2011), but which occurs independently of these events as well. Our model of neurogenesis-induced forgetting can be conceptualized as a synthesis of the additive and replacement computational models. In particular, as the continual addition of new neurons mature and form synaptic connections, they necessarily remodel (Toni et al. 2007, 2008), and potentially even replace (Yasuda et al. 2011), the synaptic circuitry upon which hippocampal memories are dependent. This remodeling of the circuit reduces the probability that a given environmental cue will reactivate the specific pattern of neural activity that mediates successful memory retrieval, resulting in forgetting (Josselyn and Frankland 2012; Frankland et al. 2013).

We recently tested the prediction that post-training hippocampal neurogenesis causes forgetting in a series of experiments involving both young and old animals of three different species: mice, guinea pigs, and degus (Akers et al. 2014) (Fig. 3). First, we showed that running increases rates of hippocampal neurogenesis in adult mice. Next, we found that post-training (but not pre-training) access to a running wheel causes forgetting of hippocampal-dependent memories (but not hippocampal-independent memories). However, voluntary wheel running alters a host of physiological phenomena outside of hippocampal neurogenesis. Thus, we next tested whether inhibiting the increase in neurogenesis would alleviate the forgetting phenotype. To do this, nestin-tk mice which express HSV-thymidine kinase (a 'cell suicide gene') driven by a neural progenitor specific promoter (nestin) were used. Activation of thymidine kinase by ganciclovir results in the death of proliferating cells (Singer et al. 2009; Niibori et al. 2012). Thus, by introducing ganciclovir into the diet of nestin-tk mice, levels of neurogenesis are drastically reduced. Using this mouse line, we found that inhibiting post-training increases in neurogenesis completely reversed the forgetting of contextual fear memory. Increasing neurogenesis pharmacologically (via memantine) also caused forgetting of contextual fear, and this effect was once again blocked using the nestin-tk mice. To determine whether genetically increasing neurogenesis would also cause forgetting of hippocampal memories, iKO-p53 mice (in which the tumor suppressor gene p53 is inducibly deleted from neural progenitors and their progeny) were used. Once again, this manipulation caused forgetting of contextual fear memory. Thus, using naturalistic, pharmacological, and genetic manipulations, we demonstrated that hippocampus neurogenesis regulates the forgetting of hippocampal-dependent memories (Akers et al. 2014).

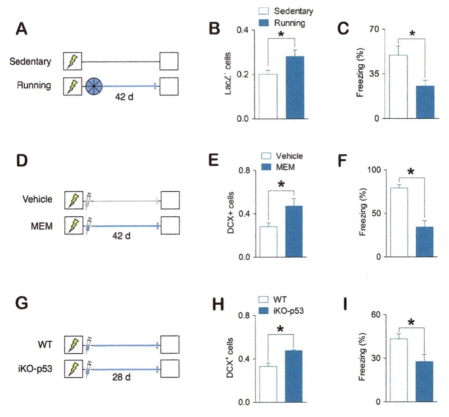

Fig. 3 Increasing neurogenesis post-learning causes forgetting in adult mice. (**a**) Mice ran following contextual fear conditioning. (**b**) Running increased rates of hippocampal neurogenesis (indicated by LacZ) and (**c**) caused forgetting of the contextual fear memory. (**d**) Mice were injected with the pharmacological agent memantine following contextual fear conditioning. (**e**) Memantine increased rates of hippocampal neurogenesis (indicated by DCX) and (**f**) caused forgetting of the contextual fear memory. (**g**) The tumor suppressor gene p53 was selectively deleted following contextual fear conditioning. (**h**) Deleting p53 increased rates of hippocampal neurogenesis (indicated by DCX) and (**i**) lead to forgetting of the contextual fear memory. Reprinted with permission from AAAS

Altricial mammalian species (including mice and humans) have high levels of neurogenesis during infancy. Interestingly, these species also exhibit more pronounced forgetting of hippocampus-dependent memories during infancy (i.e., infantile amnesia) (Howe and Courage 1993; Josselyn and Frankland 2012). To determine whether high rates of infantile neurogenesis contribute to infantile amnesia, we suppressed neurogenesis in infant mice using both genetic (nestin-tk) and pharmacological (temozolomide, a DNA alkylating agent) methodology. Both of these methods mitigated the forgetting of hippocampal-dependent memories formed during infancy (Akers et al. 2014). Next, we showed that two precocial species (guinea pigs and degus) possess significantly lower rates of hippocampal neurogenesis during infancy than mice. Importantly, and consistent with the idea that high rates of neurogenesis cause forgetting, these precocial species did not exhibit infantile amnesia. Would artificially elevating levels of neurogenesis in these precocial species induce infantile amnesia? Using both naturalistic (voluntary running) and pharmacological (memantine administration) methods in two different hippocampal-dependent tasks, this was found to be the case: infantile amnesia can be artificially induced in precocial species by increasing rates of neurogenesis (Akers et al. 2014). Together, these results suggest that neurogenesis mediates infantile amnesia in mammalian species.

How Does Neurogenesis Cause Forgetting?

Here, we outline two accounts by which neurogenesis might modulate forgetting. These explanations are not mutually exclusive, as neurogenesis has multiple effects on surrounding circuitry, many of which might contribute to forgetting.

1. *Integration-based circuit remodeling*: According to this hypothesis, the integration and synaptogenesis of adult-generated neurons remodels the neural circuitry in which hippocampal memory traces are imbedded. Given a sufficient number of newborn neurons infiltrating an established memory circuit, the end result will necessarily be the degradation of information stored in that circuit. Importantly, new neuron synaptogenesis seems to be a competitive process, with new synapses existing preferentially alongside (Toni et al. 2007, 2008), and potentially even eliminating (Yasuda et al. 2011), pre-existing synaptic connections. Thus neurogenesis results in the addition and deletion of both its efferent and afferent synaptic connections. Because successful memory retrieval is thought to rely on replaying the spatiotemporal pattern of hippocampal activity that was present at encoding, this neural rewiring would cumulatively reduce the likelihood that the same patterns of neural activity would be replayed and engaged by appropriate retrieval cues. Note that, according to this hypothesis, the hyperexcitability of newborn neurons only plays an auxiliary role (e.g., potentially playing a role in new synapses out-competing old synapses) and is not the driving factor in neurogenesis-induced forgetting per se. Thus, in contrast to theories of how neurogenesis improves memory which often emphasize the unique electrophysiological properties possessed by immature neurons, this account of neurogenesis-induced forgetting focuses instead on the structural impact the integration of new neurons has on surrounding neural circuitry.
2. *Homeostatically-induced neural silencing*: The highly excitable nature of new dentate granule neurons might result in forgetting by driving homeostatic changes in the hippocampus. The addition of excitable new neurons in the dentate gyrus would increase the excitatory drive of this region and also its upstream target CA3. Because overexcitation can be maladaptive (e.g., seizure activity), this increase in excitation would trigger homeostatic mechanisms (e.g.,

synaptic scaling) that would decrease excitability of neurons in both DG and CA3 (Desai et al. 1999; Brickley et al. 2001; Deisseroth et al. 2004; Turrigiano 2008; Restivo et al. 2015), potentially leading to silencing of some synapses (Meltzer et al. 2005; Josselyn and Frankland 2012; Frankland et al. 2013). Consistent with this possibility, we have recently shown that synaptic connections with CA3 interneurons peak when dentate granule neurons are 4 weeks old (Restivo et al. 2015), the time point when these neurons display hyperexcitability (Ge et al. 2007; Marín-Burgin et al. 2012). Because some of the synapses silenced will inevitably include those involved in memory storage and retrieval, the end result of this homeostatic response would be impaired memory retrieval.

Regardless of how neurogenesis causes forgetting, our current understanding suggests that the probability of forgetting (or retrieval failure) should incrementally increase with time (Frankland and Bontempi 2005; Frankland et al. 2013) as more and more new neurons become active and integrated into hippocampal circuitry.

Which Memories Are Forgotten?

Not all memories survive. We propose that the fate of a memory depends on the outcome of a competition between two opposing processes: memory trace consolidation (e.g., off-line memory replay (Ji and Wilson 2007; Diekelmann and Born 2010; Winocur and Moscovitch 2011) and memory trace decay (e.g., neurogenesis-induced memory destabilization (Akers et al. 2014). Conceptually, this process can be understood as a Darwinian-like competition amongst memory traces; only memories that have been sufficiently reactivated and strengthened survive (to be consolidated in the cortex), while others are gradually cleared via neurogenesis-mediated memory decay or other decay processes (Frankland et al. 2013). Our current understanding of neurogenesis-induced forgetting suggests that it is agnostic to informational content; neurogenesis impacts all hippocampus-dependent memories (Frankland et al. 2013). Factors such as the initial strength of the memory likely promote persistence. Indeed, we found that stronger memories were more resistant to neurogenesis-induced forgetting (Akers et al. 2014), perhaps because stronger events are reactivated more frequently. Thus, neurogenesis-mediated forgetting can be conceptualized as a process of unrelenting hippocampal-dependent memory decay, with the caveat that stronger (typically more significant) memories are less susceptible to information loss in response to this constant bombardment of new neurons.

Neurogenesis-Mediated Forgetting: Storage Versus Retrieval Failure?

Many questions remain unanswered regarding the role that neurogenesis plays in forgetting. One important question relates to the nature of the forgetting. In particular, is neurogenesis-mediated forgetting a deficit in memory storage, such that the memory trace is significantly degraded to the point of inaccessibility (i.e. memory erasure)? Or is the forgetting better characterized as a deficit in memory retrieval, such that the memory trace still exists in the brain, but becomes more difficult to access as a function of high levels of neurogenesis? One way of addressing this question is to give mice an environmental 'reminder' of the memory after neurogenesis-induced forgetting has taken place. If the reminder is sufficient in alleviating the forgetting, this would suggest that neurogenesis-induced forgetting is the result of a retrieval deficit. That is, some significant portion of the memory trace is still available in the brain, but the animal is unable to access it under normal, 'non-reminder' conditions. An alternative approach to providing physical reminders is to provide 'neural reminders'. That is, to take advantage of new technologies that allow for the permanent tagging of populations of cells that are active during encoding (and potentially correspond to the memory engram), and ask whether reactivating this population of tagged cells is sufficient to induce successful memory retrieval after neurogenesis-induced forgetting has taken place. If this protocol alleviates the forgetting, it suggests that the memory trace still exists in the brain, and that neurogenesis simply renders it more difficult to access. Ongoing experiments are underway in our lab to address this and other questions.

Neurogenesis-Mediated Forgetting Is Adaptive

The term 'forgetting' often has negative connotations. In the field of animal research, 'deficits' in memory retention are often understood as failures in neural processing (Kraemer and Golding 1997). Interestingly, forgetting is more readily accepted in human research as an adaptive phenomena, one which can aid in the processing and retrieval of information encountered in the future (Altmann and Gray 2002; Wixted 2004; Schooler and Hertwig 2005). This is emphasized in 'intentional (or directed) forgetting' paradigms, in which the goal of the participant is to forget some of the presented information. Consistent with this, we have recently demonstrated that neurogenesis induced forgetting provides an adaptive benefit in the acquisition of new memories. Mice that experience a post-training increase in neurogenesis show forgetting of the previously acquired hippocampus dependent memory (Epp et al. 2016). Although this forgetting may appear to be a poor outcome, we have observed that the mice that have forgotten the old information are in some cases able to acquire a new memory faster. In particular, under conditions where the new memory has a high degree of similarity to the previous memory (e.g. reversal learning) neurogenesis induced forgetting facilitates encoding of this new, conflicting memory. Therefore, neurogenesis seems to offer a mechanism that reduces the strength of old memories in order to allow for more efficient encoding of new conflicting information. In other words, neurogenesis-induced forgetting is an adaptive phenomenon that increases an animal's ability to deal with memory interference (see also: Zhang et al. 2008; Garthe et al. 2009, 2016).

Previously, there has been a large emphasis on how hippocampal neurogenesis contributes to the formation of new memories. However, our finding that neurogenesis regulates forgetting suggests that this process regulates a balance between encoding of new memories and the clearance of old memories. Clinically, impaired memory clearance may be relevant to disorders such as depression and PTSD. For example, depressed individuals display lower levels of neurogenesis (Sahay and Hen 2007; Epp et al. 2013) and tend to ruminate or focus on negative thoughts and memories (Rimes and Watkins 2005; Ciesla and Roberts 2007; Disner et al. 2011; Romero et al. 2014). Perhaps rumination on negative thoughts and memories in depressed individuals is exacerbated, at least in part, by low levels of neurogenesis. These low rates of neurogenesis would impede the clearance of negative memories, which may aggravate depressive symptoms. Consistent with this possibility, most antidepressants increase neurogenesis (Malberg et al. 2000), and in fact often depend on this increase in neurogenesis for their therapeutic effect (Santarelli et al. 2003). More recently, it has been shown that specifically targeting and increasing neurogenesis through genetic means is sufficient in alleviating depressive symptoms (open field, elevated plus, tail suspension) in mice (Hill et al. 2015). Together, the above results are consistent with forgetting being an adaptive phenomena, since the inability to forget can intensify or even be the core symptom of mental disorders such as depression and PTSD.

Summary

Why does neurogenesis continue throughout life in the dentate gyrus of the hippocampus? Perhaps neurogenesis is required for learning and memory. Yet results reviewed above suggest that developmentally generated neurons are often capable of supporting learning in the absence of newly generated neurons. Furthermore, many regions outside of the hippocampus (e.g., the amygdala) are essential in learning and yet lack neurogenesis. Perhaps 'learning' is too broad, and we should look towards the pattern separation function of the hippocampus to explain why neurogenesis occurs in the dentate gyrus but not other neural regions. But pattern separation is not unique to the hippocampus (Yassa and Stark 2011), and undoubtedly occurs in neural regions throughout the brain [e.g., amygdala, (Gilbert and Kesner 2002), perirhinal cortex (Bartko et al. 2007; Burke et al. 2011)]. Moreover, as argued elsewhere, almost any behaviour or physiological result can be interpreted as a pattern separation effect (Aimone et al. 2011). Thus, the role of neurogenesis in regulating pattern separation might not be sufficiently powerful to explain why neurogenesis persists in the adult hippocampus.

The finding that neurogenesis decreases the dependence of memories on the hippocampus (Kitamura et al. 2009) and causes forgetting (Akers et al. 2014) might shed new light on why neurogenesis persists in the hippocampus. Our evolving view is that it reflects the transient necessity of the hippocampus in memory storage. While the hippocampus is thought to automatically encode all experiences, only a fraction of these experiences become stored permanently. As we have outlined above, perhaps those that are more frequently reactivated are retained, and ultimately consolidated in the cortex while the remaining traces are eventually overwritten and forgotten as new neurons integrate into the hippocampus.

Neurogenesis-induced forgetting provides an active, mechanistic account (i.e., as oppose to passive memory decay in response to time) of how forgetting occurs across the lifespan in mammalian species. Indeed, this is one of the first biologically-relevant accounts of how information degradation occurs in the brain, and in the hippocampus in particular (see also: Shuai et al. 2010; Berry et al. 2012, 2015; Hardt et al. 2013). Future research will focus on uncovering the precise mechanism(s) through which neurogenesis causes forgetting.

References

Acsády L, Kamondi A, Sík A, Freund T, Buzsáki G (1998) GABAergic cells are the major postsynaptic targets of mossy fibers in the rat hippocampus. J Neurosci 18:3386–3403

Aimone JB, Wiles J, Gage FH (2009) Computational influence of adult neurogenesis on memory encoding. Neuron 61:187–202

Aimone JB, Deng W, Gage FH (2011) Resolving new memories: a critical look at the dentate gyrus, adult neurogenesis, and pattern separation. Eur J Neurosci 33:1160–1169

Akers KG, Martinez-Canabal A, Restivo L, Yiu AP, De Cristofaro A, Hsiang HL, Wheeler AL, Guskjolen A, Niibori Y, Shoji H, Ohira K, Richards BA, Miyakawa T, Josselyn SA, Frankland PW (2014) Hippocampal neurogenesis regulates forgetting during adulthood and infancy. Science 344:598–602

Allen E (1912) The cessation of mitosis in the central nervous system of the albino rat. Waverley Press, Baltimore

Alme CB, Buzzetti RA, Marrone DF, Leutgeb JK, Chawla MK, Schaner MJ, Bohanick JD, Khoboko T, Leutgeb S, Moser EI, Moser M-B, McNaughton BL, Barnes CA (2010) Hippocampal granule cells opt for early retirement. Hippocampus 20:1109–1123

Altman J (1962) Are new neurons formed in the brains of adult mammals? Science 135:1127–1128

Altman J (1963) Autoradiographic investigation of cell proliferation in the brains of rats and cats. Anat Rec 145:573–591

Altmann EM, Gray WD (2002) Forgetting to remember: the functional relationship of decay and interference. Psychol Sci 13:27–33

Arruda-Carvalho M, Sakaguchi M, Akers KG, Josselyn SA, Frankland PW (2011) Posttraining ablation of adult-generated neurons degrades previously acquired memories. J Neurosci 31:15113–15127

Bartko SJ, Winters BD, Cowell RA, Saksida LM, Bussey TJ (2007) Perirhinal cortex resolves feature ambiguity in configural object recognition and perceptual oddity tasks. Learn Mem 14:821–832

Berry JA, Cervantes-Sandoval I, Nicholas EP, Davis RL (2012) Dopamine is required for learning and forgetting in Drosophila. Neuron 74:530–542

Berry JA, Cervantes-Sandoval I, Chakraborty M, Davis RL (2015) Sleep facilitates memory by blocking dopamine neuron-mediated forgetting. Cell 161:1656–1667

Bizzozero G (1894) An address on the growth and regeneration of the organism: delivered before a general meeting of the XIth internationa medical congress, held in Rome, 1894. Br Med J 1:728–732

Brickley SG, Revilla V, Cull-Candy SG, Wisden W, Farrant M (2001) Adaptive regulation of neuronal excitability by a voltage-independent potassium conductance. Nature 409:88–92

Brown J, Cooper-Kuhn CM, Kempermann G, van Praag H, Winkler J, Gage FH, Kuhn HG (2003) Enriched environment and physical activity stimulate hippocampal but not olfactory bulb neurogenesis. Eur J Neurosci 17:2042–2046

Burke SN, Wallace JL, Hartzell AL, Nematollahi S, Plange K, Barnes CA (2011) Age-associated deficits in pattern separation functions of the perirhinal cortex: a cross-species consensus. Behav Neurosci 125:836–847

Cajal SRY (1888) Estructura de los centros nerviosos de las aves. Rev Trim Histol Norm Patol 1:1–10

Cajal SRY (1914) Estudios sobre la degeneración y regeneración del sistema nerviosa

Cameron HA, Woolley CS, McEwen BS, Gould E (1993) Differentiation of newly born neurons and glia in the dentate gyrus of the adult rat. Neuroscience 56:337–344

Carlén M, Cassidy RM, Brismar H, Smith GA, Enquist LW, Frisén J (2002) Functional integration of adult-born neurons. Curr Biol 12:606–608

Chambers RA, Potenza MN, Hoffman RE (2004) Simulated apoptosis/neurogenesis regulates learning and memory capabilities of adaptive neural networks. Neuropsychopharmacology 29(4):747–758

Chawla MK, Guzowski JF, Ramirez-Amaya V, Lipa P, Hoffman KL, Marriott LK, Worley PF, McNaughton BL, Barnes CA (2005) Sparse, environmentally selective expression of Arc RNA in the upper blade of the rodent fascia dentata by brief spatial experience. Hippocampus 15:579–586

Ciesla JA, Roberts JE (2007) Rumination, negative cognition, and their interactive effects on depressed mood. Emotion 7:555–565

Clelland CD, Choi M, Romberg C, Clemenson GD, Fragniere A, Tyers P, Jessberger S, Saksida LM, Barker RA, Gage FH, Bussey TJ (2009) A functional role for adult hippocampal neurogenesis in spatial pattern separation. Science 325:210–213

Coe CL, Kramer M, Czeh B, Gould E, Reeves AJ, Kirschbaum C, Fuchs E (2003) Prenatal stress diminishes neurogenesis in the dentate gyrus of juvenile rhesus monkeys. Biol Psychiatry 54:1025–1034

Creer DJ, Romberg C, Saksida LM, van Praag H, Bussey TJ (2010) Running enhances spatial pattern separation in mice. Proc Natl Acad Sci U S A 107:2367–2372

Deisseroth K, Singla S, Toda H, Monje M, Palmer TD (2004) Excitation-neurogenesis coupling in adult neural stem/progenitor cells. Neuron 42(4):535–552

Denny CA, Burghardt NS, Schachter DM, Hen R, Drew MR (2012) 4- to 6-week-old adult-born hippocampal neurons influence novelty-evoked exploration and contextual fear conditioning. Hippocampus 22:1188–1201

Desai NS, Rutherford LC, Turrigiano GG (1999) Plasticity in the intrinsic excitability of cortical pyramidal neurons. Nat Neurosci 2:515–520

Diekelmann S, Born J (2010) The memory function of sleep. Nat Rev Neurosci 11(2):114–126

Disner SG, Beevers CG, Haigh EAP, Beck AT (2011) Neural mechanisms of the cognitive model of depression. Nat Publ Group 12:467–477

Drew MR, Denny CA, Hen R (2010) Arrest of adult hippocampal neurogenesis in mice impairs single- but not multiple-trial contextual fear conditioning. Behav Neurosci 124:446–454

Dupret D, Revest J-M, Koehl M, Ichas F, De Giorgi F, Costet P, Abrous DN, Piazza PV (2008) Spatial relational memory requires hippocampal adult neurogenesis. PLoS One 3:e1959

Eckenhoff MF, Rakic P (1988) Nature and fate of proliferative cells in the hippocampal dentate gyrus during the life span of the rhesus monkey. J Neurosci 8:2729–2747

Epp JR, Spritzer MD, Galea LAM (2007) Hippocampus-dependent learning promotes survival of new neurons in the dentate gyrus at a specific time during cell maturation. Neuroscience 149:273–285

Epp JR, Haack AK, Galea LAM (2010) Task difficulty in the Morris water task influences the survival of new neurons in the dentate gyrus. Hippocampus 20:866–876

Epp JR, Haack AK, Galea LAM (2011) Activation and survival of immature neurons in the dentate gyrus with spatial memory is dependent on time of exposure to spatial learning and age of cells at examination. Neurobiol Learn Mem 95:316–325

Epp JR, Chow C, Galea LAM (2013) Hippocampus-dependent learning influences hippocampal neurogenesis. Front Neurosci 7:57

Epp JR, Silva Mera R, Kohler S, Josselyn SA, Frankland PW (2016) Neurogenesis-mediated forgetting minimizes proactive interference. Nat Commun 7:10838

Eriksson PS, Perfilieva E, Björk-Eriksson T, Alborn AM, Nordborg C, Peterson DA, Gage FH (1998) Neurogenesis in the adult human hippocampus. Nat Med 4:1313–1317

Espósito MS, Piatti VC, Laplagne DA, Morgenstern NA, Ferrari CC, Pitossi FJ, Schinder AF (2005) Neuronal differentiation in the adult hippocampus recapitulates embryonic development. J Neurosci 25:10074–10086

Faulkner RL, Jang M-H, Liu X-B, Duan X, Sailor KA, Kim JY, Ge S, Jones EG, Ming G-L, Song H, Cheng H-J (2008) Development of hippocampal mossy fiber synaptic outputs by new neurons in the adult brain. Proc Natl Acad Sci U S A 105:14157–14162

Frankland PW (2013) Neurogenic evangelism: comment on Urbach et al. (2013). Behav Neurosci 127:126–129

Frankland PW, Bontempi B (2005) The organization of recent and remote memories. Nat Rev Neurosci 6:119–130

Frankland PW, Kohler S, Josselyn SA (2013) Hippocampal neurogenesis and forgetting. Trends Neurosci 36:497–503

Fusi S, Drew PJ, Abbott LF (2005) Cascade models of synaptically stored memories. Neuron 45:599–611

Garner AR, Rowland DC, Hwang SY, Baumgaertel K, Roth BL, Kentros C, Mayford M (2012) Generation of a synthetic memory trace. Science 335:1513–1516

Garthe A, Behr J, Kempermann G (2009) Adult-generated hippocampal neurons allow the flexible use of spatially precise learning strategies. PLoS One 4:e5464

Garthe A, Roeder I, Kempermann G (2016) Mice in an enriched environment learn more flexibly because of adult hippocampal neurogenesis. Hippocampus 26:261–271

Ge S, Yang C-H, Hsu K-S, Ming G-L, Song H (2007) A critical period for enhanced synaptic plasticity in newly generated neurons of the adult brain. Neuron 54:559–566

Gilbert PE, Kesner RP (2002) The amygdala but not the hippocampus is involved in pattern separation based on reward value. Neurobiol Learn Mem 77:338–353

Gould E, Cameron HA, Daniels DC (1992) Adrenal hormones suppress cell division in the adult rat dentate gyrus. J Neurosci 12(9):3642–3650

Gould E, Beylin A, Tanapat P, Reeves A, Shors TJ (1999a) Learning enhances adult neurogenesis in the hippocampal formation. Nat Neurosci 2:260–265

Gould E, Reeves AJ, Fallah M, Tanapat P, Gross CG, Fuchs E (1999b) Hippocampal neurogenesis in adult old world primates. Proc Natl Acad Sci U S A 96:5263–5267

Gregoire C-A, Bonenfant D, Le Nguyen A, Aumont A, Fernandes KJL (2014) Untangling the influences of voluntary running, environmental complexity, social housing and stress on adult hippocampal neurogenesis. PLoS One 9:e86237

Groves JO, Leslie I, Huang G-J, McHugh SB, Taylor A, Mott R, Munafò M, Bannerman DM, Flint J (2013) Ablating adult neurogenesis in the rat has no effect on spatial processing: evidence from a novel pharmacogenetic model. PLoS Genet 9:e1003718

Gu Y, Arruda-Carvalho M, Wang J, Janoschka SR, Josselyn SA, Frankland PW, Ge S (2012) Optical controlling reveals time-dependent roles for adult-born dentate granule cells. Nat Neurosci 15:1700–1706

Han J-H, Kushner SA, Yiu AP, Cole CJ, Matynia A, Brown RA, Neve RL, Guzowski JF, Silva AJ, Josselyn SA (2007) Neuronal competition and selection during memory formation. Science 316:457–460

Han J-H, Kushner SA, Yiu AP, Hsiang H-LL, Buch T, Waisman A, Bontempi B, Neve RL, Frankland PW, Josselyn SA (2009) Selective erasure of a fear memory. Science 323:1492–1496

Hardt O, Nader K, Nadel L (2013) Decay happens: the role of active forgetting in memory. Trends Cogn Sci (Regul Ed) 17:111–120

Hastings NB, Gould E (1999) Rapid extension of axons into the CA3 region by adult-generated granule cells. J Comp Neurol 413:146–154

Henze DA, Wittner L, Buzsáki G (2002) Single granule cells reliably discharge targets in the hippocampal CA3 network in vivo. Nat Neurosci 5:790–795

Hill AS, Sahay A, Hen R (2015) Increasing adult hippocampal neurogenesis is sufficient to reduce anxiety and depression-like behaviors. Neuropsychopharmacology 40:2368–2378

Holick KA, Lee DC, Hen R, Dulawa SC (2008) Behavioral effects of chronic fluoxetine in BALB/cJ mice do not require adult hippocampal neurogenesis or the serotonin 1A receptor. Neuropsychopharmacology 33:406–417

Howe ML, Courage ML (1993) On resolving the enigma of infantile amnesia. Psychol Bull 113:305–326

Jaholkowski P, Kiryk A, Jedynak P (2009) New hippocampal neurons are not obligatory for memory formation; cyclin D2 knockout mice with no adult brain neurogenesis show learning. Learn Mem 16(7):439–451

Jedynak P, Jaholkowski P, Wozniak G, Sandi C, Kaczmarek L, Filipkowski RK (2012) Lack of cyclin D2 impairing adult brain neurogenesis alters hippocampal-dependent behavioral tasks without reducing learning ability. Behav Brain Res 227:159–166

Ji D, Wilson MA (2007) Coordinated memory replay in the visual cortex and hippocampus during sleep. Nat Neurosci 10:100–107

Josselyn SA, Frankland PW (2012) Infantile amnesia: a neurogenic hypothesis. Learn Mem 19:423–433

Kaplan MS (1981) Neurogenesis in the 3-month-old rat visual cortex. J Comp Neurol 195:323–338

Kaplan MS (1985) Formation and turnover of neurons in young and senescent animals: an electronmicroscopic and morphometric analysis. Ann N Y Acad Sci 457:173–192

Kaplan MS (2001) Environment complexity stimulates visual cortex neurogenesis: death of a dogma and a research career. Trends Neurosci 24:617–620

Kaplan MS, Hinds JW (1977) Neurogenesis in the adult rat: electron microscopic analysis of light radioautographs. Science 197:1092–1094

Kaplan MS, McNelly NA, Hinds JW (1985) Population dynamics of adult-formed granule neurons of the rat olfactory bulb. J Comp Neurol 239:117–125

Kee N, Teixeira CM, Wang AH, Frankland PW (2007) Preferential incorporation of adult-generated granule cells into spatial memory networks in the dentate gyrus. Nat Neurosci 10:355–362

Kempermann G (2002) Why new neurons? Possible functions for adult hippocampal neurogenesis. J Neurosci 22:635–638

Kempermann G, Kuhn HG, Gage FH (1997) More hippocampal neurons in adult mice living in an enriched environment. Nature 386:493–495

Kheirbek MA, Klemenhagen KC, Sahay A, Hen R (2012) Neurogenesis and generalization: a new approach to stratify and treat anxiety disorders. Nat Neurosci 15:1613–1620

Kitamura T, Saitoh Y, Takashima N, Murayama A, Niibori Y, Ageta H, Sekiguchi M, Sugiyama H, Inokuchi K (2009) Adult neurogenesis modulates the hippocampus-dependent period of associative fear memory. Cell 139:814–827

Knoth R, Singec I, Ditter M, Pantazis G, Capetian P, Meyer RP, Horvat V, Volk B, Kempermann G (2010) Murine features of neurogenesis in the human hippocampus across the lifespan from 0 to 100 years. PLoS One 5:e8809

Kornack DR, Rakic P (1999) Continuation of neurogenesis in the hippocampus of the adult macaque monkey. Proc Natl Acad Sci U S A 96:5768–5773

Kraemer PJ, Golding JM (1997) Adaptive forgetting in animals. Psychon Bull Rev 4(4):480–491

Kuhn HG, Dickinson-Anson H, Gage FH (1996) Neurogenesis in the dentate gyrus of the adult rat: age-related decrease of neuronal progenitor proliferation. J Neurosci 16:2027–2033

Leblond CP (1964) Classification of cell populations on the basis of their proliferative behavior. Natl Cancer Inst Monogr 14:119–150

Leuner B, Mendolia-Loffredo S, Kozorovitskiy Y, Samburg D, Gould E, Shors TJ (2004) Learning enhances the survival of new neurons beyond the time when the hippocampus is required for memory. J Neurosci 24:7477–7481

Liu S, Wang J, Zhu D, Fu Y, Lukowiak K, Lu YM (2003) Generation of functional inhibitory neurons in the adult rat hippocampus. J Neurosci 23:732–736

Liu X, Ramirez S, Pang PT, Puryear CB, Govindarajan A, Deisseroth K, Tonegawa S (2012) Optogenetic stimulation of a hippocampal engram activates fear memory recall. Nature 484:381–385

Malberg JE, Duman RS (2003) Cell proliferation in adult hippocampus is decreased by inescapable stress: reversal by fluoxetine treatment. Neuropsychopharmacology 28:1562–1571

Malberg JE, Eisch AJ, Nestler EJ, Duman RS (2000) Chronic antidepressant treatment increases neurogenesis in adult rat hippocampus. J Neurosci 20:9104–9110

Marín-Burgin A, Mongiat LA, Pardi MB, Schinder AF (2012) Unique processing during a period of high excitation/inhibition balance in adult-born neurons. Science 335:1238–1242

Martinez-Canabal A, Akers KG, Josselyn SA, Frankland PW (2013) Age-dependent effects of hippocampal neurogenesis suppression on spatial learning. Hippocampus 23:66–74

McDonald HY, Wojtowicz JM (2005) Dynamics of neurogenesis in the dentate gyrus of adult rats. Neurosci Lett 385:70–75

Meltzer LA, Yabaluri R, Deisseroth K (2005) A role for circuit homeostasis in adult neurogenesis. Trends Neurosci 28:653–660

Meshi D, Drew MR, Saxe M, Ansorge MS, David D (2006) Hippocampal neurogenesis is not required for behavioral effects of environmental enrichment. Nat Neurosci 9:729–731

Messier B, Leblond CP (1960) Cell proliferation and migration as revealed by radioautography after injection of thymidine-H3 into male rats and mice. Am J Anat 106:247–285

Mongiat LA, Espósito MS, Lombardi G, Schinder AF (2009) Reliable activation of immature neurons in the adult hippocampus. PLoS One 4:e5320

Niibori Y, Yu T-S, Epp JR, Akers KG, Josselyn SA, Frankland PW (2012) Suppression of adult neurogenesis impairs population coding of similar contexts in hippocampal CA3 region. Nat Commun 3:1253

Olson AK, Eadie BD, Ernst C, Christie BR (2006) Environmental enrichment and voluntary exercise massively increase neurogenesis in the adult hippocampus via dissociable pathways. Hippocampus 16:250–260

Parent JM, Yu TW, Leibowitz RT, Geschwind DH, Sloviter RS, Lowenstein DH (1997) Dentate granule cell neurogenesis is increased by seizures and contributes to aberrant network reorganization in the adult rat hippocampus. J Neurosci 17:3727–3738

Paton JA, Nottebohm FN (1984) Neurons generated in the adult brain are recruited into functional circuits. Science 225:1046–1048

Rakic P (1985a) DNA synthesis and cell division in the adult primate brain. Ann N Y Acad Sci 457:193–211

Rakic P (1985b) Limits of neurogenesis in primates. Science 227:1054–1056

Ramirez S, Liu X, Lin P-A, Suh J, Pignatelli M, Redondo RL, Ryan TJ, Tonegawa S (2013) Creating a false memory in the hippocampus. Science 341:387–391

Ramirez-Amaya V, Marrone DF, Gage FH, Worley PF, Barnes CA (2006) Integration of new neurons into functional neural networks. J Neurosci 26:12237–12241

Restivo L, Niibori Y, Mercaldo V, Josselyn SA, Frankland PW (2015) Development of adult-generated cell connectivity with excitatory and inhibitory cell populations in the hippocampus. J Neurosci 35:10600–10612

Reynolds BA, Weiss S (1992) Generation of neurons and astrocytes from isolated cells of the adult mammalian central nervous system. Science 255:1707–1710

Rimes KA, Watkins E (2005) The effects of self-focused rumination on global negative self-judgements in depression. Behav Res Ther 43:1673–1681

Romero N, Sanchez A, Vazquez C (2014) Memory biases in remitted depression: the role of negative cognitions at explicit and automatic processing levels. J Behav Ther Exp Psychiatry 45:128–135

Sahay A, Hen R (2007) Adult hippocampal neurogenesis in depression. Nat Neurosci 10:1110–1115

Sahay A, Scobie KN, Hill AS, O'Carroll CM, Kheirbek MA, Burghardt NS, Fenton AA, Dranovsky A, Hen R (2011a) Increasing adult hippocampal neurogenesis is sufficient to improve pattern separation. Nature 472:466–470

Sahay A, Wilson DA, Hen R (2011b) Pattern separation: a common function for new neurons in hippocampus and olfactory bulb. Neuron 70:582–588

Santarelli L, Saxe M, Gross C, Surget A, Battaglia F, Dulawa S, Weisstaub N, Lee J, Duman R, Arancio O, Belzung C, Hen R (2003) Requirement of hippocampal neurogenesis for the behavioral effects of antidepressants. Science 301:805–809

Santoro A (2013) Reassessing pattern separation in the dentate gyrus. Front Behav Neurosci 7:96

Saxe MD, Battaglia F, Wang J-W, Malleret G, David DJ, Monckton JE, Garcia ADR, Sofroniew MV, Kandel ER, Santarelli L, Hen R, Drew MR (2006) Ablation of hippocampal neurogenesis impairs contextual fear conditioning and synaptic plasticity in the dentate gyrus. Proc Natl Acad Sci U S A 103:17501–17506

Schmidt-Hieber C, Jonas P, Bischofberger J (2004) Enhanced synaptic plasticity in newly generated granule cells of the adult hippocampus. Nature 429:184–187

Schooler LJ, Hertwig R (2005) How forgetting aids heuristic inference. Psychol Rev 112:610–628

Shors TJ, Miesegaes G, Beylin A, Zhao M, Rydel T, Gould E (2001a) Neurogenesis in the adult is involved in the formation of trace memories. Nature 410:372–376

Shors TJ, Townsend DA, Zhao M, Kozorovitskiy Y (2001b) Neurogenesis may relate to some but not all types of hippocampal-dependent learning. Hippocampus 12(5):578–584

Shuai Y, Lu B, Hu Y, Wang L, Sun K, Zhong Y (2010) Forgetting is regulated through Rac activity in Drosophila. Cell 140:579–589

Singer BH, Jutkiewicz EM, Fuller CL, Lichtenwalner RJ, Zhang H, Velander AJ, Li X, Gnegy ME, Burant CF, Parent JM (2009) Conditional ablation and recovery of forebrain neurogenesis in the mouse. J Comp Neurol 514:567–582

Snyder JS, Cameron HA (2012) Could adult hippocampal neurogenesis be relevant for human behavior? Behav Brain Res 227:384–390

Snyder JS, Hong NS, McDonald RJ, Wojtowicz JM (2005) A role for adult neurogenesis in spatial long-term memory. Neuroscience 130:843–852

Spalding KL, Bergmann O, Alkass K, Bernard S, Salehpour M, Huttner HB, Boström E, Westerlund I, Vial C, Buchholz BA, Possnert G, Mash DC, Druid H, Frisén J (2013) Dynamics of hippocampal neurogenesis in adult humans. Cell 153:1219–1227

Stone SSD, Teixeira CM, Zaslavsky K, Wheeler AL, Martinez-Canabal A, Wang AH, Sakaguchi M, Lozano AM, Frankland PW (2011) Functional convergence of developmentally and adult-generated granule cells in dentate gyrus circuits supporting hippocampus-dependent memory. Hippocampus 21:1348–1362

Tanapat P, Galea LA, Gould E (1998) Stress inhibits the proliferation of granule cell precursors in the developing dentate gyrus. Int J Dev Neurosci 16:235–239

Tashiro A, Makino H, Gage FH (2007) Experience-specific functional modification of the dentate gyrus through adult neurogenesis: a critical period during an immature stage. J Neurosci 27:3252–3259

Toni N, Teng EM, Bushong EA, Aimone JB (2007) Synapse formation on neurons born in the adult hippocampus. Nat Neurosci. 10(6):727–734

Toni N, Laplagne DA, Zhao C, Lombardi G, Ribak CE, Gage FH, Schinder AF (2008) Neurons born in the adult dentate gyrus form functional synapses with target cells. Nat Neurosci 11:901–907

Tronel S, Belnoue L, Grosjean N, Revest J-M, Piazza PV, Koehl M, Abrous DN (2012) Adult-born neurons are necessary for extended contextual discrimination. Hippocampus 22:292–298

Trouche S, Bontempi B, Roullet P, Rampon C (2009) Recruitment of adult-generated neurons into functional hippocampal networks contributes to updating and strengthening of spatial memory. Proc Natl Acad Sci U S A 106:5919–5924

Turrigiano GG (2008) The self-tuning neuron: synaptic scaling of excitatory synapses. Cell 135:422–435

Urbach A, Robakiewicz I, Baum E, Kaczmarek L, Witte OW, Filipkowski RK (2013) Cyclin D2 knockout mice with depleted adult neurogenesis learn Barnes maze task. Behav Neurosci 127:1–8

van Praag H, Kempermann G, Gage FH (1999) Running increases cell proliferation and neurogenesis in the adult mouse dentate gyrus. Nat Neurosci 2:266–270

van Praag H, Schinder AF, Christie BR, Toni N, Palmer TD, Gage FH (2002) Functional neurogenesis in the adult hippocampus. Nature 415:1030–1034

Weisz VI, Argibay PF (2012) Neurogenesis interferes with the retrieval of remote memories: forgetting in neurocomputational terms. Cognition 125:13–25

Willner P (2005) Chronic mild stress (CMS) revisited: consistency and behavioural-neurobiological concordance in the effects of CMS. Neuropsychobiology 52:90–110

Winocur G, Moscovitch M (2011) Memory transformation and systems consolidation. J Int Neuropsychol Soc 17:766–780

Wixted JT (2004) The psychology and neuroscience of forgetting. Annu Rev Psychol 55:235–269

Yassa MA, Stark CEL (2011) Pattern separation in the hippocampus. Trends Neurosci 34:515–525

Yasuda M, Johnson-Venkatesh EM, Zhang H, Parent JM, Sutton MA, Umemori H (2011) Multiple forms of activity-dependent competition refine hippocampal circuits in vivo. Neuron 70:1128–1142

Yiu AP, Mercaldo V, Yan C, Richards B, Rashid AJ, Hsiang H-LL, Pressey J, Mahadevan V, Tran MM, Kushner SA, Woodin MA, Frankland PW, Josselyn SA (2014) Neurons are recruited to a memory trace based on relative neuronal excitability immediately before training. Neuron 83:722–735

Zhang C-L, Zou Y, He W, Gage FH, Evans RM (2008) A role for adult TLX-positive neural stem cells in learning and behaviour. Nature 451:1004–1007

Manipulating Hippocampus-Dependent Memories: To Enhance, Delete or Incept?

Hugo J. Spiers, William de Cothi, and Daniel Bendor

In a sense, he thought, all we consist of is memories. Our personalities are constructed from memories, our lives are organized around memories, our cultures are built upon the foundation of shared memories that we call history and science. But now to give up a memory, to give up knowledge, to give up the past... His entire being rebelled against the idea of forgetting.

Sphere, Michael Crichton

Abstract Memory manipulation has advanced substantially in recent years to a range of new methods available to researchers. These methods include optogenetics, transcranial stimulation, deep brain stimulation, pharmacological agents and cued reactivation of memories during sleep. Here we review and evaluate findings from these methods in relation to manipulations of hippocampus-dependent memories. In doing so we shed light on the different ways in which memories can be erased, enhanced or implanted.

In his novel *Sphere,* Michael Crichton's protagonists are faced with the dilemma of whether or not to wipe their memories in order to save others from danger. The lead character, Harry, realises in this moment just how important his memories are to him. Science fiction has continually played with memory enhancement, memory erasure and memory implantation (inception) (see Groes 2016). Recent years have seen science 'fiction' translate to science 'reality'.

The capacity to manipulate memories offers the potential for huge benefits. In the medical domain, being able to treat patients with memory problems such as Alzheimer's dementia by enhancing their memory carries the possibility for

H.J. Spiers (✉) • W. de Cothi • D. Bendor (✉)
Institute of Behavioural Neuroscience, Department of Experimental Psychology, Division of Psychology and Language Sciences, University College London, 26 Bedford Way, London WC1H 0AP, UK
e-mail: h.spiers@ucl.ac.uk; d.bendor@ucl.ac.uk

© Springer International Publishing AG 2017
D.E. Hannula, M.C. Duff (eds.), *The Hippocampus from Cells to Systems,*
DOI 10.1007/978-3-319-50406-3_5

treating their catastrophic memory loss problems. While for such patients memory enhancement is helpful, for others memory removal may be needed. Patients suffering from post-traumatic stress disorder (PTSD), phobias, or anxiety disorders suffer from memory problems that may be alleviated by dampening memory retrieval. Weighed against these potential benefits is the dark side of memory manipulation. Over the decades films have provided a continual warning about the dangers of unbridled meddling with memories (see Appendix). With the rise of new technologies a number of authors have provided careful consideration of the ethics surrounding memory manipulation (Liao and Sandberg 2008; Mohamed and Sahakian 2012; Ragan et al. 2013). Despite the need for caution, research in this domain continues apace.

In this chapter we provide an overview of recent research on memory manipulation. This review extends a recent review on this topic (Spiers and Bendor 2014). We will cover studies that manipulate memories for which the hippocampus is thought to be required, including those defined as spatial, episodic, relational, or declarative (Eichenbaum 2004; Moscovitch et al. 2006; Squire et al. 2004; Spiers 2012). Psychologists have studied memory manipulation through stimuli at length (e.g. Loftus and Palmer 1974). Here we focus on memory manipulation using invasive interventions or with cuing during sleep states. In Table 1 we summarise each of the main methods currently used to target memory, which include: optogenetics, chemogenetic tools, transcranial stimulation, deepbrain stimulation and pharmacological agents. We will also discuss results arising from recording neural activity during memory manipulation, giving an insight into the mechanisms by which the intervention may affect memory (see e.g. Bendor and Wilson 2012; Hauner et al. 2013).

Improving Memory

Much like strength is an asset for physical activities, mental tasks are facilitated by having a better memory. While generally not recommend due to the health-related side-effects, drugs such as steroids can be used to artificially accelerate the process of adding muscle tone. Is there the equivalent of "mental steroids", that can be used to artificially improve your memory?

While several putative "cognitive enhancers" have in fact been developed (e.g. Kaplan and Moore 2011; Rodríguez et al. 2013), there is simply no substitute for our brain's natural approach to memory enhancement—a good night of sleep. During sleep, memories are normally consolidated, a process whereby labile memory traces are strengthened for long-term storage in memory (Stickgold and Walker 2013; Frankland and Bontempi 2005; Squire and Alvarez 1995). Thus, through this process the brain sifts through what is to be retained and sheds the memory traces that are less behaviourally or motivationally useful. In particular, non-REM sleep plays a critical role in the consolidation of hippocampus-dependent memories, such as word pairings and spatial associations (Dudai 2004; Frankland

Table 1 A brief summary of the main methods covered in this review that are currently used to manipulate memories

Transcranial stimulation	A magnetic field generator is held externally to the head and used to stimulate brain tissue. The magnetic field passes through the skull and electromagnetically induces small, electrical currents in the regions of the brain that are within the vicinity of the field.
Deep brain stimulation	Electrodes are surgically implanted into the brain so that small electric currents can stimulate targeted brain areas via a battery pack (called a neurostimulator).
Optogenetics	Using light to stimulate *in vivo* neurons that have been genetically modified to express light-gated ion channels.
Designer receptors exclusively activated by designer drugs (DREADDs)	A chemogenetic tool that utilises G-protein-coupled-receptors to achieve spatiotemporal control over neural stimulation. A 'designer drug' (such as Clozapine-N-Oxide) is used stimulate neurons expressing a 'designer receptor' (such as hM_3D_q).
Propranolol	A medication mainly used to treat various cardiovascular conditions. It is being investigated as a potential treatment for post-traumatic stress disorder and phobias as it is thought it may block the reconsolidation of fear memories (Brunet et al. 2011b; Kindt et al. 2009).

and Bontempi 2005; Squire and Alvarez 1995; Diekelmann and Born 2010). While there are clear benefits from a good night of sleep, manipulations that have the potential to make this process more efficient could theoretically lead to further memory enhancement.

One strategy for doing this is to manipulate a number of different "brain waves", including slow wave oscillations and thalamocoritcal spindles, that occur only during non-REM sleep (Buzsaki 2009). Slow wave oscillations are large amplitude, low frequency (<1 Hz) variations in the local field potential (LFP) and are a by-product of neocortical up and down states (Buzsaki et al. 2012). Thalamocortical spindles are brief oscillations in the thalamocortical pathway (7–14 Hz)—generated by the thalamic reticular nucleus (Steriade et al. 1993). Since spindles and slow-wave oscillations are thought to be critical for memory consolidation, boosting either their quantity or amplitude during non-REM sleep could provide an avenue to strengthening memory. In order to boost slow wave oscillations, Marshall and colleagues applied a slow time-varying transcranial stimulation (0.75 Hz) to the frontal cortex of sleeping human subjects (Marshall et al. 2006). One unexpected effect of the low frequency transcranial stimulation was an increase in spindle power. Following training on a hippocampus-dependent task involving word-pair associations, subjects went to sleep and received either transcranial stimulation or sham stimulation as a control. Once the subjects had awoken, those that had received the transcranial stimulation performed better on the task than the control subjects. Since both slow-wave oscillations and spindles were affected in this experiment, the underlying mechanism (i.e. which type of oscillation) responsible for this memory enhancement is still unclear. Optogenetics may provide an approach for disambiguating the roles of slow wave oscillations and spindles during memory consolidation. Using optogenetic techniques, Halassa and colleagues artificially generated thalamocortical spindles in rodents (Halassa et al. 2011). However, whether optogenetically boosting spindle production during sleep leads to better memory consolidation has not yet been demonstrated, nor is optogenetics currently viable for human subjects.

Another type of oscillation that is observed during non-REM sleep is the sharp-wave ripple; a brief, high frequency (140–220 Hz) oscillation generated within the hippocampal complex that co-occurs with a large "sharp wave" deflection in the LFP. Sharp-wave ripples also have been observed to co-occur with the cortico-thalamic spindle oscillations (Siapas and Wilson 1998; Sirota et al. 2003). During sharp-wave ripples, sequential neural patterns linked to a previous behavioural experience reactivate spontaneously in both the hippocampus and neocortex in a phenomenon commonly referred to as "replay" (Wilson and McNaughton 1994; Lee and Wilson 2002; Ji and Wilson 2006). Replay events are a neural memory trace of a previous experience and by replaying these memory traces repeatedly, the brain could reinforce and gradually consolidate memories. Sharp-wave ripples can be suppressed by using the preceding sharp wave signal to trigger stimulation of the ventral hippocampal commissure. This disruption in replay activity leads to a memory deficit (Girardeau et al. 2009; Ego-Stengel and Wilson 2010), suggesting that memory consolidation requires hippocampal replay (or at least sharp-wave ripples). If memory consolidation depends on hippocampal sharp-wave ripples and replay, can these be manipulated to enhance memories? One approach of modifying what is replayed during a sharp-wave ripple event is to use Targeted Memory Reactivation (TMR); where a sensory cue that has previously been paired with a behavioural task is repeatedly presented to a sleeping subject. For example, after rats have received a training session for an auditory-spatial association task, playing a task-related sound cue during non-REM sleep will bias replay events towards the spatial locations associated with that cue (Bendor and Wilson 2012). Therefore, biasing replay towards reactivating a specific memory in turn strengthens the consolidation of that memory. In both rodents and humans, the presentation of task related cues during non-REM sleep improves performance in a post-nap test, compared to control conditions in which no cue is presented (Antony et al. 2012; Barnes and Wilson 2014; Rasch et al. 2007; Rudoy et al. 2009; Diekelmann et al. 2011; Rolls et al. 2013). This method of targeted memory reactivation (Oudiette and Paller 2013) is specific to non-REM sleep and presenting task related cues during either the awake state or REM sleep does not provide any improvement in memory consolidation (Rasch et al. 2007; Diekelmann et al. 2011).

Rather than directly targeting the sensory component of a memory with a cue, a second strategy for modifying memories during sleep is to target the emotional valence of an experience. For mice performing a spatial task, optogenetic stimulation of the Ventral Tegmental Area (VTA), a reward center in the brain, results in enhanced sleep replay activity and improved subsequent performance of the task (McNamara et al. 2014). Meanwhile, when electrical stimulation of the VTA in rats is precisely timed to the reactivation of a single hippocampal place cell, it results in a new place preference for the rat matching the neuron's place field (de Lavilleon et al. 2015). Thus stimulation of the VTA can be used to artificially manipulate the valence of an experience during behaviour, or of a reactivated experience during sleep, leading to an enhanced memory.

While the above examples all take advantage of the brain during non-REM sleep, recent studies have shown memory enhancement can also be achieved during wakefulness. One such approach is deep brain stimulation (DBS), where electrical current is applied to the nuclei or fibre tracks of targeted brain structures via surgically implanted electrodes. This approach has been used in multiple applications, including the treatment of Parkinson's disease, depression, severe dementias and obesity. More recently, DBS of the fornix and hypothalamus has been reported to enhance associative and episodic memory recollection (Hamani et al. 2008), and to slow down the rate of cognitive decline in patients with Alzheimer's disease (Laxton et al. 2010). Furthermore, DBS of the entorhinal cortex has been shown to affect spatial memory (Suthana et al. 2012; Jacobs et al. 2016). While DBS may provide a route to memory enhancement, a less invasive alternative could be high-frequency, repetitive transcranial magnetic stimulation (rTMS). Using rTMS to target an area of the lateral parietal cortex with strong connectivity to the hippocampus, Wang and colleagues observed a long-lasting improvement in patients' performance of an associative memory task (Wang et al. 2014), with effects lasting to 15 days (Wang and Voss 2015).

To summarise, brain stimulation and targeted memory reactivation are two different approaches that have been used to enhance the consolidation process of hippocampus-dependent memories. It is worth noting that while statistically significant, these effects are typically mild (~10 % improvement). Manipulating coordinated brain rhythms (e.g. ripple-spindle interactions) and more precisely targeting the neural circuits storing a particular memory (Liu et al. 2012) may strengthen memory consolidation even further.

Removing Unwanted Memories

Not all memories are helpful. Some memories we might want to forget. The lead characters in the film *The Eternal Sunshine of the Spotless Mind* take advantage of a new technology that can delete selected autobiographical memories from their brain. They use this to forget their unhappy relationship, however the technology turns out to be too good to be true and they face the problem of piecing their memories together. Such technology does not currently exist, and based on current evidence seems unlikely to work. While frontotemporal dementia can give rise to amnesia for personally known individuals (Thompson et al. 2004), it is highly unlikely that it would be possible to selectively erase all the memories associated with a specific person. This is because semantic memories appear to be widely distributed in the neocortex (Martin and Chao 2001; McClelland and Rogers 2003). By contrast, editing hippocampus-dependent memories for a single event or learned association is not so inconceivable. Indeed, rather than something to be feared, memory removal may prove helpful in the treatment of phobias, PTSD and anxiety disorders.

While there appear to be specific endogenous mechanisms in the brain for degrading memories (Anderson et al. 2004; Frankland et al. 2013; Hardt et al. 2013; Hulbert et al. 2016), the search for drugs that can aid this process has been a topic of recent interest. Pharmacological treatment of the persistent involuntary memory retrieval that accompanies PTSD has been explored in numerous studies (see e.g. Steckler and Risbrough 2012; de Kleine et al. 2013 for review). The unwanted memory retrieval in PTSD is highly disruptive to the patient's health. They may suffer distraction at work from involuntary flash backs and 'night terrors' while sleeping. While psychological interventions have shown impressive advancement in recent years, attempts to treat the condition with drugs has been on the rise. In both clinical and laboratory settings, a wide variety of pharmacological agents have been explored, with particular emphasis on disrupting fear-related memories (Kaplan and Moore 2011). These have focused on glucocorticoid (e.g. (de Bitencourt et al. 2013), glutamatergic (Kuriyama et al. 2013), GABAergic (Rodríguez et al. 2013) adrenergic (Kindt et al. 2009), cannabinoid (Rabinak et al. 2013), serotonergic (Zhang et al. 2013) and glycine (File et al. 1999) receptors.

In animal models the study of memory manipulation has predominately focused on Pavlovian fear conditioning in rodents, in which an electrical shock is delivered through the floor of the test cage. The dominance of this approach is due to the rapid memory formation, and the robustness of the expression of this memory in the form of freezing behaviour. 'Auditory fear conditioning' involves initial exposure to the repeated pairings of an electrical shock with a neutral tone. With time, the tone alone evokes a fear memory revealed in observed freezing behaviour (Maren 2001). In 'contextual fear conditioning' the animal is exposed to a novel environment in which it receives one or more electric shocks, eliciting a learned association between the environmental context and the potential for more shocks (Kim and Fanselow 1992). Recent contextual fear memories can be suppressed by hippocampal inactivation, but this effect is not specific to a single memory (Varela et al. 2016). However, repeated exposure to the tone or context alone leads to a natural reduction in freezing, suggesting a weakening of the memory. This is referred to as extinction. When fibroblast growth factor 2 (an agent affecting neural cell development and neurogenesis) is infused into the amygdala immediately after extinction, it strongly increases the likelihood that the fear memory will not re-surface (Graham and Richardson 2011). It has been demonstrated that the extinction of conditioned fear memories can be boosted via reactivation of the memories during non-REM sleep. For example, Hauner and colleagues conditioned humans to expect a shock when viewing certain faces, where the presentation of the faces associated with the shocks was also paired with certain odours. Subsequently, during non-REM sleep subjects were re-exposed to the odours associated with half of the feared faces. After sleep and during fMRI, conditioned responses to the faces associated with the odours that were represented during sleep were ameliorated in comparison to the faces paired with odours that were not (Hauner et al. 2013). This effect was observed in a reduced BOLD signal in the hippocampus, as well as a reorganisation of activity patterns in the amygdala when pre- and post-sleep conditioning periods were compared. Although these results might appear to go against the memory-enhancing effects of cued-reactivation during non-REM sleep (Rasch et al. 2007; Rudoy et al. 2009; Rolls et al. 2013), the extinction of a fear memory is not necessarily caused by memory removal. Contrary, it is likely that extinction involves the active suppression of a still intact fear memory by regions of the brain distinct from where the original fear memory is stored (Milad and Quirk 2002). Furthermore, recent work by Schriener and colleagues has shown that the memory benefits of cued reactivation during sleep are lost if the memory cue is immediately followed by other auditory stimulation. Sleeping patients were presented with reactivation cues in the form of Dutch vocabulary, immediately followed by either a correct or incorrect translation into German vocabulary (mother tongue), or a neutral tone. The reactivation effect caused by the initial cue was diminished by the subsequent auditory stimulus, and this was also observed via EEG as the disruption of the neural oscillations associated with learning (Schriener et al. 2015).

Applying drugs or selective cueing during sleep provides one means of disrupting memories, another approach is to manipulate the brain at a much later point in time, potentially many weeks later. Memories are thought to require restabilising after reactivation, a process known as reconsolidation (Misanin et al. 1968; Sara 2010; Dudai 2004). In an influential study by Nader and colleagues, an infusion of protein synthesis inhibitors was found to disrupt fear conditioned memories when applied during periods following the reactivation of the memory (Nader et al. 2000). Oral application of the adrenergic modulator propranolol has been used to study reconsolidation in humans, with an emphasis on preventing the reactivation of fear conditioned memories (Brunet et al. 2011; Kindt et al. 2009). It is thought that propranolol is able to block the reconsolidation of fear memories, providing a potential treatment for PTSD and phobias. However, because propranolol must be administered before the reactivation to have an effect, there has been some debate as to whether reconsolidation processes have been specifically targeted (Brunet et al. 2011) or not (Schiller and Phelps 2011).

The study of long-term potentiation (LTP) has been important for research on memory manipulation. LTP is an activity-dependent, persistent form of synaptic plasticity and provides a key model for memory storage at the cellular level (Bliss and Collingridge 1993; Malenka and Bear 2004). LTP is a complex topic beyond the scope of this review, but in a simplified model it is thought synapses that have been active during an experience become strengthened to form a memory of that experience. Whether the memory persists depends on the continued maintenance of LTP in the relevant synapses. Prior work has suggested that persistent phosphorylation by PKMζ (protein kinase M zeta) is needed for this maintenance (Pastalkova et al. 2006). An injection of synthetic ζ-pseudosubstrate inhibitory peptide (ZIP) to the hippocampus inhibits PKMζ, and consequently causes disruption to LTP (Serrano et al. 2005). One day after rats have been trained in an active place avoidance task, specific injection of ZIP into their hippocampus disrupts their performance (Pastalkova et al. 2006). Furthermore, injection of ZIP at different neuroanatomical sites can also help to delete other memory types; deletion of a taste-aversion memory stored in the insula can be achieved by ZIP injection to the insula (Shema et al. 2007). Another approach to disrupting PKMζ has been the lentivirus-induced overexpression of a dominant-negative PKMζ mutation in insular cortex. This also blocks taste-aversion memory (Shema et al. 2011). Interestingly, enhancement of taste aversion can be achieved by overexpression of PKMζ in the insular cortex (using the same lentiviral approach) (Shema et al. 2011). However, recent evidence suggests that the relationship between ZIP, PKMζ and LTP maintenance may be more complicated than previously thought. If PKMζ was essential for memory, then transgenic knockout mice lacking PKMζ should have impaired memory function, but they do not (Volk et al. 2013; Lee et al. 2013). Since ZIP is still effective in erasing memories in PKMζ null mice, ZIP does not need PKMζ to function and kinases other than PKMζ may be crucial for LTP maintenance. Indeed, a recent study found that an enzyme closely related to PKMζ, named PKCι/λ (protein kinase C iota/lambda), substitutes for PKMζ in the transgenic knockout mice (Tsokas et al. 2016) and is similarly inhibited by ZIP but at higher concentrations (Ren et al. 2013). Additionally, another recent explanation for how ZIP disrupts memories is that ZIP triggers cell death in the hippocampus (Sadeh et al. 2015). However it should be noted that Sadeh and colleagues reported the majority of these cell deaths at ZIP concentrations far higher than the doses often used to impair memory. As well as this, similar cell deaths were reported for the same concentrations of scrambled ZIP (scr-ZIP); a control peptide known not to affect long-term memory retention (Pastalkova et al. 2006; Shema et al. 2007).

Incepting Memories

In the movie *The Matrix* (see Appendix), Neo has the knowledge of kung fu "downloaded" directly into his brain. How close are we to artificially creating or "incepting" new memories into our brain? Like the sleep-specific manipulations discussed previously that can be used to enhance memories, similar approaches can be used to artificially create new memories. One approach used by Arzi and colleagues was to present paired auditory-olfactory cues (e.g. a high frequency tone with an unpleasant odour) to human subjects while they were sleeping (Arzi et al. 2012). Because larger sniff volumes are evoked by pleasant odours than unpleasant odours, the sniff volume when a sound is presented in the absence of an odour provides a proxy for the expectation of the odour (that is normally paired to the sound). After these auditory-olfactory pairings were conditioned during a pre-test, non-REM sleep, Arzi and colleagues observed that sounds associated with pleasant odours had larger sniff volumes than sounds associated with unpleasant odours. The results provide a new method for unconsciously storing new memories, albeit limited to associations between sensory cues.

Next-generation molecular-genetic methods are now being used to more directly target and manipulate the neurons encoding memories—referred to as memory engram cells. In a c-Fos-tTA transgenic mouse, the tetracycline transactivator (tTA) is under the control of the immediate early gene c-Fos, which in turn is driven by recent neural activity. Additionally, the presence of doxycycline inhibits the binding of tTA to its target. Thus, the combination of c-Fos and tTA allows the spatial and temporal restriction of gene expression to be limited to the neural circuit involved in encoding a single recent experience. Using the strategy of cFos/tTA-driven transcription with either the channelrhodopsin-2 (ChR2) or the hM_3D_q DREADD (designer receptor exclusively activated by designer drug) (Liu et al. 2012; Garner et al. 2012), mice underwent a fear conditioning protocol. After doxycycline was removed from the diet, allowing c-Fos-tTA gene transcription to function normally, mice received several mild shocks in a novel context to create in a new contextual fear memory. As a result, using either light (for ChR2 mice) or an intraperitoneal injection of Clozapine-N-Oxide (for hM_3D_q mice), activity in the neural circuit storing the newly formed fear memory could be induced, observable from the freezing behaviour of the mouse. While these two methods successfully reactivated the fear memory, it is difficult to determine if the neural circuit storing this memory was directly targeted. The ChR2 approach only targeted the dentate gyrus (Liu et al. 2012), thus it is unclear whether the actual fear memory is stored in the ChR2 expressing neurons, or if it resides further downstream in the neural cascade that produces the freezing response (e.g. CA3 and CA1 of the hippocampus). Additionally, although the hM_3D_q approach (Garner et al. 2012) targeted multiple brain regions, it is likely that the memory is only stored in a subset of the neurons expressing hM_3D_q. Therefore, it is probable that the neural circuit encoding the fear memory is not uniquely targeted by the Clozapine-N-Oxide.

To take this one step further and artificially create an entirely new memory, Ramirez and colleagues used c-Fos-tTA mice expressing ChR2 to identify memory engram cells in the hippocampus corresponding to the memory of a novel context. Using light stimulation, they then paired the reactivation of this memory with a fear conditioning (shocks) in a different, unrelated context (Ramirez et al. 2013). Upon returning the mice to the original context, the mice showed elevated freezing levels despite never having been actually shocked in this context. Hence, the mice were artificially fear conditioned by pairing the reactivation of the contextual memory with a shock, thus creating a new fear association into their memory. Extending this method even further, Redondo et al. (2014) succeeded in changing the valence of a contextual memory stored in the hippocampus of mice. By incorporating the optogenetic reactivation of a fearful engram within a rewarding context, the negative valence of this memory was decreased (Redondo et al. 2014). It is important to note that this approach, as well as the cue-pairing during sleep approach described previously, only creates a new association between previous experiences. Although we are still far away from the ability to download complex procedural memories (i.e. kung fu) into our brains, we have taken a giant step in this direction with the "inception" of new hippocampus-dependent, associative memories.

Conclusion

In this chapter, we have discussed the different approaches and methods for modifying hippocampus-dependent memories. These fall under the approaches of enhancing memories, deleting memories and implanting false memories (inception). Related to these topics is the idea that it could one day be possible 'read' peoples thoughts. Whilst seemingly deep into the realms of science fiction, it is an area of considerable interest to domains such as law and marketing. In a recent study, Uncapher and colleagues went a step closer to determining whether 'mind reading' could be viable technique in an eyewitness identification context. They conducted a study whereby participants were shown a series of previously studied and novel faces whilst undergoing fMRI scanning. Using multivariate pattern analysis (MVPA) on the fMRI data, they were able to reliably classify whether a presented face was previously known or novel to the participant. However, when the participants were asked to conceal their true memory state (i.e. pretend a novel face was known and vice versa), the ability to decode that memory state using MVPA was lost, and in some cases even reversed (Uncapher et al. 2015). Hence, it may be that mind reading techniques based on neuroimaging are never robust enough for use in a court of law.

An approach taken by many of the studies covered in this chapter is to manipulate memories during sleep, when they are more malleable (Diekelmann and Born 2010; Oudiette and Paller 2013). A second strategy has been to target specific neurons using molecular-genetic techniques, allowing control over the neural circuits regulating the encoding of a memory (Liu et al. 2012; Garner et al. 2012). Finally, a third strategy has been to manipulate the synaptic processes involved in memory maintenance (Pastalkova et al. 2006). Looking to the future, combining these three approaches may lead to a more powerful means of controlling memory. Researchers will continue to enhance, delete, and incept memories; whether one day science will be able to emulate all the concepts that science fiction has to offer remains to be seen.

Appendix: Movies About Memory Enhancement, Deletion, and Inception

The following appendix is an updated version of the appendix appearing in Spiers and Bendor (2014).

Lucy (2015): After getting overdosed with a new experimental drug that unlocks the "unused" portion of the brain, the main character develops super cognitive abilities, including telekinesis and metamorphosis. According to the movie, we use only 10 % of our brain. This is a scientific "urban legend" that is completely false. The only person that uses 10 % of their brain was perhaps the writer of this movie.

The Bourne Identity (2002): A highly-trained spy with no episodic memory, but all his procedural memory intact. Essentially James Bond with dementia and without the NHS.

Eternal Sunshine of the Spotless Mind (2004): After breaking up with his girlfriend, the main character has a procedure performed-while he sleeps, a machine zaps and deletes all the memories of his ex-girlfriend. This technology replaces more established gustatory-driven methods of recovering from a break-up, like eating several cartons of ice cream.

Inception (2010): Using a "shared dream" technology, the main character and his team attempt to implant false memories (inception) in an unsuspecting target. The larger question is how did they get all that "dream-hacking" equipment through airport security?

Limitless (2011): The main character takes a mystery pill (NZT) that substantially enhancing his cognitive abilities. The movie demonstrates some of the down-sides of "genius withdrawal".

The Manchurian Candidate (1962, 2004 (remake)): A solider captured by the enemy is "programmed" to become an assassin. After receiving the trigger (a queen of diamonds playing card), the solider unconsciously carriers out any instruction (such as assassinating a target), after which he forgets everything related to these actions. With the "queen of diamonds" as the trigger, best to avoid playing poker with this guy. . .

The Matrix Trilogy (1999, 2003): The year is 2199. After a war between humans and computers, humans now live inside a virtual reality environment called "the Matrix", where humans still think it is 1999, and are unaware of what has happened. The few humans that have managed to leave the Matrix are staging a revolution, and must re-enter the Matrix to fight the computers. As the Matrix is essentially software, computer code structured by rules, humans find that it is possible to "download" new skills and learn to bend or even break the rules of physics. The writers also decide to break the rules of physics by ignoring the first law of thermodynamics, suggesting that humans within the Matrix are used as energy sources (producing more energy than they require to survive).

Total Recall (1990): Implanting a false memory of a vacation to Mars has bizarre consequences for the main character, unlocking a supressed memory of his true identity- a secret agent. Could this movie have been the inspiration behind Newt Gingrich's plan to build a space colony on Mars?

Total Recall (2012 (remake)): A poorly done remake of the 1990 Total Recall movie. After watching this, you may want to look into some memory deletion technology (see Eternal Sunshine of the Spotless Mind)

See Baxendale (2004) for a review of movies exploring memory-related themes.

References

Anderson MC, Ochsner KN, Kuhl B, Cooper J, Robertson E, Gabrieli SW, Glover GH, Gabrieli JDE (2004) Neural systems underlying the suppression of unwanted memories. Science 303 (5655):232–235

Antony JW, Gobel EW, O'Hare JK, Reber PJ, Paller KA (2012) Cued memory reactivation during sleep influences skill learning. Nat Neurosci 15(8):1114–1116

Arzi A, Shedlesky L, Ben-Shaul M, Nasser K, Oksenberg A, Hairston IS, Sobel N (2012) Humans can learn new information during sleep. Nat Neurosci 15(10):1460–1465

Barnes DC, Wilson DA (2014) Slow-wave sleep-imposed replay modulates both strength and precision of memory. J Neurosci 34(15):5134–5142

Baxendale S (2004) Memories aren't made of this: amnesia at the movies. Br Med J 329 (7480):1480

Bendor D, Wilson MA (2012) Biasing the content of hippocampal replay during sleep. Nat Neurosci 15:1439–1444

Bliss TVP, Collingridge GL (1993) A synaptic model of memory: long-term potentiation in the hippocampus. Nature 361(6407):31–39

Brunet A, Ashbaugh AR, Saumier D, Pitman RK, Nelson M, Tremblay J, Roullet P, Birmes P (2011) Does reconsolidation occur in humans: a reply. Front Behav Neurosci 5:74

Buzsaki G (2009) Rhythms of the brain. Oxford University Press, Oxford

Buzsáki G, Anastassiou CA, Koch C (2012) The origin of extracellular fields and currents—EEG, ECoG, LFP and spikes. Nat Rev Neurosci 13(6):407–420

De Bitencourt RM, Pamplona FA, Takahashi RN (2013) A current overview of cannabinoids and glucocorticoids in facilitating extinction of aversive memories: potential extinction enhancers. Neuropharmacology 64:389–395

De Kleine RA, Rothbaum BO, van Minnen A (2013) Pharmacological enhancement of exposure-based treatment in PTSD: a qualitative review. Eur J Psychotraumatol:4. doi:10.3402/ejpt. v4i0.21626

de Lavilléon G, Lacroix MM, Rondi-Reig L, Benchenane K (2015) Explicit memory creation during sleep demonstrates a causal role of place cells in navigation. Nat Neurosci 18 (4):493–495

Diekelmann S, Born J (2010) The memory function of sleep. Nat Rev Neurosci 11(2):114–126

Diekelmann S, Büchel C, Born J, Rasch B (2011) Labile or stable: opposing consequences for memory when reactivated during waking and sleep. Nat Neurosci 14(3):381–386

Dudai Y (2004) The neurobiology of consolidations, or, how stable is the engram? Annu Rev Psychol 55:51–86

Eichenbaum H (2004) Hippocampus: cognitive processes and neural representations that underlie declarative memory. Neuron 44:109–120

Ego-Stengel V, Wilson MA (2010) Disruption of ripple-associated hippocampal activity during rest impairs spatial learning in the rat. Hippocampus 20(1):1–10

File SE, Fluck E, Fernandes C (1999) Beneficial effects of glycine (bioglycin) on memory and attention in young and middle-aged adults. J Clin Psychopharmacol 19:506–512

Frankland PW, Bontempi B (2005) The organization of recent and remote memories. Nat Rev Neurosci 6:119–130

Frankland PW, Köhler S, Josselyn SA (2013) Hippocampal neurogenesis and forgetting. Trends Cogn Sci 36(9):497–503

Garner AR, Rowland DC, Hwang SY, Baumgaertel K, Roth BL, Kentros C, Mayford M (2012) Generation of a Synthetic Memory Trace. Science 335(6075):1513–1516

Girardeau G, Benchenane K, Wiener SI, Buzsáki G, Zugaro MB (2009) Selective suppression of hippocampal ripples impairs spatial memory. Nat Neurosci 12(10):1222–1223

Graham BM, Richardson R (2011) Intraamygdala infusion of fibroblast growth factor 2 enhances extinction and reduces renewal and reinstatement in adult rats. J Neurosci 31(40):14151–14157

Groes (2016) Memory in the twenty-first century new critical perspectives from the arts, humanities, and sciences. Palgrave MacMillan, Basingstoke

Halassa MM, Siegle JH, Ritt JT, Ting JT, Feng G, Moore CI (2011) Selective optical drive of thalamic reticular nucleus generates thalamic bursts and cortical spindles. Nat Neurosci 14 (9):1118–1120

Hamani C, McAndrews MP, Cohn M, Oh M, Zumsteg D, Shapiro CM et al (2008) Memory enhancement induced by hypothalamic/fornix deep brain stimulation. Ann Neurol 63 (1):119–123

Hardt O, Nader K, Nader L (2013) Decay happens: the role of active forgetting in memory. Trends Cogn Sci 17(3):111–120

Hauner KK, Howard JD, Zelano C, Gottfried JA (2013) Stimulus-specific enhancement of fear extinction during slow-wave sleep. Nat Neurosci 16:1553–1555

Hulbert JC, Henson RN, Anderson MC (2016) Inducing amnesia through systematic suppression. Nat Commun 7:11003

Jacobs J, Miller J, Lee SA, Coffey T, Watrous AJ, Sperling MR et al (2016) Direct electrical stimulation of the human entorhinal region and hippocampus impairs memory. Neuron 92 (5):983–990. doi:10.1016/j.neuron.2016.10.062

Ji D, Wilson MA (2006) Coordinated memory replay in the visual cortex and hippocampus during sleep. Nat Neurosci 10(1):100–107

Kaplan GB, Moore KA (2011) The use of cognitive enhancers in animal models of fear extinction. Pharmacol Biochem Behav 99:217–228

Kim JJ, Fanselow MS (1992) Modality-specific retrograde amnesia of fear. Science 256 (5057):675–677

Kindt M, Soeter M, Vervliet B (2009) Beyond extinction: erasing human fear responses and preventing the return of fear. Nat Neurosci 12:256–258

Kuriyama K, Honma M, Yoshiike T, Kim Y (2013) Valproic acid but not D-cycloserine facilitates sleep-dependent offline learning of extinction and habituation of conditioned fear in humans. Neuropharmacology 64:424–431

Laxton AW, Tang-Wai DF, McAndrews MP, Zumsteg D, Wennberg R, Keren R et al (2010) A phase I trial of deep brain stimulation of memory circuits in Alzheimer's disease. Ann Neurol 68(4):521–534

Lee AK, Wilson MA (2002) Memory of sequential experience in the hippocampus during slow wave sleep. Neuron 36(6):1183–1194

Lee AM, Kanter BR, Wang D, Lim JP, Zou ME, Qiu C et al (2013) Prkcz null mice show normal learning and memory. Nature 493(7432):416–419

Liao SM, Sandberg A (2008) The normativity of memory modification. Neuroethics 1(2):85–99

Liu X, Ramirez S, Pang PT, Puryear CB, Govindarajan A, Deisseroth K, Tonegawa S (2012) Optogenetic stimulation of a hippocampal engram activates fear memory recall. Nature 484 (7394):381–385

Loftus EF, Palmer JC (1974) Reconstruction of automobile destruction: an example of the interaction between language and memory. J Verbal Learn Verbal Behav 13(5):585–589

Malenka RC, Bear MF (2004) LTP and LTD: an embarrassment of riches. Neuron 44(1):5–21

Maren S (2001) Neurobiology of Pavlovian fear conditioning. Ann Rev Neurosci 24(1):897–931

Martin A, Chao LL (2001) Semantic memory and the brain: structure and processes. Curr Opin Neurobiol 11:194–201

Marshall L, Helgadóttir H, Mölle M, Born J (2006) Boosting slow oscillations during sleep potentiates memory. Nature 444(7119):610–613

McClelland JL, Rogers TT (2003) The parallel distributed processing approach to semantic cognition. Nat Rev Neurosci 4(4):310–322

McNamara CG, Tejero-Cantero Á, Trouche S, Campo-Urriza N, Dupret D (2014) Dopaminergic neurons promote hippocampal reactivation and spatial memory persistence. Nat Neurosci 17 (12):1658–1660

Milad MR, Quirk GJ (2002) Neurons in medial prefrontal cortex signal memory for fear extinction. Nature 420(6911):70–74

Misanin JR, Miller RR, Lewis DJ (1968) Retrograde amnesia produced by electroconvulsive shock after reactivation of a consolidated memory trace. Science 160:554–555

Mohamed AD, Sahakian BJ (2012) The ethics of elective psychopharmacology. Int J Neuropsychopharmacol Off Sci J Coll Int Neuropsychopharmacol CINP 15:559–571

Moscovitch M, Nadel L, Winocur G, Gilboa A, Rosenbaum RS (2006) The cognitive neuroscience of remote episodic, semantic and spatial memory. Curr Opin Neurobiol 16:179–190

Nader K, Schafe GE, Le Doux JE (2000) Fear memories require protein synthesis in the amygdala for reconsolidation after retrieval. Nature 406(6797):722–726

Oudiette D, Paller KA (2013) Upgrading the sleeping brain with targeted memory reactivation. Trends Cogn Sci 17(3):142–149

Pastalkova E, Serrano P, Pinkhasova D, Wallace E, Fenton AA, Sacktor TC (2006) Storage of spatial information by the maintenance mechanism of LTP. Science 313(5790):1141–1144

Rabinak CA, Angstadt M, Sripada CS, Abelson JL, Liberzon I, Milad MR, Phan KL (2013) Cannabinoid facilitation of fear extinction memory recall in humans. Neuropharmacology 64:396–402

Ragan CI, Bard I, Singh I (2013) What should we do about student use of cognitive enhancers? An analysis of current evidence. Neuropharmacology 64:588–595

Ramirez S, Liu X, Lin PA, Suh J, Pignatelli M, Redondo RL et al (2013) Creating a false memory in the Hippocampus. Science 341(6144):387–391

Rasch B, Büchel C, Gais S, Born J (2007) Odour cues during slow-wave sleep prompt declarative memory consolidation. Science 315(5817):1426–1429

Redondo RL, Kim J, Arons AL, Ramirez S, Liu X, Tonegawa S (2014) Bidirectional switch of the valence associated with a hippocampal contextual memory engram. Nature 513 (7518):426–430

Ren SQ, Yan JZ, Zhang XY et al (2013) PKCλ is critical in AMPA receptor phosphorylation and synaptic incorporation during LTP. EMBO J 32(10):1365–1380

Rodríguez MLC, Campos J, Forcato C, Leiguarda R, Maldonado H, Molina VA, Pedreira ME (2013) Enhancing a declarative memory in humans: the effect of clonazepam on reconsolidation. Neuropharmacology 64:432–442

Rolls A, Makam M, Kroeger D, Colas D, de Lecea L, Heller HC (2013) Sleep to forget: interference of fear memories during sleep. Mol Psychiatry 18:1166–1170

Rudoy JD, Voss JL, Westerberg CE, Paller KA (2009) Strengthening individual memories by reactivating them during sleep. Science 326(5956):1079–1079

Sadeh N, Verbitsky S, Dudai Y, Segal M (2015) Zeta inhibitory peptide, a candidate inhibitor of protein kinase Mζ, is excitotoxic to cultured hippocampal neurons. J Neurosci 35 (36):12404–12411

Sara SJ (2010) Reactivation, retrieval, replay and reconsolidation in and out of sleep: connecting the dots. Front Behav Neurosci 4:185. doi:10.3389/fnbeh.2010.00185

Schiller D, Phelps EA (2011) Does reconsolidation occur in humans? Front Behav Neurosci 5:24. doi:10.3389/fnbeh.2011.00024

Schreiner T, Lehmann M, Rasch B (2015) Auditory feedback blocks memory benefits of cueing during sleep. Nat Commun 6:8729

Serrano P, Yao Y, Sacktor TC (2005) Persistent phosphorylation by protein kinase Mζ maintains late-phase long-term potentiation. J Neurosci 25(8):1979–1984

Shema R, Sacktor TC, Dudai Y (2007) Rapid erasure of long-term memory associations in the cortex by an inhibitor of PKMζ. Science 317(5840):951

Shema R, Haramati S, Ron S, Hazvi S, Chen A, Sacktor TC, Dudai Y (2011) Enhancement of consolidated long-term memory by overexpression of protein kinase Mζ in the neocortex. Science 331(6021):1207–1210

Siapas AG, Wilson MA (1998) Coordinated interactions between hippocampal ripples and cortical spindles during slow-wave sleep. Neuron 21(5):1123–1128

Sirota A, Csicsvari J, Buhl D, Buzsáki G (2003) Communication between neocortex and hippocampus during sleep in rodents. Proc Natl Acad Sci U S A 100(4):2065–2069

Spiers HJ (2012) Hippocampal formation. In: Ramachandran VS (ed) The encyclopedia of human behaviour, vol 2. Academic Press, New York, pp 297–304

Spiers HJ, Bendor D (2014) Enhance, delete, incept: manipulating hippocampus-dependent memories. Brain Res Bull 105:2–7

Squire LR, Alvarez P (1995) Retrograde amnesia and memory consolidation: a neurobiological perspective. Curr Opin Neurobiol 5:169–177

Squire LR, Stark CEL, Clark RE (2004) The medial temporal lobe. Annu Rev Neurosci 27:279–306

Steckler T, Risbrough V (2012) Pharmacological treatment of PTSD–established and new approaches. Neuropharmacology 62(2):617–627

Steriade M, McCormick DA, Sejnowski TJ (1993) Thalamocortical oscillations in the sleeping and aroused brain. Science 262(5134):679–685

Stickgold R, Walker MP (2013) Sleep-dependent memory triage: evolving generalization through selective processing. Nat Neurosci 16(2):139–145

Suthana N, Haneef Z, Stern J, Mukamel R, Behnke E, Knowlton B, Fried I (2012) Memory enhancement and deep-brain stimulation of the entorhinal area. N Engl J Med 366(6):502–510

Thompson SA, Graham KS, Williams G, Patterson K, Kapur N, Hodges JR (2004) Dissociating person-specific from general semantic knowledge: roles of the left and right temporal lobes. Neuropsychologia 42(3):359–370

Tsokas P, Hsieh C, Yao Y, Lesburguères E et al (2016) Compensation for PKMζ in long-term potentiation and spatial long-term memory in mutant mice. Elife 5. doi:10.7554/eLife.14846

Uncapher MR, Boyd-Meredith JT, Chow TE, Rissman J, Wagner AD (2015) Goal-directed modulation of neural memory patterns: implications for fMRI-based memory detection. J Neurosci 35(22):8531–8545

Varela C, Weiss S, Meyer R, Halassa M, Biedenkapp J, Wilson MA, Goosens KA, Bendor D (2016) Tracking the time-dependent role of the hippocampus in memory recall using DREADDs. PLoS One 11(5):e0154374

Volk LJ, Bachman JL, Johnson R, Yu Y, Huganir RL (2013) PKM-(ζ) is not required for hippocampal synaptic plasticity, learning and memory. Nature 493(7432):420–423

Wang JX, Rogers LM, Gross EZ, Ryals AR, Mehmet DE, Brandstatt KL, Hermiller MA, Voss JL (2014) Targeted enhancement of the cortical-hippocampal brain networks and associative memory. Science 346(6200):1054–1057

Wang JX, Voss JL (2015) Long-lasting enhancements of memory and hippocampal-cortical functional connectivity following multiple-day targeted noninvasive stimulation. Hippocampus 25(8):877–883

Wilson MA, McNaughton BL (1994) Reactivation of hippocampal ensemble memories during sleep. Science 265(5172):676–679

Zhang G, Ásgeirsdóttir HN, Cohen SJ, Munchow AH, Barrera MP, Stackman RW (2013) Stimulation of serotonin 2A receptors facilitates consolidation and extinction of fear memory in C57BL/6J mice. Neuropharmacology 64:403–413

Part II
Development, Aging, and Functional
Contributions to Learning and Memory

Hippocampal Development: Structure, Function and Implications

Joshua K. Lee, Elliott G. Johnson, and Simona Ghetti

Abstract There has been a substantial surge in interest in the contribution of hippocampal development to cognition in childhood. The reviewed evidence suggests that the hippocampus undergoes protracted functional and structural development after birth, and that this development is best understood in terms of cytoarchitectural and anterior–posterior functional subdivisions. The dentate gyrus may develop later than cornu ammonis 1 or the subiculum, and over development there may also be a shift towards recruiting the anterior hippocampus. Both of these changes promise new insights into the development of episodic memory and other cognitive functions, and indicate that the integration of the cytoarchitectural and anterior–posterior axes of development are necessary to address outstanding questions about the significance of hippocampal development from infancy into adulthood.

Although the hippocampus has long been considered a structure specialized for memory (Milner et al. 1968; Squire and Wixted 2011; Vargha-Khadem et al. 1997), it is becoming increasingly clear that the functional properties of this structure make it suitable for a wider range of cognitive skills and behaviors (Rubin et al. 2014), including higher-order perception (Graham et al. 2010; Monti et al. 2015; Yonelinas 2013) and language (Duff and Brown-Schmidt 2012). Consequently, understanding the development of the hippocampus has gained momentum, not only for understanding the development of episodic memory, but also that of other cognitive functions. Here, we review evidence of hippocampal development from infancy into childhood and finally adolescence, and we draw connections, when possible, about the behavioral implications of these changes. Before doing so, we note that there are a number of approaches to studying the hippocampus and that this review attempts to integrate them.

A common approach has been to treat the hippocampus as a whole (e.g. Østby et al. 2009). However, the hippocampus is not structurally homogenous (Lorente de Nó 1934; Small 2002), and other approaches have reflected this reality. One

J.K. Lee • E.G. Johnson • S. Ghetti (✉)
University of California, Davis, CA, USA
e-mail: sghetti@ucdavis.edu

© Springer International Publishing AG 2017   141
D.E. Hannula, M.C. Duff (eds.), *The Hippocampus from Cells to Systems*,
DOI 10.1007/978-3-319-50406-3_6

alternative approach, inherited from histological investigations, but increasingly used with neuroimaging research, involves partitioning the hippocampus into substructures based on its cytoarchitecture, including differences in the populations of neurons and the patterns of their intrinsic and extrinsic connectivity. These substructures, or subfields, include the dentate gyrus, cornu ammonis 3 (CA3) and 1 (CA1), as well as the subiculum complex (Amaral and Witter 1995; Lorente de Nó 1934). These subfields represent the major destinations within the primary excitatory pathway of the hippocampus; originally named the trisynaptic circuit (Andersen 1975; Ramón y Cajal 1911). The trisynaptic circuit begins in the dentate gyrus, and then proceeds thru CA3 before arriving in the CA1 subfield. This circuit has been the subject of a vast amount of research, and a number of well-developed theoretical models have been explored positing distinct operations for each of these subfields (e.g. Rolls 2007); moreover, these computational models have received empirical support from a number of investigations in rodent models and humans (Kesner 2007; Yassa and Stark 2011).

Most computational models theorize that the dentate gyrus, characterized by sparse neuronal firing patterns, supports computational transformations on inputs from the entorhinal cortex called pattern separation (Rolls 2007; Marr 1971). Pattern separation is an operation through which hippocampal inputs are orthogonalized, or differentiated to reduce interference between similar inputs. Pattern separation enables effective encoding of representations that capture the unique associations among features of an episode while reducing interference from similar cues or memory representations. For example, pattern separation processes may support discrimination of objects sharing similar perceptual features or spatial locations (Bakker et al. 2008; Gilbert et al. 1998; Gilbert et al. 2001). Complementing pattern separation in the dentate gyrus are theorized functions of binding and pattern completion in CA3 (Rolls 2007), capable of arbitrarily encoding and retrieving conjunctive relations (e.g. object–location). Finally, downstream from the dentate gyrus and CA3 is the CA1 subfield. In addition to CA3 inputs, CA1 receives direct input from the entorhinal cortex via a monosynaptic pathway and CA1 may be involved in forming context-dependent bound representations from those two primary inputs (Sheffield and Dombeck 2015). The extensive literature on subfield structure and function (e.g. Hunsaker and Kesner 2013; Moser et al. 2008) provides a strong basis to derive new hypotheses and interpret findings about the development of hippocampal subfields.

A second alternative approach to studying hippocampal substructures—one that is gaining traction—emphasizes subregions defined by their positions along the anterior–posterior axis of the structure. A common division includes a hippocampal head region anterior to the uncal apex, an intermediate hippocampal body region posterior to hippocampal head, and a tail region, posterior to hippocampal body (e.g. Daugherty et al. 2015a; DeMaster et al. 2014). Across species there is a growing body of evidence for a remarkable number of differences along the longitudinal axis including differences in gross morphology and intrinsic structural connectivity (e.g. Kondo et al. 2009) and connectivity to other brain regions (Poppenk et al. 2013). Specifically, the anterior hippocampus is preferentially connected to prefrontal networks and posterior hippocampus is preferentially connected to posterior brain networks (Blessing et al. 2016; Poppenk et al. 2013). These patterns of connectivity have been associated with different levels of cognitive flexibility, which affords the capacity to respond to current goals by manipulating, modifying or integrating mental representations. Anterior networks tend to be associated with flexible memories (Giovanello et al. 2009) and posterior networks tend to be associated with memories that include spatial or other forms of perceptually salient information reproducing the original context with higher fidelity (Ciaramelli et al. 2010; Persson and Söderlund 2015). To date the literature has explored the contributions of the hippocampal subfields and long-axis subregions independently. In this review we attempt, when possible, to begin to integrate these two levels of analysis, first in infancy and early childhood (0–2, and 3–5 years, respectively), and then in middle and late childhood, and adolescence (6–10, 11–12, and 13–17 years, respectively), and we discuss potential implications for behavioral change. Finally in our review, findings from cross-sectional studies will be described in terms of age-related differences, while findings from longitudinal studies will described as developmental changes.

Hippocampal Development in Infancy and Early Childhood

Most of what we know about early hippocampal development comes for histological evidence in non-human animals or post-mortem studies in humans. We discuss this evidence first and then draw some implications for behavioral development.

Structural Development: Evidence from Histology Substantial hippocampal development occurs during prenatal development. For example, age-related increases in raw hippocampal volumes are observed during the third trimester of gestation (Seress et al. 2001; post-mortem analysis); these increases have been primarily attributed to the cellular layers in the dentate gyrus, including a 5-fold increase in the granule layer, a 20-fold increase in the molecular layer, and a 10-fold increase in the hilar region. In a separate analysis, Seress et al. (2001) reported relatively fewer differences with age in the CA fields, suggesting that during the early stages of hippocampal development in infancy, the basic circuitry of the hippocampal CA fields is relatively more mature than that of the dentate gyrus.

Considerable age-related growth in the raw volume of the human dentate gyrus is also observed after birth in histological studies (Insausti et al. 2010; Seress et al. 2001), particularly in the granule cell layer where in non-human animals cell proliferation through processes of neurogenesis has been demonstrated (Feliciano and Bordey 2013). For example, the dentate gyrus in the macaque is the slowest developing field, with roughly 40 % of the total number of granule cells found in adults developing postnatally (Jabès and Nelson 2015; Jabès et al. 2010); the dentate gyrus still contains a significantly greater number of immature cells in the 1-year-old primate than in the 5–10 year primates (Jabès et al. 2010; Jabès and

Nelson 2015). We note that additional processes such as dendritic maturation of existing granule cells (Seress and Mrzljak 1992), and glial cell proliferation (Seress et al. 2001) may also contribute to volumetric change in the granule cell layer.

Regardless of the exact source of this volumetric increase in the dentate gyrus, it is clear that substantial change occurs postnatally. Moreover, the development of the dentate gyrus is paralleled by age-related volume increases in the CA3 subfield (Lavenex and Banta Lavenex 2013). Increases in CA3 volume are more apparent in those proximal portions of CA3 that receive input from the dentate gyrus compared to the more distal portions whose entorhinal inputs may mature somewhat earlier (Amaral and Lavenex 2007; Kondo et al. 2008, 2009), underscoring the close connection between these subfields. In contrast to these prolonged age-related increases, volume changes in the CA1 and subiculum seem to be more limited (Insausti et al. 2010). This hierarchical progression of subfield development is matched by its extrinsic connections with the entorhinal cortex. While the mono-synaptic pathway between entorhinal cortex and CA1 appears to be established prenatally in the second trimester, entorhinal connectivity to the dentate gyrus through the tri-synaptic pathway shows only limited connectivity prenatally (Hevner and Kinney 1996). Indeed the developmental time-course of entorhinal connectivity with the dentate gyrus is also protracted, with myelination of those axons occurring later in childhood and adolescence (Ábrahám et al. 2010). The reviewed histological differences have a number of behavioral implications, and these are discussed in the next section.

Implications for Behavioral Development Concurrent with early hippocampal development are rapid improvements in memory, culminating in the end of infan-tile amnesia—the inability to form lasting episodic memories—by the third year (Bauer 2004; Bauer et al. 2010; Newcombe et al. 2007), including rapid improve-ments in spatial-relational memory (Newcombe et al. 2014; Sluzenski et al. 2004), and other cognitive abilities associated with adult hippocampal function (e.g. language; Mårtensson et al. 2012). While empirical evidence establishing a causal connection between early hippocampal development and the rise of these cognitive competencies is still limited, several prominent hypotheses have been advanced. Consistent with the outcomes of histological studies outlined above, Lavenex and Banta Lavenex (2013) proposed that the hippocampus develops hierarchically, such that the monosynaptic pathway, which projects from the entorhinal cortex to the CA1 emerges earlier than the trisynaptic pathway, which projects from the entorhinal cortex to the dentate gyrus and then CA3 (Gómez and Edgin 2015; Lavenex and Banta Lavenex 2013). From this perspective, early in infancy when neither pathway is fully established, memory for events may inordi-nately rely on incremental cortical learning that is inflexible, and easily disrupted when the spatio-temporal context is altered (e.g. Gómez and Edgin 2015; Mullally and Maguire 2014).

During the second year, however, Gómez and Edgin (2015) and Lavenex and Banta Lavenex (2013) have proposed that emerging memory abilities may depend on the monosynaptic circuit, which could provide the first basic forms of episodic memory. For example, it is at this point that infants begin to be able to reproduce multi-step sequences (Bauer 1996), form transitive associations between two puppets not directly viewed together (Cuevas et al. 2006), rapidly improve in their ability to represent and form relations among multiple locations in memory, as well as begin to reliably recognize objects presented outside of their original learned context (Robinson and Pascalis 2004). The emergence of early episodic memory abilities coincident with the viability of the monosynaptic circuit is intriguing, and could offer new insights into the function of this hippocampal circuit and into the emergence of other cognitive behaviors associated with hippocampal function.

While the establishment of basic episodic memory abilities represents a major achievement of cognitive development, new episodic memory abilities continue to emerge during the preschool years, and this development may be the result of substantial changes in the trisynaptic circuit during this period (e.g. Insausti et al. 2010; Lavenex and Banta Lavenex 2013). The emergence of new abilities like allocentric spatial navigation and the ability to learn multiple item–place relations within the same spatio–temporal context (e.g. Ribordy et al. 2013), and the ability to integrate more complex temporal and associative relations (Gómez and Edgin 2015) may critically depend on operations of pattern separation in the dentate gyrus to avoid inappropriate feature integration and interference, and on operations of pattern completion in CA3 to provide representational flexibility (Jabès and Nelson 2015). Interestingly, the role of developmental neurogenesis in one of the key components of this circuit, the dentate gyrus (Frankland et al. 2013) has been proposed as a mechanism underlying infantile amnesia; while new dentate gyrus neurons might confer new memory abilities (e.g. Suárez-Pereira et al. 2015), the especially rapid proliferation during infancy may also disrupt pre-existing hippocampal memory representations (Frankland et al. 2013; Mongiat et al. 2009; Yasuda et al. 2011). However, human data addressing this possibility are not yet available. Also, empirical tests of the relation between trisynaptic circuit function and the emergence of these advanced episodic memory abilities are currently lacking.

Structural and Functional Neuroimaging Studies To our knowledge only four structural imaging studies reported analyses of age-related differences in the volume of the hippocampal formation during early childhood and none of them has examined the development of hippocampal subfields. In terms of the overall volume of the hippocampal formation, these studies have painted an inconsistent developmental picture. For example, using manual segmentation, Uematsu et al. (2012) reported rapid age-related increases in intracranial-volume (ICV)-adjusted hippocampal volume from infancy into childhood. Likewise, using the automated segmentation procedure implemented in FreeSurfer (Fischl 2012), Riggins et al. (2015) reported that 6-year-olds had marginally larger ICV-adjusted overall hippocampal volumes than 4-year-old children. In contrast to these findings, Gogtay et al. (2006) did not find changes in overall ICV-adjusted hippocampal volumes in a longitudinal study spanning five scans and ten years in children as young as four years; however the sample of very young children was small in that study. Finally, using a high accuracy segmentation method that corrects FreeSurfer segmentation errors with an algorithm trained on manual segmentations (Automatic Segmentation Adapter Tool; Wang et al. 2011), Lee et al. (2015a) examined an even younger sample of eighty-nine 2 to 4 year-olds. Contrary to expectation, age-related increases in volume were not observed (Lee et al. 2015a). The limited age range in this study could have diminished the ability to observe age-related differences during a time when the hippocampal formation may be especially susceptible to experience-dependent neurogenesis (Opendak and Gould 2015), such as experiences of maternal support (Luby et al. 2012), the result of which may have increased individual variation more than normative age-related change. In addition, Lee et al. (2015a) assessed only overall hippocampal volume, which ignores its heterogeneous substructure and can mask developmental change. Of the structural studies in early childhood, only Riggins et al. (2015) explicitly examined hippocampal subregions in childhood; however only marginally reliable age-related increases were observed in hippocampal tail in 6-year-olds compared to 4-year-olds.

The relations between hippocampal volume and cognition were also explored in early childhood in two of the above-referenced studies. Examining how a measure of episodic source memory (i.e. the relation between an item and a spatio-temporal context) is associated the volume of head, body, and tail, Riggins et al. (2015) observed a positive relation with the volume of hippocampal head in 6-year-olds, but not in 4-year-old children; moreover, this correlation was significantly greater in 6-year-olds than 4-year-olds. These associations were not found with the hippocampus as a whole underscoring that some relations with cognition may differ between hippocampal substructures. The second study examined relations between overall hippocampal volume and the development of language ability (Lee et al. 2015a) based on previous evidence of an association with foreign language learning over time (Mårtensson et al. 2012), which suggested that the hippocampus is important for vocabulary learning. One might then expect that this relation would be observed in early development where word learning is especially rapid. Initial evidence consistent with this possibility was reported in a study in which Deniz Can et al. (2013) showed that hippocampal volume assessed by voxel-based morphometry at 7 month of age predicted later expressive language ability at 12 months of age. Thus, Lee et al. (2015a) predicted that hippocampal volumes would predict language ability in early childhood. Consistent with hypothesis, age-related increases in expressive language ability were observed, and those age-related increases were larger in those with larger hippocampal volumes, bilaterally.

One final study examined how the hippocampus contributes to episodic memory development in early childhood. Riggins et al. (2016) assessed resting state functional connectivity networks in 4- and 6-year-olds associated with the anterior and posterior hippocampus. As in adults (e.g. Libby et al. 2012), young children showed different patterns of cortical connectivity in the anterior and posterior hippocampus. In addition, 4- and 6-year-olds differed in the association between hippocampal functional connectivity and their episodic memory performance. Increased connectivity from the *anterior* hippocampus to the parietal and prefrontal cortices was negatively associated with episodic memory in 4-year-olds, but positively in 6-year-olds. In contrast, increased connectivity from the *posterior* hippocampus to a region in the temporal lobe was positively associated with memory in 4-year-olds, but negatively in 6-year-olds. Critically, while 6-year-olds remembered marginally better than 4-year-olds, both remembered well over chance. These results suggest that changes in episodic memory in early childhood may depend on integration of hippocampal operations with those implemented in fronto-parietal networks.

In sum, promising theoretical accounts are being explored in brain–behavior relations between hippocampal development and the emergence of episodic memory; however, there is still limited evidence. Neuroimaging studies have found inconsistent results about the development in overall hippocampal volume in early childhood, possibly due to differences in neuroimaging and tracing methods, and the age ranges investigated. Nevertheless, initial evidence of associations between these volumetric measures and behavior including memory and language are reported, which hold promise for future investigations. Further examination of functional changes in hippocampus and regions connected to it also seems to be a promising avenue for further investigation.

Hippocampal Development in Childhood and Adolescence

There is no question that the greatest hippocampal development occurs within the first 4 years of children's life, and the idea that the hippocampus develops meaningfully after early childhood has had little traction in informing theories of cognitive development until recently. In the last several years, however, the examination of hippocampal change after early childhood has gained momentum, as indicated by the growing number of studies focusing on both structural and functional change.

Structural Development The majority of studies on hippocampal development in middle childhood and adolescence come from structural neuroimaging studies assessing overall volume of the hippocampal formation. Despite a number of well powered studies, it is still unclear if overall volume of the hippocampus continues to develop. For example, in a large cross-sectional study of 187 children, adolescents, and young adults, using FreeSurfer, Østby et al. (2009) reported non-linear age-related increases in ICV-adjusted volume from eight years onward until reaching an asymptote in adolescence. However, in the longitudinal follow-up of this sample, Tamnes et al. (2013) failed to detect within-individual changes raising questions about the reliability of the earlier cross-sectional differences. Likewise, three smaller studies using more accurate manual segmentations failed to uncover development in overall ICV-adjusted hippocampal volume (Barnea-Goraly et al. 2014; Gogtay et al. 2006; Yurgelun-Todd et al. 2003).

Several studies reporting developmental increases in overall hippocampal volume, failed to adjust for ICV. For example, two large longitudinal studies using FreeSurfer in children and adolescents reported an inverted-U shape trajectory, such that raw hippocampal volume increased from 7 years until reaching either an asymptote in adolescence (Goddings et al. 2014) or a zenith and declining thereafter (Wierenga et al. 2014). Unfortunately, unlike the other papers reviewed, by failing to account for differences in ICV we do not know whether the within-individual changes in hippocampal volume reported in these two studies reflect hippocampal development or reflect broader changes in overall brain volume.

On the other hand, some studies have revealed age-related increase in ICV-adjusted hippocampal volumes through childhood using high-quality manual segmentation (Uematsu et al. 2012). In another study using FreeSurfer, Swagerman et al. (2014) found longitudinal increases in overall ICV-adjusted hippocampal volume in a sample of monozygotic and dizygotic twins and patterns of developmental changes within the hippocampus that exhibited high genetic heritability (approximately 70 %) across sex and time point.

Overall, while the number of studies is quickly amassing, there is still a great deal of inconsistency in their findings to date. In addition to methodological differences among studies, other potential factors could provide significant sources of often unaccounted variability. For example, several studies suggest that pubertal development has substantial impacts on hippocampal development (Herting and Nagel 2012; Satterthwaite et al. 2014). Using segmentations produced by the FSL FIRST software, Satterthwaite et al. (2014) compared hippocampal volumes between pre-pubertal and post-pubertal males and females in a large sample of participants 10–20 years of age. In a model, which also accounted for age, sex, and their interaction, male puberty was associated with substantial reductions in ICV-adjusted hippocampal volumes, while female puberty was associated with much smaller reductions. Herting and Nagel (2012) also report reductions in ICV-adjusted hippocampal volume in association with puberty. Given this evidence and the considerable geographic-, socio-economic-, ethnic-, and environment-related heterogeneity in the timing of pubertal development (Parent et al. 2003), failure to account for pubertal development may hinder replicability of reported developmental trajectories. However, it is important to note that the hippocampus appears sensitive to a number of additional individual differences including, for example, diet (Baym et al. 2014; Monti et al. 2014) and cardiovascular fitness (Chaddock et al. 2010; Herting and Nagel 2012), underscoring that a full understanding of the structural development of the hippocampus faces a number of hurdles going forward, and requires an account of the myriad of factors that might influence it.

Overall, it is still unclear whether overall hippocampal volume changes in childhood and adolescence. While the hippocampus exhibits high inter-individual variability in volume that disadvantages smaller studies or those with restricted age-ranges, even large longitudinal studies have been inconsistent. Many of the reviewed studies employed the standard FreeSurfer pipeline (i.e. versions ≤ 5.1), but these subcortical segmentations have been shown to have questionable reliability, especially in child populations (Lee et al. 2015a; Schoemaker et al. 2016). Subtle developmental effects might require more precise and reliable measurements than the older methodology employed by FreeSurfer (versions ≤ 5.1) can provide. Fortunately, better automated methods have since been developed (e.g. Mendrik et al. 2015; Yushkevich et al. 2015), some of which work with existing data processing pipelines by learning to correct their systematic errors (e.g. Wang et al. 2011). Nevertheless, it is important to consider the possibility that development in the hippocampus is heterogeneous within its various subregions, with trajectories that might be obscured by only considering overall hippocampal volume. In the following sections we discuss research investigating these various possibilities.

Development of Hippocampal Subfields There is a paucity of studies examining change in hippocampal subfields in childhood. In a small post-mortem qualitative study of children, adolescents, and adults, Insausti et al. (2010) described age-related increases in raw (non-ICV-corrected) volumes of the dentate gyrus, CA3, and to lesser extent CA1 subfields, but not the subiculum complex (i.e. subiculum, presubiculum, and parasubiculum) into adulthood. Quantitative analyses in developing rhesus primates revealed protracted age-related increases in the raw volumes of the dentate gyrus and CA3 into adulthood, but not CA1 or subiculum complex (Jabès et al. 2011). Thus, these histological assessments converge to show that CA3 plus dentate gyrus follow a more protracted course of development as compared to CA1. These histological examinations have begun to be complemented *in vivo* using volumetry from structural MR imaging.

Lee et al. (2014) assessed for the first time the structural development of the hippocampal subfields using high-resolution MR imaging methods in a cross-sectional sample of children and adolescents (Fig. 1a). Age-related increases in the ICV-adjusted volume of the region comprising CA3 and the dentate gyrus (CA3/DG) subfields were observed between middle-childhood and early adolescence, consistent with the results of previous histological analyses (e.g. Insausti et al. 2010; Lavenex and Banta Lavenex 2013). In contrast to the non-human primate findings discussed in Lavenex and Banta Lavenex (2013), but potentially consistent with Insausti et al. (2010), age-related increase in ICV-adjusted CA1 volume were also observed across childhood and into early adolescence. In contrast to the histological evidence, age-related decreases from early to late adolescence were found in the CA3/DG, and also in CA1. Last, consistent with post-mortem analysis (Insausti et al. 2010), Lee et al. (2015a) failed to observe substantial differences in ICV-adjusted volume of the subiculum complex with age. Daugherty et al. (2015b) is the only additional high-resolution imaging study to have assessed the volumes of the subfields in a pediatric sample. Consistent with the age-related declines observed between early and late adolescence in Lee et al. (2014), Daugherty et al. (2015b) reported age-related declines in ICV-adjusted CA3 and dentate gyrus volume, and CA1 volume across the lifespan (albeit the timing of the decline seems to be earlier in this study), and no differences in the subiculum complex. Thus, despite some differences likely due to the extent of hippocampal

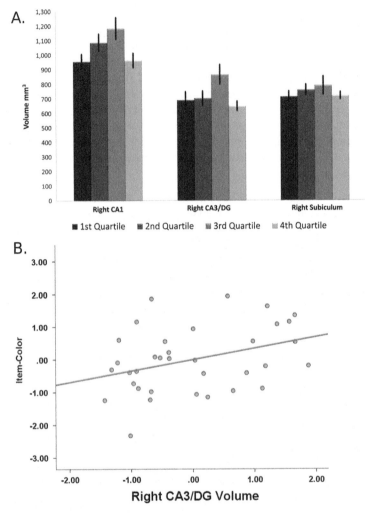

Fig. 1 (**a**) Age-related differences in subfield volume as a function of age (Quartile 1, 8–8.95 years, n = 10; Quartile 2, 8.96–10.99 years, n = 10; Quartile 3, 11.01–13.52 years, n = 10; Quartile 4, 13.53–14. 9 years, n = 9) in the right hippocampus. *Error bars* correspond to standard errors. (**b**) Plots of partial correlations between item–color memory and the volume of the right CA3/DG, controlling for age. Z-scores for each measure are plotted. Figures adapted from Lee J K, Ekstrom A, Ghetti S (2014) Volume of hippocampal subfields and episodic memory in childhood and adolescence. Neuroimage 94:162–171. doi: 10.1016/j.neuroimage.2014.03.019

body included in the volume estimates that were reported in the two studies, the two available high-resolution studies converge with the available histological evidence to suggest protracted development in CA3 and dentate gyrus and CA1, but not in the subiculum complex.

In addition to reporting age-related differences in subfield volume, Lee et al. (2014) also examined relations between subfield volumes and memory (Fig. 1b). In that study, drawings of objects paired with a red or green colored border were studied. At test, participants were asked to remember the color of the border with which studied drawings had appeared. Consistent with evidence from adult neuro-imaging (Shing et al. 2011) volume of CA3/DG positively predicted correct memory for the relation between a drawing and the color of the border, and negatively predicted the rate of false alarms to novel drawings.

It should be noted that two additional studies employed a statistical segmenta-tion method distributed with FreeSurfer (Version 5.1) to examine development of the hippocampal subfields from standard resolution images (i.e. ≥ 1 mm isotropic) (Krogsrud et al. 2014; Tamnes et al. 2013). Although results from these two studies appear somewhat consistent with those already reviewed, there are substantial concerns about the reliability and validity of their results precluding comparisons. However, this concern is not specific to Krogsrud et al. (2014) and Tamnes et al. (2014), but to the FreeSurfer hippocampal subfield segmentation method they employed (versions ≤ 5.1; Van Leemput et al. 2009). This FreeSurfer method allows the use of low–contrast T1-weighted images with voxel sizes that approx-imate the thickness of the subfields to produce segmentations not only in the hippocampal body, but also in the hippocampal head and tail. The delineation of reliable and valid subfields in the two latter sections is challenging, if not contro-versial, even when state-of-the-art high-resolution images are acquired. Moreover, the FreeSurfer delineation protocol is different from others used in the field complicating direct comparisons of the findings (Van Leemput et al. 2009; Yushkevich et al. 2015; for a comprehensive critical review of these issues see Wisse et al. (2014). While some of these criticisms have begun to be addressed (Iglesias et al. 2015), substantial validation is still needed. Given that the bound-aries of the subfields in standard T1-weighted images are not generally discernable to the eye, these boundaries must be inferred from statistical models that make distributional assumptions learned during model training; assumptions that may not be appropriate for a given imaging protocol or study population.

In sum, the available evidence is suggestive of continued structural development in the hippocampal subfields, and those subfield volumes appear to differentially relate to episodic memory. In addition, histological evidence suggests continued development in connectivity between the dentate gyrus and the cortical and sub-cortical brain. For example, fibers from the entorhinal cortex that form synapses onto the granule and molecular layers of the dentate gyrus continue to myelinate over childhood and adolescence (Insausti et al. 2010; Muftuler et al. 2012). These entorhinal fibers represent the primary cortical inputs to the hippocampal formation carrying high-level polymodal information from perirhinal cortex and parahippocampal gyrus (e.g. Burwell 2006). Thus, improvements in conduction velocity in these fibers should allow more efficient coordination with the hippo-campus during encoding and retrieval operations. Potentially equally interesting is the protracted time course of myelination of fibers in the hilus of the dentate gyrus (Ábrahám et al. 2010) that project from subcortical regions including the locus coeruleus, septal area, and the raphe nuclei (Amaral and Lavenex 2007). These subcortical connections may play pivotal roles in hippocampal function. For example, the locus coeruleus is involved in noradrenergic attentional mechanisms and contextual resets (Bouret and Sara 2005) and hippocampal response to novelty (Hagena et al. 2016), while the inter-neurons in the hilus of the dentate gyrus have been shown to selectively inhibit and enhance learning (Harley 1991, 2007; Lashgari et al. 2008; Lemon et al. 2009; Rajkumar et al. 2013; Walling and Harley 2004). Thus, these sub-cortical inputs may contribute to guiding encoding and retrieval operations within the dentate gyrus and CA3 (e.g. Gibbs et al. 2010; Hangya et al. 2009). Overall these data suggest that over childhood and adolescence there is continued development of the hippocampal subfields, particularly in the dentate gyrus, including its connections with cortical and subcortical regions.

Volumetric Development Along the Anterior–Posterior Axis About a decade ago, Gogtay et al. (2006) published a longitudinal analysis of hippocampal development across childhood into adulthood, which revealed heterogeneous non-linear changes in morphology during childhood and adolescence, despite age-invariance of total hippocampal volume. The patterns of change reported in Gogtay et al. (2006) are complex: The hippocampal body increased in volume before reaching an asymptote in early adolescence. Regions in the head, the most anterior section of the body and the tail declined in volume. Unfortunately, Gogtay et al. (2006) did not explore the factors underlying those developmental trajectories or their implications to function. Following up on Gogtay et al. (2006), DeMaster et al. (2014) sought to replicate findings in a cross-sectional sample of 8 to 11 year-olds and college-age adults, and extend them with an examination of relations between sub-regional volumes and performance on a measure of relational episodic memory. DeMaster et al. (2014) hypothesized that regions in the anterior hippocampal head and tail would become smaller, while regions in the hippocampal body would generally became larger with age. In that study, the hippocampus was segmented via the FreeSurfer pipeline and manually divided into three regions: head, body, and tail. Cross-sectional results of analyses controlling for the volume of overall hippocampus replicated Gogtay et al. (2006)'s findings, such that hippocampal head and tail were smaller in adults, while the hippocampal body was larger in adults. DeMaster et al. (2014) then associated these volumes with episodic memory performance. Results indicated that the direction of differences in volume between children and adults was consistent with the direction of its relationship with episodic memory in adulthood. That is, hippocampal head and tail were smaller in adults, and smaller head and tail volumes were associated with better memory in adults. Likewise, but in the opposite direction, hippocampal body was bigger in adults, and in adults, bigger body volumes were associated with better episodic memory. Although no other study explicitly examined changes in sub-regional volumes along the anterior-posterior axis, additional studies (Guillery-Girard et al. 2013; Hashimoto et al. 2015), using voxel-based morphometry analyses provided results about hippocampal clusters that are consistent with developmental differences along this axis, lending further support to the idea that this dimension is important to understand development.

The review of the literature thus far has suggested that structural assessments of the cytoarchitectural subfields and differences along the anterior–posterior axis may be particularly informative to understanding the trajectories and consequences of hippocampal development in childhood and adolescence. Given that the sub-fields are not uniformly distributed along the anterior–posterior axis, a natural question is whether the trajectories in head, body, and tail are reducible to an account about subfield development (e.g. DeMaster et al. 2014; Duvernoy 2005; Gogtay et al. 2006). Reductions in volume of substructures may reflect synaptic pruning (Johnson et al. 1996), while increases in volume of substructures may reflect neurogenesis and synaptic elaboration (e.g. Eckenhoff and Rakic 1991). This would suggest that the subfield subdivision is a critical unit of analysis. However, neurogenic processes differ along the longitudinal axis in non-human animals (e.g. Snyder et al. 2009); moreover, the differential connectivity of anterior and posterior hippocampus with cortical regions (e.g. Libby et al. 2012) may be particularly relevant to cognitive development. To date the anterior–posterior axis and the cytoarchitectural subfields have been examined separately. However, the anterior and posterior poles of hippocampal head and tail predominately comprise CA1, but not CA3 or dentate gyrus (Duvernoy 2005; Yushkevich et al. 2015). These are the regions exhibiting volumetric declines in hippocampal tail and head observed in Gogtay et al. (2006), and subfield-driven changes could have contrib-uted to these findings and the similar results in DeMaster et al. (2014). It will be important to ask whether these declines in anterior and posterior hippocampal volumes are indeed specific to CA1.

Functional Development Corroborating structural findings supporting the view of protracted hippocampal development beyond early childhood are findings from a handful of functional magnetic resolution imaging (fMRI) studies of episodic encoding (Ghetti et al. 2010) and retrieval (DeMaster and Ghetti 2013; DeMaster et al. 2013) that have revealed age-related activation differences along the anterior–posterior axis. Subsequent memory effects in the hippocampus were first examined in a cross-sectional sample of 8-year-olds, 10 to 11-year-olds, 14-year-olds, and college-age adults (Ghetti et al. 2010). Participants incidentally encoded line drawings of objects, which either appeared in red or green ink. At retrieval, participants were asked to recall the color of the recognized drawings. Fourteen-year-olds and adults showed selective hippocampal activation in both left and right anterior hippocampus for encoding trials in which the color was subsequently remembered correctly than when the color was not later remembered correctly or when the drawing was not later recognized. In comparison, the 8-year-old children exhibited generally stronger activation for trials in which items were subsequently remembered with their correct color detail compared to trials in which they subsequently failed to retrieve the correct color detail. Importantly, performance was well above chance and the hippocampus was strongly engaged in children, but did not discriminate between correct and incorrect source decisions. Interestingly, in 10 to 11-year-old children, subsequent memory effects appeared less reliable than either those seen in 8-year-old children or adults, suggesting that middle-childhood may be a transitional period of hippocampal contribution to memory encoding.

Only one other encoding study employed a relational memory task during encoding (Güler and Thomas 2013). Güler and Thomas (2013) failed to detect reliable hippocampal activations during encoding, much less age-differences in those activations. However due to this study's extremely limited sample and the noisy nature of fMRI data within the medial temporal lobe, type II error is very possible. Additional studies of encoding related activity have focused on item encoding instead of encoding of item-context relations (i.e. Chiu et al. 2006; Maril et al. 2010; Ofen et al. 2007; re-analyzed in Chai et al. 2010), making it challenging to interpret the results given knowledge of hippocampal role within relational processes (e.g., Cohen and Eichenbaum 1993; Eichenbaum et al. 1994; Konkel and Cohen 2009). Nevertheless, some of these studies found age differences in hippocampal activations (Chiu et al. 2006; Maril et al. 2010) and some did not (Ofen et al. 2007; Chai et al. 2010).

The development of hippocampal contributions to episodic retrieval has also been examined. For example, in DeMaster and Ghetti (2013) 8- to 11-year-old children and adults learned arbitrary relations between a drawing of an object and a red or green border. During the scanned memory test, participants indicated whether a drawing had appeared with a red or green border, or was new. In adults, the episodic memory contrast (i.e. correct retrieval of item-color relation greater than incorrect retrieval of the color) was reliable only in hippocampal head. The episodic retrieval contrasts in 8 to 11-year-old children, however, were only reliable in the hippocampal tail. Interestingly, the hippocampal body failed to reliably differentiate between correct and incorrect retrieval of item–color relations in either children or adults. These results suggest that age-related differences between adults and children in the recruitment of anterior–posterior hippocampal substructures respectively, during episodic retrieval. In a similarly designed study, DeMaster et al. (2013) examined hippocampal recruitment during the retrieval of item–space relations in a sample of 8 to 9-year-olds, 10 to 11-year-olds, and young adults. Similar to DeMaster and Ghetti (2013), episodic retrieval contrasts for item–space relations in adults were observed in the hippocampal head and body, but not in the hippocampal tail. In contrast to adults' retrieval of item-space, children did not reliably recruit the hippocampus or its substructures.

Overall, the results of these studies suggest that adults recruit anterior hippo-campal regions during episodic retrieval and this pattern is not established clearly in children. Several factors might modulate hippocampal recruitment during episodic retrieval during development. One possibility is that individual differences in memory performance in childhood are associated with the degree to which sub-structures of the hippocampus are recruited during remembering. A recent func-tional retrieval study validates this possibility. In a large sample of 8 to 9-year-old and 10 to 11-year-old children, and adults, Sastre et al. (2016) reported that the functional recruitment of the hippocampus during episodic retrieval differed between high and low episodic memory performers and age-groups (Fig. 2). In low performing adults, reliable episodic memory contrasts were found across the

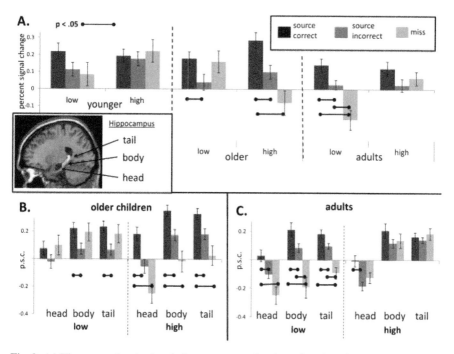

Fig. 2 (**a**) Hippocampal activation during memory retrieval as a function of age group, trial type, and performance level. (**b**) Hippocampal activation in older children as a function of trial type, performance group, and hippocampal region (i.e., head, body, and tail as shown in figure). (**c**) Hippocampal activation in adults as a function of trial type, performance group, and hippocampal region. *Error bars* depict within-subjects standard error. Figures adapted from Sastre M, Wendelken C, Lee J et al. (2016) Age- and performance-related differences in hippocampal contributions to episodic retrieval. Dev Cogn Neurosci 19:42–50. doi: 10.1016/j.dcn.2016.01.003

entire range of the hippocampal head, body, and tail, while in high-performing adults this difference was exclusively restricted to the hippocampal head. In both low and high performing 10 to 11-year-olds reliable episodic memory contrasts were observed in the hippocampal body and tail, but high-performing 10 to 11-year-olds also recruited the hippocampal head, which was similar to the pattern observed in low performing adults. Finally, no reliable differences between correct and incorrect trials were observed in any hippocampal substructure for low or high performing 8 to 9-year-old children. These results are consistent with the hypothesis that as episodic memory performance increases over middle-childhood, there is an evolution from preferential recruitment of posterior hippocampal regions towards integrating the entire hippocampal axis, and finally towards a preferential recruitment of anterior hippocampal regions.

Another potential modulator of patterns of hippocampal recruitment during retrieval is the degree to which flexible retrieval processes are required. Children have difficulties retrieving memories when the context differs between encoding and retrieval (Ackerman 1981; Ackerman 1982; Paz-Alonso et al. 2008). While developmental differences in the ability to employ and benefit from prefrontal-mediated memory strategies contribute to these difficulties, recent proposals and data implicate development in hippocampal mechanisms that may allow for increased representational flexibility (e.g. Edgin et al. 2014; Lavenex and Banta Lavenex 2013; Zeithamova and Preston 2010), including potential age-differences in the dentate gyrus and CA3 (Lee et al. 2014), and in recruitment of anterior hippocampal subregions (e.g. DeMaster et al. 2014; Giovanello et al. 2004, 2009). DeMaster et al. (2016) recently examined how hippocampal recruitment was modulated by demands on representational flexibility (Fig. 3). Participants studied pairs of objects, and at retrieval each object in the pair appeared at test in their originally encoded spatial positions (low flexibility demands), or exchanged spatial positions (high flexibility demands). Behaviorally, all age-groups responded faster on correctly recognized pairs under low flexibility demand than when under high

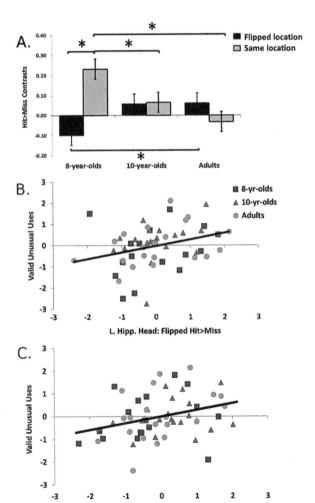

Fig. 3 (a) Left hippocampal activation across the entire sample for Hit > Miss trials as a function of age and retrieval flexibility condition, where 'same location' denotes low demand for retrieval flexibility, and 'flipped location' denotes high demand for retrieval flexibility. (b) Correlation between creative thinking (Unusual Uses Task) and activation for Hit > Miss Flipped-Location trials in the left hippocampal head and (c) in left hippocampal body. Standardized residuals are plotted corrected for age, sex, and behavioral performance (i.e., Hit-False Alarms for rearranged pairs). *Squares* indicate 8-year-olds, *triangles* indicate 10-year-olds, and *circles* indicate adults. Figures adapted from DeMaster D, Coughlin C, Ghetti S (2016) Retrieval flexibility and reinstatement in the developing hippocampus. Hippocampus 26:492–501. doi: 10.1002/hipo.22538

demand, suggesting that the manipulation was successful. Only 8-year-old's recognition memory was significantly reduced in the high flexibility condition. Likewise, only in 8-year-olds, not in 10-year-olds or adults, did demand for retrieval flexibility modulate hippocampal activation when recognized and forgotten trials were compared in a hits > miss trial contrast. A positive contrast was observed for correctly recognized pairs under low flexibility demands and a negative contrast when under high flexibility demand. Moreover, the latter contrast was substantially smaller in 8-year-olds than in other age-groups. Interestingly, the degree to which anterior hippocampal regions were recruited for successful recognition in the high-flexibility condition predicted flexible cognition beyond memory as indicated by performance in the Unusual Uses Task (i.e. list as many unusual uses of a brick). This was not the case for the hippocampal contrast in the low flexibility condition. This provides evidence that the development in the hippocampus allows flexible thinking more generally, beyond memory function (Rubin et al. 2014; see also Qin et al. 2014 for additional evidence that functional development of the hippocampus is critical for arithmetic reasoning).

Future Directions

The small, but growing body of structural and functional data support continued development of the hippocampal formation throughout childhood and into adolescence, with implications for behavioral change. However, the nature of this development is still not clear and future research should focus on the possible sources of developmental change outlined in this review.

First, future research should seek to better understand the implications of continued dentate gyrus development. This development may afford memories of finer resolution via improved pattern separation; these hypothesized improvements could be assessed with behavioral and high-resolution fMRI paradigms with tasks that manipulate the degree of stimulus difference or representational overlap. The development of the dentate gyrus may also afford more efficient and selective engagement of hippocampal encoding and retrieval mechanisms in response to task demands. There is some initial evidence that such a transition occurs during childhood and adolescence. As briefly reviewed earlier, the hilus in the dentate gyrus is a site of protracted myelination of subcortical inputs (Ábrahám et al. 2010), and these subcortical inputs activate intrinsic inhibitory circuits thought to selectively suppress or facilitate granule cell activity in the dentate gyrus in response to task demands and attentional processes, both local to the input, but also in distant regions along the anterior-posterior axis via mossy cell projections (e.g. Hendrickson et al. 2016; Myers and Scharfman 2009, 2011; Sara 2015; Scharfman and Myers 2015). The myelination of these subcortical inputs in the dentate gyrus, as well as the volumetric growth observed in the hilus (Insausti et al. 2010; Lavenex and Banta Lavenex 2013) thus suggest improvements in the selective engagement of the encoding operations in the dentate gyrus due to more efficient and effective activation of its inhibitory circuits. Consistent with this hypothesis, in a rodent study Yu et al. (2013) reported protracted developmental improvements in the selectivity and temporal precision of activity in the dentate gyrus that was specifically associated with developmental differences in dentate gyrus inhibitory circuit activity. Taken together, protracted development of the dentate gyrus suggests not only continued improvement in pattern separation, but also improvement in trisynaptic mechanisms supporting memory modulation to enhance task-relevant and to suppress task-irrelevant (or redundant) encoding of features. These possibilities should be examined during child development.

Second, future research should also focus on developmental differences in hippocampal function along the anterior–posterior axis. One hypothesis receiving attention is that a developmental shift occurs towards recruiting anterior hippocampus and away from recruiting posterior hippocampus during episodic encoding and retrieval (e.g. DeMaster and Ghetti 2013; DeMaster et al. 2014). Given accounts of hippocampal function suggesting that posterior hippocampus processes perceptual information while anterior hippocampus processes verbal, or multi-level relational information (e.g. Collin et al. 2015; Persson and Söderlund 2015), a developmental shift towards anterior hippocampal recruitment could reflect a change in the types of information submitted to the hippocampus for integration. These ideas are compatible with accounts of episodic memory development that have emphasized that improvements in true memory come from decreasing reliance on exact perceptual representations and increasing reliance on increasingly integrated complex verbal representations (e.g. gist; Brainerd and Reyna 2002; Paz-Alonso et al. 2008, 2013). One question for future research might be how developmental shifts in the division of labor between posterior and anterior hippocampus are modulated by demand for gist versus exact perceptual representations. Alternatively, hippocampal recruitment along the anterior–posterior axis may also be modulated by the type of episodic relation and content encoded and retrieved. For example, in a cross-sectional sample, Lee et al. (2015b) demonstrated heterogeneous age-related trajectories of improvement in episodic memory for item–space, item–time, and item–item relations, with memory for item–space relations maturing earlier than item–time or item–item relations. Future research should examine whether development in hippocampal substructures are differentially predictive of developmental improvements in these forms of episodic memory. Finally, overt behavior is not the only means to assess the relational component of hippocampal-dependent memory: eye movements may be particularly informative indicators of memory development and hippocampal function (Pathman and Ghetti 2016). Future research should assess how hippocampal development guides memory-related eye-movements during encoding and retrieval.

Conclusions

The hippocampus contributes to a number of cognitive abilities including episodic memory, language, and perception; thus understanding whether and how this structure develops has broad implications for cognitive development. The reviewed evidence suggests that any complete theory of hippocampal development must address the roles of cytoarchitectural and anterior–posterior axial substructures to that development. Perhaps the greatest gains in the future will be made from developmental studies that integrate both sources of possible developmental change.

References

Ábrahám H, Vincze A, Jewgenow I et al (2010) Myelination in the human hippocampal formation from midgestation to adulthood. Int J Dev Neurosci 28:401–410. doi:10.1016/j.ijdevneu.2010. 03.004

Ackerman B (1981) Encoding specificity in the recall of pictures and words in children and adults. J Exp Child Psychol 31:193–211. doi:10.1016/0022-0965(81)90012-6

Ackerman B (1982) Retrieval variability: the inefficient use of retrieval cues by young children. J Exp Child Psychol 33:413–428. doi:10.1016/0022-0965(82)90056-x

Amaral D, Lavenex P (2007) Hippocampal neuroanatomy. In: Amaral D, Andersen P, Bliss T, Morris R, O'Keefe J (eds) The hippocampus book. Oxford University Press, New York, pp 37–114

Amaral D, Witter M (1995) Hippocampal formation. In: Paxinos G (ed) The rat nervous system. Academic Press, San Diego, pp 443–494

Andersen P (1975) Organization of hippocampal neurons and their interconnections. In: Isaacson R, Pribram K (eds) The hippocampus Vol 1: structure and development. Plenum, New York, pp 155–176

Bakker A, Kirwan C, Miller M, Stark C (2008) Pattern separation in the human hippocampal CA3 and dentate gyrus. Science 319:1640–1642. doi:10.1126/science.1152882

Barnea-Goraly N, Frazier T, Piacenza L et al (2014) A preliminary longitudinal volumetric MRI study of amygdala and hippocampal volumes in autism. Prog Neuropsychopharmacol Biol Psychiatry 48:124–128. doi:10.1016/j.pnpbp.2013.09.010

Baym CL, Khan NA, Monti JM, Raine LB, Drollette ES, Moore RD et al (2014) Dietary lipids are differentially associated with hippocampal-dependent relational memory in prepubescent children. Am J Clin Nutr 99:1026–1032. doi:10.3945/□ajcn.113.079624

Bauer P (1996) What do infants recall of their lives? Memory for specific events by one- to two-year-olds. Am Psychol 51:29–41. doi:10.1037//0003-066x.51.1.29

Bauer P (2004) Getting explicit memory off the ground: steps toward construction of a neuro-developmental account of changes in the first two years of life. Dev Rev 24:347–373. doi:10. 1016/j.dr.2004.08.003

Bauer P, San Souci P, Pathman T (2010) Infant memory. Wiley Interdiscip Rev Cogn Sci 1:267–277. doi:10.1002/wcs.38

Blessing E, Beissner F, Schumann A et al (2016) A data-driven approach to mapping cortical and subcortical intrinsic functional connectivity along the longitudinal hippocampal axis. Hum Brain Mapp 37:462–476. doi:10.1002/hbm.23042

Bouret S, Sara S (2005) Network reset: a simplified overarching theory of locus coeruleus noradrenaline function. Trends Neurosci 28:574–582. doi:10.1016/j.tins.2005.09.002

Brainerd C, Reyna V (2002) Fuzzy-trace theory: dual processes in memory, reasoning, and cognitive neuroscience. Adv Child Dev Behav 28:41–100. doi:10.1016/S0065-2407(02) 80062-3

Burwell R (2006) The parahippocampal region: corticocortical connectivity. Ann N Y Acad Sci 911:25–42. doi:10.1111/j.1749-6632.2000.tb06717.x

Chaddock L, Erickson K, Prakash R et al (2010) A neuroimaging investigation of the association between aerobic fitness, hippocampal volume, and memory performance in preadolescent children. Brain Res 1358:172–183. doi:10.1016/j.brainres.2010.08.049

Chai X, Ofen N, Jacobs L, Gabrieli J (2010) Scene complexity: influence on perception, memory, and development in the medial temporal lobe. Front Hum Neurosci 4:1–10. doi:10.3389/fnhum.2010.00021

Chiu C-Y, Schmithorst V, Brown R et al (2006) Making memories: a cross-sectional investigation of episodic memory encoding in childhood using FMRI. Dev Neuropsychol 29:321–340. doi:10.1207/s15326942dn2902_3

Ciaramelli E, Grady C, Levine B et al (2010) Top-down and bottom-up attention to memory are dissociated in posterior parietal cortex: neuroimaging and neuropsychological evidence. J Neurosci 30:4943–4956. doi:10.1523/jneurosci.1209-09.2010

Cohen NJ, Eichenbaum H (1993) Memory and the hippocampal system. MIT Press, Cambridge, MA

Collin S, Milivojevic B, Doeller C (2015) Memory hierarchies map onto the hippocampal long axis in humans. Nat Neurosci 18:1562–1564. doi:10.1038/nn.4138

Cuevas K, Rovee-Collier C, Learmonth A (2006) Infants form associations between memory representations of stimuli that are absent. Psychol Sci 17:543–549. doi:10.1111/j.1467-9280. 2006.01741.x

Daugherty A, Bender A, Raz N, Ofen N (2015b) Age differences in hippocampal subfield volumes from childhood to late adulthood. Hippocampus 26:220–228. doi:10.1002/hipo.22517

Daugherty A, Yu Q, Flinn R, Ofen N (2015a) A reliable and valid method for manual demarcation of hippocampal head, body, and tail. Int J Dev Neurosci 41:115–122. doi:10.1016/j.ijdevneu. 2015.02.001

DeMaster D, Coughlin C, Ghetti S (2016) Retrieval flexibility and reinstatement in the developing hippocampus. Hippocampus 26:492–501. doi:10.1002/hipo.22538

DeMaster D, Ghetti S (2013) Developmental differences in hippocampal and cortical contributions to episodic retrieval. Cortex 49:1482–1493. doi:10.1016/j.cortex.2012.08.004

DeMaster D, Pathman T, Ghetti S (2013) Development of memory for spatial context: hippocampal and cortical contributions. Neuropsychologia 51:2415–2426. doi:10.1016/j. neuropsychologia.2013.05.026

DeMaster D, Pathman T, Lee J, Ghetti S (2014) Structural development of the hippocampus and episodic memory: developmental differences along the anterior/posterior axis. Cereb Cortex 24:3036–3045. doi:10.1093/cercor/bht160

Deniz Can D, Richards T, Kuhl P (2013) Early gray-matter and white-matter concentration in infancy predict later language skills: a whole brain voxel-based morphometry study. Brain Lang 124:34–44. doi:10.1016/j.bandl.2012.10.007

Duff MC, Brown-Schmidt S (2012) The hippocampus and the flexible use and processing of language. Front Hum Neurosci 6:69. doi:10.3389/fnhum.2012.00069

Duvernoy H (2005) The human hippocampus: functional anatomy, vascularization and serial sections with MRI. Springer Science & Business Media, New York

Eckenhoff M, Rakic P (1991) A quantitative analysis of synaptogenesis in the molecular layer of the dentate gyrus in the rhesus monkey. Dev Brain Res 64:129–135. doi:10.1016/0165-3806 (91)90216-6

Edgin J, Spanò G, Kawa K, Nadel L (2014) Remembering things without context: development matters. Child Dev 85:1491–1502. doi:10.1111/cdev.12232

Eichenbaum H, Otto T, Cohen NJ (1994) Two functional components of the hippocampal memory system. Behav Brain Sci 17:449–472. doi:10.1017/s0140525x00035391

Feliciano D, Bordey A (2013) Newborn cortical neurons: only for neonates? Trends Neurosci 36:51–61. doi:10.1016/j.tins.2012.09.004

Fischl B (2012) FreeSurfer. NeuroImage 62:774–781. doi:10.1016/j.neuroimage.2012.01.021

Frankland P, Köhler S, Josselyn S (2013) Hippocampal neurogenesis and forgetting. Trends Neurosci 36:497–503. doi:10.1016/j.tins.2013.05.002

Ghetti S, DeMaster D, Yonelinas A, Bunge S (2010) Developmental differences in medial temporal lobe function during memory encoding. J Neurosci 30:9548–9556. doi:10.1523/JNEUROSCI.3500-09.2010

Gibbs M, Hutchinson D, Summers R (2010) Noradrenaline release in the locus coeruleus modulates memory formation and consolidation; roles for α- and β-adrenergic receptors. Neuroscience 170:1209–1222. doi:10.1016/j.neuroscience.2010.07.052

Gilbert P, Kesner R, DeCoteau W (1998) Memory for spatial location: role of the hippocampus in mediating spatial pattern separation. J Neurosci 18:804–810

Gilbert P, Kesner R, Lee I (2001) Dissociating hippocampal subregions: a double dissociation between dentate gyrus and CA1. Hippocampus 11:626–636. doi:10.1002/hipo.1077

Giovanello K, Schnyer D, Verfaellie M (2009) Distinct hippocampal regions make unique contributions to relational memory. Hippocampus 19:111–117. doi:10.1002/hipo.20491

Giovanello K, Schnyer D, Verfaellie M (2004) A critical role for the anterior hippocampus in relational memory: evidence from an fMRI study comparing associative and item recognition. Hippocampus 14:5–8. doi:10.1002/hipo.10182

Goddings A, Mills K, Clasen L et al (2014) The influence of puberty on subcortical brain development. NeuroImage 88:242–251. doi:10.1016/j.neuroimage.2013.09.073

Gogtay N, Nugent T, Herman D et al (2006) Dynamic mapping of normal human hippocampal development. Hippocampus 16:664–672. doi:10.1002/hipo.20193

Gómez R, Edgin J (2015) The extended trajectory of hippocampal development: implications for early memory development and disorder. Dev Cogn Neurosci 18:57–69. doi:10.1016/j.dcn.2015.08.009

Graham K, Barense M, Lee A (2010) Going beyond LTM in the MTL: a synthesis of neuropsychological and neuroimaging findings on the role of the medial temporal lobe in memory and perception. Neuropsychologia 48:831–853. doi:10.1016/j.neuropsychologia.2010.01.001

Guillery-Girard B, Martins S, Deshayes S et al (2013) Developmental trajectories of associative memory from childhood to adulthood: a behavioral and neuroimaging study. Front Behav Neurosci 7:126. doi:10.3389/fnbeh.2013.00126

Güler O, Thomas K (2013) Developmental differences in the neural correlates of relational encoding and recall in children: an event-related fMRI study. Dev Cogn Neurosci 3:106–116. doi:10.1016/j.dcn.2012.07.001

Hagena H, Hansen N, Manahan-Vaughan D (2016) β-adrenergic control of hippocampal function: subserving the choreography of synaptic information storage and memory. Cereb Cortex 26:1349–1364. doi:10.1093/cercor/bhv330

Hunsaker M, Kesner R (2013) The operation of pattern separation and pattern completion processes associated with different attributes or domains of memory. Neurosci Biobehav Rev 37:36–58. doi:10.1016/j.neubiorev.2012.09.014

Hangya B, Borhegyi Z, Szilagyi N et al (2009) GABAergic neurons of the medial septum lead the hippocampal network during theta activity. J Neurosci 29:8094–8102. doi:10.1523/jneurosci.5665-08.2009

Harley C (1991) Noradrenergic and locus coeruleus modulation of the perforant path-evoked potential in rat dentate gyrus supports a role for the locus coeruleus in attentional and memorial processes. Prog Brain Res 88:307–321. doi:10.1016/S0079-6123(08)63818-2

Harley C (2007) Norepinephrine and the dentate gyrus. Prog Brain Res 163:299–318

Hashimoto T, Takeuchi H, Taki Y, Yokota S et al (2015) Increased posterior hippocampal volumes in children with lower increase in body mass index: a 3-year longitudinal MRI study. Dev Neurosci 37:153–160. doi:10.1159/000370064

Hendrickson P, Yu G, Song D, Berger T (2016) A million-plus neuron model of the hippocampal dentate gyrus: critical role for topography in determining spatiotemporal network dynamics. IEEE Trans Biomed Eng 63:199–209. doi:10.1109/TBME.2015.2445771

Herting M, Nagel B (2012) Aerobic fitness relates to learning on a virtual Morris Water Task and hippocampal volume in adolescents. Behav Brain Res 233:517–525. doi:10.1016/j.bbr.2012.05.012

Hevner RF, Kinney HC (1996) Reciprocal entorhinal-hippocampal connections established by human fetal midgestation. J Comp Neurol 372:384–394. doi:10.1002/(SICI)1096-9861 (19960826)372:3<384::AID-CNE4>3.0.CO;2-Z

Iglesias J, Augustinack J, Nguyen K et al (2015) A computational atlas of the hippocampal formation using ex vivo, ultra-high resolution MRI: Application to adaptive segmentation of in vivo MRI. NeuroImage 115:117–137. doi:10.1016/j.neuroimage.2015.04.042

Insausti R, Cebada-Sanchez S, Marcos P (2010) Postnatal development of the human hippocampal formation. Adv Anat Embryol Cell Biol 206:1–86

Jabès A, Lavenex P, Amaral D, Lavenex P (2010) Quantitative analysis of postnatal neurogenesis and neuron number in the macaque monkey dentate gyrus. Eur J Neurosci 31:273–285. doi:10.1111/j.1460-9568.2009.07061.x

Jabès A, Lavenex P, Amaral D, Lavenex P (2011) Postnatal development of the hippocampal formation: A stereological study in macaque monkeys. J Comp Neurol 519:1051–1070. doi:10.1002/cne.22549

Jabès A, Nelson C (2015) 20 years after "The ontogeny of human memory: a cognitive neuroscience perspective," where are we? Int J Behav Dev 39:293–303. doi:10.1177/0165025415575766

Johnson M, Perry R, Piggott M et al (1996) Glutamate receptor binding in the human hippocampus and adjacent cortex during development and aging. Neurobiol Aging 17:639–651. doi:10.1016/0197-4580(96)00064-4

Kesner R (2007) Behavioral functions of the CA3 subregion of the hippocampus. Learn Mem 14:771–781. doi:10.1101/lm.688207

Kondo H, Lavenex P, Amaral D (2009) Intrinsic connections of the macaque monkey hippocampal formation: II. CA3 connections. J Comp Neurol 515:349–377. doi:10.1002/cne.22056

Kondo H, Lavenex P, Amaral D (2008) Intrinsic connections of the macaque monkey hippocampal formation: I. dentate gyrus. J Comp Neurol 511:497–520. doi:10.1002/cne.21825

Konkel A, Cohen NJ (2009) Relational memory and the hippocampus: representations and methods. Front Neurosci 3:23. doi:10.3389/neuro.01.023.2009

Krogsrud S, Tamnes C, Fjell A et al (2014) Development of hippocampal subfield volumes from 4 to 22 years. Hum Brain Mapp 35:5646–5657. doi:10.1002/hbm.22576

Lashgari R, Khakpour-Taleghani B, Motamedi F, Shahidi S (2008) Effects of reversible inactivation of locus coeruleus on long-term potentiation in perforant path-DG synapses in rats. Neurobiol Learn Mem 90:309–316. doi:10.1016/j.nlm.2008.05.012

Lavenex P, Banta Lavenex P (2013) Building hippocampal circuits to learn and remember: insights into the development of human memory. Behav Brain Res 254:8–21. doi:10.1016/j.bbr.2013.02.007

Lee JK, Ekstrom A, Ghetti S (2014) Volume of hippocampal subfields and episodic memory in childhood and adolescence. NeuroImage 94:162–171. doi:10.1016/j.neuroimage.2014.03.019

Lee JK, Nordahl C, Amaral D et al (2015a) Assessing hippocampal development and language in early childhood: evidence from a new application of the Automatic Segmentation Adapter Tool. Hum Brain Mapp 36:4483–4496. doi:10.1002/hbm.22931

Lee JK, Wendelken C, Bunge SA, Ghetti S (2015b) A time and place for everything: developmental differences in the building blocks of episodic memory. Child Dev 87:194–210. doi:10.1111/cdev.12447

Lemon N, Aydin-Abidin S, Funke K, Manahan-Vaughan D (2009) Locus coeruleus activation facilitates memory encoding and induces hippocampal LTD that depends on beta-adrenergic receptor activation. cereb cortex 19:2827–2837. doi:10.1093/cercor/bhp065

Libby L, Ekstrom A, Ragland J, Ranganath C (2012) Differential connectivity of perirhinal and parahippocampal cortices within human hippocampal subregions revealed by high-resolution functional imaging. J Neurosci 32:6550–6560. doi:10.1523/jneurosci.3711-11.2012

Lorente de Nó R (1934) Studies on the structure of the cerebral cortex. II Continuation of the study of the ammonic system. J Psychol Neurol 45:113–177

Luby J, Barch D, Belden A et al (2012) Maternal support in early childhood predicts larger hippocampal volumes at school age. Proc Natl Acad Sci USA 109:2854–2859. doi:10.1073/pnas.1118003109

Maril A, Davis P, Koo J et al (2010) Developmental fMRI study of episodic verbal memory encoding in children. Neurology 75:2110–2116. doi:10.1212/WNL.0b013e318201526e

Marr D (1971) Simple memory: a theory for archicortex. Philos Trans R Soc Lond Ser B Biol Sci 262:23–81

Mårtensson J, Eriksson J, Bodammer N et al (2012) Growth of language-related brain areas after foreign language learning. NeuroImage 63:240–244. doi:10.1016/j.neuroimage.2012.06.043

Mendrik A, Vincken K, Kuijf H et al (2015) MRBrainS challenge: online evaluation framework for brain image segmentation in 3T MRI scans. Comp Intell Neurosci. 2015:813696

Milner B, Corkin S, Teuber H (1968) Further analysis of the hippocampal amnesic syndrome: 14-year follow-up study of H.M. Neuropsychologia 6:215–234. doi:10.1016/0028-3932(68)90021-3

Mongiat L, Espósito M, Lombardi G, Schinder A (2009) Reliable activation of immature neurons in the adult hippocampus. PLoS One 4:e5320. doi:10.1371/journal.pone.0005329

Monti JM, Baym CL, Cohen NJ (2014) Identifying and characterizing the effects of nutrition on hippocampal memory. Adv Nutr 5:337–343. doi:10.3945/an.113.005397

Monti J, Cook G, Watson P et al (2015) Relating hippocampus to relational memory processing across domains and delays. J Cogn Neurosci 27:234–245. doi:10.1162/jocn_a_00717

Moser E, Kropff E, Moser M (2008) Place cells, grid cells, and the brain's spatial representation system. Neuroscience 31:69. doi:10.1146/annurev.neuro.31.061307.090723

Muftuler L, Davis E, Buss C et al (2012) Development of white matter pathways in typically developing preadolescent children. Brain Res 1466:33–43. doi:10.1016/j.brainres.2012.05.035

Mullally S, Maguire E (2014) Learning to remember: the early ontogeny of episodic memory. Dev Cogn Neurosci 9:12–29. doi:10.1016/j.dcn.2013.12.006

Myers C, Scharfman H (2009) A role for hilar cells in pattern separation in the dentate gyrus: a computational approach. Hippocampus 19:321–337. doi:10.1002/hipo.20516

Myers C, Scharfman H (2011) Pattern separation in the dentate gyrus: a role for the CA3 backprojection. Hippocampus 21:1190–1215. doi:10.1002/hipo.20828

Newcombe NS, Balcomb F, Ferrara K, Hansen M, Koski J (2014) Two rooms, two representations? Episodic-like memory in toddlers and preschoolers. Dev Sci 17:743–756. doi:10.1111/desc.12162

Newcombe NS, Lloyd ME, Ratliff KR (2007) Development of episodic and autobiographical memory: a cognitive neuroscience perspective. In: Kail RV (ed) Advances in child development and behavior, vol 35. Elsevier, San Diego, CA, pp 37–85

Ofen N, Kao Y, Sokol-Hessner P et al (2007) Development of the declarative memory system in the human brain. Nat Neurosci 10:1198–1205. doi:10.1038/nn1950

Opendak M, Gould E (2015) Adult neurogenesis: a substrate for experience-dependent change. Trends Cogn Sci 19:151–161. doi:10.1016/j.tics.2015.01.001

Østby Y, Tamnes C, Fjell A et al (2009) Heterogeneity in subcortical brain development: a structural magnetic resonance imaging study of brain maturation from 8 to 30 years. J Neurosci 29:11772–11782. doi:10.1523/jneurosci.1242-09.2009

Parent A, Teilmann G, Juul A et al (2003) The timing of normal puberty and the age limits of sexual precocity: variations around the world, secular trends, and changes after migration. Endocr Rev 24:668–693. doi:10.1210/er.2002-0019

Pathman T, Ghetti S (2016) More to it than meets the eye: how eye movements can elucidate the development of episodic memory. Memory 24:721–736. doi:10.1080/09658211.2016.1155870

Paz-Alonso P, Gallego P, Ghetti S (2013) Age differences in hippocampus-cortex connectivity during true and false memory retrieval. J Int Neuropsychol Soc 19:1031–1041. doi:10.1017/S1355617713001069

Paz-Alonso P, Ghetti S, Donohue S et al (2008) Neurodevelopmental correlates of true and false recognition. Cereb Cortex 18:2208–2216. doi:10.1093/cercor/bhm246

Persson J, Söderlund H (2015) Hippocampal hemispheric and long-axis differentiation of stimulus content during episodic memory encoding and retrieval: an activation likelihood estimation meta-analysis. Hippocampus 25:1614–1631. doi:10.1002/hipo.22482

Poppenk J, Evensmoen H, Moscovitch M, Nadel L (2013) Long-axis specialization of the human hippocampus. Trends Cogn Sci 17:230–240. doi:10.1016/j.tics.2013.03.005

Qin S, Cho S, Chen T et al (2014) Hippocampal-neocortical functional reorganization underlies children's cognitive development. Nat Neurosci 17:1263–1269. doi:10.1038/nn.3788

Rajkumar R, Suri S, Min Deng H, Dawe G (2013) Nicotine and clozapine cross-prime the locus coeruleus noradrenergic system to induce long-lasting potentiation in the rat hippocampus. Hippocampus 23:616–624. doi:10.1002/hipo.22122

Ramón y Cajal S (1911) Histologie du système nerveux de l'homme et des vertèbrès, vol II. Maloine, Paris

Ribordy F, Jabès A, Banta Lavenex P, Lavenex P (2013) Development of allocentric spatial memory abilities in children from 18 months to 5 years of age. Cogn Psychol 66:1–29. doi:10.1016/j.cogpsych.2012.08.001

Riggins T, Blankenship S, Mulligan E et al (2015) Developmental differences in relations between episodic memory and hippocampal subregion volume during early childhood. Child Dev 86:1710–1718. doi:10.1111/cdev.12445

Riggins T, Geng F, Blankenship L, Redcay E (2016) Hippocampal functional connectivity and episodic memory in early childhood. Dev Cogn Neurosci 19:58–69. doi:10.1016/j.dcn.2016.02.002

Robinson A, Pascalis O (2004) Development of flexible visual recognition memory in human infants. Dev Sci 7:527–533. doi:10.1111/j.1467-7687.2004.00376.x

Rolls E (2007) An attractor network in the hippocampus: theory and neurophysiology. Learn Mem 14:714–731. doi:10.1101/lm.631207

Rubin R, Watson P, Duff M, Cohen N (2014) The role of the hippocampus in flexible cognition and social behavior. Front Hum Neurosci 8:742. doi:10.3389/fnhum.2014.00742

Sara S (2015) Locus coeruleus in time with the making of memories. Curr Opin Neurobiol 35:87–94. doi:10.1016/j.conb.2015.07.004

Sastre M, Wendelken C, Lee J et al (2016) Age- and performance-related differences in hippocampal contributions to episodic retrieval. Dev Cogn Neurosci 19:42–50. doi:10.1016/j.dcn.2016.01.003

Satterthwaite T, Vandekar S, Wolf D et al (2014) Sex differences in the effect of puberty on hippocampal morphology. J Am Acad Child Psychiatry 53:341–350.e1. doi:10.1016/j.jaac.2013.12.002

Scharfman H, Myers C (2015) Hilar mossy cells of the dentate gyrus: a historical perspective. Front Neural Circuits 6:106. doi:10.3389/fncir.2012.00106

Schoemaker D, Buss C, Head K et al (2016) Hippocampus and amygdala volumes from magnetic resonance images in children: assessing accuracy of FreeSurfer and FSL against manual segmentation. NeuroImage 129:1–14. doi:10.1016/j.neuroimage.2016.01.038

Seress L, Ábrahám H, Tornóczky T, Kosztolányi G (2001) Cell formation in the human hippocampal formation from mid-gestation to the late postnatal period. Neuroscience 105:831–843. doi:10.1016/s0306-4522(01)00156-7

Seress L, Mrzljak L (1992) Postnatal development of mossy cells in the human dentate gyrus: a light microscopic Golgi study. Hippocampus 2:127–141. doi:10.1002/hipo.450020205

Sheffield M, Dombeck D (2015) The binding solution? Nat Neurosci 18:1060–1062. doi:10.1038/nn.4075

Shing Y, Rodrigue K, Kennedy K et al (2011) Hippocampal subfield volumes: age, vascular risk, and correlation with associative memory. Front Aging Neurosci 3:2. doi:10.3389/fnagi.2011. 00002

Sluzenski J, Newcombe NS, Satlow E (2004) Knowing where things are in the second year of life: implications for hippocampal development. J Cogn Neurosci 16:1443–1451. doi:10.1162/ 0898929042304804

Small S (2002) The longitudinal axis of the hippocampal formation: its anatomy, circuitry, and role in cognitive function. Rev Neurosci. 13:183–194. doi:10.1515/revneuro.2002.13.2.183

Snyder J, Radik R, Wojtowicz J, Cameron H (2009) Anatomical gradients of adult neurogenesis and activity: young neurons in the ventral dentate gyrus are activated by water maze training. Hippocampus 19:360–370. doi:10.1002/hipo.20525

Squire L, Wixted J (2011) The cognitive neuroscience of human memory since H.M. Annu Rev Neurosci 34:259–288. doi:10.1146/annurev-neuro-061010-113720

Suárez-Pereira I, Canals S, Carrión Á (2015) Adult newborn neurons are involved in learning acquisition and long-term memory formation: the distinct demands on temporal neurogenesis of different cognitive tasks. Hippocampus 25:51–61. doi:10.1002/hipo.22349

Swagerman S, Brouwer R, de Geus E et al (2014) Development and heritability of subcortical brain volumes at ages 9 and 12. Genes Brain Behav 13:733–742. doi:10.1111/gbb.12182

Tamnes C, Walhovd K, Grydeland H et al (2013) Longitudinal working memory development is related to structural maturation of frontal and parietal cortices. J Cogn Neurosci 25:1611–1623. doi:10.1162/jocn_a_00434

Uematsu A, Matsui M, Tanaka C et al (2012) Developmental trajectories of amygdala and hippocampus from infancy to early adulthood in healthy individuals. PLoS One 7:e46970. doi:10.1371/journal.pone.0046970

Van Leemput K, Bakkour A, Benner T et al (2009) Automated segmentation of hippocampal subfields from ultra-high resolution in vivo MRI. Hippocampus 19:549–557. doi:10.1002/hipo. 20615

Vargha-Khadem F, Gadian D, Watkins K et al (1997) Differential effects of early hippocampal pathology on episodic and semantic memory. Science 277:376–380. doi:10.1126/science.277. 5324.376

Walling S, Harley C (2004) Locus coeruleus activation initiates delayed synaptic potentiation of perforant path input to the dentate gyrus in awake rats: a novel β-adrenergic-and protein synthesis-dependent mammalian plasticity mechanism. J Neurosci 24:598–604. doi:10.1523/ JNEUROSCI.4426-03.2004

Wang H, Das S, Suh J et al (2011) A learning-based wrapper method to correct systematic errors in automatic image segmentation: consistently improved performance in hippocampus, cortex and brain segmentation. NeuroImage 55:968–985. doi:10.1016/j.neuroimage.2011.01.006

Wierenga L, Langen M, Oranje B, Durston S (2014) Unique developmental trajectories of cortical thickness and surface area. NeuroImage 87:120–126. doi:10.1016/j.neuroimage.2013.11.010

Wisse L, Biessels G, Heringa S et al (2014) Hippocampal subfield volumes at 7T in early Alzheimer's disease and normal aging. Neurobiol Aging 35:2039–2045. doi:10.1016/j. neurobiolaging.2014.02.021

Yassa M, Stark C (2011) Pattern separation in the hippocampus. Trends Neurosci 34:515–525. doi:10.1016/j.tins.2011.06.006

Yasuda M, Johnson-Venkatesh E, Zhang H et al (2011) Multiple forms of activity-dependent competition refine hippocampal circuits in vivo. Neuron 70:1128–1142. doi:10.1016/j.neuron. 2011.04.027

Yonelinas A (2013) The hippocampus supports high-resolution binding in the service of perception, working memory and long-term memory. Behav Brain Res 254:34–44. doi:10.1016/j.bbr. 2013.05.030

Yu J, Proddutur A, Elgammal F et al (2013) Status epilepticus enhances tonic GABA currents and depolarizes GABA reversal potential in dentate fast-spiking basket cells. J Neurophysiol 109:1746–1763. doi:10.1152/jn.00891.2012

Yurgelun-Todd D, Killgore W, Cintron C (2003) Cognitive correlates of medial temporal lobe development across adolescence: a magnetic resonance imaging study. Percept Mot Skills 96:3–17. doi:10.2466/pms.2003.96.1.3

Yushkevich P, Pluta J, Wang H et al (2015) Automated volumetry and regional thickness analysis of hippocampal subfields and medial temporal cortical structures in mild cognitive impairment. Hum Brain Mapp 36:258–287. doi:10.1002/hbm.22627

Zeithamova D, Preston A (2010) Flexible memories: differential roles for medial temporal lobe and prefrontal cortex in cross-episode binding. J Neurosci 30:14676–14684. doi:10.1523/jneurosci.3250-10.2010

Age-Related Differences in the Human Hippocampus: Behavioral, Structural and Functional Measures

Cheryl L. Grady and Jennifer D. Ryan

Abstract In this chapter we review the behavioral and neuroimaging literature on age-related differences in hippocampal function. Although it is well known that older adults have reduced relational memory, which depends on the hippocampus, and that hippocampal volume is reduced in older adults, activity in this region is not uniformly lower in older than younger adults during encoding and retrieval tasks. Nevertheless, when the functional neuroimaging evidence is examined in light of current theories of the pattern separation and completion processes carried out by the hippocampus, both processes appear to be altered in older age. We conclude with some suggestions for future work in this field.

Introduction

In this chapter we review the literature on age-related differences in the hippocampus (HPC) from several viewpoints, including performance on HPC-mediated memory tasks, structure of the HPC, and functional measures, primarily from fMRI. Our focus is on memory, given the critical role of the HPC in both encoding and retrieval (Squire 1992; Nadel and Moscovitch 1997; Augustinack et al. 2014). In particular, we have focused on age-related differences in memory for single items (e.g., word, picture of an object), or some form of context (e.g., spatial, temporal) or for item-context bindings (e.g, associative pairings). In terms of the behavioral literature, we have focused primarily on memory for the bound representations of inter-item or item-context bindings, as we consider this kind of memory to be critically dependent upon function of the HPC. We have included memory for single items in our discussion of the neuroimaging literature because many of the aging studies have used items as stimuli. In addition, we have chosen to omit some aspects of memory, such as spatial memory, for which there are only a few papers, in order to include the aspects of age-related differences that are

C.L. Grady (✉) • J.D. Ryan
Rotman Research Institute at Baycrest, University of Toronto, Toronto, ON, Canada
e-mail: cgrady@research.baycrest.org

© Springer International Publishing AG 2017
D.E. Hannula, M.C. Duff (eds.), *The Hippocampus from Cells to Systems*,
DOI 10.1007/978-3-319-50406-3_7

167

supported by the most data. We also note that the goal of this review is to discuss age-related differences found in healthy older adults, so we do not cover the extensive literature on the HPC and related structures in diseases of aging such as Alzheimer's disease (e.g., Sabuncu et al. 2011; O'Brien et al. 2010; Celone et al. 2006; Sperling 2007; Jack et al. 2013).

In the following sections, we first provide a very brief overview of the anatomy of the HPC and the medial temporal lobe structures that are connected to it. This is followed by a review of behavioral studies focusing on age-related differences in HPC-mediated tasks, and how memory impairments in older adults compare to those seen in amnesic patients with HPC damage. The goal in this section is to identify similarities between older adults' performance and that of amnesics, which would presumably reflect altered HPC function in both groups, as well as differences, reflecting the involvement of other brain areas in healthy older adults. We next discuss the evidence for age-related differences in the structure of the HPC and its white matter connections, differences in HPC activation during memory tasks, and differences in functional connectivity of the HPC during memory tasks and at rest. In these sections we also consider age-related differences in other brain regions, primarily prefrontal cortex (PFC) and how these might influence HPC function in older adults. Because a number of theories have emerged recently that ascribe different functions to various subregions of the hippocampus, we attempt to interpret the structural and functional age-related differences in light of these theories. Finally, we discuss gaps in our knowledge that remain, particularly in how to reconcile the behavioural and neuroimaging literatures on the aging HPC, and propose some avenues of future research in this field.

Medial Temporal Lobe Anatomy

The HPC is in a privileged position as a memory structure due to the wide variety of cortical inputs to it via the entorhinal cortex (Fig. 1). The anterior parahippopcampal region (perirhinal cortex) receives input from visual areas in temporal cortex (such as TE and TEO) and from frontal regions, whereas visual area V4 and parietal regions project into the posterior portions of the parahippocampal gyrus (Suzuki and Amaral 1994). These areas of medial temporal lobe (MTL) cortex then project into the entorhinal region, with the perirhinal area projecting into lateral entorhinal cortex and the posterior parahippocampal region projecting into the medial entorhinal cortex (van Strien et al. 2009). The entorhinal region provides the bulk of input to the HPC via the perforant pathway; these inputs arise mainly from the superficial layers of the entorhinal cortex and project into the full long axis of the HPC (van Strien et al. 2009; Kerr et al. 2007). Within the HPC there are several subfields, including the dentate gyrus (DG), CA3 and CA4 subfields, the CA1 subfield, and the subiculum. Information flow through the HPC is mainly one way, progressing from the DG, through the CA fields and into the subiculum (van Strien et al. 2009). The CA1 area and subiculum project back into

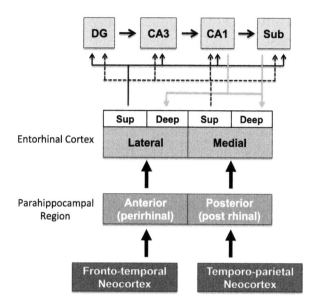

Fig. 1 This figure shows the "standard" view of inputs to the HPC and is an adaptation of a figure in Van Strien et al. (2009). The *solid lines with arrows* represent the inputs from the lateral entorhinal cortex to the HPC and the *dashed lines with arrows* represent the inputs from the medial entorhinal cortex to the HPC. The *gray lines with arrows* represent the connections from the CA1 and subiculum back to the entorhinal cortex. See text for more details regarding these connections. *DG* dentate gyrus, *CA1/CA3* cornu ammonis subfields 1 and 3, *Sub* subiculum, *Sup* superficial layers of the entorhinal area, *Deep* deep layers of the entorhinal area. For more detail on these connections see Van Strien et al. (2009) and Suzuki and Amaral (1994)

deep layers of the entorhinal cortex (Kerr et al. 2007), and hippocampal efferents via the fornix also connect to areas outside the temporal lobes.

All of these HPC and medial temporal regions are thought to carry out specific roles in memory function (Brown and Aggleton 2001; Squire et al. 2004), and they also participate in other functions such as spatial navigation (van Strien et al. 2009). The particular structure of the HPC is also thought to play a role in how memory works. For example, the unidirectional flow of information, the recurrent collaterals prominent in the CA3 region, and the specific input/ouput connections between the HPC subfields and the lateral/medial entorhinal sites, are all thought to underlie the critical functions of pattern separation and pattern completion (discussed in more detail below) that together allow for the formation and retrieval of specific and detailed memory representations (Maass et al. 2014, 2015). Therefore, the structure of the HPC and the wide variety of information that gets funnelled into it via the entorhinal cortex facilitate the role of this region in the relational binding of objects with the contexts in which they are encountered, forming the basis for what we know as episodic memory (e.g., Hsieh et al. 2014; Cohen et al. 1999; Howard and Eichenbaum 2015).

Age-Related Differences in Hippocampally-Mediated Cognitive Tasks

In this section, we focus on differences between young and older adults in performance on tasks that have been shown to critically rely on the HPC for successful performance. We contrast the performance of older adults against the performance of amnesic cases who have damage to the HPC and/or extended system (i.e., individual cases may have additional lesions in regions such as the fornix, mammillary bodies, thalamus, or additional regions within the medial temporal lobes), and where possible, against the performance of nonhuman animals who have lesions to the HPC. The findings outlined below generally focus on age-related deficits in relational memory, as tested through tasks of inter-item and item-context pairings, nonlinear problems, and future imagining.[1] Age-related differences are also observed on behavioral tasks that require discrimination among representations (of items and/or their contexts) that have considerable feature overlap. Such differences suggest that older adults have a deficit in the process of *pattern separation* (Yassa and Stark 2011), which would otherwise allow for similar inputs to be orthogonalized into distinct representations in memory. Typically, tasks of relational memory were applied first to the study of hippocampal function in non-human animals and human amnesia, and subsequently they were used to examine the nature of the memory deficit in aging. By contrast, tasks designed to tap into the specific process of pattern separation have been predominantly employed to the study of aging to examine the extent to which older adults can create distinct representations for overlapping information. Pattern separation studies of non-human animals with hippocampal lesions and with human amnesic cases are continuing to emerge and provide evidence for the engagement of specific hippocampal subregions in this process.

Relational Memory

Inter-Item and Item-Context Bindings

Research in nonhuman animals with HPC lesions led to the proposal that the HPC has a critical role in the binding of relations among distinct items and between an item and its context (for a review, see Cohen and Eichenbaum 1993). As described in detail elsewhere (e.g., Eichenbaum and Cohen 1988), research indicates that hippocampal neurons code for combinations of objects and the combination of

---

[1]Some authors (e.g., Naveh-Benjamin 2000) have used the term "associative memory" in their descriptions of these age-related memory deficits. As has been done in the past (e.g., Ryan et al. 2007), we use the term "relational memory" here. For further explanation about differences in the interpretation of these terms, readers can consult Moses et al. (2008b).

objects with specific places. Human amnesic cases with MTL damage show deficits in learning pairs of stimuli across multiple domains, and under a variety of task instructions (Cohen and Eichenbaum 1993; Eichenbaum and Cohen 2001; Moses and Ryan 2006). As well, amnesic individuals show deficits in linking items to their respective spatial and/or temporal contexts, such as knowing where an item had been previously located, or in what order items had been previously viewed (Konkel and Cohen 2009)

Behavioral research over the past 15 years has consistently showcased a deficit in aging for establishing inter-item or item-context bindings in memory that is similar to that observed in amnesia (for meta-analysis, see Old and Naveh-Benjamin 2008a). Typically, older adults have demonstrated impaired memory for pairs of stimuli, despite often showing normal or relatively preserved memory for the items themselves. Even when an age-related impairment in item memory is observed, the memory impairment for the inter-item or item-context pairings tends to be disproportionate to the item memory deficit (Fig. 2). That is, the deficit for the pairings is larger than what would be expected given the level of memory that is observed for the items. The age-related decrease in memory for stimulus pairings occurs under all manner of learning conditions (Naveh-Benjamin et al. 2009), and is reflected behaviorally in a decrease in hits, an increase in false alarms and a general

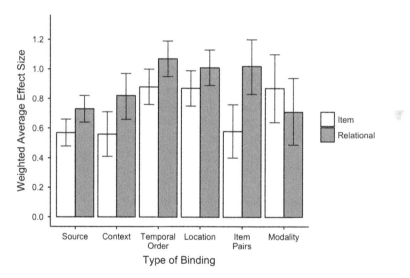

Fig. 2 Age-related relational memory decline. An adaptation of the results from a meta-analysis by Old and Naveh-Benjamin (2008a) is presented. Across studies that required different types of binding (i.e., items must be bound to either a source, context, temporal order, spatial location, another item, or a modality) and subsequently assessed memory for the items as well as their bound relations, older adults showed deficits in both item and relational memory, as indexed by the weighted effect size; however, older adults show a disproportionate decline in memory for relations. *Error bars* reflect 95% confidence intervals as reported in Old and Naveh-Benjamin (2008a)

shift towards a more liberal response bias associated with increasing age (e.g., Soei and Daum 2008; Bender et al. 2010).

Age-related memory deficits have been observed for within-domain stimulus pairings for verbal (Castel and Craik 2003; Naveh-Benjamin 2000; Light et al. 2004; Healy et al. 2005) and non-verbal visual stimuli (Naveh-Benjamin et al. 2003; Bastin and Van der Linden 2005). Deficits also have been observed for pairings of stimuli across domains (James et al. 2008; Naveh-Benjamin et al. 2004), including the pairing of objects to spatial locations (Ryan et al. 2007), odors to spatial locations (Gilbert et al. 2008), objects to orientation (Soei and Daum 2008), and even actors to actions (Kersten and Earles 2010). Additionally, older adults have shown impairments relative to their younger counterparts in remembering the details of the episode in which item information has been learned (source amnesia, Schacter et al. 1984), such as whether items were read or heard (e.g., McIntyre and Craik 1987), the gender of the person who presented an item (e.g., Bayer et al. 2011), or the list in which information was presented (e.g., Bastin and Van der Linden 2005). Source memory can be considered as another example of item-context binding (i.e., binding of an item to its source context). This large body of evidence on age-related differences in relational memory, along with relational binding deficits in amnesic individuals and lesioned animals, suggests that altered HPC function underlies these deficits in older adults.

Nonlinear Problems

Nonhuman animals with lesions to the HPC or extended HPC system (e.g., fornix damage) show impairments on tasks of nonlinear problems that require single items to be evaluated with respect to other items in order for a correct response to be generated. Successful performance on nonlinear problems cannot be achieved by merely learning that one stimulus is rewarded whereas another is not. Whether a stimulus will be rewarded on any given trial is determined by its relational context, specifically, the other item(s) with which it is presented, and these relationships among the stimuli must be learned through trial and error. Transitive inference, transverse patterning, and transitivity tasks are each nonlinear problems that have been used to examine the role of the HPC in establishing relations among overlapping pairs of items (Fig. 3), and in supporting inference decisions that require the bridging of information across existing sets of relations (Moses and Ryan 2006).

In the transitive inference task, a relational hierarchy of items ($A>B>C>D>E$) must be learned through exposure to a series of overlapping premise pairs of items (A wins over B, B wins over C, C wins over D, D wins over E). Inferences are then made by bridging across these sets of premise pairs (i.e., choose A when presented with the pair A–C, Fig. 3a). Nonhuman animals with damage to the fornix were impaired on transitive inference despite successful learning of the premise pairs (Dusek and Eichenbaum 1997) and human amnesic cases with damage to the HPC

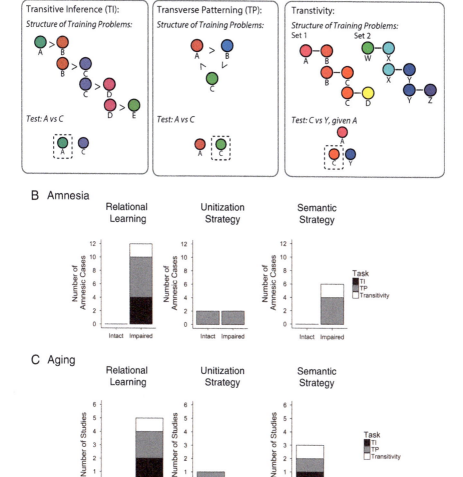

Fig. 3 Cognitive strategies that can mediate relational binding deficits in amnesia and aging. This figure summarizes the general procedures and data from the papers reviewed in the text that examined performance on nonlinear problems. In (**a**), the training and test structure is shown for the transitive inference (TI; *left*), transverse patterning (TP; *middle*) and transitivity (*right*) paradigms. In all tasks, participants are trained to select either the stimulus that 'wins' within a premise pair (denoted by >; transitive inference, transverse patterning), or the choice stimulus that 'belongs' with a given sample stimulus (transitivity; when given A, choose B, not X, when given B, choose C, not Y) such that two sets of stimuli may be learned (A,B,C,D; W,X,Y,Z). In the test phase, relational memory for inter-item bindings is tested by requiring participants to select the stimulus that either 'wins' over (TI, TP) or 'belongs' with another stimulus (transitivity). Critically, relational memory is tested by requiring memory for different contexts in which each stimulus was learned (e.g, in TP, A can either win or lose depending on the other item). Relational memory, as well as inference, is also tested by presenting a novel problem that requires the participants to bridge across sets of premise pairs in memory (TI, transitivity). The correct answer to the test questions is indicated with the *dashed line*. The number of amnesic cases who show

(Fig. 3b) were impaired for both learning of the premise pairs and subsequent transitive inference (Smith and Squire 2005). Some older adults have shown deficits in learning of the premise pairs, in subsequent inference (Fig. 3c), and in explicitly articulating the hierarchy (Ryan et al. 2009). The older adults who did perform well on transitive inference were those who were able to successfully establish the premise relations (Ryan et al. 2009) and were those who performed better on neuropsychological tests that tapped MTL function (Moses et al. 2008b).

However, the transitive inference task can be solved by learning the associative reward strength of each item (Frank et al. 2003; van Elzakker et al. 2003; Wynne et al. 1992; Wynne 1997); for instance, in the hierarchy of A>B>C>D>E, the anchor stimulus A always wins and E always loses (Fig. 3a). By virtue of the reward patterns of these anchor stimuli, appropriate responses can be made to the other stimuli, without needing to reference their relative positions within the hierarchy. Specifically, the reward value of stimulus D is de-valued due to its proximity to E, whereas B has larger associative reward strength due to its proximity to A; thus the B–D problem can be solved in the absence of memory for the relations among the items in the hierarchy (Frank et al. 2003; van Elzakker et al. 2003). A perhaps more definitive test of memory for relations among items, for which an associative strength strategy cannot be used (Sutherland and Rudy 1989), is the transverse patterning task.

The transverse patterning task is akin to the rock-paper-scissors game and requires participants to learn the relations among a set of three items (A wins over B, B wins over C, C wins over A; Fig. 3a). Importantly, whether an item wins or loses must be determined in the context of another item; an item may be 'correct' in the context of one item (e.g., A wins over B), and 'incorrect' in the context of another item (e.g., A loses to C). Thus, the prior reward history of an item is not useful for deciding whether it should be selected, as it is rewarded/ unrewarded equally often. Studies of non-human animals with HPC lesions and studies of amnesic cases (Fig. 3b) have shown that successful transverse patterning performance critically requires HPC function (Rickard and Grafman 1998; Rickard et al. 2006; Driscoll et al. 2003, 2005; Reed and Squire 1999; Moses et al. 2008a). Older adults (Fig. 3c) show deficits similar to those expressed by HPC amnesics on

---

Fig. 3 (continued) intact versus impaired performance is presented in (**b**). Amnesic cases have shown impairments for learning the relations among the items in nonlinear problems (*left*), but for at least some amnesic cases, using a unitization strategy to fuse the items into a single, blended representation through an action allows the HPC system to be bypassed, resulting in intact performance (*middle panel*). Amnesic cases do not benefit from using existing knowledge within semantic memory to support the learning of new relations (*right panel*). In (**c**), the number of studies in which older adults show intact versus impaired performance is presented. Whereas older adults show impaired relational learning that is similar to what is observed in amnesic cases (*left panel*) and can benefit from the use of unitization to support performance (*middle panel*), unlike amnesic cases, older adults can use existing semantic knowledge to boost relational learning (*right panel*), suggesting that HPC function can be supported through neocortical connections and function

transverse patterning (Ostreicher et al. 2010). However, although the transverse patterning task can be used to assess relational memory in aging, it does not allow for inference to be tested, as all of the possible pairs of stimuli have been previously studied; no novel problem sets can be presented in which information must be bridged across sets of relations.

Recently, the transitivity paradigm has been adapted from the non-human animal literature (Bunsey and Eichenbaum 1996) to assess the establishment of relations among items in memory as well as inference across those relations. Like transverse patterning, successful performance on transitivity tasks critically requires the HPC and extended system (Bunsey and Eichenbaum 1996; Ryan et al. 2016). Like the transitive inference task, transitivity tasks test for the ability to make inferences across learned pairs of relations, but, unlike the transitive inference task, learning of each stimulus' associative strength cannot support performance. In transitivity, participants learn two sets of stimuli (A–B–C–D, W–X–Y–Z), and when presented with a sample stimulus from one group (A), participants must then choose the appropriate choice stimulus that belongs to the same group (B) when that stimulus is presented alongside a choice stimulus from the other group (X). Participants learn sets of pairwise relations (when A is presented, choose B not X, when B is presented, choose C not Y; however when W is presented, choose X not B, and when X is presented, choose Y not C). During a test phase, participants must make inferences across the premise sets when presented with stimuli that had not been previously shown together during the study phase. For instance, when presented with A, the participant must select C and not Y (Fig. 3a); likewise, when presented with W, the participant must select Y and not C. Importantly, each of the two choice stimuli in any problem pair is rewarded equally often; the participants must learn to select the appropriate choice stimulus given the context of the sample stimulus.

Amnesic cases whose damage includes the HPC (DA, Ryan et al. 2016) or the fornix and thalamus (NC, D'Angelo et al. 2016b), show deficits in transitivity; they have difficulty learning the premise sets and subsequently are impaired for making inferences across the sets of relations. Similar to the amnesic cases, older adults have difficulty establishing the relations within each of the premise sets, and ultimately are impaired relative to younger adults on the inference problems (Ryan et al. 2016). The similarity between the performance of older adults and the amnesic cases suggests an age-related decline in the functioning of the HPC and/or extended system, in this case, the fornix and thalamus (Fig. 3b, c).

Although older adults show deficits on nonlinear problems similar to those seen in amnesic cases who have HPC damage, older adults and amnesic cases differ with respect to whether they can benefit from strategies to either remediate or circumvent deficits in HPC function. Older adults can rely on existing semantic knowledge to improve performance in tasks of nonlinear problems. In studies of transitive inference, when older adults were first exposed to a known hierarchy (e.g., a hierarchy of playing cards), they could use such information to support learning of a new hierarchy with previously unknown stimuli (Moses et al. 2010). Similarly, when first presented with a known structure in transverse patterning (e.g., rock-paper-scissors; playing cards Ace-King-Two in which the Ace can be treated as the high or the low card), subsequent learning of the relations among a novel set of objects is facilitated and the performance of older adults resembles that of younger adults (Ostreicher et al. 2010). In studies of transitivity, just as in transitive inference and transverse patterning, older adults benefit in the learning of arbitrary groupings when they have prior exposure to known groupings of items (e.g., sets of garden tools versus kitchen tools; Fig. 3c). Moreover, older adults are able to benefit from known pairwise relations in order to support novel inferences across non-overlapping pairs of stimuli (Ryan et al. 2016; D'Angelo et al. 2016b).

Across multiple tasks of nonlinear problems, older adults benefit from the use of prior semantic knowledge to support learning and declining function of the HPC system, however such benefits are not apparent in amnesic cases whose damage includes the HPC and extended system (Fig. 3b; Moses et al. 2008a; Ryan et al. 2013; D'Angelo et al. 2015). These findings suggest a decline in HPC function in aging, but also indicate that HPC function can be supported through the use of other neural systems, although it is not yet evident whether these additional regions upregulate HPC function itself, merely lessen the overall binding demands of the HPC by also engaging in relational processing, or support performance through some other mechanism. Nonetheless, there is evidence that regions within the PFC may permit the HPC to more effectively bind new information when such binding occurs in the context of prior knowledge, although the contribution of the HPC is still critical (Tse et al. 2007, 2011). Existing knowledge could provide a framework onto which novel relations may be mapped via coordination between the PFC and the HPC (Tse et al. 2011). Relatedly, it may be the case that older adults do not spontaneously appreciate (i.e., are not aware of) the overall task structure (e.g., Moses et al. 2008b), and the presence of prior knowledge can facilitate that understanding. A lack of appreciation for the task structure may also reflect underlying HPC decline, and awareness of the task structure could be mediated through prior knowledge and PFC function, that thereby lessens the binding demands of the HPC. Research with rodents has shown that the assimilation of new knowledge into existing schemas is associated with up-regulation of immediate early genes in the medial prefrontal cortex (Tse et al. 2011). Neuroimaging research regarding such underlying mechanisms in young adults also suggests that medial PFC is involved in memory processing using schemas (e.g., van Kesteren et al. 2010). Although it has not been tested directly, this evidence suggests that support for HPC function via schema-related activity in PFC may be one mechanism whereby older adults can maintain inferential types of relational memory. This idea also would be in line with evidence that older adults often over-recruit PFC relative to younger adults during memory tasks (e.g., Rajah and D'Esposito 2005) and that PFC function may be used to support the integration of relations in HPC (Backus et al. 2016; Schlichting et al. 2015).

Age-related deficits on tasks of nonlinear problems can also be alleviated by bypassing the HPC and MTL cortex (Ryan et al. 2013; D'Angelo et al. 2015). *Unitization* is a strategy whereby items are fused together, through an imagined action (e.g., one could imagine that one item could fit within the other or the two items could interact in some way, e.g., "the star pierces the bucket") in order to form a single representation from which the relations among the items can be derived. Older adults can successfully use unitization to support transverse patterning performance (see Fig. 3, D'Angelo et al. 2016a). However, otherwise nominally healthy older adults who have failed the Montreal Cognitive Assessment (MoCA, Nasreddine et al. 2005) do not benefit from unitization, suggesting that a certain level of cognitive function, likely the use of semantic memory and visual imagery, may be required for the unitization to successfully circumvent HPC and MTL function (D'Angelo et al. 2016a). Although the exact cognitive and neural mechanisms that support unitization remain to be fully clarified, findings from amnesic cases show that unitization can be successfully implemented despite bilateral damage to the HPC and regions of the MTL, such as the perirhinal cortex (see D'Angelo et al. (2015) for further discussion; Quamme et al. 2007; Ryan et al. 2013). Those amnesics who do not benefit from unitization have damage beyond the HPC and MTL that includes anterior temporal lobes (Ryan et al. 2013), suggesting this region may be critical for unitization. Whether amnesic cases and older adults engage similar brain regions during unitization also remains to be investigated; however, research points to the use of unitization as a viable cognitive strategy for some amnesic cases and some older adults to bypass relational binding deficits.

Future Imagining

Consistent with the evidence reviewed thus far for age-related deficits in inter-item and item-context bindings, the number of details that are recalled from past autobiographical events also is consistently reduced with age (e.g., Levine et al. 2002). These deficits in detail recollection and relational memory appear to have significant and broader consequences for other cognitive functions, in particular, thinking about, or simulating the future. A critical role for the HPC in imagining the future, in addition to recalling the past, is supported by neuropsychological studies with amnesic cases (Hassabis et al. 2007; Klein et al. 2002). Early (Tulving 1985) and more recent (Rosenbaum et al. 2005) research with the amnesic case KC particularly demonstrated that significant deficits in imagining the future were observed, even when KC was merely asked what he could imagine doing later that day. Further converging support is garnered from neuroimaging findings that demonstrate a common neural network, including the HPC, is engaged for recalling the past and imagining the future (Addis et al. 2007; Okuda et al. 2003; Szpunar et al. 2007). Age-related reductions in activity in HPC, parahippocampus and precuneus have been observed during future imagining; activity in these regions in younger adults correlates with detail generation (Addis et al. 2011). In light of the concomitant deficits for recalling personal past details and imagining future scenarios in amnesic cases (Rosenbaum et al. 2005; Klein et al. 2002), and the overlap in neural networks supporting the two functions, researchers have argued that recalling the past may be a necessary component for constructing the future

(Schacter et al. 2007). Although the HPC is engaged in both recalling the past and imagining the future, research has revealed distinct patterns of activity in the HPC that distinguish remembering from imagining (Kirwan et al. 2014), suggesting that there may be unique cognitive operations involved in imagining the future.

Research regarding age-related changes in future thinking (also termed, *episodic simulation*) was propelled by Addis et al. (2008). Using the Autobiographical Interview (Levine et al. 2002), Addis and colleagues showed that older adults generated fewer internal details (details directly related to the time, place, emotion, for the generated event) for past events as well as for future scenarios. In contrast, older adults typically generated more external details (details regarding semantic information) than the younger adults for both past and future events, indicating that the deficit for generating internal details was not merely a function of an age-related change in verbal output. The number of internal details that was generated by the older adults was correlated with their ability to remember word pairs, another indicator of relational memory, and also with backwards digit span, an indicator of executive function. A subsequent study by Addis et al. (2010) elicited details regarding people, places and objects from past memories, and then asked participants to recombine such details, across memories, to generate an imagined scenario that could have happened in the past or in the future. For the imagined scenarios, the participants were presented with the person, place, and object details that had been previously recalled for past events. The details for the to-be-imagined scenarios were either all taken from the same past event, or were randomly recombined from multiple past events. Older adults were impaired relative to younger adults for recalling episodic details from past events, and for generating episodic details when imagining a past or a future scenario, including the recombined events. Since this initial work that detailed a deficit for future imagining in aging (Addis et al. 2008, 2010), other researchers have demonstrated that older adults were more likely to misattribute generated future events, made in response to a cue word, to the past, than to misattribute past events to the future, suggesting that older adults had difficulty in retrieving features of an event memory that would characteristically define a past, autobiographical memory (McDonough and Gallo 2013). Additionally, older adults were impaired for self-generating future intentions, reflected in a reduced ability to use foresight to acquire items that would subsequently be needed to solve a problem (Lyons et al. 2014).

From these findings, Addis and colleagues (2008, 2010) suggested that future thinking requires the recall of past events, the details of which are integrated anew to form a novel future event. Older adults generally have a deficit in recalling details from the past. However, even when older adults have successfully recalled the past details, deficits are observed in integrating those past details into a coherent imagined scenario that is set in the future. This suggests that older adults have a deficit in flexibly re-binding past information in support of future event construction, above and beyond any deficit in recalling past details (i.e., constructive-episodic simulation hypothesis, Schacter and Addis 2007b, c). This conceptualization of future thinking as re-binding of the past is supported by additional investigations that continued to demonstrate that age-related deficits in the generation of internal details during future imagining are related to impairments in episodic memory, and more generally in relational memory (for a review, see Schacter et al. 2013). Age-related differences in the use of the past to imagine the future may also be related to evidence that older adults generated more past events in response to cue words, whereas younger adults generated more future events (Spreng and Levine 2006). In addition, older and younger adults generally differ with respect to their future time perspectives; older adults perceive that they have less time remaining in life compared to younger adults (Carstensen 2006). It is possible that such age-related changes in future time perspective interact with age-related differences in future thinking and detail generation. Nonetheless, the findings noted above have consistently shown age-related deficits in recalling past information and generating future scenarios, both of which engage the HPC (Addis et al. 2007; Okuda et al. 2003; Szpunar et al. 2007; Schacter et al. 2007).

Pattern Separation

Behavioral observations of age-related memory deficits as outlined above suggest that older adults have difficulty forming inter-item and item-context relational bindings in memory. This deficit may be due to an underlying deficit in HPC-mediated pattern separation for the items and/or for their associated contexts. Research suggests that the HPC is critical for pattern separation, which is the ability to orthogonalize incoming information within the dentate gyrus into separate representations that are then stored by the CA3 of the HPC (O'Reilly and Norman 2002). An age-related deficit in pattern separation would consequently result in memory deficits due to an inability to create and subsequently use separate representations for similar information. As well, during retrieval, the aging HPC may tend towards *pattern completion,* the matching of partial incoming information to stored representations, which may result in the retrieval of inappropriate information (Yassa and Stark 2011) and be related to the increased false alarms seen in older adults (e.g., Bender et al. 2010; Old and Naveh-Benjamin 2008b; Light et al. 2004). It is important to note that the terms 'pattern separation' and 'pattern completion' have been used primarily to refer to the neural computations that may orthogonalize or match, respectively, incoming information to information that is held in memory, but that these terms also have been used to characterize behavioral performance on tasks that require participants to discriminate among stimuli that are perceptually or semantically similar. There is a debate concerning whether the terms 'behavioral pattern separation' and 'behavioral pattern completion' are as appropriate as the term 'behavioral discrimination', given that there may not be perfect correspondence between behavioral separation/completion and neural separation/completion (Santoro 2013). For our purposes here, we use the terms 'behavioral pattern separation/completion' in order to specify the type of behavioral discrimination that is required, while noting that the behavioral findings may not necessarily align with neural indices of pattern separation or completion.

As well, we acknowledge, as noted more thoroughly in the later section regarding neuroimaging findings of pattern separation, that HPC regions such as the DG and CA3 may not be uniquely involved in only pattern separation or completion, but instead may contribute to both kinds of neural computations (e.g., Nakashiba et al. 2012; Leutgeb et al. 2007).

Evidence for an age-related deficit in behavioral pattern separation comes from studies showing that older adults are impaired relative to younger adults in discriminating target items from similar lures, under a variety of task instructions, on the Mnemonic Similarity Task (MST, Stark et al. 2015), which was designed specifically to tax pattern separation processes (Stark et al. 2013). Older adults also have difficulty discriminating previously viewed items (scenes) from novel items when the stimuli are perceptually degraded (Vieweg et al. 2015). Similarly, older adults have impairments in temporal pattern separation; they show more difficulty in distinguishing the relative recency of objects when the temporal lags are short, but not when temporal lags are more pronounced (Roberts et al. 2014). Interestingly, older adults who demonstrate better performance on verbal learning tests have more accurate memory performance on such discrimination tasks (Holden et al. 2012; Reagh et al. 2014).

Evidence for behavioral pattern separation deficits in aging also can be found in some of the above-cited studies regarding inter-item and item-context (source) binding deficits in aging. Some paradigms that test memory for inter-item pairings require participants to distinguish exact repetitions from pairings that are comprised of recombined stimuli. For instance, a participant may study the three pairs of stimuli, A–B, C–D, and E–F, whereby each letter denotes a single item. The presentation of the pair A–B during a subsequent test phase would represent an exact repetition, whereas the presentation of the pair C–F would represent a recombination. In such a paradigm, pairs of stimuli can only be judged as 'old' by virtue of their bindings and not on memory for the items themselves; in this example, the stimuli A, B, C and F had all been studied previously, however, the combination of C with F is novel. Typically, older adults have difficulty relative to younger adults in distinguishing true pair repetitions from recombined pairs, a relational memory impairment that may occur as a consequence of deficient pattern separation in this population (e.g., Overman and Becker 2009). One recent variant of the standard paradigm makes discrimination of intact from rearranged pairings even more difficult. In this case, participants were required to encode pairs (such as C–D and E–F) that were then recombined (e.g., C–F), and those recombined pairings were also to be learned and endorsed as 'repeated' in a later test phase. Lures were similarly comprised of recombinations of previously studied items, however those recombinations were uniquely presented in the test phase (e.g., Overman and Becker 2009). Such a paradigm may rely on pattern separation and the creation of unique representations for successful performance to occur, particularly when the representations may contain some overlapping features. On this task, older adults performed poorly across all conditions, raising the question of whether floor effects prevented the experimenters from seeing age-related differences that would have been exaggerated for the recombined stimuli that were to be endorsed as 'targets'.

In many source memory tasks, the source stimuli repeat across experimental trials; for instance, a male and a female voice are each presented on half of the trials, and the participant must subsequently decide whether a given stimulus was presented in the male or female voice. Thus, there is considerable feature overlap among the stimuli (i.e., half of stimuli are presented in the same voice). Therefore, pattern separation may be required to form unique representations of each of the studied words, and older adults may experience difficulty in forming non-overlapping representations.

Although all of this evidence suggests that older adults have behavioral deficits in pattern separation, presumably due to HPC dysfunction, to date, there has not been much research investigating behavioral pattern separation in amnesic cases. However, one study has shown that two amnesic cases with bilateral damage to the HPC showed impaired performance on a collaborative referencing paradigm when pattern separation became necessary (Duff et al. 2012). In the collaborative referencing paradigm, the amnesic individuals and their partners must develop unique labels for a set of tangrams (novel objects comprised of multiple geometric shapes) through extended interactions. Although the amnesics developed concise verbal labels for the tangrams when they were perceptually dissimilar, they had more difficulty relative to control participants for developing concise verbal labels (as measured by the number of words used) when the tangrams were visually similar.

A more recent study (Baker et al. 2015) examined behavioral pattern separation using the MST with case B.L., who presents with a rare case of selective lesions to the dentate gyrus bilaterally. As noted above, on the MST, participants must identify previously viewed objects (targets), identify objects that are similar to ones previously studied (lures), and new objects (foils). Control participants endorsed lures as 'similar' more often than foils, but B.L. tended to endorse the lures as 'targets' and consequently, was less accurate than controls for endorsing lures as 'similar', suggesting that the dentate gyrus is critical for disambiguating similar stimuli.

More data from amnesic cases is needed to thoroughly compare their performance to that of older adults, and in particular to determine whether amnesic cases show alterations in pattern completion performance. Although older adults have been shown to have a bias towards pattern completion as a result of age-related deficits in pattern separation processes (Yassa et al. 2011b), older adults have nonetheless shown deficits in pattern completion. Older adults were shown to have more difficulty identifying previously viewed scenes that were perceptually degraded compared to younger adults, even at the easiest levels of discrimination (Vieweg et al. 2015). Older adults also exhibited more false alarms, suggesting that a bias towards pattern completion may lead to incorrect recognition judgments. Additional evidence for age-related differences in pattern completion was found during a match-to-sample spatial navigation paradigm when increasing numbers of extra-maze cues were removed (Paleja and Spaniol 2013). In rats, performance on the match-to-sample paradigm declined with extra-maze cue removal when the CA3 was lesioned, suggesting that pattern completion performance may rely on this region of the HPC (Gold and Kesner 2005). Thus, this evidence for age-related differences in both behavioral pattern completion and separation indicate that older adults may have rather broad deficits in behavioral discrimination due to alterations in multiple HPC processes.

Summary

Older adults typically show deficits on memory tasks that are consistent with the deficits that are observed in individuals who have HPC damage. However, older adults also show patterns of performance that are distinct from HPC amnesic cases. Research using nonlinear problems has shown that memory performance in older adults, but not amnesic cases, can be supported through the use of existing semantic knowledge, suggesting that while there is likely some HPC dysfunction associated with aging, residual HPC function can be supported through a larger, interconnected network of regions and/or that semantic knowledge provides a framework for which older adults may appreciate the task relations. Emerging research on behavioral pattern separation and completion points to age-related differences in pattern completion (either manifested as deficient or as incorrect completion), as well as separation, suggesting a broad deficit either within the HPC or in regions beyond the HPC, although further research is needed (we return to this question in the section below on how age-related differences in structure and function can be understood in the context of neural pattern separation/completion). Additionally, although not reviewed here, older adults can show deficits in sensory processing and/or other cognitive functions that are distinct from the HPC memory deficits, suggesting dysfunction in neural regions outside of the HPC (Naveh-Benjamin and Kilb 2014; Buckner 2004).

Thus, it would be important to consider the age-related changes in hippocampally-mediated memory function within a larger context of the brain. The nature of behavioral studies permits discussion of the role of the HPC in aging to the extent that the paradigms used are drawn from those used with human and non-human lesion cases, such as many of the studies noted above. However, the nature of neuroimaging allows for a more refined examination regarding the differential structural or functional decline of the HPC and its subfields with respect to age-related changes in memory performance, while also allowing for the function of the HPC to be situated within a much broader neural context. From the consistent behavioral evidence of age reductions in the inter-item and item-context bindings of relational memory and underlying behavioral pattern separation processes, one might predict equally consistent findings of reduced HPC structure and function in older compared to younger adults. As we will see in the next sections, this is generally true for structural measurements, but functional studies have produced more variable results.

Age-Related Differences in Hippocampal Structure

The HPC has been the target of many experiments focused on age-related differences in brain structure. Cross-sectional studies have consistently found that HPC volume is reduced in older vs. younger adults (e.g., Jack et al. 1997; Fjell et al. 2014; Lemaitre et al. 2005), and longitudinal studies have shown volume declines over time in older adults (Raz et al. 2005, 2010; Resnick et al. 2003). Although other cortical and subcortical areas also show decline over time with age, in particular the PFC, the HPC shows a relatively large reduction (Raz et al. 2010). Importantly, smaller HPC volumes typically are correlated with worse performance on word recall or recognition tasks in older adults (Kramer et al. 2007; Jernigan et al. 2001; Chen et al. 2010; Ezzati et al. 2016), and on memory composite scores incorporating both verbal and nonverbal standard neuropsychological tests of episodic memory (Head et al. 2008; Rodrigue et al. 2013;Ward et al. 2015). In addition, HPC volume declines more over a 10 year period in those older adults who also show a general decline in memory performance (Persson et al. 2012).

Age-Related Differences in HPC Subregions

Some studies have examined the volumes of different sub-regions of the HPC as a function of age, for example in terms of anterior or posterior HPC. However, these studies have not reported consistent results, with some finding more atrophy in anterior portions (Chen et al. 2010; Rajah et al. 2010; Ta et al. 2012), and others reporting more robust age-related differences in the volume of posterior HPC (Driscoll et al. 2003; Malykhin et al. 2008). There is even one report of increased anterior HPC volumes with age (Kalpouzos et al. 2009). Although the jury is still out in this regard, a recent paper in a large sample of cognitively normal adults (almost 300 people) found greater age-related atrophy in anterior HPC (Gordon et al. 2013), lending support to the idea of a gradient of atrophy along the HPC axis from anterior to posterior. Although most of these studies did not explicitly examine the relationship between anterior/posterior HPC volume and memory performance, Rajah et al. (2010) failed to find a significant correlation between anterior HPC volume and memory for spatial and temporal context in older adults, providing some evidence that context memory might not be closely linked to HPC reductions in volume in healthy older adults.

Other investigators have approached age-related differences in the HPC by assessing the various subfields separately. These studies have used a variety of methods, including manual tracing of MRI scans (de Flores et al. 2015; Wisse et al. 2014; La Joie et al. 2010; Mueller and Weiner 2009; Shing et al. 2011; Doxey and Kirwan 2015; Raz et al. 2015; Daugherty et al. 2016), automatic segmentation of MRI scans (de Flores et al. 2015; Pereira et al. 2014; Ziegler et al. 2012; Voineskos et al. 2015), and cell counts of autopsy tissue (West et al. 1994; Simic et al. 1997).

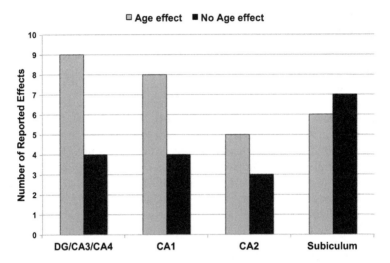

Fig. 4 This figure summarizes the data from the papers reviewed in the text that examined volumes of the different HPC subfields. Each *bar* in the figure represents the number of papers reporting an age reduction in volume (*gray bars*), or no age difference (*black bars*). The CA3 and CA4 subfields are included with the DG because they were typically combined with the DG rather than examined separately

Unfortunately, these data are complicated by the fact that the various subfields are often combined, such as DG/CA3 or CA1/CA2, and neither the CA3 or CA4 subfield has been examined on its own. For these reasons, the results are somewhat difficult to interpret, but the evidence to date does suggest that the DG and CA1 subfields are more likely to show reductions in volume with age, whereas the subiculum is less often influenced by aging (Fig. 4), as is the entorhinal cortex, at least in healthy older adults (the entorhinal cortex is affected early in Alzheimer's disease, e.g., Braak et al. 1993). In one very large study (over 500 participants) volumes were measured in adults from 20-80 years of age, and reductions were seen throughout the medial temporal lobes, with the DG and CA fields showing non-linear reductions with age, and the subiculum showing a linear effect (Ziegler et al. 2012). In the DG and CA subfields, volumes were relatively maintained until after 60 years, when marked reductions were found. Consistent with the predominance of age reductions in the DG/CA3 region, correlations have been reported between larger volumes of this region and better performance on relational memory (Shing et al. 2011) and pattern separation tasks (Doxey and Kirwan 2015). This suggests that an age-related decline in structural integrity of some HPC subfields may result in greater similarity between different memory representations (Wilson et al. 2006; Holden and Gilbert 2012) and negatively impact memory (but see Voineskos et al. 2015).

Age-Related Differences in HPC White Matter Tracts

In addition to gray matter volumes of the HPC, the white matter tracts leading into or out of the HPC also show age-related differences. These have been measured with diffusion tensor imaging, which is sensitive to the direction of motion of water molecules in white matter fibers (Johansen-Berg and Behrens 2009). The primary measure from DTI studies is known as fractional anisotropy, or FA, which is a measure of how tightly water diffusion is constrained in white matter tracts. Higher FA is thought to reflect aspects of healthy white matter (Beaulieu 2002; Lebel et al. 2012), whereas reduced FA is found in various disorders (Treit et al. 2013; Damoiseaux et al. 2009). FA also is associated with cognitive function during development (Treit et al. 2014) and during aging (Zahr et al. 2009; Engvig et al. 2012). Several studies have reported an age reduction in FA of the fornix, which is the major output tract from the HPC (Zahr et al. 2009; Bennett and Stark 2016). FA in the fornix also was correlated with word list learning in older adults (Grambaite et al. 2010). In addition to the fornix, age reductions have been found in FA of the perforant path and the cingulum bundle, which provide input to the HPC from the entorhinal cortex and posterior cingulate, respectively (Bennett and Stark 2016; Yassa et al. 2010). Stronger directional diffusion in these two input tracts predicted better behavioral pattern separation after controlling for age effects on global diffusion (Bennett and Stark 2016). FA in the cingulum bundle also was shown to predict verbal episodic memory in older adults (Ezzati et al. 2016).

Thus, the bulk of the evidence on age-related differences in HPC volume or white matter is consistent with the behavioral literature in showing reductions in older relative to younger adults, and an impact of these reductions on memory performance, including tasks of pattern separation. All HPC subfields may be reduced in volume with age, although the DG and CA1 appear to be the most vulnerable; in contrast, there is as yet no consistent evidence on age vulnerability along the anterior/posterior axis of the HPC. Next we turn to functional studies, where the picture gets considerably more complex.

Age-Related Differences in Hippocampal Function During Memory Tasks

Memory for Items

The majority of experiments assessing age-related differences in HPC activity during memory tasks can be categorized as studies of either memory for single items, or memory for the relations between items or between an item and its context. For item memory we identified ten studies that examined HPC activity during encoding, three during retrieval, and one during both phases of memory. Consistent with the behavioral literature showing relatively spared item memory in older adults, compared to relational memory (for a review, see Old and Naveh-Benjamin 2008a), most of these experiments found no age-related difference in memory performance. During encoding, younger adults were found to have more activation in the HPC than older adults for encoding of words (Dennis et al. 2007a, b; Dennis and Cabeza 2011), and scenes (Ta et al. 2012; Murty et al. 2009), and line drawings of objects (Trivedi et al. 2008b). However, equivalent HPC activation between age groups was seen in other studies involving similar kinds of encoding with multiple types of stimuli, including words (Duverne et al. 2009; Morcom et al. 2003), line drawings of objects (Grady et al. 2003), scenes (Park et al. 2013), and faces (Stevens et al. 2008). This inconsistency in whether or not age-related differences were observed during item encoding does not seem to be due to differences in how HPC activation was analyzed, because a number of these studies used a subsequent memory approach, i.e., they assessed activity during encoding for items that were subsequently correctly remembered. Also, given the variety of stimuli used, the conflicting results appear not to be related to any material specificity effect. All but two of these studies (Trivedi et al. 2008b; Stevens et al. 2008) used incidental encoding tasks, so the nature of the encoding instructions also cannot account for the inconsistency in results. However, it may be related to the part of HPC activated, as we discuss in the summary at the end of the neuroimaging section.

During old/new recognition of items, less HPC activation in older adults was reported for words (Daselaar et al. 2006; Dennis et al. 2008b) and scenes (Murty et al. 2009), whereas no age effect was reported in a study that used both objects and words as stimuli (Wang et al. 2015). Again, stimulus material does not seem to account for the different results. In addition, two of these studies assessed HPC activity specifically for recollection; one found an age-related difference (Daselaar et al. 2006) and the other did not (Wang et al. 2015).

One notable aspect of these item encoding and recognition studies is that none of them report *more* activity in the HPC in older relative to younger adults. This differs from the many experiments that have reported increased activation in older adults in other brain regions, especially but not limited to the PFC (for reviews see Grady 2012; Rajah and D'Esposito 2005), although the impact of this over-recruitment on behavior in older adults is still under debate (Grady 2012; Davis et al. 2012; de Chastelaine et al. 2011). The reported results for item memory studies (including behavioral studies) suggest that the encoding, storage or retrieval of single stimuli involves computations in the HPC that can be successfully carried out in older adults much of the time, even with less overall activation, as long as there is little demand on relational memory and/or the older adults maintain a reasonable level of cognitive health. This latter factor may be critical, as HPC *hyperactivity* has been reported in older individuals with mild cognitive impairment during encoding of single items (Kircher et al. 2007; Trivedi et al. 2008a) compared to cognitively normal older adults. Thus, increased activity in the HPC during item memory in an older adult, above that seen in younger adults, may be a marker of declining function.

Relational Memory

FMRI studies of memory involving some type of contextual relations, such as autobiographical or source memory, mostly find age reductions in performance, in line with the behavioral literature (Naveh-Benjamin 2000; Glisky et al. 2001). Seven studies assessing encoding of context were identified, five assessing memory for stimulus pairs and two that studied source memory. One of the source memory experiments and a study involving encoding of face-name pairs found greater HPC activity in younger than in older adults (Salami et al. 2012; Dennis et al. 2008a). Interestingly, a second source memory experiment found greater HPC activity during encoding in older adults, specifically in a posterior portion of the left HPC (Dulas and Duarte 2011). The remaining four studies examined encoding of word pairs (de Chastelaine et al. 2011; Addis et al. 2014), face-name pairs (Miller et al. 2008), or object pairs (Leshikar et al. 2010), and found no age-related differences in HPC activity. In all seven of these encoding studies younger adults showed better memory for the context than older adults, suggesting that the level of HPC activity during context encoding in older adults, relative to that of young adults, is not strongly related to differences in subsequent memory for the context.

Eleven studies reporting HPC activation in younger and older adults during context retrieval were identified, nine of which found age-related differences in activity levels. Young adults had more HPC activity than older adults during retrieval of word pairs (Giovanello et al. 2010; Giovanello and Schacter 2012), recognition of names and job titles associated with faces (Tsukiura et al. 2011), and retrieval of detailed autobiographical memories (St Jacques et al. 2012; Addis et al. 2011). An additional study of cued source recognition found that young adults activated the HPC during the cue period, whereas the older group activated the HPC during actual retrieval of the source information (Dew et al. 2012). This finding is consistent with the suggestion that an age-related failure to engage in proactive processes early during memory retrieval limits the constraints that can be used to guide retrieval (Velanova et al. 2007), thus necessitating more processing later on (Braver et al. 2009) and hampering relational memory in older adults. A few studies found more HPC activity in older adults compared to younger adults during source retrieval (Morcom et al. 2007; Duverne et al. 2008), and autobiographical retrieval (Maguire and Frith 2003). However this over-recruitment of the HPC in older adults did not seem to aid their performance, as the young out-performed the old on at least some aspect of the task in all three experiments. The two studies that found no age-related difference in HPC activity involved retrieval of face-name pairs (Persson et al. 2011), on which the age groups performed the same, and autobio-graphical retrieval (Martinelli et al. 2013), where young adults retrieved more details than the older adults.

It would thus appear that studies of context retrieval more consistently report age-related differences in HPC activation than studies of context encoding (4/8) or item memory (6/14). The majority of studies of context retrieval (9/11) found an age-related difference, and most of these found a difference in favor of the young group (6/9). This suggests that HPC activity is likely to differ between older and younger adults during retrieval of memories involving contextual details, consistent with their reduced retrieval of these details, but the direction of this difference is less predictable.

Summary and Implications for Memory Performance

In summary, both item and relational memory experiments tend to find more HPC activity in younger adults or equivalent activity in young and older adults, with only a few studies of relational memory finding more activity in older adults. As noted above, this differs from other brain regions that are frequently found to be more active in older adults. For example, some have found that increased PFC activity in older adults is related to less activity elsewhere in the brain, e.g. occipital cortex, suggesting that older adults may compensate for reduced processing effectiveness by increasing cognitive control (e.g., Davis et al. 2008). This over-recruitment of PFC in older adults may reflect an additional demand on control during memory processing, given that memory performance regardless of age depends to some extent on PFC and other areas of cortex, in addition to the HPC (e.g., Brewer et al. 1998; Wagner et al. 1998; de Chastelaine et al. 2011; Jenkins and Ranganath 2010; Kim 2011; Gottlieb and Rugg 2011). Indeed, in some of the memory studies reviewed above, the older adults showed more activity in PFC compared to young adults (Dennis et al. 2007a, b; Morcom et al. 2003; Murty et al. 2009; de Chastelaine et al. 2011; Miller et al. 2008; Leshikar et al. 2010; Giovanello et al. 2010), consistent with the aging literature in general (for reviews see Rajah and D'Esposito 2005; Grady 2012; Park and Reuter-Lorenz 2009). Others reported more activity in older adults in rhinal cortex (Daselaar et al. 2006) or retrosplenial cortex (Dennis et al. 2008a). Most of these experiments did not report correlations between PFC activity and memory performance, but in one study this over-recruitment was associated with better task performance (Murty et al. 2009). However, in three other experiments PFC over-recruitment was only found in those older adults with poorer memory (de Chastelaine et al. 2011; Duverne et al. 2009; Persson et al. 2011), suggesting that if increased engagement of cognitive control occurs in response to reduced function in the HPC, it may not sufficiently compensate for this loss. In contrast, those studies reporting robust behavioral correlations with HPC activation found that this activity was positively associated with performance in both young and older adults (Daselaar et al. 2006; Salami et al. 2012; Mander et al. 2014; Dew et al. 2012; Addis et al. 2011). It would seem then that HPC activation can support memory success regardless of age, although there is still controversy about the benefit of activation elsewhere, particularly in PFC, for memory performance in older adults.

We should note that the somewhat inconsistent results regarding age-related differences in HPC activation in the item and relational memory studies reviewed here could be due to a number of reasons. For example, neuroimaging studies have been hampered over the years by relatively small sample sizes, a problem which has only started to be remedied in the past few years. Small sample sizes, along with variability in how carefully one rules out disorders common in older adults, such as hypertension, may have influenced whether or not a given study reported age-related differences in HPC activity. Another issue is that the presence of various risk factors for Alzheimer's disease, such as APOE4 genotype and amyloid deposition, also can influence brain structure/function and memory in otherwise healthy older adults (e.g., Sheline et al. 2010; Kennedy et al. 2012; Mormino et al. 2012; Brier et al. 2016), and the extent of this influence is typically unknown in most cognitive aging studies. A third possibility is that different subregions of the HPC may be differentially involved in specific memory processes as well as differentially vulnerable to aging (as noted above). This could tax the limits of typical fMRI scanning parameters and analyses, as well as add variability across experiments. This is a possibility that we return to below (in the section on relating age-related differences in structure and function to the role of the HPC in pattern separation/completion). Finally, it may be that mean task-related levels of activity per se in the HPC do not adequately reflect this region's role in either encoding or recognition, and that assessing patterns of activity across the HPC (e.g., using multivoxel pattern analysis as in Carp et al. 2011) or measuring the way in which the HPC is functionally connected to other brain regions may provide a more accurate and consistent picture of age-related differences.

Age-Related Differences in Hippocampal Functional Connectivity

Connectivity During Memory Tasks

Functional connectivity refers to the correlations that exist between activity in a specific brain area, such as the HPC, and other regions under a given experimental condition, or measured at rest. As such, it is thought to reflect communication among brain areas in the service of cognition and is one way of defining spatially and temporally coherent networks (McIntosh 1999; Bressler and Menon 2010). Consistent with the idea that functional connectivity may be more sensitive than mean activity levels to changes with age, all of the experiments examining functional connectivity of the HPC during either encoding or retrieval have found age-related differences. One study found greater connectivity between the HPC and occipital cortex in younger adults during associative encoding of visually complex stimuli despite no age-related differences in HPC activation during the task (Leshikar et al. 2010). Another also found greater HPC-occipital connectivity in young adults during encoding in a specific region of the HPC that showed lower activation in older adults (Dennis et al. 2008a). Interestingly, in both of these studies older adults performed more poorly on subsequent memory tasks, suggesting that reduced communication between visual processing regions and the HPC might have led to poorer representations of the stimuli in memory. A third encoding study (Grady et al. 2003) found greater functional connectivity in young adults between HPC and inferior frontal regions that are important for encoding (Wagner et al. 1998; Grady et al. 1995), whereas the older group had greater connectivity between HPC and dorsolateral PFC, a region of PFC usually thought to underlie cognitive control (e.g., Vincent et al. 2008; Badre and D'Esposito 2007). In addition, performance on subsequent memory tasks was correlated with activity in these age-unique connectivity patterns, suggesting that different HPC connectivity patterns support successful memory as a function of age. The results from these encoding studies suggest that there are age-related changes in the functional connectivity between HPC and other regions that are specific to the encoding task at hand. Importantly, this alteration in functional connectivity may contribute to memory performance in older adults, for better or worse, even if there are no age-related differences in mean levels of HPC activity.

Four studies of HPC functional connectivity during retrieval have reported age-related differences in the functional connectivity of the HPC with prefrontal regions. One study (Mander et al. 2013) examined verbal paired-associate memory after a delay of several hours, and found greater connectivity in young compared to older adults between HPC and ventromedial prefrontal cortex (vmPFC), a region thought to be important as memories become more consolidated over time (Winocur et al. 2010). A unique aspect of this study was that participants learned the material prior to sleeping but were tested after they awoke the next morning. More slow wave activity during sleep was associated with stronger functional connectivity between the HPC and vmPFC, which in turn was correlated with better retention of the material learned the night before, suggesting an interesting link between sleep and memory in older adults via the HPC. A second experiment examined autobiographical retrieval and found that connections between ventral PFC and the HPC were modulated by the richness of the episodic detail that could be retrieved, and that this modulation was greater in young adults (St Jacques et al. 2012). The third experiment (Dew et al. 2012) found that functional connectivity between the HPC and PFC depended on both age and the timing of cued source retrieval; young adults showed stronger connectivity during the cue phase and older adults showed stronger connectivity during subsequent retrieval. This finding is similar to the delay in HPC activation in the older adults reported in this study and mentioned above, and indicates that delayed PFC activity in older adults reported by others in memory tasks (Velanova et al. 2007) may be due to a delay in information coming from the HPC. Finally, young adults have been reported to show stronger functional connectivity between HPC and parietal cortex, consistent with their superior recollective ability, whereas older adults have stronger functional connectivity between perirhinal cortex and PFC, in line with their maintained familiarity (Daselaar et al. 2006). All of these studies, as well as the encoding studies, indicate that there are robust age-related differences in how the HPC interacts with other brain regions during memory tasks, and that these differences influence memory ability. These studies also highlight the utility of assessing functional connectivity of the HPC, in addition to activation levels during memory tasks, in providing a fuller picture of how this region changes with age.

Connectivity During Rest

Measures of intrinsic functional connectivity obtained when participants are at rest have received considerable attention in recent years (e.g., Yeo et al. 2011; Allen et al. 2014). Many studies of intrinsic connectivity have focused on large scale networks such as the default mode network (DMN), which is involved in a number of cognitive processes, such as memory retrieval, self-reference and theory of mind (e.g., Buckner et al. 2008; Raichle et al. 2001; Grigg and Grady 2010; Spreng and Grady 2010; Schacter and Addis 2007a; Andrews-Hanna et al. 2014). The DMN includes vmPFC, posterior cingulate cortex, angular gyri (in the inferior parietal cortex) and the parahippocampal gyrus. The HPC is also often functionally connected with the DMN, although not as consistently as other regions, and evidence suggests that its connection is through the parahippocampal gyrus (Ward et al. 2014). Despite this inconsistent connection between the HPC and the DMN, this connectivity can be important for memory function, as shown in a recent study highlighting this relation (Salami et al. 2014). Salami et al showed that the resting functional connections between the HPC and other DMN nodes were reduced with age in a large longitudinal sample of adults from 20 to 80 years of age. In contrast, connectivity between the right and left HPC was increased with age. This right/left HPC connectivity was particularly enhanced in those older adults who showed typical age-related cognitive decline over several years, and further was associated with reductions in HPC activation during an episodic memory task and in performance on the task (Salami et al. 2014). This finding, along with studies showing reductions of intrinsic HPC-DMN functional connectivity in people with Alzheimer's disease (Greicius et al. 2004), suggests that age-related alterations in intrinsic functional connectivity of the HPC can have important implications for cognitive decline.

Other work also has shown that DMN connectivity is generally reduced in older compared to younger adults and associated with reductions in performance on memory tasks (Andrews-Hanna et al. 2007; Wang et al. 2010; Ferreira and Busatto 2013; Grady et al. 2016). The specific connections involving the HPC with the DMN that are reduced with age include those with posterior cingulate cortex and the angular gyri (Andrews-Hanna et al. 2007), and with vmPFC and anterior cingulate cortex nodes (Wang et al. 2010). Efficiency of information flow through the HPC to other brain regions also is reduced with age (Achard and Bullmore 2007). Perhaps paradoxically, there is some evidence that long range functional connections between the MTL and non-DMN regions are increased with age (Tomasi and Volkow 2012), in contrast to the well-documented reduction with age in connectivity between the HPC and the DMN. This increase is consistent with the intrinsic hyper-connectivity between right and left HPC mentioned above (Salami et al. 2014), which was related to reduced recruitment during encoding and poorer memory performance. Increased limbic connectivity also is consistent with reports of increased variability in the BOLD signal in HPC and parahippocampal gyrus during working memory tasks in older vs younger adults (Garrett et al. 2010; Guitart-Masip et al. 2016) and associated with slower responding on the tasks. Thus, there is increasing evidence that alterations in the dynamic functional connections between the HPC and DMN, as well as those involving the HPC and other brain regions outside the DMN, are important potential mechanisms underlying memory reductions in older adults.

Mapping Age-Related Differences in Structure and Function onto Ideas About the Role of the HPC in Pattern Separation and Completion

The evidence that we have reviewed thus far indicates that behavioral and structural brain measures often show reductions in older compared to younger adults, but that functional activation of the HPC shows a more variable pattern. Nevertheless, if the activation findings are considered in the context of current ideas of HPC processing during memory, some commonalities might emerge. As mentioned above, one prominent hypothesis about HPC function is that it is involved in pattern separation and completion processes, both of which contribute to memory. Pattern separation reflects a lack of overlap in representations of stored inputs, and could signal novelty, whereas pattern completion reflects the "filling-in" of incomplete cue representations by comparing them to previously stored representations, thus facilitating integration of information (Yassa and Stark 2011; Norman and O'Reilly 2003). Some researchers have suggested that the anterior HPC enables integration of different streams of information, or pattern completion, whereas the posterior HPC subserves separate representations of individual items/events via pattern separation (Poppenk et al. 2013; Schlichting et al. 2015). Others have suggested that specific subfields of the HPC carry out pattern separation and completion. One such idea is that the inputs from the DG to CA3 are critical for pattern separation and novelty, whereas outputs from CA3 to CA1, and then to extra-hippocampal structures, are important for pattern completion and more closely related to memory retrieval (Maass et al. 2014; Bakker et al. 2008; Yassa and Stark 2011; Kesner and Rolls 2015).

To take these in turn, we first address potential anterior/posterior age-related differences. Figure 5 summarizes the studies reviewed above to see how age-related differences in HPC activity, or lack of such differences, are expressed along the long axis of the HPC and whether age effects in item or relational memory cluster in anterior or posterior HPC. For item memory there are more reports of age differences in the posterior HPC, all of which are age reductions, whereas for the anterior HPC there are more studies reporting equivalent HPC activity in young and older adults, suggesting that age-related differences in item memory are more prominent in posterior HPC. In contrast, for relational memory the number of reported anterior

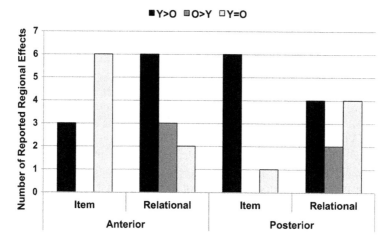

Fig. 5 This figure summarizes the data from the papers reviewed in the text that examined functional activation in the HPC. Each *bar* in the figure represents the total number of times (across all the reviewed papers) a particular effect was reported, whether it was an age-related difference in activity (and which group had the higher activity), or no age-related difference. Effects are reported separately for item and relational memory tasks, and are subdivided by anterior or posterior HPC location. An area of activity was considered anterior if its Y coordinate (in the MNI coordinate system) was equal or anterior to $Y = -20$; the region was considered posterior if its Y coordinate was posterior to -20 (after Poppenk et al. 2013). If a paper reported multiple HPC regions, all were included

HPC regions where young adults have more activity than older adults is twice that with the opposite effect and only a few studies have reported no age-related difference. In posterior HPC the picture is more mixed, with roughly equivalent numbers of relational memory papers reporting the presence or absence of age-related differences. One conclusion that can be drawn from this pattern of results is that if the anterior HPC carries out pattern completion and there tend to be age reductions in anterior HPC during relational memory tasks, then altered pattern completion may impact relational memory in older adults. In contrast, if posterior HPC is involved in pattern separation, then the predominance of age reductions here during item memory could adversely affect formation of distinct item representations in older adults. This tentative conclusion is in line with our suggestion above that aging is accompanied by alterations in both pattern separation and completion processes in the HPC, although the results to date leave many questions unanswered about how these processes are influenced by older age.

To specifically examine HPC subfields in relation to pattern separation some researchers have devised ways of delineating these subfields on structural MRIs and applying this categorization to functional data. These studies have shown that older adults have less activity in the DG/CA3 during tasks designed to tap into the ability to separate similar lures from target stimuli, along with poorer performance on the task (Yassa et al. 2011a, b). In addition, DG volume is related to pattern separation ability in both younger and older adults (Doxey and Kirwan 2015). Similarly, studies comparing the volumes of various HPC subfields in young and older adults suggest that the DG, often combined with CA3 in these studies, is more likely to show age reductions than other subfields, with the possible exception of CA1 (see Fig. 4). A study of HPC function in old monkeys and rats also identified the DG as the HPC subfield with the greatest vulnerability to aging (Small et al. 2004). These studies, taken together, support the intriguing idea that there might be subfield and process specific vulnerabilities in the HPC to advanced age in humans. A recent model of HPC dysfunction in aging proposes that changes are more prominent in DG and CA1 than CA3, and that this, coupled with a decrease in cholinergic input that normally inhibits CA3, causes a relative increase of activity in the auto-associative fibers in CA3, thus tilting the balance to pattern completion at the expense of pattern separation (Wilson et al. 2006). Although the behavioral data, particularly increased false alarms in older adults, do seem to provide support for this idea of overactive pattern completion in older adults, the neuroimaging data reviewed here do not provide any clear evidence of an imbalance between pattern separation and completion, favoring completion, at the neural level in older humans. Such evidence might consist of more HPC activity in older than in young adults in those regions involved in pattern completion (e.g., anterior HPC), but this type of effect is not clearly seen in the data published to date. Future research into the interaction of these two processes would be of considerable interest to the field.

Thus, when looked at in the context of regional function, the neuroimaging data on age-related differences in HPC activity provide some evidence that human aging is associated with functional reductions in regions of the HPC that are important for pattern separation, but also may impact pattern completion. That is, Figs. 4 and 5 suggest that the DG and posterior HPC, both of which have been linked to pattern separation, are vulnerable to age-related changes particularly when representations of single items must be constructed and distinguished from other items that may be similar perceptually. However, pattern completion processes occurring in anterior HPC and CA3/CA1, which have been linked to this process, are also affected (see Figs. 4 and 5) and seem to impact relational memory in particular, perhaps because memory for complex contextual details depends on the ability to correctly "complete the pattern" and integrate information across time and space. Therefore, the safest conclusion to date is that age-related differences in memory are not limited to one portion of the HPC or one function subserved by this region, but to a combination of effects distributed across the long axis of the HPC.

Future Directions

The bulk of the evidence from behavioral and structural MRI studies on the HPC provides a fairly consistent picture of age-related differences in favor of younger adults relative to older adults, i.e., better performance on most hippocampally-mediated tasks, larger volumes of the HPC and many of its subfields, and better measures of diffusion in white matter structures. In contrast, the activation data from functional neuroimaging experiments are more complex, and measures of functional connectivity in particular may hold promise for providing useful insights into the ways in which HPC function can impact memory in older adults. Our view is that at least part of the inconsistency in studies of functional activation in the HPC can be explained by the concept of degeneracy in brain function. Degeneracy is the idea that any given cognitive process can be brought about by multiple patterns of activity in the brain when one looks across individuals (Noppeney et al. 2004). From this point of view, one older adult could have atrophy in the HPC and poor episodic memory, but not necessarily reduced activation of the HPC during an encoding or retrieval task; if the functional connections of the HPC with other memory-related regions, or activity in those other regions themselves were dysfunctional, this could underlie the memory deficit. A second older adult could have smaller HPC volumes and reduced activity in the HPC, but greater engagement of other brain areas, such as PFC, or strong functional connectivity between the HPC and PFC, and show very little memory deficit. The main point is that the functional activity of the brain is highly plastic and adaptable into older age, so that a memory deficit, or lack of one, could come about because of a number of different scenarios of activity involving the HPC and other brain regions. This of course is what the literature to date suggests, making it difficult to predict the outcome of any one fMRI memory experiment in older adults.

However, we should not give up hope of ever finding a clear answer to the question of how age changes in the HPC affect memory in older adults. Indeed, the studies reviewed here suggest that it would be productive to compare tasks tapping into pattern separation and completion in the same participants, to examine whether there is an imbalance in these processes in older adults, accompanied by too much or too little activity in the HPC regions thought to underlie these processes. Additionally, it would be productive to adapt tasks that have been used in the nonhuman animal and behavioral literatures for use in imaging, in order to programmatically understand the cognitive processes that are engaged by other neural regions to modulate HPC function. For example, tasks involving nonlinear problems have not been studied with neuroimaging in older adults, but could provide valuable information regarding HPC dysfunction as well as the neural systems that allow older adults to perform these tasks with the right kind of semantic support. Finally, predictive modeling approaches to fMRI data that have emerged in recent years (such as MVPA) have proven to be useful in defining patterns of brain activity associated with specific memories (Bonnici et al. 2012), and have shown some age-related differences in these specific patterns, although not yet in the HPC (St-Laurent et al. 2014). However, we recently used predictive modeling to show that a pattern of brain activity, including the parahippocampal gyrus, linked to the encoding of pairs of items was less predictable in older than younger adults, and that this ability to predict the encoding pattern in the brain was related to better associative memory for the item pairs (Saverino et al. 2016), indicating that this approach may well be useful for understanding age-related differences in relational memory. Clearly, we still need quite a bit of additional work to understand how functional activity and connectivity in the HPC are modulated during episodic memory tasks in older adults, and to delineate the influence of other brain regions. In short, we need to explore the variety of neural systems and networks that can support memory as a function of age.

Finally, the field needs to consider more fully the range of factors that might influence HPC function in older adults. One potentially critical influence is sleep quality, which declines in older adults (Scullin and Bliwise 2015). As mentioned above, there is evidence that sleep can influence HPC activity and functional connectivity in older adults. In particular, sleep spindles during non-REM sleep are thought to represent a coordinated mechanism that facilitates next day HPC-dependent learning (Walker 2009), and there is evidence that aging reduces sleep spindles in frontal regions, which adversely impacts HPC activity during learning, thus impairing learning ability (Mander et al. 2014). Neurotransmitter systems also change with age, e.g., the density of receptors in the dopamine (Kaasinen et al. 2000) and serotonin systems (Dillon et al. 1991; Meltzer et al. 1998). Both of these neurotransmitters can influence HPC activity (Chowdhury et al. 2012) or volume (O'Hara et al. 2007) and memory in older adults, suggesting that future work will need to identify the impact of age reductions in neurotransmitters on hippocampally-mediated memory function.

In conclusion, there is substantial evidence that episodic memory depends heavily on the HPC, and that at least some older adults have reduced structure and function in the HPC that is associated with poorer memory compared to that seen in younger adults. There is also behavioral and neuroimaging evidence that other brain regions, and the cognitive processes they support, can be utilized by older adults to aid memory function, although we still do not understand the best way to encourage these compensatory mechanisms or which neural systems are responsible for the observed improvements in performance. The major challenge ahead for this field, in addition to broadening our knowledge of the brain mechanisms underlying memory failures and successes in older adults, is to determine how best to harness this knowledge to facilitate the development of improved memory rehabilitation strategies.

References

Achard S, Bullmore E (2007) Efficiency and cost of economical brain functional networks. PLoS Comput Biol 3:174–183

Addis DR, Wong AT, Schacter DL (2007) Remembering the past and imagining the future: common and distinct neural substrates during event construction and elaboration. Neuropsychologia 45:1363–1377

Addis DR, Wong AT, Schacter DL (2008) Age-related changes in the episodic simulation of future events. Psychol Sci 19:33–41

Addis DR, Musicaro R, Pan L, Schacter DL (2010) Episodic simulation of past and future events in older adults: evidence from an experimental recombination task. Psychol Aging 25:369–376

Addis DR, Roberts RP, Schacter DL (2011) Age-related neural changes in autobiographical remembering and imagining. Neuropsychologia 49:3656–3669

Addis DR, Giovanello KS, Vu MA, Schacter DL (2014) Age-related changes in prefrontal and hippocampal contributions to relational encoding. NeuroImage 84:19–26

Allen EA, Damaraju E, Plis SM, Erhardt EB, Eichele T, Calhoun VD (2014) Tracking whole-brain connectivity dynamics in the resting state. Cereb Cortex 24:663–676

Andrews-Hanna JR, Smallwood J, Spreng RN (2014) The default network and self-generated thought: component processes, dynamic control, and clinical relevance. Ann N Y Acad Sci 1316:29–52

Andrews-Hanna JR, Snyder AZ, Vincent JL, Lustig C, Head D, Raichle ME, Buckner RL (2007) Disruption of large-scale brain systems in advanced aging. Neuron 56:924–935

Augustinack JC, van der Kouwe AJ, Salat DH, Benner T, Stevens AA, Annese J, Fischl B, Frosch MP, Corkin S (2014) H.M.'s contributions to neuroscience: a review and autopsy studies. Hippocampus 24:1267–1286

Backus AR, Schoffelen JM, Szebenyi S, Hanslmayr S, Doeller CF (2016) Hippocampal-prefrontal theta oscillations support memory integration. Curr Biol 26:450–457

Badre D, D'Esposito M (2007) Functional magnetic resonance imaging evidence for a hierarchical organization of the prefrontal cortex. J Cogn Neurosci 19:2082–2099

Baker SW, Gao FQ, Vieweg P, Wolbers T, Black SE, Gilboa A, Rosenbaum RS (2015) A necessary role for the dentate gyrus of the hippocampus in pattern separation in humans. Poster presented at the 23rd Annual Cognitive Neuroscience Society Meeting, April 2–5, 2015, New York, NY

Bakker A, Kirwan CB, Miller M, Stark CE (2008) Pattern separation in the human hippocampal CA3 and dentate gyrus. Science 319:1640–1642

Bastin C, Van der Linden M (2005) Memory for temporal context: effects of ageing, encoding instructions, and retrieval strategies. Memory 13:95–109

Bayer ZC, Hernandez RJ, Morris AM, Salomonczyk D, Pirogovsky E, Gilbert PE (2011) Age-related source memory deficits persist despite superior item memory. Exp Aging Res 37:473–480

Beaulieu C (2002) The basis of anisotropic water diffusion in the nervous system – a technical review. NMR Biomed 15:435–455

Bender AR, Naveh-Benjamin M, Raz N (2010) Associative deficit in recognition memory in a lifespan sample of healthy adults. Psychol Aging 25:940–948

Bennett IJ, Stark CE (2016) Mnemonic discrimination relates to perforant path integrity: an ultra-high resolution diffusion tensor imaging study. Neurobiol Learn Mem 129:107–112

Bonnici HM, Chadwick MJ, Lutti A, Hassabis D, Weiskopf N, Maguire EA (2012) Detecting representations of recent and remote autobiographical memories in vmPFC and hippocampus. J Neurosci 32:16982–16991

Braak H, Braak E, Bohl J (1993) Staging of Alzheimer-related cortical destruction. Eur Neurol 33:403–408

Braver TS, Paxton JL, Locke HS, Barch DM (2009) Flexible neural mechanisms of cognitive control within human prefrontal cortex. Proc Natl Acad Sci USA 106:7351–7356

Bressler SL, Menon V (2010) Large-scale brain networks in cognition: emerging methods and principles. Trends Cogn Sci 14:277–290

Brewer JB, Zhao Z, Desmond JE, Glover GH, Gabrieli JDE (1998) Making memories: brain activity that predicts how well visual experience will be remembered. Science 281:1185–1187

Brier MR, Gordon B, Friedrichsen K, McCarthy J, Stern A, Christensen J, Owen C, Aldea P, Su Y, Hassenstab J, Cairns NJ, Holtzman DM, Fagan AM, Morris JC, Benzinger TL, Ances BM (2016) Tau and Abeta imaging, CSF measures, and cognition in Alzheimer's disease. Sci Transl Med 8:338ra366

Brown MW, Aggleton JP (2001) Recognition memory: what are the roles of the perirhinal cortex and hippocampus? Nat Rev Neurosci 2:51–61

Buckner RL (2004) Memory and executive function in aging and AD: multiple factors that cause decline and reserve factors that compensate. Neuron 44:195–208

Buckner RL, Andrews-Hanna JR, Schacter DL (2008) The brain's default network: anatomy, function, and relevance to disease. Ann N Y Acad Sci 1124:1–38

Bunsey M, Eichenbaum H (1996) Conservation of hippocampal memory function in rats and humans. Nature 379:255–257

Carp J, Park J, Polk TA, Park DC (2011) Age differences in neural distinctiveness revealed by multi-voxel pattern analysis. NeuroImage 56:736–743

Carstensen LL (2006) The influence of a sense of time on human development. Science 312:1913–1915

Castel AD, Craik FI (2003) The effects of aging and divided attention on memory for item and associative information. Psychol Aging 18:873–885

Celone KA, Calhoun VD, Dickerson BC, Atri A, Chua EF, Miller SL, DePeau K, Rentz DM, Selkoe DJ, Blacker D, Albert MS, Sperling RA (2006) Alterations in memory networks in mild cognitive impairment and Alzheimer's disease: an independent component analysis. J Neurosci 26:10222–10231

Chastelaine M, Wang TH, Minton B, Muftuler LT, Rugg MD (2011) The effects of age, memory performance, and callosal integrity on the neural correlates of successful associative encoding. Cereb Cortex 21:2166–2176

Chen KH, Chuah LY, Sim SK, Chee MW (2010) Hippocampal region-specific contributions to memory performance in normal elderly. Brain Cogn 72:400–407

Chowdhury R, Guitart-Masip M, Bunzeck N, Dolan RJ, Duzel E (2012) Dopamine modulates episodic memory persistence in old age. J Neurosci 32:14193–14204

Cohen NJ, Eichenbaum H (1993) Memory, amnesia and the hippocampal system. MIT Press, Cambridge

Cohen NJ, Ryan J, Hunt C, Romine L, Wszalek T, Nash C (1999) Hippocampal system and declarative (relational) memory: summarizing the data from functional neuroimaging studies. Hippocampus 9:83–98

D'Angelo MC, Kacollja A, Rabin JS, Rosenbaum RS, Ryan JD (2015) Unitization supports lasting performance and generalization on a relational memory task: evidence from a previously undocumented developmental amnesic case. Neuropsychologia 77:185–200

D'Angelo MC, Smith VM, Kacollja A, Zhang F, Binns MA, Barense MD, Ryan JD (2016a) The effectiveness of unitization in mitigating age-related relational learning impairments depends on existing cognitive status. Neuropsychol Dev Cogn B Aging Neuropsychol Cogn 23:667–690

D'Angelo MC, Rosenbaum RS, Ryan JD (2016b) Impaired inference in the case of developmental amnesia. Hippocampus 26(10):1291–1302

Damoiseaux JS, Smith SM, Witter MP, Sanz-Arigita EJ, Barkhof F, Scheltens P, Stam CJ, Zarei M, Rombouts SA (2009) White matter tract integrity in aging and Alzheimer's disease. Hum Brain Mapp 30:1051–1059

Daselaar SM, Fleck MS, Dobbins IG, Madden DJ, Cabeza R (2006) Effects of healthy aging on hippocampal and rhinal memory functions: an event-related fMRI study. Cereb Cortex 16:1771–1782

Daugherty AM, Bender AR, Raz N, Ofen N (2016) Age differences in hippocampal subfield volumes from childhood to late adulthood. Hippocampus 26:220–228

Davis SW, Dennis NA, Daselaar SM, Fleck MS, Cabeza R (2008) Que PASA? The posterior-anterior shift in aging. Cereb Cortex 18:1201–1209

Davis SW, Kragel JE, Madden DJ, Cabeza R (2012) The architecture of cross-hemispheric communication in the aging brain: linking behavior to functional and structural connectivity. Cereb Cortex 22:232–242

Dennis NA, Cabeza R (2011) Age-related dedifferentiation of learning systems: an fMRI study of implicit and explicit learning. Neurobiol Aging 32(2318):e2317–e2330

Dennis NA, Daselaar S, Cabeza R (2007a) Effects of aging on transient and sustained successful memory encoding activity. Neurobiol Aging 28:1749–1758

Dennis NA, Kim H, Cabeza R (2007b) Effects of aging on true and false memory formation: an fMRI study. Neuropsychologia 45:3157–3166

Dennis NA, Hayes SM, Prince SE, Madden DJ, Huettel SA, Cabeza R (2008a) Effects of aging on the neural correlates of successful item and source memory encoding. J Exp Psychol Learn Mem Cogn 34:791–808

Dennis NA, Kim H, Cabeza R (2008b) Age-related differences in brain activity during true and false memory retrieval. J Cogn Neurosci 20:1390–1402

Dew IT, Buchler N, Dobbins IG, Cabeza R (2012) Where is ELSA? The early to late shift in aging. Cereb Cortex 22:2542–2553

Dillon KA, Gross-Isseroff R, Israeli M, Biegon A (1991) Autoradiographic analysis of serotonin 5-HT1A receptor binding in the human brain postmortem: effects of age and alcohol. Brain Res 554:56–64

Doxey CR, Kirwan CB (2015) Structural and functional correlates of behavioral pattern separation in the hippocampus and medial temporal lobe. Hippocampus 25:524–533

Driscoll I, Hamilton DA, Petropoulos H, Yeo RA, Brooks WM, Baumgartner RN, Sutherland RJ (2003) The aging hippocampus: cognitive, biochemical and structural findings. Cereb Cortex 13:1344–1351

Driscoll I, Howard SR, Prusky GT, Rudy JW, Sutherland RJ (2005) Seahorse wins all races: hippocampus participates in both linear and non-linear visual discrimination learning. Behav Brain Res 164:29–35

Duff MC, Warren DE, Gupta R, Vidal JP, Tranel D, Cohen NJ (2012) Teasing apart tangrams: testing hippocampal pattern separation with a collaborative referencing paradigm. Hippocampus 22:1087–1091

Dulas MR, Duarte A (2011) The effects of aging on material-independent and material-dependent neural correlates of contextual binding. NeuroImage 57:1192–1204

Dusek JA, Eichenbaum H (1997) The hippocampus and memory for orderly stimulus relations. Proc Natl Acad Sci USA 94:7109–7114

Duverne S, Habibi A, Rugg MD (2008) Regional specificity of age effects on the neural correlates of episodic retrieval. Neurobiol Aging 29:1902–1916

Duverne S, Motamedinia S, Rugg MD (2009) The relationship between aging, performance, and the neural correlates of successful memory encoding. Cereb Cortex 19:733–744

Eichenbaum H, Cohen NJ (1988) Representation in the hippocampus: what do hippocampal neurons code? Trends Neurosci 11:244–248

Eichenbaum H, Cohen NJ (2001) From conditioning to conscious recollection: memory systems of the brain. Oxford University Press, New York, NY

Engvig A, Fjell AM, Westlye LT, Moberget T, Sundseth O, Larsen VA, Walhovd KB (2012) Memory training impacts short-term changes in aging white matter: a longitudinal diffusion tensor imaging study. Hum Brain Mapp 33:2390–2406

Ezzati A, Katz MJ, Lipton ML, Zimmerman ME, Lipton RB (2016) Hippocampal volume and cingulum bundle fractional anisotropy are independently associated with verbal memory in older adults. Brain Imaging Behav 10(3):652–659

Ferreira LK, Busatto GF (2013) Resting-state functional connectivity in normal brain aging. Neurosci Biobehav Rev 37:384–400

Fjell AM, McEvoy L, Holland D, Dale AM, Walhovd KB (2014) What is normal in normal aging? Effects of aging, amyloid and Alzheimer's disease on the cerebral cortex and the hippocampus. Prog Neurobiol 117:20–40

Flores R, Joie R, Landeau B, Perrotin A, Mezenge F, Sayette V, Eustache F, Desgranges B, Chetelat G (2015) Effects of age and Alzheimer's disease on hippocampal subfields: comparison between manual and FreeSurfer volumetry. Hum Brain Mapp 36:463–474

Frank MJ, Rudy JW, O'Reilly RC (2003) Transitivity, flexibility, conjunctive representations, and the hippocampus. II. A computational analysis. Hippocampus 13:341–354

Garrett DD, Kovacevic N, McIntosh AR, Grady CL (2010) Blood oxygen level-dependent signal variability is more than just noise. J Neurosci 30:4914–4921

Gilbert PE, Pirogovsky E, Ferdon S, Brushfield AM, Murphy C (2008) Differential effects of normal aging on memory for odor-place and object-place associations. Exp Aging Res 34:437–452

Giovanello KS, Schacter DL (2012) Reduced specificity of hippocampal and posterior ventrolateral prefrontal activity during relational retrieval in normal aging. J Cogn Neurosci 24:159–170

Giovanello KS, Kensinger EA, Wong AT, Schacter DL (2010) Age-related neural changes during memory conjunction errors. J Cogn Neurosci 22:1348–1361

Glisky EL, Rubin SR, Davidson PS (2001) Source memory in older adults: an encoding or retrieval problem? J Exp Psychol Learn Mem Cogn 27:1131–1146

Gold AE, Kesner RP (2005) The role of the CA3 subregion of the dorsal hippocampus in spatial pattern completion in the rat. Hippocampus 15:808–814

Gordon BA, Blazey T, Benzinger TL, Head D (2013) Effects of aging and Alzheimer's disease along the longitudinal axis of the hippocampus. J Alzheimers Dis 37:41–50

Gottlieb LJ, Rugg MD (2011) Effects of modality on the neural correlates of encoding processes supporting recollection and familiarity. Learn Mem 18:565–573

Grady C (2012) The cognitive neuroscience of ageing. Nat Rev Neurosci 13:491–505

Grady CL, McIntosh AR, Horwitz B, Maisog JM, Ungerleider LG, Mentis MJ, Pietrini P, Schapiro MB, Haxby JV (1995) Age-related reductions in human recognition memory due to impaired encoding. Science 269:218–221

Grady CL, McIntosh AR, Craik FI (2003) Age-related differences in the functional connectivity of the hippocampus during memory encoding. Hippocampus 13:572–586

Grady CL, Sarraf S, Saverino C, Campbell KL (2016) Age differences in the functional interactions among the default, frontoparietal control and dorsal attention networks. Neurobiol Aging 41:159–172

Grambaite R, Stenset V, Reinvang I, Walhovd KB, Fjell AM, Fladby T (2010) White matter diffusivity predicts memory in patients with subjective and mild cognitive impairment and normal CSF total tau levels. J Int Neuropsychol Soc 16:58–69

Greicius MD, Srivastava G, Reiss AL, Menon V (2004) Default-mode network activity distinguishes Alzheimer's disease from healthy aging: evidence from functional MRI. Proc Natl Acad Sci USA 101:4637–4642

Grigg O, Grady CL (2010) The default network and processing of personally relevant information: converging evidence from task-related modulations and functional connectivity. Neuropsychologia 48:3815–3823

Guitart-Masip M, Salami A, Garrett D, Rieckmann A, Lindenberger U, Backman L (2016) BOLD variability is related to dopaminergic neurotransmission and cognitive aging. Cereb Cortex 26:2074–2083

Hassabis D, Kumaran D, Vann SD, Maguire EA (2007) Patients with hippocampal amnesia cannot imagine new experiences. Proc Natl Acad Sci USA 104:1726–1731

Head D, Rodrigue KM, Kennedy KM, Raz N (2008) Neuroanatomical and cognitive mediators of age-related differences in episodic memory. Neuropsychology 22:491–507

Healy MR, Light LL, Chung CS (2005) Dual-process models of associative recognition in young and older adults: evidence from receiver operating characteristics. J Exp Psychol Learn Mem Cogn 31:768–788

Holden HM, Gilbert PE (2012) Less efficient pattern separation may contribute to age-related spatial memory deficits. Front Aging Neurosci 4:9. doi:10.3389/fnagi.2012.00009

Holden HM, Hoebel C, Loftis K, Gilbert PE (2012) Spatial pattern separation in cognitively normal young and older adults. Hippocampus 22:1826–1832

Howard MW, Eichenbaum H (2015) Time and space in the hippocampus. Brain Res 1621:345–354

Hsieh LT, Gruber MJ, Jenkins LJ, Ranganath C (2014) Hippocampal activity patterns carry information about objects in temporal context. Neuron 81:1165–1178

Jack CR Jr, Petersen RC, Xu YC, Waring SC, O'Brien PC, Tangalos EG, Smith GE, Ivnik RJ, Kokmen E (1997) Medial temporal atrophy on MRI in normal aging and very mild Alzheimer's disease. Neurology 49:786–794

Jack CR Jr, Knopman DS, Jagust WJ, Petersen RC, Weiner MW, Aisen PS, Shaw LM, Vemuri P, Wiste HJ, Weigand SD, Lesnick TG, Pankratz VS, Donohue MC, Trojanowski JQ (2013) Tracking pathophysiological processes in Alzheimer's disease: an updated hypothetical model of dynamic biomarkers. Lancet Neurol 12:207–216

James LE, Fogler KA, Tauber SK (2008) Recognition memory measures yield disproportionate effects of aging on learning face-name associations. Psychol Aging 23:657–664

Jenkins LJ, Ranganath C (2010) Prefrontal and medial temporal lobe activity at encoding predicts temporal context memory. J Neurosci 30:15558–15565

Jernigan TL, Ostergaard AL, Fennema-Notestine C (2001) Mesial temporal, diencephalic, and striatal contributions to deficits in single word reading, word priming, and recognition memory. J Int Neuropsychol Soc 7:63–78

Johansen-Berg H, Behrens T (eds) (2009) In: Diffusion MRI: from quantitative measurement to in-vivo neuroanatomy. Academic, London

Joie R, Fouquet M, Mezenge F, Landeau B, Villain N, Mevel K, Pelerin A, Eustache F, Desgranges B, Chetelat G (2010) Differential effect of age on hippocampal subfields assessed using a new high-resolution 3T MR sequence. NeuroImage 53:506–514

Kaasinen V, Vilkman H, Hietala J, Nagren K, Helenius H, Olsson H, Farde L, Rinne J (2000) Age-related dopamine D2/D3 receptor loss in extrastriatal regions of the human brain. Neurobiol Aging 21:683–688

Kalpouzos G, Chetelat G, Baron JC, Landeau B, Mevel K, Godeau C, Barre L, Constans JM, Viader F, Eustache F, Desgranges B (2009) Voxel-based mapping of brain gray matter volume and glucose metabolism profiles in normal aging. Neurobiol Aging 30:112–124

Kennedy KM, Rodrigue KM, Devous MD Sr, Hebrank AC, Bischof GN, Park DC (2012) Effects of beta-amyloid accumulation on neural function during encoding across the adult lifespan. NeuroImage 62:1–8

Kerr KM, Agster KL, Furtak SC, Burwell RD (2007) Functional neuroanatomy of the parahippocampal region: the lateral and medial entorhinal areas. Hippocampus 17:697–708

Kersten AW, Earles JL (2010) Effects of aging, distraction, and response pressure on the binding of actors and actions. Psychol Aging 25:620–630

Kesner RP, Rolls ET (2015) A computational theory of hippocampal function, and tests of the theory: new developments. Neurosci Biobehav Rev 48:92–147

Kim H (2011) Neural activity that predicts subsequent memory and forgetting: a meta-analysis of 74 fMRI studies. NeuroImage 54:2446–2461

Kircher TT, Weis S, Freymann K, Erb M, Jessen F, Grodd W, Heun R, Leube DT (2007) Hippocampal activation in patients with mild cognitive impairment is necessary for successful memory encoding. J Neurol Neurosurg Psychiatry 78:812–818

Kirwan CB, Ashby SR, Nash MI (2014) Remembering and imagining differentially engage the hippocampus: a multivariate fMRI investigation. Cogn Neurosci 5:177–185

Klein SB, Loftus J, Kihlstrom JF (2002) Memory and temporal experience: the effects of episodic memory loss on an amnesic patient's ability to remember the past and imagine the future. Soc Cogn 20:353–379

Konkel A, Cohen NJ (2009) Relational memory and the hippocampus: representations and methods. Front Neurosci 3:166–174

Kramer JH, Mungas D, Reed BR, Wetzel ME, Burnett MM, Miller BL, Weiner MW, Chui HC (2007) Longitudinal MRI and cognitive change in healthy elderly. Neuropsychology 21:412–418

Lebel C, Gee M, Camicioli R, Wieler M, Martin W, Beaulieu C (2012) Diffusion tensor imaging of white matter tract evolution over the lifespan. NeuroImage 60:340–352

Lemaitre H, Crivello F, Grassiot B, Alperovitch A, Tzourio C, Mazoyer B (2005) Age- and sex-related effects on the neuroanatomy of healthy elderly. NeuroImage 26:900–911

Leshikar ED, Gutchess AH, Hebrank AC, Sutton BP, Park DC (2010) The impact of increased relational encoding demands on frontal and hippocampal function in older adults. Cortex 46:507–521

Leutgeb JK, Leutgeb S, Moser MB, Moser EI (2007) Pattern separation in the dentate gyrus and CA3 of the hippocampus. Science 315:961–966

Levine B, Svoboda E, Hay J, Winocur G, Moscovitch M (2002) Aging and autobiographical memory: dissociating episodic from semantic retrieval. Psychol Aging 17:677–689

Light LL, Patterson MM, Chung C, Healy MR (2004) Effects of repetition and response deadline on associative recognition in young and older adults. Mem Cognit 32:1182–1193

Lyons AD, Henry JD, Rendell PG, Corballis MC, Suddendorf T (2014) Episodic foresight and aging. Psychol Aging 29:873–884

Maass A, Schutze H, Speck O, Yonelinas A, Tempelmann C, Heinze HJ, Berron D, Cardenas-Blanco A, Brodersen KH, Stephan KE, Duzel E (2014) Laminar activity in the hippocampus and entorhinal cortex related to novelty and episodic encoding. Nat Commun 5:5547

Maass A, Berron D, Libby L, Ranganath C, Duzel E (2015) Functional subregions of the human entorhinal cortex. Elife 4:1–20

Maguire EA, Frith CD (2003) Lateral asymmetry in the hippocampal response to the remoteness of autobiographical memories. J Neurosci 23:5302–5307

Malykhin NV, Bouchard TP, Camicioli R, Coupland NJ (2008) Aging hippocampus and amygdala. Neuroreport 19:543–547

Mander BA, Rao V, Lu B, Saletin JM, Lindquist JR, Ancoli-Israel S, Jagust W, Walker MP (2013) Prefrontal atrophy, disrupted NREM slow waves and impaired hippocampal-dependent memory in aging. Nat Neurosci 16:357–364

Mander BA, Rao V, Lu B, Saletin JM, Ancoli-Israel S, Jagust WJ, Walker MP (2014) Impaired prefrontal sleep spindle regulation of hippocampal-dependent learning in older adults. Cereb Cortex 24:3301–3309

Martinelli P, Sperduti M, Devauchelle AD, Kalenzaga S, Gallarda T, Lion S, Delhommeau M, Anssens A, Amado I, Meder JF, Krebs MO, Oppenheim C, Piolino P (2013) Age-related changes in the functional network underlying specific and general autobiographical memory retrieval: a pivotal role for the anterior cingulate cortex. PLoS One 8:e82385

McDonough IM, Gallo DA (2013) Impaired retrieval monitoring for past and future autobiographical events in older adults. Psychol Aging 28:457–466

McIntosh AR (1999) Mapping cognition to the brain through neural interactions. Memory 7:523–548

McIntyre JS, Craik FIM (1987) Age differences in memory for item and source information. Can J Psychol 41:175–192

Meltzer CC, Smith G, Price JC, Reynolds CF 3rd, Mathis CA, Greer P, Lopresti B, Mintun MA, Pollock BG, Ben-Eliezer D, Cantwell MN, Kaye W, DeKosky ST (1998) Reduced binding of [18F]altanserin to serotonin type 2A receptors in aging: persistence of effect after partial volume correction. Brain Res 813:167–171

Miller SL, Celone K, DePeau K, Diamond E, Dickerson BC, Rentz D, Pihlajamaki M, Sperling RA (2008) Age-related memory impairment associated with loss of parietal deactivation but preserved hippocampal activation. Proc Natl Acad Sci USA 105:2181–2186

Morcom AM, Good CD, Frackowiak RS, Rugg MD (2003) Age effects on the neural correlates of successful memory encoding. Brain 126:213–229

Morcom AM, Li J, Rugg MD (2007) Age effects on the neural correlates of episodic retrieval: increased cortical recruitment with matched performance. Cereb Cortex 17:2491–2506

Mormino EC, Brandel MG, Madison CM, Marks S, Baker SL, Jagust WJ (2012) Abeta deposition in aging is associated with increases in brain activation during successful memory encoding. Cereb Cortex 22:1813–1823

Moses SN, Ryan JD (2006) A comparison and evaluation of the predictions of relational and conjunctive accounts of hippocampal function. Hippocampus 16:43–65

Moses SN, Ostreicher ML, Rosenbaum RS, Ryan JD (2008a) Successful transverse patterning in amnesia using semantic knowledge. Hippocampus 18:121–124

Moses SN, Villate C, Binns MA, Davidson PSR, Ryan JD (2008b) Cognitive integrity predicts transitive inference performance bias and success. Neuropsychologia 46:1314–1325

Moses SN, Ostreicher ML, Ryan JD (2010) Relational framework improves transitive inference across age groups. Psychol Res 74:207–218

Mueller SG, Weiner MW (2009) Selective effect of age, Apo e4, and Alzheimer's disease on hippocampal subfields. Hippocampus 19:558–564

Murty VP, Sambataro F, Das S, Tan HY, Callicott JH, Goldberg TE, Meyer-Lindenberg A, Weinberger DR, Mattay VS (2009) Age-related alterations in simple declarative memory and the effect of negative stimulus valence. J Cogn Neurosci 21:1920–1933

Nadel L, Moscovitch M (1997) Memory consolidation, retrograde amnesia and the hippocampal complex. Curr Opin Neurobiol 7:217–227

Nakashiba T, Cushman JD, Pelkey KA, Renaudineau S, Buhl DL, McHugh TJ, Rodriguez Barrera V, Chittajallu R, Iwamoto KS, McBain CJ, Fanselow MS, Tonegawa S (2012) Young dentate granule cells mediate pattern separation, whereas old granule cells facilitate pattern completion. Cell 149:188–201

Nasreddine ZS, Phillips NA, Bedirian V, Charbonneau S, Whitehead V, Collin I, Cummings JL, Chertkow H (2005) The Montreal Cognitive Assessment, MoCA: a brief screening tool for mild cognitive impairment. J Am Geriatr Soc 53:695–699

Naveh-Benjamin M (2000) Adult age differences in memory performance: tests of an associative deficit hypothesis. J Exp Psychol Learn Mem Cogn 26:1170–1187

Naveh-Benjamin M, Kilb A (2014) Age-related differences in associative memory: the role of sensory decline. Psychol Aging 29:672–683

Naveh-Benjamin M, Hussain Z, Guez J, Bar-On M (2003) Adult age differences in episodic memory: further support for an associative-deficit hypothesis. J Exp Psychol Learn Mem Cogn 29:826–837

Naveh-Benjamin M, Guez J, Kilb A, Reedy S (2004) The associative memory deficit of older adults: further support using face-name associations. Psychol Aging 19:541–546

Naveh-Benjamin M, Shing YL, Kilb A, Werkle-Bergner M, Lindenberger U, Li S-C (2009) Adult age differences in memory for name-face associations: the effects of intentional and incidental learning. Memory 17:220–232

Noppeney U, Friston KJ, Price CJ (2004) Degenerate neuronal systems sustaining cognitive functions. J Anat 205:433–442

Norman KA, O'Reilly RC (2003) Modeling hippocampal and neocortical contributions to recognition memory: a complementary-learning-systems approach. Psychol Rev 110:611–646

O'Brien JL, O'Keefe KM, LaViolette PS, DeLuca AN, Blacker D, Dickerson BC, Sperling RA (2010) Longitudinal fMRI in elderly reveals loss of hippocampal activation with clinical decline. Neurology 74:1969–1976

O'Hara R, Schroder CM, Mahadevan R, Schatzberg AF, Lindley S, Fox S, Weiner M, Kraemer HC, Noda A, Lin X, Gray HL, Hallmayer JF (2007) Serotonin transporter polymorphism, memory and hippocampal volume in the elderly: association and interaction with cortisol. Mol Psychiatry 12:544–555

O'Reilly RC, Norman KA (2002) Hippocampal and neocortical contributions to memory: advances in the complementary learning systems framework. Trends Cogn Sci 6:505–510

Okuda J, Fujii T, Ohtake H, Tsukiura T, Tanji K, Suzuki K, Kawashima R, Fukuda H, Itoh M, Yamadori A (2003) Thinking of the future and past: the roles of the frontal pole and the medial temporal lobes. NeuroImage 19:1369–1380

Old SR, Naveh-Benjamin M (2008a) Differential effects of age on item and associative measures of memory: a meta-analysis. Psychol Aging 23:104–118

Old SR, Naveh-Benjamin M (2008b) Memory for people and their actions: further evidence for an age-related associative deficit. Psychol Aging 23:467–472

Ostreicher ML, Moses SN, Rosenbaum RS, Ryan JD (2010) Prior experience supports new learning of relations in aging. J Gerontol B Psychol Sci Soc Sci 65B:32–41

Overman AA, Becker JT (2009) The associative deficit in older adult memory: recognition of pairs is not improved by repetition. Psychol Aging 24:501–506

Paleja M, Spaniol J (2013) Spatial pattern completion deficits in older adults. Front Aging Neurosci 5:3. doi:10.3389/fnagi.2013.00003

Park DC, Reuter-Lorenz P (2009) The adaptive brain: aging and neurocognitive scaffolding. Annu Rev Psychol 60:173–196

Park H, Kennedy KM, Rodrigue KM, Hebrank A, Park DC (2013) An fMRI study of episodic encoding across the lifespan: changes in subsequent memory effects are evident by middle-age. Neuropsychologia 51:448–456

Pereira JB, Valls-Pedret C, Ros E, Palacios E, Falcon C, Bargallo N, Bartres-Faz D, Wahlund LO, Westman E, Junque C (2014) Regional vulnerability of hippocampal subfields to aging measured by structural and diffusion MRI. Hippocampus 24:403–414

Persson J, Kalpouzos G, Nilsson LG, Ryberg M, Nyberg L (2011) Preserved hippocampus activation in normal aging as revealed by fMRI. Hippocampus 21:753–766

Persson J, Pudas S, Lind J, Kauppi K, Nilsson LG, Nyberg L (2012) Longitudinal structure-function correlates in elderly reveal MTL dysfunction with cognitive decline. Cereb Cortex 22:2297–2304

Poppenk J, Evensmoen HR, Moscovitch M, Nadel L (2013) Long-axis specialization of the human hippocampus. Trends Cogn Sci 17:230–240

Quamme JR, Yonelinas AP, Norman KA (2007) Effect of unitization on associative recognition in amnesia. Hippocampus 17:192–200

Raichle ME, MacLeod AM, Snyder AZ, Powers WJ, Gusnard DA, Shulman GL (2001) A default mode of brain function. Proc Natl Acad Sci USA 98:676–682

Rajah MN, D'Esposito M (2005) Region-specific changes in prefrontal function with age: a review of PET and fMRI studies on working and episodic memory. Brain 128:1964–1983

Rajah MN, Kromas M, Han JE, Pruessner JC (2010) Group differences in anterior hippocampal volume and in the retrieval of spatial and temporal context memory in healthy young versus older adults. Neuropsychologia 48:4020–4030

Raz N, Lindenberger U, Rodrigue KM, Kennedy KM, Head D, Williamson A, Dahle C, Gerstorf D, Acker JD (2005) Regional brain changes in aging healthy adults: general trends, individual differences and modifiers. Cereb Cortex 15:1676–1689

Raz N, Ghisletta P, Rodrigue KM, Kennedy KM, Lindenberger U (2010) Trajectories of brain aging in middle-aged and older adults: regional and individual differences. NeuroImage 51:501–511

Raz N, Daugherty AM, Bender AR, Dahle CL, Land S (2015) Volume of the hippocampal subfields in healthy adults: differential associations with age and a pro-inflammatory genetic variant. Brain Struct Funct 220:2663–2674

Reagh ZM, Roberts JM, Ly M, DiProspero N, Murray E, Yassa MA (2014) Spatial discrimination deficits as a function of mnemonic interference in aged adults with and without memory impairment. Hippocampus 24:303–314

Reed JM, Squire LR (1999) Impaired transverse patterning in human amnesia is a special case of impaired memory for two-choice discrimination tasks. Behav Neurosci 113:3–9

Resnick SM, Pham DL, Kraut MA, Zonderman AB, Davatzikos C (2003) Longitudinal magnetic resonance imaging studies of older adults: a shrinking brain. J Neurosci 23:3295–3301

Rickard TC, Grafman J (1998) Losing their configural mind: amnesic patients fail on transverse patterning. J Cogn Neurosci 10:509–524

Rickard TC, Verfaellie M, Grafman J (2006) Transverse patterning and human amnesia. J Cogn Neurosci 18:1723–1733

Roberts JM, Ly M, Murray E, Yassa MA (2014) Temporal discrimination deficits as a function of lag interference in older adults. Hippocampus 24:1189–1196

Rodrigue KM, Daugherty AM, Haacke EM, Raz N (2013) The role of hippocampal iron concentration and hippocampal volume in age-related differences in memory. Cereb Cortex 23:1533–1541

Rosenbaum RS, Köhler S, Schacter DL, Moscovitch M, Westmacott R, Black SE, Gao F, Tulving E (2005) The case of K.C.: contributions of a memory-impaired person to memory theory. Neuropsychologia 43:989–1021

Ryan JD, Leung G, Turk-Browne NB, Hasher L (2007) Assessment of age-related changes in inhibition and binding using eye movement monitoring. Psychol Aging 22:239–250

Ryan JD, Moses SN, Villate C (2009) Impaired relational organization of propositions, but intact transitive inference, in aging: implications for understanding underlying neural integrity. Neuropsychologia 47:338–353

Ryan JD, Moses SN, Barense M, Rosenbaum RS (2013) Intact learning of new relations in amnesia as achieved through unitization. J Neurosci 33:9601–9613

Ryan JD, D'Angelo MC, Kamino D, Ostreicher M, Moses SN, Rosenbaum RS (2016) Relational learning and transitive expression in aging and amnesia. Hippocampus 26:170–184

Sabuncu MR, Desikan RS, Sepulcre J, Yeo BT, Liu H, Schmansky NJ, Reuter M, Weiner MW, Buckner RL, Sperling RA, Fischl B, Alzheimer's Disease Neuroimaging I (2011) The dynamics of cortical and hippocampal atrophy in Alzheimer disease. Arch Neurol 68:1040–1048

Salami A, Eriksson J, Nyberg L (2012) Opposing effects of aging on large-scale brain systems for memory encoding and cognitive control. J Neurosci 32:10749–10757

Salami A, Pudas S, Nyberg L (2014) Elevated hippocampal resting-state connectivity underlies deficient neurocognitive function in aging. Proc Natl Acad Sci USA 111:17654–17659

Santoro A (2013) Reassessing pattern separation in the dentate gyrus. Front Behav Neurosci 7:95. doi:10.3389/fnbeh.2013.00096

Saverino C, Fatima Z, Sarraf S, Oder A, Strother SC, Grady CL (2016) The associative deficit in aging: insights from functional neuroimaging. J Cogn Neurosci. 28(9):1331–1344

Schacter DL, Addis DR (2007a) The cognitive neuroscience of constructive memory: remembering the past and imagining the future. Philos Trans R Soc Lond Ser B Biol Sci 362:773–786

Schacter DL, Addis DR (2007b) The ghosts of past and future: a memory that works by piecing together bits of the past may be better suited to simulating future events than one that is a store of perfect records. Nature 445:27

Schacter DL, Addis DR (2007c) On the constructive episodic simulation of past and future events. Behav Brain Sci 30:331–332

Schacter DL, Harbluk JL, McLachlan DR (1984) Retrieval without recollection: an experimental analysis of source amnesia. J Verbal Learn Verbal Behav 23:593

Schacter DL, Addis DR, Buckner RL (2007) Remembering the past to imagine the future: the prospective brain. Nat Rev Neurosci 8:657–661

Schacter DL, Gaesser B, Addis DR (2013) Remembering the past and imagining the future in the elderly. Gerontology 59:143–151

Schlichting ML, Mumford JA, Preston AR (2015) Learning-related representational changes reveal dissociable integration and separation signatures in the hippocampus and prefrontal cortex. Nat Commun 6:8151

Scullin MK, Bliwise DL (2015) Sleep, cognition, and normal aging: integrating a half century of multidisciplinary research. Perspect Psychol Sci 10:97–137

Sheline YI, Morris JC, Snyder AZ, Price JL, Yan Z, D'Angelo G, Liu C, Dixit S, Benzinger T, Fagan A, Goate A, Mintun MA (2010) APOE4 allele disrupts resting state fMRI connectivity in the absence of amyloid plaques or decreased CSF Abeta42. J Neurosci 30:17035–17040

Shing YL, Rodrigue KM, Kennedy KM, Fandakova Y, Bodammer N, Werkle-Bergner M, Lindenberger U, Raz N (2011) Hippocampal subfield volumes: age, vascular risk, and correlation with associative memory. Front Aging Neurosci 3:2. doi:10.3389/fnagi.2011.00002

Simic G, Kostovic I, Winblad B, Bogdanovic N (1997) Volume and number of neurons of the human hippocampal formation in normal aging and Alzheimer's disease. J Comp Neurol 379:482–494

Small SA, Chawla MK, Buonocore M, Rapp PR, Barnes CA (2004) Imaging correlates of brain function in monkeys and rats isolates a hippocampal subregion differentially vulnerable to aging. Proc Natl Acad Sci USA 101:7181–7186

Smith C, Squire LR (2005) Declarative memory, awareness, and transitive inference. J Neurosci 25:10138–10146

Soei E, Daum I (2008) Course of relational and non-relational recognition memory across the adult lifespan. Learn Mem 15:21–28

Sperling R (2007) Functional MRI studies of associative encoding in normal aging, mild cognitive impairment, and Alzheimer's disease. Ann N Y Acad Sci 1097:146–155

Spreng RN, Grady CL (2010) Patterns of brain activity supporting autobiographical memory, prospection, and theory of mind, and their relationship to the default mode network. J Cogn Neurosci 22:1112–1123

Spreng RN, Levine B (2006) The temporal distribution of past and future autobiographical events across the lifespan. Mem Cogn 34:1644–1651

Squire LR (1992) Memory and the hippocampus: a synthesis from findings with rats, monkeys, and humans. Psychol Rev 99:195–231

Squire LR, Stark CE, Clark RE (2004) The medial temporal lobe. Annu Rev Neurosci 27:279–306

St Jacques PL, Rubin DC, Cabeza R (2012) Age-related effects on the neural correlates of autobiographical memory retrieval. Neurobiol Aging 33:1298–1310

Stark SM, Yassa MA, Lacy JW, Stark CE (2013) A task to assess behavioral pattern separation (BPS) in humans: data from healthy aging and mild cognitive impairment. Neuropsychologia 51:2442–2449

Stark SM, Stevenson RA, Wu CC, Rutledge S, Stark CE (2015) Stability of age-related deficits in the mnemonic similarity task across task variations. Behav Neurosci 129:257–268

Stevens WD, Hasher L, Chiew K, Grady CL (2008) A neural mechanism underlying memory failure in older adults. J Neurosci 28:12820–12824

St-Laurent M, Abdi H, Bondad A, Buchsbaum BR (2014) Memory reactivation in healthy aging: evidence of stimulus-specific dedifferentiation. J Neurosci 34:4175–4186

Sutherland RJ, Rudy JW (1989) Configural association theory: the role of the hippocampal formation in learning, memory, and amnesia. Psychobiology 17:129–144

Suzuki WA, Amaral DG (1994) Perirhinal and parahippocampal cortices of the macaque monkey: cortical afferents. J Comp Neurol 350:497–533

Szpunar KK, Watson JM, McDermott KB (2007) Neural substrates of envisioning the future. Proc Natl Acad Sci USA 104:642–647

Ta AT, Huang SE, Chiu MJ, Hua MS, Tseng WY, Chen SH, Qiu A (2012) Age-related vulnerabilities along the hippocampal longitudinal axis. Hum Brain Mapp 33:2415–2427

Tomasi D, Volkow ND (2012) Aging and functional brain networks. Mol Psychiatry 17:549–558

Treit S, Lebel C, Baugh L, Rasmussen C, Andrew G, Beaulieu C (2013) Longitudinal MRI reveals altered trajectory of brain development during childhood and adolescence in fetal alcohol spectrum disorders. J Neurosci 33:10098–10109

Treit S, Chen Z, Rasmussen C, Beaulieu C (2014) White matter correlates of cognitive inhibition during development: a diffusion tensor imaging study. Neuroscience 276:87–97

Trivedi MA, Murphy CM, Goetz C, Shah RC, Gabrieli JD, Whitfield-Gabrieli S, Turner DA, Stebbins GT (2008a) fMRI activation changes during successful episodic memory encoding and recognition in amnestic mild cognitive impairment relative to cognitively healthy older adults. Dement Geriatr Cogn Disord 26:123–137

Trivedi MA, Schmitz TW, Ries ML, Hess TM, Fitzgerald ME, Atwood CS, Rowley HA, Asthana S, Sager MA, Johnson SC (2008b) fMRI activation during episodic encoding and metacognitive appraisal across the lifespan: risk factors for Alzheimer's disease. Neuropsychologia 46:1667–1678

Tse D, Langston RF, Kakeyama M, Bethus I, Spooner PA, Wood ER, Witter MP, Morris RGM (2007) Schemas and memory consolidation. Science 316:76–82

Tse D, Takeuchi T, Kakeyama M, Kajii Y, Okuno H, Tohyam C, Bito H, Morris RGM (2011) Schema-dependent gene activation and memory encoding in neocortex. Science 333:891–895

Tsukiura T, Sekiguchi A, Yomogida Y, Nakagawa S, Shigemune Y, Kambara T, Akitsuki Y, Taki Y, Kawashima R (2011) Effects of aging on hippocampal and anterior temporal activations during successful retrieval of memory for face-name associations. J Cogn Neurosci 23:200–213

Tulving E (1985) Memory and consciousness. Can Psychol 26:1–12

van Elzakker M, O'Reilly RC, Rudy JW (2003) Transitivity, flexibility, conjunctive representations, and the hippocampus. I. An empirical analysis. Hippocampus 13:334–340

van Kesteren MT, Fernandez G, Norris DG, Hermans EJ (2010) Persistent schema-dependent hippocampal-neocortical connectivity during memory encoding and postencoding rest in humans. Proc Natl Acad Sci USA 107:7550–7555

van Strien NM, Cappaert NL, Witter MP (2009) The anatomy of memory: an interactive overview of the parahippocampal-hippocampal network. Nat Rev Neurosci 10:272–282

Velanova K, Lustig C, Jacoby LL, Buckner RL (2007) Evidence for frontally mediated controlled processing differences in older adults. Cereb Cortex 17:1033–1046

Vieweg P, Stangl M, Howard LR, Wolbers T (2015) Changes in pattern completion—a key mechanism to explain age-related recognition memory deficits? Cortex 64:343–351

Vincent JL, Kahn I, Snyder AZ, Raichle ME, Buckner RL (2008) Evidence for a frontoparietal control system revealed by intrinsic functional connectivity. J Neurophysiol 100:3328–3342

Voineskos AN, Winterburn JL, Felsky D, Pipitone J, Rajji TK, Mulsant BH, Chakravarty MM (2015) Hippocampal (subfield) volume and shape in relation to cognitive performance across the adult lifespan. Hum Brain Mapp 36:3020–3037

Wagner AD, Schacter DL, Rotte M, Koutstaal W, Maril A, Dale AM, Rosen BR, Buckner RL (1998) Building memories: remembering and forgetting of verbal experiences as predicted by brain activity. Science 281:1188–1191

Walker MP (2009) The role of sleep in cognition and emotion. Ann N Y Acad Sci 1156:168–197

Wang L, Laviolette P, O'Keefe K, Putcha D, Bakkour A, Van Dijk KR, Pihlajamaki M, Dickerson BC, Sperling RA (2010) Intrinsic connectivity between the hippocampus and posteromedial cortex predicts memory performance in cognitively intact older individuals. NeuroImage 51:910–917

Wang WC, Dew IT, Cabeza R (2015) Age-related differences in medial temporal lobe involvement during conceptual fluency. Brain Res 1612:48–58

Ward AM, Schultz AP, Huijbers W, Van Dijk KR, Hedden T, Sperling RA (2014) The parahippocampal gyrus links the default-mode cortical network with the medial temporal lobe memory system. Hum Brain Mapp 35:1061–1073

Ward AM, Mormino EC, Huijbers W, Schultz AP, Hedden T, Sperling RA (2015) Relationships between default-mode network connectivity, medial temporal lobe structure, and age-related memory deficits. Neurobiol Aging 36:265–272

West MJ, Coleman PD, Flood DG, Troncoso JC (1994) Differences in the pattern of hippocampal neuronal loss in normal ageing and Alzheimer's disease. Lancet 344:769–772

Wilson IA, Gallagher M, Eichenbaum H, Tanila H (2006) Neurocognitive aging: prior memories hinder new hippocampal encoding. Trends Neurosci 29:662–670

Winocur G, Moscovitch M, Bontempi B (2010) Memory formation and long-term retention in humans and animals: convergence towards a transformation account of hippocampal-neocortical interactions. Neuropsychologia 48:2339–2356

Wisse LE, Biessels GJ, Heringa SM, Kuijf HJ, Koek DH, Luijten PR, Geerlings MI (2014) Hippocampal subfield volumes at 7T in early Alzheimer's disease and normal aging. Neurobiol Aging 35:2039–2045

Wynne CDL (1997) Pigeon transitive inference: tests of simple accounts of a complex performance. Behav Processes 39:95–112

Wynne CD, von Fersen L, Staddon JE (1992) Pigeons' inferences are transitive and the outcome of elementary conditioning principles: a response. J Exp Psychol Anim Behav Process 18:313–315

Yassa MA, Stark CE (2011) Pattern separation in the hippocampus. Trends Neurosci 34:515–525
Yassa MA, Muftuler LT, Stark CE (2010) Ultrahigh-resolution microstructural diffusion tensor imaging reveals perforant path degradation in aged humans in vivo. Proc Natl Acad Sci USA 107:12687–12691
Yassa MA, Lacy JW, Stark SM, Albert MS, Gallagher M, Stark CE (2011a) Pattern separation deficits associated with increased hippocampal CA3 and dentate gyrus activity in nondemented older adults. Hippocampus 21:968–979
Yassa MA, Mattfeld AT, Stark SM, Stark CE (2011b) Age-related memory deficits linked to circuit-specific disruptions in the hippocampus. Proc Natl Acad Sci USA 108:8873–8878
Yeo BT, Krienen FM, Sepulcre J, Sabuncu MR, Lashkari D, Hollinshead M, Roffman JL, Smoller JW, Zollei L, Polimeni JR, Fischl B, Liu H, Buckner RL (2011) The organization of the human cerebral cortex estimated by intrinsic functional connectivity. J Neurophysiol 106:1125–1165
Zahr NM, Rohlfing T, Pfefferbaum A, Sullivan EV (2009) Problem solving, working memory, and motor correlates of association and commissural fiber bundles in normal aging: a quantitative fiber tracking study. NeuroImage 44:1050–1062
Ziegler G, Dahnke R, Jancke L, Yotter RA, May A, Gaser C (2012) Brain structural trajectories over the adult lifespan. Hum Brain Mapp 33:2377–2389

Physical Activity and Cognitive Training: Impact on Hippocampal Structure and Function

Rachel Clark, Christopher Wendel, and Michelle W. Voss

Abstract This chapter will review the current state of knowledge on the effects of physical and mental (cognitive) training on hippocampal structure and function. We will primarily focus on normal aging and patient populations, though some relevant examples with young adults will also be described. Where possible, we will briefly review relevant research with animal models, in order to discuss potential mechanisms for beneficial effects of physical activity and cognitive training on hippocampal health.

Introduction

Normal age-related cognitive decline occurs for most individuals (Park and Reuter-Lorenz 2009; Salthouse 2010) and this can have a negative impact on quality of life and independence. In addition, age-related neurological and neurodegenerative conditions such as Alzheimer's Disease are associated with enormous societal cost in terms of morbidity, mortality, loss of independence, loss of employment, and caregiving costs (Barnes and Yaffe 2011; Hurd et al. 2013). Furthermore, because the risk of these disorders increases with age, this problem will likely surge as our population becomes older. The hippocampus is central to these public health concerns because normal aging is known to affect hippocampal structure and function (Nyberg et al. 2012). Furthermore, the hippocampus is a central region where pathology develops in Alzheimer's Disease (Jack et al. 2013), the leading age-related neurodegenerative disorder affecting older adults (Brookmeyer et al. 2011). This increase in the aging population and age-related diseases that affect the hippocampus raises questions about the extent to which such changes in the brain with aging and disease are inevitable or whether they can be prevented, delayed, or

R. Clark
Interdisciplinary Graduate Program in Neuroscience, University of Iowa, Iowa City, IA, USA

C. Wendel • M.W. Voss (✉)
Department of Psychological and Brain Sciences, University of Iowa, Iowa City, IA, USA
e-mail: michelle-voss@uiowa.edu

© Springer International Publishing AG 2017 209
D.E. Hannula, M.C. Duff (eds.), *The Hippocampus from Cells to Systems*,
DOI 10.1007/978-3-319-50406-3_8

improved (e.g., Hertzog et al. 2009; Norton et al. 2014). Questions like these have been historically conceptualized under theories of a lifespan view of plasticity (Lindenberger 2014; Lövdén et al. 2010). In general, plasticity can be viewed as the capacity for change within the individual at multiple levels of analysis (e.g., cellular, systems, behavioral), and the concept of lifespan plasticity extends this capacity of modifiability throughout the lifespan (Lerner 1984; Raz and Lindenberger 2013; Reuter-Lorenz and Park 2014).

Recently Lövden and colleagues proposed a theoretical framework for the study of adult cognitive plasticity that synthesized historical views of lifespan plasticity and asserts several predictions for its realization and measurement (Lövdén et al. 2010). First, they operationalize *plasticity* as the capacity for reactive change, where reactive means in response to a stimulus such as experience or brain injury. A distinction is made between the stimulus and the response to this stimulus. For example, with age-related neurodegeneration or brain injury, the damage itself is the stimulus whereas the actual response of the organism (e.g., repair, compensation) reflects plasticity. With learning new skills, the perceptual demands and the representations during practice would not constitute plasticity, but rather it is the secondary response (improved performance, structural and functional brain alterations) that reflect the plasticity of the organism. Thus, plasticity processes are one component of successful aging because they enable an adaptive response to the brain aging process. Another concept related to successful aging is *flexibility*, which denotes the capacity for the organism to meet the demands of the current context using existing structural and functional resources, limited by existing constraints in the system (Lövdén et al. 2010). The concept of flexibility acknowledges the dynamic range of brain processes that meet everyday demands of cognitive function, while reserving the notion of adaptive plasticity for when this range of performance and functioning has been increased. Finally, in addition to plasticity and flexibility, an equally important process for successful aging is *brain mainte-nance*, whereby the primary stimulus of brain aging is delayed or slowed and therefore this concept focuses more on postponing age-related changes and pathology rather than on how the brain copes with their presence (Lindenberger 2014; Nyberg et al. 2012).

Given that the hippocampus is a region highly vulnerable to the effects of aging, yet shows tremendous variability in age-related decline in structure and function (suggesting maintenance is possible), and has been shown to increase in size or function in response to interventions (suggesting plasticity is possible), the region represents a valuable test case for determining what types of environmental and lifestyle factors optimize lifespan plasticity (Lövdén et al. 2010; Nyberg et al. 2012). Within this context, the goal of this chapter is to review the current state of knowledge on the effects of physical and mental (cognitive) training on hippo-campal structure and function. We primarily focus on normal aging and patient populations, though some relevant examples with young adults are also described. With respect to brain maintenance, we will evaluate whether the evidence supports the preservation of more youth-like brain structure and function. With respect to plasticity processes, we will evaluate whether evidence supports the restoration of structural or functional circuits known to be vulnerable to aging or supports a compensatory response (i.e., creating new circuits in response to primary aging-related losses of structure and function). This theoretical foundation will create the basis for evaluating the methodological strengths and weaknesses of the empirical literature and for identifying important directions for future research.

Furthermore, for brain maintenance and plasticity to have their broadest impact, it is ideal for lifestyle and intervention-related changes to affect a wide range of cognitive abilities, rather than only skills and abilities that were the targets of training. This type of wide-ranging impact is known as *transfer*, where the target (s) of improvement for training have the capacity to extend to a wide range of functional skills and abilities beyond only the trained capacity (Lustig et al. 2009; Noack et al. 2014). The hippocampus has tremendous potential as an intervention target that could engender wide transfer because of its vulnerabilities to aging (e.g., Nyberg et al. 2012; Persson et al. 2012) and its role in a broad range of skills and abilities that require forming new relational memories that can be flexibly reconfigured (Konkel et al. 2008; Shohamy and Turk-Browne 2013). In general, the concept of aiming interventions at brain regions that could serve as a center for overlap with a broad range of cognitive abilities has been proposed. This proof of concept was demonstrated with the striatum in the context of working memory training (Dahlin et al. 2008, 2009). Therefore, we will evaluate whether there is also evidence supporting the proposal that the hippocampus is a powerful target of interventions because of its potential for broad transfer of training. The easiest case for this concept is non-cognitive training such as physical exercise training that induces change in hippocampal structure and function that is then beneficial for forming new relational memories (e.g., Erickson et al. 2011; Maass et al. 2014; Pereira et al. 2007).

Finally, where possible, we will integrate relevant research with animal models, in order to discuss potential *cellular and molecular mechanisms* for beneficial effects of physical activity and cognitive training on hippocampal health. Such discussion is important for bridging between basic and cognitive mechanisms of *how* interventions with great translational potential engender maintenance, plasticity, and transfer (Voss et al. 2013b).

Cognitive Training

Cognitive training protocols are designed to challenge cognitive functions in order to cause a mismatch between supply and demand in the brain that stimulates structural or functional brain plasticity (Lövdén et al. 2010). Training typically consists of scheduled routines of adaptive mental exercises or games that are either delivered in person or on a computer. Strategy training involves teaching the participant a particular strategy in order to improve performance on a task (e.g. training method of loci to improve memory performance). Process training, on the other hand, involves targeting specific cognitive processes, without explicit strategy training. Process training programs typically include many tasks that place heavy demands on a particular process (e.g. working memory).

As described above, a crucial aspect of cognitive training is the potential for transfer of benefits to untrained tasks. Transfer may depend on overlapping activation across multiple functions, such that when a region underlying the trained task experiences increased volume or altered activation, other functions that engage the same region(s) also demonstrate benefits. Depending on the level of similarity to the trained task, transfer can be defined as near or far. Near transfer is demonstrated when the training of a particular task affects performance on a task that is similar in stimuli, strategy, or outcome. Far transfer, on the other hand, is demonstrated when training of a particular task affects performance on a task that has few common elements with the trained task. Similarity can span many domains (such as knowledge, physical context, temporal context, functional context, social context, and modality), each separately influencing the distinction of near or far transfer (Zelinski 2009). A relevant analogy of near transfer might be training of a forehand swing in tennis followed by improved performance on backhand tennis swings. There are many common elements across dimensions of stimulus-response cues and the movements required to strike the ball. Alternatively, far transfer in this case would be transferring common elements across sports such as applying basics of ball striking to a golf or baseball swing or applying lateral movements from tennis to the same type of movements in basketball (Perkins and Salomon 1992). Similarly, many tasks across varied cognitive domains include demands for rapid and flexible relational binding, such as acquisition of complex task sets, and this could be a basis for far transfer.

An important emerging market for cognitive training is the aging population. Because normal aging is accompanied by deterioration of brain structure (Hedden et al. 2014; Raz et al. 2005) and function (Rieckmann et al. 2011; Shaw et al. 2015), and a broad range of corresponding cognitive abilities (Park and Reuter-Lorenz 2009; Salthouse 2010), cognitive training may offer an especially helpful tool for maintaining or improving mental processes in this population. Within this context, as introduced above, the hippocampus is an important target for training given its central role in Alzheimer's Disease and its capacity for plasticity across the lifespan (Jack et al. 2013). In order to evaluate the extent to which cognitive training has been shown to affect the hippocampus, below we only describe cognitive training studies that included analysis of brain structure or function. We describe evidence from studies where training was focused on functions known to depend on the hippocampus and we consider studies that have focused on multi-domain cognitive training (such as playing video games) or process-specific training of cognitive functions such as working memory training.

Changes in Hippocampal Structure Following Cognitive Training

Although many studies have demonstrated behavioral improvement on both trained and untrained tasks (near transfer) in response to memory training (for review see

Lustig et al. 2009), fewer studies have examined how the brain changes in response to some type of memory training in healthy older adults. However, a few seminal studies of experience-induced brain plasticity should be considered as they found evidence of changes in the hippocampus.

A series of studies investigated the changes induced by visuomotor training in both young and older adults. Draganski et al. (2004) found that in 24 healthy college aged adults, 3 months of juggling practice (to reach juggling fluency), compared to no juggling practice, increased grey matter in the bilateral mid-temporal area and the left posterior intraparietal sulcus. These volumetric changes were positively related to juggling performance and were also transient, such that the grey matter increase vanished after 3 months of no further practice. In a larger study of older adults (n = 50) using the same 3-month juggling training protocol, Boyke et al. (2008) found similar training-related transient grey matter increases in mid-temporal area, but in addition they found increased volume in the left hippo-campus and bilateral nucleus accumbens. However, the increase in volume was not related to post-training juggling performance or time spent practicing the skill, unlike the positive relationship observed in young adults in other parts of the brain. Further, a much lower percentage of older adults than younger adults became proficient in juggling following training (16% for old, 100% for young), possibly indicating lower limits of potential plasticity. The authors state that hippocampal changes could have resulted from the motor movements, learning and/or spatial skills associated with juggling. Although the hippocampus was not, at the time of these studies, predicted to be involved with juggling, current understanding of hippocampal involvement in motor learning supports and may help explain this finding (Doyon et al. 2009; Schendan et al. 2003).

Based on the involvement of the hippocampus in learning and memorizing abstract and declarative information (Eichenbaum 2004; Squire et al. 2004), Draganski et al. (2006) predicted that extensive studying of medical information in young adults would induce volumetric increases in the parietal lobe and hippo-campus. The authors found that 3 months of daily studying in medical students, compared to 3 months during which the control cohort of dental students was not studying for exams, resulted in significant hippocampal grey matter increases. Further, significant increase in hippocampal volume was also found at a 3-month follow up. Given that much less studying presumably occurred between the exam and follow-up date, this result demonstrated consistent plasticity even after the learning period. While hippocampal change was not related to performance on the exam, the authors did not control for additional variables such as IQ, learning strategies, or workload, which may account for some variation in performance.

This series of studies begins to characterize how long it might take to modify the hippocampus and how long that change may last. Although the studies were performed with different populations, with both juggling and extensive studying, 3 months was sufficient to induce and measure changes in human hippocampal volume with MRI methods for measuring brain structure.

More recent studies have better examined the impact of experience with cogni-tive functions primarily thought to be hippocampal-dependent. Based on cross-sectional findings that healthy adult London taxi cab drivers had larger hippocampi than comparison individuals, Woollett and Maguire (2011) found that extensive spatial learning over a 3–4 year period resulted in not only increased volume in the posterior hippocampi, but also improved performance on a spatial relations test. In addition, compared to a control sample that either studied but did not pass examinations or had no spatial training over the 3–4 year period, individuals with spatial training performed worse on a complex figure test, a measure of free recall of spatial material after a short delay. Authors explained this pattern of results as a possible trade-off between different spatial abilities. Indeed, the increased volume of the posterior hippocampi was accompanied by decreased volume of the anterior hippocampi (Woollett and Maguire 2011). Although this study lacked a true experimental design, the brain imaging data prior to and following spatial training provide evidence for training-related hippocampal plasticity. Though unfortunately aside from the complex figure test, no other cognitive functions were assessed, which prevents an evaluation of far transfer of training.

Further, it has also been shown that after retiring from taxi driving, hippocampal volume and spatial memory tend to return to normal levels (Woollett et al. 2009). Given that structural plasticity is thought to be dependent on continued use of the hippocampus, this re-normalization is not surprising, and supports that real-life experiences can trigger transient changes in hippocampal volume. This result aligns with the finding that juggling practice in older adults led to increased hippocampal grey matter, followed by decreased volume after termination of practice (Draganski et al. 2004).

As a brief methodological note, the studies summarized above all used Voxel-Based Morphometry (VBM), a technique designed to evaluate local changes in brain structure as a function of training and which allows testing for changes across the whole brain. The advantages of VBM are that it is in principle fully automated and therefore has perfect repeatability, and it does not require time-consuming manual tracing of anatomical structures (Kennedy et al. 2009). However, there are also some limitations to VBM analyses, which include vulnerabilities to bias in areas that commonly have image artifacts and registration errors such as the medial temporal lobes (for methodological reviews, see: Bookstein 2001; Davatzikos 2004; Thomas et al. 2009).

Instead of looking across the whole brain for training-induced change in volume, another approach is to evaluate change in volume of a defined anatomical structure (for methodological reviews, see: Morey et al. 2009; Mulder et al. 2014; Schoemaker et al. 2016; Wenger et al. 2014). For example, a study by Lövdén et al. (2012) used techniques for manually tracing the hippocampus from the rest of the brain and predicted that extensive, long-term engagement of the hippocampus through virtual environment spatial navigation training would modify hippocampal volumes more than an active, no spatial navigation training, control group. Forty-four healthy young and 47 healthy older men were included in the analyses. Individuals in the spatial navigation group learned to navigate through a virtual zoo to learn and memorize the locations of certain animals, while the control group walked comfortably on a treadmill. Both groups completed 42 50-min sessions across a 4-month period. To ensure hippocampal engagement, the spatial navigation task required numerous functions thought to rely on the hippocampus: allocentric processes, associative memory, encoding of novel information, and consolidation of information. The authors' predictions included improved training-related increases in spatial navigation performance and transfer of training to untrained tasks requiring allocentric spatial processing.

Findings revealed that, compared to the control group, men that underwent spatial navigation training improved more on the navigation task at post-testing. Results also revealed a significant effect of age, such that older adults' performance was worse than young adults' on the navigation task prior to training. However, at post-test, older adults that completed spatial navigation training performed equal to young adults in the control group, demonstrating training-related cognitive plasticity for older adults. Further, performance improvements related to training were partially maintained following 4 months without training. The authors also predicted that integrity (measured by volume and mean diffusivity (MD)) of the hippocampus would increase following spatial training in both young and older adults. Note that both younger and older men in the control group experienced a natural decline in hippocampal volume (0.75% for the left and 1.59% decrease for the right) from the beginning of the study to post-testing. However, both young and older men in the navigation-training group displayed stable volumes across the training period and also throughout the 4 months post-training. Given that the training did not appear to result in increased volume, this may provide evidence for hippocampal maintenance that deters normal age-related decrease in hippocampal volume. These results are interesting, as they reveal maintenance for brain structure (in the face of age-related natural declines), but also cognitive plasticity for performance on the spatial navigation task.

Notably, no main effect of age for hippocampal volume was observed, but there was a main effect of age for MD, such that older men presented with higher MD than young adults. Higher MD is thought to reflect lower structural integrity, possibly by quantifying the density of membranes within a particular region (for review, see Assaf and Pasternak 2008). Within the training group, hippocampal MD decreased during training and then returned to baseline during the 4 months post-training. The control group did not demonstrate significant changes in MD across time. Although the functional relevance of microstructural changes in the hippocampus is still unknown, the evaluation of measures aside from volume is critical, as some experience-dependent changes may take place on a smaller scale than would affect overall volume.

A novel aspect of this study was their investigation of the transfer of navigation training benefits on a variety of other tasks. However, a trend toward better performance for the training group was observed only on a task of spatial orientation. No other tasks showed signs of transfer, including tasks of intelligence (Raven's progressive matrices), mental rotations, vocabulary, processing speed (Digit-Symbol Substitution), route memory, location memory, object-position memory, numerical memory updating, numerical and figural comparison, spatial 2-back, word-list recall, and number-noun pairs.

Thus far the evidence suggests that extensive spatial navigation training can result in structural hippocampal changes and improved spatial abilities, but these spatial ability benefits may not transfer to other tasks, even for tasks that would also seem to evoke hippocampal processing. Greater transfer may occur when individuals are trained on an activity that demands a wide array of cognitive and motor functions, such as playing video games. Given the broad involvement of cognitive processes, this type of activity may have more potential to convey benefits to other tasks. Videogame playing has previously been found to correlate with attention, perception, and executive control abilities (Green and Bavelier 2003), and many have suggested that videogame experience may simultaneously train multiple skills (Basak et al. 2008). Kühn and colleagues (2014) studied young adults' brains before and after 2 months of playing a SuperMario videogame daily. In contrast to the methods used by Lövdén et al. (2012), this study employed VBM to determine training-related group differences in volume. Grey matter increases were observed for the training group, in comparison to the control group who performed no tasks during the training time, in the right hippocampus, right dorsolateral prefrontal cortex, and cerebellum. Interestingly, greater increases in hippocampal volume were associated with the participants' tendency toward an allocentric orientation strategy, suggesting that structural plasticity in the hippocampus was functionally relevant for strategy choice on the SuperMario game.

To summarize, in healthy adults, there are a few studies of interventions that target the hippocampus by training functions like memory and spatial abilities. Changes in the hippocampus have also been seen following other types of training (juggling, abstract learning, video game playing), but evidence is lacking for transfer of benefits. There are many behavioral studies demonstrating improved memory following memory training, but the evidence has not yet shown whether hippocampal plasticity underlies observed benefits.

Because individuals with subjective memory impairment (SMI) or mild cognitive impairment (MCI) are at high risk for developing Alzheimer's Disease (Jessen et al. 2010), studies have also designed interventions for such individuals. Theoretically, bolstering the hippocampus through memory training might decrease the risk or rate of disease progression. In the first of a series of papers by Engvig et al. (2012), the authors found that memory training that employed method of loci to enhance verbal recall in older adults with SMI resulted in increased memory performance, and change in memory was associated with pre-training left hippocampal volume. Following this finding, Engvig et al. (2014) tested for training-related changes in brain structure in the same 19 individuals with SMI from the previous study, along with 42 healthy older adult controls. The authors found that after 2 months of episodic memory training both healthy and impaired older adult individuals experienced improved memory. In particular, a larger training-related effect size for memory improvement was seen in the SMI training group. Both healthy and SMI training groups experienced increased cortical grey matter volume, but only healthy older adults experienced training-related increases in the left hippocampus. The SMI training group showed slightly less (though not statistically significant) negative change in the left hippocampus than the no-training healthy controls. Notably, change in left hippocampal volume was positively correlated with change in free recall only in the SMI training group. As the authors discuss, the lack of significant structural hippocampal plasticity in the SMI training group could be due to lower potential for plasticity in the impaired hippocampus. However, another likely explanation could be that the training mitigated even greater expected atrophy in SMI individuals, an example of brain maintenance. Unfortunately, the study lacked a no-training SMI control group, so we are unable to discern how much hippocampal atrophy might have occurred during the intervention period in a SMI no-training group.

Changes in Hippocampal Functional MRI Outcomes Following Cognitive Training

Given the relative recency of functional neuroimaging techniques, cognitive training studies in healthy older adults that have demonstrated functional outcomes are limited. Unlike structural changes, which can clearly reflect plasticity, functional changes may reflect change in a variety of processes, including response to familiar stimuli, reallocation of available resources, altering representations, and switching between existing cognitive states (for methodological reviews, see: Kelly et al. 2006; Poldrack 2000). Given these possibilities, it would be misleading to conclude that all functional changes represent experience-dependent plasticity because some training-related patterns of change may stem from more transient adaptive mechanisms associated with flexibility rather than structural changes (Lövdén et al. 2010).

As a first example, a study of 14 healthy older adults by Kirchhoff and colleagues (2012) found that change in hippocampal activity, as measured by functional MRI, during memory retrieval following 2 days of memory training was positively associated with training-related improvements in memory encoding and retrieval. Greater hippocampal activity during memory retrieval may represent greater neural activity or recruitment of more neuronal groups within the hippocampus in response to the previously encoded stimuli. Although this training paradigm included only two training sessions across a 2-week period and the sample size was small relative to many other studies, the positive correlation between activation change and memory performance suggests that the hippocampus in older adults can be a key target for improving memory abilities.

Voss et al. (2012) employed a complex skill-learning videogame-based program and examined the effects of various training programs on interactions among multiple neural networks of interest in healthy young adults. Twenty-nine individuals completed 10 2-h training sessions across 2–3 weeks and were randomly assigned to be instructed to focus either on the entirety of the game (fixed priority) or on sub-components of the game (variable priority). Previous studies showed that variable priority training leads to more transfer (Kramer et al. 1995, 1999b), and this study examined the neural mechanisms of this pattern using functional imaging. For instance, variable priority training may recruit neural networks that are involved with higher-order and flexible goals and actions, such as fronto-parietal networks and regions associated with the declarative memory system such as the hippocampus. In contrast, fixed priority training is thought to invoke a more rigid learning style that is more dependent on procedural learning systems. Such a pattern of systems-level change as a function of training strategy may help explain why variable priority training has been shown to lead to more transfer compared to fixed priority training. Indeed results showed that as expected, the functional connectivity of the hippocampus with regions in a fronto-parietal brain network associated with orienting of attention was most influenced by variable priority training, compared to fixed priority training which increased functional connectivity between the caudate nucleus and the same fronto-parietal brain network. Importantly, the interaction between the fronto-parietal network and the hippocampus was related to a faster learning rate on the videogame across training sessions only for the variable priority group, and connectivity between the caudate and the fronto-parietal system was not associated with learning rate across training sessions for either group. The relationships with learning rate are consistent with the idea that functional plasticity occurred rather than only flexibility of the learning systems. Overall, while the results are promising with regard to understanding the role of the hippocampus in transfer of learning, unfortunately there was little evidence of transfer to other cognitive abilities in this study (Boot et al. 2010). Therefore, it will be important for future studies to replicate this result and examine the role of these network interactions in transfer of learning.

Some studies have also explored functional brain plasticity using techniques other than fMRI (Langer et al. 2013). While techniques such as electroencephalography (EEG) are informative of cortical brain networks, they typically are unable to measure signals within subcortical structures such as the hippocampus. However, Langer et al. (2013) did find that working memory training in 34 young adults (20 30-min sessions across 4 weeks), compared to 32 young adults in the active control group, resulted in increased theta oscillations, which have been tied to functions involving the hippocampus, such as spatial navigation and working memory.

In an attempt to quantify plasticity across multiple time points and multiple structural and functional systems, Lampit et al. (2015) measured a variety of brain outcomes at two times points during and following a randomized, 12-week group-based multi-domain computerized cognitive training program. Participants were 12 older adults from a neuroimaging subsample of a larger cognitive training program in older adults with at least one risk factor for dementia (Lampit et al. 2014). This program consisted of 36 60-min sessions involving exercises of memory, attention, response speed, executive functions, and language. The active control group viewed National Geographic videos and answered multiple-choice questions about the videos, for the same amount and duration of time as the training group. The authors examined training-related changes using structural measures, including VBM, vertex-based analysis, diffusion tensor imaging, and MR spectroscopy (metabolite signals), as well as fMRI. Results revealed a significant effect of the 3-month training on global cognition (measured as a composite of memory, information processing speed, and executive function). Between baseline and the first follow-up (after 9 h of training), the training group demonstrated increased functional connectivity (FC) between the right hippocampus and the left superior temporal gyrus, whereas the control group showed decreased FC between these regions in that same time period. Interestingly, greater FC increases from baseline to the first follow-up were related to greater increases in global cognition from baseline to the second follow-up (after 36 h of training spread across 3 months). While this may suggest that functional changes occur early on in training and at least partially predict later cognitive changes, the group differences in FC found at the first follow-up were not found after the full 3 months of cognitive training. However, it is critical to note that this study consisted of a relatively small sample (12 participants) and, more importantly, the cognitive training group consisted almost exclusively of women, and the active control group consisted exclusively of men. Therefore, while the results can help guide future hypotheses and continued examination of these outcome measures, there is a chance that the findings described are largely due to sex differences across the groups. This is especially concerning for the functional findings, given that the authors found baseline group differences in whole-brain FC maps for the selected ROIs (right hippocampus and posterior cingulate). Despite the limitations of this pilot study, Lampit and colleagues were able to demonstrate findings of differential training-related effects in early and later stages of training, in addition to semi-converging structural findings between VBM and vertex-based analysis (though these findings did not include the hippocampus).

Functional outcomes of cognitive training have also been examined in patient populations, most often in MCI and Alzheimer's Disease (AD). Most studies focus on memory-based training in an attempt to increase memory abilities and mitigate some symptoms of the diseases. For example, Belleville et al. (2011) employed a training program targeting episodic memory in an older adult population in which half of the participants had a diagnosis of Mild Cognitive Impairment (MCI). The other participants were healthy, age-matched adults. The training program was previously shown to improve delayed word recall, face-name memory, and self-reported daily memory functioning (Belleville et al. 2006). Though effects of training were examined between healthy adults and those with MCI, no training-control groups were used for either population. A key aspect of the study design (Belleville et al. 2011) included two MRI scans prior to the cognitive training. This design helped control for practice effects in the functional MRI measures. The group-based memory training lasted for 12 h spaced across 6 weeks. Neuroimaging outcomes for this study included brain activation during a memory-encoding and retrieval task. No group differences in hippocampal activation during encoding were found prior to training. Healthy adults experienced a training-related decrease in hippocampal activation during encoding, and a training-related increase in hippocampal activation during retrieval processes. In contrast, while training-related increases in cortical activation were found for MCI subjects, no training-related changes in activation were found in the hippocampus. Although the authors did not find hippocampal-specific plasticity, this study did demonstrate plasticity across a wide network of cortical areas even within this diseased state.

A similar study by Rosen et al. (2011) specifically hypothesized that a training program from Posit Science that included adaptive games would influence the hippocampus and improve memory performance in older adult participants with MCI. Interestingly, the games were designed to improve speed and accuracy in auditory processing, not necessarily to improve memory, but this program has previously been found to improve memory performance in both healthy older adults and those with MCI (Mahncke et al. 2006). Each participant completed the program for 100 min per day for 5 days until criterion performance was reached (asymptotic performance over several days or completion of 80% of the training material), with training lasting approximately 2 months. The control group (also individuals with MCI) completed non-adaptive cognitive engagement tasks (reading, listening to audio books, playing a visuospatial computer game) for 90 min per day, 5 days per week. Consistent with predictions, the adaptive training group did show a significantly greater increase in performance on neuropsychological tests of memory ability than control participants. The control participants experienced a decline in memory ability throughout the intervention, which is not surprising since all participants had been diagnosed with MCI. Brain function was measured during an auditory decision-making task, during which participants chose whether words were concrete or abstract. A small increase in brain activation was found in the left hippocampus exclusively for individuals in the adaptive training group. There was also a trending, though non-significant, positive correlation between increase in activation and change in neuropsychological performance from pre-training to post-training. Although this study consisted of a small number of participants (6 per training/control group), the results suggest that adaptive game training mitigated the negative effects that were seen in the control MCI group.

Finally, Hampstead et al. (2011) and (2012) focused on rehabilitating cognitive impairment using mnemonic strategy training for face-name associations (2011) or object-location associations (2012). In the 2011 study, the training procedure was fairly short (five sessions within a 2 week period), the number of participants was small (6 individuals with MCI), and no placebo training control group or comparison individuals were used. Instead, within-subject comparisons were made across trained and untrained stimuli. Although the mnemonic training resulted in increased performance on laboratory memory tasks, widespread activation increases across the cortex, and increased connectivity within neural networks, this study did not show training-related changes in hippocampal activity or connectivity. In the 2012 study, a healthy older adult group of 16 individuals was included as comparison to 18 MCI patients, and all individuals were randomized to mnemonic strategy training or exposure-matched control training. Critically, group (MCI versus healthy individuals) differences in brain activation were observed prior to training, such that individuals with MCI showed lower encoding and retrieval-related hippocampal activity than healthy comparisons in the head, body and tail of the hippocampus bilaterally. Note that the hippocampus was the main region of interest, so analyses were limited to manually traced bilateral hippocampal regions.

Both MCI and healthy individuals in the mnemonic strategy training group showed greater improvement in object-location memory following the training than the individuals in the exposure-matched control group. This finding was accompanied by greater training-related increases in activation for MCI compared to healthy comparisons, particularly in the hippocampus. These results suggest that the mnemonic strategy training worked to partially restore hippocampal activation during encoding.

Overall, studies of training-related functional changes tend to suggest that, in healthy adults, certain types of cognitive training can result in decreased brain activation during tasks, and this pattern of results may reflect increased efficiency of neural processing (c.f., Poldrack 2015). Evidence also demonstrates that memory training can result in increased activation during memory retrieval and recollection. These differing patterns of training-related changes may depend on the processes involved at the time of measurement. There is little consensus on the type or duration of training that best improves neural function and achieves successful transfer to other tasks. In patient populations, cognitive training may work to restore or mitigate the declines observed due to disease, though the patterns of functional response may differ from healthy individuals and are difficult to interpret unless closely linked to behavioral changes.

Possible Mechanisms Based on Animal Models

Animal and human literature has presented many possibilities for age-related changes in the brain (López-Otín et al. 2013; Thomas et al. 2012; Voss et al. 2013b). Some of these changes include alterations between synaptic connections, cortical thinning and reduced brain volume due to neuronal loss and less efficient neuromodulatory processes. Structural plasticity in the hippocampus can manifest as increased structural volume, which may reflect increased neurogenesis, proliferation of glial cells and astrocytes, increased synaptic density and vascular density, or prolonged cell survival (Kempermann et al. 1998, 2002; Opendak and Gould 2015; van Praag et al. 1999; for review, see Voss et al. 2013b).

In animal studies, increased neurogenesis in the dentate gyrus/CA1 region has been found to relate to better cognitive function, though it is unclear whether neurogenesis itself is the direct cause of improved performance (van Praag et al. 1999; for review, see Voss et al. 2013b). Because it is quite difficult to directly measure neurogenesis in human studies, volume is considered the primary structural adaptation measurable by MRI. Supporting evidence for this is that hippocampal volume has been found to positively relate to some types of learning (Herting and Nagel 2012; Konishi and Bohbot 2013) and memory (Chaddock et al. 2010; Erickson et al. 2009). However, because it is still unknown precisely how microscale cellular, molecular, and synaptic changes translate into macroscale age-related changes in brain structure and function observable with human neuroimaging, further translational research in these areas is needed to improve our ability to understand whether and how cognitive training might work against the basic mechanisms of brain aging.

Review of Literature for Physical Activity and Fitness Training

Exercise training has also been shown in both animal and human studies to significantly impact the hippocampus. Many studies have now demonstrated evidence for a positive relationship between cardiorespiratory fitness or fitness training and varied cognitive abilities including processing speed, visuospatial processing, attention, and executive function (Colcombe and Kramer 2003; for review, see Erickson et al. 2009, 2014; Kramer et al. 1999a; Smith et al. 2010). Protection from age-related cognitive decline has also been associated with greater self-reported (Benedict et al. 2012; Yaffe et al. 2001) and objectively sensor-based measurements of physical activity (Makizako et al. 2014). Interventions typically aim to increase cardiorespiratory fitness through structured, aerobically challenging exercise sessions. However, some interventions aim primarily to increase physical activity level (particularly if the activity takes place outside the laboratory). Without monitoring heart rate and effort it can be difficult for an intervention to increase cardiorespiratory fitness in all participants. Physical activity is defined as movement that increases the body's energy expenditure beyond resting levels, while exercise is the structured process of completing movement for the purpose of increasing fitness. These distinctions are important when considering whether activity needs to improve fitness in order to improve brain and cognitive health. Further, while most studies use aerobic training, some have examined the effects of other types of training, such as resistance and coordinative training.

Changes in Hippocampal Structure Following Physical Exercise Training

Colcombe et al. (2006) examined changes in fitness and brain volume in 59 healthy but sedentary older adults following a 6-month, 3 times per week, aerobic (walking) exercise-training program. While this study did not find specific volume increases in the hippocampus, this exercise regimen did result in both grey matter and white matter increases in the frontal and temporal lobe. This study used VBM, which, as noted previously, can be vulnerable to bias in areas such as the medial temporal lobes.

Measuring training-related changes in cognition is critical for understanding the relevance of exercise's impact on the brain in everyday life, specifically for cognitive functions that are known to decline with age. To this end, Erickson et al. (2011) employed a 1-year randomized control trial using an aerobic (walking) exercise training paradigm in 120 healthy older adults and measured hippocampal volume and spatial memory before and after training. Spatial memory is related to hippocampal volume (Erickson et al. 2009) and because hippocampal volume experiences significant decline during aging, training-related changes in the hippo-campus may result in improved spatial memory. Significant increases in hippo-campal volume were observed for individuals in the training group, compared to control group individuals who experienced decline in hippocampal volume over the course of the year. The increase in volume was found exclusively for the anterior hippocampus. Further, greater increases in hippocampal volume in both the right and left hemispheres were related to greater improvements in fitness, suggesting that increasing fitness may impact the hippocampus in a continuous manner. This study also examined serum BDNF levels, the results of which are discussed below in the possible mechanisms section. Finally, the authors found increases in spatial memory for both the exercise and the control groups. Although this lack of a time x group interaction effect was unexpected, the authors did find that cardiorespiratory fitness both before and after the intervention was positively correlated with perfor-mance on the spatial memory task. Further, increased hippocampal volume in the aerobic exercise group was correlated with improvements in spatial memory. These brain and behavior relationships increase the possibility that it was specifically cardiovascular fitness gains that caused plasticity in the hippocampus, here mea-sured by increased hippocampal volume and spatial memory performance.

In addition to volumetric changes of grey or white matter, structural plasticity of the hippocampus has also been measured by vascular plasticity. Maass et al. (2014) conducted a 3-month intervention with 40 healthy, sedentary older adults that were pseudo-randomized to either a program of thrice weekly treadmill interval training or a stretching/relaxation control program that met for the same frequency and duration. The authors found that change in fitness was positively related to change in hippocampal perfusion, a measure of localized blood flow, which may indirectly reflect increased neural activity. However, results also showed that vascular plastic ity in the hippocampus might be age-dependent, as fitness-related improvement in perfusion was negatively associated with age. The authors also found a positive correlation between change in fitness and change in hippocampal volume in only the head of the hippocampus, as well as positive correlations between hippocampal volume, perfusion, and memory performance (early recall and recognition on a verbal list learning task). Overall, the results suggest that fitness-related changes in hippocampal volume underlie the subsequent changes in memory performance. Importantly, the relationships that the authors found did not apply to whole-brain measures, and instead appeared hippocampal-specific.

Extending the work on aerobic exercise training, evidence also suggests that training of movements that require extensive motor control (balance, obstacle avoidance, speeded reactions) may facilitate processes in the brain separate from those of cardiovascular training. Niemann et al. (2014) studied 91 healthy older adults and examined whether a 1-year, thrice weekly, group-based coordinative training program would also increase the size of the hippocampus differently from that of cardiovascular training. The authors found that both cardiovascular and coordinative training, compared to the stretching and relaxation training control group, resulted in increased hippocampal volume. In particular, the right hippocampus was more responsive to coordination training than to cardiovascular training. Interestingly, volume changes could be detected as early as 6 months in the cardiovascular training group, but not until 12 months in the coordination-training group. Although more work is needed to confirm and extend findings of differential time courses for various types of training, this finding may reflect the involvement of both rapid and more stable and longer-lasting neuroplasticity mechanisms that depend on exercise modality.

Exercise training may impact the hippocampus in individuals with certain diseases, particularly diseases that have previously been shown to have hippocampal abnormalities, such as MCI, Alzheimer's Disease and schizophrenia. Reduced hippocampal volume is a well-known feature of schizophrenia. In a randomized, controlled study, Pajonk et al. (2010) randomized 24 schizophrenic patients to either a 3-month aerobic cycling program or a non-aerobic condition of playing table-top football. Further, the effect of exercise training for patients was compared to a group of 8 matched healthy control participants who also did exercise training. Results showed that exercise increased hippocampal volume for both patients and healthy controls, and there was no effect of non-aerobic training on volume. For both patients and controls that did exercise training, greater increases in fitness were related to greater increases in hippocampal volume. Additionally, the authors found a greater training-related increase in a neuronal metabolite marker for schizophrenic exercise group than for the healthy control exercise group. While interesting, this metabolite change was not correlated to the increase in hippocampal volume, which limits the likelihood that this mechanism underlies exercise-related increases in brain volume. Cognitively, only the schizophrenic exercise group experienced increases in memory ability. Within the entire schizophrenic group (exercise and non-exercise), this change in memory was associated with hippocampal volume change, but this correlation was not significant in each separate group. While this study suggests hippocampal plasticity is present in schizophrenia, it is still unknown how the hippocampus differs between the diseased and healthy states and whether those differences alter the capacity for plasticity in any situations.

In a study of 86 women with probable MCI, ten Brinke et al. (2015) found that a twice-weekly, 2-month aerobic exercise program increased hippocampal volume relative to a resistance training or stretching control. However, average change in hippocampal volume was not significantly different from zero, supporting that participants in the aerobic group maintained their hippocampal volume compared to atrophy shown by the other two exercise groups. Unexpectedly, across all participants, while controlling for group membership, greater increase in hippocampal volume was associated with a change towards poorer recall performance on the Rey Auditory Verbal Learning Task (RAVLT). The authors suggested that this may be in part due to other important moderators of the relationship between change in hippocampal volume and memory performance in an MCI population, such as white matter degeneration. Thus, future studies will be needed to more robustly tie exercise-induced changes in hippocampal volume to improvements in hippocampal-mediated cognitive functions.

Changes in Hippocampal Functional MRI Outcomes Following Physical Exercise Training

The profound effect of exercise training on hippocampal structure has led many researchers to explore how exercise may impact the function of the hippocampus, including measures of blood flow change, indirect measures of brain activation, and degree of connectivity within and between the hippocampus and other brain regions.

Based on evidence of exercise-induced neurogenesis from animal studies, Pereira et al. (2007) examined angiogenesis in mice and humans by measuring cerebral blood volume (CBV) with high resolution MR imaging of the hippocampus. Through a unique comparison of in vivo mouse CBV (accompanied by post-mortem histology data) with human CBV, Pereira and colleagues examined if exercise training would increase CBV in the dentate gyrus for both animal and human participants. Results showed that dentate gyrus CBV increased in 11 healthy young adult humans (9 females, 2 males) following a 12-week (4 times weekly) aerobic exercise intervention and that the CBV increases in the dentate gyrus were related to the increase in cardiorespiratory fitness across individuals. Further, increases in fitness and CBV were related to improvement in the early learning and free recall of a list of words (RAVLT). These correlations were selective for the early learning trial and not other performance measures, such as delayed recall, recognition, and source memory. These findings paralleled the results from the animal model, where rodents that exercised voluntarily for 2 weeks showed increased CBV in the dentate gyrus that corresponded with a measure of increased neurogenesis for the exercise group. Limitations of the study include a relatively small sample for the human exercise training group (N = 11) and no control group for either humans or animals to rule out confounds such as increased social enrichment from the exercise program or other lifestyle changes that could have improved fitness. Even despite these limitations, this study provides initial evidence that CBV may be a meaningful outcome measure corresponding to underlying mechanisms related to exercise training.

In the first study to investigate the effect of exercise on functional brain networks in healthy older adults, Voss et al. (2010) employed a 1-year exercise-training program that compared thrice weekly walking to a control group of light stretching and toning. The brain network known as the default mode network (DMN) includes the hippocampus and has been shown to have lower functional connectivity in older compared to young adults (Andrews-Hanna et al. 2007; Buckner et al. 2008; Damoiseaux et al. 2008; Fox et al. 2005). Given that these changes correspond to declines in certain cognitive functions (Andrews-Hanna et al. 2007; Damoiseaux et al. 2008), this network is thought to play an important role in healthy aging. After

12 months of training, Voss and colleagues found that the walking group had greater connectivity within the temporal lobe and between the medial temporal lobe and lateral occipital cortex and prefrontal regions of the DMN. Greater change in DMN functional connectivity was also related to greater change in executive function performance across all participants. Further, exercise training increased the negative association between prefrontal regions and an anterior left hippocampus region, suggesting training-related differentiation between functional networks, an outcome that made older adults' FC patterns look more similar to healthy, young adults. Interestingly, in this study FC outcomes were generally not reliably different after 6 months of training suggesting that functional brain outcomes may continue to increase with greater training durations.

To better understand the training-related changes in the hippocampus, Burdette et al. (2010) examined functional changes in 11 older adults (aged 70–85) by measuring hippocampal blood flow following 4 months of thrice weekly aerobic exercise training compared to a healthy aging educational training program. Of note, all older adults in this single-blinded randomized control pilot trial were at risk for cognitive decline based on age and self-report memory loss. Eligible individuals were randomized into the exercise training (n = 6) or healthy aging educational training (n = 5) program. Following training, greater hippocampal cerebral blood flow was found in the exercise-training individuals relative to the control group. While there were no overall differences in global brain network metrics for the exercise and control groups, the exercise-training group had higher connectivity between the hippocampus and the rest of the brain after training. In particular, for the exercise-training group, the hippocampus showed higher connectivity with the anterior cingulate cortex (ACC) and anterior medial prefrontal cortex compared to the control group. Finally, greater connectivity between the hippocampus and the medial prefrontal cluster was related to greater perfusion (blood flow) in the hippocampus. However, it is critical to note that all brain measures were assessed only after training. In order to clarify cause and effect, these results will need to be replicated with an assessment of change between before and after an exercise-training program.

Using another yearlong intervention program, Voelcker-Rehage et al. (2011) explored whether the type of training would have a significant effect on outcomes. Evidence from animal and human literature suggest that different types of training may affect brain metabolism and molecular cascades in various ways. This was further supported by the fact that the "control" (flexibility, toning and balance) group in Voss et al. (2010) experienced increased functional connectivity within brain networks, indicating that the non-cardiovascular activities may have beneficially impacted the brain. The use of other types of interventions is also supported by cross-sectional evidence that motor fitness, defined by movement speed, balance, coordination, and flexibility, is associated with greater cognitive ability (Voelcker-Rehage et al. 2011). The interventions used by Voelcker-Rehage et al. (2011) consisted of both a cardiovascular training (n = 17) and a coordinative training group (n = 16), along with a control group (n = 11). The total sample studied included 44 healthy older adults. The authors found that cognitive performance on the Flanker test was improved over the 12-month study interval for both cardiovascular and coordinative groups, but not the control group. Similarly, while the control group experienced increased activation in cortical regions as well as the parahippocampus following the intervention, the intervention groups experienced no or very little activation change in these regions. Overall, based on changes in cortical activation patterns, the authors concluded that both cardiovascular and coordination training result in brain activation in older adults that is more similar to young adults. The authors also demonstrated that improvements continued to rise across the intervention duration in a linear fashion, suggesting the improvements do not trail off with longer training durations.

Some studies have also investigated whether resistance training can exert a similar beneficial effect as aerobic training. There is some evidence of training-related improvements in cognitive functions, including associative memory, (Liu-Ambrose et al. 2012; Nagamatsu et al. 2012). Such changes were related to changes in functional brain activation patterns in cortical regions, but not the hippocampus. To our knowledge, no studies have shown resistance-training-related structural or functional changes within the human hippocampus.

Possible Mechanisms Based on Animal Models

Evidence from animal studies supports that exercise increases the rate of neurogenesis in the hippocampus (Kronenberg et al. 2003; for review, see van Praag 2008; van Praag et al. 1999). Further, as mentioned in the previous section, a neurogenesis marker was positively correlated with dentate gyrus CBV in exercising mice, and controlling for neurogenesis abolished effects on CBV (Pereira et al. 2007). These observations suggest that exercise-induced increases in neurogenesis may induce increases in hippocampal blood flow. However, more translational studies of this nature are needed to test whether there are other molecular cascades and effects related to exercise-induced increases in cerebral blood volume/flow. It is also possible that the beneficial effect of exercise on angiogenesis may vary with age, as some studies have shown that exercise training can increase neurogenesis and learning in aged rodents without marked changes in angiogenesis (Creer et al. 2010; van Praag et al. 2005).

Training or experience-related neuroplasticity likely occurs in part through a series of growth factor and protein cascades (Nishijima et al. 2015; for review see Voss et al. 2013b). There are two well-known growth factors that are up-regulated in response to exercise. Brain derived neurotrophic factor (BDNF) and vascular endothelial growth factor (VEGF) have been shown in animal studies to increase following exercise training (Fabel et al. 2003; Neeper et al. 1996; Vaynman et al. 2004) and there is evidence that BDNF increases with exercise in humans as well (Pereira et al. 2007; Rojas Vega et al. 2006). Further, there is some evidence that BDNF may play a role in exercise-induced brain and cognitive changes in humans (Ferris et al. 2007; Pereira et al. 2007; Rasmussen et al. 2009; Voss et al. 2013a).

Although Erickson et al. (2011) did not find a difference in training-related change in circulating BDNF between the exercise and control groups, they did find within the exercise group that greater increases in serum BDNF were related to greater increases in hippocampal volume. It is possible that other lifestyle activities could also up-regulate BDNF, which may account for the lack of an exercise-related group difference. Nevertheless, this result suggests that BDNF's response to exercise may be related to the structural plasticity observed in the human adult brain.

Multimodal Interventions that Combine Cognitive and Physical Training

Given the potentially unique effects of cognitive and exercise training, programs combining various types of training may be the most promising for widespread cognitive increases. In these studies, improvements in cognitive function can be described by the cognitive-enrichment theory, in which improving general cognitive capacity overall can affect many different abilities based on greater capacity and resources (Hertzog et al. 2009). Further, multiple reviews of the literature have suggested that multimodel interventions might take advantage of many different molecular pathways (Bamidis et al. 2014; Dhami et al. 2015). Fissler et al. (2013) propose synergistic effects on multiple mechanisms when combining difficult, novel tasks with physical activity components. Such components may interact through "guided plasticity facilitation," such that cognitive activity guides spatial and temporal characteristics of changes in the brain, while physical activity acts to facilitate, or enhance, these changes. Figure 1 visualizes this interaction and points to relevant papers for each relationship. Further, this physical activity-induced facilitation might be the most effective immediately after a bout of exercise, as it has also been shown that BDNF increases acutely following exercise (Knaepen et al. 2010). Fissler and colleagues conclude that across a population of highly

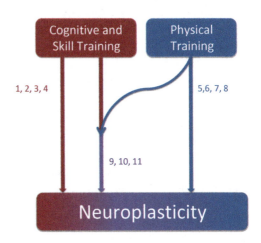

Fig. 1 Training leads to neuroplasticity that engenders changes in cognitive function and brain health. References denote key papers that demonstrate evidence for each type of relationship in healthy adults (see Table 1)

Table 1 Studies referenced in Fig. 1

# in figure	Author (First; Last)	Year	Type of training	Experimental Training Modality (control group)	Population Details unless otherwise stated, controls match population of intervention group
1	Boyke; May	2008	Cognitive and Skill training	Juggling (no-contact)	25 healthy older adults; 25 controls
2	Lövdén; Bäckman	2012	Cognitive and Skill Training	Virtual environment spatial navigation (walk-time yoked, no navigation)	23 healthy young adults; 21 young controls; 23 healthy older adults; 24 older controls
3	Kühn; Gallinat	2014	Cognitive and Skill Training	Video game (no-contact)	23 healthy young adults; 25 controls
4	Lampit; Valenzuela	2015	Cognitive and Skill Training	Multidomain computerized cognitive training (video watching and quizzes)	7 older adults with at least one dementia risk factor; 5 active controls
5	Colcombe; Kramer	2006	Physical activity	Walking (light stretching and toning)	30 sedentary healthy older adults; 29 active controls
6	Erickson; Kramer	2011	Physical activity	Walking (light stretching and toning)	60 healthy older adults; 60 active controls
7	Voss; Kramer	2012	Physical activity	Walking (light flexibility, toning, and balance)	30 sedentary healthy older adults; 25 active controls; 32 young adult controls
8	Maass; Düzel	2014	Physical activity	Treadmill walking (muscle relaxation/stretching)	21 sedentary healthy older adults; 19 active controls
9	Anderson-Hanley; Zimmerman	2012	Multimodal	Cybercycle with virtual reality display (stationary bike)	38 healthy older adults; 41 active controls
10	Li; Li	2014	Multimodal	Cognitive training, Tai Chi & group counseling (lectures on health and aging)	26 healthy older adults; 19 healthy older adult controls
11	Carlson; Fried	2015	Multimodal	Experience Corps (wait-list)	58 healthy older adults; 53 healthy older adult controls

variable individuals, interventions that combine cognitive, social, and physical activities are the most likely to be effective. Although the literature is still sparse concerning brain outcomes from multimodal interventions, we briefly highlight a few promising studies.

In a recent study, Carlson et al. (2015) report the beneficial effects of an intergenerational social health promotion program (Experience Corps) on cortical and hippocampal volume and memory, as measured by the Rey Auditory Verbal Learning Test. The Experience Corps program is a randomized control trial involving older adults volunteering as mentors and tutors at Baltimore area elementary schools for 2 years, which increased their physical, cognitive, and social activity. The Experience Corps group was compared to a wait-list control group of healthy older adults who were referred to lower-intensity volunteer opportunities. A subset of Experience Corps participants were randomized to a Brain Health Substudy (N = 111). Compared to the control group, the experimental group showed intervention-related increases in cortical and hippocampal volume after 2 years, though this relationship was found only in men. Importantly, following the 2 years, male control participants demonstrated age-related atrophy in cortical and hippocampal volume. Thus, the intervention appeared to not only mitigate expected declines, but also facilitate positive plasticity throughout the brain, at least in male participants. For both women and men, greater benefits in the hippocampus were observed after 2 years of the intervention than after just 1 year of participation (although for women the improvements did not reach statistical significance).

Although changes in hippocampal volume were not significantly related to improvements in memory, cortical volume changes were positively associated with changes in recall ability for men in the intervention group. In addition, benefits of the Experience Corps program were found for executive function (Carlson et al. 2008) and neural activity (Carlson et al. 2009), such that greater intervention-related increases in activity in the regions associated with executive function were related to greater behavioral improvement on the Flanker task. The Experience Corps program was proposed as targeting the hippocampus, based on hippocampal plasticity from enriched environments that simultaneously combine physical, mental, and social stimulation (van Praag et al. 2000). Varma et al. (2015) did demonstrate an increase in physical activity in the women from a similar cohort of participants from the Brain Health Substudy of the Experience Corps program (total N = 114), as well as a positive relationship between physical activity and hippocampal volume (Varma et al. 2014). This relationship was observed only for women in the Brain Health Substudy. Thus, overall, results suggest this type of volunteer and community-based multimodal intervention is beneficial for hippocampal structure but more research is needed to link these changes to improved hippocampal function.

In a study of 39 otherwise healthy young individuals with schizophrenia, Malchow et al. (2015) examined the effects of a 3-month exercise intervention combined with cognitive training, compared to a non-exercise table soccer intervention that included cognitive training. Compared to schizophrenic patients in the control group there was not a significant effect of endurance exercise + cognitive training on hippocampal volume. This null effect may be a result of the lack of appropriate control (table soccer requires some level of both movement and cognitive engagement). However, the authors did find a significant decrease in hippocampal volume in schizophrenic participants in the non-exercise control group at a follow-up 3 months following the end of the intervention. If nothing else, these data should encourage individuals, particularly those with expected hippocampal decline, to engage in some type of physical or cognitive activity.

In addition, a recent paper by Li et al. (2014) demonstrated functional plasticity following a 6-week multimodal intervention in 26 healthy older adults. Because the prefrontal cortex and the hippocampus are especially sensitive to aging, the authors examined hippocampal-prefrontal connectivity before and after 6 weeks of a multimodal intervention, which was a combination of cognitive training (mnemonic and executive function training), Tai Chi, and group counseling. Of the 45 total participants, 19 individuals were randomly assigned to the control group; these participants received instructions to attend 2 lectures on health and aging. Cognitively, the intervention group improved more on paired associates learning than the control group. The intervention group also showed a greater increase in FC between the prefrontal cortex (PFC) and the left parahippocampal cortex (PHC) than the control group. In an exploratory examination of functional connections with the medial prefrontal cortex, the authors found that the greatest functional connectivity was found with regions in the DMN including the anterior hippocampus. Importantly, in the intervention group, the increase in connectivity between the PFC and the PHC was related to increased improvement in the Category Fluency Test. Across all individuals, post-training FC between the two regions was also positively related to performance on the Trail Making Test, a measure of executive function. This again demonstrates the potential of functional plasticity in the older adult brain, and suggests that some regions and/or networks may be more responsive than others to cognitive and physical intervention training.

Although the literature concerning mechanisms underlying multimodal intervention is sparse, some evidence can be gleaned from Anderson-Hanley and colleagues' study of cybercycling (2012) in healthy older adults from retirement communities. A 3-month cybercycling intervention was compared to a 3-month stationary cycling program that did not have a cognitive component. The authors found that cybercycling increased plasma BDNF more than exercise alone. This may suggest that combining cognitive and physical activity boosts BDNF-mediated enhancements in brain structure and function. However, this does not explain why the exercise-only group did not also show increases in BDNF given the extent of evidence that supports BDNF up-regulation in response to exercise (Knaepen et al. 2010). It is also still unclear how plasma BDNF relates to BDNF in the hippocampus.

Summary and Outstanding Questions

We have shown that experience-dependent brain plasticity has been demonstrated within the hippocampus in healthy adults and selected patient populations with neurological diseases that affect the hippocampus. Structural changes that take place in response to cognitive or physical training, or their combination, and which correspond with improved cognitive functions reflect the characteristics of plasticity described by Lövdén and colleagues (2010). We detailed many studies that demonstrated these plasticity processes in humans in response to a variety of training programs, with durations ranging from 8 weeks (Engvig et al. 2012, 2014; Kühn et al. 2014) to 2 years (Carlson et al. 2015). These data are consistent with another current theory of plasticity (Walhovd et al. 2015), which proposes that the medial temporal lobe, including the hippocampus, possesses many characteristics that provide a premise for plasticity, including low evolutionary expansion, low genetic correlation or heritability, and high variability in change of cortical thickness and myelin content. As a result, the medial temporal lobe is uniquely positioned for experience-dependent change, in both negative (e.g., neurological disease) and positive (e.g., neurogenesis) ways. However, there are surprisingly few studies that target hippocampal plasticity through the training of specific functions thought to depend on the hippocampus, and which include MR imaging to evaluate changes in hippocampal structure and function in relation to training-related changes in cognitive performance. This is an important gap to address for empirically testing the role of the hippocampus in the broad cognitive decline associated with aging and its responsiveness to experiential stimulation from cognitive and physical training.

While studies that demonstrate training-induced changes in brain structure can be classified as demonstrating plasticity, it is often more difficult to determine whether changes in functional imaging outcomes reflect plasticity or flexibility. Again, when thinking about functional outcomes, flexibility refers to the behavioral repertoire and functional activity within the range of currently available resources, whereas plasticity refers to changes in behavior that are the result of rejuvenation or reorganization in brain structures and their connections. Flexibility reflects the brain's need for ongoing rapid, transient reconfigurations to support interactions between internal goal states and continually changing external demands. Unless these flexible processes are accompanied by physical changes (such as synaptic connections, density of neurotransmitter receptors, neurogenesis, etc.), they may not indicate training-induced plasticity. An example of a change in functional brain activity or connectivity due to flexibility would be a change in strategy from pre- to post-training that could alter the functional activation observed during performance without altering the structure or functional characteristics of brain regions that give rise to performance.

Given this possible scenario, we can think of several (not mutually exclusive) strategies for studies to examine functional plasticity in the hippocampus with MR imaging that have rarely been applied thus far. First, investigators could include measures of self-reported strategy that are acquired following any behavioral tasks administered during functional imaging. Information about change in strategy would provide context for interpreting change in activation as plasticity or flexibility. In addition, studies could include assessment of changes in the resting FC of hippocampal networks that are known to overlap with networks evoked during behavioral performance, and it would be unlikely that there would be a strategy change in response to instructions to rest quietly in the scanner. Further, studies including either task or resting state functional imaging could more directly test for links between training-induced changes in structure and function. While changes in structure and function do not necessarily occur on the same timescales, this should not be a prerequisite for observing an association between the two and this analytic approach would ground changes in functional outcomes more strongly to plasticity than functional outcomes alone.

Another strategy would be exploring the similarities between various types of learning in animals and humans using cross-species investigations (Mishra and Gazzaley 2015). Animal studies have been able to investigate many variables that cannot be easily measured in humans such as micro-scale tissue changes like dendrite density or receptor density, genetic interactions, or electrophysiology of hippocampal neurons. Therefore, experimental designs that enable connecting training-induced changes in these cellular and molecular outcomes to changes in MR imaging outcomes could provide deep insight into how to measure brain plasticity with human imaging. For example, one study (Sagi et al. 2012) determined that in both humans and rats, as little as 2 h of spatial learning and memory training improved performance and decreased MD in the hippocampus and other nearby regions, suggesting training increased hippocampal tissue density.

Supporting evidence from the animal study provides a conceptual replication and strengthens knowledge of possible cellular changes underlying the change observed with MRI in humans. This study also highlights just how short the "training" period can be before observing structural changes, and may also point out that we need to consider whether and when it is necessary to conceptually distinguish between learning and cognitive training. In any case, future studies comparing effects of cognitive training in animals and humans may help clarify the time-scale of different types of experience-induced changes in hippocampal structure and function and their relationship to changes in performance. Similar to this approach with cognitive training, animal models can provide a parallel foundation of knowledge of basic mechanisms underlying exercise-related plasticity (Voss et al. 2013b). While animal models may not fully translate to humans because of the larger repertoire of human behaviors and experiences that could impact the experience and effectiveness of training, greater understanding of how basic processes studied with animal models relate to human outcomes could inspire interventions that are driven more by an understanding of the mechanisms for change.

We also evaluated examples of brain maintenance, wherein cognitive or physical training delayed decline in hippocampal structure or function, or both. This was most notably demonstrated by Lövdén et al. (2012) in their study of virtual environment spatial navigation training in young and older men. Compared to a group that slowly walked on a treadmill with no spatial navigation training, the authors observed a training ability-related maintenance of hippocampal volume and an increase in spatial navigation. Brain maintenance can also be inferred in studies of individuals with MCI or SMI (Engvig et al. 2012, 2014), when training seems to mitigate greater expected atrophy. However, studies need to include appropriate control and comparison groups to clarify the precise effects of training (similar age groups, cognitive health, and active control conditions). Importantly, ideally plasticity, flexibility, and brain maintenance are each evaluated in the context of accompanying behavioral and cognitive changes thought to rely on the underlying brain regions affected by training.

As alluded to above, a conceptual weakness of this literature is the lack of appropriate control groups that limit the extent to which we can conclude effects of training are specific to the training program (Boot et al. 2013; Thomas and Baker 2013). For cognitive training, researchers should include active control groups that differ from the cognitive training group in one key way. For example, the intervention group may undergo training of various cognitive functions (episodic memory, spatial navigation, task-switching) by completing tasks of increasing difficulty (adaptive training). In this case, the control program could include training of those same functions, but with tasks of fixed difficulty. At a bare minimum, the control group should engage in an active program that acts as a placebo to safeguard against differential expectations and motivation at post-testing. Similarly, with physical training, control programs should be matched on every aspect possible except for the characteristic of interest. For instance, control groups should be matched on social interaction. If training participants are meeting multiple times a week with trainers and other participants, control participants should have the same schedule.

Many exercise studies have used stretching and toning as control exercises, in comparison to aerobic exercises. However, there could be multiple factors that differ between the experimental aerobic and control stretching/toning programs. The level of engagement with trainers, involvement of spatial navigation, or one's own physiologic response (such as monitoring heart rate and reaching heart rate goals) may all vary between aerobic and stretching/toning groups. As an alternative, for maximal experimental control, the control group should experience similar stimulation as the experimental group, with only the variable of interest eliminated or modified. As evident by our review, numerous studies have already demonstrated general benefits of training compared to no training, therefore control groups should now be used to help clarify the critical ingredients for a training program to be effective.

Further, studies should more carefully consider whether the claim of "training-induced" plasticity is appropriately tested and whether sufficient statistics are used to estimate power and control for multiple comparisons (Thomas and Baker 2013). Strong evidence for training-induced change requires an interaction between group and time variables on the outcome measure (either brain or cognitive measures). Some studies have instead reported a main effect of time within the training group, but not the control group, without reporting the interaction effects, and this raises questions about the strength of the finding. Regarding specificity of training-induced change, another conceptual limitation of many studies is the use of targeted analyses of the hippocampus. Such analyses include segmenting or defining the hippocampus as an isolated region of interest without also evaluating other regions for comparison. Although determining the effects of training on comparison regions is not necessary for detecting an effect on the hippocampus, comparison regions are critical for determining the specificity of change in relation to change in performance and for narrowing in on possible mechanistic explanations for change. As discussed more below, limiting analyses to only the hippocampus also limits the ability to evaluate how training impacts the structural and functional connections of the hippocampus with other brain regions. This type of systems-approach will be especially important for understanding how training drives changes in behavior and cognitive performance.

Overall, while many studies did demonstrate changes in the hippocampus, other studies that anticipated hippocampal changes did not find support for this prediction. This mixed support across different studies may in part be driven by the complex patterns of anatomical and functional gains and losses in normal aging (Ballesteros et al. 2015), and the wide variability seen across individuals in the amount and rate of change each person experiences (Baltes et al. 2005). To overcome these intrinsic challenges in the data, the field will benefit from evaluating promising training effects with designs that have enough statistical power to compare change across sub-groups based on baseline demographic, health, cognitive, or brain characteristics in order to identify profiles for responders and non-responders to different interventions. It may also be that timeseries designs would be a valuable complement to reports of change averaged over groups of individuals in traditional clinical trial designs.

We also found little empirical evidence for far transfer of benefits from cognitive training studies that either targeted hippocampal-dependent cognitive functions or which showed change in structure or function of the hippocampus. On the other hand, improvement in hippocampal structure and function from exercise programs may be evidence of far transfer, as improvements in cognitive function are seen in executive function and memory domains and it is unclear how exercise would directly train these processes. As inspiration from this pattern, one strategy for understanding the mechanisms of transfer for cognitive training may be to adopt a more systems-level perspective, rather than focusing on isolated brain regions, as "targets" for transfer. For instance, one reason exercise is proposed to have such broad influence on cognition is its capacity to affect the functional integration of the hippocampus in brain networks known to degrade with age and neurological disease (Voss et al. 2010, 2016). One such network is the DMN, which is also thought to interact with many other systems during cognition, and this interaction may serve as a point of transfer to a broad range of cognitive abilities. Similarly, to better understand the widespread effects of training across the brain, Taya et al. (2015) emphasized the value in characterizing training-related brain outcomes using graph theoretical approaches. In general, a graph approach examines the brain as a complex system of interacting regions, and can reveal which brain regions have the most influence on how well many other regions interact with each other for effective information processing. Therefore, similar to the idea of transfer occurring due to brain regions that have multiple functions and are affected by training, another promising mechanism for transfer could be training-induced plasticity in regions that have broad influence in the overall function of the complex (brain) network by simultaneously affecting many overlapping sub-networks involved in abilities outside of abilities that were explicitly trained (Taya et al. 2015).

In sum, while there is good evidence the hippocampus remains plastic into late life and that cognitive and physical activity can stimulate this plasticity, there is still much to learn in order to optimize the application of this knowledge. One important unknown apparent from our review is the time-scale of various types of training or how much those timescales vary across training type. Although current technology limits the spatial resolution of human imaging with which we can measure structural and functional changes, it is still possible to further define the time course and duration of training effects on brain structure and function across hours, days, weeks, months, and years. This coupled with parallel animal models promises to enhance our understanding of how to drive plasticity over extended periods of time. Determining individual time courses for training gains and change in outcomes for training programs will also be critical for effectively combining interventions for maximum benefit. In addition, current theories of cognitive enhancement and exercise-induced brain plasticity do not yet specify with great detail which aspects of training programs are most effective for engendering primary change and transfer of training. This may occur as more data are collected that allow us to better understand the mechanisms through which experience physically changes the brain and through which transfer occurs. Ultimately, designing interventions based on components that deliver maximum benefit becomes especially important in the context of the public arena, as certain ingredients of evidence-based interventions will be critical to preserve when implementing any program into realistic guidelines for the public. In this way, studies examining mechanisms in humans with MR imaging are positioned well to increase our understanding of how to take advantage of the natural plasticity of the hippocampus in order to translate training studies from the lab to theoretically-based training programs that are fun, effective, and easily accessible to the broader community.

References

Anderson-Hanley C, Arciero P, Brickman A, Nimon J, Okuma N, Westen S et al (2012) Exergaming and older adult cognition: A cluster randomized clinical trial. Am J Prev Med 42(2): 109–119. doi:10.1016/j.amepre.2011.10.016

Andrews-Hanna J, Snyder A, Vincent J, Lustig C, Head D, Raichle M, Buckner R (2007) Disruption of large-scale brain systems in advanced aging. Neuron 56(5):924–935. doi:10. 1016/j.neuron.2007.10.038

Assaf Y, Pasternak O (2008) Diffusion tensor imaging (DTI)-based white matter mapping in brain research: a review. J Mol Neurosci 34(1):51–61. doi:10.1007/s12031-007-0029-0

Ballesteros S, Kraft E, Santana S, Tziraki C (2015) Maintaining older brain functionality: a targeted review. Neurosci Biobehav Rev 55:453–477. doi:10.1016/j.neubiorev.2015.06.008

Baltes P, Freund A, Li S-C (eds) (2005) The psychological science of human ageing. Cambridge University Press, Cambridge

Bamidis P, Vivas A, Styliadis C, Frantzidis C, Klados M, Schlee W et al (2014) A review of physical and cognitive interventions in aging. Neurosci Biobehav Rev 44:206–220. doi:10. 1016/j.neubiorev.2014.03.019

Barnes D, Yaffe K (2011) The projected effect of risk factor reduction on Alzheimer's disease prevalence. Lancet Neurol 10(9):819–828. doi:10.1016/S1474-4422(11)70072-2

Basak C, Boot W, Voss M, Kramer A (2008) Can training in a real-time strategy video game attenuate cognitive decline in older adults? Psychol Aging 23(4):765–777. doi:10.1037/ a0013494

Belleville S, Gilbert B, Fontaine F, Gagnon L, Ménard E, Gauthier S (2006) Improvement of episodic memory in persons with mild cognitive impairment and healthy older adults: evidence from a cognitive intervention program. Dement Geriatr Cogn Disord 22(5-6):486–499 Retrieved from http://www.karger.com/DOI/10.1159/000096316

Belleville S, Clement F, Mellah S, Gilbert B, Fontaine F, Gauthier S (2011) Training-related brain plasticity in subjects at risk of developing Alzheimer's disease. Brain 134(Pt 6):1623–1634. doi:10.1093/brain/awr037

Benedict C, Brooks S, Kullberg J, Nordenskjöld R, Burgos J, Le Grevès M et al (2012) Association between physical activity and brain health in older adults. Neurobiol Aging 34:83–90. doi:10. 1016/j.neurobiolaging.2012.04.013

Bookstein F (2001) "Voxel-based morphometry" should not be used with imperfectly registered images. NeuroImage 14(6):1454–1462. doi:10.1006/nimg.2001.0770

Boot W, Basak C, Erickson K, Neider M, Simons D, Fabiani M et al (2010) Transfer of skill engendered by complex task training under conditions of variable priority. Acta Psychol 135(3):349–357. doi:10.1016/j.actpsy.2010.09.005

Boot W, Simons D, Stothart C, Stutts C (2013) The pervasive problem with placebos in psychology: why active control groups are not sufficient to rule out placebo effects. Perspect Psychol Sci 8(4):445–454. doi:10.1177/1745691613491271

Boyke J, Driemeyer J, Gaser C, Büchel C, May A (2008) Training-induced brain structure changes in the elderly. J Neurosci 28(28):7031–7035. doi:10.1523/JNEUROSCI.0742-08.2008

Brookmeyer R, Evans D, Hebert L, Langa K, Heeringa S, Plassman B, Kukull W (2011) National estimates of the prevalence of Alzheimer's disease in the United States. Alzheimers Dement 7(1):61–73. doi:10.1016/j.jalz.2010.11.007

Buckner R, Andrews-Hanna J, Schacter D (2008) The brain's default network: anatomy, function, and relevance to disease. Ann N Y Acad Sci 1124(1):1–38. doi:10.1196/annals.1440.011

Burdette J, Laurienti P, Espeland M, Morgan A, Telesford Q, Vechlekar C et al (2010) Using network science to evaluate exercise-associated brain changes in older adults. Front Aging Neurosci 2(June):23–23. doi:10.3389/fnagi.2010.00023

Carlson M, Saczynski J, Rebok G, Seeman T, Glass T, McGill S et al (2008) Exploring the effects of an "everyday" activity program on executive function and memory in older adults: experience corps. The Gerontologist 48(6):793–801. doi:10.1093/geront/48.6.793

Carlson M, Erickson K, Kramer A, Voss M, Bolea N, Mielke M et al (2009) Evidence for neurocognitive plasticity in at-risk older adults: the experience corps program. J Gerontol Ser A Biol Med Sci 64A(12):1275–1282. doi:10.1093/gerona/glp117

Carlson M, Kuo J, Chuang Y, Varma V, Harris G, Albert M et al (2015) Impact of the baltimore experience corps trial on cortical and hippocampal volumes. Alzheimers Dement:1–9. doi:10.1016/j.jalz.2014.12.005

Chaddock L, Erickson K, Prakash R, Kim J, Voss M, Vanpatter M et al (2010) A neuroimaging investigation of the association between aerobic fitness, hippocampal volume, and memory performance in preadolescent children. Brain Res 1358:172–183. doi:10.1016/j.brainres.2010.08.049

Colcombe S, Kramer A (2003) Fitness effects on the cognitive function of older adults: a meta-analytic study. Psycholog Sci 14(2):125–130 Retrieved from http://www.ncbi.nlm.nih.gov/pubmed/12661673 http://pss.sagepub.com/content/14/2/125.full.pdf

Colcombe S, Erickson K, Scalf P, Kim J, Prakash R, McAuley E et al (2006) Aerobic exercise training increases brain volume in aging humans. J Gerontol A Biol Sci Med Sci 61(11):1166–1170. doi:10.1093/gerona/61.11.1166

Creer D, Romberg C, Saksida L, van Praag H, Bussey T (2010) Running enhances spatial pattern separation in mice. PNAS 107(5):2367–2372. doi:10.1073/pnas.0911725107

Dahlin E, Neely A, Larsson A, Backman L, Nyberg L (2008) Transfer of learning after updating training mediated by the striatum. Science 320(5882):1510–1512. doi:10.1126/science.1155466

Dahlin E, Backman L, Neely A, Nyberg L (2009) Training of the executive component of working memory: subcortical areas mediate transfer effects. Restor Neurol Neurosci 27(5):405–419. doi:10.3233/RNN-2009-0492

Damoiseaux J, Beckmann C, Arigita E, Barkhof F, Scheltens P, Stam C et al (2008) Reduced resting-state brain activity in the "default network" in normal aging. Cereb Cortex 18(8):1856–1864. doi:10.1093/cercor/bhm207

Davatzikos C (2004) Why voxel-based morphometric analysis should be used with great caution when characterizing group differences. NeuroImage 23(1):17–20. doi:10.1016/j.neuroimage.2004.05.010

Dhami P, Moreno S, DeSouza J (2015) New framework for rehabilitation – fusion of cognitive and physical rehabilitation: the hope for dancing. Front Psychol 5(January):1–15. doi:10.3389/fpsyg.2014.01478

Doyon J, Bellec P, Amsel R, Penhune V, Monchi O, Carrier J et al (2009) Contributions of the basal ganglia and functionally related brain structures to motor learning. Behav Brain Res 199(1):61–75. doi:10.1016/j.bbr.2008.11.012

Draganski B, Gaser C, Busch V, Schuierer G, Bogdahn U, May A (2004) Changes in grey matter induced by training newly honed juggling skills show up as a transient feature on a brain-imaging scan. Nature 427:311–312. doi:10.1038/427311a

Draganski B, Gaser C, Kempermann G, Kuhn H, Winkler J, Büchel C, May A (2006) Temporal and spatial dynamics of brain structure changes during extensive learning. J Neurosci 26(23):6314–6317. doi:10.1523/JNEUROSCI.4628-05.2006

Eichenbaum H (2004) Hippocampus: cognitive processes and neural representations that underlie declarative memory. Neuron 44(1):109–120. doi:10.1016/j.neuron.2004.08.028

Engvig A, Fjell A, Westlye L, Skaane N, Sundseth Ø, Walhovd K (2012) Hippocampal subfield volumes correlate with memory training benefit in subjective memory impairment. Neuro-Image 61(1):188–194. doi:10.1016/j.neuroimage.2012.02.072

Engvig A, Fjell A, Westlye L, Skaane N, Dale A, Holland D et al (2014) Effects of cognitive training on gray matter volumes in memory clinic patients with subjective memory impairment. J Alzheimers Dis 41(3):779–791. doi:10.3233/JAD-131889

Erickson K, Prakash R, Voss M, Chaddock L, Hu L, Morris K et al (2009) Aerobic fitness is associated with hippocampal volume in elderly humans. Hippocampus 19(10):1030–1039. doi:10.1002/hipo.20547

Erickson K, Voss M, Prakash R, Basak C, Szabo A, Chaddock L et al (2011) Exercise training increases size of hippocampus and improves memory. Proc Natl Acad Sci USA 108(7):3017–3022. doi:10.1073/pnas.1015950108

Erickson K, Leckie R, Weinstein A (2014) Physical activity, fitness, and gray matter volume. Neurobiol Aging 35(Suppl 2):S20–S28. doi:10.1016/j.neurobiolaging.2014.03.034

Fabel K, Fabel K, Tam B, Kaufer D, Baiker A, Simmons N et al (2003) VEGF is necessary for exercise-induced adult hippocampal neurogenesis. Eur J Neurosci 18(10):2803–2812. doi:10.1046/j.1460-9568.2003.03041.x

Ferris L, Williams J, Shen C (2007) The effect of acute exercise on serum brain-derived neurotrophic factor levels and cognitive function. Med Sci Sports Exerc 39(4):728–734. doi:10.1249/mss.0b013e31802f04c7

Fissler P, Kuster O, Schlee W, Kolassa I (2013) Novelty interventions to enhance broad cognitive abilities and prevent dementia: synergistic approaches for the facilitation of positive plastic change. Prog Brain Res 207:403–434

Fox M, Snyder A, Vincent J, Corbetta M, Essen D, Raichle M (2005) The human brain is intrinsically organized into dynamic, anticorrelated functional networks. Proc Natl Acad Sci USA 102:9673–9678

Green C, Bavelier D (2003) Action video game modi es visual selective attention. Nature 423 (May):3–6

Hampstead B, Stringer A, Stilla R, Deshpande G, Hu X, Moore A, Sathian K (2011) Activation and effective connectivity changes following explicit-memory training for face-name pairs in patients with mild cognitive impairment: a pilot study. Neurorehabil Neural Repair 25(3):210–222. doi:10.1177/1545968310382424

Hampstead B, Stringer A, Stilla R, Giddens M, Sathian K (2012) Mnemonic strategy training partially restores hippocampal activity in patients with mild cognitive impairment. Hippocampus 22(8):1652–1658. doi:10.1002/hipo.22006

Hedden T, Schultz A, Rieckmann A, Mormino E, Johnson K, Sperling R, Buckner R (2014) Multiple brain markers are linked to age-related variation in cognition. Cereb Cortex:1–13. doi:10.1093/cercor/bhu238

Herting M, Nagel B (2012) Aerobic fitness relates to learning on a virtual Morris Water Task and hippocampal volume in adolescents. Behav Brain Res 233(2):517–525. doi:10.1016/j.bbr.2012.05.012

Hertzog C, Kramer A, Wilson R, Lindenberger U (2009) Enrichment effects on adult cognitive development can the functional capacity of older adults be preserved and enhanced? Psychol Sci Public Interest 9(1):1–65. doi:10.1111/j.1539-6053.2009.01034.x

Hurd M, Martorell P, Delavande A, Mullen K, Langa K (2013) Monetary costs of dementia in the United States. N Engl J Med 368(14):1326–1334. doi:10.1056/NEJMsa1204629

Jack CJ, Knopman D, Jagust W, Petersen R, Weiner M, Aisen P et al (2013) Tracking pathophysiological processes in Alzheimer's disease: an updated hypothetical model of dynamic biomarkers. Lancet Neurol 12(2):207–216. doi:10.1016/S1474-4422(12)70291-0

Jessen F, Wiese B, Bachmann C, Eifflaender-Gorfer S, Haller F, Kolsch H et al (2010) Prediction of dementia by subjective memory impairment. Arch Gen Psychiatry 67(4):414–422 Retrieved from http://archpsyc.jamanetwork.com/data/Journals/PSYCH/5293/yoa90083_414_422.pdf

Kelly C, Foxe J, Garavan H (2006) Patterns of normal human brain plasticity after practice and their implications for neurorehabilitation. Arch Phys Med Rehabil 87(Suppl. 12):S20–S29. doi:10.1016/j.apmr.2006.08.333

Kempermann G, Brandon E, Gage F (1998) Environmental stimulation of 129/SvJ mice causes increased cell proliferation and neurogenesis in the adult dentate gyrus. Curr Biol 8(16): 939–944. doi:10.1016/S0960-9822(07)00377-6

Kempermann G, Gast D, Gage F (2002) Neuroplasticity in old age: sustained fivefold induction of hippocampal neurogenesis by long-term environmental enrichment. Ann Neurol 52(2): 135–143. doi:10.1002/ana.10262

Kennedy K, Erickson K, Rodrigue K, Voss M, Colcombe S, Kramer A et al (2009) Age-related differences in regional brain volumes: a comparison of optimized voxel-based morphometry to manual volumetry. Neurobiol Aging 30(10):1657–1676. doi:10.1016/j.neurobiolaging.2007. 12.020

Kirchhoff B, Anderson B, Smith S, Barch D, Jacoby L (2012) Cognitive training-related changes in hippocampal activity associated with recollection in older adults. NeuroImage 62(3): 1956–1964. doi:10.1016/j.neuroimage.2012.06.017

Knaepen K, Goekint M, Heyman E, Meeusen R (2010) Neuroplasticity – exercise-induced response of peripheral brain-derived neurotrophic factor: a systematic review of experimental studies in human subjects. Sports Med 40(9):765–801. doi:10.2165/11534530-000000000-00000

Konishi K, Bohbot V (2013) Spatial navigational strategies correlate with gray matter in the hippocampus of healthy older adults tested in a virtual maze. Front Aging Neurosci 5 (February):1–8. doi:10.3389/fnagi.2013.00001

Konkel A, Warren D, Duff M, Tranel D, Cohen N (2008) Hippocampal amnesia impairs all manner of relational memory. Front Hum Neurosci 2:15. doi:10.3389/neuro.09.015.2008

Kramer A, Larish J, Strayer D (1995) Training for attentional control in dual task settings: a comparison of young and old adults. J Exp Psychol Appl 1(1):50–76. doi:10.1037/1076-898X.1.1.50

Kramer A, Hahn S, Cohen N, Banich M, McAuley E, Harrison C et al (1999a) Ageing, fitness and neurocognitive function. Nature. 400(6743):418–419 Retrieved from http://www.nature.com/nature/journal/v400/n6743/abs/400418a0.html

Kramer A, Larish J, Weber T, Bardell L (eds) (1999b) Training for executive control: task coordination strategies and aging. MIT Press, Cambridge, MA

Kronenberg G, Reuter K, Steiner B, Brandt M, Jessberger S, Yamaguchi M, Kempermann G (2003) Subpopulations of proliferating cells of the adult hippocampus respond differently to physiologic neurogenic stimuli. J Comp Neurol 467(4):455–463. doi:10.1002/cne.10945

Kühn S, Gleich T, Lorenz R, Lindenberger U, Gallinat J (2014) Playing Super Mario induces structural brain plasticity: gray matter changes resulting from training with a commercial video game. Mol Psychiatry 19(2):265–271. doi:10.1038/mp.2013.120

Lampit A, Hallock H, Moss R, Kwok S, Rosser M, Lukjanenko M et al (2014) The timecourse of global cognitive gains from supervised computer-assisted cognitive training: a randomised, active-controlled trial in elderly with multiple dementia risk factors. J Prev Alzheimers Dis 1(1):33–39

Lampit A, Hallock H, Suo C, Naismith S, Valenzuela M (2015) Cognitive training-induced short-term functional and long-term structural plastic change is related to gains in global cognition in healthy older adults: a pilot study. Front Aging Neurosci 7(14):1–13. doi:10.3389/fnagi.2015. 00014

Langer N, von Bastian C, Wirz H, Oberauer K, Jäncke L (2013) The effects of working memory training on functional brain network efficiency. Cortex 49(9):2424–2438. doi:10.1016/j.cortex. 2013.01.008

Lerner R (1984). The life-span view of human development: philisophical, historical, and substantive bases. In: On the nautre of human plasticity. Cambridge University Press, Cambridge, 22–31

Li R, Zhu X, Yin S, Niu Y, Zheng Z, Huang X et al (2014) Multimodal intervention in older adults improves resting-state functional connectivity between the medial prefrontal cortex and medial temporal lobe. Front Aging Neurosci 6(39):1–13. doi:10.3389/fnagi.2014.00039

Lindenberger U (2014) Human cognitive aging: corriger la fortune? Science 346(6209):572–578. doi:10.1126/science.1254403

Liu-Ambrose T, Nagamatsu L, Voss M, Khan K, Handy T (2012) Resistance training and functional plasticity of the aging brain: a 12-month randomized controlled trial. Neurobiol Aging 33(8):1690–1698. doi:10.1016/j.neurobiolaging.2011.05.010

López-Otín C, Blasco M, Partridge L, Serrano M, Kroemer G (2013) The hallmarks of aging. Cell 153(June):1194–1217. doi:10.1016/j.cell.2013.05.039

Lövdén M, Backman L, Lindenberger U, Schaefer S, Schmiedek F (2010) A theoretical framework for the study of adult cognitive plasticity. Psychol Bull 136(4):659–676. doi:10.1037/a0020080

Lövdén M, Schaefer S, Noack H, Bodammer N, Kuhn S, Heinze H et al (2012) Spatial navigation training protects the hippocampus against age-related changes during early and late adulthood. Neurobiol Aging 33(3):620.e9–620.e22. doi:10.1016/j.neurobiolaging.2011.02.013

Lustig C, Shah P, Seidler R, Reuter-Lorenz P (2009) Aging, training, and the brain: a review and future directions. Neuropsychol Rev 19(4):504–522. doi:10.1007/s11065-009-9119-9

Maass A, Duzel S, Goerke M, Becke A, Sobieray U, Neumann K et al (2014) Vascular hippocampal plasticity after aerobic exercise in older adults. Mol Psychiatry. doi:10.1038/mp.2014.114

Mahncke H, Connor B, Appelman J, Ahsanuddin O, Hardy J, Wood R et al (2006) Memory enhancement in healthy older adults using a brain plasticity-based training program: a randomized, controlled study. Proc Natl Acad Sci USA 103(33):12523–12528. doi:10.1073/pnas.0605194103

Makizako H, Liu-Ambrose T, Shimada H, Doi T, Park H, Tsutsumimoto K et al (2014) Moderate-intensity physical activity, hippocampal volume, and memory in older adults with mild cognitive impairment. J Gerontol A Biol Sci Med Sci:1–7. doi:10.1093/gerona/glu136

Malchow B, Keeser D, Keller K, Hasan A, Rauchmann B, Kimura H et al (2015) Effects of endurance training on brain structures in chronic schizophrenia patients and healthy controls. Schizophr Res. doi:10.1016/j.schres.2015.01.005

Mishra J, Gazzaley A (2015) Cross-species approaches to cognitive neuroplasticity research. NeuroImage. doi:10.1016/j.neuroimage.2015.09.002

Morey R, Petty C, Xu Y, Pannu Hayes J, Wagner H, Lewis D et al (2009) A comparison of automated segmentation and manual tracing for quantifying hippocampal and amygdala volumes. NeuroImage 45(3):855–866. doi:10.1016/j.neuroimage.2008.12.033

Mulder E, de Jong R, Knol D, van Schijndel R, Cover K, Visser P et al (2014) Hippocampal volume change measurement: quantitative assessment of the reproducibility of expert manual outlining and the automated methods FreeSurfer and FIRST. NeuroImage 92:169–181. doi:10.1016/j.neuroimage.2014.01.058

Nagamatsu L, Handy T, Hsu C, Voss M, Liu-Ambrose T (2012) Resistance training promotes cognitive and functional brain plasticity in seniors with probable mild cognitive impairment. Arch Intern Med 172:666–668. doi:10.1001/archinternmed.2012.379

Neeper S, Gómez-Pinilla F, Choi J, Cotman C (1996) Physical activity increases mRNA for brain-derived neurotrophic factor and nerve growth factor in rat brain. Brain Res 726(1–2):49–56 Retrieved from http://www.ncbi.nlm.nih.gov/pubmed/8836544

Niemann C, Godde B, Voelcker-Rehage C (2014) Not only cardiovascular, but also coordinative exercise increases hippocampal volume in older adults. Front Aging Neurosci 6:1–12. doi:10.3389/fnagi.2014.00170

Nishijima T, Kawakami M, Kita I (2015) A bout of treadmill exercise increases matrix metalloproteinase-9 activity in the rat hippocampus. Neurosci Lett 594:144–149. doi:10.1016/j.neulet.2015.03.063

Noack H, Lövdén M, Schmiedek F (2014) On the validity and generality of transfer effects in cognitive training research. Psychol Res 78(6):773–789. doi:10.1007/s00426-014-0564-6

Norton S, Matthews F, Barnes D, Yaffe K, Brayne C (2014) Potential for primary prevention of Alzheimer's disease: an analysis of population-based data. Lancet Neurol 13(8):788–794. doi:10.1016/S1474-4422(14)70136-X

Nyberg L, Lövdén M, Riklund K, Lindenberger U, Bäckman L (2012) Memory aging and brain maintenance. Trends Cogn Sci 16(5):292–305. doi:10.1016/j.tics.2012.04.005

Opendak M, Gould E (2015) Adult neurogenesis: a substrate for experience-dependent change. Trends Cogn Sci 19(3):151–161. doi:10.1016/j.tics.2015.01.001

Pajonk F, Wobrock T, Gruber O, Scherk H, Berner D, Kaizl I et al (2010) Hippocampal plasticity in response to exercise in schizophrenia. Arch Gen Psychiatry 67(2):133–143. doi:10.1001/archgenpsychiatry.2009.193

Park D, Reuter-Lorenz P (2009) The adaptive brain: aging and neurocognitive scaffolding. Annu Rev Psychol 60:173–196. doi:10.1146/annurev.psych.59.103006.093656

Pereira A, Huddleston D, Brickman A, Sosunov A, Hen R, McKhann G et al (2007) An in vivo correlate of exercise-induced neurogenesis in the adult dentate gyrus. Proc Natl Acad Sci USA 104(13):5638–5643. doi:10.1073/pnas.0611721104

Perkins DN, Salomon G (1992) Transfer of learning. Int enc educ 2:6452–6457

Persson J, Pudas S, Lind J, Kauppi K, Nilsson L, Nyberg L (2012) Longitudinal structure-function correlates in elderly reveal MTL dysfunction with cognitive decline. Cereb Cortex 22(10):2297–2304. doi:10.1093/cercor/bhr306

Poldrack R (2000) Imaging brain plasticity: conceptual and methodological issues—a theoretical review. NeuroImage 12:1–13. doi:10.1006/nimg.2000.0596

Poldrack R (2015) Is "efficiency" a useful concept in cognitive neuroscience? Dev Cogn Neurosci 11:12–17. doi:10.1016/j.dcn.2014.06.001

Rasmussen P, Brassard P, Adser H, Pedersen M, Leick L, Hart E et al (2009) Evidence for a release of brain-derived neurotrophic factor from the brain during exercise. Exp Physiol 94(10):1062–1069. doi:10.1113/expphysiol.2009.048512

Raz N, Lindenberger U (2013) Life-span plasticity of the brain and cognition: from questions to evidence and back. Neurosci Biobehav Rev 37(9 Pt B):2195–2200. doi:10.1016/j.neubiorev.2013.10.003

Raz N, Lindenberger U, Rodrigue K, Kennedy K, Head D, Williamson A et al (2005) Regional brain changes in aging healthy adults: general trends, individual differences and modifiers. Cereb Cortex 15(11):1676–1689. doi:10.1093/cercor/bhi044

Reuter-Lorenz P, Park D (2014) How does it STAC up? Revisiting the scaffolding theory of aging and cognition. Neuropsychol Rev 24(3):355–370. doi:10.1007/s11065-014-9270-9

Rieckmann A, Karlsson S, Karlsson P, Brehmer Y, Fischer H, Farde L et al (2011) Dopamine D1 receptor associations within and between dopaminergic pathways in younger and elderly adults: links to cognitive performance. Cereb Cortex 21(9):2023–2032. doi:10.1093/cercor/bhq266

Rojas Vega S, Strüder H, Vera Wahrmann B, Schmidt A, Bloch W, Hollmann W (2006) Acute BDNF and cortisol response to low intensity exercise and following ramp incremental exercise to exhaustion in humans. Brain Res 1121(1):59–65. doi:10.1016/j.brainres.2006.08.105

Rosen A, Sugiura L, Kramer J, Whitfield-Gabrieli S, Gabrieli J (2011) Cognitive training changes hippocampal function in mild cognitive impairment: a pilot study. J Alzheimers Dis 26(Suppl 3):349–357. doi:10.3233/JAD-2011-0009

Sagi Y, Tavor I, Hofstetter S, Tzur-Moryosef S, Blumenfeld-Katzir T, Assaf Y (2012) Learning in the fast lane: new insights into neuroplasticity. Neuron 73(6):1195–1203. doi:10.1016/j.neuron.2012.01.025

Salthouse T (2010) Selective review of cognitive aging. J Int Neuropsychol Soc 16(5):754–760. doi:10.1017/S1355617710000706

Schendan H, Searl M, Melrose R, Stern C (2003) An fMRI study of the role of the medial temporal lobe in implicit and explicit sequence learning. Neuron 37(6):1013–1025. doi:10.1016/S0896-6273(03)00123-5

Schoemaker D, Buss C, Head K, Sandman C, Davis E, Chakravarty M et al (2016) Hippocampus and amygdala volumes from magnetic resonance images in children: assessing accuracy of FreeSurfer and FSL against manual segmentation. NeuroImage 129:1–14. doi:10.1016/j.neuroimage.2016.01.038

Shaw E, Schultz A, Sperling R, Hedden T (2015) Functional connectivity in multiple cortical networks is associated with performance across cognitive domains in older adults. Brain Connect 5(8):505–516. doi:10.1089/brain.2014.0327

Shohamy D, Turk-Browne N (2013) Mechanisms for widespread hippocampal involvement in cognition. J Exp Psychol Gen 142(4):1159–1170. doi:10.1037/a0034461

Smith P, Blumenthal J, Hoffman B, Cooper H, Strauman T, Welsh-Bohmer K et al (2010) Aerobic exercise and neurocognitive performance: a meta-analytic review of randomized controlled trials. Psychosom Med 72(3):239–252. doi:10.1097/PSY.0b013e3181d14633

Squire L, Stark C, Clark R (2004) The medial temporal lobe. Annu Rev Neurosci 27:279–306. doi:10.1146/annurev.neuro.27.070203.144130

Taya F, Sun Y, Babiloni F, Thakor N, Bezerianos A (2015) Brain enhancement through cognitive training: a new insight from brain connectome. Front Syst Neurosci 9(April):1–19. doi:10.3389/fnsys.2015.00044

ten Brinke L, Bolandzadeh N, Nagamatsu L, Hsu C, Davis J, Miran-Khan K, Liu-Ambrose T (2015) Aerobic exercise increases hippocampal volume in older women with probable mild cognitive impairment: a 6-month randomised controlled trial. Br J Sports Med 49(4):248–254. doi:10.1136/bjsports-2013-093184

Thomas C, Baker C (2013) Teaching an adult brain new tricks: a critical review of evidence for training-dependent structural plasticity in humans. NeuroImage 73:225–236. doi:10.1016/j.neuroimage.2012.03.069

Thomas A, Marrett S, Saad Z, Ruff D, Martin A, Bandettini P (2009) Functional but not structural changes associated with learning: an exploration of longitudinal Voxel-Based Morphometry (VBM). NeuroImage 48(1):117–125. doi:10.1016/j.neuroimage.2009.05.097

Thomas A, Dennis A, Bandettini P, Johansen-Berg H (2012) The effects of aerobic activity on brain structure. Frontiers in Psychology 3(Mar):1–9. doi:10.3389/fpsyg.2012.00086

van Praag H (2008) Neurogenesis and exercise: past and future directions. NeuroMolecular Med 10(2):128–140. doi:10.1007/s12017-008-8028-z

van Praag H, Christie B, Sejnowski T, Gage F (1999) Running enhances neurogenesis, learning, and long-term potentiation in mice. Proc Natl Acad Sci USA 96(23):13427–13431 Retrieved from http://www.pubmedcentral.nih.gov/articlerender.fcgi?artid=23964&tool=pmcentrez&rendertype=abstract http://www.ncbi.nlm.nih.gov/pmc/articles/PMC23964/pdf/pq013427.pdf

van Praag H, Kempermann G, Gage F (2000) Neural consequences of environmental enrichment. Nature reviews. Neuroscience 1(December):191–198. doi:10.1038/35044558

van Praag H, Shubert T, Zhao C, Gage F (2005) Exercise enhances learning and hippocampal neurogenesis in aged mice. J Neurosci 25(38):8680–8685. doi:10.1523/JNEUROSCI.1731-05.2005

Varma V, Chuang Y, Harris G, Tan E, Carlson M (2014) Low-intensity daily walking activity is associated with hippocampal volume in older adults. Hippocampus 25(5):605–615. doi:10.1002/hipo.22397

Varma V, Tan E, Gross A, Harris G, Romani W, Fried L et al (2015) Effect of community volunteering on physical activity. Am J Prev Med:1–5. doi:10.1016/j.amepre.2015.06.015

Vaynman S, Ying Z, Gomez-Pinilla F (2004) Hippocampal BDNF mediates the efficacy of exercise on synaptic plasticity and cognition. Eur J Neurosci 20(10):2580–2590. doi:10.1111/j.1460-9568.2004.03720.x

Voelcker-Rehage C, Godde B, Staudinger U (2011) Cardiovascular and coordination training differentially improve cognitive performance and neural processing in older adults. Front Hum Neurosci 5:1–12. doi:10.3389/fnhum.2011.00026

Voss M, Prakash R, Erickson K, Basak C, Chaddock L, Kim J et al (2010) Plasticity of brain networks in a randomized intervention trial of exercise training in older adults. Front Aging Neurosci 2(August):1–17. doi:10.3389/fnagi.2010.00032

Voss M, Prakash R, Erickson K, Boot W, Basak C, Neider M et al (2012) Effects of training strategies implemented in a complex videogame on functional connectivity of attentional networks. NeuroImage 59(1):138–148. doi:10.1016/j.neuroimage.2011.03.052

Voss M, Erickson K, Shaurya R, Chaddock L, Kim J, Alves H et al (2013a) Neurobiological markers of exercise-related brain plasticity in older adults. Brain Behav Immun 28:90–99

Voss M, Vivar C, Kramer A, van Praag H (2013b) Bridging animal and human models of exercise-induced brain plasticity. Trends Cogn Sci 17(10):525–544. doi:10.1016/j.tics.2013.08.001

Voss MW, Weng TB, Burzynska AZ, Wong CN, Cooke GE, Clark R, Fanning J, Awick E, Gothe NP, Olson EA, McAuley E, Kramer AF (2016) Fitness, but not physical activity, is related to functional integrity of brain networks associated with aging. NeuroImage 131:113–125

Walhovd K, Westerhausen R, Glasø de Lange A, Bråthen A, Grydeland H, Engvig A, Fjell A (2015) Premises of plasticity – and the loneliness of the medial temporal lobe. NeuroImage. doi:10.1016/j.neuroimage.2015.10.060

Wenger E, Mårtensson J, Noack H, Bodammer N, Kühn S, Schaefer S et al (2014) Comparing manual and automatic segmentation of hippocampal volumes: reliability and validity issues in younger and older brains. Hum Brain Mapp 35(8):4236–4248. doi:10.1002/hbm.22473

Woollett K, Maguire E (2011) Acquiring "the Knowledge" of London's layout drives structural brain changes. Curr Biol 21(24):2109–2114. doi:10.1016/j.cub.2011.11.018

Woollett K, Spiers H, Maguire E (2009) Talent in the taxi: a model system for exploring expertise. Philos Trans R Soc Lond B Biol Sci 364(1522):1407–1416. doi:10.1098/rstb.2008.0288

Yaffe K, Barnes D, Nevitt M, Lui L, Covinsky K (2001) A prospective study of physical activity and cognitive decline in elderly women: women who walk. Arch Intern Med 161(14): 1703–1708 Retrieved from http://www.ncbi.nlm.nih.gov/pubmed/11485502 http://archinte.jamanetwork.com/data/Journals/INTEMED/11989/ioi00861.pdf

Zelinski E (2009) Far transfer in cognitive training of older adults. Restor Neurol Neurosci 27(5): 455–471. doi:10.3233/RNN-2009-0495

Hippocampal Contributions to Declarative Memory Consolidation During Sleep

James W. Antony and Ken A. Paller

Abstract The human brain faces a fundamental information storage challenge—forming useful new memories while not over-writing important old ones. Memory consolidation, and the corresponding interplay between the hippocampus and neocortex, is a protracted process to adjudicate between these two competing factors. Converging evidence from behavioral, cellular, and systems neuroscience strongly implicates a special role for sleep in stabilizing new declarative memories. In this chapter, we review evidence that during sleep the reactivation of newly acquired neuronal traces has lasting implications for memory transformation and stabilization. We first summarize relevant theoretical issues in memory research and then outline the physiological properties of sleep that may allow for this reactivation. We consider many factors that affect spontaneous memory reactivation, and we highlight research showing that memories can be selectively targeted and modified using learning-related stimuli. Ultimately, the ability to rescue otherwise fleeting episodes from oblivion plays a vital role in human life. Research elucidating this ability will also be critical for understanding how memory breaks down in aging and disease.

During a young scientist's graduate school interviews, a senior researcher told her that she would not cut it in such a competitive field. At each major junction of her life—her first publication, first tenure-track job, a named professorship, and a lifetime achievement award—she remembered the researcher's exact words, his dismissive tone, and the seeds of doubt he planted about her career path as vividly as the day it happened.

Most learning requires repetition. A barrage of visual experience in early life is required for plasticity within the visual system (Wiesel and Hubel 1963). Hundreds to thousands of hours of practice are required to form expert procedural skills. So

J.W. Antony
Princeton Neuroscience Institute, Princeton University, Princeton, NJ 08544, USA

K.A. Paller (✉)
Department of Psychology, Northwestern University, Evanston, IL 60208, USA
e-mail: kap@northwestern.edu

© Springer International Publishing AG 2017   245
D.E. Hannula, M.C. Duff (eds.), *The Hippocampus from Cells to Systems*,
DOI 10.1007/978-3-319-50406-3_9

how can an event that occurred just once and took less than 3 s be stored in the connections of the brain for a full lifetime?

The answer appears to lie in the unique physiological properties of the hippocampus and its relationship with the neocortex. A personally important memory, though played out in the world only once, becomes repeatedly replayed after that one unique event by the networks of neurons involved in its formation and storage. A key feature of this ability lies in how much occurs outside of the agent's consciousness. Whereas the young scientist's memory of being told that she would never succeed in science likely returned to her consciousness when revisiting the memory in her mind or recounting the story to a friend, it seems that forming this lasting memory trace required nothing like the number of hours of experience or practice required for a highly refined skill. Thus, while most long-lasting experience-dependent changes in the brain require numerous repetitions to drive requisite changes in synaptic weights, episodic memories must become embedded in the brain and replayed on their own, without extensive efforts to re-live the experience over and over again.

This is not to say offline changes do not play a role in other types of memories. On the contrary, sensory and procedural memories benefit from offline processes, including sleep (Brawn et al. 2010b; Mednick et al. 2003). Additionally, there is evidence that the hippocampus may play a role in types of learning previously deemed to be hippocampal-independent (Albouy et al. 2013).

Nevertheless, something unique must occur that allows for lasting episodic memory traces. The following discussion will focus mostly on changes that occur to a memory trace after its initial formation, with consideration for how various factors operative at encoding might alter this process. We will take a historical perspective on the concept of memory consolidation and then consider the role by which memory reactivation influences consolidation. Although consolidation certainly occurs to some extent during wake, we will focus on the physiological properties present during sleep that create unique conditions for interactions between the hippocampus and neocortex.

Memory Consolidation

Brief Historical Perspective

In this section, we will discuss two major advancements in the concept of memory consolidation, specifically Müller and Pilzecker's (1900) original study that precipitated the creation of the concept and Scoville and Milner's (1957) research with patient H.M. (Fig. 1). We will finish by outlining what researchers theorize about the hippocampus and neocortex in explaining consolidation.

In a series of studies, Müller and Pilzecker (1900) taught participants lists of nonsense syllables and tested them after a delay of typically a few hours (Lechner et al. 1999). Between encoding and testing, they introduced other lists at various

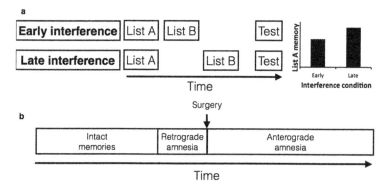

Fig. 1 Historical concepts in memory consolidation. (**a**) After encoding of a list of syllables (List A), introducing interfering information (List B) impairs List A memory when introduced sooner than later (Müller and Pilzecker 1900). A period of time without interference after learning is therefore beneficial for long-term memory stabilization. (**b**) Following medial temporal damage (such as when patient H.M. underwent surgery), the patient not only becomes unable to form new memories going forward (anterograde amnesia), but also cannot recall memories from an earlier time period (retrograde amnesia). Older memories remain largely intact. Therefore, during a prolonged period after learning, some declarative memories become independent of the medial temporal lobe

times within the delay interval and found that the later they gave the new lists, the better participants typically remembered the original information. Thus, they proposed a prolonged process eventually termed "consolidation" by which memories become increasingly resistant to interference. Their seminal finding remains a consistent and fundamental aspect of memory research today.

The next large piece of evidence about memory consolidation came from one of the most famous patients in the history of neuroscience known as H.M. (Scoville and Milner 1957). After H.M. had lived into his 20s with a form of epilepsy that resisted all other forms of treatment, the neurosurgeon William Scoville proposed performing brain surgery to remove the part of H.M.'s brain from which his seizures originated: a large portion of the medial temporal lobe, including the hippocampus. Little was known about this brain structure at the time, but the severity and frequency of his seizures seemed to warrant attempting such an experimental surgery. After removal of the hippocampus on both sides of his brain, the neuropsychologist Brenda Milner found H.M. could perform a whole host of mental functions and he retained many old memories. However, he forgot ones lasting up to a few years before his surgery and, critically, could no longer acquire new memories. He was thereafter, in the words of Suzanne Corkin (2013), trapped in a "permanent present tense."

H.M.'s impairment, while devastating, radicalized how researchers understood memory. The emerging theoretical picture suggested that when an event is learned, a confluence of information concurrently processed in the brain becomes bound together. This information enters the hippocampus from multiple neocortical streams specialized for processing highly detailed sensory inputs or thoughts as well as the spatial and temporal context in which the information arrives. The hippocampus then rapidly binds together these distinct components. Initially highly susceptible to interference, new links reach a stable form in the neocortex only after a period of time. The relevant steps may require a certain number of reactivations rather than a certain period of time, per se. In either case, consolidation can progress such that networks within the neocortex are sufficient for retrieval—although whether some memories, or the full re-experience of some memories, then depend on only neocortical networks or on both neocortical and hippocampal networks remains a hotly debated issue (see Moscovitch et al. (2005) for one perspective and Squire and Bayley (2007) for another). In either case, converging data across many amnesic patients and animal models has forged agreement among researchers upon this basic conceptualization of consolidation.

Consolidation = Reactivation?

We will begin this section by outlining how memory retrieval acts to reactivate, strengthen, and reorganize a memory trace. We will then argue how this process resembles what occurs spontaneously in the gradual process by which memories become stabilized.

Cognitive psychological models of retrieval have long stressed the role of context on memory retrieval (Godden and Baddeley 1975; Jensen et al. 1971). For instance, psychology students often learn that studying information in the room where the test eventually occurs produces better results. An analogous mental trick is to try to create a mental context that allows for the successful retrieval of a memory. A series of neuroimaging studies have investigated how this phenomenon manifests itself in the brain. Unsurprisingly, neural reactivation patterns at retrieval tend to match those at encoding (Buchsbaum et al. 2012; Gelbard-Sagiv et al. 2008; Johnson and Rugg 2007; Nyberg et al. 2000; Polyn et al. 2005). Moreover, better matches predict better memory (Johnson et al. 2009; Manning et al. 2012; St-Laurent et al. 2014; Wing et al. 2015). A straightforward explanation of these findings follows from the reinstatement hypothesis (Tulving and Thomson 1973); the better the learning context is reinstated, the more likely a memory will be successfully remembered.

A vast psychological literature shows that memory retrieval does not simply involve finding a memory from storage and placing it back unaffected, as in retrieving a book from a library and later returning it in the same condition. Rather, successful retrieval produces better long-term enhancement than re-exposure to the material itself, a phenomenon known as the testing effect (for a thorough review, see Roediger and Karpicke 2006). During retrieval or during repeated study, stored information can be reactivated, leading in theory to superior storage. A straightforward prediction follows that better neural reactivation of an encoding context would produce better long-term memory at a later test. This prediction has been supported both during repeated study (Newman and Norman 2010; Xue et al. 2010) and retrieval (Kuhl et al. 2010, 2012a, b; Vilberg and Davachi 2013; though see Karlsson Wirebring et al. 2015).

The above studies show how neural reactivation can counteract forgetting; we will now discuss how this relates to memory consolidation. We mentioned earlier how remote memories eventually become independent of the hippocampus. The gradual changes in activation from the hippocampus to the neocortex in animal data (Lesburguères et al. 2011; Maviel et al. 2004) and human data (Takashima et al. 2009) offer a mechanism by which this transfer occurs. Even stronger support comes from two studies showing that repeated reactivations render memories more quickly resistant to disruption following hippocampal damage (Lehmann and McNamara 2011; Lehmann et al. 2009). In these studies, rodents underwent contextual fear conditioning and then were (or were not) re-exposed to the learning context for 5 days after initial learning. After 5 days, they received sham or hippocampal damage. Without re-exposure, hippocampal damage strongly impaired memory, showing the memory was hippocampal-dependent. However, with repeated exposures, hippocampal damage had no effect on memory. This suggests that memory reactivation, and not time per se, causes memories to become hippocampal-independent. The overarching idea is that newly formed memories initially rely on the hippocampus and neocortex, whereas reorganization through repeated hippocampal-neocortical interaction produces neocortical networks that are sufficient for retrieval (Lesburguères et al. 2011; Redondo and Morris 2011; Squire et al. 2015; Tse et al. 2011).

In these studies reported by Lehmann and colleagues, animals were placed back into the original context, which was inferred to produce memory reactivation. However, behavioral (Craig et al. 2015a; b; Dewar et al. 2012) and neural evidence (Staresina et al. 2013; Tambini et al. 2010) suggests memories also undergo a stabilization process in the absence of overt retrieval. In studies that controlled for overt retrieval, increased resting functional connectivity patterns correlated with the amount of previous learning (Peigneux et al. 2006) and post-learning changes in hippocampal-cortical connectivity correlated with subsequent memory retention (Tambini et al. 2010). To strengthen this idea, it was even found that post-learning changes predicted memory during a non-learning control task (Staresina et al. 2013—but see Dewar et al. 2012, for evidence that rest benefits memories more than does performing non-learning control tasks). Additionally, strongly encoded items become the most preferentially reactivated (Tambini et al. 2010). Finally, it may be especially important that hippocampal processes occur immediately after learning. In humans, breaks between short video clips benefit memory, and allow for the onset of a strong post-clip hippocampal response that predicts memory (Ben-Yakov et al. 2013). Accordingly in rodents, replay occurs during learning (Davidson et al. 2009; Karlsson and Frank 2009) and correlates with memory measures (Dupret et al. 2010; Jadhav et al. 2012). Together, these studies provide a mechanism by which learning produces neural changes that in turn provide for stabilizing newly formed neural traces, presumably by assisting in hippocampal-neocortical transfer. In other words, in considering the findings of Müller and

Pilzecker over 100 years ago, these studies suggest consolidation may occur simply because of reactivation and the processes may be indistinguishable.

We have thus far outlined evidence for how wakeful memory reactivation contributes to stabilization. However, reactivation occurring during sleep plays an important and unique role as well. The following sections will outline the basics of sleep physiology and how it contributes to memory processing.

Characteristics of Sleep Physiology

The emergence of electroencephalographic investigations of sleep in the middle of the twentieth century showed that sleeping was an active process. Throughout the night, the brain progresses through cycles of distinctive stages of brain activity, each cycle lasting approximately 90 min. The physiological features of these stages provide clues about the functions of sleep in homeostasis and memory consolidation.

Typically, wakefulness transitions into stages of light sleep, known as stage-1 and stage-2, followed by the stage of deep sleep known as slow-wave sleep (SWS). To complete a full cycle, there will be a subsequent return to lighter stages, followed by rapid eye movement (REM) sleep. Stages of light and deep sleep are also known as non-rapid eye movement or NREM sleep (with Stage-1, Stage-2, and SWS sometimes termed N1, N2, and N3). In general, the character of these stages is influenced by factors such as circadian rhythms, such that cycles early in the night contain longer SWS periods, whereas later ones contain longer REM periods.

Physiological signals during wakefulness are chaotic. Fast, low-amplitude rhythms predominate in the EEG, muscle tension varies from moderate to high levels, and wakeful rest with the eyes closed produces a prominent occipital alpha rhythm, though with larger amplitudes in some people than in others. The alpha rhythm is most strongly observed over occipital regions, particularly during periods of rest with the eyes closed.

Stage-1 sleep appears at sleep onset with the appearance of quick vertex waves, rolling eye movements, and the decline of the alpha rhythm. This stage is considered to be the bridge between sleep and wakefulness, characterized by brief hallucinations and a low arousal threshold, meaning subjects can be easily awoken by external stimuli.

During stage-2 sleep, the predominant EEG rhythm is theta (4–8 Hz) with occasional *K-complexes*, which are high-amplitude deflections at approximately 0.8 Hz. Stage-2 also includes *sleep spindles*, which are short bursts of sigma activity at 11–16 Hz. Slow oscillations appear to originate in the frontal cortex (Cash et al. 2009), whereas spindles begin in the thalamus and continue as a series of reverberating thalamocortical oscillations (Morison and Bassett 1945). Arousal thresholds increase, which may connect to the findings that K-complexes and sleep spindles both coincide with reduced stimulus processing (Cote et al. 2000; Dang-Vu et al. 2010, 2011; Schabus et al. 2012). Stimuli can additionally elicit K-complexes and spindles (Cash et al. 2009; Sato et al. 2007), and K-complexes are especially prominent after emotional or personally-relevant stimuli (e.g. one's own name) (Brain 1958; Bremer 1954; Oswald et al. 1960). For these reasons, K-complexes and spindles have been proposed to protect sleep by preventing unnecessary arousals from occurring. Because K-complexes in stage-2 sleep resemble slow oscillations in SWS in frequency, amplitude, and origin (Cash et al. 2009), stage-2 sleep can be considered a transitional bridge to SWS.

No naturally occurring brain state differs more from wakefulness than SWS. Slow, high-amplitude oscillations functionally segregate neuronal firing into discrete time bins, acting as the orchestrator of large-scale hyperpolarization and depolarization across the brain. Each oscillation has a down-state, during which there is a large-scale bias towards hyperpolarization and low neuronal firing, as well as an up-state, when there is bias towards depolarization and high neuronal firing (Mölle et al. 2011). Spindles persist into this stage, beginning most frequently during the slow oscillation up-state. Activation from neuromodulator systems prevalent during wake, such as those mediated by acetylcholine (ACh) and cortisol, wanes greatly during SWS (Diekelmann and Born 2010). Finally, arousal thresholds are highest in this stage (Rechtschaffen and Kales 1968).

After SWS, the brain progresses back towards lighter stages, and then to REM. Physiologically, REM resembles wakefulness in a number of ways. The EEG shows high-frequency, low-amplitude activity and neuromodulator levels for ACh and cortisol resemble their waking levels (Diekelmann and Born 2010). Subjectively, REM coincides with dreaming episodes more than any other stage (though dreaming also occurs in other stages). Despite these similarities between REM and wakefulness, there are obviously major differences. Muscle activity is nearly completely suppressed during REM. Brain areas involved in self-monitoring show dramatically lowered activity, whereas emotional areas reach higher levels than wake (Nir and Tononi 2010), likely corresponding to the emotionality and lack of self-awareness during dreams. Finally, arousal thresholds during REM vary widely, although the dreaming brain's ability to incorporate and re-interpret information coming from the outside world has been known at least since the days of Freud and postulated at least since Aristotle (Freud 1913).

In the following sections, we will discuss sleep's unique role in learning and memory processes.

Sleep as an Ideal State for Memory Reactivation

As with many important findings in psychology, studies on the role of sleep in memory began by accident (Jenkins and Dallenbach 1924). The story of this accident begins with Hermann Ebbinghaus, the German psychologist who pioneered experimental research on human memory in the late nineteenth and early twentieth centuries. His studies mostly consisted in presenting auditory strings of nonsense syllables and measuring how well they were remembered at various retention intervals. Arguably his most influential finding came from what is known as the forgetting curve (Ebbinghaus 1885). Not only does memory fade over time, it does so in a systematic and mathematically predictable way. Forgetting occurs rapidly just after learning and less and less so over time, resulting in the curve he championed. However, in creating this curve an anomaly repeatedly crept in: forgetting was drastically lessened when the intervals included sleep than when they did not. Ebbinghaus largely ignored this, perhaps because he had no plausible explanation for it, but Jenkins and Dallenbach (1924) famously followed up the anomaly. Indeed, their extensive study showed that sleep, as compared to wake, benefits memory.

Approaches to Sleep Research and Multiple Types of Memory

Researchers have implemented three general approaches to isolating the importance of sleep: (1) testing memory retention across sleep versus wake intervals, (2) restricting sleep, either to particular parts of the night or to specific stages, (3) manipulating the conditions of intact sleep using pharmacology, sensory, or electrical stimulation. The first approach is effective for testing whether a task could be sleep-dependent; however, it produces rather limited conclusions given that sleep intervals can produce different arousal levels than wake intervals and that wake intervals can entail more interference than sleep intervals. Greater confidence can be reached using the second and third approaches. Indeed, these three approaches can provide increasingly more convincing evidence towards establishing causal relationships between sleep and memory.

Since Jenkins and Dallenbach's (1924) landmark study showing sleep benefits for declarative memories, sleep has been shown to affect nearly every type of memory. Well-established research paradigms have been used to show that sleep benefits (1) motor sequence learning (Barakat et al. 2011; Brawn et al. 2010a; Cohen et al. 2005; Fischer et al. 2002; Fogel and Smith 2006; Gulati et al. 2014; Korman et al. 2007; Kuriyama et al. 2004; Manoach et al. 2009; Maquet et al. 2000; Morin et al. 2008; Nishida and Walker 2007; Rasch et al. 2009; Robertson et al. 2004; Song and Cohen 2014; Walker et al. 2002b, 2003, 2005; Wamsley et al. 2012); (2) procedural memory (Huber et al. 2004; Landsness et al. 2009; Plihal and Born 1997; Smith and MacNeill 1994; Tamaki et al. 2008); (3) visual perceptual learning (Frank et al. 2001; Gais et al. 2000; Karni et al. 1994; Mednick et al. 2002, 2003, 2008; Stickgold et al. 2000); and (4) auditory perceptual learning (Brawn et al. 2010b, 2013; Fenn et al. 2003; Gaab et al. 2004; Shank and Margoliash 2009). Other aspects of cognition that show improvement across sleep include anagram problem-solving (Walker et al. 2002b), statistical learning (Batterink et al. 2014; Batterink and Paller 2015; Durrant et al. 2011), language abstraction in infants (Gómez et al. 2006; Lany and Gómez 2008), and creative insight (Wagner et al. 2004; Yordanova et al. 2012). And of course, sleep replenishes attention, processing speed, rational decision-making, and many other cognitive processes that go beyond the scope of this chapter.

Understanding the mechanisms underlying these findings depends on elucidating how memory relates to hallmarks of sleep physiology. It has often been tempting to seek simplistic assignments between sleep stages and memory types, as if there were always one-to-one relationships. Varying methods, contradictory findings, and lumping disparate tasks into a single category have accounted for much of the confusion on this point. Countless other factors have likely had blurring effects as well—task difficulty and extent of learning influence sleep, which likely influences sleep's role in retention (Gais et al. 2002; Kuriyama et al. 2004); species differences in their learning aptitudes and sleep physiology (Buzsáki et al. 2013); human population and individual differences (Fenn and Hambrick 2012); the time between learning and sleep (Benson and Feinberg 1977); circadian differences between nap sleep and nocturnal sleep (Payne et al. 2008, 2012). However, a few general patterns have emerged, and relationships between sleep stages and learning may actually add nuances to how we understand differences between various learning categories.

Early on in sleep/memory investigations, the research focus rested entirely on REM sleep. This focus made intuitive sense, as the benefits of memory rehearsal were established, and a reasonable assumption would be that if a sleep benefit existed it would most likely come about through the reactivation of memories during dreams. Indeed, dreams were seen as a necessary phenomenon, as depriving REM sleep caused more subsequent entrances into it, as if it were a homeostatic need (Dement 1960). (The same homeostatic pressure can be observed for SWS). Accordingly, REM boosts were reported in the context of procedural learning prior to sleep, including avoidance conditioning (Smith et al. 1980), Morse code learning (Mandai et al. 1989), trampolining (Buchegger and Meier-Koll 1988), and other types of procedural learning (see Smith 2001 for a more extensive review). Later studies showed that REM sleep deprivation negatively affected learning on avoidance learning (Fishbein 1971), operant conditioning (Smith and Wong 1991), complex problem-solving tasks (Smith 1995), and visual discrimination (Karni et al. 1994). Furthermore, playing learning-related cues during subsequent REM sleep was found to strengthen complex procedural learning tasks such as Morse code learning (Guerrien et al. 1989), a complex logic task (Smith and Weeden 1990) and fear conditioning (Hars et al. 1985). These findings implicated a strong role for REM in procedural learning tasks.

This one-to-one relationship between procedural memory and REM sleep seemed to provide a clear and simple principle for sleep/memory theorizing, but it eventually broke down. For example, simpler motor tasks, such as explicit motor sequence learning and visual rotor pursuit, relied on NREM stages, especially stage 2 (Nishida and Walker 2007; Smith and MacNeill 1994; Tamaki et al. 2008; Walker et al. 2002a; though see Fischer et al. 2002, where performance in an identical task correlates with REM). In addition, pharmacological REM suppression boosted, rather than impaired, this type of learning (Rasch et al. 2009). To preserve some sort of REM sleep mapping, one way to potentially account for these findings was to invoke the idea that simpler procedural memory tasks rely on NREM sleep, whereas more complex ones rely on REM sleep (Smith et al. 2004).

On the other hand, the story grew yet more complex with new evidence implicating SWS in procedural tasks. For example, performance in a motor adaptation task correlated with measures of SWS (Huber et al. 2004), and impairments were found after SWS deprivation (Landsness et al. 2009). Also, some procedural knowledge acquired with awareness of learning can be modulated by learning-related cues during SWS (Antony et al. 2012; Cousins et al. 2014; Schönauer et al. 2014).

Finally, the preponderance of extant evidence indicates that NREM sleep underlies the consolidation of declarative memories. This is presented with a caveat that declarative memories are nearly universally studied now as item or paired associations. Studies investigating SWS and REM deprivation separately showed that, while REM deprivation had no effect on simple associations, it impaired declarative memory for full stories (Empson and Clarke 1970; Scrima 1982; Tilley and Empson 1978). Full stories are arguably more complex than associations, which accords with the role of REM in other complex cognitive tasks, such as complex procedural learning (see above), creativity in the remote associates task (Cai et al. 2009), solving anagrams (Walker et al. 2002a), Tower of Hanoi problems (Smith and Smith 2003), and categorical probabilistic learning (Djonlagic et al. 2009). Therefore, the role of REM sleep in declarative memory could be understated by the choice in tasks typically employed in these studies.

We will now delve deeper into the role of sleep, particularly NREM sleep, in declarative memory processing. We will again take a historical perspective and cover a wide range of converging evidence using different methods.

Passive vs. Active Role for Sleep in Declarative Memory

In the early part of the century, retrograde interference was one of the better-known characteristics of memory (Müller and Pilzecker 1900). As a result, Jenkins and Dallenbach (1924) ascribed sleep only a passive role in memory, in providing a temporary reprieve from constant encoding during wake. This hypothesis was plausible; indeed, an alternative view only became prominent a half-century later.

The first counter-evidence came with a pair of studies in the 1970s. First, Yaroush et al. (1971) measured memory retention over three 4-h intervals: the first half of the night, containing large amounts of deep NREM sleep; the second half of the night, containing large amounts of REM; and during 4 h of daytime wakefulness. Retention over the first 4 h of sleeping consistently trumped that for the other two conditions, indicating there may be something to NREM sleep physiology that specifically reduced forgetting. However, it was possible the effects could be explained by circadian factors, contributions from less predominant stages, or that NREM offered more of a release from interference, especially as the EEG during REM more resembled wakefulness and possibly interference from dreams. This partial sleep restriction method has also been used successfully (Plihal and Born 1997; Smith et al. 2004) to replicate the link between early, NREM-rich sleep and declarative memory while showing that procedural memories benefit more from the later, REM-rich part of the night. The second 1970s study investigated memory retention with an equivalent amount of sleep and wakefulness, but allowed for sleep to come immediately or later on in the 24-h interval (Benson and Feinberg 1977). Better memory after immediate sleep showed that the total amount of overall interfering wakefulness could not alone explain the role of sleep in memory [a finding replicated by Gais et al. (2006), Payne et al. (2012), and Talamini et al. (2008)].

Converging evidence for memory replay during sleep accrued from neuronal reactivation in cellular physiology in the 1980s–1990s and human behavioral manipulations performed in the 2000s. The findings from cellular physiology will be discussed in the next section. First, it is helpful to cover how the behavioral studies have unfolded.

In Ellenbogen et al. (2006), Ellenbogen and colleagues set out to test whether sleep helped stabilize a memory by experimentally inducing interference. Participants learned A–B paired associates and then experienced a 12-h sleep, 12-h wake, or 24-h sleep-then-wake interval (among other conditions). Subsequently, they returned to the lab and learned A–C associates before tests on the original A–B pairs. As predicted, comparing 12-h conditions revealed superior memory for sleep over wake. However, if sleep protected memories from interference, the investigators argued, participants in the 24-h sleep-then-wake interference condition should perform better than the 12-h wake condition, even though the interval was longer and they had more time awake. This is indeed what they found.

Additional evidence for memory replay during sleep that included data on sleep physiology was produced by directly manipulating the conditions of NREM sleep. Gais and Born (2004) studied the role of low ACh levels during NREM sleep. They found that administering ACh agonists to prevent these low levels interfered with retention. Along with evidence that cholinergic activity suppresses output from the hippocampus to extrahippocampal regions (Chrobak and Buzsáki 1994; Hasselmo and McGaughy 2004), these findings suggest the low ACh levels during NREM sleep create an important state for consolidation. Using a novel approach to link SWS with memory processing, Marshall et al. (2006) directly manipulated slow oscillations using transcranial direct current stimulation at 0.8 Hz. This oscillating current, compared with sham stimulation, significantly boosted both slow oscillations and declarative memory, and thus strongly linked slow oscillations to memory consolidation.

These slow oscillations are not the only facet of sleep physiology apparently playing a role in memory processing. Despite the abundance of divergent theories about sleep spindles, as described above, they have emerged as a key physiological factor in memory consolidation. Numerous studies have demonstrated correlations between spindles and motor memory consolidation (Barakat et al. 2011; Kurdziel et al. 2013; Nishida and Walker 2007; Rasch et al. 2009; Tamaki et al. 2008; Walker et al. 2002a), as well as with declarative memory consolidation (Clemens et al. 2005, 2006; Cox et al. 2012; Schabus et al. 2004; Studte et al. 2015; van der Helm et al. 2011). However, these correlations are complicated by another line of research showing that spindles positively correlate with general cognitive abilities (Bódizs et al. 2009; Fenn and Hambrick 2015; Schabus et al. 2006, 2008), leaving open the possibility that any memory effects are merely secondary to general cognitive effects that indirectly influence learning.

Evidence for a causal role for spindles in memory consolidation has slowly accumulated. Many researchers have employed intra-subject measures comparing learning and non-learning (control) sleep to control for individual differences. Gais et al. (2002) first showed that learning boosted spindle density during subsequent sleep. Spindle density correlated with memory at both pre- and post-nap tests, but not memory change across the nap. Schabus et al. (2004, 2008) found no boost from learning, but did find a correlation between experimental-control group density and memory retention, meaning that individuals with increased spindles showed better improvements. Schmidt et al. (2006) found that difficult learning (though not easy learning) boosted and correlated with sleep spindles, suggesting that spindles may effect changes depending on cognitive demands. Finally, learning-related spindle boosts arise in the rodent EEG (Eschenko et al. 2006) and in human epilepsy patients undergoing novel training on a brain-computer interface (Johnson et al. 2012). In a heroic effort, Bergmann et al. (2012) showed using combined fMRI-EEG that spindle amplitude increased in a specific set of brain regions related to learning but not in other areas unrelated to learning. However, this increase correlated with pre-sleep memory but not with memory change across the nap.

Another indirect line of support for spindles comes from methods aimed at boosting slow oscillations. Marshall et al. (2006) could not measure spindle activity during stimulation due to artifacts caused by the current, but did find enhanced slow oscillatory activity between successive 5-min stimulation periods. Intriguingly, during these 1-min non-stimulation periods, slow spindle power was enhanced. Bolstering these findings, a later study found that playing two auditory noise bursts in time with slow oscillation up-states can similarly boost slow oscillatory power and memory (Ngo et al. 2013). This auditory stimulation protocol also boosted fast spindles, which positively correlated with memory retention. Using a different variation on this general methodology, Ong et al. (2016) also showed that acoustic stimulation to increase slow oscillations also produced an increase in fast spindles. In another follow-up experiment, Ngo et al. (2015) showed that playing four auditory bursts during up-states had no greater effect on memory than the two-burst condition, and did not elicit a further boost in fast spindle power. Altogether, these studies show that spindles may represent an essential factor mediating the effect of increased slow oscillatory power on memory enhancement.

Using a very different approach, Mednick et al. (2013) delivered the most convincing causal evidence that spindles benefit memory to date. They found that delivering zolpidem (Ambien) boosts spindle density without increasing slow oscillation power. Furthermore, spindle density increases under zolpidem predicted within-subject memory retention improvements. Although zolpidem increased time in SWS, neither this SWS measure nor slow oscillatory power predicted memory improvements under zolpidem. In a follow-up study, a similar effect of zolpidem was found on emotional memories (Kaestner et al. 2013).

In summary, there is good evidence that slow oscillations and spindles benefit memory. We will now discuss how these measures of cortical activity fit together with events occurring in the hippocampus, where newly formed memories stir.

Memory Replay During Sleep

By the 1980s, it was clear the hippocampus played an important role in forming new memories and that memories underwent a period of consolidation. However, the specific hippocampal mechanisms responsible for driving this process were largely unknown. György Buzsáki et al. (1983) described a physiological process consisting of a sharp wave in the local field potential followed by a high frequency burst (150–250 Hz, termed a ripple) occurring uniquely in the hippocampus. Intriguingly, these events were most prevalent during NREM sleep (Buzsaki 1986; Hartse et al. 1979). Buzsaki (1989) prophetically proposed that they played a role in memory consolidation. However, it was difficult to corroborate this view at the time, as evidence linking sharp-wave/ripples (SWRs) to specific memory traces was lacking.

The evidence for replay came in steps. First, Pavlides and Winson (1989) showed that hippocampal place cells with enhanced activation during wake continued to show enhanced activation during sleep. While intriguing, there remained the possibility that each cell simply kept firing on its own as a homeostatic mechanism without any relation to other cells. Wilson and McNaughton's (1994) seminal study put this concern to rest and largely legitimized future studies on sleep and memory relationships. They recorded from numerous hippocampal place cells in the hippocampus before, during, and after a rat explored a novel spatial environment. Remarkably, cell pairs that fired together while the rat explored the environment similarly fired together during post-learning sleep. Because these cells did not fire together during pre-learning sleep, the post-learning results can be attributed to learning rather than merely a function of neurons that were already highly connected.

Wilson and McNaughton's findings inspired numerous studies that expanded upon how and when replay occurred. Not only do previously co-activated place cells correlate during post-learning sleep, they fire in the same order (though with less fidelity), as if one could read out the spatial location the rat was traversing during sleep (Louie and Wilson 2001; Skaggs and McNaughton 1996). Replay of place-cell firing patterns occur most commonly during hippocampal SWRs (Dupret et al. 2010; Kudrimoti et al. 1999; O'Neill et al. 2006, 2008; Pennartz et al. 2004; Peyrache et al. 2009). Enhanced co-firing of cell pairs during wake increases replay during sleep (O'Neill et al. 2008). In relation to activity in other parts of the brain, SWRs overlap with and slightly precede sleep spindle events in other areas such as the ventral striatum (Lansink et al. 2009) and neocortex (Siapas and Wilson 1998). Moreover, wakeful hippocampal-neocortical (Ji and Wilson 2007) and neocortical-neocortical (Hoffman and McNaughton 2002) firing patterns replay during sleep.

Other details about the speed, conditions, and timing of replay have also been uncovered. Place-cell replay patterns become compressed by a factor of 6–7x during sleep (Euston et al. 2007), occur for extended events spanning as long as 10 m of track over 60 s (replaying at approximately 8 m/s) across multiple SWRs (Davidson et al. 2009), and also occur, though to a lesser extent, during wake (Carr et al. 2011; Diba and Buzsáki 2007; Dupret et al. 2010; Karlsson and Frank 2009).

Though ensemble reactivations occur most frequently during SWRs, one could argue that the large-scale synchrony that encompassed SWR events reflected previous neuronal firing without relating to memory. However, a few findings speak against this idea. First, learning (Eschenko et al. 2008; Ramadan et al. 2009) and LTP (Behrens et al. 2005; Buzsaki 1984) boost SWRs during subsequent NREM sleep. Second, reactivation events are specific to learning-related ensembles (Dupret et al. 2010; Peyrache et al. 2009) and correlate with memory retention (Dupret et al. 2010). Third, and most definitively, manipulating SWRs alters memory. Imposing replay with artificial stimulation during SWS SWRs enhances fear memory (Barnes and Wilson 2014). Suppressing SWRs impairs memory both when done during learning in a spatial working memory task (Jadhav et al. 2012) and during subsequent sleep when rodents learn a maze over a series of days (Ego-Stengel and Wilson 2009; Girardeau et al. 2009). These studies provide a crucial causal link to the role of hippocampal SWRs and memory consolidation. Since hippocampal replay occurs most frequently during SWRs, they constitute strong indirect evidence for the role of replay in memory consolidation.

Early neuroimaging studies using positron emission tomography gave the first and currently most illustrative evidence of reactivation on a systems-level. Maquet et al. (2000) showed learning-specific activation of brain areas during REM sleep that were previously activated by motor-sequence learning. In a similar vein, Peigneux et al. (2004) showed learning-specific hippocampal activation after learning a novel spatial environment, and this activation correlated with memory improvement. However, these studies showed enhanced activity over a long time scale. In contrast, one recent study (Deuker et al. 2013) enlisted multivariate methods to decode whether newly formed memories were reactivated during sleep and wake after learning. Possible reactivation patterns were observed, though puzzlingly only during stage-1 sleep. One shortcoming of this study relates to the difficulty participants had reaching deep sleep, but the presence of an effect nevertheless offers encouragement for pursuing these sorts of approaches.

Across a wide range of neuroscientific techniques, there is strong evidence that replay reflects learning. However, not all learning events are remembered in the long-term, so there must be a mechanism by which memories become differentiated over time. Below, we will cover how various factors influence spontaneous memory reactivation and how these influences may play a role in determining which memories endure.

Factors Influencing Spontaneous Reactivation

Humans form far more episodic memories than they can remember after some time passes, suggesting there are computational limits on the hippocampal-neocortical system. Thus, human memory is cluttered with competition among memories, forcing the system to devise a mechanism by which it can keep the memories deemed to be of the greatest future use, even if it comes at the expense of other, less relevant memories. Over the last decade, it has become increasingly clear that sleep plays a role in this prioritization (Fischer and Born 2009; Rauchs et al. 2011; Saletin et al. 2011; van Dongen et al. 2012b; Wilhelm et al. 2011; though see Baran et al. 2013).

A prominent theory suggests that memories of higher importance become "tagged" during wake by the hippocampus to undergo further consolidation during sleep (Morris 2006; Redondo and Morris 2011). The "synaptic tagging and capture" hypothesis (Morris 2006) suggests there are at least two steps in the consolidation process: an early-LTP process occurring at encoding that rapidly decays and a late-LTP process that involves hippocampal-neocortical dialogue and results in a relatively persistent neural trace. Central to this idea is that a molecular "tag" influenced by the early process (albeit not deterministically) signals the late process to enact enduring changes that occur during offline periods like sleep.

At the molecular level, some relationships have been worked out between early- and late-LTP, memory persistence, and NMDA- and dopamine receptor-dependence within the hippocampus (Wang et al. 2010). The amount of cell co-firing within 50 ms of learning during spatial exploration resulted in enhanced SWRs (O'Neill et al. 2008), and we previously mentioned links between learning and LTP on SWRs. Therefore, replay of tagged memories could provide a mechanism by which memories differentially persist.

Electrophysiological oscillations during wake that group neuronal activity across regions may play a role in the tagging process. At the cellular level, prefrontal neuron assemblies producing high theta coherence during learning were preferentially replayed during sleep SWRs (Benchenane et al. 2010). Similarly, a recent study in humans (Heib et al. 2015) showed theta power during word-pair encoding mediated the positive relationship between fast spindles and memory retention. Thus, theta power may reflect an effective encoding process that tags memories to undergo further consolidation.

Further evidence for differential memory tagging comes from experimental manipulations that alter the future relevance, reward, or emotional content of various items. Sleep benefits memory items that participants are directed to remember at encoding (Rauchs et al. 2011; Saletin et al. 2011), directed to bring to mind (Fischer et al. 2011), told later would be important (van Dongen et al. 2012), or would even be tested at all (Wilhelm et al. 2011). Importantly, sleep physiology appears to become biased in conjunction with this memory prioritization. Participants who expected to be tested showed a pronounced increase in slow oscillatory power and the number of spindles in relation to a control night (Wilhelm et al.

2011), and sleep spindles correlated positively with memory change for to-be-remembered items and also negatively with that for to-be-forgotten items (Saletin et al. 2011). Using a similar paradigm with fMRI, Rauchs et al. (2011) found that hippocampal activity at encoding predicted overnight changes in memory only for to-be-remembered items. Crucially, this activity failed to predict overnight changes for a separate group of subjects who were sleep-deprived. These results support the idea that memories tagged during wake undergo further processing during sleep (Morris 2006).

In rodents, a few studies found replay during SWRs occurs more frequently for memory traces that are motivationally relevant, as assessed by the presence or absence of reward (Lansink et al. 2008, 2009; Peyrache et al. 2009). Moreover, one study showed reactivation does not occur when rewarded locations are cued and no learning is required (Dupret et al. 2010), showing it relates directly with memory importance. Dopaminergic (DA) fiber bundles emanating from the ventral tegmental area (VTA) heavily control reward processes. The VTA contains fiber bundles that innervate the hippocampus and these have been shown to affect long-term potentiation within the hippocampus (Bethus et al. 2010; Lisman and Grace 2005). Additionally, one study found activity in the VTA and hippocampus predicts memory for high-reward cues (Wittmann et al. 2005), and another found functional interactions between these regions predicts long-term memory formation (Adcock et al. 2006). Therefore, DA may modulate hippocampal reactivation as a function of reward during offline periods such as sleep.

Indeed, a recent study showed direct links between VTA-hippocampal stimulation, neuronal reactivation, and memory retention (McNamara et al. 2014). The authors found that VTA neurons increased their firing rate while rats explored a novel environment. Optogenetic stimulation of VTA-hippocampal fibers increased subsequent reactivation of related memory traces, which could be blocked by administering DA antagonists before learning. Finally, this optogenetic stimulation improved memory retention. Another recent study showed that new memories could be implanted by pairing VTA-hippocampal firing with spontaneous place cell reactivation (de Lavilléon and Lacroix 2015). First, the authors separately found hippocampal place cells while rats explored a spatial environment and stimulations of VTA ensembles that animals found rewarding. Next, during offline periods of wake or sleep, spontaneous place cell reactivation was assessed online and paired with rewarding VTA-hippocampal fiber stimulation. When rats were re-introduced into the environment, they spent more time immediately in the location represented by the place cell undergoing co-activation with VTA-hippocampal fibers. Therefore, interactions between DA inputs to the hippocampus appear to strongly influence reactivation and subsequent memory retention (Atherton et al. 2015; but see Berry et al. 2015).

Two pharmacological studies support this idea by specifically highlighting the role dopamine plays in reward-enhanced consolidation. Wang et al. (2010) found that strong rewards induced persistent memory when weak rewards did not, and this enhancement could be blocked with dopamine antagonists. Additionally, in a study with human subjects, Feld et al. (2014) showed participants a number of objects preceded by a high or low reward while administering either a dopamine agonist or placebo to participants. Under the placebo condition, high reward items were remembered better after sleep, but this difference was eliminated when participants received the dopamine agonist. These results suggest that high reward items would receive preferential processing during sleep under normal conditions, but the presence of the dopamine agonist made low items receive further processing.

Another biologically adaptive way some memories will be given priority comes from their level of emotional content (Richter-Levin and Akirav 2003). In some experimental paradigms, sleep appears to play a role in this prioritization. For instance, Payne et al. (2008) showed participants a series of pictures with a neutral or emotional central image against a neutral background (e.g., an undamaged car or a wrecked car, respectively, against a city backdrop). They tested participants after encoding and then again after 30 min, 12 h of wake, or 12 h of sleep. They found that sleep resulted in an enhanced selective benefit for emotional over neutral items relative to both the 30-min and 12-h wake intervals, demonstrating specificity for sleep in prioritizing maintenance of emotional material. The same research group has repeatedly replicated this effect (Bennion et al. 2015; Payne et al. 2012, 2015; Payne and Kensinger 2011), and other paradigms have produced comparable evidence supporting a role for sleep in emotional memory (Hu et al. 2006; Nishida et al. 2009; Wagner et al. 2001), with differential sleep effects lasting up to at least 4 years (Wagner et al. 2006).

Current conceptions of the mechanisms underlying emotional memory consolidation accord with this idea. Generally for emotional memories, amygdala activation leads to enhanced memory retention (Cahill et al. 1996; Canli et al. 2000), and enhanced amygdala-hippocampal interactions at encoding leads to better memory (Dolcos et al. 2004). Furthermore, elevated levels of the stress hormone cortisol predict higher levels of amygdala activity for negative pictures (van Stegeren et al. 2005) and predict enhanced levels of selective memory enhancement for negative stimuli after sleep (Bennion et al. 2015). Thus, a plausible mechanism is that cortisol modulates amygdala activity, which, via interactions with the hippocampus, tags memories for further rehearsal during sleep (Bennion et al. 2015).

Emotional memory consolidation presents an intriguing case for the role of emotions in REM sleep. In dream reports, REM sleep is frequently associated with greater emotional content than other stages (Fosse et al. 2001) and REM sleep also shows higher levels of amygdala activity than NREM sleep and wakefulness (Maquet et al. 1996). Accordingly, several studies have found that emotional memory retention correlates with REM sleep (Nishida et al. 2009; Payne et al. 2012; Wagner et al. 2001). These findings may, however, appear surprising, as the emotional information would certainly be categorized as an example of declarative memory, which otherwise is linked with the involvement of NREM sleep. Indeed, unlike the case with nocturnal sleep, when participants take afternoon naps they typically only attain NREM sleep, and emotional memory retention correlates with SWS measures during the nap (Payne et al. 2015). There may be more to decipher about these disparate findings, as they hint at the complexity of competing and/or complementary processes operative during the various stages of sleep.

Illuminating findings have also arisen in the context of investigations showing that sleep does not selectively benefit emotional memories. Baran et al. (2012) and Lewis et al. (2011) found no interaction between emotional versus neutral information and sleep versus wake. Additionally, Atienza and Cantero (2008) found that sleep deprivation hurt memory for emotional items less than for neutral information, suggesting that in some paradigms emotional memories are simply less susceptible to interference and may remain robust with reactivation processes that occur during wake.

These apparently contradictory findings can be reconciled. The interaction of emotion and memory is complex, and the general assertion that emotions enhance memory is by no means universal (Mather and Sutherland 2011). For instance, memories can become enhanced or inhibited depending on numerous factors at encoding, such as whether they occur before, after, or simultaneously with emotional information, their level of association with the emotional content, the level of perceptual contrast, or the relevance of the information for current goals (see Mather and Sutherland 2011, for an extensive review). Furthermore, a complex set of hormonal factors (McGaugh 2000), interactions between the amygdala and the hippocampus (Dolcos et al. 2004) or vmPFC (Bennion et al. 2015), and emotional learning during consolidation (Dunsmoor et al. 2015) can affect memory beyond the time of encoding. The type of molecular tag, including the strength of the tag and what does or does not become tagged alongside emotional information likely differs based on experimental paradigm, and this will affect what role offline processes play in memory. This research topic deserves much further attention, as it could aid treatment for disorders such as depression and post-traumatic stress disorder (LaBar and Cabeza 2006).

In this section, we have outlined some major factors that naturally influence memory reactivation. In the next section, we will discuss a relatively new method that involves artificially targeting memories for reactivation at specific times during sleep.

Targeted Declarative Memory Reactivation

As with many aspects of science and human thought, speculations and unexplained findings supported the idea that TMR could work long before it became part of established and accepted ways of thinking. In the late 1980s and early 1990s, a few studies showed altered memory after linking a stimulus with a learning episode and re-administering the stimulus as a memory cue during EEG-verified sleep. In some of the first successful studies employing TMR, Hars et al. (1985) enhanced active avoidance conditioning in rats by cueing during REM, whereas Hennevin and Hars (1987) impaired the same type of learning by cueing during SWS. Hars and Hennevin (1990) again found an effect for REM stimulation impairing spatial memories. In human participants, Smith and Weeden (1990) enhanced Morse code learning by re-playing learning-related auditory clicks during REM sleep.

To understand why these studies were largely ignored, it is important to note these studies preceded much of the reactivation literature that grounded sleep and memory at the neuronal level. Furthermore, the mechanisms at work during the various sleep stages were largely unknown. To be sure, the mechanisms for each of these effects are still somewhat mysterious, though with current knowledge of neuronal reactivation mechanisms in TMR, it is easier to envision their workings.

In a seminal study, Rasch et al. (2007) revived TMR and bolstered the idea that memories are actively reprocessed during sleep. Participants learned pairs of pictures on a spatial grid akin to a memory game, all while smelling a rose odor. Next, they slept in the lab, and some subjects received the rose odor again upon entering SWS. After waking up, those receiving the rose odor remembered significantly more pairs than those who did not. This method failed to boost memory when the rose cue was delivered during wake, or during REM, or when the rose odor was delivered during SWS but was not present during learning, demonstrating the specificity of reactivation of the learning episode.

Rudoy et al. (2009) followed up on this finding to investigate its specificity for individual memories (Fig. 2a). Participants learned 50 object-location associations against a background grid and a semantically related sound cue played concurrently with each visual object presentation (e.g. cat image—"meow" sound). During a subsequent afternoon nap, Rudoy used 25 sounds to cue half of the object locations during SWS. After the nap, participants recalled locations more accurately for objects associated with those sounds, in comparison to objects associated with sounds not presented during sleep, showing that TMR can be used to reactivate individual memories.

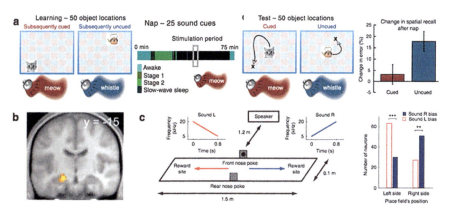

Fig. 2 Mechanisms of targeted declarative memory reactivation during sleep. (**a**) After unique auditory cues were paired with individual items, presenting those cues during subsequent SWS significantly enhanced memory (Rudoy et al. 2009). (**b**) Targeting memories using olfactory stimuli resulted in enhanced activity in the hippocampus (Rasch et al. 2007). (**c**) After unique cues were paired with different locations in a rectangular grid, presenting the cues during sleep resulted in enhanced firing of corresponding place cells (Bendor and Wilson 2012). Such biased cellular firing patterns presents a plausible mechanism by which targeting memories results in differential memory performance

Numerous other studies have subsequently shown TMR benefits and begun to elucidate the corresponding neural mechanisms. TMR enhances spatial memories via specific odors (Rihm et al. 2014), otherwise forgotten low-priority memories (Oudiette et al. 2013), memories of moderate initial strength (Creery et al. 2014), emotional memories (Cairney et al. 2015), and vocabulary words with the words as cues (Schreiner and Rasch 2014; Schreiner et al. 2015). It reduces subsequent retroactive interference (Diekelmann et al. 2011) and accelerates the consolidation process (Diekelmann et al. 2012). Additionally, stimulation boosts spindle power over learning-related regions (Cox et al. 2014) and enhances parahippocampal-mPFC connectivity (van Dongen et al. 2012a), implicating spindles and dialogue with the neocortex as possible underlying mechanisms.

In line with the expectation that targeted memory reactivation should resemble spontaneous memory reactivation, four studies suggest a role for replay in TMR. The first showed that cueing a bird's own newly-learned song during post-learning sleep elicited replay of neurons involved in forming the memory (Dave and Margoliash 2000). The second showed with fMRI that odor presence enhanced activity in the hippocampus relative to its absence (Rasch et al. 2007; Fig. 2b). The third involved cueing different sounds as rats explored the two sides of a rectangular environment (Bendor and Wilson 2012; Fig. 2c). Upon subsequent sleep, the sounds elicited corresponding place cell activity for each respective side of the grid, suggesting the cues directly activated the neurons involved in forming those memories. The fourth showed that patients with bilateral hippocampal lesions did not benefit from TMR and that memory benefits from TMR correlated inversely with amount of hippocampal damage (Fuentemilla et al. 2013), offering causal evidence that the hippocampus plays an important role in TMR.

Finally, TMR has been shown to influence other types of cognition such as creativity (Ritter et al. 2012), procedural memories (Antony et al. 2012; Cousins et al. 2014; Schönauer et al. 2014), fear memories (Barnes and Wilson 2014; Hauner et al. 2013; Rolls et al. 2013), and learning to reduce implicit social biases (Hu et al. 2015).

Basic Model of Sleep Reactivation and Major Open Questions

The aforementioned lines of evidence can be integrated into a basic model of declarative memory consolidation (Fig. 3). Hippocampal SWRs time-lock to slow wave up-states (Mölle et al. 2006), neocortical spindles time-lock to slow wave up-states (Mölle et al. 2011), and SWRs time-lock to spindle down-states during the slow-wave up-state (Ayoub et al. 2012; Siapas and Wilson 1998; Staresina et al. 2015). This scheme suggests that slow waves coordinate reactivation in the form of hippocampal-neocortical dialogue, like a conductor leading an orchestra (Mölle and Born 2011). However, there are major open questions and a few possible contradictions about the processes underlying reactivation.

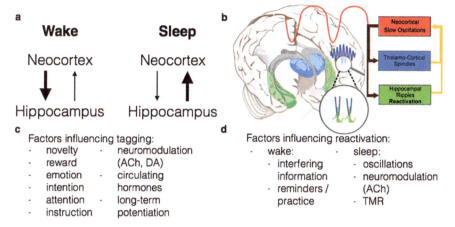

Fig. 3 Overview of sleep-dependent declarative memory consolidation. (**a**) Newly-encoded neural traces formed in the hippocampus become reactivated and consolidated via interactions with the neocortex. (**b**) Prominent theories suggest slow oscillations group hippocampal SWRs and thalamocortical spindles to coordinate hippocampal-neocortical transfer (from Born and Wilhelm 2012). (**c**) Numerous factors influence the filtering process determining which memories become later reactivated. (**d**) Reactivation processes operate on tagged neural traces to influence the later stability of the memory trace. Many of the factors in (**c**) and (**d**) are highly interrelated concepts from different levels of analysis

One major question involves the role of various parts of NREM sleep in declarative memory consolidation. A recent review (Genzel et al. 2013) argues there is likely a difference between lighter NREM sleep (stage-2 sleep and early SWS) and deep NREM sleep (late SWS). Slow oscillations are more frequently global in nature during light NREM sleep (Nir et al. 2011), which likely improves coordination between disparate brain areas such as the hippocampus and neocortex. Furthermore, SWRs (Clemens et al. 2007) and spindles (de Gennaro and Ferrara 2003) occur more frequently during lighter NREM stages, though this could be due to analysis issues with respect to identifying spindles by eye when they are superimposed on predominant slow oscillations (Cox et al. 2012). Efforts to boost memory via inducing slow oscillations with stimulation have begun in stage-2 sleep (Marshall et al. 2006; Ngo et al. 2013), creating the possibility that changes crucial for memory occurred in early NREM stages. Additional confusion might have arisen due to differences in sleep-stage terminology in humans and animal models, whereby SWS has been used as a term for all of NREM sleep (Genzel et al. 2013). Altogether, considering physiological differences in the hippocampus and neocortex, the distinction of early and late NREM for declarative memory is plausible and well worth further investigation.

The open issue above highlights another set of troubling complexities with respect to the role of spindles. As reviewed above, much of the evidence on spindles to date could be attributed to correlation rather than causation. Despite this, converging evidence from four sources implicates a direct role of spindles in stabilizing memories: (1) methods showing causal roles for slow oscillations in memory also boost spindle power, suggesting that spindles could mediate the effect of slow oscillations on memory (Marshall et al. 2006; Ngo et al. 2013, 2015), (2) TMR induces spindle amplitude enhancements for learning-specific brain regions (Cox et al. 2014), (3) spindles have specifically been shown to enhance long-term potentiation between synapses *in vitro* (Rosanova and Ulrich 2005), and (4) a pharmacological method to induce spindles enhances memory (Kaestner et al. 2013; Mednick et al. 2013). However, none of these findings show that spindles benefit memory directly. For example, the pharmacological results could reflect changes in other underlying processes (e.g., hippocampal SWRs) that indirectly influence spindles. Additionally, real-time evidence for the role of spindles in reactivation remains obscure, and questions remain about whether neural measures of reactivation precede, become embedded in, or follow the spindle itself.

Other key questions arise about the nature of reactivation. On the cellular level, there is at least a basic understanding that hippocampal replay re-emerges during post-learning sleep and affects later memory retention (Ego-Stengel and Wilson 2009; Girardeau et al. 2009). To date, researchers investigating relationships between replay and behavior have understandably focused on the fidelity of offline reactivation to learning episodes. However, in addition to memory, sleep aids the generalization and abstraction of information, which may rely on reactivation (Stickgold and Walker 2013). In this light, findings in rodents (Karlsson and Frank 2009) that replay has higher fidelity during wake than sleep may prove illuminating. It would be interesting to discover if these abilities rely not on high-fidelity replay, but on some intermediate level of replay fidelity that allows for incorporating the trace into other semantic networks or statistically similar episodes.

A different line of research has begun to outline the molecular mechanisms required for long-term plasticity (Takeuchi et al. 2014). However, it is currently unknown how cellular reactivation interacts with plasticity on the single neuron level. Specifically, would blocking reactivation (for instance, using optogenetics) prevent placticity? Or vice versa, would blocking plasticity, as with protein synthesis inhibitors and/or post-translational modification regulators (Routtenberg and Rekart 2005) reduce reactivation?

On the systems level, there is no clear picture for what constitutes reactivation. Over a full night of sleep, learning-related neural activity becomes enhanced (Deuker et al. 2013; Maquet et al. 2000; Peigneux et al. 2004) and correlates with memory retention (Deuker et al. 2013; Peigneux et al. 2004), and TMR/fMRI studies have also implicated enhanced activity in medial temporal lobe structures (Rasch et al. 2007; van Dongen et al. 2012a). However, we currently lack solid real-time evidence of systems-level reactivation to correspond with results from neuronal reactivation, although methods such as EEG or MEG hold promise in this regard. Also, it remains unclear exactly how reactivation events are connected to hallmarks of sleep physiology such as slow oscillations and spindles.

Finally, the timescale for systematic changes in the neural locus of memories is not understood. Mander et al. (2011) showed that sleep, as opposed to wake, can promote the acquisition of new declarative memories, and this improvement correlates with spindle activity. Sleep may therefore act as a way to "refresh" the hippocampus to learn anew the next day, which accords with findings that sleep enhances activation in the neocortex while reducing it in the hippocampus (Takashima et al. 2009). However, discrepancies exist between this model and the effects of hippocampal damage, which typically cause retrograde amnesia for memories formed over much longer time periods. Critically, there is scant experimental evidence on the extent to which reactivation occurs for memories more than a single day old. How long does it take for memories to become independent of the hippocampus? And does sleep reactivation continue to play a role beyond even the first day of memory formation?

Concluding Thoughts

In 2005, *Science* magazine released a list of the 125 biggest questions the field of science had yet to answer. Among them was, "Why do we sleep?" and "Why do we dream?" Both remain perplexing. Why we spend a third of our lives in near-complete inactivity has thus far eluded scientists. This is likely because, as with many solutions to environmental pressures during evolution, there is no singular purpose but rather a series of co-opted adaptations that best fit ecological niches.

Some lines of evidence suggest that sleep protects an agent from predators (Siegel 2009) and aids brain metabolism and restoration (Silva et al. 2004; Vyazovskiy et al. 2008, 2009, 2011; Xie et al. 2013; see Vyazovskiy and Harris 2013 and Tononi and Cirelli 2014 for helpful reviews). More pertinent to this chapter, a recent theory posited that certain types of brain plasticity may only become possible after the organism becomes detached from the environment, so sleep may be the price paid for plasticity (Tononi and Cirelli 2014). Considering the presence of circadian rhythms and sleep patterns in organisms without our complex system of memory (Cirelli and Tononi 2008), the argument that sleep evolved originally and primarily for memory is not strong.

However, that sleep plays a unique role in learning and memory has gradually become an irrefutable position. Throughout evolution, many organs and networks of cells originally evolved for one purpose and have later been used for another. The inner ear originally evolved in early vertebrates for balance, but later became involved in hearing (Torres and Giraldez 1998). The brain itself evolved to coordinate movement, but has clearly taken on numerous other abilities. Thus, it seems highly plausible that sleep originally evolved for purposes other than plasticity, but became co-opted later as new selection pressures incentivized the need for greater plasticity.

Behavioral, cellular, and systems level evidence suggests NREM sleep plays a special role, though perhaps not an exclusive role, in consolidating declarative memories. Reactivation is instrumental to our ability to retain information from a single, unique episode. One could easily envision a world in which no moment effectively lived beyond the present. Humans forget most of their life's episodes, as the natural, entropic fate of any episode is oblivion. However, offline reactivation can rescue otherwise fleeting episodes, especially those of high priority like the experience of the young scientist on her interview outlined at the beginning of this chapter. That the hippocampal-neocortical system has evolved a way to solidify experiences that were formed only once is a marvel, and that it co-opted natural sleep processes to effect its end is another testament to nature's ability to find unique solutions to adaptation challenges.

Acknowledgements Parts of this chapter were adapted from James Antony's Ph.D. thesis. We offer special thanks to Paul Reber for his insightful comments and Aryeh Routtenberg for his willingness to discuss these topics in detail. This work was supported by NIMH grant F31MH100958 and Princeton University's C.V. Starr Fellowship to JWA and NSF grants NSF BCS-1461088 and BCS-1533512 to KAP.

References

Adcock RA, Thangavel A, Whitfield-Gabrieli S, Knutson B, Gabrieli JDE (2006) Reward-motivated learning: mesolimbic activation precedes memory formation. Neuron 50(3): 507–517. doi:10.1016/j.neuron.2006.03.036

Albouy G, King BR, Maquet P, Doyon J (2013) Hippocampus and striatum: dynamics and interaction during acquisition and sleep-related motor sequence memory consolidation. Hippocampus 23(11):985–1004. doi:10.1002/hipo.22183

Antony JW, Gobel EW, O'Hare JK, Reber PJ, Paller KA (2012) Cued memory reactivation during sleep influences skill learning. Nat Neurosci 15(8):1114–1116. doi:10.1038/nn.3152

Atherton LA, Dupret D, Mellor JR (2015) Memory trace replay: the shaping of memory consolidation by neuromodulation. Trends Neurosci 38(9):560–570. doi:10.1016/j.tins.2015.07.004

Atienza M, Cantero JL (2008) Modulatory effects of emotion and sleep on recollection and familiarity. J Sleep Res 17:285–294. doi:10.1111/j.1365-2869.2008.00661.x

Ayoub A, Mölle M, Preissl H, Born J (2012) Grouping of MEG gamma oscillations by EEG sleep spindles. NeuroImage 59(2):1491–1500. doi:10.1016/j.neuroimage.2011.08.023

Barakat M, Doyon J, Debas K, Vandewalle G, Morin A, Poirier G et al (2011) Fast and slow spindle involvement in the consolidation of a new motor sequence. Behav Brain Res 217(1): 117–121

Baran B, Pace-Schott EF, Ericson C, Spencer RMC (2012) Processing of emotional reactivity and emotional memory over sleep. J Neurosci 32(3):1035–1042. doi:10.1523/JNEUROSCI.2532-11.2012

Baran B, Daniels D, Spencer RMC (2013) Sleep-dependent consolidation of value-based learning. PLoS One 8(10):e75326. doi:10.1371/journal.pone.0075326

Barnes DC, Wilson DA (2014) Slow-wave sleep-imposed replay modulates both strength and precision of memory. J Neurosci 34(15):5134–5142. doi:10.1523/JNEUROSCI.5274-13.2014

Batterink LJ, Paller KA (2015) Sleep-based memory processing facilitates grammatical generalization: evidence from targeted memory reactivation. Brain Lang. Available online Oct 3, 2015 [Epub ahead of print]. doi:10.1016/j.bandl.2015.09.003

Batterink LJ, Oudiette D, Reber PJ, Paller KA (2014) Sleep facilitates learning a new linguistic rule. Neuropsychologia 65(2014):169–179. doi:10.1016/j.neuropsychologia.2014.10.024

Behrens CJ, van den Boom LP, de Hoz L, Friedman A, Heinemann U (2005) Induction of sharp wave-ripple complexes in vitro and reorganization of hippocampal networks. Nat Neurosci 8(11):1560–1567. doi:10.1038/nn1571

Benchenane K, Peyrache A, Khamassi M, Tierney PL, Gioanni Y, Battaglia FP, Wiener SI (2010) Coherent theta oscillations and reorganization of spike timing in the hippocampal-prefrontal network upon learning. Neuron 66(6):921–936. doi:10.1016/j.neuron.2010.05.013

Bendor D, Wilson MA (2012) Biasing the content of hippocampal replay during sleep. Nat Neurosci 15(10):1439–1444. doi:10.1038/nn.3203

Bennion KA, Mickley Steinmetz KR, Kensinger EA, Payne JD (2015) Sleep and cortisol interact to support memory consolidation. Cereb Cortex 25(3):646–657

Benson K, Feinberg I (1977) The beneficial effect of sleep in an extended Jenkins and Dallenbach paradigm. Psychophysiology 14(4):375–384

Ben-Yakov A, Eshel N, Dudai Y (2013) Hippocampal immediate poststimulus activity in the encoding of consecutive naturalistic episodes. J Exp Psychol Gen 142(4):1255–1263. doi:10.1037/a0033558

Bergmann TO, Mölle M, Diedrichs J, Born J, Siebner HR (2012) Sleep spindle-related reactivation of category-specific cortical regions after learning face-scene associations. NeuroImage 59(3):2733–2742

Berry JA, Cervantes-Sandoval I, Chakraborty M, Davis RL (2015) Sleep facilitates memory by blocking dopamine neuron-mediated forgetting. Cell 161(7):1656–1667. doi:10.1016/j.cell.2015.05.027

Bethus I, Tse D, Morris RGM (2010) Dopamine and memory: modulation of the persistence of memory for novel hippocampal NMDA receptor-dependent paired associates. J Neurosci 30(5):1610–1618. doi:10.1523/JNEUROSCI.2721-09.2010

Bódizs R, Körmendi J, Rigó P, Lázár AS (2009) The individual adjustment method of sleep spindle analysis: methodological improvements and roots in the fingerprint paradigm. J Neurosci Methods 178(1):205–213

Born J, Wilhelm I (2012) System consolidation of memory during sleep. Psychol Res 76(2):192–203

Brain R (1958) The physiological basis of consciousness: a critical review. Brain 81(3):426–455. doi:10.1093/brain/81.3.426

Brawn TP, Fenn KM, Nusbaum HC, Margoliash D (2010a) Consolidating the effects of waking and sleep on motor-sequence learning. J Neurosci 30(42):13977–13982. doi:10.1523/JNEUROSCI.3295-10.2010

Brawn TP, Nusbaum HC, Margoliash D (2010b) Sleep-dependent consolidation of auditory discrimination learning in adult starlings. J Neurosci 30(2):609–613. doi:10.1523/JNEUROSCI.4237-09.2010

Brawn TP, Nusbaum HC, Margoliash D (2013) Sleep consolidation of interfering auditory memories in starlings. Psychol Sci 24(4):439–447. doi:10.1177/0956797612457391

Bremer F (1954) The neurophysiological problem of sleep. In: Adrian E, Bremer F, Japser H (eds) Brain mechanisms and consciousness. Blackwell, Oxford, pp 137–162

Buchegger J, Meier-Koll A (1988) Motor learning and ultradian sleep cycle: an electroencephalo-graphic study of trampoliners. Percept Mot Skills 67(1972):635–645

Buchsbaum BR, Lemire-Rodger S, Fang C, Abdi H (2012) The neural basis of vivid memory is patterned on perception. J Cogn Neurosci 24(9):1867–1883

Buzsaki G (1984) Long-term changes of hippocampal sharp-waves following high frequency afferent activation. Brain Res 300(1):179–182. doi:10.1016/0006-8993(84)91356-8

Buzsaki G (1986) Hippocampal sharp waves: their origin and significance. Brain Res 398:242–252

Buzsaki G (1989) Two-stage model of memory trace formation: a role for "noisy" brain states. Neuroscience 31(3):551–570

Buzsáki G, Lai-Wo SL, Vanderwolf CH (1983) Cellular bases of hippocampal EEG in the behaving rat. Brain Res Rev 6(2):139–171. doi:10.1016/0165-0173(83)90037-1

Buzsáki G, Logothetis N, Singer W (2013) Scaling brain size, keeping timing: evolutionary preservation of brain rhythms. Neuron 80(3):751–764. doi:10.1016/j.neuron.2013.10.002

Cahill L, Haiert RJ, Fallon J, Alkirei MT, Tangii C, Keatorii D et al (1996) Amygdala activity at encoding correlated with long-term, free recall of emotional information. Proc Natl Acad Sci USA 93:8016–8021

Cai DJ, Mednick SA, Harrison EM, Kanady JC, Mednick SC (2009) REM, not incubation, improves creativity by priming associative networks. Proc Natl Acad Sci 106(25):10130–10134

Cairney SA, Durrant SJ, Power R, Lewis PA (2015) Complementary roles of slow-wave sleep and rapid eye movement sleep in emotional memory consolidation. Cereb Cortex 25(6): 1565–1575. doi:10.1093/cercor/bht349

Canli T, Zhao Z, Brewer J, Gabrieli JD, Cahill L (2000) Event-related activation in the human amygdala associates with later memory for individual emotional experience. J Neurosci 20(19):RC99 20004570 [pii]

Carr MF, Jadhav SP, Frank LM (2011) Hippocampal replay in the awake state: a potential substrate for memory consolidation and retrieval. Nat Neurosci 14(2):147–153. doi:10.1038/nn.2732

Cash SS, Halgren E, Dehghani N, Rossetti AO, Thesen T, Wang C et al (2009) The human K-complex represents an isolated cortical down-state. Science 324(5930):1084–1087. doi:10.1126/science.1169626

Chrobak JJ, Buzsáki G (1994) Selective activation of deep layer (V-VI) retrohippocampal cortical neurons during hippocampal sharp waves in the behaving rat. J Neurosci 14(10):6160–6170

Cirelli C, Tononi G (2008) Is sleep essential? PLoS Biol 6(8):1605–1611

Clemens Z, Fabó D, Halász P (2005) Overnight verbal memory retention correlates with the number of sleep spindles. Neuroscience 132(2):529–535

Clemens Z, Fabó D, Halász P (2006) Twenty-four hours retention of visuospatial memory correlates with the number of parietal sleep spindles. Neurosci Lett 403(1-2):52–56

Clemens Z, Mölle M, Eross L, Barsi P, Halász P, Born J (2007) Temporal coupling of parahippocampal ripples, sleep spindles and slow oscillations in humans. Brain 130: 2868–2878

Cohen DA, Pascual-Leone A, Press DZ, Robertson EM (2005) Off-line learning of motor skill memory: a double dissociation of goal and movement. Proc Natl Acad Sci USA 102(50): 18237–18241. doi:10.1073/pnas.0506072102

Corkin S (2013) Permanent present tense. Basic Books, New York, NY

Cote K, Epps T, Campbell K (2000) The role of the spindle in human information processing of high-intensity stimuli during sleep. J Sleep Res 9:19–26

Cousins JN, El-Deredy W, Parkes LM, Hennies N, Lewis PA (2014) Cued memory reactivation during slow-wave sleep promotes explicit knowledge of a motor sequence. J Neurosci 34(48): 15870–15876. doi:10.1523/JNEUROSCI.1011-14.2014

Cox R, Hofman WF, Talamini LM (2012) Involvement of spindles in memory consolidation is slow wave sleep-specific. Learn Mem 19(7):264–267

Cox R, Hofman WF, de Boer M, Talamini LM (2014) Local sleep spindle modulations in relation to specific memory cues. NeuroImage 99:103–110 Retrieved from http://www.ncbi.nlm.nih.gov/pubmed/24852461

Craig M, Dewar M, Della Sala S, Wolbers T (2015a) Rest boosts the long-term retention of spatial associative and temporal order information. Hippocampus 25:1017–1027. doi:10.1002/hipo.22424

Craig M, Dewar M, Harris MA, Della Sala S, Wolbers T (2015b) Wakeful rest promotes the integration of spatial memories into accurate cognitive maps. Hippocampus. doi:10.1002/hipo.22502

Creery JD, Oudiette D, Antony JW, Paller KA (2014) Targeted memory reactivation during sleep depends on prior learning. Sleep 38(5):755–763 Retrieved from http://www.ncbi.nlm.nih.gov/pubmed/25515103

Dang-Vu TT, McKinney SM, Buxton OM, Solet JM, Ellenbogen JM (2010) Spontaneous brain rhythms predict sleep stability in the face of noise. Curr Biol 20(15):R626–R627

Dang-Vu TT, Bonjean M, Schabus M, Boly M, Darsaud A, Desseilles M et al (2011) Interplay between spontaneous and induced brain activity during human non-rapid eye movement sleep. Proc Natl Acad Sci USA 108(37):15438–15443

Dave A, Margoliash D (2000) Song replay during sleep and computational rules for sensorimotor vocal learning. Science 290(5492):812–816. doi:10.1126/science.290.5492.812

Davidson TJ, Kloosterman F, Wilson MA (2009) Hippocampal replay of extended experience. Neuron 63(4):497–507. doi:10.1016/j.neuron.2009.07.027

de Gennaro L, Ferrara M (2003) Sleep spindles: an overview. Sleep Med Rev 7(5):423–440

de Lavilléon G, Lacroix MM (2015) Explicit memory creation during sleep demonstrates a causal role of place cells in navigation. Nat Neurosci 18:493–495. doi:10.1038/nn.3970

Dement W (1960) The effect of dream deprivation. Science 131(3415):1705–1707. doi:10.1126/science.131.3415.1705

Deuker L, Olligs J, Fell J, Kranz TA, Mormann F, Montag C et al (2013) Memory consolidation by replay of stimulus-specific neural activity. J Neurosci 33(49):19373–19383. doi:10.1523/JNEUROSCI.0414-13.2013

Dewar M, Alber J, Butler C, Cowan N, Della Sala S (2012) Brief wakeful resting boosts new memories over the long term. Psychol Sci 23(9):955–960. doi:10.1177/0956797612441220

Diba K, Buzsáki G (2007) Forward and reverse hippocampal place-cell sequences during ripples. Nat Neurosci 10(10):1241–1242. doi:10.1038/nn1961

Diekelmann S, Born J (2010) The memory function of sleep. Nat Rev Neurosci 11(2):114–126

Diekelmann S, Büchel C, Born J, Rasch B (2011) Labile or stable: opposing consequences for memory when reactivated during waking and sleep. Nat Neurosci 14(3):381–386. doi:10.1038/nn.2744

Diekelmann S, Biggel S, Rasch B, Born J (2012) Offline consolidation of memory varies with time in slow wave sleep and can be accelerated by cuing memory reactivations. Neurobiol Learn Mem 98(2):103–111. doi:10.1016/j.nlm.2012.07.002

Djonlagic I, Rosenfeld A, Shohamy D, Djonlagic I, Rosenfeld A, Shohamy D, Myers C, Gluck M, Stickgold R (2009) Sleep enhances category learning. Learn Mem 16:751–755

Dolcos F, Labar KS, Cabeza R (2004) Interaction between the amygdala and the medial temporal lobe memory system predicts better memory for emotional events. Neuron 42:855–863

Dunsmoor JE, Murty VP, Davachi L, Phelps EA (2015) Emotional learning selectively and retroactively strengthens memories for related events. Nature 520:345–348. doi:10.1038/nature14106

Dupret D, O'Neill J, Pleydell-Bouverie B, Csicsvari J (2010) The reorganization and reactivation of hippocampal maps predict spatial memory performance. Nat Neurosci 13(8):995–1002. doi:10.1038/nn.2599

Durrant SJ, Taylor C, Cairney S, Lewis PA (2011) Sleep-dependent consolidation of statistical learning. Neuropsychologia 49(5):1322–1331. doi:10.1016/j.neuropsychologia.2011.02.015

Ebbinghaus H (1885) Memory: a contribution to experimental psychology. Columbia University Press, New York

Ego-Stengel V, Wilson MA (2009) Disruption of ripple-associated hippocampal activity during rest impairs spatial learning in the rat. Hippocampus 20(1):1–10. doi:10.1002/hipo.20707

Ellenbogen JM, Hulbert JC, Stickgold R, Dinges DF, Thompson-Schill SL (2006) Interfering with theories of sleep and memory: sleep, declarative memory, and associative interference. Curr Biol 16(13):1290–1294. doi:10.1016/j.cub.2006.05.024

Empson J, Clarke P (1970) Rapid eye movements and remembering. Nature 227:287–288

Eschenko O, Mölle M, Born J, Sara SJ (2006) Elevated sleep spindle density after learning or after retrieval in rats. J Neurosci 26(50):12914–12920

Eschenko O, Ramadan W, Mölle M, Born J, Sara S (2008) Sustained increase in hippocampal sharp-wave ripple activity during slow-wave sleep after learning. Learn Mem 15:222–228

Euston DR, Tatsuno M, McNaughton BL (2007) Fast-forward playback of recent memory sequences in prefrontal cortex during sleep. Science 318(5853):1147–1150. doi:10.1126/science.1148979

Feld GB, Besedovsky L, Kaida K, Münte TF, Born J (2014) Dopamine D2-like receptor activation wipes out preferential consolidation of high over low reward memories during human sleep. J Cogn Neurosci 26(10):2310–2320. doi:10.1162/jocn

Fenn KM, Hambrick DZ (2012) Individual differences in working memory capacity predict sleep-dependent memory consolidation. J Exp Psychol Gen 141(3):404–410. doi:10.1037/a0025268

Fenn KM, Hambrick DZ (2015) General intelligence predicts memory change across sleep. Psychon Bull Rev 22(3):791–799. doi:10.3758/s13423-014-0731-1

Fenn KM, Nusbaum HC, Margoliash D (2003) Consolidation during sleep of perceptual learning of spoken language. Nature 425:614–616. doi:10.1038/nature01971.1

Fischer S, Born J (2009) Anticipated reward enhances offline learning during sleep. J Exp Psychol Learn Mem Cogn 35(6):1586–1593. doi:10.1037/a0017256

Fischer S, Hallschmid M, Elsner AL, Born J (2002) Sleep forms memory for finger skills. Proc Natl Acad Sci USA 99(18):11987–11991

Fischer S, Diekelmann S, Born J (2011) Sleep's role in the processing of unwanted memories. J Sleep Res 20(2):267–274. doi:10.1111/j.1365-2869.2010.00881.x

Fishbein W (1971) Disruptive effects of rapid eye movement sleep deprivation on long-term memory. Physiol Behav 6(1):279–282

Fogel SM, Smith CT (2006) Learning-dependent changes in sleep spindles and Stage 2 sleep. J Sleep Res 15:250–255

Fosse R, Stickgold R, Hobson JA (2001) The mind in REM sleep: reports of emotional experience. Sleep 24(8):947–955 Retrieved from http://www.ncbi.nlm.nih.gov/pubmed/11766165

Frank MG, Issa NP, Stryker MP (2001) Sleep enhances plasticity in the developing visual cortex. Neuron 30:275–287

Freud S (1913) The interpretation of dreams. Macmillan, New York, NY

Fuentemilla L, Miró J, Ripollés P, Vilà-Balló A, Juncadella M, Castañer S et al (2013) Hippocampus-dependent strengthening of targeted memories via reactivation during sleep in humans. Curr Biol 23(18):1769–1775. doi:10.1016/j.cub.2013.07.006

Gaab N, Paetzold M, Becker M, Walker MP, Schlaug G (2004) The influence of sleep on auditory learning: a behavioral study. NeuroReport 15(4):731–734. doi:10.1097/01.wnr.0000113532.32218.d6

Gais S, Born J (2004) Declarative memory consolidation: mechanisms acting during human sleep. Learn Mem 11(6):679–685. doi:10.1101/lm.80504

Gais S, Plihal W, Wagner U, Born J (2000) Early sleep triggers memory for early visual discrimination skills. Nat Neurosci 3(12):1335–1339

Gais S, Mo M, Helms K, Born J (2002) Learning-dependent increases in sleep spindle density. J Neurosci 22(15):6830–6834

Gais S, Lucas B, Born J (2006) Sleep after learning aids memory recall. Learn Mem 13(3):259–262. doi:10.1101/lm.132106

Gelbard-Sagiv H, Mukamel R, Harel M, Malach R, Fried I (2008) Internally generated reactivation of single neurons in human hippocampus during free recall. Science 322(5898):96–101. doi:10.1126/science.1164685

Genzel L, Kroes MCW, Dresler M, Battaglia FP (2013) Light sleep versus slow wave sleep in memory consolidation: a question of global versus local processes? Trends Neurosci 37(1):1–10. doi:10.1016/j.tins.2013.10.002

Girardeau G, Benchenane K, Wiener SI, Buzsáki G, Zugaro MB (2009) Selective suppression of hippocampal ripples impairs spatial memory. Nat Neurosci 12(10):1222–1223. doi:10.1038/nn.2384

Godden D, Baddeley A (1975) Context-dependent memory in two natural environments: on land and underwater. Br J Psychol 66(3):325–331

Gómez RL, Bootzin RR, Nadel L (2006) Naps promote abstraction in language-learning infants. Psychol Sci 17(8):670–674. doi:10.1111/j.1467-9280.2006.01764.x

Guerrien A, Dujardin K, Mandai O, Sockeel P, Leconte P (1989) Enhancement of memory by auditory stimulation during postlearning REM sleep in humans. Physiol Behav 45:947–950

Gulati T, Ramanathan DS, Wong CC, Ganguly K (2014) Reactivation of emergent task-related ensembles during slow-wave sleep after neuroprosthetic learning. Nat Neurosci 17(8):1107–1113. doi:10.1038/nn.3759

Hars B, Hennevin E (1990) Reactivation of an old memory during sleep and wakefulness. Anim Learn Behav 18(4):365–376

Hars B, Hennevin E, Pasques P (1985) Improvement of learning by cueing during postlearning paradoxical sleep. Behav Brain Res 18:241–250

Hartse KM, Eisenhart SF, Bergmann BM, Rechtschaffen A (1979) Ventral hippocampus spikes during sleep, wakefulness, and arousal in the cat. Sleep 1(3):231–246

Hasselmo ME, McGaughy J (2004) High acetylcholine levels set circuit dynamics for attention and encoding and low acetylcholine levels set dynamics for consolidation. Prog Brain Res 145:207–231. doi:10.1016/S0079-6123(03)45015-2

Hauner KK, Howard JD, Zelano C, Gottfried JA (2013) Stimulus-specific enhancement of fear extinction during slow-wave sleep. Nat Neurosci 16(11):1553–1555. doi:10.1038/nn.3527

Heib DPJ, Hoedlmoser K, Anderer P, Gruber G, Zeitlhofer J, Schabus M (2015) Oscillatory theta activity during memory formation and its impact on overnight consolidation: a missing link? J Cogn Neurosci 27(8):1648–1658. doi:10.1162/jocn

Hennevin E, Hars B (1987) Is increase in post-learning paradoxical sleep modified by cueing? Behav Brain Res 24(3):243–249

Hoffman K, McNaughton B (2002) Coordinated reactivation of distributed memory traces in primate neocortex. Science 297:2070–2073

Hu P, Stylos-Allan M, Walker MP (2006) Sleep facilitates consolidation of emotional declarative memory. Psychol Sci 17(10):891–898

Hu X, Antony JW, Creery JD, Vargas IM, Bodenhausen GV, Paller KA (2015) Unlearning implicit social biases during sleep. Science 348(6238):1013–1015

Huber R, Ghilardi MF, Massimini M, Tononi G (2004) Local sleep and learning. Nature 430: 78–81

Jadhav SP, Kemere C, German PW, Frank LM (2012) Awake hippocampal sharp-wave ripples support spatial memory. Science 336(6087):1454–1458. doi:10.1126/science.1217230

Jenkins J, Dallenbach K (1924) Obliviscence during sleep and waking. Am J Psychol 35(4): 605–612

Jensen LC, Harris K, Anderson DC (1971) Retention following a change in ambient contextual stimuli for six age groups. Dev Psychol 4(3):394–399. doi:10.1037/h0030957

Ji D, Wilson MA (2007) Coordinated memory replay in the visual cortex and hippocampus during sleep. Nat Neurosci 10(1):100–107. doi:10.1038/nn1825

Johnson JD, Rugg MD (2007) Recollection and the reinstatement of encoding-related cortical activity. Cereb Cortex 17(11):2507–2515. doi:10.1093/cercor/bhl156

Johnson JD, McDuff SGR, Rugg MD, Norman KA (2009) Recollection, familiarity, and cortical reinstatement: a multivoxel pattern analysis. Neuron 63(5):697–708. doi:10.1016/j.neuron.2009.08.011

Johnson L, Blakely T, Hermes D, Hakimian S, Ramsey N, Ojemann J (2012) Sleep spindles are locally modulated by training on a brain-computer interface. Proc Natl Acad Sci USA 109(45): 18583–18588

Kaestner EJ, Wixted JT, Mednick SC (2013) Pharmacologically increasing sleep spindles enhances recognition for negative and high-arousal memories. J Cogn Neurosci 25(10): 1597–1610

Karlsson MP, Frank LM (2009) Awake replay of remote experiences in the hippocampus. Nat Neurosci 12(7):913–918. doi:10.1038/nn.2344

Karlsson Wirebring L, Wiklund-Hornqvist C, Eriksson J, Andersson M, Jonsson B, Nyberg L (2015) Lesser neural pattern similarity across repeated tests is associated with better long-term memory retention. J Neurosci 35(26):9595–9602. doi:10.1523/JNEUROSCI.3550-14.2015

Karni A, Tanne D, Rubenstein B, Askenasy J, Sagi D (1994) Dependence on REM sleep of overnight improvement of a perceptual skill. Science 265:679–682

Korman M, Doyon J, Doljansky J, Carrier J, Dagan Y, Karni A (2007) Daytime sleep condenses the time course of motor memory consolidation. Nat Neurosci 10(9):1206–1213. doi:10.1038/nn1959

Kudrimoti HS, Barnes CA, Mcnaughton BL (1999) Reactivation of hippocampal cell assemblies: effects of behavioral state, experience, and EEG dynamics. J Neurosci 19(10):4090–4101

Kuhl BA, Shah AT, DuBrow S, Wagner AD (2010) Resistance to forgetting associated with hippocampus-mediated reactivation during new learning. Nat Neurosci 13(4):501–506. doi:10. 1038/nn.2498

Kuhl BA, Bainbridge WA, Chun MM (2012a) Neural reactivation reveals mechanisms for updating memory. J Neurosci 32(10):3453–3461. doi:10.1523/JNEUROSCI.5846-11.2012

Kuhl BA, Rissman J, Wagner AD (2012b) Multi-voxel patterns of visual category representation during episodic encoding are predictive of subsequent memory. Neuropsychologia 50(4): 458–469. doi:10.1016/j.neuropsychologia.2011.09.002

Kurdziel L, Duclos K, Spencer RMC (2013) Sleep spindles in midday naps enhance learning in preschool children. Proc Natl Acad Sci USA 110(43):17267–17272. doi:10.1073/pnas. 1306418110

Kuriyama K, Stickgold R, Walker MP (2004) Sleep-dependent learning and motor-skill complexity. Learn Mem 11:705–713. doi:10.1101/lm.76304.appears

LaBar KS, Cabeza R (2006) Cognitive neuroscience of emotional memory. Nat Rev Neurosci 7(1):54–64. doi:10.1038/nrn1825

Landsness EC, Crupi D, Hulse BK, Peterson MJ, Huber R, Ansari H et al (2009) Sleep-dependent improvement in visuomotor learning: a causal role for slow waves. Sleep 32(10):1273–1284

Lansink CS, Goltstein PM, Lankelma JV, Joosten RNJMA, McNaughton BL, Pennartz CMA (2008) Preferential reactivation of motivationally relevant information in the ventral striatum. J Neurosci 28(25):6372–6382. doi:10.1523/JNEUROSCI.1054-08.2008

Lansink CS, Goltstein PM, Lankelma JV, McNaughton BL, Pennartz CMA (2009) Hippocampus leads ventral striatum in replay of place-reward information. PLoS Biol 7(8):e1000173. doi:10. 1371/journal.pbio.1000173

Lany J, Gómez RL (2008) Twelve-month-old infants benefit from prior experience in statistical learning. Psychol Sci 19(12):1247–1252. doi:10.1111/j.1467-9280.2008.02233.x

Lechner H, Squire L, Byrne J (1999) 100 years of consolidation—remembering Müller and Pilzecker. Learn Mem 6:77–88

Lehmann H, McNamara KC (2011) Repeatedly reactivated memories become more resistant to hippocampal damage. Learn Mem 18(3):132–135. doi:10.1101/lm.2000811

Lehmann H, Sparks FT, Spanswick SC, Hadikin C, McDonald RJ, Sutherland RJ (2009) Making context memories independent of the hippocampus. Learn Mem 16(7):417–420. doi:10.1101/ lm.1385409

Lesburguères E, Gobbo OL, Alaux-Cantin S, Hambucken A, Trifilieff P, Bontempi B (2011) Early tagging of cortical networks is required for the formation of enduring associative memory. Science 331(6019):924–928. doi:10.1126/science.1196164

Lewis PA, Cairney S, Manning L, Critchley HD (2011) The impact of overnight consolidation upon memory for emotional and neutral encoding contexts. Neuropsychologia 49:2619–2629. doi:10.1016/j.neuropsychologia.2011.05.009

Lisman JE, Grace AA (2005) The hippocampal-VTA loop: controlling the entry of information into long-term memory. Neuron 46(5):703–713. doi:10.1016/j.neuron.2005.05.002

Louie K, Wilson MA (2001) Temporally structured replay of awake hippocampal ensemble activity during rapid eye movement sleep. Neuron 29(1):145–156. doi:10.1016/S0896-6273 (01)00186-6

Mandai O, Guerrien A, Sockeel P, Dujardin K, Leconte P (1989) REM sleep modifications following a Morse code learning session in humans. Physiol Behav 46(4):639–642. doi:10. 1016/0031-9384(89)90344-2

Mander BA, Santhanam S, Saletin JM, Walker MP (2011) Wake deterioration and sleep restoration of human learning. Curr Biol 21(5):R183–R184. doi:10.1016/j.cub.2011.01.019

Manning JR, Sperling MR, Sharan A, Rosenberg EA, Kahana MJ (2012) Spontaneously reactivated patterns in frontal and temporal lobe predict semantic clustering during memory search. J Neurosci 32(26):8871–8878. doi:10.1523/JNEUROSCI.5321-11.2012

Manoach DS, Thakkar KN, Stroynowski E, Ely A, McKinley SK, Wamsley E et al (2009) Reduced overnight consolidation of procedural learning in chronic medicated schizophrenia is related to specific sleep stages. J Psychiatr Res 44(2):112–120. doi:10.1016/j.jpsychires. 2009.06.011

Maquet P, Péters J, Aerts J, Delfiore G, Degueldre C, Luxen A, Franck G (1996) Functional neuroanatomy of human rapid-eye-movement sleep and dreaming. Nature 383(6596):163–166. doi:10.1038/383163a0

Maquet P, Laureys S, Peigneux P, Fuchs S, Petiau C, Phillips C et al (2000) Experience-dependent changes in cerebral activation during human REM sleep. Nat Neurosci 3(8):831–836

Marshall L, Helgadóttir H, Mölle M, Born J (2006) Boosting slow oscillations during sleep potentiates memory. Nature 444(7119):610–613

Mather M, Sutherland MR (2011) Arousal-biased competition in -perception and memory. Perspect Psychol Sci 6:114–133. doi:10.1177/1745691611400234

Maviel T, Durkin TP, Menzaghi F, Bontempi B (2004) Sites of neocortical reorganization critical for remote spatial memory. Science 305(5680):96–99. doi:10.1126/science.1098180

McGaugh JL (2000) Memory—a century of consolidation. Science 287(5451):248–251. doi:10. 1126/science.287.5451.248

McNamara CG, Tejero-Cantero Á, Trouche S, Campo-Urriza N, Dupret D (2014) Dopaminergic neurons promote hippocampal reactivation and spatial memory persistence. Nat Neurosci 17: 1658–1660. doi:10.1038/nn.3843

Mednick SC, Nakayama K, Cantero JL, Atienza M, Levin AA, Pathak N, Stickgold R (2002) The restorative effect of naps on perceptual deterioration. Nat Neurosci 5(7):677–681

Mednick S, Nakayama K, Stickgold R (2003) Sleep-dependent learning: a nap is as good as a night. Nat Neurosci 6(7):697–698

Mednick SC, Drummond SPA, Arman AC, Boynton GM (2008) Perceptual deterioration is reflected in the neural response: fMRI study of nappers and non-nappers. Perception 37(7): 1086–1097. doi:10.1068/p5998

Mednick S, McDevitt E, Walsh J, Wamsley E, Paulus M, Kanady J, Drummond S (2013) The critical role of sleep spindles in hippocampal-dependent memory: a pharmacology study. J Neurosci 33(10):4494–4504

Mölle M, Born J (2011) Slow oscillations orchestrating fast oscillations and memory consolidation. Prog Brain Res 193:93–110

Mölle M, Yeshenko O, Marshall L, Sara S, Born J (2006) Hippocampal sharp wave-ripples linked to slow oscillations in rat slow-wave sleep. J Neurophysiol 96(1):62–70

Mölle M, Bergmann T, Marshall L, Born J (2011) Fast and slow spindles during the sleep slow oscillation: disparate coalescence and engagement in memory processing. Sleep 34(10): 1411–1421

Morin A, Doyon J, Dostie V, Barakat M, Tahar AH, Korman M et al (2008) Motor sequence learning increases sleep spindles and fast frequencies in post-training sleep. Sleep 31(8): 1149–1156

Morison R, Bassett D (1945) Electrical activity of the thalamus and basal ganglia in decorticate cats. J Neurophysiol 8(5):309–314

Morris RGM (2006) Elements of a neurobiological theory of hippocampal function: the role of synaptic plasticity, synaptic tagging and schemas. Eur J Neurosci 23(11):2829–2846. doi:10. 1111/j.1460-9568.2006.04888.x

Moscovitch M, Rosenbaum RS, Gilboa A, Addis DR, Westmacott R, Grady C et al (2005) Functional neuroanatomy of remote episodic, semantic and spatial memory: a unified account based on multiple trace theory. J Anat 207(1):35–66. doi:10.1111/j.1469-7580.2005.00421.x

Müller G, Pilzecker A (1900) Experimental contributions to the theory of memory. Z Psychol Ergänzungsband 1:1–300

Newman EL, Norman KA (2010) Moderate excitation leads to weakening of perceptual representations. Cereb Cortex 20(11):2760–2770. doi:10.1093/cercor/bhq021

Ngo HV, Martinetz T, Born J, Mölle M (2013) Auditory closed-loop stimulation of the sleep slow oscillation enhances memory. Neuron 78(3):545–553

Ngo H, Miedema A, Faude I, Martinetz T, Molle M, Born J (2015) Driving sleep slow oscillations by auditory closed-loop stimulation—a self-limiting process. J Neurosci 35(17):6630–6638. doi:10.1523/JNEUROSCI.3133-14.2015

Nir Y, Tononi G (2010) Dreaming and the brain: from phenomenology to neurophysiology. Trends Cogn Sci 14(2):88–100. doi:10.1016/j.tics.2009.12.001

Nir Y, Staba RJ, Andrillon T, Vyazovskiy VV, Cirelli C, Fried I, Tononi G (2011) Regional slow waves and spindles in human sleep. Neuron 70(1):153–169

Nishida M, Walker MP (2007) Daytime naps, motor memory consolidation and regionally specific sleep spindles. PLoS One 2(4):e341

Nishida M, Pearsall J, Buckner RL, Walker MP (2009) REM sleep, prefrontal theta, and the consolidation of human emotional memory. Cereb Cortex 19:1158–1166. doi:10.1093/cercor/bhn155

Nyberg L, Habib R, McIntosh AR, Tulving E (2000) Reactivation of encoding-related brain activity during memory retrieval. Proc Natl Acad Sci USA 97(20):11120–11124. doi:10.1073/pnas.97.20.11120

O'Neill J, Senior T, Csicsvari J (2006) Place-selective firing of CA1 pyramidal cells during sharp wave/ripple network patterns in exploratory behavior. Neuron 49(1):143–155. doi:10.1016/j.neuron.2005.10.037

O'Neill J, Senior TJ, Allen K, Huxter JR, Csicsvari J (2008) Reactivation of experience-dependent cell assembly patterns in the hippocampus. Nat Neurosci 11(2):209–215. doi:10.1038/nn2037

Ong JL, Lo JC, Chee NIYN, Santostasi G, Paller KA, Zee PC, Chee MWL (2016) Effects of phase-locked acoustic stimulation during a nap on EEG spectra and declarative memory consolidation. Sleep Med 20:88–97

Oswald I, Taylor A, Treisman M (1960) Discriminative responses to stimuli during human sleep. Brain 83(3):440–453. doi:10.1093/brain/83.3.440

Oudiette D, Antony JW, Creery JD, Paller KA (2013) The role of memory reactivation during wakefulness and sleep in determining which memories endure. J Neurosci 33(15):6672–6678. doi:10.1523/JNEUROSCI.5497-12.2013

Pavlides C, Winson J (1989) Influences of hippocampal place cell firing in the awake state on the activity of these cells during subsequent sleep episodes. J Neurosci 9(8):2907–2918

Payne JD, Kensinger EA (2011) Sleep leads to changes in the emotional memory trace: evidence from fMRI. J Cogn Neurosci 23(6):1285–1297. doi:10.1162/jocn.2010.21526

Payne JD, Stickgold R, Swanberg K, Kensinger EA (2008) Sleep preferentially enhances memory for emotional components of scenes. Psychol Sci 19(8):781–788. doi:10.1111/j.1467-9280.2008.02157.x

Payne JD, Chambers AM, Kensinger EA (2012) Sleep promotes lasting changes in selective memory for emotional scenes. Front Integr Neurosci 6:108. doi:10.3389/fnint.2012.00108

Payne JD, Kensinger EA, Wamsley EJ, Spreng RN, Alger SE, Gibler K et al (2015) Napping and the selective consolidation of negative aspects of scenes. Emotion 15(2):176–186. doi:10.1037/a0038683

Peigneux P, Laureys S, Fuchs S, Collette F, Perrin F, Reggers J et al (2004) Are spatial memories strengthened in the human hippocampus during slow wave sleep? Neuron 44:535–545

Peigneux P, Orban P, Balteau E, Degueldre C, Luxen A, Laureys S, Maquet P (2006) Offline persistence of memory-related cerebral activity during active wakefulness. PLoS Biol 4(4):e100. doi:10.1371/journal.pbio.0040100

Pennartz CMA, Lee E, Verheul J, Lipa P, Barnes CA, McNaughton BL (2004) The ventral striatum in off-line processing: ensemble reactivation during sleep and modulation by hippocampal ripples. J Neurosci 24(29):6446–6456. doi:10.1523/JNEUROSCI.0575-04.2004

Peyrache A, Khamassi M, Benchenane K, Wiener SI, Battaglia FP (2009) Replay of rule-learning related neural patterns in the prefrontal cortex during sleep. Nat Neurosci 12(7):919–926. doi:10.1038/nn.2337

Plihal W, Born J (1997) Effects of early and late nocturnal sleep on declarative and procedural memory. J Cogn Neurosci 9(4):534–547

Polyn SM, Natu VS, Cohen JD, Norman KA (2005) Category-specific cortical activity precedes retrieval during memory search. Science 310(5756):1963–1966. doi:10.1126/science.1117645

Ramadan W, Eschenko O, Sara SJ (2009) Hippocampal sharp wave/ripples during sleep for consolidation of associative memory. PLoS One 4(8):1–9. doi:10.1371/journal.pone.0006697

Rasch B, Büchel C, Gais S, Born J (2007) Odor cues during slow-wave sleep prompt declarative memory consolidation. Science:1426–1429

Rasch B, Pommer J, Diekelmann S, Born J (2009) Pharmacological REM sleep suppression paradoxically improves rather than impairs skill memory. Nat Neurosci 12(4):396–397. doi:10.1038/nn.2206

Rauchs G, Feyers D, Landeau B, Bastin C, Luxen A, Maquet P, Collette F (2011) Sleep contributes to the strengthening of some memories over others, depending on hippocampal activity at learning. J Neurosci 31(7):2563–2568. doi:10.1523/JNEUROSCI.3972-10.2011

Rechtschaffen A, Kales A (1968) A manual of standardized terminology, techniques and scoring system of sleep stages in human subjects. University of California, Los Angele

Redondo RL, Morris RGM (2011) Making memories last: the synaptic tagging and capture hypothesis. Nat Rev Neurosci 12(1):17–30. doi:10.1038/nrn2963

Richter-Levin G, Akirav I (2003) Emotional tagging of memory formation—in the search for neural mechanisms. Brain Res Rev 43:247–256. doi:10.1016/j.brainresrev.2003.08.005

Rihm JS, Diekelmann S, Born J, Rasch B (2014) Reactivating memories during sleep by odors: odor specificity and associated changes in sleep oscillations. J Cogn Neurosci 26(8):1806–1818. doi:10.1162/jocn_a_00579

Ritter SM, Strick M, Bos MW, van Baaren RB, Dijksterhuis A (2012) Good morning creativity: task reactivation during sleep enhances beneficial effect of sleep on creative performance. J Sleep Res 21(6):643–647. doi:10.1111/j.1365-2869.2012.01006.x

Robertson EM, Pascual-Leone A, Press DZ (2004) Awareness modifies the skill-learning benefits of sleep. Curr Biol 14(3):208–212. doi:10.1016/j.cub.2004.01.027

Roediger HL, Karpicke JD (2006) The power of testing memory: basic research and implications for educational practice. Perspect Psychol Sci 1(3):181–210

Rolls A, Makam M, Kroeger D, Colas D, de Lecea L, Heller HC (2013) Sleep to forget: interference of fear memories during sleep. Mol Psychiatry 18(11):1166–1170. doi:10.1038/mp.2013.121

Rosanova M, Ulrich D (2005) Pattern-specific associative long-term potentiation induced by a sleep spindle-related spike train. J Neurosci 25(41):9398–9405

Routtenberg A, Rekart JL (2005) Post-translational protein modification as the substrate for long-lasting memory. Trends Neurosci 28(1):12–19. doi:10.1016/j.tins.2004.11.006

Rudoy J, Voss J, Westerberg C, Paller K (2009) Strengthening individual memories by reactivating them during sleep. Science 326:1079

Saletin JM, Goldstein AN, Walker MP (2011) The role of sleep in directed forgetting and remembering of human memories. Cereb Cortex 21(11):2534–2541

Sato Y, Fukuoka Y, Minamitani H, Honda K (2007) Sensory stimulation triggers spindles during sleep Stage 2. Sleep 30(4):511–518

Schabus M, Gruber G, Parapatics S, Sauter C, Klösch G, Anderer P et al (2004) Sleep spindles and their significance for declarative memory consolidation. Sleep 27(8):1479–1485

Schabus M, Hödlmoser K, Gruber G, Sauter C, Anderer P, Klösch G et al (2006) Sleep spindle-related activity in the human EEG and its relation to general cognitive and learning abilities. Eur J Neurosci 23(7):1738–1746

Schabus M, Kerstin H, Thomas P, Anderer P, Gruber G, Parapatics S et al (2008) Interindividual sleep spindle differences and their relation to learning-related enhancements. Brain Res 29:127–135

Schabus M, Dang-Vu T, Heib D, Boly M, Desseilles M, Vandewalle G et al (2012) The fate of incoming stimuli during NREM sleep is determined by spindles and the phase of the slow oscillation. Front Neurol 3:40

Schmidt C, Peigneux P, Muto V, Schenkel M, Knoblauch V, Münch M et al (2006) Encoding difficulty promotes postlearning changes in sleep spindle activity during napping. J Neurosci 26(35):8976–8982. doi:10.1523/JNEUROSCI.2464-06.2006

Schönauer M, Geisler T, Gais S (2014) Strengthening procedural memories by reactivation in sleep. J Cogn Neurosci 26(1):143–153. doi:10.1162/jocn_a_00471

Schreiner T, Rasch B (2014) Boosting vocabulary learning by verbal cueing during sleep. Cereb Cortex 25(11):4169–4179. doi:10.1093/cercor/bhu139

Schreiner T, Göldi M, Rasch B (2015) Cueing vocabulary during sleep increases theta activity during later recognition testing. Psychophysiology 52(11):1538–1543. doi:10.1111/psyp.12505

Scoville WB, Milner B (1957) Loss of recent memory after bilateral hippocampal lesions. J Neurol Neurosurg Psychiatry 20:11–21

Scrima L (1982) Isolated REM sleep facilitates recall of complex associative information. Psychophysiology 19(3):252–259. doi:10.1111/j.1469-8986.1982.tb02556.x

Shank SS, Margoliash D (2009) Sleep and sensorimotor integration during early vocal learning in a songbird. Nature 458(7234):73–77. doi:10.1038/nature07615

Siapas AG, Wilson MA (1998) Coordinated interactions between hippocampal ripples and cortical spindles during slow-wave sleep. Neuron 21:1123–1128

Silva RH, Abílio VC, Takatsu AL, Kameda SR, Grassl C, Chehin AB et al (2004) Role of hippocampal oxidative stress in memory deficits induced by sleep deprivation in mice. Neuropharmacol 46(6):895–903

Skaggs WE, McNaughton BL (1996) Replay of neuronal firing sequences in rat hippocampus during sleep following spatial experience. Science 271(5257):1870–1873. doi:10.1126/science.271.5257.1870

Smith C (1995) Sleep states and memory processes. Behav Brain Res 69:137–145

Smith C (2001) Sleep states and memory processes in humans: procedural versus declarative memory systems. Sleep Med Rev 5(6):491–506. doi:10.1053/smrv.2001.0164

Smith C, MacNeill C (1994) Impaired motor memory for a pursuit rotor task following stage 2 sleep loss in college students. J Sleep Res 3:206–213

Smith CT, Smith D (2003) Ingestion of ethanol just prior to sleep onset impairs memory for procedural but not declarative tasks. Sleep 26(2):185–191

Smith C, Weeden K (1990) Post training REMs coincident auditory stimulation enhances memory in humans. Psychiatr J Univ Ott 15(2):85–90

Smith C, Wong PT (1991) Paradoxical sleep increases predict successful learning in a complex operant task. Behav Neurosci 105(2):282–288. doi:10.1037//0735-7044.105.2.282

Smith C, Young J, Young W (1980) Prolonged increases in paradoxical sleep during and after avoidance-task acquisition. Sleep 3(1):67–81

Smith C, Aubrey J, Peters K (2004) Different roles for REM and Stage-2 sleep in motor learning: a proposed model. Psychol Belg 44:79–102

Song S, Cohen LG (2014) Practice and sleep form different aspects of skill. Nat Commun 5:3407. doi:10.1038/ncomms4407

Squire LR, Bayley PJ (2007) The neuroscience of remote memory. Curr Opin Neurobiol 17(2):185–196. doi:10.1016/j.conb.2007.02.006

Squire LR, Genzel L, Wixted JT, Morris RG (2015) Memory consolidation. Cold Spring Harb Perspect Biol 7(8):a021766

Staresina BP, Alink A, Kriegeskorte N, Henson RN (2013) Awake reactivation predicts memory in humans. Proc Natl Acad Sci USA 110(52):21159–21164. doi:10.1073/pnas.1311989110

Staresina BP, Bergmann TO, Bonnefond M, van der Meij R, Jensen O, Deuker L et al (2015) Hierarchical nesting of slow oscillations, spindles and ripples in the human hippocampus during sleep. Nat Neurosci 18:1679–1686. doi:10.1038/nn.4119

Stickgold R, Walker MP (2013) Sleep-dependent memory triage: evolving generalization through selective processing. Nat Neurosci 16(2):139–145. doi:10.1038/nn.3303

Stickgold R, Whidbee D, Schirmer B, Patel V, Hobson JA (2000) Visual discrimination task improvement: a multi-step process occurring during sleep. J Cogn Neurosci 12(2):246–254

St-Laurent M, Abdi H, Buchsbaum BR (2014) Distributed patterns of reactivation predict vividness of recollection. J Cogn Neurosci 27(10):2000–2018. doi:10.1162/jocn

Studte S, Bridger E, Mecklinger A (2015) Nap sleep preserves associative but not item memory performance. Neurobiol Learn Mem 120:84–93. doi:10.1016/j.nlm.2015.02.012

Takashima A, Nieuwenhuis ILC, Jensen O, Talamini LM, Rijpkema M, Fernández G (2009) Shift from hippocampal to neocortical centered retrieval network with consolidation. J Neurosci 29(32):10087–10093. doi:10.1523/JNEUROSCI.0799-09.2009

Takeuchi T, Duszkiewicz AJ, Morris RGM (2014) The synaptic plasticity and memory hypothesis: encoding, storage and persistence. Philos Trans R Soc Lond B Biol Sci 369(1633):20130288. doi:10.1098/rstb.2013.0288

Talamini LM, Nieuwenhuis ILC, Takashima A, Jensen O (2008) Sleep directly following learning benefits consolidation of spatial associative memory. Learn Mem 15(4):233–237. doi:10.1101/lm.771608

Tamaki M, Matsuoka T, Nittono H, Hori T (2008) Fast sleep spindle (13–15 Hz) activity correlates with sleep-dependent improvement in visuomotor performance. Sleep 31(2):204–211

Tambini A, Ketz N, Davachi L (2010) Enhanced brain correlations during rest are related to memory for recent experiences. Neuron 65(2):280–290. doi:10.1016/j.neuron.2010.01.001

Tilley A, Empson J (1978) REM sleep and memory consolidation. Biol Psychol 6(4):293–300. doi:10.1016/0301-0511(78)90031-5

Tononi G, Cirelli C (2014) Sleep and the price of plasticity: from synaptic and cellular homeostasis to memory consolidation and integration. Neuron 81(1):12–34. doi:10.1016/j.neuron.2013.12.025

Torres M, Giraldez F (1998) The development of the vertebrate inner ear. Mech Dev 71:5–21

Tse D, Takeuchi T, Kakeyama M, Kajii Y, Okuno H, Tohyama C et al (2011) Schema-dependent gene activation and memory encoding in neocortex. Science 333:891–895

Tulving E, Thomson D (1973) Encoding specificity and retrieval processes in episodic memory. Psychol Rev 80(5):352–373

van der Helm E, Gujar N, Nishida M, Walker MP (2011) Sleep-dependent facilitation of episodic memory details. PLoS One 6(11):e27421. doi:10.1371/journal.pone.0027421

van Dongen EV, Takashima A, Barth M, Zapp J, Schad LR, Paller KA, Fernández G (2012a) Memory stabilization with targeted reactivation during human slow-wave sleep. Proc Natl Acad Sci USA 109(26):10575–10580. doi:10.1073/pnas.1201072109

van Dongen EV, Thielen J-W, Takashima A, Barth M, Fernández G (2012b) Sleep supports selective retention of associative memories based on relevance for future utilization. PLoS One 7(8):e43426. doi:10.1371/journal.pone.0043426

van Stegeren AH, Goekoop R, Everaerd W, Scheltens P, Barkhof F, Kuijer JPA, Rombouts SARB (2005) Noradrenaline mediates amygdala activation in men and women during encoding of emotional material. NeuroImage 24:898–909. doi:10.1016/j.neuroimage.2004.09.011

Vilberg KL, Davachi L (2013) Perirhinal-hippocampal connectivity during reactivation is a marker for object-based memory consolidation. Neuron 79(6):1232–1242. doi:10.1016/j.neuron.2013.07.013

Vyazovskiy VV, Harris KD (2013) Sleep and the single neuron: the role of global slow oscillations in individual cell rest. Nat Rev Neurosci 14(6):443–451

Vyazovskiy VV, Cirelli C, Pfister-Genskow M, Faraguna U, Tononi G (2008) Molecular and electrophysiological evidence for net synaptic potentiation in wake and depression in sleep. Nat Neurosci 11(2):200–208

Vyazovskiy VV, Olcese U, Lazimy YM, Faraguna U, Esser SK, Williams JC et al (2009) Cortical firing and sleep homeostasis. Neuron 63(6):865–878

Vyazovskiy VV, Olcese U, Hanlon EC, Nir Y, Cirelli C, Tononi G (2011) Local sleep in awake rats. Nature 472(7344):443–447

Wagner U, Gais S, Born J (2001) Emotional memory formation is enhanced across sleep intervals with high amounts of rapid eye movement sleep. Learn Mem 8:112–119. doi:10.1101/lm.36801.sleep

Wagner U, Gais S, Haider H, Verleger R, Born J (2004) Sleep inspires insight. Nature 427(22): 352–355

Wagner U, Hallschmid M, Rasch B, Born J (2006) Brief sleep after learning keeps emotional memories alive for years. Biol Psychiatry 60:788–790. doi:10.1016/j.biopsych.2006.03.061

Walker MP, Brakefield T, Morgan A, Hobson JA, Stickgold R (2002a) Practice with sleep makes perfect: sleep-dependent motor skill learning. Neuron 35:205–211

Walker MP, Liston C, Hobson JA, Stickgold R (2002b) Cognitive flexibility across the sleep–wake cycle: REM-sleep enhancement of anagram problem solving. Cogn Brain Res 14(3):317–324. doi:10.1016/S0926-6410(02)00134-9

Walker MP, Brakefield T, Hobson JA (2003) Dissociable stages of human memory consolidation and reconsolidation. Nature 425:616–620. doi:10.1038/nature01951.1

Walker MP, Stickgold R, Alsop D, Gaab N, Schlaug G (2005) Sleep-dependent motor memory plasticity in the human brain. Neuroscience 133(4):911–917. doi:10.1016/j.neuroscience.2005.04.007

Wamsley EJ, Tucker MA, Shinn AK, Ono KE, McKinley SK, Ely AV et al (2012) Reduced sleep spindles and spindle coherence in schizophrenia: mechanisms of impaired memory consolidation? Biol Psychiatry 71(2):154–161. doi:10.1016/j.biopsych.2011.08.008

Wang S-H, Redondo RL, Morris RGM (2010) Relevance of synaptic tagging and capture to the persistence of long-term potentiation and everyday spatial memory. Proc Natl Acad Sci 107(45):19537–19542

Wiesel TN, Hubel DH (1963) Effects of visual deprivation on morphology and physiology of cells in the cat's lateral geniculate body. J Neurophysiol 26:978–993 Retrieved from http://hubel.med.harvard.edu/papers/HubelWiesel1963Jneurophysiol3.pdf. http://www.ncbi.nlm.nih.gov/pubmed/14084170

Wilhelm I, Diekelmann S, Molzow I, Ayoub A, Mölle M, Born J (2011) Sleep selectively enhances memory expected to be of future relevance. J Neurosci 31(5):1563–1569. doi:10.1523/JNEUROSCI.3575-10.2011

Wilson M, McNaughton B (1994) Reactivation of hippocampal ensemble memories during sleep. Science 265:676–679

Wing EA, Ritchey M, Cabeza R (2015) Reinstatement of individual past events revealed by the similarity of distributed activation patterns during encoding and retrieval. J Cogn Neurosci 27(4):679–691. doi:10.1162/jocn

Wittmann BC, Schott BH, Guderian S, Frey JU, Heinze HJ, Düzel E (2005) Reward-related fMRI activation of dopaminergic midbrain is associated with enhanced hippocampus-dependent long-term memory formation. Neuron 45:459–467. doi:10.1016/j.neuron.2005.01.010

Xie L, Kang H, Xu Q, Chen MJ, Liao Y, Thiyagarajan M et al (2013) Sleep drives metabolite clearance from the adult brain. Science 342:373–377

Xue G, Dong Q, Chen C, Lu Z, Mumford JA, Poldrack RA (2010) Greater neural pattern similarity across repetitions is associated with better memory. Science 330(6000):97–101. doi:10.1126/science.1193125

Yaroush R, Sullivan MJ, Ekstrand BR (1971) Effect of sleep on memory: II. Differential effect of the first and second half of the night. J Exp Psychol 88(3):361–366. doi:10.1037/h0030914

Yordanova J, Kolev V, Wagner U, Born J, Verleger R (2012) Increased alpha (8–12 Hz) activity during slow wave sleep as a marker for the transition from implicit knowledge to explicit insight. J Cogn Neurosci 24(1):119–132

Beyond Long-Term Declarative Memory: Evaluating Hippocampal Contributions to Unconscious Memory Expression, Perception, and Short-Term Retention

Deborah E. Hannula, Jennifer D. Ryan, and David E. Warren

Abstract Contributions made by the hippocampus to long-term declarative memory are well established, but recent work compels reconsideration of the perspective that performance in other cognitive domains is independent of hippocampal function. In this chapter, we review literature that points to a role for the hippocampus in three additional domains—namely, perception, short-term or working memory, and unconscious expressions of memory. Counterevidence against claims for this broader reach are considered along with methodological challenges in each domain, and questions that remain to be addressed in future work are proposed. In the end, we argue that while there is much to be done, evidence strongly suggests that the reach of the hippocampus extends well beyond long-term declarative memory.

It is well established that the hippocampus and adjacent medial temporal lobe (MTL) cortical structures are necessary for long-term declarative (conscious) memory, but investigators continue to cast a wider net, suggesting a considerably broader reach for these structures than standard perspectives have proposed. The objective of this chapter is to explore possible contributions made by the hippocampus to perception, short-term or working memory, and expressions of memory in the absence of conscious awareness. Questions about whether and how the hippocampus supports processing in these domains have garnered a good deal of interest in recent years, and healthy debate about the viability of claims that have been made in the literature is ongoing (cf. Eichenbaum 2013; Squire and Dede

D.E. Hannula (✉)
Department of Psychology, University of Wisconsin, Milwaukee, WI, USA
e-mail: hannula@uwm.edu

J.D. Ryan
Rotman Research Institute, Baycrest, Toronto, ON, Canada

Department of Psychology, University of Toronto, Toronto, ON, Canada

D.E. Warren
Department of Neurological Sciences, College of Medicine, University of Nebraska, Omaha, NE, USA

© Springer International Publishing AG 2017
D.E. Hannula, M.C. Duff (eds.), *The Hippocampus from Cells to Systems*,
DOI 10.1007/978-3-319-50406-3_10

2015). In the sections that follow, background context for each of these domains is provided followed by select empirical findings that hint at possible hippocampal contributions to cognition beyond long-term declarative memory. Along the way, dissenting viewpoints and methodological hurdles are considered alongside alternative accounts for key findings. As will be seen, it is not always the case that we advocate for a particular perspective, but we do make efforts to be even-handed in our treatment of the literature. In the end, we conclude by attempting to identify questions that remain unresolved and offer some suggestions about how ongoing controversies might be reconciled in future work.

Some Context: The Medial Temporal Lobe Memory System

As is often the case when MTL function is considered, it is appropriate to begin with a brief discussion of Henry Molaison (H.M.), who participated in research for decades following bilateral MTL surgical resection in 1953 (c.f. Corkin 2002; Eichenbaum 2013; Squire 2009). In early descriptions, and subsequent empirical work, it was immediately clear that H.M.'s long-term memory (LTM) was severely compromised—indeed, he was said to "forget the incidents of . . . daily life as fast as they occur[ed]" (p. 15, Scoville and Milner 1957). Nonetheless, as reported by Scoville and Milner (1957), he could retain three digit numbers and unrelated word pairs for several minutes in the absence of distraction, and his performance on a battery of tests that tapped perception, abstract thinking, and reasoning ability was preserved.

Subsequent studies of MTL function largely confirmed these initial observations and set the stage for decades of research that has been squarely focused on questions about how exactly structures in the MTL contribute to LTM. Especially important for our purposes, this work has led to claims for the dissociation of declarative (consciously accessible, reportable) LTM, which arguably depends critically on MTL integrity, and non-declarative (consciously inaccessible) LTM, said to be independent of these structures. Once again, some of the earliest evidence in favor of this dissociation originated with H.M. For example, severe impairments were evident on standardized tests of LTM, and it was noted that "once he had turned to a new task the nature of the preceding one could no longer be recalled, nor the test recognized if repeated" (Scoville and Milner 1957, p. 108). Nevertheless, H.M. did acquire new skills. For example, like healthy controls, his ability to trace within the outline of a star using only the reflection from a mirror improved with practice (Milner 1962; Milner et al. 1998; Gabrieli et al. 1993). What made this observation so striking was that his memories for the *experiences* associated with skill acquisition (e.g., the testing apparatus, task, and experimenter) were lost despite clear evidence for long-lasting gains in performance (for review, see Hannula and Greene 2012).

A major effort of contemporary research has been to determine whether the hippocampus contributes to LTM in a qualitatively different way than surrounding MTL cortical structures (i.e., perirhinal, parahippocampal, and entorhinal cortices). While general consensus has not yet been achieved, and perspectives continue to evolve (see chapter "Dynamic Cortico-Hippocampal Networks Underlying Memory and Cognition: The PMAT Framework" by Inhoff & Ranganath), many theories seem to share some version of the view that the hippocampus, which sits at the top of the MTL processing hierarchy, is ideally positioned to bind together converging inputs (e.g., Cohen and Eichenbaum 1993; Davachi 2006; Diana et al. 2007; Eichenbaum et al. 1994; Montaldi and Mayes 2010). As described in detail elsewhere, the resulting relational memory representations permit us to retrieve rich, multifaceted episodic memories of objects that co-occur in space and time (cf. Eichenbaum and Cohen 2014). It is in this context that questions have often been posed about whether or not the role of the hippocampus in cognition might go further than had originally been appreciated. Indeed, it was only in this context that we could begin to address these questions because tasks had to be developed that would tax the processing and representational affordances that are unique to the hippocampus. In so doing, it seems that the reach of the hippocampus is indeed broader than standard textbook descriptions would have us believe; research outcomes consistent with this claim are considered in the sections that follow, along with associated counterevidence that has been reported in the literature.

Unconscious or Implicit Memory

That the hippocampus contributes critically to consciously accessible, or declarative, memory is not subject to debate. Indeed, all previous and current memory systems theories acknowledge a connection between hippocampal function and conscious awareness—specifically, explicit memory. This position is based on indisputable evidence showing that amnesic individuals with hippocampal lesions have impaired conscious appreciation for prior learning episodes (Squire 1992; Cohen and Eichenbaum 1993; Moses and Ryan 2006; Henke 2010; Moscovitch 1992). These effects were first observed anecdotally in informal interaction with H.M. who was described by Scoville and Milner (1957) as being unable to remember where he had been, or what he had done, just hours after events had transpired. The same observations have been made in formal testing conducted with H.M. and other amnesic patients on tasks that require recall or recognition of materials presented during an encoding phase (Squire and Wixted 2011). However, it is important to note that even amnesic individuals who have severe memory impairments and widespread damage that goes beyond the hippocampus and surrounding MTL structures have conscious appreciation for the present moment (e.g., amnesic case K.C.—Rosenbaum et al. 2005; amnesic case E.P.—Insausti et al. 2013). That is, amnesic individuals can understand the current contextual setting, engage in conversation appropriately, follow instructions and perform tasks, etc. Therefore, the hippocampus does not appear to be critical for conscious experience, *per se*. Instead, it is conscious access to information experienced in the past minutes, hours, days, or years (i.e. prior learning episodes) that is severely compromised. Consequently, it is the position of declarative memory theory (and others) that the contents of hippocampus-mediated representations must be within conscious apprehension during encoding and that conscious awareness is part and parcel of the retrieval of such representations (e.g. Squire 2004; Graf and Schacter 1985; Moscovitch 1992). However, if conscious awareness were indeed a fundamental property of hippocampal processing and/or hippocampus-dependent representations, it would be difficult to imagine how amnesic individuals retain conscious appreciation for what is happening in the present moment (see also the Perception section).

In recent years, alternative theories of MTL function have emerged that suggest the primary role of the hippocampus in memory has less to do with conscious awareness than with the nature of the information that is retained in memory. In other words, it is the representational affordances and/or processing capabilities of MTL structures that set them apart from other brain regions. Specifically, relational memory theory posits that the hippocampus is critical for binding relations among distinct objects, and that these relational memory representations can be encoded, retrieved, and subsequently used in service of ongoing cognition. This is the case whether information is available to conscious access or not (Eichenbaum and Cohen 2001; Cohen and Eichenbaum 1993; Ryan et al. 2000). Similarly, the binding of items in context (BIC) model suggests that there may not be a one-to-one mapping of MTL structures and explicit memory; instead, the relationship between regions of the MTL and explicit memory processes may depend on task demands (Diana et al. 2007). Finally, this position has perhaps been articulated most strongly by Henke (2010), who indicates that "... hippocampal damage will impair the rapid associative encoding of compositional and flexible associations irrespective of consciousness of encoding and retrieval" (p. 530). In general, the prediction from these models is that the hippocampus is critical for fully-intact performance whenever the information processing demands of a task require representation of relational (or item-in-context) bindings whether or not that information is subject to conscious access. Consistent with this possibility, there are several reports in the literature of unconscious, implicit, memory that is indeed hippocampus-dependent. Many of these findings were reviewed in detail by Hannula and Greene (2012), and therefore, in this section of the chapter, we highlight just a few recent examples. Before turning to these studies, however, it is important to acknowledge that the number of examples is far fewer than reports linking the hippocampus to explicit memory; this is likely due, at least in part, to challenges associated with conducting studies that deal with conscious awareness—some of these pitfalls are referenced in the text below.

Neuropsychological Investigations of Implicit, Unconscious, Memory

Evidence that the hippocampus is critically involved in implicit, unconscious, memory began with two neuropsychological studies conducted with MTL amnesic patients. First, Chun and Phelps (1999) demonstrated that control participants were faster to search for and identify a target among distractors when search arrays were repeated (versus novel) across blocks, an outcome known as the contextual cuing effect. Effects of contextual cuing occurred even though participants could not explicitly identify or recognize the displays that had been repeated. Amnesic individuals whose damage included the hippocampus showed response time facilitation across blocks, demonstrating spared skill learning. However, compared to controls, these individuals were not differentially faster for repeated displays, suggesting that they were unable to create, and benefit from, the requisite memories in which the target could be located in reference to the relative positions of corresponding distractors.

The second study to provide evidence in favor of hippocampus-dependent memory expression absent awareness was reported by Ryan et al. (2000). Using eye tracking, these investigators demonstrated that both control participants and amnesic patients showed a decrease in the number of fixations that were made to repeated, as compared to novel, scenes. As above, this result suggests that basic reprocessing, or fluency, effects are intact in amnesia. However, only the control participants showed eye movements that were differentially attracted to changed regions within scenes. No evidence for this preferential viewing effect was evident in patient data, suggesting that the MTL, and the hippocampus specifically (see Ryan and Cohen 2004), was critical for binding the spatial relations among items that were embedded in previously studied pictures. Importantly, these eye-movement-based relational memory effects were absent from the viewing patterns of amnesic patients even though the same effects were observed in control data when awareness for what had been altered in the scenes was absent. In other words, eye movements were sensitive to relational memory even in the absence awareness, but not when individuals with hippocampal damage were tested. Whether or not viewing patterns index memory without awareness has been subject to some debate in the literature (Smith et al. 2006). However, the same investigators who have reported null outcomes in past work recently found that these effects are sensitive to instructional manipulations (Smith and Squire 2015). This is discussed in more detail below, but is mentioned here because it seems that discrepancies in the literature may come down to experiment-specific implementation details.

In the time since publication of these initial reports, Henke and colleagues have made great strides in this domain, reporting in several studies that the hippocampus contributes to unconscious encoding. In one of these studies (Duss et al. 2014), amnesic patients and matched controls were presented with pairs of unrelated words (e.g., rain-screw, coffee-tango) embedded in a visual masking sequence. Subsequently, pairs of words were presented supraliminally, and participants were asked to indicate whether the words in each pair were a good fit. Notably, all of these visible word pairs were novel (i.e. had not been presented subliminally), and were either semantically related to a previously encoded pair (intact pair: snow-nail), or not (broken pair: hail-waltz). Results indicated that intact pairs were endorsed more often by controls as a 'fit' than broken pairs. This outcome was said to reflect the influence of memory for the relations among subliminally presented word pairs on subsequent performance, and was reduced in the amnesic sample. Notably, some of the amnesic patients performed at levels comparable to the control group on the unconscious encoding/retrieval task, but were impaired when memory was tested directly. Neuroimaging data indicated that these individuals recruited spared tissue in the hippocampus during task performance. The authors conclude that the hippocampus has a role in both conscious and unconscious encoding/retrieval, and that based on functional connectivity results, a larger network of the hippocampal-anterior thalamic axis and neocortical connections may be required to support conscious access. Considered together, the above studies demonstrate that awareness is not an absolute requirement for hippocampus-supported memory.

Early Information Processing Is Modified Following Hippocampal Damage

While it is clear that amnesic patients have deficits in *conscious* access to remembered content, evidence also suggests that there are important changes in how information is processed by these individuals well before explicit memory decisions might be made. One possibility then is that these early processing abnormalities occur outside of conscious awareness. For example, in past work we have reported that eye movements index memory for learned scene-face relationships during a test trial within 500–750 ms of display onset, and as much as 1–1.5 s in advance of explicit recognition responses (Hannula et al. 2007); the same effect is completely absent from viewing patterns of amnesic patients with hippocampal damage. Based on this outcome, it was proposed that this eye-movement-based prioritization occurs in advance of, and may contribute to the development of conscious awareness for the associate (see also Hannula and Ranganath 2009). Studies outlined below suggest that in addition to the absence of changes in viewing that precede conscious reports, the manner by which hippocampal amnesic patients engage in basic processing is altered in the earliest moments of stimulus exploration (also see the Perception section).

Changes in online processing are particularly well illustrated by an experiment that required amnesic patients and control participants to study an array of objects for a subsequent memory test (Voss et al. 2011). Critically, the objects used in this experiment were not revealed simultaneously; rather, the participant's eye movements were used to reveal the objects through a moving window. During exploration, control participants would occasionally revisit previously inspected objects/locations, however, this "spontaneous revisitation" effect was nearly absent in the amnesic data. Further, results from control participants indicated that revisitation predicted subsequent memory and was associated with hippocampal activity as revealed with functional magnetic resonance imaging (fMRI). This study illustrates the utility of converging research methods (eye tracking with neuropsychological cases, functional neuroimaging of healthy individuals), and it provides initial evidence for the online influence of hippocampal processing on the manner by which information is extracted from the external world.

A recent study from Olsen et al. (2015) with the developmental amnesic patient H.C. echoes the findings described above (Voss et al. 2011) and provides yet another compelling example of changes in online information processing as a consequence of hippocampal damage. H.C. presents with hippocampal volume loss and abnormal development of the extended hippocampal system (Rosenbaum et al. 2014). However, the volume of H.C.'s MTL cortical structures are similar to those of controls. In this study, when faces were presented during an incidental encoding phase, H.C. directed significantly more viewing to the eyes, and less viewing to other face features, compared to the control participants (see Fig. 1). Furthermore, H.C. had a lower transition-to-fixation ratio than controls. Consistent with past reports (e.g., Bird and Burgess 2008; Mayes et al. 2002), H.C. showed relatively intact recognition for faces that were presented from the same viewpoint during study and test, but was impaired when the viewpoint at test was different from corresponding study exposures, or when faces had been presented from different viewpoints across individual study trials. These outcomes suggest that the manner in which the faces are studied and tested (i.e., same versus different viewpoint) can considerably impact recognition performance in amnesic patients, and that deficiencies in how materials are processed (as indexed by eye movement behavior) may contribute to this outcome. Consistent with descriptions in the STM section below, results from this experiment suggest a role for the hippocampus in intra-item feature binding when a high fidelity representation of encoded information is required for successful task performance. In other words, the focus of the hippocampus can be relatively wide, encompassing several objects embedded in an episodic context, or narrow (i.e., limited to relationships among item features), depending on task demands. Whether non-normative viewing patterns are responsible for compromised binding or vice-versa cannot be determined based on the outcomes of this work, but we suspect that the relationship is bi-directional (i.e., ongoing binding deficits change viewing patterns that are, in turn, non-optimal for binding; see Olsen et al. 2012).

Importantly, online processing, as indexed by eye movement behavior in studies described briefly above, is likely to be outside the domain of conscious experience. For example, while the externally presented face in the study conducted with H.C. was certainly subject to conscious apprehension, it is unlikely that participants in this experiment were completely aware of their particular eye movement

Fig. 1 This figure illustrates differences in the distribution of fixations to a face that was presented during encoding. Data from a representative control participant can be seen on the *left* and data from developmental amnesic patient H.C. can be seen on the *right (top)*. The proportion of fixations directed to the nose and the mouth is reduced in H.C. relative to control participants; in contrast, more fixations were directed by H.C. to the eyes. Figure adapted from Olsen et al. (2015) and reproduced with permission according to the Creative Commons License agreement with the Journal of Neuroscience

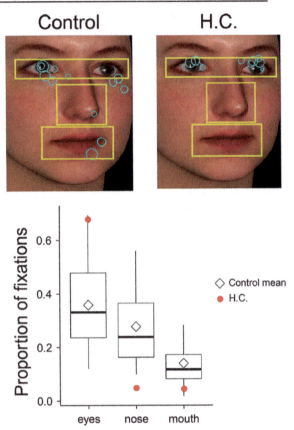

patterns, or the specifics of ongoing processing. Consistent with this possibility, participants perform poorly when they attempt to distinguish their own fixations patterns from those of other participants (Foulsham and Kingstone 2013). Furthermore, while participants can reasonably introspect about the placement of their own eye movements during a visual search task (Marti et al. 2015), introspection was not perfect. Indeed, reported gaze position was frequently inaccurate and false fixations were reported as well. More generally, there was an effect on eye movement placement that was related to the task of introspection itself, suggesting that asking people to report the position of their eye movements changes at least some aspects of eye movement behavior. Finally, while this was not tested directly, it may have been difficult for participants to distinguish instances of covert (attention in the absence of direct fixation) from overt attention (attention coincident with a direct fixation); evidence for this kind of misattribution error has been reported previously by Hollingworth et al. (2008). In sum, evidence suggests that participants have poor insight into their fixation patterns, and this may reflect a lack of insight into online processing. Questions about how exactly patterns of "free viewing" (absent search requirements or any other specific instruction) are related to conscious awareness (or introspection) have yet to be addressed.

One final bit of evidence that suggests eye movements and conscious experience can be dissociated comes from a study conducted by Spering et al. (2011). Specifically, these authors reported that the trajectory of eye movements can be separated from the conscious percept of a presented stimulus. Participants in this experiment were presented with two horizontally (90°) or vertically (0°) oriented sine-wave gratings that drifted orthogonally to their orientation. One of the gratings was adapted to one eye, and then re-presented to the same eye as the other grating was presented to the other eye simultaneously. Whereas eye movement trajectories responded to the integrated motion of the two gratings (the diagonal), the conscious percept of the participants was typically in the direction of the un-adapted grating, or of two separate motions (one weak, one strong). Together with findings outlined above, this work indicates that the link between eye movements and continuous, accurate conscious apprehension is tenuous at best. Thus, what is observed in H.C., and other amnesic patients, is a change in behavior that is not likely to be fully within conscious apprehension. In short, the hippocampus may contribute information that supports conscious awareness of remembered content, but consciousness may not be bound up in the representation itself (Hannula and Greene 2012).

Neuroimaging Investigations of Implicit, Unconscious, Memory

Neuroimaging investigations have provided additional support for hippocampal contributions to memory in the absence of explicit knowledge for prior learning experiences. For instance, Reber et al. (2016) presented participants who were undergoing intracranial electroencephalography (iEEG) with sequences of word pairs, some of which contained a common associate (e.g., "winter-red", "red-cat"), and asked participants to judge the goodness of fit of each pair. Although participants were not aware of the indirect relationships that linked distinct pairs (e.g., the word "red" in our example above), an ERP difference recorded from the hippocampus was observed 400 ms following the onset of the second word pair during encoding (e.g., "red-cat") when the match was present. Subsequently, a test pair was presented that combined the words that were related indirectly by virtue of their shared associate (e.g., "winter-cat"), however there were no ERP differences that distinguished these pairs from others. Based on these outcomes, the authors proposed that relational learning occurred during encoding, even in the absence of awareness.

Work from Ryals et al. (2015) has indicated that hippocampal engagement during retrieval is sensitive to memory absent awareness as well. Their findings are similar to other recent reports, which show that hippocampus-dependent eye movement effects can be dissociated from explicit behavioral responses (e.g., Hannula and Ranganath 2009; Ryan et al. 2000; Nickel et al. 2015). Using eye movement monitoring and fMRI, Ryals et al. (2015) presented participants with scenes that were either new, or configurally similar to scenes that had been previously studied. Participants were asked to identify scenes that they felt were a configural match to (i.e. had the same global layout as) previously encoded exemplars. Results indicated that there was significant overlap in eye-movement-based exploration of configurally similar and previously studied scenes, and that this viewing effect was related to hippocampal activity. Furthermore, and especially important in the context of this section, eye-movement-based exploration effects were correlated with activity differences in the hippocampus even though performance (i.e., explicit identification of configurally similar scenes) was at chance.

Finally, a recent study that combined event-related potentials (ERPs) with patient testing indicated that a neural signature of recognition memory, evident in control data irrespective of awareness, was absent from patient data (Addante 2015). In this experiment, participants were presented with several words, and for each exemplar, made either an animacy or manmade judgment. In an unexpected, subsequent memory test, participants indicated whether individual words were old or new, and specified what kind of source judgment had been made earlier. Results indicated that both explicit item recognition and source memory decisions were impaired in amnesic patients. Additionally, amnesic individuals failed to show a neural signature in posterior regions that, in control participants, distinguished between previously studied and novel words, and was independent from explicit recognition reports. Once again, and much like studies described above, this outcome suggests that consciousness may be orthogonal to hippocampal function.

Early Information Processing Engages the Hippocampus

As indicated above, effects of memory on eye movement behavior are evident shortly after stimulus onset and precede explicit recognition responses (Hannula et al. 2007; see also Ryan et al. 2007); the same effects are absent from the viewing patterns of amnesic patients. One possibility suggested by this observation is that *early* recruitment of the hippocampus (not measured in the cited studies) indexes pattern completion processes and corresponding retrieval of memory representations that are then used in service of conscious awareness. That is, hippocampal representations may not be the seat of consciousness itself (Voss et al. 2012), but rather, may support the subsequent experience of conscious awareness (Hannula and Ranganath 2009; Ranganath 2010). This possibility is consistent with a two-stage model of conscious recollection (Moscovitch 2008; Sheldon and Moscovitch 2010), which states that the hippocampus supports automatic and obligatory retrieval of encoded content during stage one, and contributes to conscious appreciation of retrieved content (perhaps via interactions with the PFC), subsequently, in stage two.

Consistent with the two-stage model, it has been reported that activity differences in the hippocampus during presentation of a scene cue predicted eye-movement-based prioritization of a learned associate when a test display was presented (Hannula and Ranganath 2009). Because these activity differences were evident even when explicit recognition responses were incorrect, it was proposed that this outcome corresponds to stage one of the two-stage model (i.e. automatic, or obligatory retrieval of encoded content). Also consistent with the model, functional coupling of the hippocampus with PFC, identified in a connectivity analysis of data collected during test display presentation, was associated with successful explicit recognition memory performance. While these results suggest that the hippocampus contributes to obligatory retrieval of relational memory representations as indexed by early viewing patterns, nothing can be said about the time-course of hippocampal recruitment because fMRI methods were used. Only by using other neuroimaging approaches with much finer temporal resolution (e.g., iEEG, magnetoencephalography) can questions like these be addressed.

Consistent with the proposal that hippocampal engagement can occur early in processing, Riggs et al. (2009) reported that hippocampal theta responses, indexed with magnetoencephalography (MEG), distinguished old from new scenes during performance of a recognition task within just 250 ms of stimulus onset. This outcome suggests that the hippocampus may be obligatorily engaged during perceptual processing, well before explicit recognition would occur, when a task requires the comparison of external stimuli to internal representations of encoded content. Work from Olsen et al. (2013) complements this report by demonstrating that hippocampal theta responses index binding requirements when information must be integrated across time. In this experiment, objects were presented sequentially and participants were required to encode their relative visuospatial positions. Importantly, because items appeared one at a time, the amount of stimulus information visible from moment to moment remained the same over the course of the trial. Presumably, given that the participants were aware of the task demands (i.e. to integrate the objects), which remained constant as well, conscious experience was not appreciably different across time. Binding demands, however, did increase, as more elements had to be integrated into the existing memory representation as the trial progressed. Results indicated that hippocampal theta responses tracked binding demands, increasing with the introduction of each new item. As such, ongoing modulation of hippocampal responses seems to be especially sensitive to binding operations, rather than conscious experience per se. Of course, any strong claim in this regard would require evaluation of hippocampal theta oscillations absent awareness, perhaps by rendering materials invisible at encoding via masking, or by binning trials based on recognition accuracy.

Several additional studies provide converging evidence in favor of early hippocampal engagement. For example, hippocampal replay (i.e. reinstatement of neural activity patterns evident at encoding) has been reported within 500 ms of memory cue onset (Jafarpour et al. 2014). Additionally, Horner et al. (2012) recorded neural responses with MEG in a group of younger (predominantly developmental amnesic) patients of varying etiologies with a range of hippocampal volumes, as well as control participants. All of the participants were required to study words (items) superimposed on scenes (context). Patients' item memory did not differ from controls, but context memory (selection of a scene from a three alternative forced choice) was impaired and the magnitude of this impairment was correlated with hippocampal volume (Horner et al. 2012). Control participants showed a frontotemporal MEG effect between 350–400 ms following stimulus onset that reflected item memory and an effect between 500–600 ms that distinguished context hits from misses; such effects were absent in the patient data suggesting they were hippocampus-dependent. These findings point to a role for the hippocampus in both item and context memory, but importantly for discussion here they showcase the early engagement of the hippocampus or brain regions that are connected to—i.e., depend upon information processing supported by—the hippocampus.

Perhaps most notable, was a report that provided specific information about the timing of hippocampal responses relative to explicit recognition decisions in a recent iEEG investigation (Staresina et al. 2012; See Fig. 2). In this experiment, recordings taken directly from the hippocampus in pre-surgical epilepsy patients indicated that there was a significant effect of successful source memory retrieval within 250–750 ms of stimulus onset during test. This source effect was followed by a sustained response sensitive to new (i.e. not studied) items. The late onset of this item-based response suggested that it might reflect the engagement in post-retrieval processing. Consistent with this possibility, a response-locked analysis of the data indicated that item-specific responses in the hippocampus were only evident after explicit recognition decisions had been made, and may therefore have reflected incidental encoding of new items into memory. Critically, source-specific responses were evident in hippocampal recordings before explicit recognition decisions were made. While the authors do not discuss this outcome in terms of conscious access, it aligns well with eye movement studies described above, and with the idea that the role of the hippocampus in conscious experience may be secondary to, and emerge from, its primary role in supporting a particular type of representation—here, bound representations of item and source.

Challenges for Evaluating Unconscious Memory

Any study that points to a role for the hippocampus in memory function outside of conscious awareness must consider whether there is potential contamination from explicit remembering. That is, a person may not explicitly report remembered content because they have adopted a strict response criterion, or because perceived task demands preclude them from disclosing awareness (for review, see Simons et al. 2007). However, counterarguments must be considered as well when linking hippocampal function to explicit memory. For instance, it is possible that responses

A) Task and Implantation

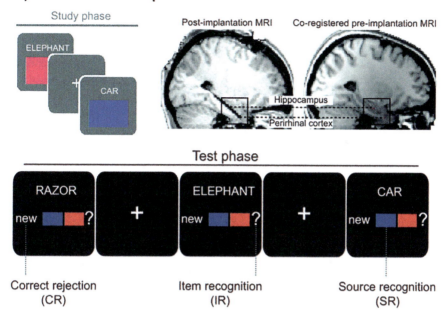

B) iEEG Results - Hippocampus

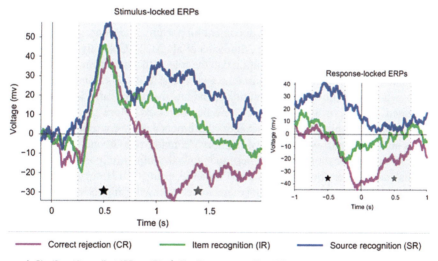

Fig. 2 Hippocampal iEEG recordings during performance of a source memory task. (**a**) Illustration of the experimental task and iEEG electrode placement. During the study phase, participants attempted to encode associations between concrete nouns and corresponding colors, indicating whether the combination was plausible. During the test phase, a concrete noun was presented at the *top* of the screen and participants indicated whether the word was "new", old and they remembered the source (indicated via color selection), or old but the source was forgotten ("?"). (**b**) iEEG

that appear to index conscious knowledge of formed/stored information were influenced by use of a liberal response criterion, or accurate guessing. Regardless, there are increasing numbers of studies that point to hippocampus-dependent memory effects outside of conscious awareness that have used careful methodological approaches in order to minimize contributions from explicit memory. Examples include subliminal masking procedures that, when effective, render a stimulus invisible as confirmed by strict forced-choice testing procedures (cf. Henke et al. 2003a; Nickel et al. 2015), and task designs that preclude the use of effortful retrieval strategies or strategic processing (Carlesimo et al. 2005). These methods should be considered in future studies that attempt to address questions about when and how the hippocampus contributes to unconscious expressions of memory.

Summary and Conclusion: Awareness

Evidence in support of the view that the hippocampus contributes to implicit, unconscious memory comes from four lines of work—namely, studies that report hippocampus-dependent encoding when materials are masked from view (e.g., Henke et al. 2003a, b), studies that indicate learning is impaired in the face of hippocampal damage, even when improvements in performance occur without awareness in controls (e.g., Chun and Phelps 1999; Smyth and Shanks 2008), studies that link hippocampal integrity or function to the expression of implicit eye-movement-based memory effects at retrieval (Hannula and Ranganath 2009; Ryan et al. 2000), and studies that document hippocampal responses in advance of explicit recognition decisions (e.g., Staresina et al. 2012). Collectively, these outcomes make reasonable the proposition that the role of the hippocampus in memory is outside of, or orthogonal to, conscious awareness. With this in mind, questions about when and how hippocampus-dependent memories are formed and/or expressed outside of awareness can now be addressed. More generally, studies might attempt to pin down how exactly the hippocampus contributes to unconscious and conscious expressions of memory (Hannula and Greene 2012).

Relevant to questions about when and how the hippocampus contributes to unconscious expressions of memory, Verfaellie et al. (2012) suggest that some forms of implicit relational memory are intact following hippocampal damage, while others are compromised. The authors used a category exemplar task in which participants read a word pair (e.g., mall-rain), heard a sentence that used the two

Fig. 2 (continued) results locked to stimulus onset (*left*) and to behavioral responses (*right*). iEEG responses were greater for correct source recognition responses than for correct rejections and item recognition shortly after the presentation of the test trial and in advance of button press responses. Figure adapted from Staresina et al. (2012) and reproduced with permission from the Nature Publishing Group and Copyright Clearance Center

words, and rated the plausibility of the sentence. In an indirect testing condition, participants saw one word from the pair (e.g., the context word—mall) spelled backwards, and were asked to list the first four words that came to mind given a particular category descriptor (e.g., weather pattern). Amnesic patients with hippocampal damage generated the associates of context words at rates similar to controls when the context-category pairs from study were reinstated (vs. recombined) at test. However, when participants were asked to explicitly report the target words, in a direct test of memory, amnesic individuals failed to show the normative reinstatement benefit. The authors suggested that while direct expressions of memory required the hippocampus, indirect, and perhaps implicit, expressions of verbal relational memory were not compromised in hippocampal amnesia. It remains to be determined whether performance could have been supported by strategies that do not depend on the hippocampus (e.g., unitization), but the findings raise important questions about the role of the hippocampus in unconscious versus conscious memory.

Furthermore, and as indicated earlier, recent work suggests that task demands influence whether or not expressions of memory require conscious awareness (e.g., Smith and Squire, submitted). Specifically, it has been reported that whether or not the expression of eye-movement-based repetition effects (i.e., decreases in the number of fixations for previously viewed versus novel stimuli) depends on conscious awareness is influenced by task instructions. When participants were told that their memory would be tested, the repetition effect was only observed with concomitant conscious awareness of having previously viewed the scenes. However, when participants were simply instructed to view the scenes, the repetition effect was observed whether participants recognized the scenes as studied or not. Under free viewing conditions, the repetition (or reprocessing) effect was evident in viewing patterns of amnesic patients, as has been reported previously (Althoff and Cohen 1999; Ryan et al. 2000). These results indicate that changes in task instructions can dictate whether or not the same metric of memory is associated with conscious access.

Perception

Like unaware expressions of memory, perception is among the putative new roles that has been ascribed to the hippocampus (Bussey and Saksida 2007; Graham et al. 2010; Suzuki and Baxter 2009). In this section, we present empirical findings relevant to this topic, but first, we anchor our discussion by considering what constitutes perception and how it differs from other cognitive processes. With this information in mind, we briefly revisit a small subset of studies described above (Unconscious Memory section) in service of evaluating whether a hippocampal contribution to perception is feasible based on how quickly information is available for processing. Finally, we summarize the significant empirical literature from neuropsychological and neuroimaging studies that informs whether the human hippocampus contributes to perception, consider whether any such contributions are necessary, and present interim conclusions on this matter. To foreshadow that commentary, we will suggest that while the hippocampus contributes to ongoing cognition beyond long-term memory, it is not clear whether the term perception best describes those contributions.

Perception: Dissociations and Definitions

In the literature describing hippocampal function, the term perception has often been used without elaboration (e.g., Bussey and Saksida 2007; Graham et al. 2010; Suzuki and Baxter 2009), leaving interpretation to individual readers. Resulting differences in how this term is understood may therefore drive some of the debate over hippocampal involvement in perceptual processes. Differences of interpretation are not difficult to understand because perception interacts extensively with other cognitive processes, and these interactions are necessary for the integration and interpretation of information. For example, perception of external stimuli overlaps significantly with later stages of sensation, and a clear delineation between perceptual and sensory processes may be impossible (Lezak 2012).

Similarly, perception also interacts with memory in ways ranging from simple maintenance of current neural activity (STM), to processing of the contents of short-term memory (working memory, WM), and the ability to encode, store, and retrieve preexisting memories (LTM). Consider the example of a typical visual scene such as an office desk decorated with multiple complex objects arranged in a three-dimensional spatial configuration. Although a gist-level perceptual representation of this scene might be available with only a very brief exposure (Thorpe et al. 1996), elaboration and maintenance of the objects comprising the scene might require serial attention to multiple locations reflected in many fixations of the eyes spread across several seconds (Henderson and Hollingworth 1999). Furthermore, perception of the individual objects as such must rely to some extent on previous experience (i.e., memory). As with sensation, strict separation of perception from memory—especially short-term or working memory—may not be possible.

In addition to lying at the interface of other cognitive domains, perception is an ongoing process. That is, perception does not deliver a single, final product but instead provides a succession of interpretations that evolves over time in response to input from external sources and feedback from internal sources. An ambiguous part of a jigsaw puzzle may be resolved by finding an edge; motion may cause a roadside shrub to be re-evaluated as a half-seen deer; and extended viewing of a Necker cube will flip the observer's perspective. These scenarios illustrate the difficulty of deciding when perceptual processes have finished. Further, they illustrate the challenge of strictly distinguishing between perception and other cognitive processes, and they raise important questions about the nature of perception. In the example of the desk from earlier, at what moment has the desk scene been perceived? Is conscious awareness of the scene necessary for perception? At what instant should we expect a necessary contribution of memory processes for inter-pretation of information from the scene? To what extent does information have to be actively maintained to determine whether two percepts (e.g., of complex scenes) are exact copies or slightly different from one another?

In part, perception overlaps with other cognitive processes because our models of these processes are imperfect descriptions of complex, highly interactive systems that operate in parallel (Lezak 2012). Should concerns about whether perception is fundamentally dissociable from other cognitive processes influence our discussion? Or more simply, do these considerations obviate this section of the current chapter? We believe not. While perception clearly overlaps with other aspects of cognition, it has been established as a partially dissociable process that can be separately tested and uniquely impaired. In the spirit of decades of neuropsychological and cognitive neuroscience research studying brain-behavior relationships, we believe that it is perfectly appropriate to investigate whether the neural correlates of perception include the hippocampus. However, we hope that by noting the substantial inter-activity between these different processes we might inform future discussions of whether the hippocampus could reasonably be said to contribute, for example, jointly to memory and perception rather than solely to memory. Isolating densely intertwined cognitive processes is difficult even in controlled laboratory tasks; conclusive dissociation of their neural correlates presents an even greater challenge.

To have any hope of distinguishing perception from other cognitive processes, careful definition of terms is important. For the purposes of this chapter, we will consider perception to be a set of cognitive processes representing the interaction of ongoing elementary sensory experience with top-down influences by other cogni-tive processes including memory, attention, and executive functions. For example, in the case of an external stimulus, perception is preceded by sensation, which involves the transduction of physical energy into neural signals, and can be succeeded by various other cognitive processes that may lead, for example, to encoding of stimulus information into lasting memory representations. A leading neuropsychological text describes perception as follows:

> Perception involves active processing of the continuous torrent of sensations …. This processing comprises many successive and interactive stages. The simplest physical or sensory characteristics, such as color, shape, or tone, come first … and serve as foundations for the more complex 'higher' levels of processing that integrate sensory stimuli with one another and with past experience (Lezak 2012, p. 26).

We will rely on this description and consider perception to be a process that supports interpretation of the most recent several seconds of sensory experience through the lens of existing knowledge and that has hierarchical as well as parallel aspects.

We note one further caveat here, which is that our consideration will focus almost exclusively on alleged hippocampal contributions to *visual* perception because that modality has received the most attention from researchers. Although we will speculate that hippocampal contributions generalize across many modali-ties, further research will be necessary to address this important issue.

The Timecourse of Hippocampal Involvement in Cognitive Processes

Perception is an active, ongoing cognitive process, which may place greater demands on speed than would be expected of cognitive processes historically associated with the hippocampus such as memory. Speed of processing is relevant to the current discussion because if the hippocampus is to contribute meaningfully to perception, it must be capable of receiving, processing, and transmitting information quickly. In this section, we briefly revisit a subset of the empirical findings that were described above (Unconscious Memory section) in service of evaluating whether the hippocampus might reasonably be expected to contribute to perception.

Results from studies that have examined the latency of hippocampal responses suggest that this structure can be engaged quickly, within a time window that begins as early as 250 ms following stimulus onset (e.g., Riggs et al. 2009; Staresina et al. 2012). Furthermore, response-locked analyses, based on iEEG recordings, indicate that hippocampal responses, sensitive to source memory, are evident before explicit recognition responses have been made by the participants (Staresina et al. 2012). Research studies have also indicated that individuals with hippocampal amnesia process visual stimulus information, as indexed by eye movement behavior, in qualitatively different ways than neurologically healthy controls (Voss et al. 2011; Olsen et al. 2015). They fail, for example, to distribute viewing among face features, which seems, in turn to affect recognition memory performance when faces are seen from different perspectives at study and test. Outcomes like these, particularly the latency data, indicate that the hippocampus does indeed respond quickly when stimuli are in view, although these activity differences were associated with memory rather than perception.

In short, the intervals in question are sufficiently brief that the hippocampus could reasonably be expected to respond to and influence activity in other brain regions within the scope of our working definition of perception (i.e., as a process that interprets the most recent several seconds of sensory experience). By comparison, other brain regions that have been less controversially associated with perception for complex stimuli such as faces are similarly situated in or near ventral temporal cortex and receive, process, and transmit information with similar latencies (Schmolesky et al. 1998). This prompts us to note that many brain regions would of course respond to visual stimuli at least as quickly as the hippocampus and therefore potentially contribute to perception. These non-hippocampal contributions to perception are no doubt critical, but they do not affect our main point, viz., the latency of hippocampal responses to external stimuli is not so long that the structure would be prevented from contributing to perception simply by virtue of its connectivity. In short, hippocampal processing is rapid enough to actively influence ongoing cognition rather than simply responding and recording.

Empirical Findings

In the following sections, we discuss key empirical findings in the domain of perception and the human hippocampus. The dependence of perception on hippocampal function has perhaps most often been evaluated using an oddity task (e.g., Lee et al. 2005b; Behrmann et al. 2016). In this case, participants view an array of stimuli (e.g. colors, simple shapes, faces, scenes) and must select, from the alternatives that are present, the stimulus that is different from the remainders (i.e. the 'odd-one-out'). Other tasks require participants to select the exemplar from two or more alternatives that is most like a simultaneously presented sample stimulus (e.g., Sidman et al. 1968; Lee et al. 2005a; Hartley et al. 2007; Warren et al. 2010), to name/identify objects that are degraded or overlap in space (Warren et al. 2012), or to determine whether or not two pictures, presented simultaneously, are an exact match (e.g., Aly et al. 2013). In the text that follows, neuropsychological studies that have provided critical insight regarding the necessity of hippocampus for perception are described and neuroimaging studies that have informed debate about how the hippocampus is functionally involved in putatively perceptual tasks are summarized. Notably, while there is also an extensive literature documenting MTL and hippocampal contributions to perception from animal models including rodents and non-human primates, a description of that work is beyond the scope of the current chapter (Graham and Gaffan 2005). Instead, we focus on the rich scholarship describing relevant work in human participants.

Neuropsychological Studies of Perception

As was outlined briefly at the outset of this chapter, the hippocampus and surrounding MTL structures have been associated with LTM since the seminal report of Scoville and Milner (1957). Generally, damage to the medial temporal lobes or hippocampus has been reported to leave perception and STM intact (Cave and Squire 1992; Drachman and Arbit 1966; Warrington and Baddeley 1974; Wickelgren 1968). However, as described briefly below, the large literature based on work conducted with amnesic individuals has long included hints that the MTL and/or hippocampus might contribute to cognitive processes beyond LTM. The decades-long absence of research on this topic may seem odd in hindsight, but when interacting with individuals who have amnesia the severe memory deficit is obvious while any perceptual deficits are relatively subtle. Nevertheless, careful contemporary experimentation has revealed reliable performance deficits attributed to impaired perceptual processing among patients with broader MTL damage (e.g., Barense et al. 2007, 2012), as well as patients with more focal hippocampal damage (e.g., Lee et al. 2005a, b).

Without a doubt, the pattern of impaired declarative and spared non-declarative (or procedural) memory found in patient H.M. transformed theories of memory

(Scoville 1968; Scoville and Milner 1957; Cohen and Squire 1980), but deficits were also evident in his ability to maintain or perceive visual information. Although H.M.'s STM for many types of familiar, verbalizable stimuli was relatively normal, studies of the durability and quality of his non-verbal visual representations indicated impairment. In particular, Sidman et al. (1968) tested H.M.'s ability to perceive and maintain simple visual stimuli—ellipses of varying eccentricity—over intervals ranging from 0 to 32 s. With no delay, H.M. was as accurate as healthy control participants when choosing from a selection of related alternatives, but the accuracy of his responses decreased as a function of the maintenance interval until they were essentially at chance after 32 s; in contrast, the performance of control participants remained nearly unchanged even at the longest delay. This unexpected finding went largely unremarked when it was published, but suggested that either perception or maintenance processes were altered by H.M.'s MTL damage.

This early example of impairment in the representation of visual information at short intervals is important supporting evidence for more recent observations described below and we offer the speculative suggestion that results from additional non-published studies may have also pointed to a role for the hippocampus in short-lived representations but suffered from the "file drawer problem" (Rosenthal 1979). One piece of evidence potentially supporting this notion can be found in the doctoral dissertation work of Prisko (1963) which included findings similar to those reported by Sidman et al. (1968) but was never published in a peer-reviewed format. Formal analysis of this file-drawer suggestion is beyond the scope of this review, but the prospect is intriguing and may be worth further investigation.

Returning to results reported by Sidman et al. (1968), the impaired ability of patient H.M. to maintain hard-to-verbalize visual information for short periods of time was potentially attributable to deficits in at least two distinct abilities: visual perception or visual STM. From our perspective, evidence which supports the proposition that the hippocampus is involved in perception should rely on tasks that meet two key criteria: (1) very limited maintenance demands; and (2) relatively low memory load. Failure to meet either criterion would allow critics to suggest that LTM processes might have been recruited in service of task performance (Hales et al. 2015; Jeneson and Squire 2012; Jeneson et al. 2012; Squire and Wixted 2011).

Initial observations that may meet these criteria were reported by Lee et al. (2005a, b, 2006) who observed impairments of perception in patients with focal hippocampal damage when they were asked to perform visual discrimination tasks using complex, three-dimensional scene stimuli. In one such experiment (Lee et al. 2005b; see Fig. 3), participants were presented with a sample stimulus (e.g. a face, object, scene, art, or color swatch) at the top of the screen and had to choose the exemplar from two alternatives presented below that most resembled that item. The choice stimuli were blended exemplars of two baseline objects, one of which served as the sample. Use of this blending procedure meant that choice stimuli were more or less similar to each other and to the sample stimulus across trials. Consequently, selection of the closer match could not be achieved based on a single diagnostic feature, particularly when the level of blending was high. Results indicated that

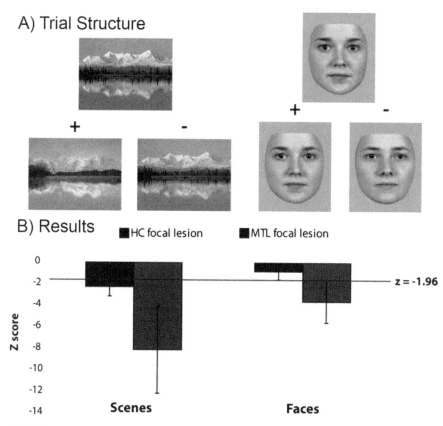

Fig. 3 Perceptual identification task. (**a**) Participants were to select the image from two alternatives that was the best match of a picture (scene or face, depending on the trial type) presented simultaneously at the *top* of the screen. (**b**) Patients with focal hippocampal lesions were impaired on the scene matching task, but performed within the normative range for faces. Figure adapted from Lee et al. (2012) and reproduced with permission according to the Creative Commons License agreement with Frontiers

patients with damage limited to the hippocampus (based on visual inspection of scans) had trouble distinguishing between alternatives when scenes were presented, but only when there was considerable feature overlap due to high levels of blending. Discrimination was intact for faces, objects, art, and color swatches.

Converging evidence for deficient scene perception among hippocampal amnesic patients has been reported using the oddity task (Lee et al. 2005a; Behrmann et al. 2016). In these experiments deficits in identification of the 'odd-exemplar-out' were only evident when scenes in a choice array were rendered from different, as compared to the same, viewpoints (see also Hartley et al. 2007). Much like the blending procedure, this manipulation places high demands on perceptual discrimination processes, as there is considerable ambiguity across array exemplars. This very specific outcome is consistent with a proposed role for the hippocampus in the processing and representation of arbitrary relationships among items embedded in scene contexts and taxes representational flexibility, a key property of hippocampal function (cf. Cohen and Eichenbaum 1993). Recent work also points to a potential role for the fornix, the main output pathway for the hippocampus, in perceptual disambiguation (Lech et al. 2016; Postans et al. 2014), but additional testing is required to determine how to best characterize these contributions.

While evidence consistent with a role for the hippocampus in perception has been reported with increasing frequency over the last decade, a significant volume of counter-evidence has also been published. Most frequently, this evidence has come in the form of failures to replicate relevant behavioral findings in samples of amnesic patients. In one early example, Stark and Squire (2000) adapted the methods employed in a study of non-human primates (Buckley et al. 2001). The original work suggested that perirhinal cortex played a role in object perception, but Stark and Squire did not observe behavioral impairments consistent with this account in a sample of patients with MTL damage (including hippocampus and perirhinal cortex). A potential explanation for this ambiguity was proposed by Aly et al. (2013) who have found that perceptual discrimination of complex scenes can be supported by two distinct processes. According to this work, identification of specific details that permit disambiguation of perceptually similar inputs depends upon a *state-based process*, whereas a general sense of relational (mis)match used to the same end depends on a *strength-based process* (see Aly and Yonelinas (2012) for details). Critically, these processes were expected to show dissociable dependence on the hippocampus. Specifically, it was predicted that estimates of strength-based relational processing would be significantly reduced when hippocampal amnesics were tested, but that state-based processing, which might depend on the size or position of an individual scene feature, would be preserved. Indeed, this was the observed pattern when hippocampal amnesic patients were tested. Furthermore, converging evidence from an fMRI task, conducted with healthy young participants, confirmed that hippocampal activity differences were sensitive to the strength of the relational mismatch between scenes, as indexed by subjective confidence judgments. Based on these outcomes, the authors proposed that inconsistencies in the literature might reflect differences in the use of state- as compared to strength-based processing strategies when tasks require discrimination of perceptually similar complex pictures.

Other evidence fitting the criteria for perceptual impairment outlined above were reported by Warren et al. (2012) who found that patients with focal hippocampal damage were impaired on tasks requiring visual discrimination or recognition of complex objects based on partial information. These latter findings contrast to some extent with results from other labs which indicate that broader MTL damage (particularly to perirhinal cortex) may be necessary to impair object (as opposed to scene) discrimination performance (Barense et al. 2007, 2010; Lee et al. 2006). Critically though, this discrepancy does not diminish the most important implication of these findings which is that the hippocampus contributes to the representation of information even when stimuli are continuously present.

While the just-described findings seem to meet the minimum criteria that we established for excluding substantial contributions of LTM to performance (i.e., by limiting maintenance demands and memory load), labelling the underlying deficit as one of perception remains controversial (Hales et al. 2015; Shrager et al. 2006; Squire et al. 2006; Suzuki 2009). In our view, the ongoing debate over how to interpret these data reflects the complexity of disentangling cognitive abilities such as visual perception and visual STM. For example, while tasks such as visual discrimination or recognition based on partial information do not explicitly require maintenance (because all of the materials are presented simultaneously), there are still implicit demands on participants to maintain some amount of information while developing their response (cf. Olsen et al. 2012). When discriminating one complex scene from another, participants must examine the first scene (scene 1) and then maintain enough information about that scene to discriminate it from another (scene 2) (Barense et al. 2007, 2010; Lee et al. 2005a, b, 2006; Aly et al. 2014). Even if the intervals between examining scenes 1 and 2 are very short (i.e., on the order of hundreds of milliseconds for attentional shifts and saccadic eye movements) there is still an implicit maintenance demand for visual or conceptual information sufficient to discriminate the two stimuli. Against this, it has been argued that the eye movements of amnesic patients do not differ from control participants during visual comparison or search tasks (e.g., Erez et al. 2013), but others have shown differences in eye-movement or related behaviors during search, comparison, or study tasks (Warren et al. 2011; Lee and Rudebeck 2010a; Olsen et al. 2015; Voss et al. 2011).

Finally, although delays of hundreds of milliseconds may seem trivial, there is evidence that damage to the MTL or hippocampus is sufficient to impair maintenance of very simple visual information (i.e., color or shape) over intervals as short as 1 s (Warren et al. 2014). Furthermore, it has been shown that amnesic patients (those with limited hippocampal damage and more extensive lesions) can successfully perform the oddity task when they are allowed to draw lines linking exact matches, which was "intended to reduce the burden on working memory" (Knutson et al. 2013, p. 609). In short, use of this memory aid meant that after identifying a match, that set of items could be completely disregarded. Collectively then, these findings suggest that the hippocampus is necessary for maintaining information over brief delays with the implication that even visual discrimination tasks that do not explicitly require LTM may still rely on hippocampus-dependent maintenance processes. This is consistent with the perspective that the hippocampus is necessary for normal visual experience. Whether the underlying deficit is best described as one of perception will be considered in more depth later.

Neuroimaging Studies of Perception

Studies using functional neuroimaging methods such as fMRI to investigate whether the hippocampus is involved in on-line cognition have found evidence which is consistent with that perspective. More specifically, fMRI studies testing perception have shown correlations between performance of perceptual tasks and hippocampal activity. Notably, many of the same caveats and concerns that were raised in the context of neuropsychological findings discussed above will also be relevant here.

Use of functional neuroimaging to investigate perception has motivated the adaptation of tasks previously used in neuropsychological studies (Lee et al. 2005a, 2006). In particular, scene discrimination tasks that are difficult for patients with hippocampal damage also evoke hippocampal activation in neurological healthy adults (Barense et al. 2010b; Lech and Suchan 2014; Lee et al. 2013; Lee and Rudebeck 2010b). These neuroimaging findings show a correlation between on-line scene discrimination performance and hippocampal activity which converges with neuropsychological findings (Lee et al. 2005a, 2006), and the originating authors suggest that the underlying deficit is perceptual. Additionally, one of the neuroimaging publications had the promising goal of—as suggested by the title— "Investigating the interaction between spatial perception and working memory in the human medial temporal lobe" (Lee and Rudebeck 2010b) which is highly relevant to this chapter. The authors used 2 × 2 design to cross working memory load (1- or 2-back task) with item complexity (simple shapes vs. complex scenes) in a within-subjects design that required participants to perform these task conditions while fMRI data were collected. Analysis of this data revealed an interaction between working memory load and stimulus type in the right posterior hippocampus and parahippocampal cortex such that activation increased with working memory load in the complex-item condition but not the simple item condition. As such, this report is most consistent with a role for the hippocampus in perception and working memory rather than one or the other exclusively.

Notably, several of these studies have included measures intended to control for potentially confounding influences of incidental LTM or STM processes (Lee et al. 2013; Lee and Rudebeck 2010b; Zeidman et al. 2015). In one typical example, Lee et al. (2013) asked participants to perform an oddity-detection task while fMRI data were collected, and later administered a surprise recognition task testing memory for the oddity task materials. They reported increased hippocampal activity associated with correct oddity responses irrespective of later recognition performance for the same items. These and similar findings are suggestive of a unique hippocampal contribution to scene discrimination or perception over and above activation related to LTM processes. Finally, one finding is intriguingly consistent with a perceptual role for the hippocampus but would extend that role beyond scenes to include faces and other complex but non-scenic stimuli (Barense et al. 2011) which would be consistent with other neuropsychological findings (Warren et al. 2012). Briefly, Barense et al. (2011) collected fMRI data from healthy participants while they performed a perceptual discrimination task that crossed two types of visual stimuli (faces and objects) with two levels of familiarity (familiar and unfamiliar). Object and face stimuli increased activity in the hippocampus and perirhinal cortex relative to a baseline condition, and a main effect for familiarity was evident in the same regions. These activity differences were orthogonal to subsequent memory, suggesting that hippocampus (and perirhinal cortex) may contribute to object perception.

Recent developments in fMRI data analysis, which test the predictive accuracy derived from patterns of brain activity have also produced results that bear on a perceptual role for the hippocampus. Specifically, Lee et al. (2013) followed up their univariate analysis—described earlier—by applying multi-voxel pattern analysis (MVPA) to the same fMRI data collected from participants who were performing an oddity judgment task. The authors found that functional data from the regions of interest including the hippocampus or the parahippocampal cortex were sufficient to predict accurate performance of individual oddity judgment trials significantly better than chance (~57 % correct predictions) irrespective of later recognition memory performance for the test materials. A second MVPA analysis showed that the same functional data was also sufficient to predict subsequent recognition performance significantly better than chance (~53 %) irrespective of oddity judgment performance. Following on their findings from a univariate analysis in which hippocampal activation was more strongly related to oddity judgment than subsequent recognition, the authors produced new results consistent with their account that the hippocampus contributes to perception in addition to memory.

To summarize the neuroimaging findings, there is fMRI evidence that is consistent with the perspective that the hippocampus contributes to visual perception (Barense et al. 2010, 2011; Lech and Suchan 2014; Lee et al. 2013; Lee and Rudebeck 2010b; Zeidman et al. 2015). However, as described in the section describing neuropsychological studies, the tasks used in neuroimaging studies cannot definitively be said to be process-pure; that is, these tasks cannot exclude the possibility that the observed associations between hippocampal activation and visual discrimination performance are due to other processes (e.g., maintenance). This concern is tempered to some extent by studies that control for subsequent memory effects (Barense et al. 2011; Lee et al. 2013; Zeidman et al. 2015), but that approach cannot entirely mitigate potential memory-related activity because subsequent memory is not perfectly related to hippocampal activity. Again, much like the neuropsychological evidence, neuroimaging data are suggestive of and consistent with a hippocampal role in perception, but not conclusive.

Summary and Conclusions: Perception

As evidenced by studies discussed earlier in this section, the hippocampus appears to make necessary contributions to ongoing cognitive processes that may include perception. Although these findings have sometimes been critiqued on the grounds that hippocampal involvement may be related to coincident LTM processes (Hales et al. 2015; Jeneson et al. 2010, 2012; Jeneson and Squire 2012; Squire and Wixted 2011) several studies reviewed earlier addressed this issue in design and/or analysis have still found significant evidence of hippocampal contributions to cognitive processes over short intervals (Barense et al. 2012; Lee et al. 2013; Warren et al. 2012, 2014; Zeidman et al. 2015).

If—for the sake of discussion—we accept that the hippocampus makes necessary contributions to ongoing cognitive processes, a key question remains: does the hippocampus contribute to perception *per se*? To respond, we return to our working definition, which described perception as a "process that supports interpretation of the most recent several seconds of sensory experience through the lens of existing knowledge and that has hierarchical as well as parallel aspects". Considering these characteristics in turn, the findings reviewed here fall within the interval described in the definition, and existing knowledge appears to be exercised in support of task performance when available (e.g., real-world knowledge of spatial layouts). The critical remaining characteristic is "interpretation" and its meaning in this context. That is, "interpretation" could mean full appreciation of a stimulus in all its complexity simultaneously; "interpretation" could also mean understanding the broad nature of a stimulus without understanding it completely. For example, when the stimulus is a complex scene, the scene could be said to be perceived (i.e., interpreted) at any of the following stages: when its presence influences responses to other materials; when it is known to be a scene; when the type of scene is known; when objects in the scene have been identified; when the scene is recognized as previously viewed; etc. This nuance is important because—from our perspective—findings that could arbitrate questions of hippocampal necessity for perception hinge on exactly this issue. Therefore, we suggest that consensus on the theoretical issues at stake in this debate depend first on achieving consensus on what is meant by the "interpretation" of a stimulus during perception.

Other definitional issues related to a hippocampal role in perception also require further consideration. First, our understanding of perception as a separable cognitive construct may be an imperfect reflection of the underlying cognitive processes or neural representations. Second, it is not clear whether perception necessarily requires conscious awareness. Third, defining the timeline for an ongoing process such as perception is challenging; does failure to interpret a stimulus before a deadline constitute a perceptual failure? Fourth, defining the success or failure of perception is challenging because perception always yields a product whether accurate, normative, or otherwise. Finally, while the end result may be an adequate interpretation of the current environment and be sufficient for accurate performance of a perceptual test, the manner in which this outcome is achieved may be quite different across individuals or after brain injury. While these concerns are also important components of an expanded understanding of perception, we believe that a clear operational definition of perceptual "interpretation" remains most critical for understanding the role of the hippocampus in the prevailing ontology of cognition and for drawing strong conclusions about hippocampal contributions to perception.

Despite our inability to draw strong conclusions about whether hippocampus is necessary for normal perception based on empirical data, we suggest that an interim conclusion can be derived by drawing on the literature of neurology and neuropsychology for descriptions of alternative perceptual deficits. For example, remaining in the realm of visual perception, we can consider the examples of object agnosia, alexia, and prosopagnosia. In each of these three examples, relatively focal brain injury or dysfunction can produce a severe, selective cognitive deficit in the perception of objects, orthography, or faces, respectively. The severity of these deficits stands in stark contrast to the deficits in, for example, scene perception reported in patients with bilateral hippocampal damage or fornix disconnection (Lech et al. 2016; Lee et al. 2005a, 2006). In everyday life, patients with bilateral hippocampal damage are typically able to navigate through space relatively well, localize objects without noticeable difficulty, copy complicated shapes accurately, and describe complex scenes comprehensibly. In fact, these patients often perform less well than expected only when tested with challenging spatial tasks such as discrimination of very similar scenes (Lech et al. 2016; Lee et al. 2005a, 2006). Meanwhile, patients with object agnosia have famously mistaken their spouses for headgear (Sacks 1998), patients with pure alexia have gross deficits in the ability to perceive written language (Damasio and Damasio 1983), and patients with severe prosopagnosia are often unable to recognize individual faces with any success (Moscovitch et al. 1997; Newcombe et al. 1994). Returning to our earlier discussion, these perceptual deficits illustrate obvious failures of the ability to interpret sensory input normally.

The severity of these alternative examples of widely recognized visual perceptual deficits provide context for putative perceptual deficits in patients with hippocampal damage. While the latter findings are statistically significant, reported impairments in perception among patients with hippocampal damage present with much less urgency than the memory deficits of those patients, and with much less salience than the perceptual deficits experienced by patients with non-hippocampal brain injuries. Notably, impairments in scene discrimination performance are hardly unique in this regard—many non-LTM deficits reported in patients with hippocampal damage are statistically significant but modest relative to the patients' LTM deficits. Therefore, as an interim conclusion on this matter, we suggest that theories of hippocampal involvement in perception describe phenomena that are real and important, but that it is not clear whether perception is an appropriate descriptor. With that in mind, we consider, in a final section, whether the hippocampus might be reasonably said to contribute to short-term or working memory.

Short-Term or Working Memory

Short-term memory is a repository for information that is being kept active or in mind and, as is the case with perception, recent findings challenge claims that STM is completely independent of hippocampal function. The term STM is often used interchangeably with working memory in the literature, but the two are not synonymous. This is because working memory involves not only active retention, but also manipulation of content that is currently being represented. Here, we frequently refer to STM, as many investigations that have addressed questions about hippocampal contributions to these processes have required active retention, but not manipulation of stimulus information over the course of a brief delay. Furthermore, while early work evaluating the dependence of STM on hippocampal function emphasized active retention of verbal materials—e.g., unrelated word pairs, a string of digits (see Olsen et al. 2012)—contemporary studies have most often used visual stimuli and efforts have been made to minimize the influence or effectiveness of verbal rehearsal strategies. It is in this context that neuropsychological deficits in STM have most often been reported and hippocampal activity differences in neuroimaging investigations are evident.

A fundamental characteristic of STM is its limited capacity. While chunking can increase the capacity of STM tremendously (Miller 1956), the standard view has been that a small, fixed number of simple elements or items can be actively retained over the short term (Luck and Vogel 1997). This view has considerable appeal and a good deal of empirical support in the literature, but some researchers have recently come to endorse a different model of STM capacity that is based on a finite amount of available *resource* (Alvarez and Cavanagh 2004). In this case, there is not a fixed item-based STM capacity limit. Instead, capacity is determined by the complexity of to-be-retained information and the precision with which that information must be represented in order to meet task demands. In short, this perspective suggests that there is a tradeoff between the number of items that can be actively retained and the fidelity with which key features are represented. We revisit this important issue later in this section.

Much of the time, questions about the defining characteristics of STM are addressed using a *change detection task* (Luck 2008) though match- or non-match-to-sample protocols, n-back tasks, and delayed alternation tasks are also common. In a standard visual STM change detection task, participants attempt to actively retain information presented during a sample phase (e.g., a scene, a face, a set of simple objects) over the course of a short delay. At the end of the trial, when a test display appears, participants indicate yes or no, whether anything in the display has changed (e.g., the identity of a cued object). In some experiments, the number of items in the sample display is manipulated across trials or blocks so that investigators can evaluate changes in accuracy as a function of load and obtain STM capacity estimates based on participant performance (cf. Cowan 2001). Recent adaptations of the standard change detection task permit investigators to address more nuanced questions about the representational precision, or fidelity, of STM. In this case, participants are required to report specific information using a continuous scale about a characteristic feature of an item that was presented during the sample phase (e.g. color, orientation; Wilken and Ma 2004; Zhang and Luck 2008). This approach provides more precise insights into *why* forgotten information was not successfully retained—i.e., (1) because the representation is simply gone, or (2) because the representation became degraded and imprecise due to high memory loads or when longer delays were imposed. As described in more detail below, a handful of investigators have adapted this new testing procedure to address questions about the quality of visual STM in amnesic patients with hippocampal damage.

Finally, before describing relevant empirical findings, it is important to note that models of short-term or WM are currently in a state of flux (cf. Jonides et al. 2008). In contrast to the long-standing view that short- and long-term memory depend upon strictly dissociable systems, recent models propose that short-term retention is best characterized by states of representational accessibility that are mediated by interactions between attention and LTM. As summarized by LaRocque et al. (2014), state-based models conceptualize STM as activated (or currently relevant) representations from the long-term store. A small subset of information is prioritized and immediately accessible (e.g., in the "focus of attention") and additional information is either held in a "region of direct access", or remains in a heightened state ("activated LTM") by virtue of its recent prioritization (cf. Cowan 1993; Oberauer 2002). In short, there is a notable shift underway from systems- to state-based models in the STM literature, which is consistent with a broader movement in the cognitive neuroscience community pointing to association (rather than dissociation) of short- and long-term memory (cf. Ranganath and Blumenfeld 2005; Olsen et al. 2012). It is in this context that it becomes increasingly clear the time is ripe to re-evaluate claims for the complete independence of short-term retention from the hippocampus—we do so below based on recent empirical reports from the neuropsychological and neuroimaging literatures.

Neuropsychological Investigations of STM

As was outlined briefly in the section on perceptual processing, results from some of the earliest neuropsychological studies that evaluated whether, and to what extent, simple visual materials could be actively retained over the course of a short delay are difficult to reconcile with standard views of MTL function. For example, H.M.'s performance on a task that required identification of an ellipse that exactly matched the eccentricity of a sample stimulus was increasingly compromised as the retention interval between sample and test was lengthened. Indeed, performance was impaired even when the imposed delay was no more than 5 s long, suggesting that active maintenance was deficient (e.g., Sidman et al. (1968); see Ranganath and Blumenfeld (2005), Olsen et al. (2012) for more information about early work). However, reports of intact amnesic performance on STM tests (e.g. Cave and Squire 1992; Warrington and Baddeley 1974; Wickelgren 1968), combined with scores in the normative range on standardized neuropsychological tests (e.g., digit span; cf. Cave and Squire 1992; Rose et al. 2011) led to general consensus that STM does not depend on the integrity of MTL structures, including the hippocampus. Furthermore, as has been argued by some investigators, use of a short retention interval does not obviate instantiation of LTM processes (for review see Jeneson and Squire 2011). Whether these processes simply occur coincident with active retention or are necessary for fully intact performance on STM tests has been difficult to pin down.

Short-Term Retention of Inter-item and Item-Context Bindings

Doubts about a role for the hippocampus limited to LTM were raised in recent neuropsychological studies when deficits were documented using tasks that encouraged active retention of inter-item and item-context relationships (i.e., spatial positions of objects embedded in scenes, scene-face pairings, and object-location associations; Hannula et al. 2006; Olson et al. 2006b). For example, results reported by Olson et al. (2006b) indicated that amnesic patients, including a subset with damage limited to the hippocampus, were impaired on tests that required active retention of just three object-location associations over the course of 1 (experiment 2) or 8 (experiments 1 and 2) second delays. The deficits in these investigations were quite specific as the same hippocampal amnesic patients who performed poorly on tests of relational memory *could* successfully distinguish old from new scenes, old from new objects, and previously filled from empty locations (see also Cashdollar et al. 2009). A peculiarity, perhaps, of our work (Hannula et al. 2006) was the use of a lag-based design in which corresponding sample and test stimuli (i.e. rendered scenes) were not always presented in immediate succession. While this design choice meant that we could determine whether task performance did in fact depend critically on the hippocampus (i.e. chance performance at long lags), it also meant that we could not conclusively rule out potential contributions of LTM to performance when sample and test displays were presented consecutively. This is because participants may not have used an active retention strategy and because the interleaved lag-based trial structure meant that information about several scenes had to be stored simultaneously for upcoming test trials (see also Jeneson et al. 2011).

The above concerns were addressed recently in two new experiments that examined memory for the locations of items embedded in scenes (Yee et al. 2014; see Fig. 4). Several design changes were made, among them use of a standard delay-based change detection protocol. Replicating previous findings, results indicated that patients were impaired on the basic change detection decision, but perhaps more compelling was the finding that patients frequently failed to identify an object that had been displaced (via forced-choice response) despite having successfully indicated that a change was present. This result suggests that the memory representation was incomplete or degraded. Especially important for our purposes, marked deficits were documented despite performance among control participants that was near ceiling (experiment 2), and were evident even in a patient with confirmed volume reductions limited to the hippocampus, sparing adjacent MTL structures, parietal, frontal, and other temporal lobe regions.

Additional, complementary evidence for hippocampal contributions to active retention of relational memory representations has been reported recently in the literature. For example, impairments have been reported on tests that required short-term retention of inter-object bindings (van Geldorp et al. 2014), simple color-location associations (Finke et al. 2008, 2013; Braun et al. 2008, 2011), and color-shape associations (Parra et al. 2015). In this last example color patches and

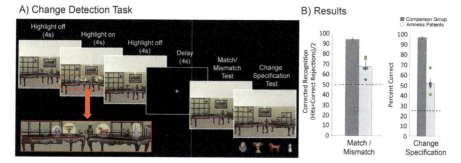

Fig. 4 Illustration of the change detection task and corresponding results from Yee et al. (2014). (**a**) Participants were presented with a scene during the sample phase of each trial. Four objects were highlighted briefly while the scene was in view and one might be displaced when the test picture was presented. Following a brief delay, participants indicated whether any of the objects had changed locations (match/mismatch test), and then attempted to identify the displaced item from four alternatives. (**b**) Results from amnesic patients and matched controls for the match/mismatch and change specification tests. Amnesic patients were significantly impaired on both tests, and frequently failed to specify the change correctly, even when a correct mismatch response had been made (shown here). Figure adapted from Yee et al. (2014) and reproduced with permission according to the Creative Commons License agreement with Frontiers

corresponding shapes were presented side by side to discourage unitization, and in each experiment, patients had lateralized MTL damage (e.g., due to stroke, tumor resection, or temporal lobe epilepsy). Impaired performance has also been reported on a test that required active retention of rendered topographical landscapes, but here participants were patients with damage limited to the hippocampus (Hartley et al. 2007). In this experiment, scenes in the choice array were shown from different perspectives than the sample, which meant that successful identification of the match (from four alternatives) required flexible representation of relative positions amongst key landscape features. Relevant to the Perception section, results indicated that two of four patients were impaired even when choice arrays were presented simultaneously with the sample, but all four patients were impaired when a delay was imposed. As above, short-term retention of other visual information in each of these studies—e.g., non-spatial components of the rendered landscapes (Hartley et al. 2007), individual colors or locations (e.g., Finke et al. 2008), object-color associations when color was a feature of the object, encouraging unitization (Parra et al. 2015; see also van Geldorp et al. 2014)—was intact. It seems then that one could conclude the hippocampus contributes to STM when participants must bind objects with context or with other objects (inter-item bindings; e.g., faces with scenes, objects with color patches), but not when single objects or fused/unitized associations (intra-item bindings; e.g. a green shoe) are to be maintained. Indeed, similar dissociations have been reported in the LTM literature (cf. Davachi 2006; Diana et al. 2007); however, as we shall see, findings summarized below suggest that this conclusion may require some modification.

Identification of impairments like those described above ultimately led investigators to question whether anything more specific could be said about the *kinds* of errors made by amnesic patients on STM tests. In one experiment (Watson et al. 2013), participants were presented with an array of two, three, four, or five objects. Subsequent to exposure, the objects were cleared to one side of the table, and after an eyes-closed delay of approximately 4 s, participants attempted to replace the objects in their previous locations. Several metrics were used to examine performance (e.g. misplacement distance, changes in overall configuration or shape, presence of swap errors) and amnesic patients were impaired on all of these measures relative to a healthy control group. Furthermore, with just one exception (i.e., the global configuration metric), the magnitude of reported impairment was unaffected by memory load. Critically, careful analysis revealed that patients made one kind of error far more often than others—namely, a "swap" error. This error was observed even during trials that required active retention of just two objects, and the same mistake was rarely made by control participants. Deficits on a similar task were also reported for some patients at low loads (i.e. 1–4 items) by Jeneson et al. (2010) when participants were required to minimize displacement errors to reach a criterion level of performance, but this modest low load impairment was deemphasized relative to a sharp discontinuity in displacement error among patients when four, five, or six objects had been presented. This sudden high-load performance change was not evident in results reported by Watson et al., and what drove the between-study differences is not clear. Procedural details, including the use of just four trials per condition and systematic increases in memory load across trials, may have rendered deficits at low loads less robust in the task reported by Jeneson, but because similar information was not reported by Watson, this is merely speculation. Nonetheless, results from these studies converge with findings described above, and implicate the hippocampus in short-term retention of memory representations; here, especially when mappings of objects to specific, previously filled, spatial locations had to be retained.

Precision of STM Representations

Efforts to better characterize STM deficits that have been reported in hippocampal amnesia continue to gain traction in the literature, and a handful of studies have approached this issue in terms of the representational precision or fidelity of information retained over the short term. In one early example (Warren et al. 2010; see also Ezzyat and Olson 2008), participants had to determine whether a target was present among foils created so that their resemblance to the corresponding sample stimulus varied parametrically. This manipulation meant that successful performance required retention of precise information about a tested feature (e.g. shape, luminance, line tilt, spatial frequency). The task was difficult for both patients and control participants, with performance near chance levels whether a delay was imposed or not and it was in this context that eye movements, which were recorded along with button press responses, proved particularly informative. Eye tracking results showed that when the sample stimulus was present simultaneously with the choice array, both groups of participants spent more time fixating foils that most resembled the sample. However, when a 6-s delay separated sample from test, the visual-similarity-based preferential viewing effect was attenuated in patient data; the basic effect persisted, but the correspondence between visual similarity and fixation time was reduced. This outcome suggests that representations were degraded, but had not been completely lost (see also Warren et al. 2011), and a potential mechanistic explanation for this pattern of performance is deficient hippocampus-supported pattern separation—a process that establishes orthogonalized representations of similar or confusable inputs (e.g., Yassa and Stark 2011).

Eye movement methods are notable because they provide researchers with a continuous index of cognitive processing while stimulus materials are being viewed (cf. Hannula et al. 2010). New behavioral testing procedures that use continuous rather than binary response metrics also permit investigators to address increasingly specific questions about the fidelity of STM representations, and recent neuropsychological studies have adapted these methods (Pertzov et al. 2013; Warren et al. 2014). In general, participants in these experiments attempt to identify a key feature (e.g., color, orientation) of one object from the sample array. This target object appears at test, stripped of critical information, and participants choose from a continuous range of options (e.g., on a color wheel, by manipulating the orientation of a colored bar) the feature value that provides the most precise fit (e.g. a specific shade of blue, a 45° angle). In two experiments, Pertzov et al. (2013) found that patients with amnesia secondary to a specific subtype of limbic encephalitis were impaired on STM tests that used continuous reporting metrics, but that their mistakes were due to swap errors. For example, when patients attempted to drag a fractal to its previously occupied location, they were just as likely as controls to get it near one of the locations occupied during the sample phase, but were more likely than controls to place it closest to a location previously filled by a different exemplar. Similarly, when patients attempted to specify the studied orientation of a colored bar, they oftentimes matched orientation to a different colored line presented prior to the delay, an effect that was evident even when the imposed load was just two object-orientation associations. In both of these examples, the fidelity or precision of memory for sample features (e.g., orientation) was intact, but the binding of objects to feature values or spatial location was compromised.

Much like Pertzov et al. (2013), Warren et al. (2014) reported that the fidelity of feature-based memory representations was comparably stable in amnesic patients and matched control participants across very short intervals (900 ms). In a task of color-location associations (Zhang and Luck 2008), responses made by selecting a remembered color from a color wheel were similarly accurate for healthy control participants and patients with hippocampal damage. In contrast to other studies though, the amnesic patients' memory representations were more likely to be completely lost than control participants' when the imposed memory load was three or six color values (see Fig. 5). Furthermore, follow-up analyses confirmed that this forgetting was not due to relational memory (or "swap") errors.

These outcomes are difficult to reconcile. While results from some studies suggest that STM representations in amnesia are degraded or lack fidelity (Warren et al. 2010, 2011; Yee et al. 2014), others suggest a very systematic pattern of

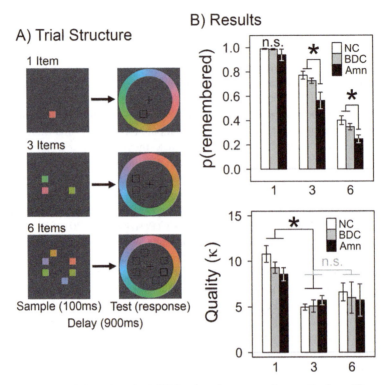

Fig. 5 Illustration of the color-wheel STM task and corresponding results from Warren et al. (2014). (**a**) Representative examples of 1, 3, and 6 item sample displays. On every trial, a sample stimulus was presented for 100 ms, followed by a brief delay (900 ms), and finally the appearance of the color wheel. One of the *squares* was marked as the target (thicker black outline) and participants attempted to specify the color of that exemplar. (**b**) Results indicated that the complete loss of information was more common among amnesic patients (amn) than normal controls (nc) and brain damaged controls (bdc) for sample sizes 3 and 6 (*top*). In contrast, the quality of retained color information was well-matched across groups (*bottom*). Figure adapted from Warren et al. (2014) and reproduced with permission according to the Creative Commons License agreement with Cold Spring Harbor Laboratory Press

mistakes—namely, swap errors (Pertzov et al. 2013; Watson et al. 2014). Furthermore, one study provides evidence for abnormally elevated levels of lost representations, even at low loads and in the context of a standard STM feature specification task (Warren et al. 2014). Notably, many of these studies have been conducted with the same group of well-characterized patients (e.g. Warren et al. 2010, 2011, 2014; Watson et al. 2013; Yee et al. 2014), which discounts the possibility that discrepant results are due to patient-specific qualities like differences in age, lesion location or extent, etc. across experiments. This suggests then, that properties of the tasks— e.g., the instructions, the duration of trial events, the materials—are driving reported differences. Consistent with this idea, event timing was considerably shorter in Warren et al. (2013) than other studies. As is standard (Zhang and Luck 2008), the sample array in this experiment, which consisted of one, three, or six colored squares was in view for just 100 ms, and was followed by a 900 ms delay. By comparison, sample arrays used by Pertzov et al. (2013), Watson et al. (2013), and others were in view for at least 1 s, and often several seconds more; furthermore, imposed delays were seconds, rather than milliseconds, long. With this in mind, one possibility is that amnesic patients require more time to establish (or vulcanize; Luck 2008) mental representations of the sample stimulus and that, even with more time, representational precision or relational mappings remain below normal levels. These possibilities could be tested in future work.

STM for Items

There are some exceptions to what have become fairly standard reports of impaired amnesic performance on tests that require retention of inter-item and item-context bindings across short delays. First, and perhaps most notable, there is some compelling evidence for deficits on tasks that seem not to have the same kinds of binding requirements as studies outlined above. For example, several reports indicate that active retention of a single face is deficient in amnesia (Ezzyat and Olson 2008; Nichols et al. 2006; Olson et al. 2006a; Race et al. 2013; Rose et al. 2011). These impairments have been documented at delays of just 1 s, although the faces in that case were artificial, rendered without hair, and morphed to obtain a range of foils for test that were more or less similar to the sample (Ezzyat and Olson 2008). To the extent that the hippocampus contributes to pattern separation, these relatively homogenous faces may have become nearly indistinguishable when presented in sequence. Nevertheless, amnesic patients could successfully indicate whether pairs of faces presented simultaneously were a match or not—that is, impairment was only evident when the delay was imposed.

Results like these seem to be at odds with claims that hippocampal contributions to STM are limited to situations that require inter-item or item-context binding, but are compatible with other observations in the literature. For example, as described above, deficits have been reported on tests that require short-term retention of complex novel objects (Warren et al. 2011), and are evident even when STM for simple features is tested provided that items in the choice array resemble the sample stimulus (Warren et al. 2010, 2014).

Second, two additional recent studies (Olson et al. 2006a; Piekema et al. 2007) have reported impairments on tests that require active retention of simple features (e.g., spatial locations, colors) absent high-fidelity testing protocols, but deficits may have been a consequence of more extensive MTL damage. In fact, it was proposed recently that even the reported deficits in active face retention are a consequence of broader MTL lesions. Race et al. (2013) tested two groups of patients—individuals with limited hippocampal damage and those with more extensive MTL lesions—and performance was only impaired when lesions went beyond the hippocampus. As indicated by the investigators, some caution is warranted in the interpretation of this outcome because patients with extensive damage also had greater volume reduction in the hippocampus itself; this is especially notable in light of neuroimaging findings summarized below. Collectively, however, these findings suggest that some reconsideration of our original conclusion about hippocampal contributions to STM might be needed. While there is good reason to expect hippocampal involvement when tasks require representation of inter-item or item-context bindings, there is also a growing body of evidence that points to hippocampal involvement when choice arrays require more precise representation of intra-item bindings. In other words, STM tasks that require high resolution bound representations of object features, object combinations, or objects and contexts may depend on processing that is supported by the hippocampus (Yonelinas 2013).

Evidence Against Hippocampus-Supported Short-Term Retention

The literature also contains evidence that runs counter to the observations summarized above (Allen et al. 2014; Baddeley et al. 2010, 2011; Jeneson et al. 2010, 2011, 2012; Shrager et al. 2008). It is possible that performance in some of these studies was intact because tasks required active retention of simple or unitized items/features and did not use testing protocols that would be expected to require representation of high-resolution bindings. We consider just one representative example. Jeneson et al. (2012) reported that estimates of STM capacity derived from performances of hippocampal amnesic patients on a standard STM change detection task were within normal limits at short delays. Critically though, the test displays in this experiment, which required short-term retention of a small collection of colored squares, did not tap memory for color-location bindings. When a change was present, the target object (specified with a bounding box), was always a new color that had not been presented in the sample array. Indeed, as reported by the authors, "the task was to decide whether a new color had been introduced, not whether a color that was present in the first array was now presented in a new location" (p. 3585). More generally, the colors themselves were perceptually distinctive (e.g., red, green, blue, yellow), effectively ruling out any requirement for high fidelity representation of the critical feature value. Another potential obstacle concerns the patients themselves. Recent work has indicated that the neural correlates of STM for object-location associations may be subject to considerable reorganization among patients treated surgically for epilepsy versus the presence of a tumor (Finke et al. 2013; see also Braun et al. 2008). Epilepsy patients often perform normally on object-location change detection tasks and show compensatory recruitment of contralesional hippocampus and STM network structures (e.g. DLPFC) relative to a healthy control group. Tumor patients, who have a much abbreviated disease history with very little time for neural reorganization are impaired on the same task, and do not show increased recruitment of these structures. In this context, it is notable that several of the published studies in which STM deficits have not been forthcoming were based on work conducted with Jon (Allen et al. 2014; Baddeley et al. 2010, 2011), a developmental amnesic patient in whom the possibility of neural reorganization seems not to have been explored.

Final Considerations from the Neuropsychological Literature

STM research has shown that there are documented tradeoffs between representational fidelity and stimulus complexity (Alvarez and Cavanagh 2004). If a stimulus is particularly complex and/or discrimination at test depends on high quality representation or differentiation of feature-specific minutiae, then the number of items stored in STM may go down. Results described above suggest that these reductions may be more pronounced following hippocampal damage. A particularly vexing problem, one that permeates the perception literature as well, concerns the potential impact of LTM on performance. While amnesic patients can effectively leverage preexisting knowledge (e.g. semantic information) to improve their performance on STM tasks (Race et al. 2015), they cannot encode durable LTM representations of new information. As such, it is possible that at least a subset of impairments reported in the literature reflect deficiencies in LTM, not STM. In light of these concerns, any resolution of questions about the boundary conditions and characteristics of STM deficits following hippocampal damage will require systematic consideration of these factors. This is particularly challenging because, in our opinion (as outlined below), definitive procedures for disambiguating the contributions of LTM and STM to performance have yet to be described.

The premise behind one such approach is as follows—if healthy control performance is disrupted by the introduction of interference during a STM delay period, active retention must have been required. In this case, the argument is that new information has displaced the active memory representation and because a more durable LTM trace was never established, response accuracy is reduced (Shrager et al. 2008). Consequently, one benchmark for concluding that the hippocampus does indeed contribute critically to short-term retention is impairment in an amnesic sample on the very same test where control performance drops in the face of interference. In principle, this seems like a reasonable suggestion, but in practice, there are problems that impact the viability and interpretation of reported outcomes. For example, as we have described in detail elsewhere (Yee et al. 2014), it is not clear to what extent control performance must drop for investigators to say conclusively that active retention was driving task performance. In the original work outlining this procedure, control performance was significantly reduced on a test of memory for six object-location bindings in the face of interference; amnesic patients were impaired on this test as well. It seems then, that this meets the definition of evidence for hippocampal contributions to STM. Instead, however, it was indicated that the drop in control performance, while significant, was insufficient for making these claims. More generally, as described in detail by Race et al. (2013), the kind of interference matters. In other words, failures to document interference effects in control performance may simply mean that representational requirements and/or processing demands of the interference task were orthogonal to task features or insufficiently taxing to displace represented content. Until these issues are addressed, any claims about disambiguation of short- from long-term memory contributions to task performance based on this method seem premature.

As we will see, questions about STM-LTM interactions have figured prominently in the neuroimaging literature as well, which is summarized next.

Neuroimaging Investigations of STM

Consistent with the neuropsychological literature, a great deal of effort has been made in the neuroimaging community to determine whether and under what circumstances the hippocampus (and adjacent MTL structures) might contribute to STM. This work provides important insights that are not afforded by studies conducted with amnesic individuals, including observations of sustained delay period activity, information about contributions of specific hippocampal subfields to active retention, and insights into STM/LTM interactions. Furthermore, recent advances in neuroimaging analyses permit investigators to decode the representational content of delay period activity. As is described elsewhere (cf. Norman et al. 2006) and below, these multivariate statistical approaches are sensitive when univariate outcomes are inconclusive, and therefore, promise to be informative in future work.

Three fMRI investigations reported some of the earliest evidence for hippocampal activity differences during the performance of STM tasks (Mitchell et al. 2000; Ranganath and D'Esposito 2001; Stern et al. 2001; see also Curtis et al. 2000). One of these experiments (Mitchell et al. 2000) was a procedural match to the neuropsychological study conducted by Olson et al. (2006b), described above. Participants, who were older and younger adults, were presented with three object-location associations and were either instructed to maintain information about the items, the filled locations, or the pairs in anticipation of a probe display. When this display was presented, participants indicated yes or no, whether the presented information had been seen during the sample phase. Most important for our purposes, activity was significantly greater in anterior hippocampus when young participants attempted to retain object-location bindings as compared to objects or locations alone; this activity pattern was absent from the fMRI data of older adults. This result is consistent with reported impairments of hippocampal amnesic patients on the same task, and with the claim that the hippocampus contributes to STM when relational memory representations are required for accurate performance. In contrast to subsequent investigations, individual trial components (i.e., sample, delay, and test) were not modeled separately here; instead, activity differences were modeled using timing parameters that collapsed across the sample phase and the early delay.

In two subsequent studies (Ranganath and D'Esposito, 2001; Stern et al. 2001), the common denominator was hippocampal recruitment associated with active retention of novel, trial-unique materials (i.e., faces or scenes; see also Schon et al. 2013). In one of these experiments (Ranganath and D'Esposito, 2001) activity differences were evaluated for subcomponents of the STM trial, and results indicated that short-term retention of novel but not familiar faces was correlated with sustained delay period activity in the hippocampus. A control experiment conducted with the same materials and timing parameters confirmed that this outcome could not be reproduced when participants engaged in intentional LTM encoding, and a subsequent report based on reanalysis of this data set indicated that there was significant functional coupling between an FFA seed region and the hippocampus (along with PFC, parietal, and occipital sites) during the delay period (Gazzaley et al. 2004). These observations provide compelling evidence in favor of the view that hippocampal contributions go beyond LTM, here, when active retention of object-location associations, single faces, or a set of complex scenes was required.

Delay Period Activity in the Hippocampus Predicts Subsequent Memory

In the wake of these early studies, a number of investigators reported that hippocampal engagement during the delay period of a STM test predicted subsequent recognition memory performance (e.g., Axmacher et al. 2008; Nichols et al. 2006; Ranganath et al. 2005; Schon et al. 2004). Indeed, this was an outcome that had been anticipated early on, as Ranganath and D'Esposito (2001) had proposed that hippocampal delay period activity may serve two purposes—namely, active retention *and* incidental encoding. One example of evidence in favor of this dual-purpose role came from an experiment that required active retention of novel complex objects (Ranganath et al. 2005). In this experiment, hippocampal activity was evident early, but not late in the delay, and recruitment predicted performance on a surprise subsequent memory test. Corresponding results from a behavioral study indicated that LTM for actively retained objects was reduced when processing was disrupted via interference early, but not mid- or late-delay. Based on these observations, it was inferred that the hippocampus is a key site of STM-LTM interactions, and that incidental encoding operations supported by the hippocampus build a representation that is then reconstructed and retained late in the delay by specialized neocortical processing sites. Notably, despite robust effects of early distraction on LTM performance, active retention was not compromised. This may seem like evidence against hippocampal contributions to STM, but the authors point out that interference in the behavioral task was unlikely to affect retention of low-level features (e.g., a simple shape in the upper right corner of a complex object), and that this kind of detail could be used to rescue STM performance even when a bound high-resolution representation of the whole object had been lost. In contrast, because subsequent recognition required disambiguation of 120 complex, novel objects that were likely to share these elementary features, the same low-level information could not support accurate performance on the LTM test.

Consistent with the perspective outlined above, recent evidence suggests that delay period activity in the hippocampus may be particularly important for establishing and retaining orthogonalized representations of objects that are characterized by high levels of feature overlap (Newmark et al. 2013). In this high-resolution neuroimaging study, hippocampal subfields DG/CA3 and CA1 were engaged disproportionately when feature overlap was high (versus low) during the sample phase of the STM task. Furthermore, these activity differences were sustained in CA1 (and adjacent MTL cortical structures) over the course of the delay.[1] As above, these outcomes align well with recent results from neuropsychological studies that were described earlier.

The same group (Nauer et al. 2015) has also reproduced the finding that hippocampal activity is robust early in the delay, but decays with time. Notably, use of high-resolution neuroimaging methods meant that the effects could be localized to specific hippocampal subfields (i.e. DG/CA3 and CA1). Skeptics might argue that it sounds suspiciously like delay period activity in these studies is merely a carryover effect associated with processing the sample stimulus, but Nauer et al. (2015) reported that a model based strictly on timing of the sample stimulus, excluding the ensuing delay period, was a poor fit to the raw fMRI data. Consistent with observations in the rodent literature (Knauer et al. 2013), it was suggested that activity differences early in the delay may be an indirect index of persistent neuronal spiking in hippocampal subfields (and elsewhere in the MTL). In sum then, results from both the neuroimaging and neuropsychological literatures coalesce by implicating the hippocampus in active retention of object representations when performance depends on the integrity of high-fidelity bound representations of intra-item features. Neuroimaging experiments go further though, as they permit investigators to evaluate correlations between delay period activity and subsequent LTM, and to examine hippocampal recruitment at the subfield level.

Activity Differences in Hippocampus Are Sensitive to STM Performance

Much of the time, STM task performance has been near ceiling in neuroimaging investigations, but there are a few reports in the literature that suggest activity differences and/or activity patterns in the hippocampus predict successful STM performance. In a difficult matching-to-sample task, for example, Olsen et al. (2009) had participants attempt to identify the face from two alternatives that had been presented during the sample phase. To make the task challenging, sample displays consisted of two faces and both faces had to be retained in anticipation of the test display. A small set of male faces, cropped to remove the hair, was used repeatedly across trials, and pre-exposure to the faces meant that, in contrast to studies above, the materials were not novel. Despite these changes in protocol, delay period activity differences were evident in anterior hippocampus (and other MTL structures); these activity differences distinguished high confidence correct STM responses from low confidence and incorrect responses, and persisted for the entire duration of the 30 s delay period. Much like results reported by Nauer et al.

---

[1]For more information about hippocampal anatomy, including the designation of specific subfields, readers should consult chapter "The Nonhuman Primate Hippocampus: Neuroanatomy and Patterns of Cortical Connectivity" of this book.

(2015), this outcome confirms that delay period activity was not a consequence of carry-over effects from the sample stimulus. Furthermore, this work demonstrates that novel, trial-unique materials are not required to drive hippocampal activity up during the delay. Instead, activity differences were said to reflect requirements to bind specific, known face exemplars to the temporal context of a particular trial, insulating the representations from proactive interference.

Another approach that has been used to evaluate STM success effects incorporates a surprise LTM test (Bergmann et al. 2012, 2015, 2016). In these experiments, activity differences associated with successful (versus failed) short-term retention were identified for the subset of trials with incorrect long-term recognition responses. This meant that STM activity differences were unlikely to be a spurious consequence of LTM encoding and storage, although any absence of activity differences might reflect failed delay-period retention even if recognition (upon visual presentation of the test materials) was ultimately successful. Results across studies, two that required active retention of four face-house pairs and one that required active retention of the relative positions of objects embedded in scene contexts, varied. For example, it was found in the first study (face-house pairs) that hippocampal activity during the sample phase predicted subsequent long-term recognition, but not success on the test of STM; activity differences associated with delay and test could not be evaluated. The next study, a slightly modified version of the original, was run to permit separate analysis of STM trial components. In contrast to predictions, no suprathreshold activation in the MTL or anywhere else in the brain predicted STM accuracy during the delay period. The most recent experiment (Bergmann et al. 2016) required participants to retain information about the locations of four objects embedded in a rendered scene. Accuracy effects associated with "pure" STM contrasts (i.e. when LTM responses were incorrect) were evident in bilateral hippocampus during the test phase of STM trials (see also Hannula and Ranganath 2008), but there were no suprathreshold activations during the delay period. Furthermore, there were no activity differences anywhere in the brain that predicted LTM outcomes. Results from these experiments are surprising because amnesic patients with hippocampal damage are impaired on tasks like these when short delays are imposed (e.g., Hannula et al. 2006; Hartley et al. 2007; Yee et al. 2014), and because robust delay period activity in the hippocampus has been reported in several other experiments. One potential explanation for null outcomes, based on the report by Olsen et al. (2009) above, is that collapsing across correct STM responses without considering confidence obscured hippocampal activity differences. As a reminder, contrasts performed by Olsen et al. distinguished high confidence correct responses from low confidence and inaccurate responses. It seems worthwhile then for future studies to include subjective confidence ratings or remember/know judgments, and to subdivide STM trials on this basis.

Finally, null delay period effects were also reported by Hannula and Ranganath (2008) on a test of spatial relational working memory. In this study, on every trial, participants were presented with four objects (from a set of nine), each in one of nine possible spatial locations in a 3×3 rendered grid. Over the course of the delay, participants attempted to mentally rotate the encoded sample stimulus so that they could detect, and discriminate among, changes in object-location bindings when the test display, which was presented from a different viewpoint, appeared on the screen. This was a very difficult test, and while univariate contrasts indicated that activity differences in the hippocampus were greater for correct than for incorrect WM responses during presentation of the sample and test displays, there was no evidence for above-threshold accuracy effects anywhere in the brain during the delay. As above, it is possible that incorporating a measure of memory strength would change reported outcomes, but here, because participants had to disambig-uate test displays based not only on the presence (or absence) of a position change, but also the specific type of change that was in play, correct responses likely required precise memory representations. While we prefer not to place too much stock in null findings, alternative explanations for the lack of delay period activity are worth considering. First, the absence of differential recruitment could reflect efforts to retain and work with whatever had been encoded, whether those repre-sentations were complete or not. Second, it is possible that univariate BOLD signal contrasts were insufficiently sensitive to neural differences that are correlated with successful retention in this task. Consistent with this second possibility, recent re-evaluation of this data set using multivariate representational similarity analysis (RSA) showed that activity patterns during encoding and delay were correlated when participants successfully identified relational matches or manipulations on the WM test (Libby et al. 2014). A more compelling approach might have looked at delay period pattern similarity across trials with shared relational content (i.e. that required representations of the same bindings), but the experiment had been designed so that object-to-space and inter-object-to-space bindings were always trial unique. Therefore, decoding of specific relational representations could not be performed. Nevertheless, this outcome does suggest that representations of rela-tional information persisted from encoding into the delay period when participants made correct responses, and more importantly, speaks to the sensitivity of multi-variate approaches to fMRI data analysis. We are not aware of any other studies that have used multivariate (MVPA, RSA) techniques specifically to evaluate when and how the *hippocampus* supports active short-term retention (but see Lewis-Peacock et al. (2012) for an example of a promising paradigm that could be adapted for this purpose)—future studies could be performed with this in mind.

Effects of Memory Load and Task Demands on Hippocampal Recruitment

We end with a brief discussion of the potential impact of memory load and task demands on hippocampal recruitment during the performance of STM tasks. These issues are important to consider, as it is possible that STM capacity is exceeded when the imposed load is high or tasks become especially difficult. Consequently, any resulting activity differences in the hippocampus may reflect a shift to depen-dence on LTM mechanisms. Evidence that suggests this may be the case comes from a study that has shown a trade-off in functional connectivity with the fusiform face area (FFA) between the inferior frontal gyrus (IFG) and the hippocampus when the number of to-be-retained faces is systematically manipulated across trials (Rissman et al. 2008). Specifically, functional coupling between FFA and hippocampus increased linearly with sample size (i.e. 1–4 faces); IFG showed the opposite pattern. Much like this result, greater hippocampal recruitment for high (four symbols) versus low (one symbol) load working memory trials was also reported by Axmacher et al. (2009). However, in this experiment, participants were also required to encode a face that was presented during the delay period. Subsequent face recognition and hippocampal activity associated with face encoding were both down when the concurrent WM load was high. In contrast to the load-dependent proposal, but consistent with conclusions drawn in past work (e.g. Ranganath et al. 2004; see above), it was suggested that this outcome points to a dual-purpose role for the hippocampus in active retention *and* LTM encoding. When the hippocampus is recruited to support *active retention* of four symbols, it is less available for face *encoding* and subsequent recognition is compromised. The authors acknowledge that activity differences associated with high load WM trials may be a consequence of exceeding the capacity limits of STM (i.e. in the high load condition symbols may have been encoded into LTM and then subsequently retrieved), but raise several counterpoints against this interpretation of the data.

Finally, as proposed by Zanto et al. (2015), challenging tasks may elicit hippocampal recruitment because LTM is required. To test this possibility, they had participants attempt to retain a single face over the course of a delay, but in three experimental conditions, this basic task was made more difficult. Across conditions, a task-irrelevant distractor face was presented during the delay, the delay was lengthened, or foil faces at test were purposely selected based on their visual similarity to the sample. In each case, including a baseline condition (face retention absent the above challenges), participants knew what to expect (i.e. whether the task would be relatively easy or difficult). Relative to baseline, hippocampal engagement increased during the performance of "challenge" tasks when the sample face was in view and during the delay period there was increased functional coupling of hippocampus (and other structures in the parahippocampal gyrus) with FFA for the same contrast. Interestingly though, there were also significant activity differences in the hippocampus during the sample phase greater for baseline trials than for passive viewing trials. Delay period activity differences and connectivity patterns seem not to have been evaluated in a baseline/passive viewing contrast. Another key outcome of the study concerns the pattern of behavioral performance, which included assessment of face recognition on a surprise test of LTM. Specifically, there was a significant tradeoff between STM and LTM performance across tasks—short-term change detection was better in the baseline condition and LTM was better in the challenge conditions. This likely had to do with expectations participants had about task difficulty (based on instruction) and corresponding efforts to encode/retain the sample more effectively. Notably though, activity differences and connectivity patterns with the hippocampus during performance of "challenge" tasks were not correlated with subsequent LTM performance. While it was concluded that results provide strong evidence in favor of the standard view that dissociable brain systems support short- and long-term memory, the presence of hippocampal activity differences in the baseline condition and the lack of correspondence between hippocampal recruitment/connectivity and subsequent recognition performance leave room for alternative interpretation.

In sum, the results summarized above indicate that delay period hippocampal engagement is only present when load is high or tasks are especially challenging, but these outcomes do not jibe with reports of hippocampal recruitment during tasks that require active retention of just one item (e.g. a single face or object—Ranganath and D'Esposito 2001; Ranganath et al. 2004). More generally, there is important counterevidence to the LTM-based load argument that bears consideration. For instance, von Allmen et al. (2013) reported that set-size dependent hippocampal recruitment was evident within the capacity limits of visual STM when participants were required to retain color-location associations, and that these activity differences actually collapsed when capacity limits were exceeded. Furthermore, research that is based on recently proposed models of STM that distinguish between the "focus of attention", a "region of direct access", and the "activated part of LTM" (more accessible by virtue of its recent use) has consistently reported hippocampal recruitment (Nee and Jonides 2013, 2014; Öztekin et al. 2009, 2010). For example, Nee and Jonides (2013) have shown that hippocampal activity during presentation of a test stimulus is evident when decisions are being made about information that was held in the "region of direct access"; notably, and consistent with von Allmen et al. (2013), these activity differences were evident below the individual STM capacity limits of tested participants. It is difficult to reconcile these observations with the view that hippocampal engagement is only evident when LTM has to be engaged to support performance (i.e., because capacity limits were exceeded). In short, there is solid evidence in the literature consistent with a role for the hippocampus in short-term retention (see also Soto et al. 2012).

Summary and Conclusions: Short-Term Memory

There is a long-held tradition in cognitive neuroscience to view memory from a systems perspective. Especially notable here is what has been considered unambiguous dissociation of systems that support short- and long-term memory. The strongest evidence in favor of this perspective came from work with amnesic patients who had severely compromised LTM, but remained quite capable of retaining a limited amount of information in mind over the short term. Anecdotally, this divide is apparent in interactions with individuals who have hippocampal damage. While they can engage in basic conversation without difficulty, it soon becomes clear that their narrative is not anchored in the context of events that transpired even moments earlier. They can carry on though, as long as the topic of conversation stays on course. This is why, when short-term retention is tested, it has been so important to develop tasks that tap hippocampus-dependent representations. Only under these circumstances has it become clear that performance is not on par with healthy control participants, even when imposed delays are on the order of hundreds of milliseconds (e.g., Warren et al. 2014).

The neuroimaging literature has provided additional insights and converging evidence for a hippocampal stake in STM, and goes further than patient work, as specific questions about the delay period (e.g., whether activity differences persist) and STM-LTM interactions can be examined. While mechanism was not discussed here, human neuroimaging (particularly magnetoencephalography; Cashdollar et al. 2009; Olsen et al. 2013), and intracranial recording (e.g., Leszczynski et al. 2015) studies are a source of compelling evidence for *how* short-term retention may be achieved and/or mediated by the hippocampus. Based on these observations, it seems that some serious reconsideration of the memory systems perspective is required.

That said, there remains much to do in this domain. For example, it is increasingly apparent that brain injury may give rise to significant reorganization of the neural correlates of STM depending on the time-course of the underlying disease process (e.g., Finke et al. 2013). Studies that combine functional neuroimaging and patient work have great potential to provide new insights into how and when brain function is reorganized subsequent to damage, and whether reorganization affects performance on STM tasks. Work is also needed in the neuropsychological literature that makes inroads with contemporary STM models. For example, as proposed by LaRocque et al. (2014) and consistent with fMRI outcomes (Nee and Jonides 2013, 2014), one might expect that amnesic patients would be impaired on simple short-term memory tasks when information is being held in the "region of direct access", but not the "focus of attention". Alternatively, impairments might even be evident for information in the focus of attention depending on the representational demands of the task. In turn, fMRI studies that take advantage of multivariate analysis techniques might provide important new insights into what exactly is represented by the hippocampus over the course of a delay period. This kind of work could serve to test claims made here, and elsewhere (Yonelinas 2013), that the hippocampus is likely to support or contribute to STM when tasks require active retention of inter-item and item-context bindings, or when the testing procedure requires representation of high-resolution object details that distinguish the target on the current trial, from one seen several trials earlier, or from similar foils in the test array. In short, it seems reasonable to conclude that the hippocampus contributes to STM. Consistent with conclusions drawn elsewhere (e.g., Ranganath and Blumenfeld 2005), this is important because intact performance on STM tasks following hippocampal damage was considered linchpin evidence for separate short- and long-term memory systems.

General Conclusions

In this chapter, we reviewed three current topics related to hippocampal function, each of which is addressed by a distinct portion of the literature. However, perception, short-term retention, and conscious awareness are linked by a common, historical exclusion: according to long-held views of hippocampal function (cf. Squire and Dede 2015), none depends on the hippocampus. Recent work has prompted our field to reconsider this widely-held perspective by suggesting that perception, short-term retention, and memory expression absent awareness may in fact require and recruit the hippocampus. The field's acknowledgement of broader hippocampal contributions is evident in the proliferation of new theories (or refocusing of existing theories) to describe a synthesis between recent findings and the established role of the hippocampus in LTM processes. We close by (re-) considering a few theoretical accounts related to the topics we reviewed.

Relational memory theory (Eichenbaum and Cohen 2001, 2014) and related proposals (Davachi and Dobbins 2008; Ranganath 2010) have indicated that the hippocampus supports the binding together of arbitrarily related stimuli at encoding, and supports part-cued retrieval of associated content during a temporally-extended consolidation process. Empirical support for the predictions of relational memory theory in LTM is considerable, but a key theoretical question for this chapter has been to what extent the hippocampus contributes this kind of relational processing to other cognitive operations. For example, when binding is required by tests that do not tap long-term declarative memories, is a hippocampal contribution required? Much of the evidence that we have reviewed here is consistent with this possibility. As such, the relational memory theory continues to make important and accurate predictions more than two decades after its debut.

Despite the continued success of relational memory theory, findings that imply a hippocampal role in perception could constitute something of a challenge. As implied by its name, the perceptual-mnemonic theory (PMT) of MTL and hippocampal function suggests that these structures contribute to (at least) two distinct cognitive domains, namely perception and memory (Bussey and Saksida 2007; Graham et al. 2010; Graham and Gaffan 2005; Lee et al. 2012). A key concept in PMT is that the hierarchical organization of the ventral stream (Mishkin et al. 2000) is preserved and extended in the MTL (Bussey and Saksida 2007). PMT is appealing because it tackles recent findings for hippocampal involvement in cognitive processes over short intervals head on, and because it extends an established model of hierarchical visual representation in the brain (Mishkin et al. 2000). Befitting the apical position of the hippocampus in the ventral visual stream, PMT suggests that this structure is uniquely capable of contributing to the perception of complex scenes (Bussey and Saksida 2007; Graham et al. 2010; Graham and Gaffan 2005; Lee et al. 2012). Some findings described in this chapter support this claim (reviewed by Douglas and Lee 2015; Lee et al. 2012), but it is not yet clear whether the scope of PMT is sufficient to encompass the entire breadth of hippocampal contributions to cognition. Despite this uncertainty, the originators deserve great credit for proposing a theory with solid empirical foundations that is capable of generating empirically testable hypotheses.

Another recent account that explicitly attempts to address a potential dual role for the hippocampus in perception and memory is the high-resolution binding theory (HRBT) (Yonelinas 2013). HBRT suggests that the hippocampus supports ". . . the generalization and utilization of complex high-resolution bindings that link together the qualitative aspects that make up an event" (p. 34). HRBT incorporates key components of the declarative and relational memory theories to address hippocampal contributions to memory as well as portions of perceptual-mnemonic theory to account for recent perception-oriented findings. The claims of HRBT are broadly consistent with contemporary data although certain findings of hippocampal involvement in the maintenance of relatively simple stimuli over short intervals or specific relational failures may not be addressed (Race et al. 2013; Warren et al. 2010, 2014; Watson et al. 2013). A more thorough evaluation of HRBT may require the accumulation of new data to test whether its impressive explanatory power will be matched by the quality of its novel predictions (e.g., evidence of high-resolution bindings operating in recollection, language, and other cognitive processes).

Finally, a model proposed recently by Henke (2010) takes an aggressive stance on the consciousness issue. This model shares a number of key tenets with the relational memory theory (Cohen and Eichenbaum 1993) and related proposals that have made increasingly specific claims about the role of MTL cortical structures (particularly perirhinal and parahippocampal cortices) in memory (e.g., Davachi 2006; Eichenbaum et al. 2007; Diana et al. 2007). This model holds tight to proposed divisions between long-term memory systems (e.g., episodic, semantic, procedural), but suggests that the differences among them come down to processing speed and flexibility of the resulting memory representations, rather than consciousness. There is a good deal of existing empirical support for this model, and it suggests a number of hypotheses that can be tested to further evaluate the viability of claims that have been made. It does not seem, however, to directly consider hippocampal contributions to cognitive function outside the domain of LTM (e.g., perception and short-term retention), though it seems possible that the same basic principles would apply.

In conclusion, our summary finds the literature describing hippocampal contributions to cognition at a moment of significant change that prompts fundamental questions about the nature of conscious memory access, perception, and representation of information over the short-term. For example, an important constraint on hippocampal involvement in cognitive processes beyond LTM may be the representational and/or processing demands of a particular task. Much recent work was initiated in the context of theories that have proposed a role for the hippocampus in relational binding and representation. As indicated above, this view implicates the hippocampus in the encoding, subsequent retrieval, and flexible use of representations that contain information about items bound together in space and time. In turn, this new work, including several of the studies that were summarized here, has led to important observations that compel reconsideration of some key tenets of established theories. For example, in each of the three domains that were examined, it seems to be the case that the hippocampus contributes not only to binding of items and context or inter-item binding, but also to feature binding when task demands require detailed intra-item information for successful performance. As we have suggested, it may be the case that the lens of the hippocampus can be dynamically adjusted, so that the "focus" of this structure targets items in broader contextual settings, or is optimized to process features within an item, depending on task demands. For example, when face recognition depends critically on high-fidelity representation of the component parts, because it has been viewed from several different perspectives during encoding, flexible representation of the relationships among face features may be required to support successful performance. Similarly, when an ellipse, tilted 45° from vertical, has to be distinguished from similar exemplars in a test display, or insulated from other similar exemplars across trials, a bound representation of *that* item (i.e. its features) to specific temporal context might be required for successful performance. In sum, it seems that the reach of the hippocampus does indeed go beyond long-term declarative memory; now, investigators must begin to address questions about the specific characteristics of these contributions.

References

Addante RJ (2015) A critical role of the human hippocampus in an electrophysiological measure of implicit memory. *Neuroimage 109*:515–528

Allen RJ, Vargha-Khadem F, Baddeley AD (2014) Item-location binding in working memory: is it hippocampus-dependent? *Neuropsychologia 59*:74–84

Althoff RR, Cohen NJ (1999) Eye-movement based memory effect: a reprocessing effect in face perception. *J Exp Psychol 25*:997–1010

Alvarez GA, Cavanagh P (2004) The capacity of visual short-term memory is set both by visual information load and by number of objects. *Psychol Sci 15*:106–111

Aly M, Yonelinas AP (2012) Bridging consciousness and cognition in memory and perception: evidence for both state and strength processes. *PLoS One 7*(1):e30231

Aly M, Ranganath C, Yonelinas AP (2013) Detecting changes in scenes: the hippocampus is critical for strength-based perception. *Neuron 78*:1127–1137

Aly M, Ranganath C, Yonelinas AP (2014) Neural correlates of state- and strength-based perception. *J Cogn Neurosci 26*:1–18

Axmacher N, Schmitz DP, Wagner T, Elger CE, Fell J (2008) Interactions between medial temporal lobe, prefrontal cortex, and inferior temporal regions during visual working memory: a combined intracranial EEG and functional magnetic resonance imaging study. *J Neurosci 28* (29):7304–7312

Axmacher N, Haupt S, Cohen MX, Elger CE, Fell J (2009) Interference of working memory load with long-term memory formation. Eur J Neurosci 29:1501–1513

Baddeley AD, Allen RJ, Vargha-Khadem F (2010) Is the hippocampus necessary for visual and verbal binding in working memory? *Neuropsychologia 48*:1089–1095

Baddeley AD, Jarrold C, Vargha-Khadem F (2011) Working memory and the hippocampus. *J Cogn Neurosci 23*:3855–3861

Barense MD, Gaffan D, Graham KS (2007) The human medial temporal lobe processes online representations of complex objects. *Neuropsychologia 45*:2963–2974

Barense MD, Groen IIA, Lee ACH, Yeung LK, Brady SM, Gregori M, Kapur N, TJ B, LM S, RN H (2012) Intact memory for irrelevant information impairs perception in amnesia. *Neuron* 75:157–167

Barense MD, Henson RNA, Graham KS (2011) Perception and conception: temporal lobe activity during complex discriminations of familiar and novel faces and objects. *J Cogn Neurosci* 23:3052–3067

Barense MD, Henson RNA, Lee ACH, Graham KS (2010) Medial temporal lobe activity during complex discrimination of faces, objects, and scenes: effects of viewpoint. *Hippocampus* 20:389–401

Behrmann M, Lee ACH, Geskin JZ, Graham KS, Barense MD (2016) Temporal lobe contribution to perceptual function: a tale of three patient groups. *Neuropsychologia* 90:33–45

Bergmann HC, Daselaar SM, Beul SF, Rijpkema M, Gernandez G, Kessels RPC (2015) Brain activation during associative short-term memory maintenance is not predictive for subsequent retrieval. Front Hum Neurosci 9, article 479

Bergmann HC, Daselaar SM, Fernandez G, Kessels RP (2016). Neural substrates of successful working memory and long-term memory formation in a relational spatial memory task. Cogn Process, epub ahead of print

Bergmann HC, Rijpkema M, Fernández G, Kessels RP (2012) Distinct neural correlates of associative working memory and long-term memory encoding in the medial temporal lobe. *Neuroimage* 63(2):989–997

Bird CM, Burgess N (2008) Report: the hippocampus supports recognition memory for familiar words, but not unfamiliar faces. *Curr Biol* 18:1932–1936

Braun M, Finke C, Ostendorf F, Lehmann TN, Hoffmann KT, Ploner CJ (2008) Reorganization of associative memory in humans with long-standing hippocampal damage. *Brain* 131:2742–2750

Braun M, Weinrich C, Finke C, Ostendorf F, Lehmann TN, Ploner CJ (2011) Lesions affecting the right hippocampal formation differentially impair short-term memory of spatial and nonspatial associations. *Hippocampus* 21:309–318

Buckley MJ, Booth MC, Rolls ET, Gaffan D (2001) Selective perceptual impairments after perirhinal ablation. *J Neurosci* 21:9824–9836

Bussey TJ, Saksida LM (2007) Memory, perception, and the ventral visual-perirhinal-hippocampal stream: thinking outside of the boxes. *Hippocampus* 17:898–908

Carlesimo GA, Perri R, Costa A, Serra L, Caltagirone C (2005) Priming for novel between-word associations in patients with organic amnesia. *J Int Neuropsychol Soc* 11(5):566–573

Cashdollar N, Malecki U, Rugg-Gunn FJ, Duncan JS, Lavie N, Düzel E (2009) Hippocampus-dependent and -independent theta-networks of active maintenance. *Proc Natl Acad Sci* 106:20493–20498

Cave CB, Squire LR (1992) Intact verbal and nonverbal short-term memory following damage to the human hippocampus. *Hippocampus* 2:151–163

Chun MM, Phelps EA (1999) Memory deficits for implicit contextual information in amnesic subjects with hippocampal damage. *Nat Neurosci* 2(9):844–847

Cohen NJ, Eichenbaum H (1993) Memory, amnesia, and the hippocampal system. The MIT Press, Cambridge, MA

Cohen NJ, Squire LR (1980) Preserved learning and retention of pattern-analyzing skill in amnesia: dissociation of knowing how and knowing that. *Science* 210:207–210

Corkin S (2002) What's new with the amnesic patient H.M.? *Nat Rev Neurosci* 3(2):153–160

Cowan N (1993) Activation, attention, and short-term memory. *Mem Cognit* 21:162–167

Cowan N (2001) The magical number 4 in short-term memory: a reconsideration of mental storage capacity. *Behav Brain Sci* 24:87–185

Curtis CE, Zald DH, Lee JT, Pardo JV (2000) Object and spatial alternation tasks with minimal delays activate the right anterior hippocampus proper in humans. *Neuroreport* 11 (10):2203–2207

Damasio AR, Damasio H (1983) The anatomic basis of pure alexia. *Neurology* 33(12):1573–1573

Davachi L (2006) Item, context and relational episodic encoding in humans. *Curr Opin Neurobiol* *16*:693–700

Davachi L, Dobbins IG (2008) Declarative memory. *Curr Direct Psychol Sci 17*:112–118

Diana RA, Yonelinas AP, Ranganath C (2007) Imaging recollection and familiarity in the medial temporal lobe: a three-component model. *Trends Cogn Sci 11*:379–386

Douglas D, Lee ACH (2015) Medial temporal lobe contributions to memory and perception. In: Addis DR, Barense MD, Duarte A (eds) *The Wiley handbook on the cognitive neuroscience of memory*. Wiley-Blackwell, Hoboken, NJ, pp 190–217

Drachman DA, Arbit J (1966) Memory and the hippocampal complex II. Is memory a multiple process? *Arch Neurol 15*:52–61

Duss SB, Reber TP, Hänggi J, Schwab S, Wiest R, Müri RM, Brugger P, Gutbrod K, Henke K (2014) Unconscious relational encoding depends on hippocampus. *Brain 137*(12):3355–3370

Eichenbaum H (2013) What H.M. taught us. *J Cogn Neurosci 25*:14–21

Eichenbaum H, Cohen NJ (2001) *From conditioning to conscious recollection: memory systems of the brain*. Oxford University Press, New York, NY

Eichenbaum H, Cohen NJ (2014) Can we reconcile the declarative memory and spatial navigation views on hippocampal function? *Neuron 83*:764–770

Eichenbaum H, Yonelinas AP, Ranganath C (2007) The medial temporal lobe and recognition memory. Annu Rev Neurosci 30:123–152

Eichenbaum H, Otto T, Cohen NJ (1994) Two functional components of the hippocampal memory system. *Behav Brain Sci 17*(03):449–472

Erez J, Lee ACH, Barense MD (2013) It does not look odd to me: perceptual impairments and eye movements in amnesic patients with medial temporal lobe damage. *Neuropsychologia 51*:168–180

Ezzyat Y, Olson IR (2008) The medial temporal lobe and visual working memory: comparisons across tasks, delays, and visual similarity. *Cogn Affect Behav Neurosci 8*:32–40

Finke C, Braun M, Ostendorf F, Lehmann TN, Hoffman KT, Kopp U, Ploner CJ (2008) The human hippocampal formation mediates short-term memory of colour-location associations. *Neuropsychologia 46*:614–623

Finke C, Bruehl H, Düzel E, Heekeren HR, Ploner CJ (2013) Neural correlates of short-term memory reorganization in humans with hippocampal damage. *J Neurosci 33*:11061–11069

Foulsham T, Kingstone A (2013) Where have eye been? Observers can recognise their own fixations. *Perception 42*(10):1085–1089

Gabrieli JD, Corkin S, Mickel SF, Growdon JH (1993) Intact acquisition and long-term retention of mirror-tracing skill in Alzheimer's disease and in global amnesia. *Behav Neurosci 107*:899–910

Gazzaley A, Rissman J, D'Esposito M (2004) Functional connectivity during working memory maintenance. *Cogn Affect Behav Neurosci 4*:580–599

Graf P, Schacter DL (1985) Implicit and explicit memory for new associations in normal and amnesic subjects. *J Exp Psychol Learn Mem Cogn 11*(3):501–518

Graham KS, Gaffan D (2005) The role of the medial temporal lobe in memory and perception: evidence form rats, nonhuman primates, and humans. *Quart J Exp Psychol Sect B 58*:193–201

Graham KS, Barense MD, Lee AC (2010) Going beyond LTM in the MTL: a synthesis of neuropsychological and neuroimaging findings on the role of the medial temporal lobe in memory and perception. *Neuropsychologia 48*:831–853

Hales JB, Broadbent NJ, Velu PD, Squire LR, Clark RE (2015) Hippocampus, perirhinal cortex, and complex visual discrimination in rats and humans. *Learn Mem 22*:83–91

Hannula DE, Greene AJ (2012) The hippocampus reevaluated in unconscious learning and memory: at a tipping point? Front Hum Neurosci 6, article 80

Hannula DE, Ranganath C (2008) Medial temporal lobe activity predicts successful relational memory binding. *J Neurosci 28*:116–124

Hannula DE, Ranganath C (2009) The eyes have it: hippocampal activity predicts expression of memory in eye movements. *Neuron 63*:592–599

Hannula DE, Althoff RR, Warren DE, Riggs L, Cohen NJ, Ryan JD (2010) Worth a glance: using eye movements to investigate the cognitive neuroscience of memory. Front Hum Neurosci 4, article 166

Hannula DE, Tranel D, Cohen NJ (2006) The long and the short of it: relational memory impairments in amnesia, even at short lags. J Neurosci 26:8352–8359

Hannula DE, Tranel D, Cohen NJ (2007) Rapid onset relational memory effects are evident in eye movement behavior, but not in hippocampal amnesia. J Cogn Neurosci 19(10):1690–1705

Hartley T, Bird CM, Chan D, Cipolotti L, Husain M, Vargha-Khadem F, Burgess N (2007) The hippocampus is required for short-term topographical memory in humans. Hippocampus 17:34–48

Henderson JM, Hollingworth A (1999) High-level scene perception. Annu Rev Psychol 50:243–271

Henke K (2010) A model for memory systems based on processing modes rather than consciousness. Nat Rev Neurosci 11(7):523–532

Henke K, Mondadori CRA, Treyer V, Nitsch RM, Buck A, Hock C (2003a) Nonconscious formation and reactivation of semantic associations by way of the medial temporal lobe. Neuropsychologia 41(8):863–876

Henke K, Treyer V, Nagy ET, Kneifel S, Düsteler M, Nitsch RM, Buck A (2003b) Active hippocampus during nonconscious memories. Conscious Cogn 12(1):31–48

Hollingworth A, Richard AM, Luck SJ (2008) Understanding the function of visual short-term memory: transaccadic memory, object correspondence, and gaze correction. JEP Gen 137:163–181

Horner AJ, Gadian DG, Fuentemilla L, Jentschke S, Vargh-Khadem F, Duzel E (2012) A rapid, hippocampus-dependent, item-memory signal that initiates context memory in humans. Curr Biol 22(24):2369–2374

Insausti R, Annese J, Amaral DG, Squire LR (2013) Human amnesia and the medial temporal lobe illuminated by neuropsychological and neurohistological findings for patient E.P. Proc Natl Acad Sci 110:E1953–E1962

Jafarpour A, Fuentemilla L, Horner AJ, Penny W, Duzel E (2014) Replay of very early encoding representations during recollection. J Neurosci 34(1):242–248

Jeneson A, Squire LR (2012) Working memory, long-term memory, and medial temporal lobe function. Learn Mem 19:15–25

Jeneson A, Mauldin KN, Hopkins RO, Squire LR (2011) The role of the hippocampus in retaining relational information across delays: the importance of memory load. Learn Mem 18:301–305

Jeneson A, Squire LR (2011) Working memory, long-term memory, and medial temporal lobe function. Learn Mem 19:15–25

Jeneson A, Mauldin KN, Squire LR (2010) Intact working memory for relational information after medial temporal lobe damage. J Neurosci 30:13624–13629

Jeneson A, Wixted JT, Hopkins RO, Squire LR (2012) Visual working memory capacity and the medial temporal lobe. J Neurosci 32:3584–3589

Jonides J, Lewis RL, Nee DE, Lustig CA, Berman MG, Moore KS (2008) The mind and brain of short-term memory. Annu Rev Psychol 59:193–224

Knauer B, Jochems A, Valero-Aracama MJ, Yoshida M (2013) Long-lasting intrinsic persistent firing in rat CA1 pyramidal cells: a possible mechanism for active maintenance of memory. Hippocampus 23(9):820–831

Knutson AR, Hopkins RO, Squire LR (2013) A pencil rescues impaired performance on a visual discrimination task in patients with medial temporal lobe lesions. Learn Mem 20:607–610

LaRocque JJ, Lewis-Peacock JA, Postle BR (2014) Multiple neural states of representation in short-term memory? It's a matter of attention. Front Hum Neurosci 8, article 5

Lech RK, Suchan B (2014) Involvement of the human medial temporal lobe in a visual discrimination task. Behav Brain Res 268:22–30

Lech RK, Koch B, Schwarz M, Suchan B (2016) Fornix and medial temporal lobe lesions lead to comparable deficits in complex visual perception. Neurosci Lett 620:27–32

Lee AC, Rudebeck SR (2010a) Human medial temporal lobe damage can disrupt the perception of single objects. J Neurosci 30:6588–6594

Lee ACH, Rudebeck SR (2010b) Investigating the interaction between spatial perception and working memory in the human medial temporal lobe. J Cogn Neurosci 22:2823–2835

Lee AC, Buckley MJ, Pegman SJ, Spiers H, Scahill VL, Gaffan D, Bussey TJ, Davies RR, Kapur N, Hodges JR, Graham KS (2005a) Specialization in the medial temporal lobe for processing of objects and scenes. Hippocampus 15:782–797

Lee AC, Bussey TJ, Murray EA, Saksida LM, Epstein RA, Kapur N, Hodges JR, Graham KS (2005b) Perceptual deficits in amnesia: challenging the medial temporal lobe mnemonic view. Neuropsychologia 43:1–11

Lee ACH, Brodersen KH, Rudebeck SR (2013) Disentangling spatial perception and spatial memory in the hippocampus: a univariate and multivariate pattern analysis fMRI study. J Cogn Neurosci 25:534–546

Lee AC, Buckley MJ, Gaffan D, Emery T, Hodges HR, Graham KS (2006) Differentiating the roles of the hippocampus and perirhinal cortex in processes beyond long-term declarative memory: a double dissociation in dementia. J Neurosci 26:5198–5203

Lee ACH, Yeung LK, Barense MD (2012) The hippocampus and visual perception. Front Hum Neurosci 6:91

Leszczynski M, Fell J, Axmacher N (2015) Rhythmic working memory activation in the human hippocampus. Cell Rep 13:1272–1282

Lewis-Peacock JA, Drysdale AT, Oberauer K, Postle BR (2012) Neural evidence for a distinction between short-term memory and the focus of attention. J Cogn Neurosci 24:61–79

Lezak MD (ed) (2012) Chapter 02: Basic concepts. In: Neuropsychological assessment, 5th edn. Oxford University Press, Oxford

Libby LA, Hannula DE, Ranganath C (2014) Medial temporal lobe coding of item and spatial information during relational binding in working memory. J Neurosci 34:14233–14242

Luck SJ (2008) Visual short-term memory. In: Luck SJ, Hollingworth A (eds) Visual memory: advances in visual cognition. Oxford University Press, New York, NY, pp 43–86

Luck SJ, Vogel EK (1997) The capacity of visual working memory for features and conjunctions. Nature 390:279–281

Marti S, Bayet L, Dehaene S (2015) Subjective report of eye fixations during serial search. Conscious Cogn 33:1–15

Mayes AR, Holdstock JS, Isaac CL, Hunkin NM, Roberts N (2002) Relative sparing of item recognition in a patient with adult-onset damage limited to the hippocampus. Hippocampus 12:325–340

Miller GA (1956) The magical number seven, plus or minus two: some limits on our capacity for processing information. Psychol Rev 63:81–97

Milner B (1962) Les troubles de la memorie accompagnant les lesions hippocampiques bilaterales. In: Physiologie de l'Hippocampe. Colloques Internationaux, No. 107. C.N.R.S., Paris, pp 257–272

Milner B, Squire LR, Kandel ER (1998) Cognitive neuroscience and the study of memory. Neuron 20:445–468

Mishkin M, Ungerleider LG, Macko KA (2000) Object vision and spatial vision: two cortical pathways. Brain Behav Crit Concepts Psychol 4:1209

Mitchell KJ, Johnson MK, Raye CL, D'Esposito M (2000) fMRI evidence of age-related hippocampal dysfunction in feature binding in working memory. Cogn Brain Res 10(1):197–206

Montaldi D, Mayes AR (2010) The role of recollection and familiarity in the functional differentiation of the medial temporal lobes. Hippocampus 20(11):1291–1314

Moscovitch M (1992) A neuropsychological model of memory and consciousness. In: Neuropsychology of memory, 2nd edn. Guilford Press, New York, NY, pp 5–22

Moscovitch M (2008) The hippocampus as a "stupid," domain-specific module: implications for theories of recent and remote memory, and of imagination. Can J Exp Psychol/Revue Canadienne De Psychologie Expérimentale 62(1):62–79

Moscovitch M, Winocur G, Behrmann M (1997) What is special about face recognition? Nineteen experiments on a person with visual object agnosia and dyslexia but normal face recognition. *J Cogn Neurosci 9*:555–604

Moses SN, Ryan JD (2006) A comparison and evaluation of the predictions of relational and conjunctive accounts of hippocampal function. *Hippocampus 16*(1):43–65

Nauer RK, Whiteman AS, Dunne MF, Stern CE, Schon K (2015) Hippocampal subfield and medial temporal cortical persistent activity during working memory reflects ongoing encoding. *Front Syst Neurosci 9*:30

Nee DE, Jonides J (2013) Neural evidence for a 3-state model of visual short-term memory. *Neuroimage 74*:1–11

Nee DE, Jonides J (2014) Frontal-medial temporal interactions mediate transitions among representational states in short-term memory. *J Neurosci 34*:7964–7975

Newcombe F, Mehta Z, de Haan EHF (1994) Category specificity in visual recognition. In: Farah MJ, Ratcliff G (eds) The neuropsychology of high-level vision: collected tutorial essays. Lawrence Erlbaum Associates, Hillsdale, NJ

Newmark RE, Schon K, Ross RS, Stern CE (2013) Contributions of the hippocampal subfields and entorhinal cortex to disambiguation during working memory. *Hippocampus 23*(6):467–475

Nichols EA, Kao YC, Verfaellie M, Gabrieli JD (2006) Working memory and long-term memory for faces: evidence from fMRI and global amnesia for involvement of the medial temporal lobes. *Hippocampus 16*:604–616

Nickel AE, Henke K, Hannula DE (2015) Relational memory is evident in eye movement behavior despite use of subliminal testing methods. *PLoS One 10*(10):e0141677

Norman KA, Polyn SM, Detre GJ, Haxby JV (2006) Beyond mind-reading: multi-voxel pattern analysis of fMRI data. *Trends Cogn Sci 10*:424–430

Oberauer K (2002) Access to information in working memory: exploring the focus of attention. *J Exp Psychol Learn Mem Cogn 28*:411–421

Olsen RK, Lee Y, Kube J, Rosenbaum RS, Grady CL, Moscovitch M, Ryan JD (2015) The role of relational binding in item memory: evidence from face recognition in a case of developmental amnesia. *J Neurosci 35*(13):5342–5350

Olsen RK, Moses SN, Riggs L, Ryan JD (2012) The hippocampus supports multiple cognitive processes through relational binding and comparison. *Front Hum Neurosci 6*:146

Olsen RK, Nichols EA, Chen J, Hunt JF, Glover GH, Gabrieli JD, Wagner AD (2009) Performance-related sustained and anticipatory activity in human medial temporal lobe during delayed match-to-sample. *J Neurosci 29*(38):11880–11890

Olsen RK, Palombo DJ, Rabin JS, Levine B, Ryan JD, Rosenbaum RS (2013) Volumetric analysis of medial temporal lobe subregions in developmental amnesia using high-resolution magnetic resonance imaging. *Hippocampus 23*(10):855–860

Olson IR, Moore KS, Stark M, Chatterjee A (2006a) Visual working memory is impaired when the medial temporal lobe is damaged. *J Cogn Neurosci 18*:1807–1897

Olson IR, Page K, Moore KS, Chatterjee A, Verfaellie M (2006b) Working memory for conjunctions relies on the medial temporal lobe. *J Neurosci 26*:4596–4601

Öztekin I, Davachi L, McElree B (2010) Are representations in working memory distinct from representations in long-term memory? Neural evidence in support of a single store. Psychol Sci 21(8):1123–1133

Öztekin I, McElree B, Staresina BP, Davachi L (2009) Working memory retrieval: contributions of the left prefrontal cortex, the left posterior parietal cortex, and the hippocampus. *J Cogn Neurosci 21*(3):581–593

Parra MA, Fabi K, Luzzi S, Cubelli R, Valdes Hernandez M, Della Sala S (2015) Relational and conjunctive binding functions dissociate in short-term memory. *Neurocase 21*:55–56

Pertzov Y, Miller TD, Gorgoraptis N, Caine D, Schott JM, Butler C, Husain M (2013) Binding deficits in memory following medial temporal lobe damage in patients with voltage-gated potassium channel complex antibody-associated limbic encephalitis. *Brain 136*:2474–2485

Piekema C, Fernández G, Postma A, Hendriks MP, Wester AJ, Kessels RP (2007) Spatial and non-spatial contextual working memory in patients with diencephalic or hippocampal dysfunction. *Brain Res 1172*:103–109

Postans M, Hodgetts CJ, Mundy ME, Jones DK, Lawrence AD, Graham KS (2014) Interindividual variation in fornix microstructure and macrostructure is related to visual discrimination accuracy for scenes but not faces. *J Neurosci 34*(36):12121–12126

Prisko L (1963) Short-term memory in focal cerebral damage. Unpublished doctoral dissertation. McGill University, Montreal

Race E, LaRocque KF, Keane MM, Verfaellie M (2013) Medial temporal lobe contributions to short-term memory for faces. *J Exp Psychol Gen 142*:1309–1322

Race E, Palombo DJ, Cadden M, Burke K, Verfaellie M (2015) Memory integration in amnesia: prior knowledge supports verbal short-term memory. *Neuropsychologia 70*:272–280

Ranganath C (2010) A unified framework for the functional organization of the medial temporal lobes and the phenomenology of episodic memory. *Hippocampus 20*:1263–1290

Ranganath C, Blumenfeld RS (2005) Doubts about double dissociations between short- and long-term memory. *Trends Cogn Sci 9*:374–380

Ranganath C, D'Esposito M (2001) Medial temporal lobe activity associated with active maintenance of novel information. *Neuron 31*:865–873

Ranganath C, Cohen MX, Brozinsky CJ (2005) Working memory maintenance contributes to long-term memory formation: neural and behavioral evidence. *J Cogn Neurosci 17*(7):994–1010

Ranganath C, Yonelinas AP, Cohen MX, Dy CJ, Tom SM, D'Esposito M (2004) Dissociable correlates of recollection and familiarity within the medial temporal lobes. *Neuropsychologia 42*(1):2–13

Reber TP, Do Lam ATA, Axmacher N, Elger CE, Helmstaedter C, Henke K, Fell J (2016) Intracranial eeg correlates of implicit relational inference within the hippocampus. Hippocampus 26(1):54–66

Riggs L, Moses SN, Bardouille T, Herdman AT, Ross B, Ryan JD (2009) A complementary analytic approach to examining medial temporal lobe sources using magnetoencephalography. *Neuroimage 45*(2):627–642

Rissman J, Gazzaley A, D'Esposito M (2008) Dynamic adjustments in prefrontal, hippocampal, and inferior temporal interactions with increasing visual working memory load. *Cereb Cortex 18*:1618–1629

Rose NS, Olsen RK, Craik FI, Rosenbaum RS (2011) Working memory and amnesia: the role of stimulus novelty. *Neurospsychologia 50*:11–18

Rosenbaum RS, Gao F, Honjo K, Raybaud C, Olsen RK, Palombo DJ, Levine B, Black SE (2014) Congenital absence of the mammillary bodies: a novel finding in a well-studied case of developmental amnesia. *Neuropsychologia 65*:82–87

Rosenbaum RS, Kohler S, Schacter DL, Moscovitch M, Westmacott R, Black SE, Gao F, Tulving E (2005) The case of K.C.: contributions of a memory-impaired person to memory theory. *Neurospsychologia 43*:989–1021

Rosenthal R (1979) The file drawer problem and tolerance for null results. *Psychol Bull 86*:638

Ryals AJ, Wang JX, Polnaszek KL, Voss JL (2015) Hippocampal contribution to implicit configuration memory expressed via eye movements during scene exploration. *Hippocampus 25*(9):1028–1041

Ryan JD, Cohen NJ (2004) Processing and short-term retention of relational information in amnesia. *Neuropsychologia 42*:497–511

Ryan JD, Althoff RR, Whitlow S, Cohen NJ (2000) Amnesia is a deficit in relational memory. *Psychol Sci 11*(6):454–461

Ryan JD, Hannula DE, Cohen NJ (2007) The obligatory effects of memory on eye movements. *Memory 15*(5):508–525

Sacks O (1998) *The man who mistook his wife for a hat and other clinical tales*. Simon & Schuster, New York, NY

Schmolesky MT, Wang Y, Hanes DP, Thompson KG, Leutgeb S, Schall JD, Leventhal AG (1998) Signal timing across the macaque visual system. J Neurophysiol 79:3272–3278

Schon K, Hasselmo ME, LoPresti ML, Tricarico MD, Stern CE (2004) Persistence of parahippocampal representation in the absence of stimulus input enhances long-term encoding: a functional magnetic resonance imaging study of subsequent memory after a delayed match-to-sample task. *J Neurosci* 24(49):11088–11097

Schon K, Ross RS, Hasselmo ME, Stern CE (2013) Complementary roles of medial temporal lobes and mid-dorsolateral prefrontal cortex for working memory for novel and familiar trial-unique visual stimuli. *Eur J Neurosci* 37(4):668–678

Scoville WB (1968) Amnesia after bilateral mesial temporal-lobe excision: introduction to case H.M. *Neuropsychologia* 6:211–213

Scoville WB, Milner B (1957) Loss of recent memory after bilateral hippocampal lesions. *J Neurol Neurosurg Psychiatry* 20:11–21

Sheldon SAM, Moscovitch M (2010) Recollective performance advantages for implicit memory tasks. *Memory* 18(7):681–697

Shrager Y, Gold JJ, Hopkins RO, Squire LR (2006) Intact visual perception in memory-impaired patients with medial temporal lobe lesions. *J Neurosci* 26:2235–2240

Shrager Y, Levy DA, Hopkins RO, Squire LR (2008) Working memory and the organization of brain systems. *J Neurosci* 28:4818–4822

Sidman M, Stoddard LT, Mohr JP (1968) Some additional quantitative observations of immediate memory in a patient with bilateral hippocampal lesions. *Neuropsychologia* 6:245–254

Simons DJ, Hannula DE, Warren DE, Day SW (2007) Behavioral, neuroimaging, and neuropsychological approaches to implicit perception. In: Zelazo P, Moscovitch M, Thompson E (eds) *Cambridge handbook of consciousness.* Cambridge University Press, New York, NY, pp 207–250

Smith CN, Squire LR (2015) Differential eye movements for old and new scenes under free viewing conditions are unrelated to awareness and are hippocampus-independent. Poster presented at the Annual Meeting of the Society for Neuroscience

Smith CN, Hopkins RO, Squire LR (2006) Experience-dependent eye movements, awareness, and hippocampus-dependent memory. *J Neurosci* 26:11304–11312

Smyth AC, Shanks DR (2008) Awareness in contextual cuing with extended and concurrent explicit tests. *Mem Cognit* 36(2):403–415

Soto D, Greene CM, Kiyonaga A, Rosenthal CR, Egner T (2012) A parieto-medial temporal pathway for the strategic control over working memory biases in human visual attention. *J Neurosci* 32(49):17563–17571

Spering M, Pomplun M, Carrasco M (2011) Tracking without perceiving: a dissociation between eye movements and motion perception. *Psychol Sci* 22(2):216–225

Squire LR (1992) Declarative and nondeclarative memory: multiple brain systems supporting learning and memory. *J Cogn Neurosci* 4(3):232–243

Squire LR (2004) Memory systems of the brain: a brief history and current perspective. *Neurobiol Learn Mem* 82:171–177

Squire LR (2009) The legacy of patient H.M. for neuroscience. *Neuron* 61:6–9

Squire LR, Dede AJ (2015) Conscious and unconscious memory systems. *Cold Spring Harb Perspect Biol* 7:a021667

Squire LR, Wixted JT (2011) The cognitive neuroscience of memory since patient H.M. *Annu Rev Neurosci* 34:259–288

Squire LR, Shrager Y, Levy DA (2006) Lack of evidence for a role of medial temporal lobe structures in visual perception. *Learn Mem* 13:106–107

Staresina BP, Fell J, Lam D, Anne TA, Axmacher N, Henson RN (2012) Memory signals are temporally dissociated in and across human hippocampus and perirhinal cortex. *Nat Neurosci* 15(8):1167–1173

Stark CEL, Squire LR (2000) Intact perceptual discrimination in humans in the absence of perirhinal cortex. *Learn Mem* 7:273–278

Stern CE, Sherman SJ, Kirchhoff BA, Hasselmo ME (2001) Medial temporal and prefrontal contributions to working memory tasks with novel and familiar stimuli. *Hippocampus 11* (4):337–346

Suzuki WA (2009) Perception and the medial temporal lobe: evaluating the current evidence. *Neuron 61*:657–666

Suzuki WA, Baxter MG (2009) Memory, perception, and the medial temporal lobe: a synthesis of opinions. *Neuron 61*(5):678–679

Thorpe S, Fize D, Marlot C (1996) Speed of processing in the human visual system. Nature 381:520–522

van Geldorp B, Bouman Z, Hendriks MPH, Kessels RPC (2014) Different types of working memory binding in epilepsy patients with unilateral anterior temporal lobectomy. *Brain Cogn 85*:231–238

Verfaellie M, LaRocque KF, Keane MM (2012) Intact implicit verbal relational memory in medial temporal lobe amnesia. *Neuropsychologia 50*(8):2100–2106

von Allmen DY, Wurmitzer K, Martin E, Klaver P (2013) Neural activity in the hippocampus predicts individual visual short-term memory capacity. *Hippocampus 23*:606–615

Voss J, Lucas HD, Paller KA (2012) More than a feeling: pervasive influences of memory without awareness of retrieval. *Cogn Neurosci 3*(3-4):193–207

Voss JL, Warren DE, Gonsalves BD, Federmeier KD, Tranel D, Cohen NJ (2011) Spontaneous revisitation during visual exploration as a link among strategic behavior, learning, and the hippocampus. *Proc Natl Acad Sci 108*:E402–E409

Warren DE, Duff MC, Jensen U, Tranel D, Cohen NJ (2012) Hiding in plain view: lesions of the medial temporal lobe impair online representation. *Hippocampus 22*:1577–1588

Warren DE, Duff MC, Tranel D, Cohen NJ (2010) Medial temporal lobe damage impairs representation of simple stimuli. Front Hum Neurosci 4, article 35

Warren DE, Duff MC, Tranel D, Cohen NJ (2011) Observing degradation of visual representations over short intervals when medial temporal lobe is damaged. *J Cogn Neurosci 23*:3862–3873

Warren DE, Duff MC, Tranel D, Cohen NJ (2014) Hippocampus contributes to the maintenance but not the quality of visual information over time. *Learn Mem 22*:6–10

Warrington EK, Baddeley AD (1974) Amnesia and memory for visual location. *Neuropsychologia 12*:257–263

Watson PD, Voss JL, Warren DE, Tranel D, Cohen NJ (2013) Spatial reconstruction by patients with hippocampal damage is dominated by relational memory errors. Hippocampus 23:570–580

Wickelgren WA (1968) Sparing of short-term memory in amnesia: implications for strength theory of memory. *Neuropsychologia 6*:235–244

Wilken P, Ma WJ (2004) A detection theory account of change detection. *J Vis 4*:1120–1135

Yassa MA, Stark CEL (2011) Pattern separation in the hippocampus. *Trends Neurosci 34*:515–525

Yee LT, Hannula DE, Tranel D, Cohen NJ (2014) Short-term retention of relational memory in amnesia revisited: accurate performance depends on hippocampal integrity. Front Hum Neurosci 8, article 16

Yonelinas AP (2013) The hippocampus supports high-resolution binding in the service of perception, working memory, and long-term memory. *Behav Brain Res 254*:34–44

Zanto TP, Clapp WC, Rubens MT, Karlsson J, Gazzaley A (2015) Expectations of task demands dissociate working memory and long-term memory systems. *Cereb Cortex 26*:1176–1186

Zeidman P, Mullally SL, Maguire EA (2015) Constructing, perceiving, and maintaining scenes: hippocampal activity and connectivity. *Cereb Cortex 25*:3836–3855

Zhang W, Luck SJ (2008) Discrete fixed-resolution representations in visual working memory. *Nature 453*:233–235

Memory, Relational Representations, and the Long Reach of the Hippocampus

Rachael D. Rubin and Neal J. Cohen

Abstract What is memory and what is it for? How does the brain represent our experiences and use stored information to guide future behavior? In this chapter, we explore the contributions of the hippocampus to memory and to the many aspects of behavior it supports. We start with discussion of the view that hippocampal-dependent memory is a fundamentally relational and compositional representation system, supporting the binding of even arbitrary or accidental relations among constituent elements of experience into memory, representing experience via links that maintain the separate representational integrity of those elements rather than fusing them into wholistic or unitized memories, and supporting the subsequent reactivation of these relational representations. The major portion of the chapter discusses the surprisingly long reach of hippocampal influence, due to the relational and flexible nature of its representations. Hippocampal influence extends both across all manner of relations, including spatial, temporal, and associative relations, and across timescales, including not only the classically described role in long-term memory, but also on the timescale of short-term or working memory and even in moment-to-moment processing. Furthermore, the long reach of hippocampal influence is seen in the flexible use of relational representations in service of a rich variety of behavioral repertoires, helping critically in guiding flexible and adaptive choices—uses of hippocampal representations that clearly stretch the classical definition of memory. Finally, we consider the debt owed to clinical studies in providing insights about the nature of relational representations and about the extent of hippocampal influence.

R.D. Rubin
Beckman Institute for Advanced Science and Technology, University of Illinois, 405 N. Mathews Ave., Urbana, IL 61801, USA

Carle Neuroscience Institute, Carle Foundation Hospital, Urbana, IL, USA
e-mail: rrubin2@illinois.edu

N.J. Cohen (✉)
Beckman Institute for Advanced Science and Technology, University of Illinois, 405 N. Mathews Ave., Urbana, IL 61801, USA
e-mail: njc@illinois.edu

© Springer International Publishing AG 2017
D.E. Hannula, M.C. Duff (eds.), *The Hippocampus from Cells to Systems*,
DOI 10.1007/978-3-319-50406-3_11

Introduction

What did you do yesterday? Have you ever seen this place before? When did you wake up today? Who was the last person you had a conversation with and when and where did that occur? How do you get to home from here? The ability to answer these questions of course relies on memory mechanisms of the brain that record information about our experiences, recent and long ago, and that permit the use of this stored information now and into the future. But, these memory mechanisms provide more than the immensely powerful ability to recall information from the many, many events and experiences from our past; as we will argue, memory mechanisms also provide the key to flexible and adaptive behavior, guiding choice behavior in the moment to solve the many cognitive and social challenges we confront in navigating life.

Different kinds of memory rely on different brain systems, which are used for processing and representing different kinds of information. One well-known description of the taxonomy of memory systems, based on decades of converging human and animal research, includes the distinction between declarative memory, which supports remembering of facts and events derived from our experiences, and relies on the hippocampus and surrounding medial temporal lobe (MTL) structures, in interaction with various neocortical regions, versus a procedural or non-declarative memory system, which supports the acquisition and expression of new skills, and relies on structures such as the striatum and basal ganglia, in conjunction with various neocortical regions (Cohen and Eichenbaum 1993; Eichenbaum and Cohen 2001; see below).

These systems, in turn, support different behavioral repertoires, dependent critically on the nature of the representations to which the different memory systems give rise (Cohen and Eichenbaum 1993; Konkel and Cohen 2009). In this chapter, we focus on the nature of the representations that characterize the hippocampal-dependent declarative memory system, emphasizing that they are fundamentally relational. We then go on to highlight research findings that reveal the range of behavioral repertoires that engage and rely upon relational representations, across timescales and domains, illustrating the very long reach of hippo-campal influence in a way that stretches the very definition of memory.

Relational Memory

Our characterization of the functional contribution of the hippocampus to memory and behavior, a project of very long standing for one of the authors (NJC), has attempted to integrate literatures across species and methodologies, as the human and animal (especially rodent) literatures have had very different traditions over the past several decades. Rather than aiming to model the role of the hippocampus in any one specific task, whether it be performance on tests of recognition memory in humans or performance on the Morris water maze or other tests of spatial memory and navigation in rodents, our theorizing has aimed to capture the nature of representations supported by the hippocampus and the role of these representations in supporting various behavioral repertoires (Cohen 2015; Cohen and Eichenbaum 1993; Eichenbaum and Cohen 2001, 2014).

A little historical context may be helpful here. In the human literature, theorizing about hippocampal function was initially guided largely by findings from patients with amnesia consequent to damage to the hippocampus, starting with the neurological patient H.M. (Scoville and Milner 1957), which revealed a critical dissociation between the impairment in forming (and subsequently retrieving on demand) new long-term memories of facts and events versus preserved ability to acquire and express new skills (Cohen and Eichenbaum 1993; Cohen and Squire 1980; Corkin 1968; Eichenbaum and Cohen 2001; Milner 1962; Milner et al. 1968). Such findings led to the notion of multiple memory systems in the brain and the selective role of the hippocampus and related structures in one form of memory, leaving other forms of memory intact (e.g., Cohen and Eichenbaum 1993; Cohen and Squire 1980; Eichenbaum and Cohen 2001, 2014; Norman and O'Reilly 2003; Poldrack and Packard 2003; Squire 2004; Tranel et al. 1994; Schacter 1987; Schacter and Tulving 1994; Tulving and Schacter 1990). Even while ideas about multiple memory systems also began to emerge in the animal literature, for a time much theorizing in the human literature focused on the role of the hippocampus specifically in explicit remembering and conscious recollection, phenomena that are difficult to relate to animal models (although see work by Eichenbaum, elsewhere in this book).

A separate literature on animal, primarily rodent, studies centered around the discovery of place cells in the hippocampus and, later, of grid cells in the entorhinal cortex, identifying a critical role for the hippocampus, entorhinal cortex, and related structures in mapping and navigating spatial environments (e.g., Hafting et al. 2005; Moser et al. 2008; Muller et al. 1987; O'Keefe and Burgess 1996; O'Keefe and Dostrovsky 1971; O'Keefe and Nadel 1978; Shapiro et al. 1997). Such findings led to the awarding of the Nobel Prize in 2014 to John O'Keefe, Edvard Moser, and May-Britt Moser. Relating these findings to the human neuropsychological data in patients with hippocampal damage is challenging, however, in light of abundant evidence that hippocampal amnesia impairs domains of memory and its uses beyond spatial memory and spatial navigation. Instead, hippocampal amnesia is widely seen to be modality- and domain-general, affecting memory for facts and events (declarative memory) very broadly (Cohen and Squire 1980; Cohen and Eichenbaum 1993).

In our efforts to integrate these separate threads we have long emphasized the critical role of the hippocampus in capturing the fundamentally relational character of declarative memory. We see it as involving the binding of relations among the constituent elements of experience into compositional relational representations, and, upon subsequently retrieving or reactivating these representations, being able to use them flexibly across a broad range of tasks, contexts, and conditions, both in humans and animals (Cohen 2015; Cohen and Eichenbaum 1993; Cohen et al. 1997; Eichenbaum and Cohen 2001, 2014).

The binding and subsequent reactivation/retrieval of relational representations permit us to: encode memories of the (often arbitrary or accidental) co-occurrences of people, places, and things, along with the spatial, temporal, and interactional relations among them; permit us to appreciate and make use of past experiences, in all their overlapping and non-overlapping complexity; and use these representations in service of an immense range of behaviors, including when challenged in novel circumstances. Relational representations permit us to learn the names of new acquaintances (completely arbitrary relations that cannot be appreciated *a priori* or inferred upon initial encounter), to remember where and at what time to go for a new appointment (relying on representations previously created from past learning experiences and flexibly using them prospectively in service of a new goal), and, as discussed later, even to facilitate new learning of information (guiding on-line choice behaviors so as to adaptively devote processing resources to optimize new binding).

In rodents, relational representations permit learning and remembering of the spatial relations among environmental cues, acquired across experiences, supporting the ability of the animal to appreciate its location in space relative to the location of other objects and boundaries, and thereby to successfully navigate the environment even when it requires novel paths through the environment. They support, as well, the learning and remembering of what objects were encountered in which locations in space and in what order, and thereby to guide choice behavior to optimize reward. It isn't only spatial relations that are encoded by hippocampal neurons and reflected in "place cell" activity. Rather, hippocampal neurons also represent temporal relations ("time cells"), collectively representing temporal maps of event structure just as they collectively represent spatial maps of event structure, and, moreover, hippocampal neuronal activity reflects various conjunctions of objects, space, and time (see Eichenbaum 2004, 2014; Eichenbaum et al. 1999; McKenzie et al. 2014)—i.e., hippocampal neurons can be seen to code the binding of relations among the constituent elements of experience, which we take to be the

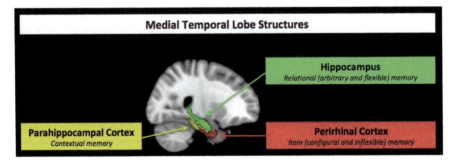

Fig. 1 Medial temporal lobe structures. Reprinted from Rubin and Cohen (2014)

very definition of relational memory. Accordingly, we have long seen hippocampal neurons as being "relational cells" (Eichenbaum and Cohen 1988).

Anatomically, the hippocampus is uniquely positioned to support relational memory binding of experience, consistent with our functional characterization, above. The hippocampus is situated in the MTL (see Fig. 1), alongside the MTL cortices (perirhinal cortex and the parahippocampal cortex), which are the recipients of converging input from the various high-level neocortical processing and association areas, and that input is in turn directed to the hippocampus proper. There follow several stages of serial and parallel processing within the hippocampal formation (i.e., sub-regions CA1–CA3, dentate gyrus, and subiculum) and then output is directed back to the neocortical regions of origin.

The neocortical processing and association areas parse the experienced world into the objects or items for which they are specialized, and thus, upon sending their converging input into the MTL cortices and thence to the hippocampus, they provide the streams of faces, objects, landmarks, words, and so on, as they are encountered in time and space, which are then bound into relational representations capable of capturing the who, what, where, and when of events.

Accordingly, in this chapter, when we offer views about the role of the hippocampus in relational memory we always mean the hippocampus in interaction with cortical processing and storage areas operating in the context of larger-scale brain networks, i.e., a hippocampal-cortical system. Research into the details of those interactions, identification of the various brain networks in which the hippocampus may play a part, and the characterization of the separate contributions of each of the participating brain regions is still very much in its infancy. We leave more detailed consideration of the functional parcellation and characterization of the components of the hippocampal-cortical system to others, except for a few observations below that bear importantly on our representational claims. Likewise, the functional anatomy of the hippocampus itself, involving characterization of the functional roles of and possible interactions among the various hippocampal subfields, is an important emerging area of research in both the animal and human literatures that we largely leave to others, instead focusing here on function at the level of the hippocampus as a whole embedded in its larger brain networks.

There is a critical distinction we have drawn among elements of the hippocampal-cortical network, specifically between the flexible and compositional relational representations supported by the hippocampus and the representations supported by the MTL cortical regions feeding into the hippocampus (Eichenbaum et al. 1994; Cohen et al. 1997). On our view, the hippocampus provides relational representations of experience via links among the item representations of the constituent elements of an event (e.g., of a dinner out, each individual item—the faces and people, the food, the table and restaurant, the conversation, etc.—in its respective cortical processing region) that can maintain the separate representational integrity of those elements (reflecting "compositionality"). This would be in contrast to fusing them into wholistic or unitized memories (as in a flat "snapshot" of the face-scene compound item). The property of compositionality provides a natural way for memories to be linked across events via their common elements (e.g., the multiple conversations between the authors of this chapter at different times in different places) in a rich "memory space" or relational database (Eichenbaum et al. 1999; Konkel and Cohen 2009), permitting the ability to distinguish among the events while nonetheless abstracting and generalizing across them. We have argued (Eichenbaum et al. 1994) that, by contrast, while some amount of associative processing and memory can be accomplished in cortical areas, including perirhinal cortex, it is done in a manner that results in inflexibly bound, unitized representations of perceptual experience.

Consider, for example, cortical areas involved in face-processing, visual object processing, or word processing. All are capable of processing and storing the configural relations among the constituent elements of the items for which they are specialized (the universal features of faces, the distinct parts of objects, and the individual letters of words, respectively). But they do so in service of identifying the whole face, the whole object, or the whole word, and rearrangements of the elements at input will cause the parsing of the information stream into wholly different object representations (e.g., faces with different configural relations among the eyes, nose, and mouth are recognized as different people; likewise different arrangement of the same letters can give rise to different words, as any player of the game Scrabble can attest). Contrast that with a hippocampal-supported relational representation of say, Scarlett, Angelina, and Tom touring the sites of Boston one day. If on a different day they toured the sites in a different order, or if across subsequent days they toured different cities, you would have no trouble remembering which people were touring, which places they toured, regardless of order, and so forth.

An emerging body of work is now available that confirms these representational notions, in that various associations can indeed be learned by cortical regions independently of the hippocampus, but then the acquired representations are highly inflexible, and when conditions or instructions lead participants to create unitized representations, for example by treating pairs of words as compound items, then perirhinal cortex is capable of supporting those representations consistent with its support of other item representations.

Another emerging body of work has been successfully illuminating distinct roles in memory for different MTL regions (Davachi 2006; Davachi and Preston, 2014; Diana et al. 2008; Eichenbaum et al. 2007). Although not the province of the current chapter, a comment about one thread of this work would seem to be in order here. The BIC (binding of item and context) theory (Diana et al. 2007; Eichenbaum et al. 2007) distinguishes among the hippocampus, the perirhinal cortex, and the parahippocampal cortex in terms of their respective mnemonic roles, with perirhinal cortex encoding item information, the parahippocampal cortex encoding background contextual information, and the hippocampus mediating the binding of items with their temporal and/or spatial context into memory. This theory consti-tutes a clear advance in explicating some of the functional neuroanatomy of the MTL in supporting recognition memory, tying the activity of and interactions among different MTL regions both to performance and to familiarity and recollec-tion processes. We are pleased to see that the role it ascribes to the hippocampus remains relational memory binding.

Hippocampus Supports the Binding of Arbitrary Spatial, Temporal, and Associative Relations and Their Re-activation

In remembering the rich tapestry of real-life events, all manner of relational information must be represented in order to capture the who, what, where, and when of events, as well as the common elements across events. It is our contention that the hippocampus, together with its input-output relations with various cortical processors, mediates this capability via relational representations of the converging sensory inputs as well as between current inputs and reactivated relational memories (Eichenbaum and Cohen 2014; Konkel and Cohen 2009). This section outlines some of the critical evidence that the hippocampus does indeed support memory for all manner of arbitrary relations.

Before proceeding, a quick point about memory for arbitrary or accidental relations. The critical capability that only the hippocampus (or hippocampal-cortical system) provides is the ability to create new bindings in memory between previously unrelated items upon experiencing them together and to use those representations to support performance. But, tasks that are intended to challenge the learning of relational information presented in a study trial could easily be solved by other brain systems using various sources of information, if they allow previous knowledge, prior exposure, or statistical learning of regularities among items to permit the answer to be derived on the fly rather than needing to retrieve the representations created at study time. For example, if at study the materials are pairings of faces and occupations, and they include some already known pairings (e.g., Lebron James and "athlete"), then of course there is no need to do relational memory binding at study and retrieve the new representations at test time. Similarly, if at study the materials are sequences of items, and the sequences happen to be constructed using some rule structure that provides the ability to make accurate predictions about the next item, then successful performance would not necessarily require relational memory binding at study and retrieval of the new representations at test time.

By contrast, arbitrary or accidental relations, such as the real-life circumstances of who you happen to bump into while walking down an unfamiliar street (reminiscent of the famous line spoken by Humphrey Bogart in the movie *Casablanca*: "Of all the gin joints in all the towns in all the world, she walks into mine"), require relational representation capability and the hippocampus. The conditions of testing arbitrary or accidental relations are easily met in the laboratory, where it is possible for the experimenter to carefully control the assignment of items such that any item could be paired with or related at study to any other previously unrelated item. In this way, *from the perspective of the participant*, the relations are entirely arbitrary or accidental, and there is no way to derive the answer, thereby requiring relational memory binding and the subsequent retrieval of the new relational representations.

All Manner of Arbitrary Relations

Now, we can consider the evidence for the role of the hippocampus in various relational representations. As noted above, real-life events typically entail a combination of different kinds of to-be-remembered relational information, which may be difficult to disentangle, at least through introspective means. Empirical studies are good at isolating for investigation one or another kind of relation among items. Using such an approach, evidence from patient, neuroimaging, eye tracking, and animal studies implicate the hippocampus in memory for spatial, temporal, and associative relations; these are discussed, in turn, in sections "Spatial Relations", "Temporal Relations" and "Associative Relations".

But, in this section we detail findings from one study showing the contribution of hippocampus to all of the above types of relations, assessed in unconfounded manner all in the same task with the same set of stimuli (Konkel et al. 2008). Participants included patients with damage largely circumscribed to the hippocampus, patients with both hippocampal and more extensive MTL damage that included perirhinal cortex, and matched comparison participants. The study involved presentation of computer-generated stimuli constructed to be odd shapes of different colors, all roughly the same size, but with different textual patterns, so each one was clearly distinguishable from the others (see Fig. 2a). Shapes were presented one at a time, sequentially, each in one of three consistent locations on the screen (upper left, upper right, and bottom center). Any given shape could appear in any of the three temporal slots and any of the three spatial locations. A blank screen was presented in between sets of three stimuli to encourage the temporal grouping of stimuli into triplets. Study consisted of sets of several triplets, followed by corresponding test displays in which triplets were presented in different layouts to allow for the separate and unconfounded assessment of either spatial relations (correct assignment of the three stimuli to their studied spatial locations), or temporal relations (correct assignment of the three stimuli to their studied temporal order), or associative relations (correct assignment of the three stimuli as having co-occurred in the same studied triplet, independent of spatial location or temporal order) (see Fig. 2b–d). Importantly, participants were encouraged to represent all types of relation at study, as the type of relation to be assessed at test time was only apparent upon presentation of the test display. Additional test displays presented individual stimuli for the assessment of item memory, in order to disambiguate any observed deficits in memory for relations among items from any deficits in memory for the items themselves.

Consistent with predictions of relational memory theory, findings (see Fig. 2e) showed that patients with damage to the hippocampus, whether relatively circumscribed or as part of more extensive MTL damage, were impaired for each of the types of relations (spatial, temporal, and associative) relative to matched comparison participants. Moreover, the two groups of patients differed, such that patients with circumscribed hippocampal damage were disproportionately impaired on memory for all manner of relations, compared to memory for individual items,

a)

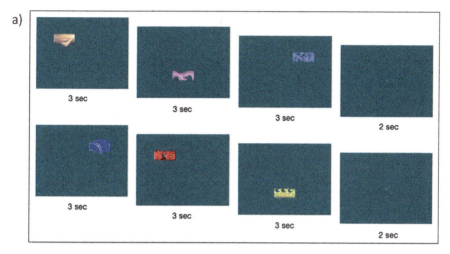

b)

Fig. 2 Memory for all manner of relations. (**a**) Stimuli presentation. (**b**) The spatial test presented three stimuli in each of the three repeating locations with the question—"Same Positions?"—isolating the spatial relations for assessment. (**c**) The temporal test presented each of the three stimuli in the center of the screen, one followed by the other, with the question—"Same Sequence"—isolating the temporal relations for assessment. (**d**) The associative test presented the three stimuli all next to each other, in the middle of the screen, with the question—"Studies Together?"—isolating the associative relations for assessment. (**e**) Performance on each task by group. *Dots* represent individual participants' scores. Comparison = matched comparison participants. Hipp = patients with damage to hippocampus. MTL = patients with more extensive damage to medial temporal lobe. Reprinted from Konkel et al. (2008)

whereas patients with MTL damage that included the perirhinal cortex additionally showed impairment on item memory relative to the other groups. Together, these findings provide strong support for the differential contribution of hippocampus to relational memory, across all manner of relations, and of perirhinal cortex to item memory, consistent with many other studies described below (but see Gold et al. 2006; Stark and Squire 2003).

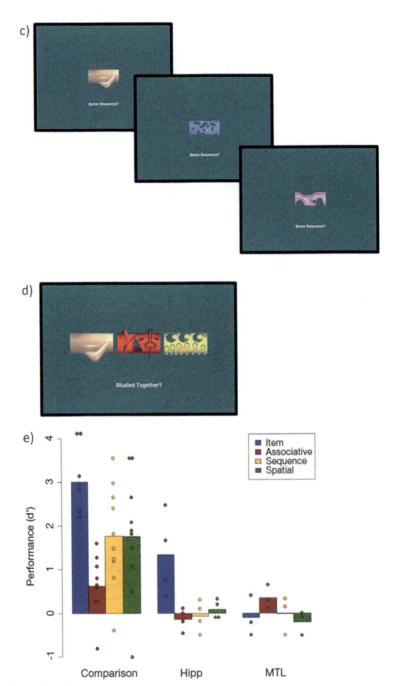

Fig. 2 (continued)

Spatial Relations

The contribution of hippocampus to memory for spatial relations has been demonstrated in studies of patients with hippocampal or more extensive MTL damage, including the following examples showing memory impairment for arbitrary object locations (Crane and Milner 2005; Holdstock et al. 2002, 2005), spatial arrangements among items in scenes (Hannula et al. 2006, 2007; Ryan et al. 2000), spatial lay-outs of residences occupied since the onset of amnesia (Bayley and Squire 2005), and spatial navigation of complex large-scale environments (Maguire et al. 2006). Consistent with these findings, functional neuroimaging data in healthy participants reveal MTL activation in areas including the hippocampus on tasks requiring memory for spatial relations (e.g., Epstein 2008; Hannula and Ranganath 2008; Hartley et al. 2003; Howard et al. 2014; Pine et al. 2002; Spiers and Maguire 2006, 2007, 2008; Zhang and Ekstrom 2013) (for reviews see Cohen et al. 1999; Davachi 2006; Konkel and Cohen 2009; Ranganath 2010). Another well-known finding relating hippocampus to memory for spatial relations is from structural MRI, showing that the volume of posterior hippocampus is greater in London taxi drivers, increasing in size with more years of experience navigating through the city among a wide array of possible locations, but no effect in hippocampus in London bus drivers with similar years of service driving along a fixed route (Maguire et al. 2006).

Temporal Relations

Among the evidence for the contribution of hippocampus to temporal relations are neuroimaging findings in healthy participants, revealing hippocampal activation on tasks requiring memory for pictures of objects grouped in temporal order (Kumaran and Maguire 2006) and sequences of numbers and related response-button mappings on a serial reaction time task (Schendan et al. 2003). Hippocampal activity patterns have been shown to reflect the temporal positions of objects in sequences, explain individual differences in sequence learning, and support representations that differentiate the same object in different sequence contexts (Hsieh et al. 2014). Further, dissociable functional networks in the brain have been identified that support the retrieval of spatial and temporal order source information, both involving the hippocampus, suggesting the hippocampus contributes to representing both kinds of relations (Ekstrom and Bookheimer 2007; Ekstrom et al. 2011). High-resolution fMRI has also recently been employed to examine the contribution of various hippocampal subfields to memory for spatial and temporal relations, finding the dentate gyrus/CA3 region of the hippocampus to be engaged in processing both temporal and spatial relational information (Azab et al. 2014). Finally, a growing animal literature implicates the hippocampus in representations of temporal contexts and sequential memory (e.g., Fortin et al. 2002; MacDonald et al. 2011;

Manns et al. 2007; Schapiro et al. 2012, 2014), including work that links CA3 to sequence memory for both spatial and nonspatial material (Farovik et al. 2010) and other work that suggests separate roles for the CA1 for temporal information and the dentate gyrus for sequential memory (Gilbert et al. 2001).

Associative Relations

Evidence for the contribution of hippocampus to memory for associative relations comes from a variety of paradigms employing a range of methodologies. For example, patients with hippocampal damage demonstrated impairments on tasks requiring memory for the arbitrary association of pairs of words (Giovanello et al. 2003), faces (Kroll et al. 1996; Turriziani et al. 2004), and faces and words (Turriziani et al. 2004). Functional neuroimaging studies have found hippocampal activations in healthy adults on associative or relational memory tasks involving word triplets (Davachi and Wagner 2002) or word–color mental imagery (Staresina and Davachi 2006), as well as pairings of words (Giovanello et al. 2004), of faces with scenes (Henke et al. 1997; also see below), of faces with occupations (Degonda et al. 2005; Henke et al. 2003), and of words with fonts (Prince et al. 2005), along with other associative and source memory tasks (e.g., Davachi et al. 2003; Ranganath et al. 2004).

One paradigm from our laboratory assessing associative relations, involving memory for arbitrary pairings of faces and scenes, has been widely employed in studies using patient, neuroimaging, and eye tracking approaches to reveal hippocampal involvement in associative relational memory (see Fig. 3). Patients with hippocampal damage were impaired at this face-scene binding task relative to matched comparison participants on behavioral testing, falling to chance performance levels with delays of just a few minutes (Hannula et al. 2006). Other testing employed eye tracking to assess relational memory via evidence of disproportionate viewing to the "target" face (the face that had been paired with that scene during study) from among equally familiar faces appearing on the scene in a 3-face test display. Healthy participants showed disproportionate viewing of the target face relative to competitor faces, an effect that emerged automatically and very rapidly, within 500–750 ms of the presentation of the test display, when they had the opportunity to preview the test scene briefly before the 3-face test display was superimposed on it; however, patients with hippocampal damage failed to show preferential viewing of the target face (Hannula et al. 2007). In patients with schizophrenia, reported in some studies to have modest hippocampal abnormalities, the usual preferential viewing effect was markedly reduced in size and delayed in latency (Williams et al. 2010).

In other studies, neuroimaging and eye tracking methods have been combined to investigate more thoroughly the reactivation of relational representations for the face-scene pairings. In an fMRI study (Hannula and Ranganath 2009), hippocampal activation elicited by the scene preview was observed during the delay between

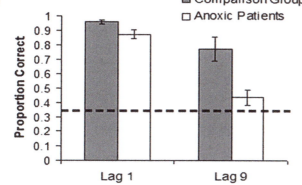

Fig. 3 Face-scene associations. (**a**) Three-face test displays superimposed on previously viewed scenes probed memory for face-scene relations, at either short (lag 1) or long (lag 9) lags. *Red boxes* are for illustrative purposes only. (**b**) Deficit in relational memory in amnesia. The proportion of correct relational memory judgments for the comparison group (*white bars*) and the amnesic patients (*gray bars*) are shown. SE bars are plotted around the means. The *dashed line* represents chance performance. Modified from Hannula et al. (2006)

scene preview and the onset of the superimposed 3-face test display, and was found to predict subsequent preferential viewing of the target face. This effect occurred whether or not the participant ended up making the correct behavioral choice of the target face. These findings suggest that relational representations of studied face-scene pairings are re-activated when previewing the scene, prior to the presentation of the test face; that the reactivation involves the engagement of the hippocampus; and that this relational memory effect can drive eye movements even without successful explicit remembering or conscious recollection. Finally, in a pair of imaging studies using the event-related optical signal (EROS) to investigate cortical re-activation of face-scene relational memory (Walker et al. 2014, 2016), increased activity was found in the face processing area STS (superior temporal sulcus) in response to the scene preview, prior to the presentation of the test face, for scenes that had been previously paired with faces as compared to novel scenes; and individual differences in the size of the relational memory effect in STS was significantly predicted by individual differences in the structural volume of the hippocampus. These findings provide further evidence of re-activation of relational representations of face-scene pairings prompted just by the scene, and of the involvement of the hippocampus in this effect. In addition, they show the engagement in relational memory reactivation of the cortical processor that is likely part of the relational memory itself.

Role of Hippocampus and Relational Representations Across Timescales

The previous sections described the role of the hippocampus in relational memory, showing its critical involvement in relational representations for all manner of arbitrary relations. We emphasize that whether the task involves the use of spatial, temporal, or associative relations, the need to bind, create, maintain, and make use of relational representations places demands on the hippocampus. In this and the next section we discuss the surprisingly long reach of hippocampal influence, showing its involvement in aspects of performance that challenge classical views of memory and its major systems.

We start with recent discoveries that have changed our understanding of the timescale over which the hippocampus acts in mediating relational memory. Rather than being exclusively the province of long-term memory, as has been the classical view, hippocampal involvement in relational memory is now known to occur for tasks on the timescale of short term or working memory, and even for moment-to-moment processing.

The classical distinction between short-term or working memory and long-term memory relied for some of its support on findings from patients with hippocampal amnesia, such as the patient H.M., who could readily track conversations and manage interactions in the immediate present but whose memory was grossly impaired with the imposition of significant delays. But, as findings began to accumulate regarding the critical role of the hippocampus in relational memory as opposed to item memory, we decided to revisit the status of short-term or working memory in amnesia, because the formal, experimental findings on this topic were from an earlier period of the history of the field, and they focused on memory for items rather than on relational memory.

Accordingly, we turned to testing relational memory at short timescales in patients with hippocampal amnesia, and found impairment in some eye movement measures of change detection between successive pairs of scenes involving relational changes (Ryan and Cohen 2004), as well as deficits in behavioral measures of memory for both relations among items in scenes and for arbitrary face-scene pairings, even when study and test trials were separated by only a few seconds and with no intervening items (Hannula et al. 2006). We concluded that the hippocampus contributes to online representations, and not just to long-term memory (Ryan and Cohen 2004), and that "characterizing the memory functions of the hippocampus may have less to do with any distinction between long-term and short-term (or working) memory than it has to do with the distinction between relational memory and memory for items" (Hannula et al. 2006).

As already noted, such findings were a significant departure from the classical view that hippocampus only contributes to long-term memory formation, but they conform with findings from neuroimaging studies that the hippocampus is activated even for certain tasks that require memory only over a short timescale, and there is now a considerable and growing number of studies showing evidence of impairment in patients with hippocampal damage in tasks that place a high demand on relational representations even when there are very short delays or no experimenter-imposed delay (e.g., see Barense et al. 2007; Cabeza et al. 2002; Hannula and Ranganath 2008; Henke 2010; Olsen et al. 2012; Öztekin et al., 2001; Ranganath and D'Esposito 2001; Rubin et al. 2011; Staresina and Davachi 2009; Warren et al. 2011; Watson et al. 2013; Yonelinas 2013). Perhaps the most surprising of these findings concerns the contribution of hippocampus and relational memory to the on-line processing of information. These findings and their implications are discussed in some detail in the next section, in the context of hippocampal influence on in-the-moment processing supporting aspects of social processing, active exploration and navigation of the environment, and creative thinking.

One last note here, before moving to online processing. Some of the neuropsychological findings implicating hippocampus over short delays involve only modest deficits in patients with hippocampal amnesia, compared to the very profound deficits seen over longer delays. But another study from our laboratory tells a different story (Watson et al. 2013). This study investigated the performance of hippocampal patients on a spatial reconstruction task in which participants were instructed to study a set of objects (such as a pen, button, toy car, etc.), varying in set size from 2 to 5, that were placed on a table, and then were to reconstruct the arrangement of objects after a brief delay of about 4 s (see Fig. 4a).

Previous studies involving spatial reconstruction tasks assessed performance by measuring change in position, or misplacement, of each of the objects in participants' reconstructions compared to the originally studied positions—an assessment that emphasizes memory for positions of items individually. But, given our view that the critical role of the hippocampus is in memory for the relations among items, even over short timescales, new measures were included of relational memory that proved to be particularly revealing. The critical finding came from a measure called "swap errors", reflecting the occurrence of errors in which the reconstructed positions of two or more objects were swapped with one other in the reconstruction, while maintaining the original x,y coordinates of filled positions. (A real-life example might be the following: if asked to reconstruct the seating positions of John and Mary at a table for 4, the error would be swapping the positions of John and Mary in the reconstruction, but placing them in seats that had in fact been occupied rather than mistakenly shifting their positions to seats that had been unoccupied). In one condition of the study, patients with hippocampal amnesia were 40 times more likely than comparison participants to make this swap error, even at a set size of only two items and a delay of only 4 s (see Fig. 4b); comparison participants almost never made swap errors at such small set sizes with such short delays. This finding underscores the need for relational memory representations to

Fig. 4 Spatial reconstruction. (**a**) Example display of study configuration with three objects, depicted here as a *star*, *triangle*, and *hexagon*. After a brief delay of a few seconds, participants were instructed to place the objects back in the same locations. (**b**) Disproportionately high swap errors in patients. The ratios of mean patient performance to mean comparison performance are provided for each of the five performance metrics. Error bars indicate SE, calculated by error propagation. Modified from Watson et al. (2013)

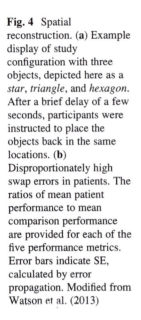

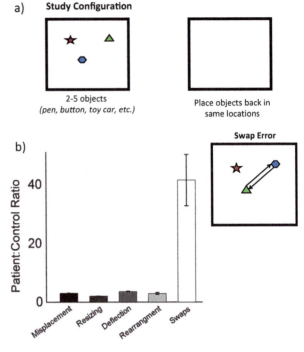

capture the arbitrary bindings of particular objects to particular positions, even on the timescale of short term or working memory.

Role of Hippocampus and Relational Representations Across Domains

Having shown that the influence of the hippocampus extends beyond the province of long-term memory, playing a critical role when relational representations are required for successful performance regardless of timescale, in this section we emphasize the contribution of the hippocampus and relational representations to performances in various domains of cognition and behavior that do not seem like memory tasks and that, as a result, seem to stretch the very definition of memory.

In our earliest writings on memory and the hippocampus we emphasized that the nature of the representations supported by the hippocampus makes them promiscuously accessible to various processing systems of the brain, and not just to the systems originally engaged in acquiring the information (Cohen 1984). We have noted, as well, that successful real-world behavior often relies upon actively acquiring and representing information about the environment and people, as well as manipulating and using those acquired representations to act optimally in and on the world around us (Rubin et al. 2014). If hippocampal representations are indeed promiscuously accessible to various brain processors, and if indeed the hippocampus contributes to performance whenever relational representations are required for optimal performance, whether in the real world or in the laboratory, then we should not be surprised to find any number of tasks outside of the usual province of official memory tests that show hippocampal (and relational memory) dependence. And indeed this has been a major focus of our research program over the last decade or more.

There are now numerous studies, from our laboratory and many others, that demonstrate the contribution of the hippocampus to performance not only on what are obviously memory tasks but also on tasks as diverse as navigation, exploration, imagination, creativity, decision-making, character judgments, establishing and maintaining social relationships, empathy, social discourse, and language use (as reviewed in Rubin et al. (2014) and described in more detail throughout Part III of this book). Here, we highlight just a small set of findings in patients with hippocampal amnesia on tasks of active exploration and navigation of the environment, character judgments, and creativity, as an illustration of the breadth of hippocampal contributions to flexible cognition and behavior and to the ability "to navigate life in all its beautiful complexity" (Cohen 2015).

Active Exploration and Navigation of the Environment

The role of the hippocampus in actively using memory to explore and navigate the environment has traditionally been appreciated in animal models of hippocampal function (e.g., Burgess et al. 2002; O'Keefe and Nadel 1978). Indeed, most animal studies are designed to require the active use of memory to navigate the spatial environment, imposing a much more realistic demand on the animal, whereas most human studies of memory instruct participants to passively view the to-be-remembered study materials. Thus, we wondered about how much more would be demanded of hippocampal function in humans when the task allows, indeed demands, the same use of active memory by the human participant. We have suggested the active use of memory is required to make adaptive cognitive and/or behavioral choices to navigate spatial and non-spatial environments alike, a process achieved through automatic, obligatory, and at times covert (unconscious), action-memory simulation, relying on interactions between the hippocampus and PFC to quickly provide multiple simulations of potential outcomes used to evaluate possible choices (Wang et al. 2015). Evidence in support of this proposal comes from both animal and human studies and provides an account of why hippocampus has been shown to make critical contributions to the short-term, adaptive control of behavior (e.g., Song et al. 2005; Voss et al. 2011a, b; Yee et al. 2014). Likewise, a number of studies indicate the involvement of hippocampus in human studies of spatial navigation, finding evidence from functional neuroimaging (Ghaem et al.

1997; Hartley et al. 2003; Kumaran and Maguire 2005; Maguire et al. 1997, 1998; Spiers and Maguire 2006), as well as hippocampal patients (Maguire et al. 2006).

In particular, we review a set of human studies that demonstrate the contribution of hippocampus to the active control of memory, revealing a benefit for active learning in healthy participants, but not in patients with hippocampal damage (Voss et al. 2011a, b). In these studies, participants studied an array of common objects arranged on a grid, viewing one object at a time through a small moving window. Participants studied the objects under two viewing conditions, one with self-initiated active control of the window position using a computer mouse or joystick (i.e., the volitional condition) and the other a passive condition. Importantly, the passive condition was the self-controlled, active movements of a previous particip-ipant, recorded and played back, as the passive condition for the next participant.

Subsequently, memory was tested for object identity on a recognition memory test and for object position on a spatial recall test. Across both measures, in healthy adults volitional control benefited memory performance relative to passive study (an effect that could not be attributed either to motor control or to facilitated perception). Furthermore, neuroimaging evidence from the same task linked the active control benefits to a brain network centered on the hippocampus, suggesting that volitional control optimizes interactions among specialized neural systems via the hippocampus (Voss et al. 2011a). But participants with hippocampal damage did not demonstrate a benefit from volition control (Voss et al. 2011a), nor did they exhibit and benefit from the "spontaneous revisitation" of recently seen objects during visual exploration (Voss et al. 2011b). Spontaneous revisitation, the back-and-forth viewing of previously studied objects, is interpreted as a series of advantageous choices concerning what to study to enhance performance on a later memory test, since in healthy adults the behavior correlates with better subsequent memory performance and covaries linearly with activation in hippo-campus and PFC (Voss et al. 2011b). Thus, this task ties hippocampus and rela-tional memory to the ability to bind and re-activate the constituent elements of experience, across and within episodes, automatically, obligatorily, and even covertly, to guide adaptive behavior in the moment.

Character Judgments

The ability to learn new information that is tied to a specific event or experience is a characteristic feature of hippocampal-dependent relational memory. Moreover, the information may be about a person, or ourselves, and thus contribute to the ability to form relationships with others, influence behavior towards others, and affect judg-ments and perceptions of others. Indeed, relational representations enable us to access multiple lines of associated information, often remote in time and space, and flexibly integrate the information with new experiences (Cohen and Eichenbaum 1993; Eichenbaum and Cohen 2001). Here, we describe the impact of relational memory on social interactions, notably in terms of character judgments, such that information about the way people have behaved in the past will influence the way we evaluate their character and expectations for future behavior.

One recent study of note investigated the contribution of the hippocampus to forming and updating character judgments on a task in which participants rated unfamiliar persons, before and after the presentation of scenarios in which the person was shown engaging in morally good, bad, or neutral behaviors (Croft et al. 2010). The ability to incorporate new information about the unfamiliar person and update their associated representation was measured by the change in the rating of the unfamiliar person from before the presentation of the scenario to after the presentation of the scenario. The study compared the performance of hippocampal patients to patients with damage to vmPFC (a brain region associated with processing emotional salience and moral information), as well as other brain damaged control participants (see Fig. 5). Patients with vmPFC damage demonstrated the least amount of change in character judgments, likely due to their impairment in processing emotional information. The hippocampal patients, on the other hand, demonstrated the greatest amount of change in character judgments, ultimately rating the unfamiliar person as either extremely good or extremely bad depending on the respective scenario to which they were exposed.

These findings suggest the hippocampus is important to establish an appropriate context in which to evaluate new information and make sensible character judgments, requiring the flexible binding of information across multiple experiences into an integrated representation. Without such ability, patients with hippocampal damage overvalue the present event and make more polarized judgments, as we had previously observed in a non-social context on the Iowa Gambling Task (Gupta et al. 2009). Furthermore, the findings are consistent with a growing literature that provides evidence for the contribution of the hippocampus to other aspects of social cognition and behavior (e.g., Beadle et al. 2013; Davidson et al. 2012; Duff et al.

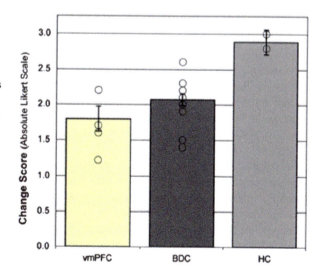

Fig. 5 Character judgments. Moral updating for valenced scenarios as a function of group. This figure shows the group changes in moral judgments (in absolute Likert scale units) for morally good and bad (valenced) scenarios. Group means represent adjusted values after taking into account the effects of the covariate. Individual raw data points are plotted as *open circles*. Error bars represent SEM. Reprinted from Croft et al. (2010)

2007, 2008a, b, 2009; Johnson et al. 1985; Tavares et al. 2015; Todorov and Olson 2008; Tranel and Damasio 1993), as well as those discussed in chapter "The Hippocampus and Social Cognition".

Creativity

Our last example concerns the influence of hippocampus in supporting creativity or creative thinking. Creativity is a capacity most often attributed to PFC, requiring the ability to rapidly combine and recombine existing mental representations (whether ones created long ago or recently) in order to generate novel ideas and ways of thinking (Damasio 2001; Bristol and Viskontas 2006). Yet, we have suggested that such tasks place a demand on hippocampus, both in terms of re-activating relational representations of previous experience as well as flexibly recombining their elements (critically dependent on the property of compositionality, which is fundamental to relational representations) to generate creative thought and behavior.

In a recent study, we evaluated the performance of patients with hippocampal damage on a well-validated, standardized measure of creativity, the Torrance Tests of Creative Thinking (TTCT) (Duff et al. 2013). The TTCT is comprised of both verbal and figural measures of creativity, in which a written prompt is provided and then participants have between 5 and 10 min to generate the most creative responses they can imagine, either by writing or drawing on the respective forms. An example from the verbal form consisted of participants using written language to generate creative uses for cardboard boxes during a 10 min period, whereas an example from the figural form consisted of participants being presented with a filled-in oval shape and asked to draw a picture, adding new ideas, to make the picture tell as interesting and exciting a story as possible, also within a 10 min period.

Both quantitative and qualitative differences between the patients with hippocampal damage and the matched comparison participants were stiking. On formalized scoring measures, in which a score of 100 is standardized to be average performance for that age group (as was observed in the comparison participants), patients with hippocampal damage scored significantly below average on both the verbal and figural forms, scoring several standard deviations worse than comparisons on the verbal form (see Fig. 6a). The patients' performance on the figural form was also notably less creative, with less richness of the picture constructions and fewer associated contextual details produced, than in comparison participants; this is exemplified in a side-by-side comparison of the responses from a matched comparison participant and a hippocampal patient in Fig. 6b. The deficit in creativity is consistent with the role of the hippocampus in representational flexibility and resonates with previous findings in patients in which hippocampal damage disrupted verbal play and the creative use of language in social interaction (Duff et al. 2009 and also see chapter "Hippocampal Contributions to Language Use and Processing").

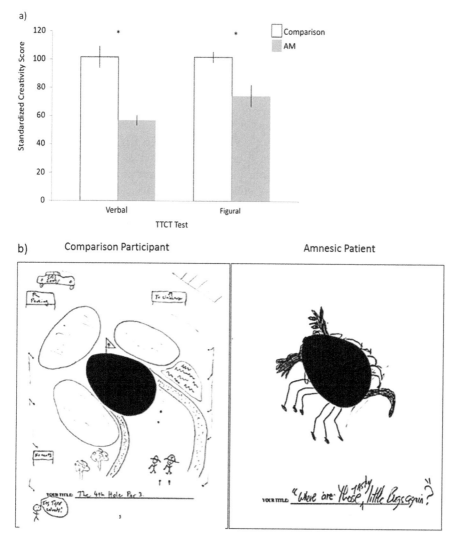

Fig. 6 Verbal and figural performance on creativity task. (**a**) Means and standard error; *asterisk* indicates significant (*p* < 0.05) differences between groups. TTCT Test = Torrance Tests of Creative Thinking. AM = amnesic participants. Comparison = Comparison participants. (**b**) Figural form example: picture construction from oval stimulus. *Left*: Comparison participant— title: The 4th Hole Par 3; notations read from *upper left* clockwise: To parking; To clubhouse; Its Tiger Woods!; No carts; *Right*: Amnesic patient—title: Where are those tasty little buggers? Modified from Duff et al. (2013)

Furthermore, these findings are consistent with a growing body of literature in which the hippocampus has been shown to contribute to imagination and mental depictions of possible future events (as described more in chapter "Physical Activity and Cognitive Training: Impact on Hippocampal Structure and Function").

Examples are found using neuroimaging methods in healthy participants (e.g., Addis et al. 2007; Addis and Schacter 2008; Buckner and Carroll 2007; Hassabis et al. 2007a; Schacter and Addis 2007, 2009; Schacter et al. 2007; Szpunar and McDermott 2008). As well, studies of hippocampal patients have shown their responses to be poorer in overall quality, more fragmented, and contain fewer episodic and semantic details than those of comparison participants (e.g., Hassabis et al. 2007b; Kwan et al. 2010; Race et al. 2013).

Taken together, these findings tie hippocampus and relational representations both to imagination and to creativity. We would suggest that the dependence of both of these classes of performance on the contribution of the hippocampus is related to the requirement for dynamically recombining previously formed compositional representations in service of creative new thoughts and behavior.

Clinical Studies

We would be remiss if we did not point out the debt owed to patient studies in providing insights about the aspects of memory supported by the hippocampus, and the influence of the hippocampus and relational memory on aspects of performance that challenges traditional views about memory. Work with patient H.M. and other hippocampal amnesic patients was central in developing the theoretical framework for understanding the nature of memory representations supported by the hippocampus and its associated cortical networks, a framework that has continued to be refined with contributions from other methodologies including advances in neuroimaging of intact human participants. These converging methods offer another level of precision to our understanding of and ability to measure hippocampal function.

However, it was the findings from patient studies of hippocampal amnesia, many of which were reviewed above, that have been critical in revealing the surprisingly long (and seemingly ever-growing!) reach of the hippocampus into aspects of cognitive and social performances that push the boundaries of what we take memory to be. Additionally, we note that for our lab's work on deficits in hippocampal amnesic patients for performances outside of the classic tests and domains of memory, we had specifically designed the tasks and predicted the deficits based on our theoretical commitment to the relational memory account of hippocampal function; hence they provide additional support for that account.

Even as "big data" is increasing touted as the future of medicine and clinical science, let us remember that many of the advances in our understanding of human learning and memory have clear roots in patient studies of hippocampal amnesia that examine a relatively small number of individuals with theory-driven measures. This approach to investigation has contributed to and will continue to drive robust advances in our field, even while it continues to benefit from insights of complementary methods as well as the development of ever more sensitive and specific measures of hippocampal function. The more we examine the role of hippocampus and relational memory in domains such as interactive language use, social behaviors, creativity, and active exploration and choice behavior, aspects vital to the human experience, the more important it is that we maintain a commitment to studying amnesic patients, and to do so in increasingly real-world contexts, in order to reveal other potentially "surprising" deficits in other non-obvious examples of memory in action.

There is also an argument to be made for using what has been learned from the patient studies to, in turn, inform cognitive and social therapies that aim to improve function in such patients. Given the number of diseases that present with some form of hippocampal abnormality or dysfunction, with structural changes and related functional impairments of hippocampus, manifesting to varying degrees in patients with neurological conditions such as traumatic brain injury and Alzheimer's disease, and with psychiatric conditions such as schizophrenia, post-traumatic stress disorder, depression, anxiety, and autism (Heckers et al. 1998; Nelson et al. 1998; Campbell and MacQueen 2004; Schumann et al. 2004; Shin et al. 2006; Etkin and Wager 2007), there is plenty of opportunity to make an impact.

For example, we are encouraged by observations that task demands that manipulate the degree to which relational representations are required determine the success vs failure of hippocampal amnesic patients to develop mutually shared knowledge with a communication partner (i.e., common ground) (Rubin and Cohen 2014). Thus, developing task procedures that minimize relational memory demands should be able to maximize the performance of which patients with compromised hippocampal systems may be capable, across all domains.

Conclusion

In this chapter, we set out to consider the contributions of the hippocampus to memory and to the many aspects of behavior it supports. The characteristic features of the hippocampus, namely its ability to form representations of arbitrary relations and to permit the flexible use of such representations, regardless of timescale, provide for an extraordinarily powerful system capable not just of encoding and retrieving enduring memories but of guiding behavioral choices and action in a complex world.

A tremendous corpus of empirical findings across methodologies, species, timescales, and domains of cognition and behavior converge in emphasizing the long reach of relational memory representations supported by hippocampal function, in service of a rich variety of behavioral repertoires, helping critically in guiding flexible and adaptive choices—uses of hippocampal representations that clearly stretch the classical definition of memory.

The findings discussed here also emphasize the role of the hippocampus in the larger network of structures important for executing such behaviors. Whereas previous accounts of these complex behaviors have emphasized the contribution of prefrontal cortex (PFC) and associated working memory capabilities, we would argue that it is the interconnection of the hippocampus with PFC, and other cortical systems, that supports the flexible use of information in general, relying on both old and recently formed relational representations (see Cohen and Eichenbaum 1993; Eichenbaum and Cohen 2001; Rubin et al. 2014).

References

Addis DR, Schacter DL (2008) Constructive episodic simulation: temporal distance and detail of past and future events modulate hippocampal engagement. Hippocampus 18:227–237

Addis DR, Wong AT, Schacter DL (2007) Remembering the past and imagining the future: common and distinct neural substrates during event construction and elaboration. Neuropsychologia 45:1363–1377

Azab M, Stark SM, Stark CE (2014) Contributions of the human hippocampal subfields to spatial and temporal pattern separation. Hippocampus 24:293–302

Barense MD, Gaffan D, Graham KS (2007) The human medial temporal lobe processes online representations of complex objects. Neuropsychologia 45:2963–2974

Bayley PJ, Squire LR (2005) Failure to acquire new semantic knowledge in patients with large medial temporal lobe lesions. Hippocampus 15:273–280

Beadle JN, Tranel D, Cohen NJ, Duff MC (2013) Empathy in hippocampal amnesia. Front Psychol 4:1–12

Bristol A, Viskontas I (2006) Dynamic processes within associative memory stores. In: Kaufman J, Baer J (eds) Creativity and reason in cognitive development. Cambridge University Press, Cambridge, pp 60–80

Buckner RL, Carroll DC (2007) Self-projection and the brain. Trends Cogn Sci 11:49–57

Burgess N, Maguire EA, O'Keefe J (2002) The human hippocampus and spatial and episodic memory. Neuron 35:625–641

Cabeza R, Dolcos F, Graham R, Nyberg L (2002) Similarities and differences in the neural correlates of episodic memory retrieval and working memory. Neuroimage 16:317–330

Campbell S, MacQueen G (2004) The role of the hippocampus in the pathophysiology of major depression. J Psychiatry Neurosci 29:417–426

Cohen NJ (1984) Preserved learning capacity in amnesia: evidence for multiple memory systems. In: Butters N, Squire LR (eds) Neuropsychology of memory. Guilford Press, San Diego, CA, pp 83–103

Cohen NJ (2015) Navigating life. Hippocampus 25:704–708

Cohen NJ, Eichenbaum H (1993) Memory, amnesia, and the hippocampal system. MIT Press, Cambridge, MA

Cohen NJ, Squire LR (1980) Preserved learning of pattern-analyzing skill in amnesia: dissociation of knowing how and knowing that. Science 210:207–210

Cohen NJ, Poldrack RA, Eichenbaum H (1997) Memory for items and memory for relations in the procedural/declarative memory framework. Memory 5:131–178

Cohen NJ, Ryan J, Hunt C, Romine L, Wszalek T, Nash C (1999) Hippocampal system and declarative (relational) memory: summarizing the data from functional neuroimaging studies. Hippocampus 9:83–98

Corkin S (1968) Acquisition of motor skill after bilateral medial temporal-lobe excision. Neuropsychologia 6:255–265

Crane J, Milner B (2005) What went where? Impaired object-location learning in patients with right hippocampal lesions. Hippocampus 15:216–231

Croft KE, Duff MC, Kovach CK, Anderson SW, Adolphs R, Tranel D (2010) Detestable or marvelous? Neuroanatomical correlates of character judgments. Neuropsychologia 48:1789–1801

Damasio H (2001) Neural basis of language disorders. In: Chapey R (ed) Language intervention strategies in aphasia and related neurogenic communication disorders, 4th edn. Lippincott William and Wilkins, Philadelphia, PA, pp 18–36

Davachi L (2006) Item, context and relational episodic encoding in humans. Curr Opin Neurobiol 16:693–700

Davachi L, Preston AR (2014) The medial temporal lobe and memory. In: Gazzaniga MS, Mangun GR (eds) The cognitive neurosciences, 5th edn. MIT Press, Cambridge, MA, pp 539–546

Davachi L, Wagner AD (2002) Hippocampal contributions to episodic encoding: insights from relational and item-based learning. J Neurophysiol 88(2):982–990

Davachi L, Mitchell JP, Wagner AD (2003) Multiple routes to memory: distinct medial temporal lobe processes build item and source memories. Proc Natl Acad Sci U S A 100(4):2157–2162

Davidson PSR, Drouin H, Kwan D, Moscovitch M, Rosenbaum RS (2012) Memory as social glue: close interpersonal relationships in amnesic patients. Front Psychol 3:531

Degonda N, Mondadori CRA, Brossard S, Schmidt, C.F., Besieger, P., Nitsch, R.M., Hock, C., & Henke, K. (2005). Implicit associative learning engages the hippocampus and interacts with explicit associative learning. Neuron 46(3):505-520.

Diana RA, Yonelinas AP, Ranganath C (2007) Imaging recollection and familiarity in the medial temporal lobe: a three-component model. Trends Cogn Sci 11(9):379–386

Diana RA, Yonelinas AP, Ranganath C (2008) High-resolution multi-voxel pattern analysis of category selectivity in the medial temporal lobes. Hippocampus 18(6):536–541

Duff MC, Hengst JA, Tranel D, Cohen NJ (2007) Talking across time: using reported speech as a communicative resource in amnesia. Aphasiology 21:702–716

Duff MC, Hengst JA, Tengshe C, Krema A, Tranel D, Cohen NJ (2008a) Hippocampal amnesia disrupts the flexible use of procedural discourse in social interaction. Aphasiology 22:866–880

Duff MC, Hengst JA, Tranel D, Cohen NJ (2008b) Collaborative discourse facilitates efficient communication and new learning in amnesia. Brain Lang 106:41–54

Duff MC, Hengst JA, Tranel D, Cohen NJ (2009) Hippocampal amnesia disrupts verbal play and the creative use of language in social interaction. Aphasiology 23:926–939

Duff MC, Kurczek J, Rubin R, Cohen NJ, Tranel D (2013) Hippocampal amnesia disrupts creative thinking. Hippocampus 23:1143–1149

Eichenbaum H (2004) Hippocampus: cognitive processes and neural representations that underlie declarative memory. Neuron 44:109–120

Eichenbaum H (2014) Time cells in the hippocampus: a new dimension for mapping memories. Nat Rev Neurosci 15:732–744

Eichenbaum H, Cohen NJ (1988) Representation in the hippocampus: what do hippocampal neurons code? Trends Neurosci 11(6):244–248

Eichenbaum H, Cohen NJ (2001) From conditioning to conscious recollection: memory systems of the brain. Oxford University Press, New York, NY

Eichenbaum H, Cohen NJ (2014) Can we reconcile the declarative memory and spatial navigation views on hippocampal function? Neuron 83:764–770

Eichenbaum H, Dudchenko P, Wood E, Shapiro M, Tanila H (1999) The hippocampus, memory, and place cells: is it spatial memory or a memory space? Neuron 23(2):209–226

Eichenbaum H, Otto T, Cohen NJ (1994) Two distinctions of hippocampal-dependent memory processing. Behav Brain Sci 17:449–517

Eichenbaum H, Yonelinas AP, Ranganath C (2007) The medial temporal lobe and recognition memory. Annu Rev Neurosci 30:123–152

Ekstrom AD, Bookheimer SY (2007) Spatial and temporal episodic memory retrieval recruit dissociable functional networks in the human brain. Learn Mem 14:645–654

Ekstrom AD, Copara MS, Isham EA, Wang W, Yonelinas AP (2011) Dissociable networks involved in spatial and temporal order source retrieval. Neuroimage 56:1803–1813

Epstein RA (2008) Parahippocampal and retrosplenial contributions to human spatial navigation. Trends Cogn Sci 12:388–396

Etkin A, Wager T (2007) Functional neuroimaging of anxiety: a meta-analysis of emotional processing in PTSD, social anxiety disorder and specific phobia. Am J Psychiatry 164:1476–1488

Farovik A, Dupont LM, Eichenbaum H (2010) Distinct roles for dorsal CA3 and CA1 in memory for nonspatial sequential events. Learn Mem 17:801–806

Fortin NJ, Agster KL, Eichenbaum H (2002) Critical role of the hippocampus in memory for sequences of events. Nat Neurosci 5:458–462

Ghaem O, Mellet E, Crivello F, Tzourio N, Mazoyer B, Berthoz A, Dennis M (1997) Mental navigation along memorized routes activates the hippocampus, precuneus and insula. Neuroreport 8:739–744

Gilbert PE, Kesner RP, Lee I (2001) Dissociating hippocampal subregions: double dissociation between dentate gyrus and CA1. Hippocampus 11:626–636

Giovanello KS, Verfaellie M, Keane MM (2003) Disproportionate deficit in associative recognition relative to item recognition in global amnesia. Cogn Affect Behav Neurosci 3:186–194

Giovanello KS, Schnyer DM, Verfaellie M (2004) A critical role for the anterior hippocampus in relational memory: evidence from an fMRI study comparing associative and item recognition. Hippocampus 14:5–8

Gold JJ, Smith CN, Bayley PJ, Shrager Y, Brewer JB, Stark CE, Hopkins RO, Squire LR (2006) Item memory, source memory, and the medial temporal lobe: concordant findings from fMRI and memory-impaired patients. Proc Natl Acad Sci U S A 103:9351–9356

Gupta R, Duff MC, Denburg NL, Cohen NJ, Bechara A, Tranel D (2009) Declarative memory is critical for sustained advantageous complex decision-making. Neuropsychologia 47:1686–1693

Hafting T, Fyhn M, Molden S, Moser M-B, Moser EI (2005) Microstructure of a spatial map in the entorhinal cortex. Nature 436:801–806

Hannula DE, Ranganath C (2008) Medial temporal lobe activity predicts successful relational memory binding. J Neurosci 28:116–124

Hannula DE, Ranganath C (2009) The eyes have it: hippocampal activity predicts expression of relational memory in eye movements. Neuron 63:592–599

Hannula DE, Tranel D, Cohen NJ (2006) The long and the short of it: relational memory impairments in amnesia, even at short lags. J Neurosci 26:8352–8359

Hannula DE, Ryan JD, Tranel D, Cohen NJ (2007) Rapid onset relational memory effects are evident in eye movement behavior, but not in hippocampal amnesia. J Cogn Neurosci 19:1690–1705

Hartley T, Maguire EA, Spiers HJ, Burgess N (2003) The well-worn route and the path less traveled: distinct neural bases of route following and way finding in humans. Neuron 37:877–888

Hassabis D, Kumaran D, Maguire EA (2007a) Using imagination to understand the neural basis of episodic memory. J Neurosci 27:14365–14374

Hassabis D, Kumaran D, Vann SD, Maguire EA (2007b) Patients with hippocampal amnesia cannot imagine new experiences. Proc Natl Acad Sci U S A 104:1726–1731

Heckers S, Rauch SL, Goff D, Savage CR, Schacter DL, Fischman AJ, Alpert NM (1998) Impaired recruitment of the hippocampus during conscious recollection in schizophrenia. Nat Neurosci 1:318–323

Henke K (2010) A model for memory systems based on processing modes rather than consciousness. Nat Rev Neurosci 11:523–532

Henke K, Buck A, Weber B, Wieser HG (1997) Human hippocampus establishes associations in memory. Hippocampus 7:249–256

Henke K, Mondadori CRA, Treyer V, Nitsch RM, Buck A, Hock C (2003) Nonconscious formation and reactivation of semantic associations by way of the medial temporal lobe. Neuropsychologia 41(10):863–876

Holdstock JS, Mayes AR, Roberts N, Cezayirli E, Isaac CL, O'Reilly RC (2002) Under what conditions is recognition spared relative to recall after selective hippocampal damage in humans. Hippocampus 12(3):341–351

Holdstock JS, Mayes AR, Gong Q, Roberts N, Kapur N (2005) Item recognition is less impaired than recall and associative recognition in a patient with selective hippocampal damage. Hippocampus 15:203–215

Howard M, MacDonald CJ, Tiganj Z, Shankar KH, Du Q, Hasselmo ME, Eichenbaum H (2014) A unified mathematical framework for coding time, space, and sequences in the medial temporal lobe. J Neurosci 34:4692–4707

Hsieh LT, Gruber MJ, Jenkins LJ, Ranganath C (2014) Hippocampal activity patterns carry information about objects in temporal context. Neuron 81:1165–1178

Johnson MK, Kim JK, Risse G (1985) Do alcoholic Korsakoff's syndrome patients acquire affective reactions? J Exp Psychol Learn Mem Cogn 11:22–36

Konkel A, Cohen NJ (2009) Relational memory and the hippocampus: representations and methods. Front Neurosci 3:166–174

Konkel A, Warren DE, Duff MC, Tranel DN, Cohen NJ (2008) Hippocampal amnesia impairs all manner of relational memory. Front Hum Neurosci 2:15

Kroll NEA, Knight RT, Metcalfe J, Wolf ES, Tulving E (1996) Cohesion failure as a source of memory illusions. J Mem Lang 35(10):176–196

Kumaran D, Maguire EA (2005) The human hippocampus: cognitive maps or relational memory? J Neurosci 25:7254–7259

Kumaran D, Maguire EA (2006) The dynamics of hippocampal activation during encoding of overlapping sequences. Neuron 49:617–629

Kwan D, Carson N, Addis DR, Rosenbaum RS (2010) Deficits in past remembering extend to future imagining in a case of developmental amnesia. Neuropsychologia 48:3179–3186

MacDonald CJ, Lepage KQ, Eden UT, Eichenbaum H (2011) Hippocampal "time cells" bridge the gap in memory for discontiguous events. Neuron 71:737–749

Maguire EA, Frackowiak RS, Frith CD (1997) Recalling routes around London: activation of the right hippocampus in taxi drivers. J Neurosci 17:7103–7110

Maguire EA, Burgess N, Donnett JG, Frackowiak RS, Frith CD, O'Keefe J (1998) Knowing where and getting there: a human navigation network. Science 280:921–924

Maguire EA, Nannery R, Spiers HJ (2006) Navigation around London by a taxi driver with bilateral hippocampal lesions. Brain 129:2894–2907

Manns JR, Howard M, Eichenbaum H (2007) Gradual changes in hippocampal activity support remembering the order of events. Neuron 56:530–540

McKenzie S, Frank AJ, Kinsky NR, Porter B, Riviere PD, Eichenbaum E (2014) Hippocampal representation of related and opposing memories develop with distinct, hierarchically organized neural schemas. Neuron 83:202–215

Milner B (1962) Les troubles de la memoire accompagnant des lesions hippocampiques bilaterales [memory problems with bilateral hippocampal lesions]. In Physiologie de L'hippocampe. Centre de la Recherche Scientifique, Paris, pp 257–272

Milner B, Corkin S, Teuber H-L (1968) Further analysis of the hippocampal amnesic syndrome: 14-year follow-up study of H.M. Neuropsychologia 6:215–234

Moser EI, Kropff E, Moser MB (2008) Place cells, grid cells, and the brain's spatial representation system. Neuroscience 31(1):69

Muller RU, Kubie JL, Ranck JB Jr (1987) Spatial firing patterns of hippocampal complex spike cells in a fixed environment. J Neurosci 7:1935–1950

Nelson MD, Saykin AJ, Flashman LA, Riordan HJ (1998) Hippocampal volume reduction in schizophrenia as assessed by magnetic resonance imaging: a meta-analytic study. Arch Gen Psychiatry 55:433–440

Norman KA, O'Reilly RC (2003) Modeling hippocampal and neocortical contributions to recognition memory: a complementary learning systems approach. Psychol Rev 110:611–646

O'Keefe J, Burgess N (1996) Geometric determinants of the place fields of hippocampal neurons. Nature 381(6581):425–428

O'Keefe J, Dostrovsky J (1971) The hippocampus as a spatial map: preliminary evidence from unit activity in the freely-moving rat. Brain Res 34(1):171–175

O'Keefe J, Nadel L (1978) The hippocampus as a cognitive map. Oxford University Press, New York, NY

Olsen RK, Moses SN, Riggs L, Ryan JD (2012) The hippocampus supports multiple cognitive processes through relational binding and comparison. Front Hum Neurosci 6:1–13

Pine DS, Grun J, Maguire EA, Burgess N, Zarahn E, Koda V, Fyer A, Szeszko PR, Bilder RM (2002) Neurodevelopmental aspects of spatial navigation: a virtual reality fMRI study. Neuroimage 15(10):396–406

Poldrack RA, Packard MG (2003) Competition among multiple memory systems: converging evidence from animal and human brain studies. Neuropsychologia 41:245–251

Prince SE, Daselaar SM, Cabeza R (2005) Neural correlates of relational memory: successful encoding and retrieval of semantic and perceptual associations. J Neurosci 25:1203–1210

Race E, Keane MM, Verfaellie M (2013) Losing sight of the future: impaired semantic prospection following medial temporal lobe lesions. Hippocampus 23:268–277

Ranganath C (2010) A unified framework for the functional organization of the medial temporal lobes and the phenomenology of episodic memory. Hippocampus 20:1263–1290

Ranganath C, D'Esposito M (2001) Medial temporal lobe activity associated with active maintenance of novel information. Neuron 31:865–873

Ranganath C, Yonelinas AP, Cohen MX, Dy CJ, Tom SM, D'Esposito M (2004) Dissociable correlates for recollection and familiarity within the medial temporal lobes. Neuropsychologia 42(1):2–13

Rubin RD, Cohen NJ (2014) Insights into hippocampal dependent declarative memory: recent findings and implications. SIG 2 Perspect Neurophysiol Neurogenic Speech Lang Disord 24:34–42

Rubin RD, Brown-Schmidt S, Duff MC, Tranel D, Cohen NJ (2011) How do I remember that I know you know that I know? Psychol Sci 22:1574–1582

Rubin RD, Watson PD, Duff MC, Cohen NJ (2014) The role of the hippocampus in flexible cognition and social behavior. Front Hum Neurosci 8:742

Ryan JD, Cohen NJ (2004) The nature of change detection and on-line representations of scenes. J Exp Psychol Hum Percept Perform 30:988–1015

Ryan JD, Althoff RR, Whitlow S, Cohen NJ (2000) Amnesia is a deficit in relational memory. Psychol Sci 11:454–461

Schacter DL (1987) Implicit memory: history and current status. J Exp Psychol Learn Mem Cogn 13:501–518

Schacter DL, Addis DR (2007) The cognitive neuroscience of constructive memory: remembering the past and imagining the future. Philos Trans R Soc Lond B Biol Sci 362:773–786

Schacter DL, Addis DR (2009) On the nature of medial temporal lobe contributions to the constructive simulation of future events. Philos Trans R Soc Lond B Biol Sci 364:1245–1253

Schacter DL, Tulving E (eds) (1994) Memory systems. MIT Press, Cambridge, MA, pp. 1–38

Schacter DL, Addis DR, Buckner RL (2007) Remembering the past to imagine the future: the prospective brain. Nat Rev Neurosci 8:657–661

Schapiro AC, Kustner LV, Turke-Browne NB (2012) Shaping of object representations in the human medial temporal lobe based on temporal regularities. Curr Biol 22:1622–1627

Schapiro AC, Gregory E, Landau B, McCloskey M, Turk-Browne NB (2014) The necessity of the medial temporal lobe for statistical learning. J Cogn Neurosci 26(8):1736–1747

Schendan HE, Searl MM, Melrose RJ, Stern CE (2003) An fMRI study of the role of the medial temporal lobe in implicit and explicit sequence learning. Neuron 37:1013–1025

Schumann CM, Hamstra J, Goodlin-Jones BL, Lotspeich LJ, Kwon H, Buonocore MH, Lammers CR, Reiss AL, D.G. A (2004) The amygdala is enlarged in children but not adolescents with autism; the hippocampus is enlarged at all ages. J Neurosci 24:6392–6401

Scoville WB, Milner B (1957) Loss of recent memory after bilateral hippocampal lesions. J Neurol Neurosurg Psychiatry 20:11–21

Shapiro ML, Tanila H, Eichenbaum H (1997) Cues that hippocampal place cells encode: dynamic and hierarchical representation of local and distal stimuli. Hippocampus 7:624–642

Shin LM, Rauch SL, Pitman RK (2006) Amygdala, medial prefrontal cortex and hippocampal function in PTSD. Ann N Y Acad Sci 1071:67–79

Song EY, Kim YB, Kim YH, Jung MW (2005) Role of active movement in place-specific firing of hippocampal neurons. Hippocampus 15:8–17

Spiers HJ, Maguire EA (2006) Thoughts, behaviour and brain dynamics during navigation in the real world. Neuroimage 31:1826–1840

Spiers HJ, Maguire EA (2007) A navigational guidance system in the human brain. Hippocampus 17:618–626

Spiers HJ, Maguire EA (2008) The dynamic nature of cognition during wayfinding. J Environ Psychol 28(3):232–249

Squire LR (2004) Memory systems of the brain: a brief history and current perspective. Neurobiol Learn Mem 82:171–177

Staresina BP, Davachi L (2006) Differential encoding mechanisms for subsequent associative recognition and free recall. J Neurosci 26:9162–9172

Staresina BP, Davachi L (2009) Mind the gap: binding experiences across space and time in the human hippocampus. Neuron 63:267–276

Stark CE, Squire LR (2003) Hippocampal damage equally impairs memory for single items and memory for conjunctions. Hippocampus 13:281–292

Szpunar KK, McDermott KB (2008) Episodic future thought and its relation to remembering: evidence from ratings of subjective experience. Conscious Cogn 17:330–334

Tavares RM, Mendelsohn A, Grossman Y, Williams CH, Shapiro M, Trope Y, Schiller D (2015) A map for social navigation in the human brain. Neuron 87(1):231–243

Todorov A, Olson IR (2008) Robust learning of affective trait associations with faces when the hippocampus is damaged, but not when the amygdala and temporal pole are damaged. Soc Cogn Affect Neurosci 3:195–203

Tranel D, Damasio AR (1993) The covert of affective valence does not require structures in hippocampal system or amygdala. J Cogn Neurosci 5:79–88

Tranel D, Damasio AR, Damasio H, Brandt JP (1994) Sensorimotor skill learning in amnesia: additional evidence for the neural basis of nondeclarative memory. Learn Mem 1:165–179

Tulving E, Schacter DL (1990) Priming and human memory systems. Science 247:301–306

Turriziani P, Fadda L, Caltagirone C, Carlesimo GA (2004) Recognition memory for single items and for associations in amnesic patients. Neuropsychologia 42:426–433

Voss JL, Gonsalves BD, Federmeier KD, Tranel D, Cohen NJ (2011a) Hippocampal brain-network coordination during volitional exploratory behavior enhances learning. Nat Neurosci 14:115–120

Voss JL, Warren DE, Gonsalves BD, Federmeier KD, Tranel D, Cohen NJ (2011b) Spontaneous revisitation during visual exploration as a link among strategic behavior, learning and the hippocampus. Proc Natl Acad Sci U S A 108:E402–E409

Walker JA, Low KA, Cohen NJ, Fabiani M, Gratton G (2014) When memory leads the brain to take scenes at face value: face areas are reactivated at test by scenes that were paired with faces at study. Front Hum Neurosci 8:18

Walker JA, Low KA, Fletcher MA, Cohen NJ, Gratton G, Fabiani M (2016) Hippocampal structure predicts cortical indices of reactivation of related items. Neuropsychologia. Advance online publication. doi:10.1016/j.cortex.2016.09.015

Wang WC, Dew ITZ, Cabeza R (2015) Age-related differences in medial temporal lobe involvement during conceptual fluency. Brain Res 1612:48–58

Warren DE, Duff MC, Tranel D, Cohen NJ (2011) Observing degradation of visual representations over short intervals when medial temporal lobe is damaged. J Cogn Neurosci 23 (12):3862–3873

Watson PD, Voss JL, Warren DE, Tranel D, Cohen NJ (2013) Spatial reconstruction by patients with hippocampal damage is dominated by relational memory errors. Hippocampus 23:570–580

Williams LE, Must A, Avery S, Woolard A, Woodward ND, Cohen NJ, Heckers S (2010) Eye-movement behavior reveals relational memory impairment in schizophrenia. Biol Psychiatry 68:617–624

Yee LTS, Warren DE, Voss JL, Duff MC, Tranel D, Cohen NJ (2014) The hippocampus uses information just encountered to guide efficient ongoing behavior. Hippocampus 24:154–164

Yonelinas AP (2013) The hippocampus supports high-resolution binding in the service of perception, working memory and long-term memory. Behav Brain Res 254:34–44

Zhang H, Ekstrom AD (2013) Human neural systems underlying rigid and flexible forms of allocentric spatial representation. Hum Brain Mapp 34:1070–1087

Part III
Beyond Memory: Contributions to Flexible Cognition

How Hippocampal Memory Shapes, and Is Shaped by, Attention

Mariam Aly and Nicholas B. Turk-Browne

Abstract Attention has historically been studied in the context of sensory systems, with the aim of understanding how information in the environment affects the deployment of attention and how attention in turn affects the perception of this information. More recently, there has been burgeoning interest in how long-term memory can serve as a cue for attention, and ways in which attention influences long-term memory encoding and retrieval. In this chapter, we highlight this emerging body of human behavioral, neuroimaging, and neuropsychological work that elucidates these bidirectional interactions between attention and memory. Special emphasis will be given to recent findings on how the quintessential "memory system" in the brain—the hippocampus—influences and is influenced by attention.

Introduction

At one time or another, we have all puzzled over why some things are easily remembered and others are frustratingly forgotten. This question is not just one of casual introspection, but also one that has intrigued and stumped cognitive neuroscientists for decades. Studies of memory behavior have long established that the way we direct our attention strongly determines what we encode into memory. Yet, how attention influences mnemonic processes in the brain has only been investigated more recently. In fact, despite its clear importance for the encoding of new memories, research on how attention modulates the hippocampus is only just beginning. These efforts have coincided with growing interest in how

M. Aly (✉)
Princeton Neuroscience Institute, Princeton University, Peretsman-Scully Hall 320, Princeton, NJ 08544, USA
e-mail: aly@princeton.edu

N.B. Turk-Browne
Princeton Neuroscience Institute, Princeton University, Peretsman-Scully Hall 320, Princeton, NJ 08544, USA

Department of Psychology, Princeton University, Princeton, NJ 08544, USA

© Springer International Publishing AG 2017
D.E. Hannula, M.C. Duff (eds.), *The Hippocampus from Cells to Systems*,
DOI 10.1007/978-3-319-50406-3_12

369

memory in turn influences attention. This work has led to the exciting discovery that hippocampal memories can have powerful effects on how we move our eyes and orient our attention. That hippocampal representations can influence attentional processing in this way provides a compelling demonstration of the reach of the hippocampus beyond explicit memory. In this chapter, we provide a review of these bidirectional interactions between hippocampal memory and attention. In the first section, we discuss how attention affects memory encoding and retrieval at the behavioral and neural levels, and how attention modulates the hippocampus in the absence of demands on long-term memory. Then, we turn to how hippocampal memories guide attentional allocation and eye movements during visual exploration, highlighting the influence of both explicit and implicit long- and short-term memories. We end by discussing future directions for research on the interplay between attention and memory, including studies of network connectivity, neuropsychology, neurofeedback, and neuromodulation.

How Does Attention Influence Hippocampal Memory?

Behavioral Studies

Memory comes in different forms. Imagine someone says "hi!" to you in a local coffee shop, and you subsequently try to remember if you've seen this person before. You can make that decision based on different types of information. In some cases, you may be able to *recollect* specific, qualitative details about who this person is or when you last saw them—e.g., that this person is your new neighbor, who you met last week. Alternatively, you may be unable to bring to mind details about who the person is, but they nevertheless seem *familiar*—you have seen them somewhere before, but you do not remember where or when. These different types of memory differentially tax hippocampal processing: recollection, but not familiarity, is critically dependent on the hippocampus (for review, see Yonelinas et al. 2010).

A rich body of literature on behavioral expressions of memory has shown that dividing attention impairs performance primarily on those types of memory that are dependent on the hippocampus, such as recollection (Chun and Turk-Browne 2007). For example, dividing attention during encoding—by having participants make judgments on the pitch of auditory tones while encoding a list of visually presented words—impairs subsequent memory judgments made on the basis of episodic recollection, but not memory based on a general feeling of familiarity (Gardiner and Parkin 1990). Subsequent research confirmed that divided attention at encoding produces large impairments in hippocampally-mediated forms of memory (e.g., Craikm et al. 1996; Fernandes and Moscovitch 2000; for reviews, see Craik 2001; Yonelinas 2002).

Although divided attention at encoding impairs memory, early studies suggested that divided attention at retrieval is less detrimental (Craik et al. 1996; Craik 2001). Later studies, however, found that divided attention does interfere with memory retrieval when the concurrent task depends on the same representations (e.g., verbal distracting task and verbal memory retrieval; Fernandes and Moscovitch 2000). Moreover, a review of the literature suggested that divided attention at retrieval produces impairments in recollection-based, but not familiarity-based, memory (Yonelinas 2002). Indeed, the mere presence of task-irrelevant, distracting information can impair episodic memory (Wais et al. 2010), even when the distracting information is in a different sensory modality (Wais and Gazzaley 2011).

In contrast to the extensive literature on divided attention and memory, relatively little work has been done on how *selective* attention influences memory. In divided attention studies, attention is split between the memory task and an unrelated secondary task, both of which must be performed concurrently. In selective attention studies, attention must be used to select one stimulus for further processing amongst other stimuli that need to be ignored. An early example is the dichotic listening paradigm (Cherry 1953), in which participants verbally shadowed one of two auditory streams, each presented to one ear. Participants had essentially no memory for information in the unattended auditory channel (Moray 1959), showing that selective attention strongly gates what is encoded into memory. Selective attention can also apply to different representations of the same stimulus—for example, the meaning versus sound of words. Studies that encourage participants to direct attention selectively to one characteristic of a stimulus while ignoring others have found effects on memory: When the task at retrieval orients participants to the sound of words, memory is better when sound was attended during encoding; in contrast, attention to the meaning of words during encoding produces better memory in a standard recognition task, which is assumed to rely on word meaning (Morris et al. 1977).

Selective attention is especially important when stimuli are in strong competition with one another. For example, with composite stimuli that consist of superimposed faces and scenes, participants show above-chance memory only when the tested aspect of the composite stimulus (e.g., a scene) was selectively attended during encoding (Yi and Chun 2005). Finally, memory is superior when to-be-encoded objects appear in spatial locations at which attention has been selectively directed, compared to unexpected or neutral locations (Turk-Browne et al. 2013; Uncapher et al. 2011).

Attentional Modulation of the Hippocampus

Despite the abundant evidence that attention influences behavioral expressions of episodic memory (also see Hardt and Nadel 2009), how this modulation occurs in the brain is only just starting to be understood (Posner and Rothbart 2014). There are at least two potential routes by which attention might modulate memory. The first, and perhaps prevailing, view is that attention modulates hippocampal memory as a downstream consequence of its effects on sensory representations. According to biased competition models (Desimone 1996), information that is selected by attention is more robustly represented in sensory systems, and thus fares better in competition with unattended information for downstream processing. This biased competition is often manifest as higher levels of activity in visual regions that code for attended features or locations, or sharper, more precise representations of attended information (Gilbert and Li 2013; Kastner and Ungerleider 2000; Maunsell and Treue 2006; Sprague et al. 2015). Thus, according to this framework, strengthened sensory representations are more likely to be transmitted downstream to the hippocampus for further processing, either as items to be encoded or as retrieval cues for existing memories.

A different potential route is that attention directly modulates the hippocampus itself. However, in contrast to the robust effects of attention on sensory cortex, there has been little evidence of attentional modulation in the hippocampus in tasks with no overt demands on long-term memory. Indeed, studies that have manipulated attention to locations (Yamaguchi et al. 2004) and stimulus categories (Dudukovic et al. 2010) while participants underwent functional neuroimaging have failed to observe attentional modulation of the hippocampus. Instead of concluding that there are no direct effects of attention on the hippocampus outside of memory tasks, we recently suggested that these effects exist but were missed in prior studies because of how attention was manipulated and measured (Aly and Turk-Browne 2016a).

The traditional way of studying neural effects of attention is to manipulate attention to relatively simple features or locations, and to measure the effects on the representation of those features or locations in the brain (Kastner and Ungerleider 2000; Maunsell and Treue 2006). For example, participants might be cued to pay attention to the left or right side of fixation, while neuroimaging is used to measure brain activity in areas of early visual cortex that respond selectively to the left or right side of space. Such an approach is sufficient for studying sensory cortex but may be inadequate for studying the hippocampus, whose representations are fundamentally relational and contextual, consisting of (often multimodal) associations between items and the spatial and temporal contexts in which they occur (Brown and Aggleton 2001; Bussey and Saksida 2005; Cohen and Eichenbaum 1993; Davachi 2006; Graham et al. 2010; Ranganath 2010; Yonelinas 2013). Thus, in order to study attentional modulation of the hippocampus, one might have to study the types of relational information that it represents, rather than simple features or locations.

The signature of attention may also be different in cortex vs. hippocampus. In sensory areas, the primary measure of attentional modulation has been the overall level of activity, whether measured with single-cell recordings in animals or functional neuroimaging in humans (Gilbert and Li 2013; Kastner and Ungerleider 2000; Maunsell and Treue 2006). In the hippocampus, however, attentional effects may more strongly manifest as changes in *representational stability*. That is, attention may modulate the reliability of activity patterns in the hippocampus, as opposed to the overall strength of a scalar signal (cf. Dudukovic et al. 2010; Yamaguchi et al. 2004). This would produce distinct patterns of activity for different attentional states: Distributed patterns of activity in the hippocampus would be more similar to each other (or more stable) across multiple instances of the same attentional state than across different attentional states. Evidence in support of this hypothesis came first from animal studies (Fenton et al. 2010; Jackson and Redish 2007; Kelemen and Fenton 2010; Kentros et al. 2004; Muzzio et al. 2009b; for reviews, see Muzzio et al. 2009a; Rowland and Kentros 2008) and was subsequently observed by us in functional neuroimaging studies in humans (Aly and Turk-Browne 2016a, b).

In animal models, representational stability is realized as a change in the reliability of cell firing in the hippocampus as a function of the task relevance of particular aspects of the environment. For example, place cells in the hippocampus—which fire when an animal is in a particular location (Ekstrom et al. 2003; O'Keefe and Dostrovsky 1971)—show increases in the reliability of firing as the task-relevance of spatial cues increases (Kentros et al. 2004; Muzzio et al. 2009b). Conversely, hippocampal cells that respond to odor fire more reliably when olfactory information is task-relevant (Muzzio et al. 2009b). Such representational stability is also observed at the level of networks of cells—for example, different cell assembles consistently activate for different spatial reference frames (Jackson and Redish 2007; Kelemen and Fenton 2010; see also Fenton et al. 2010). Insofar as the environmental cues that animals are orienting to influence, or reflect, their attentional state, this work suggests that attention-like processes may modulate hippocampal representational stability.

Inspired by this work in animal models, we used high-resolution functional MRI to explore the idea of representational stability in human hippocampus—i.e., the notion that attention creates stable and distinct patterns of activity for different attentional states (Fig. 1; Aly and Turk-Browne 2016a). Keeping with the intuition that attentional modulation of the hippocampus might be observed only if attention is oriented to relational information, we designed a novel "art gallery" task in which participants were cued to attend to high-level relations. The stimuli consisted of 3D-rendered rooms, each with a unique configuration of walls and furniture, and a single painting. On each trial, participants were cued to attend either to the paintings (art state) or to the layout of the rooms (room state), as they viewed a stream of rooms with art. On art-state trials, they were to attend to the artistic style of the paintings, in order to identify paintings that could have been painted by the same artist. These paintings were similar in style (e.g., use of color, brushstroke, detail) but not necessarily content. On room-state trials, participants were to attend to the furniture and wall arrangements, in order to identify rooms with the same spatial layout from a different perspective. These rooms had the same configuration of walls and furniture, but different wall colors and furniture exemplars (e.g., a chair would be swapped for a different chair). At the end of the trial, participants had to respond "yes" or "no" as to whether they had found a match. Thus, these tasks emphasized higher-order relations—of abstract artistic style and spatial geometry, respectively—rather than individual features. Importantly, the same stimuli were

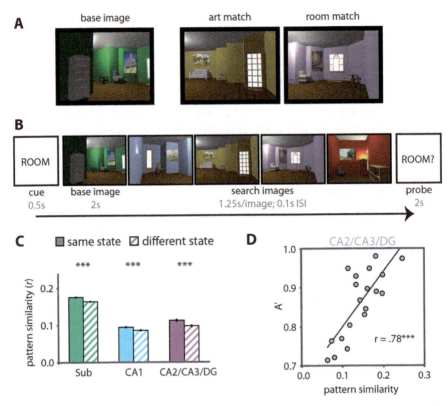

Fig. 1 Attentional modulation of the hippocampus. Attention induces representational stability in the hippocampus, with distinct patterns of activity for different attentional states. (**a**) Stimuli were rooms with a unique layout of walls and furniture and a single painting. For any given "base image", an "art match" was a room containing a painting that was painted by the same artist as the painting in the base image, and a "room match" was a room with the same spatial layout as the base image but viewed from a different perspective. (**b**) On each trial, participants were cued to attend either to the art or to the room. They then viewed a base image followed by a search set of four images. On art trials, participants had to examine the search set for an art match to the base image; on room trials, they looked for a room match to the base image. Finally, they were probed as to whether they found a match, and had to respond yes or no. (**c**) Each hippocampal subfield showed attentional state representations: activity patterns across voxels were more highly correlated for trials of the same (i.e., art/art and room/room) vs. different (i.e., art/room) state. (**d**) Individual differences in the stability of activity patterns in the room attentional state correlated with attentional behavior in the room task. This correlation was selective to the CA2/CA3/DG region of interest, and not observed anywhere else in the brain. ***$p < .001$. Figure adapted from Aly and Turk-Browne (2016a)

used in both tasks, allowing us to isolate the neural effects of top-down attention from those related to bottom-up stimulation.

Consistent with the representational stability hypothesis, we found that attention induced distinct and reliable activity patterns for the two attentional states in each hippocampal subfield: Activity patterns in each hippocampal subfield were more highly correlated for trials of the same attentional state (i.e., art/art and room/room), compared to trials of different states (i.e., art/room). Such representational stability may reflect enhanced processing of the information that is relevant in each state. That is, distinct activity patterns for different attentional states may be a result of prioritizing those hippocampal representations that are necessary for goal-directed behavior in the current attentional state. This prioritization may in turn have consequences for attentional behavior as well as the encoding of goal-relevant information into long-term memory (see section "Attentional Modulation of Hippocampal Encoding").

Indeed, in one hippocampal subfield—comprising subfields CA2/3 and dentate gyrus—individual differences in representational stability for the room state were correlated with attentional behavior on the room task, highlighting the behavioral relevance of attentional states in the hippocampus for online task performance. This brain/behavior correlation was highly selective—it was not found in any other hippocampal subfield, medial temporal lobe cortical region, or anywhere else in the brain. Insofar as attention modulates what we remember, and memory encoding has been linked to CA2/3 and dentate gyrus (e.g., Eldridge et al. 2005; Suthana et al. 2011, 2015; Wolosin et al. 2013; Zeineh et al. 2003), this finding suggests that these subfields may mediate the effect of attention on memory via the creation of state-dependent activity patterns that prioritize goal-relevant information.

We also found that modulation of representational stability was dissociable from modulation of overall activity levels in the hippocampus in a number of ways: For example, overall activity was not correlated with behavior, and voxels with both high activity and low activity contributed to the stability of activity patterns in the hippocampus. Attention also had distinct effects on the hippocampus and medial temporal lobe cortex: Modulation of representational stability in medial temporal lobe cortex was in part due to increases in overall activity. Thus, cortical state-dependent "patterns" differed from those in hippocampus, where a balance of activation and deactivation together produced representational stability. Also, as mentioned above, only attentional modulation of the hippocampus predicted behavior.

These findings provide initial evidence that attention can modulate representational stability in the human hippocampus, and in a way that is relevant for attention behavior. They also suggest that modulation of representational stability might be a means by which attention enhances hippocampally-mediated memory (see section "Attentional Modulation of Hippocampal Encoding").

Attentional Modulation of Hippocampal Encoding

In contrast to the relatively small body of work on attentional modulation in the hippocampus without overt demands on long-term memory, several studies have investigated how attention modulates hippocampal signals related to long-term memory encoding. The dominant signal of interest has been the overall level of activity during encoding, as a function of memory performance on a later test. A *subsequent memory effect* is observed if differential activity at encoding is observed for subsequently remembered vs. forgotten information (Brewer et al. 1998; Wagner et al. 1998). Thus, these studies examine whether univariate subsequent memory effects in the hippocampus are modulated by attention at encoding.

The findings from these studies are mixed: many, but not all, find evidence of attentional modulation of hippocampal encoding. At least some of the null effects might arise from the use of paradigms and methods that are not ideally suited for detecting modulation of hippocampal subsequent memory effects. For example, one early study found no difference in hippocampal activity for full vs. divided attention at encoding (Iidaka et al. 2000). However, this was a PET study, and the slow temporal resolution of this method does not allow isolation of brain activity associated with encoding of individual items that are subsequently remembered vs. forgotten. Indeed, later studies utilizing fMRI—which allows measurement of brain activity related to the processing of individual items—found that divided attention during encoding reduced hippocampal subsequent memory effects (Kensinger et al. 2003; Uncapher and Rugg 2008). Methodological considerations alone do not account for all discrepancies in the literature. For example, an easy vs. hard secondary task at encoding did not modulate hippocampal subsequent memory effects in an event-related fMRI study (Uncapher and Rugg 2005). Another line of work has manipulated the level (or type) of attention by having participants encode items with either a deep (e.g., semantic) or shallow (e.g., phonological) task. Again, data are inconsistent, with some (Otten et al. 2001; Strange and Dolan 2001) but not all (Fletcher et al. 2003; Schott et al. 2013) findings suggesting that hippocampal encoding is sensitive to the attentional depth of processing.

Other studies have more precisely manipulated selective attention at encoding, and have generally observed attentional modulation of hippocampal memory signals. For example, hippocampal activity predicts subsequent memory for words encoded in a relational manner (i.e., when encoding required the formation of inter-item associations), but not those encoded in an item-based manner (Davachi and Wagner 2002; also see Henke et al. 1997, 1999). Moreover, when attention is oriented to one of two contextual pieces of information at encoding—either the location of an object or the color surrounding it—hippocampal activity predicts subsequent memory for the attended, but not the unattended, contextual information (Fig. 2; Uncapher and Rugg 2009).

Selective spatial attention also modulates hippocampal encoding. For example, the hippocampus shows subsequent memory effects for objects that appear in expected, but not unexpected, locations (Uncapher et al. 2011). Moreover, a recent study found that hippocampal subfields CA1 and subiculum showed an interaction between attention at encoding and subsequent memory: Subsequent memory effects were found when participants attended to the distinctiveness of faces at encoding, but not when they attended to their similarities (Carr et al. 2013). In contrast, a combined region of interest for subfields CA2/3 and dentate gyrus showed subsequent memory effects that were comparable for both tasks. These data suggest that

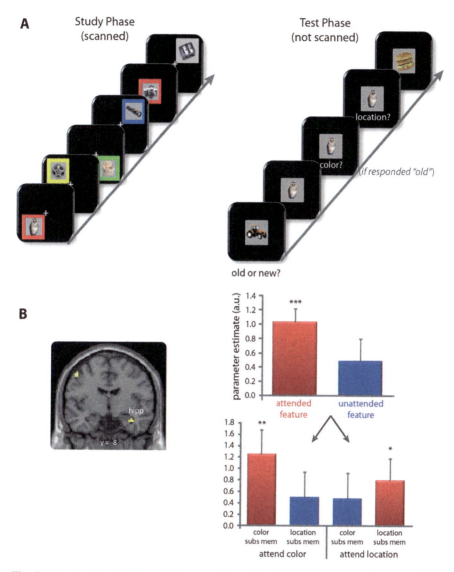

Fig. 2 Attentional modulation of hippocampal encoding. The hippocampus shows a reliable subsequent memory effect only for contextual information that was selectively attended at encoding. (**a**) During encoding, participants attended either to the location of objects on the screen or the color of the border surrounding them. Memory was then tested for the items they encoded as well as the attended and unattended contextual information. (**b**) The overall level of hippocampal activity at encoding was examined based on whether color or location was attended and whether color or location was subsequently remembered. Hippocampal activity at encoding predicted memory (i.e., showed a subsequent memory effect) for attended, but not unattended, contextual information. $*p < .05$, $**p < .005$, $***p < .0001$. Figure adapted from Uncapher and Rugg (2009)

the mnemonic benefit conferred to distinctive items might arise because of the robust recruitment of the entire hippocampal system.

Thus, studies that manipulate selective attention more consistently find effects on hippocampal encoding than those that divide attention or otherwise manipulate processing resources (e.g., the depth of processing). Why might this be the case? The hippocampus may obligatorily encode information that is consciously apprehended, that is, information in the focus of attention (Moscovitch 2008; Moscovitch et al. 2016). Divided attention manipulations reduce the amount of attention directed toward to-be-remembered information, but may not reduce attention enough to interfere consistently with automatic hippocampal encoding. On the other hand, selective attention manipulations entail processing one aspect of the environment while filtering out others, and this ignored information may not reach the threshold for conscious apprehension necessary for hippocampal encoding.

There is also evidence that voluntary control over attention at encoding confers benefits to memory, and that this effect is dependent on the hippocampus (Voss et al. 2010). Participants memorized objects arranged in a grid by moving a window around the screen that enabled them to view one object at a time, while the rest were obscured. In one condition, participants had volitional control over the movement of the window; in the other condition, they passively viewed the movements made by another participant. Thus, pairs of participants viewed the same objects in the same order, but for one learning phase, they had control over the order in which the objects were viewed; in the other learning phase, they did not. Volitional control over the trajectory of attention during encoding conferred benefits to subsequent memory for the objects as well as their spatial locations. Moreover, hippocampal activity was elevated during volitional vs. passive encoding, and patients with hippocampal damage failed to show the mnemonic benefits of volitional attention. Thus, control over the trajectory of attention is beneficial to memory encoding, and this effect requires the hippocampus.

Together, these findings largely suggest that univariate measures of hippocampal encoding are modulated by attention. However, the reason for inconsistent effects needs to be explored in future studies. One possibility is that in order to observe attentional modulation of hippocampal encoding, attention must be focused on relational information, which is a key component of hippocampal processing (Cohen and Eichenbaum 1993). Indeed, studies that manipulate attention to different types of relations (e.g., Carr et al. 2013; Uncapher and Rugg 2009) or compare relational and item-based processing (Davachi and Wagner 2002; Henke et al. 1997, 1999; cf. Uncapher and Rugg 2006), consistently find effects of attention on hippocampal encoding.

Another possibility is that attentional modulation of hippocampal memory encoding may be more robustly observed when representational stability, rather than the level of activity, is the dependent variable (see Aly and Turk-Browne 2016a and section "Attentional Modulation of the Hippocampus"). Support for this hypothesis comes from a place cell study in rodents, which measured both the rate of firing of place cells as well as the stability of their spatial firing patterns (measured as the correlation between firing rate maps in sequential sessions, where a firing rate map indicates where and how highly a cell fired in a spatial environment). Place fields were more stable when mice engaged in a task that put heavy demands on spatial processing, and this stability correlated with spatial memory (Kentros et al. 2004). In contrast, there were no differences in overall place cell firing rates for tasks that involved high vs. low demands on spatial processing. Moreover, another study found that when rats engaged in "attentive scanning" of a particular environmental location, a place field subsequently formed at that location on the very next pass through it (Monaco et al. 2014), an effect reminiscent of single-shot encoding of a new episodic memory for attended information. Spatial attention therefore modulates the formation and stability of spatial representations in the rodent hippocampus, and predicts the formation and retention of spatial memories.

We recently investigated how hippocampal representational stability during encoding influences episodic memory formation in humans (Aly and Turk-Browne 2016b). Inspired by the rodent studies mentioned above, and our own work showing that attention modulates representational stability in human hippocampus (Aly and Turk-Browne 2016a), we predicted that goal-relevant information would be more likely encoded into long-term memory if the attentional state of the hippocampus during encoding prioritized that type of information. That is, given a particular behavioral goal, attention should serve to focus hippocampal processing on goal-relevant aspects of the environment; to the extent that the pattern of activity in the hippocampus is indicative of being in the goal-relevant attentional state, information pertaining to those goals should be prioritized with respect to online processing as well as transformation into a durable long-term memory.

As in our previous study, we also explored the roles of different hippocampal subfields. We predicted that the attentional state of CA2/3 and dentate gyrus should be most closely linked to successful memory formation, based on studies linking activity and pattern similarity in these subfields to memory encoding (e.g., Eldridge et al. 2005; Suthana et al. 2011, 2015; Wolosin et al. 2013; Zeineh et al. 2003) and based on our finding that representational stability in these subfields predicted attentional behavior (Aly and Turk-Browne 2016a). Thus, the attentional state of these subfields may be particularly important for the attentional modulation of memory.

To test these hypotheses, we designed a three-part study that allowed us to identify attentional state representations in the hippocampus—that is, patterns of activity that are stable across multiple instances of the same attentional state—and then test whether more evidence for the goal-relevant attentional state during encoding predicted subsequent long-term memory (Fig. 3).

While undergoing high-resolution fMRI, participants first completed the "art gallery" task we used in our prior study (Aly and Turk-Browne 2016a) and discussed in a previous section ("Attentional Modulation of the Hippocampus"). On different trials, they attended either to the artistic style of paintings or to the layouts of rooms. We used the neuroimaging data from this part of the experiment

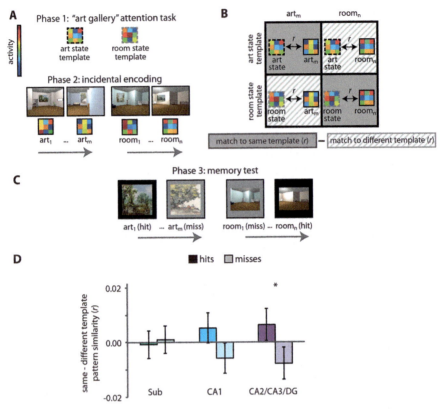

Fig. 3 Attentional modulation of hippocampal encoding via creation of state-dependent activity patterns. The fidelity of state-dependent activity patterns in the hippocampus during encoding predicts memory for goal-relevant information. (**a**) Participants first performed the "art gallery" attention task described in Fig. 1: They viewed images of rooms with art and attended to the artistic style of the paintings or the layouts of the rooms on different trials. From that task, "template" activity patterns for the art and room states were obtained in each hippocampal subfield by averaging activity patterns across all trials of the respective state. Participants then performed an incidental encoding task, viewing trial-unique rooms with paintings and attending to the paintings or the rooms in different blocks. The activity pattern for each encoding trial was extracted from each region of interest. (**b**) These trial-specific encoding patterns were correlated with the goal-relevant attentional state template (e.g., art encoding trial and art state template) and the goal-irrelevant attentional state template (e.g., art encoding trial and room state template). The difference of these correlations measures the extent to which the hippocampus was in the goal-relevant attentional state during encoding, and was the dependent measure of interest. (**c**) These correlation values were binned according to memory in the recognition test that was subsequently completed; that is, each encoding trial was back-sorted as a subsequent hit (remembered) or a miss (forgotten). (**d**) In the CA2/3 and dentate gyrus region of interest, there was greater pattern similarity with the goal-relevant vs. -irrelevant attentional state template during the encoding of items that were subsequently remembered vs. forgotten. This effect was not statistically significant in subiculum or CA1. $*p < .05$. Figure adapted from Aly and Turk-Browne (2016b)

to identify patterns of activity in each hippocampal subfield that corresponded to each of those attentional states; below, we refer to these activity patterns as "templates" for the art state and room state. Participants then completed an incidental encoding task with trial-unique images (rooms with art), attending to artistic style in one block and room layouts in the other. We obtained trial-specific activity patterns in each hippocampal subfield during encoding, and correlated these encoding activity patterns with the attentional state "templates" from the first part of the study. This allowed us to measure the extent to which the activity pattern in the hippocampus on any given encoding trial more closely resembled the goal-relevant vs. -irrelevant attentional state. Finally, participants were tested on their memory for the goal-relevant aspects of the images from the encoding phase: art from the art task and room layouts from the room task.

Consistent with our hypothesis, we found that successful episodic encoding was associated with a better attentional state in CA2/3 and dentate gyrus. That is, during encoding, activity patterns in these subfields more closely resembled the goal-relevant (vs. -irrelevant) attentional state when goal-relevant information was subsequently remembered vs. forgotten. This effect was selective to the hippocampus, and not found in medial temporal lobe cortex or object- and scene-selective regions in ventral temporal cortex (Aly and Turk-Browne 2016b). Together, these data shed light on the mechanisms by which attention transforms what we perceive into what we remember: Attention creates state-dependent patterns of activity in the hippocampus, which serve to prioritize the processing of goal-relevant aspects of the environment and create durable memory traces.

Attentional Modulation of Hippocampal Retrieval

Only a few studies have investigated how attention during retrieval modulates hippocampal memory signals. The initial studies used divided attention paradigms, and—as with the studies of divided attention during encoding—showed mixed results. For example, the PET study mentioned earlier with respect to divided attention during encoding (Iidaka et al. 2000) also found null effects during memory retrieval: Hippocampal activity was not different for full vs. divided attention. In contrast, an fMRI study—also using a blocked design, with no separation of brain activity for particular items as a function of memory success—found a reduction in hippocampal activity for divided vs. full attention during retrieval (Fernandes et al. 2005).

However, studies that have manipulated selective attention by instructing participants about which aspects of a stimulus to attend have consistently found modulation of hippocampal retrieval. For example, one study found evidence that novelty signals (enhanced activity for novel vs. familiar stimuli) in anterior hippocampus are sensitive to attention (Hashimoto et al. 2012). Participants were shown objects in a memory test that were either identical to ones that had been encoded earlier ("Same" items), perceptually different but in the same semantic category

("Similar" items; i.e., if a dog had been studied, a different dog would be included in the memory test), or entirely new ("New" items). Attention at test was oriented to either perceptual or semantic information. For perceptual attention, participants had to respond "old" if an object was perceptually identical to one they had studied, and "new" otherwise. For semantic attention, participants had to respond "old" if an object was perceptually or semantically identical to one they had studied, and "new" otherwise. Thus, a Similar item was called "new" in the perceptual task but "old" in the semantic task. Hippocampal activity for Similar items was comparable to New items in the perceptual attention task, and activity was higher than for Same items. In contrast, for the semantic attention task, Same and Similar items were associated with comparable hippocampal activity, and less activity than for New items. Thus, attention to perceptual vs. semantic information at retrieval modulates what is considered "novel" by the hippocampus.

Another study found converging evidence that novelty signals in the hippocampus are modulated by attention: Posterior hippocampus showed greater activity for correct vs. incorrect memory judgments when participants assessed the relative recency of items, while anterior hippocampus showed greater activity for correct memory judgments when participants assessed their novelty (Dudukovic and Wagner 2007). Moreover, attention to object vs. spatial information during retrieval modulated the response of hippocampal subfield CA1 to the amount of change in a probe item as compared to a similar studied item (Fig. 4; Duncan et al. 2012).

The capacity for the hippocampus to retrieve memories can also be voluntarily suppressed. That is, we can try to control the extent to which a retrieved memory comes to mind by selectively directing attention toward or away from retrieving that memory. Attempts to suppress memory retrieval do in fact worsen memory, and these suppression events are associated with reductions in hippocampal activity (Anderson et al. 2004; Anderson and Levy 2009; Hulbert et al. 2016). Subsequent research has investigated the dynamics that underlie our ability to selectively retrieve information while inhibiting competing information (e.g., Hulbert et al. 2016; Kuhl et al. 2011). In one such study (Wimber et al. 2015), participants learned associations between word cues and two images (e.g., the word "antique" paired with Albert Einstein, and, later, the word "antique" paired with goggles). They then selectively retrieved, in as much detail as possible, the first learned associate given the word cue (i.e., they would have to recall Einstein given "antique"). Presumably, during retrieval, selective attention is directed toward retrieving the target (Einstein), and away from the competitor (goggles). As a result of competitive retrieval, representations of the sought-for memory were strengthened in the hippocampus, while representations of the interfering competitor were weakened.

Another form of competition can come from distracting information. Indeed, the mere presence of task-irrelevant information during retrieval can reduce memory-related signals in the hippocampus (Wais et al. 2010). Bottom-up distraction from irrelevant stimulation can therefore interfere with the ability of the hippocampus to support episodic memory retrieval, perhaps by impairing our ability to selectively attend to task-relevant information.

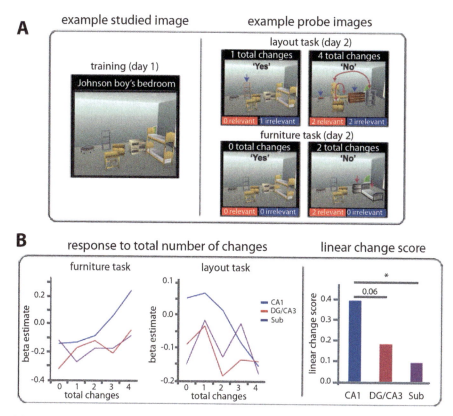

Fig. 4 Attentional modulation of hippocampal retrieval. The response of hippocampal subfield CA1 to associative mismatch at retrieval is modulated by attention. (**a**) Participants learned a set of distinctive rooms with unique furniture layouts. On each test trial, participants were presented with a probe image that corresponded to one they had studied, and performed one of two tasks. For the layout task, they had to indicate if the layout of the room was the same as what they had studied, ignoring any changes to the visual details of the furniture (e.g., if a bookcase was swapped for a different bookcase). On furniture trials, they had to indicate if the furniture was the same as what they had studied, ignoring any changes to the layout. The total number of changes in the probe image consisted of both task-relevant changes (e.g., layout changes on layout trials) and task-irrelevant changes (e.g., furniture changes on layout trials). (**b**) During retrieval, activity in CA1 was monotonically modulated by the number of changes in the probe image. The nature of this relationship differed as a function of attention to furniture vs. layout at retrieval: an increasing trend for the furniture task and a decreasing one for the layout task. *$p < .05$. Figure adapted from Duncan et al. (2012)

Thus, as with encoding, studies that manipulate selective attention more consistently find effects on hippocampal retrieval than those that divide attention. And, as with encoding, this pattern of results may be related to aspects of hippocampal memory that are relatively automatic. Some stages of recollection are presumed to be obligatory once a cue is consciously apprehended (Moscovitch 2008; Moscovitch et al. 2016). Divided attention studies may not reduce processing resources enough to prevent memory cues from being registered by the brain and triggering rapid hippocampal recollection. Conversely, selective attention, which focuses processing on one aspect of a stimulus and filters out others, changes what information is consciously apprehended, perhaps leaving some information below threshold for rapid hippocampal retrieval.

Another possible reason for why some studies fail to find effects of attention on hippocampal retrieval is that, in at least some situations, attention only has transient effects on hippocampal activity (Vilberg and Rugg 2012; also see Vilberg and Rugg 2014). In one study, participants studied word-picture associations and later had to remember the picture given the word as a cue. They were told to maintain the picture in mind over a delay, until a prompt appeared indicating which of three judgments had to be made about the remembered picture. Thus, the delay period served as time during which attention had to be focused on the contents retrieved from memory. Hippocampal activity related to successful recollection was transient—it did not persist during the delay, but was momentarily elevated after the appearance of the word cue. In contrast, elevated activity related to recollection was sustained over the delay in the intraparietal sulcus and angular gyrus, among other regions. Thus, the effects of maintaining attention to retrieve information from memory may only transiently engage the hippocampus, perhaps reflecting an initial, rapid recollection process (Moscovitch 2008; Moscovitch et al. 2016), while parietal cortical activity may be sustained because it indexes the amount of retrieved information (Vilberg and Rugg 2007).

How Do Hippocampal Memories Guide Attention?

Episodic Memory and Attention

We now turn to the other side of the story relating the hippocampus and attention—how hippocampal memories affect attentional orienting (Hutchinson and Turk-Browne 2012). We begin with studies showing that episodic memories can serve as guides for the allocation of attention during visual search and visual change detection (Hollingworth 2006).

In classic visual search paradigms, participants look for a particular, pre-defined target and respond as quickly as they can when they find it. One way of studying the influence of memory on target detection is by comparing search times for targets in new contexts to search times for targets in familiar contexts (Chun and Jiang 1998). For example, participants are faster at detecting targets in a fixed location within a real-world scene that is repeated vs. novel, with responses getting progressively faster across scene repetitions (Brockmole and Henderson 2006). This facilitation of visual search is accompanied by episodic memory for the repeated scenes (i.e., above chance recognition accuracy), as well as accurate recall of the specific target position within the scenes. Moreover, "previewing" a scene before performing a visual search task facilitates the detection of a target object (compared to a no-preview baseline), whether or not the previewed scene actually contains the target object (Hollingworth 2009). Thus, visual search is facilitated both by memory for specific goal-relevant object-location associations as well as memory for the general context.

This facilitation of visual search by long-term memory has been linked to the hippocampus (Fig. 5; Stokes et al. 2012; Summerfield et al. 2006; cf. Rosen et al. 2015). In these studies, participants have to search for a particular target—e.g., a key—in a visual scene. Information about where the key might be is provided either by memory or by perception on different trials. On memory-cued trials, participants had, on the previous day, learned the location of the key for that particular scene. Thus, they could rely on long-term memory in order to guide attention to the previously learned location of the key. Memory could also be uninformative, however—some scenes, although studied on the previous day, had never contained a key. On perception-cued trials, a box was presented on the screen around the location of the key, so that this visual cue could be used to direct attention. This visual cue could also be uninformative, however—on some trials, it could be presented at the center of the screen, and not around the key.

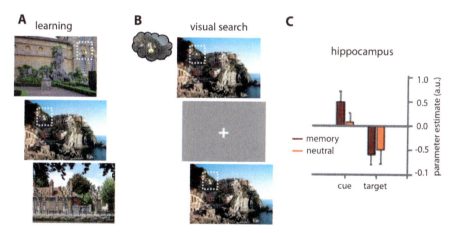

Fig. 5 Attentional guidance by episodic memory. The hippocampus is recruited for visual search cued by explicit long-term memory. (**a**) In the first phase of the study, participants explored a set of scenes, looking for a key in each. During this learning phase, therefore, associations were formed between particular scenes and the location of the key. Other scenes did not contain a key. (**b**) The next day, participants performed a visual search task with scenes that had been encoded the previous day, detecting the brief appearance of a key in those scenes (the key was present 50 % of the time). On memory-cued trials, the scene had contained a key when it was encoded on the previous day. On neutral trials, the scene had not contained a key during encoding. The scenes were shown first without a key (i.e., just the scene cue), and then, on target present trials, the key was superimposed. When present on memory-cued trials, the target always appeared in the learned location. Displayed here is an example memory-cued trial with target present. (**c**) The hippocampus was recruited by memory-guided attention, with greater activity for memory-cued vs. neutral trials, specifically during the cue (vs. target) period of the trial. Figure adapted from Stokes et al. (2012)

Both forms of cuing—memory-based and perception-based—conferred benefits to visual search: Participants were faster at responding when those cues were informative vs. uninformative (e.g., Summerfield et al. 2006). Critically, hippo-campal activity was higher for trials in which memory provided predictive infor-mation about the location of the target vs. trials in which memory was uninformative. Informative perceptual cues, however, were not associated with more hippocampal activity than uninformative ones. In addition, hippocampal activity was more strongly correlated with behavioral benefits from memory cueing than those from perceptual cuing. Because participants were able to recall the target locations, these findings implicate the hippocampus in visual search cued by explicit memory in particular.

Memory for item-context associations is just one way in which memory can guide visual search. Another role for memory is in the maintenance of item representations. That is, even in a novel or changing context, memory for the target(s) of search plays an important role. This is studied in paradigms in which there are many potential targets, and visual search therefore draws upon memory for the set of possible targets. Such tasks are meant to model real-world search situations in which, for example, you go to a soccer game with a group of friends, get separated, and can rapidly scan the crowd for any one or more of them. The visual characteristics of those targets (i.e., your friends) are stored in memory, and the number of friends you are searching for can be thought of as the memory "set size". This can be contrasted to the size of the crowd, which is the perceptual set size. Perceptual set size has a much greater cost for search efficiency than memory set size: As perceptual set size increases, response times increase linearly, whereas as memory set size increases, response times increase logarithmically (Wolfe 2012). Follow-up studies have linked this efficient search process to flexible memory representations: The memories are flexible in that they do not have to perceptually match the sought-for target, and search remains efficient even with few experiences with the item stored in memory (Guild et al. 2014). Concretely, you would still be incredibly efficient at searching a crowd for several people even if you had only been given descriptions of what they looked like, or if you'd seen them before but only once or twice. This efficient search is thought to be mediated by memories retrieved via a rapid form of recollection argued to be an obligatory, unconscious first stage of hippocampal retrieval (Moscovitch 2008; Moscovitch et al. 2016).

The benefits of long-term memory for attentional behavior are not limited to visual search. Memory also facilitates visual change detection—that is, the identi-fication of perceptual changes in scenes. In change detection paradigms, partici-pants view two versions of a scene in alternation (either several times or just once each), and have to identify the difference between the two (Rensink et al. 1997). In one such study, participants first had to detect the addition of a particular object to a scene (Becker and Rasmussen 2008). They were then shown the scenes again, and had to detect the addition of a new object in a new location, the old target object in a new location, or a new object in the old location. Change detection was faster for new objects in old locations, and old objects in new locations, compared to new objects in new locations. Thus, memory for previously goal-relevant objects and locations facilitates visual change detection. Moreover, a recent study found that long-term memory facilitates change detection performance even more when multiple different locations are made goal-relevant by prior experience (Rosen et al. 2014).

Finally, items studied at a particular spatial location (e.g., left vs. right side of a computer screen) subsequently bias attention toward that spatial location, even when they are centrally presented (Ciaramelli et al. 2009). This attentional bias facilitated the detection of dot probes that appeared on the side of the screen associated with the centrally presented item in memory. Furthermore, this facilitation of target detection by memory was correlated with subjective reports of recollection. These results offer further evidence that the contents of episodic memory can automatically, and rapidly, affect the spatial deployment of attention.

Another type of long-term associative memory that can affect the allocation of attention is semantic memory, or general knowledge about the world. For example, knowledge of what objects are typically found in a kitchen can guide how we move our eyes (and attention) as we search for a particular kitchen item (Torralba et al. 2006) and can facilitate the identification of objects that are expected in a kitchen context (Bar 2004). Semantic knowledge is not always helpful, and can even interfere with performance: Visual search is impaired by the presence of distractors that are semantically related to the target (Moores et al. 2003). Moreover, the allocation of attention to semantically related information can be automatic, occurring even when that information is completely irrelevant to the task at hand (Seidl-Rathkopf et al. 2015).

An unexplored possibility is that some effects of semantic memory on attention are at least in part linked to episodic memory. For example, when using memory to guide visual search for a particular kitchen item, you may rely on episodic memory for the last time you were in your kitchen rather than semantic memory of kitchens in general. Whether, and how, semantic and episodic memory interact in guiding attention is unclear. Moreover, whether the effects of semantic memory on attention are ever mediated by the hippocampus—perhaps when access to semantic knowledge is bolstered by episodic memory (Sheldon and Moscovitch 2012)—is currently unknown.

Implicit Learning and Attention

There is evidence that more unconscious knowledge of prior experience can guide attention during visual search, and that such implicit learning might be linked to the hippocampus. When the spatial context in a visual search task consists of a repeated (vs. novel) configuration of simple letters or shapes, rather than a real-world scene as in the studies above, recognition memory is at chance but visual search is still facilitated (Chun and Jiang 1998). That is, targets that appear at fixed locations (or have fixed identities) within repeating configurations of distractor locations

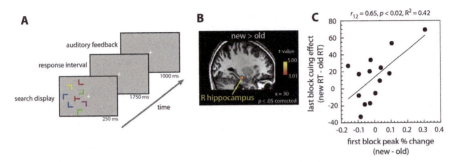

Fig. 6 Attentional guidance by implicit contextual memory. The hippocampus is recruited for visual search cued by implicit long-term memory. (**a**) Participants performed a visual search task in which they had to locate a target (the letter T) amongst distractors (the letter L) and indicate whether the T was oriented 90° clockwise or counter-clockwise. Some search displays were entirely novel ("new" contexts) while others were repeated several times over the experiment ("old" contexts). (**b**) Hippocampal activity was modulated by the type of display, with greater activity for new vs. old contexts. (**c**) Across participants, modulation of hippocampal activity by new vs. old contexts early in learning was correlated with the magnitude of the RT benefit for old vs. new contexts by the end of learning. Figure adapted from Giesbrecht et al. (2013)

(or identities) are more quickly detected than targets in new configurations (Endo and Takeda 2004).

This *contextual cuing* effect seems to depend on the hippocampus. It was impaired in patients with damage to the hippocampus and surrounding medial temporal lobe cortex (Chun and Phelps 1999; Chun 2000; Manns and Squire 2001). It was also impaired following administration of midazolam, which produces temporary amnesia (Park et al. 2004). Moreover, in healthy adults, hippocampal activity measured with fMRI was lower for repeated (vs. novel) configurations, despite chance performance on an explicit recognition task for those configurations, and this activity was inversely related to search response time (Greene et al. 2007). This overall effect was replicated and extended in subsequent studies (Fig. 6; Giesbrecht et al. 2013; Goldfarb et al. 2016; also see Kasper et al. 2015).

However, studies of the role of the hippocampus in contextual cuing are not entirely consistent. One study found a link between hippocampal activity and explicit memory for repeated contexts, rather than search facilitation, but the reverse pattern of results in adjacent perirhinal and entorhinal cortices (Preston and Gabrieli 2008). A potential reason for the discrepancy between studies is explicit recognition memory for the repeated configurations: In the only study that linked the hippocampus to explicit memory rather than implicit configural learning (Preston and Gabrieli 2008), participants showed above-chance recognition memory; this was not the case for the studies that linked the hippocampus to implicit search facilitation (Giesbrecht et al. 2013; Goldfarb et al. 2016; Greene et al. 2007; Kasper et al. 2015). Perhaps when individuals form episodic memories for the repeated configurations, retrieval of these explicit memories overshadows or prevents more implicit hippocampal memories from guiding visual search, especially when the episodic memories are for the configurations per se and not the target locations (cf. Stokes et al. 2012).

Contextual cuing shows that prior experience can facilitate visual search. Experience can confer other processing benefits as well: By learning what types of things should be attended, salient but irrelevant distractors can be better ignored. For example, training of a particular attentional set can carry over to another task, reducing susceptibility to interference from distraction (Leber and Egeth 2006; Leber et al. 2009). This might result from associating the attentional set with the current context (Cosman and Vecera 2013a). If so, then the hippocampus and/or medial temporal lobe cortex—critical for representing contexts and linking items to the contexts in which they occurred (Cohen and Eichenbaum 1993; Davachi 2006; Ranganath 2010)—might mediate this effect. Indeed, amnesic patients with medial temporal lobe damage failed to show it: The patients were able to overcome distraction in the training task, but this beneficial attentional set was not carried over to a subsequent task in the same experimental context (Cosman and Vecera 2013b).

These examples suggest that implicit contextual learning supported by the hippocampus can facilitate attentional behavior when familiar contexts are re-encountered. Another example comes from studies of statistical learning, which refers to our ability to extract structure from the environment and use it to anticipate likely upcoming events (Schapiro and Turk-Browne 2015). Such structure can occur in space (e.g., items that are typically found next to each other in a grocery store) and time (e.g., phonemes that typically follow each other in a particular language). Participants show sensitivity to statistical regularities on a number of implicit measures (e.g., faster reaction times to predicted vs. unpredicted stimuli), but are usually not explicitly aware of the underlying structure (e.g., Turk-Browne et al. 2005, 2009).

Attention is biased toward information streams that contain statistical regularities, suggesting one way that implicit statistical learning can guide the allocation of attention (Yu and Zhao 2015; Zhao et al. 2013). For example, if several task-irrelevant streams of shapes are presented simultaneously in different locations on a screen in between visual search trials, with one stream containing regularities and the others not, search targets are detected more quickly at the location that had contained regularities (Zhao et al. 2013). This attentional bias also exists for features: During visual search, attention is captured by the color of a (task-irrelevant) structured vs. random information stream. These biases can be long-lasting, persisting even if structure is no longer present (Yu and Zhao 2015). Finally, statistical structure can also guide perception and attention during development: Infants look longer at moderately predictable (vs. completely random or overly repetitive) visual and auditory sequences (Kidd et al. 2012, 2014).

Interestingly, the hippocampus seems to be involved in statistical learning. For example, hippocampal activity is enhanced for blocks of stimuli that contain temporal regularities (Turk-Browne et al. 2009) and for individual stimuli that license a prediction about what should appear next based on past exposure to regularities (Turk-Browne et al. 2010). Beyond overall activity, representations in the hippocampus are shaped by statistical learning: Hippocampal activity patterns elicited by objects that are part of the same regularity become more similar to one another (Schapiro et al. 2012). Moreover, damage to the hippocampus and medial temporal lobe cortex impairs statistical learning (Schapiro et al. 2014). However, because this was a single case study and the patient had extensive medial temporal lobe damage, future studies with selective hippocampal lesion patients will be important. Nevertheless, these initial studies linking the hippocampus to statistical learning suggest an additional way in which hippocampal mechanisms can influence attention—by setting up predictions that both facilitate processing of expected stimuli and highlight unexpected stimuli for additional processing (Hindy et al. 2016).

Implicit Memory and the Guidance of Eye Movements

When we move our attention, we also often move our eyes. Thus, eye tracking provides a powerful method to unobtrusively assess where people are directing their attention. Moreover, eye movements provide insight into cognitive operations that are not accessible to subjective awareness and thus to explicit reports (Hannula et al. 2010).

An emerging body of research suggests that hippocampal memories guide eye movements even when those memories are not conscious (for review, see Hannula et al. 2010; Meister and Buffalo 2016). An initial study of this type examined how healthy individuals and amnesic patients moved their eyes when viewing novel scenes, scenes they had viewed previously, and manipulated versions of previously viewed scenes (Ryan et al. 2000). Scene manipulations consisted of the addition, removal, or positional shift of an object. These changes alter the relations among scene components by disrupting the overall configuration of objects in the scene. Eye movements were used to assess memory for items and for relations: Implicit relational memory was measured by the modulation of eye movements to relational changes in scenes, and implicit item memory (where the "item" is the entire scene) was measured by the modulation of eye movements to repeated vs. novel scenes (see also West Channon and Hopfinger 2008). Healthy individuals made fewer fixations to repeated vs. novel scenes, and this eye movement marker of item memory was preserved in amnesic patients. In addition, healthy individuals who were not explicitly aware of relational changes in scenes made more fixations to the altered portions of those scenes. This eye movement marker of relational memory was not present in amnesic patients. Although these findings did not directly implicate the hippocampus in the guidance of eye movements by past experience (although all patients were amnesic, their etiologies were diverse), they inspired further research into how implicit forms of hippocampal memories might guide eye movements, specifically in situations that call for relational processing.

One such study tested amnesic patients with medial temporal lobe damage, most of whom had disproportionate damage to the hippocampus (Hannula et al. 2007).

Rather than measuring eye movements to changes in spatial relations, this study assessed whether hippocampal associative memory in the form of item-context bindings can bias the way people move their eyes. Participants encoded face-scene associations, and were subsequently presented with three equally familiar faces superimposed on a studied scene. Healthy participants spent more time viewing the face that had been studied with the scene, an effect that emerged rapidly—well before any explicit responses were made. This pattern of eye movements, an implicit manifestation of relational memory, critically depended on the hippocampus, as it was not found in the patients. A later study with the same paradigm found that hippocampal activity during the scene cue—before any faces were presented—was higher for trials in which participants subsequently fixated the correct face (Hannula and Ranganath 2009). Amazingly, this effect was observed even when explicit memory failed.

More evidence for implicit effects of hippocampal memories on eye movements came from a study in which participants viewed scenes that were configurally similar, but featurally dissimilar, to scenes previously encoded (Fig. 7; Ryals et al. 2015). These configurally similar scenes were behaviorally indistinguishable from entirely new scenes, in that participants' overt recognition judgments did not differ. Yet, eye movements tended to explore overlapping regions of space for the configurally similar and old scenes, and hippocampal activity correlated with this exploration overlap. This provides additional evidence that implicit memory for spatial configuration, a type of memory often supported by the hippocampus, can influence how attention is allocated, as indexed by eye movement behavior.

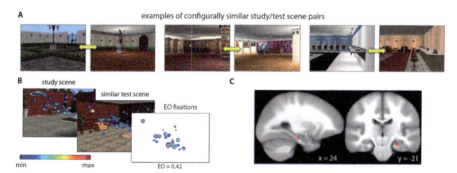

Fig. 7 Guidance of eye movements by implicit configural memory. Hippocampal activity is correlated with eye-movement expressions of implicit configural memory. (**a**) Participants encoded a set of images, and at test were presented with another set of images, half of which were entirely new and the remainder which were configurally similar (but featurally dissimilar) to the previously encoded scenes. Shown here are examples of studied scenes with their configurally-similar matches. (**b**) Example old (studied) scene and its configurally-similar test scene, overlaid with the mean heat map indicating where participants fixated their eyes. Exploration overlap (EO) is a measure of how much fixations overlapped between the studied and similar scenes. (**c**) Hippocampal activity was positively correlated with exploration overlap. Figure adapted from Ryals et al. (2015)

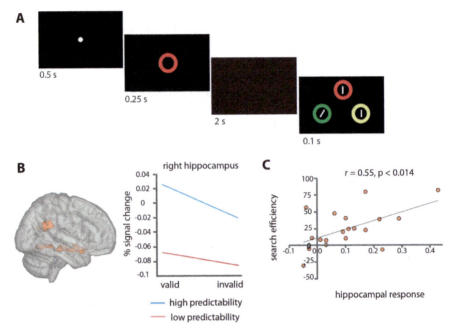

Fig. 8 Attentional guidance by working memory. Hippocampal activity is correlated with the modulation of visual search by the predictability of overlap with working memory contents. (**a**) Participants performed a visual search task in which they had to find the tilted line amongst vertical distractors, and indicate whether it was tilted to the left or the right. Prior to the search display, they viewed a circle—the memory cue—whose color they had to maintain in working memory (on some trials, memory for the color was tested). On valid trials, the circle whose color matched that of the memory cue contained the visual search target. On invalid trials, it contained a distractor. Thus, shown here is an example of an invalid trial. In high predictability blocks, the memory cue was either always valid or always invalid. In low predictability blocks, the cue was valid half the time, and thus could not be reliably used to guide search. (**b**) Hippocampal activity was enhanced for high- vs. low-predictability blocks. (**c**) Across participants, this enhancement was correlated with predictability-related modulation of visual search efficiency. Figure adapted from Soto et al. (2012)

Together, these data raise the possibility that, when a memory cue is presented, the hippocampus retrieves associated information, and these associations in turn guide eye movements (and attention) even when the memory contents do not reach conscious awareness. This view converges with the proposal that hippocampal memory retrieval consists of two stages: The first is rapid and outside of awareness, but can nevertheless affect behavior, while the second is slower, with its output accessible to conscious report (Moscovitch 2008; Moscovitch et al. 2016). These findings suggest that the first stage of hippocampal retrieval has far-reaching effects, directing the movement of our eyes and attention.

An issue for future investigation concerns the role of the hippocampus and eye movements in the facilitation of visual search and change detection by long-term memory (discussed in section "Episodic Memory and Attention"). Long-term memory improves the detection of changes to previously relevant objects and locations in natural scenes, and this is associated with more direct eye movements to those previously relevant objects and locations (Becker and Rasmussen 2008). Likewise, relatively direct eye movements to visual targets are made for familiar scenes in a visual search task (Summerfield et al. 2006). Whether hippocampal activity mediates the relationship between memory, visual search/change detection, and eye movements is an open question.

Working Memory and Attention

Thus far, we have focused on how hippocampal long-term learning and memory influence attention. Recent research has also highlighted a role for the hippocampus in *working* memory—that is, memory over several seconds (for review, see Olsen et al. 2012; Yonelinas 2013). The hippocampus plays a role in working memory for relational or associative (as compared to item) information, and may be particularly important for the maintenance of precise, high-resolution representations (Yonelinas 2013). For example, patients with hippocampal damage are impaired on working memory tasks that require the maintenance of relational information in scenes (i.e., where different scene components are with respect to one another), and this impairment increases with working memory load and retention interval (e.g., Hannula et al. 2006; Jeneson et al. 2011).

Information retained in working memory can attract attention to visually or semantically related items (e.g., Downing 2000; Huang and Pashler 2007; for review, see Soto et al. 2008), raising the possibility that hippocampally mediated working memory can bias attention. Evidence in support of this possibility came from a study that manipulated whether the contents of working memory overlapped with the target of attention, and whether this overlap was predictable or not (Fig. 8; Soto et al. 2012; also see Soto et al. 2007). On each trial, participants were presented with a working memory cue (a colored circle) to maintain over a delay before the brief appearance of a search display of three colored circles, one of which matched the color maintained in working memory. The matching circle contained the search target (a tilted line) or a distractor (a vertical line). Thus, the contents of working memory could either facilitate detection of the target or hinder performance, depending on whether the matching colored circle was around the target or a distractor. In different blocks, the relationship between working memory contents and the search target was manipulated: In predictable blocks, the search target was either always or never in the circle that matched the color in working memory. In unpredictable blocks, the matching circle contained the search target or a distractor with equal probability. Search times were faster when the search target was in a circle whose color matched the contents of working memory. Moreover, search times were faster when the relationship between working memory contents and the search target was predictable. Finally, hippocampal activity was enhanced for predictable vs. unpredictable blocks, and this enhancement correlated with the modulation of behavioral search efficiency by predictability.

One surprising aspect of this study is that the hippocampus is generally not required for working memory when the stimuli consist of simple features (such as colors and shapes) but is more important for working memory tasks that require retention of relational or associative information (Ranganath and Blumenfeld 2005; Yonelinas 2013). How hippocampal damage affects attentional guidance on this task is thus an interesting question for future research. Additionally, comparison of hippocampal involvement for this task vs. tasks that require working memory for more complex relations will be informative.

Conclusions and Future Directions

In this chapter, we discussed several ways in which attention and memory interact in the hippocampus. Attention influences the encoding and retrieval of hippocampally mediated episodic memories. Moreover, attention creates state-dependent patterns of activity in the hippocampus, and these state-dependent patterns predict online attentional behavior as well as long-term memory for goal-relevant aspects of experience. In turn, many forms of hippocampal memories influence attention: Explicit episodic, implicit contextual, and working memories can serve as a cue for attention and guide eye movements. All of these areas are relatively nascent and so will benefit from additional work. We conclude by highlighting four particular methodological approaches that could provide mechanistic insight in future investigations: studies of network connectivity, neuropsychology, neurofeedback, and neuromodulation.

Although our focus has been on the hippocampus, investigations of the interplay between attention and memory would benefit from consideration of the cortical networks with which the hippocampus interacts (Ranganath and Ritchey 2012). The hippocampus receives input—via medial temporal lobe cortex—from occipital, temporal, and parietal cortical regions (Felleman and Van Essen 1991; Lavenex and Amaral 2000), which may be an important means by which cortical perceptual and attentional signals interface with those in the hippocampus, and a means by which hippocampal memories can in turn influence perception and attention. Indeed, hippocampal activity at rest spontaneously fluctuates with that in lateral and medial parietal cortex, lateral and medial temporal cortex, and medial prefrontal cortex (Buckner et al. 2008; Kahn et al. 2008; Libby et al. 2012; Vincent et al. 2008). This connectivity may allow for the exchange of attentional, perceptual, and mnemonic signals in the brain. It will be informative for future studies to investigate how hippocampal-cortical interactions subserve attentional modulation of memory and the mnemonic modulation of attention. For example, coupling between the attentional states of the hippocampus and retrosplenial cortex has been linked to the modulation of memory by attentional states (Aly and Turk-Browne 2016b).

Another line of investigation for future studies is neuropsychological approaches. Patient studies will continue to yield important insights into the necessity of the hippocampus for attentional modulation of memory and for the guidance of attention by memory. Such studies have already made important contributions (e.g., Chun and Phelps 1999; Cosman and Vecera 2013b; Hannula et al. 2007; Ryan et al. 2000; Schapiro et al. 2014), but many of these studies have relied on patients with damage that extends beyond the hippocampus, making inferences about the specific role of the hippocampus difficult. On the other hand, patients with selective hippocampal lesions often only have partial damage, posing additional interpretational challenges if functions are preserved. The key will be to obtain results across a range of patients, and to analyze behavior as a function of the etiology, extent, and precise location of lesions. Moreover, many recent discoveries in the field of attention and memory have relied on functional neuroimaging, and have yet to be tested in any patient population. Insofar as multiple (potentially non-hippocampal) memory systems can influence attention (Hutchinson and Turk-Browne 2012; also see Hutchinson et al. 2016), it remains an important question whether hippocampal damage will eliminate some forms of attentional guidance, or if other systems can support performance.

Another way that causal inferences might be made about interactions between hippocampal memories and attention is via neurofeedback with real-time fMRI (Sulzer et al. 2013). These studies involve giving participants moment-by-moment feedback about overall activity (or the presence of an activity pattern) in a given brain region. This can be done by, for example, showing a participant a dial on the screen and having them try to move it to the left or right based on the activity in a brain region of interest. In this way, researchers can train participants to exert control over, and thus influence the state of, a given brain region. A potentially more powerful approach than using a participant's brain state to move a dial, however, is to have the participant's brain state change the stimulus that is the target of their behavioral goals (deBettencourt et al. 2015). For example, if a participant is making decisions on faces, the pattern of activity in that participant's brain—which is affected by the quality of their attentional state—could be used to degrade or clarify a perceptually noisy face. Such *closed-loop* designs—in which the state of the brain determines the content or timing of stimulus presentation, which in turn influences the state of the brain, then the next stimulus, and so on—provide an enticing method for pseudo-causal investigations with fMRI by manipulating the activity of brain regions hypothesized to be involved in a task. For example, by comparing the effects on attention and memory of real-time neurofeedback from the hippocampus to the effects of neurofeedback from a control region, conclusions can be made about the specific contributions of the hippocampus. That is, by exerting control over activity in the hippocampus, we can more confidently assess whether that activity is necessary for a particular cognitive function. With standard fMRI techniques, one can only say whether a particular type of brain activity is correlated with that function.

Finally, studies of neuromodulatory systems can elucidate the mechanisms by which hippocampal memories and attention influence each other. The hippocampus is modulated by all of the main neurotransmitter systems implicated in attention, including cholinergic, dopaminergic, and noradrenergic systems (Muzzio et al.

2009a; Rowland and Kentros 2008). These systems have strong influences on hippocampal representations of space and on hippocampal memories (Lisman and Grace 2005; Newman et al. 2012; Parent and Baxter 2004). For example, manipulations of acetylcholine and dopamine alter place field stability in the hippocampus (Brazhnik et al. 2003; Kentros et al. 2004), raising the possibility that these neurotransmitters mediate the effects of attention on hippocampal representational stability. Acetylcholine enhances the influence of environmental input on hippocampal processing by amplifying afferent signals and suppressing excitatory recurrent connections in CA3 (Hasselmo 2006; Newman et al. 2012), providing a potential mechanism by which attention can modulate activity patterns in the hippocampus. These and other neuromodulatory influences can be studied with a variety of methods, including magnetic resonance spectroscopy and pharmacological interventions (e.g., administration of neurotransmitter agonists or antagonists) in humans and in animal models. Neuromodulatory studies would be particularly informative because they could shed light on the physiological mechanisms by which attention creates, shapes, and maintains hippocampal representations. For example, if cholinergic modulation is essential for representational stability in the hippocampus, this would suggest that such stability arises as a result of enhancing the influence of the external environment (via afferent signals from entorhinal cortex) and suppressing memory retrieval (via recurrent connections in CA3).

Despite these exciting future opportunities, existing work has already convincingly demonstrated that hippocampal functions cannot be fully described without consideration of attentional processes, and in turn, that our understanding of attention is illuminated and expanded by considering the influence of the hippocampus. This body of literature also convincingly demonstrates the broad reach of the hippocampus beyond explicit memory, showing that its influence pervades even our moment-to-moment attentional behavior.

References

Aly M, Turk-Browne NB (2016a) Attention stabilizes representations in the human hippocampus. Cereb Cortex 26:783–796

Aly M, Turk-Browne NB (2016b) Attention promotes episodic encoding by stabilizing hippocampal representations. Proc Natl Acad Sci 113:E420–E429

Anderson MC, Levy BJ (2009) Suppressing unwanted memories. Curr Direct Psychol Sci 18:189–194

Anderson MC, Ochsner KN, Kuhl B, Cooper J, Robertson E, Gabrieli SW, Glover GH, Gabrieli JDE (2004) Neural systems underlying the suppression of unwanted memories. Science 303:232–235

Bar M (2004) Visual objects in context. Nat Rev Neurosci 5:617–629

Becker MW, Rasmussen IP (2008) Guidance of attention to objects and locations by long-term memory of natural scenes. J Exp Psychol Learn Mem Cogn 34:1325–1338

Brazhnik ES, Muller RU, Fox SE (2003) Muscarinic blockade slows and degrades the location-specific firing of hippocampal pyramidal cells. J Neurosci 23:611–621

Brewer JB, Zhao Z, Desmond JE, Glover GH, Gabrieli JDE (1998) Making memories: brain activity that predicts how well visual experience will be remembered. Science 281:1185–1187

Brockmole JR, Henderson JM (2006) Using real-world scenes as contextual cues for search. Vis Cogn 13:99–108

Brown MW, Aggleton JP (2001) Recognition memory: what are the roles of the perirhinal cortex and hippocampus? Nat Rev Neurosci 2:51–61

Buckner RL, Andrews-Hanna JR, Schacter DL (2008) The brain's default network: anatomy, function, and relevance to disease. Ann N Y Acad Sci 1124:1–38

Bussey TJ, Saksida LM (2005) Object memory and perception in the medial temporal lobe: an alternative approach. Curr Opin Neurobiol 15:730–737

Carr VA, Engel SA, Knowlton BJ (2013) Top-down modulation of hippocampal encoding activity as measured by high-resolution functional MRI. Neuropsychologia 51:1829–1837

Cherry EC (1953) Experiments on the recognition of speech with one and two ears. J Acoust Soc Am 25:975

Chun MM (2000) Contextual cueing of visual attention. Trends Cogn Sci 4:170–178

Chun MM, Jiang Y (1998) Contextual cuing: implicit learning and memory of visual context guides spatial attention. Cogn Psychol 36:28–71

Chun MM, Phelps EA (1999) Memory deficits for implicit contextual information in amnesic subjects with hippocampal damage. Nat Neurosci 2:844–847

Chun MM, Turk-Browne NB (2007) Interactions between attention and memory. Curr Opin Neurobiol 17:177–184

Ciaramelli E, Lin O, Moscovitch M (2009) Episodic memory for spatial context biases spatial attention. Exp Brain Res 192:511–520

Cohen NJ, Eichenbaum H (1993) Memory, amnesia, and the hippocampal system. MIT Press, Cambridge, MA

Cosman JD, Vecera SP (2013a) Context-dependent control over attentional capture. J Exp Psychol Hum Percept Perform 39:836–848

Cosman JD, Vecera SP (2013b) Learned control over distraction is disrupted in amnesia. Psychol Sci 24:1585–1590

Craik FI (2001) Effects of dividing attention on encoding and retrieval processes. In: Roediger HL, Nairne JS, Neath I (eds) The nature of remembering: essays in honor of Robert G. Crowder. American Psychological Association, Washington, DC, pp 55–68

Craik FI, Govoni R, Naveh-Benjamin M, Anderson MD (1996) The effects of divided attention on encoding and retrieval processes in human memory. J Exp Psychol Gen 125:159–180

Davachi L (2006) Item, context and relational episodic encoding in humans. Curr Opin Neurobiol 16:693–700

Davachi L, Wagner AD (2002) Hippocampal contributions to episodic encoding: insights from relational and item-based learning. J Neurophysiol 88:982–990

deBettencourt MT, Cohen JD, Lee RF, Norman KA, Turk-Browne NB (2015) Closed-loop training of attention with real-time brain imaging. Nat Neurosci 18:47–475

Desimone R (1996) Neural mechanisms for visual memory and their role in attention. Proc Natl Acad Sci 93:13494–13499

Downing PE (2000) Interactions between visual working memory and selective attention. Psychol Sci 11:467–473

Dudukovic NM, Wagner AD (2007) Goal-dependent modulation of declarative memory: neural correlates of temporal recency decisions and novelty detection. Neuropsychologia 45:2608–2620

Dudukovic NM, Preston AR, Archie JJ, Glover GH, Wagner AD (2010) High-resolution fMRI reveals match enhancement and attentional modulation in the human medial temporal lobe. J Cogn Neurosci 23:670–682

Duncan K, Ketz N, Inati SJ, Davachi L (2012) Evidence for area CA1 as a match/mismatch detector: a high-resolution fMRI study of the human hippocampus. Hippocampus 22:389–398

Ekstrom AD, Kahana MJ, Caplan JB, Fields TA, Isham EA, Newman EL, Fried I (2003) Cellular networks underlying human spatial navigation. Nature 425:184–187

Eldridge LL, Engel SA, Zeineh MM, Bookheimer SY, Knowlton BJ (2005) A dissociation of encoding and retrieval processes in the human hippocampus. J Neurosci 25:3280–3286

Endo N, Takeda Y (2004) Selective learning of spatial configuration and object identity in visual search. Percept Psychophys 66:293–302

Felleman DJ, Van Essen DC (1991) Distributed hierarchical processing in the primate cerebral cortex. Cereb Cortex 1:1–47

Fenton AA, Lytton WW, Barry JM, Lenck-Santini PP, Zinyuk LE, Kubík S, Bureš J, Poucet B, Muller RU, Olypher AV (2010) Attention-like modulation of hippocampus place cell discharge. J Neurosci 30:4613–4625

Fernandes MA, Moscovitch M (2000) Divided attention and memory: evidence of substantial interference effects at retrieval and encoding. J Exp Psychol Gen 129:155–176

Fernandes MA, Moscovitch M, Ziegler M, Grady C (2005) Brain regions associated with successful and unsuccessful retrieval of verbal episodic memory as revealed by divided attention. Neuropsychologia 43:1115–1127

Fletcher PC, Stephenson CME, Carpenter TA, Donovan T, Bullmore ET (2003) Regional brain activations predicting subsequent memory success: an event-related fMRI study of the influence of encoding tasks. Cortex 39:1009–1026

Gardiner JM, Parkin AJ (1990) Attention and recollective experience in recognition. Mem Cognit 18:579–583

Giesbrecht B, Sy JL, Guerin SA (2013) Both memory and attention systems contribute to visual search for targets cued by implicitly learned context. Vision Res 85:80–89

Gilbert CD, Li W (2013) Top-down influences on visual processing. Nat Rev Neurosci 14:350–363

Goldfarb EV, Chun MM, Phelps EA (2016) Memory-guided attention: independent contributions of the hippocampus and striatum. Neuron 89:317–324

Graham KS, Barense MD, Lee ACH (2010) Going beyond LTM in the MTL: a synthesis of neuropsychological and neuroimaging findings. Neuropsychologia 48:831–853

Greene AJ, Gross WL, Elsinger CL, Rao SM (2007) Hippocampal differentiation without recognition: an fMRI analysis of the contextual cueing task. Learn Mem 14:548–553

Guild EB, Cripps JM, Anderson ND, Al-Aidroos N (2014) Recollection can support hybrid visual memory search. Psychon Bull Rev 21:142–148

Hannula DE, Ranganath C (2009) The eyes have it: hippocampal activity predicts expression of memory in eye movements. Neuron 63:592–599

Hannula DE, Tranel D, Cohen NJ (2006) The long and short of it: relational memory impairments in amnesia, even at short lags. J Neurosci 26:8352–8359

Hannula DE, Ryan JD, Tranel D, Cohen NJ (2007) Rapid onset relational memory effects are evident in eye movement behavior, but not in hippocampal amnesia. J Cogn Neurosci 19:1690–1705

Hannula DE, Althoff RR, Warren DE, Riggs L, Cohen NJ, Ryan JD (2010) Worth a glance: using eye movements to investigate the cognitive neuroscience of memory. Front Hum Neurosci 4:1–16, Article 166

Hardt O, Nadel L (2009) Cognitive maps and attention. In: Srinivasan N (ed) Progress in brain research, vol 176. Elsevier, The Netherlands, pp 181–194

Hashimoto R, Abe N, Ueno A, Fujii T, Takahashi S, Mori E (2012) Changing the criteria for old/new recognition judgments can modulate activity in the anterior hippocampus. Hippocampus 23:141–148

Hasselmo ME (2006) The role of acetylcholine in learning and memory. Curr Opin Neurobiol 16:710–715

Henke K, Buck A, Weber B, Wieser GH (1997) Human hippocampus establishes associations in memory. Hippocampus 7:249–256

Henke K, Weber B, Kneifel S, Wieser HG, Buck A (1999) Human hippocampus associates information in memory. Proc Natl Acad Sci 96:5884–5889

Hindy NC, Ng FY, Turk-Browne NB (2016) Linking pattern completion in the hippocampus to predictive coding in visual cortex. Nat Neurosci. doi:10.1038/nn.4284

Hollingworth A (2006) Visual memory for natural scenes: evidence from change detection and visual search. Vis Cogn 14:781–807

Hollingworth A (2009) Two forms of scene memory guide visual search: memory for scene context and memory for the binding of target object to scene location. Vis Cogn 17:273–291

Huang L, Pashler H (2007) Working memory and the guidance of visual attention: consonance-driven orienting. Psychon Bull Rev 14:148–153

Hulbert JC, Henson RN, Anderson MC (2016) Inducing amnesia through systemic suppression. Nat Commun 7(11003):1–9

Hutchinson JB, Turk-Browne NB (2012) Memory-guided attention: control from multiple memory systems. Trends Cogn Sci 16:576–579

Hutchinson JB, Pak SS, Turk-Browne NB (2016) Biased competition during long-term memory formation. J Cogn Neurosci 28:187–197

Iidaka T, Anderson ND, Kapur S, Cabeza R, Craik FIM (2000) The effect of divided attention on encoding and retrieval in episodic memory revealed by positron emission tomography. J Cogn Neurosci 12:267–280

Jackson J, Redish AD (2007) Network dynamics of hippocampal cell-assemblies resemble multiple spatial maps within single tasks. Hippocampus 17:1209–1229

Jeneson A, Mauldin KN, Hopkins RO, Squire LR (2011) The role of the hippocampus in retaining relational information across short delays: the importance of memory load. Learn Mem 18:301–305

Kahn I, Andrews-Hanna JR, Vincent JL, Snyder AZ, Buckner RL (2008) Distinct cortical anatomy linked to subregions of the medial temporal lobe revealed by intrinsic functional connectivity. J Neurophysiol 100:129–139

Kasper RW, Grafton ST, Eckstein MP, Giesbrecht B (2015) Multimodal neuroimaging evidence linking memory and attention systems during visual search cued by context. Ann N Y Acad Sci 1339:176–189

Kastner S, Ungerleider LG (2000) Mechanisms of visual attention in the human cortex. Annu Rev Neurosci 23:315–341

Kelemen E, Fenton AA (2010) Dynamic grouping of hippocampal neural activity during cognitive control of two spatial frames. PLoS Biol 8(e1000403):1–14

Kensinger EA, Clarke RJ, Corkin S (2003) What neural correlates underlie successful encoding and retrieval? A functional magnetic resonance imaging study using a divided attention paradigm. J Neurosci 23:2407–2415

Kentros CG, Agnihotri NT, Streater S, Hawkins RD, Kandel ER (2004) Increased attention to spatial context increases both place field stability and spatial memory. Neuron 42:283–295

Kidd, C., Piantadosi, S.T., Aslin, R.N. (2012). The Goldilocks effect: human infants allocate attention to visual sequences that are neither too simple nor too complex. PLoS One, 7, e36399. doi: 10.1371/journal.pone.0036399.

Kidd C, Piantadosi ST, Aslin RN (2014) The Goldilocks effect in infant auditory attention. Child Dev 85:1795–1804

Kuhl BA, Rissman J, Chun MM, Wagner AD (2011) Fidelity of neural reactivation reveals competition between memories. Proc Natl Acad Sci 108:5903–5908

Lavenex P, Amaral DG (2000) Hippocampal-neocortical interaction: a hierarchy of associativity. Hippocampus 10:420–430

Leber AB, Egeth HE (2006) It's under control: top-down search strategies can override attentional capture. Psychon Bull Rev 13:132–138

Leber AB, Kawahara JI, Gabari Y (2009) Long-term abstract learning of attentional set. J Exp Psychol Hum Percept Perform 35:1385–1397

Libby LA, Ekstrom AD, Ragland JD, Ranganath C (2012) Differential connectivity of perirhinal and parahippocampal cortices within human hippocampal subregions revealed by high-resolution functional imaging. J Neurosci 32:6550–6560

Lisman JE, Grace AA (2005) The hippocampal-VTA loop: controlling the entry of information into long-term memory. Neuron 46:703–713

Manns JR, Squire LR (2001) Perceptual learning, awareness, and the hippocampus. Hippocampus 11:776–782

Maunsell JHR, Treue S (2006) Feature-based attention in visual cortex. Trends Neurosci 29:317–322

Meister MLR, Buffalo EA (2016) Getting directions from the hippocampus: the neural connection between looking and memory. Neurobiol Learn Mem. doi:10.1016/j.nlm.2015.12.004

Monaco JD, Rao G, Roth ED, Knierim JJ (2014) Attentive scanning behavior drives one-trial potentiation of hippocampal place fields. Nat Neurosci 17:725–731

Moores E, Laiti L, Chelazzi L (2003) Associative knowledge controls deployment of visual selective attention. Nat Neurosci 6:182–189

Moray N (1959) Attention in dichotic listening: affective cues and the influence of instructions. Q J Exp Psychol 11:56–60

Morris CD, Bransford JD, Franks JJ (1977) Levels of processing versus transfer appropriate processing. J Verb Learn Verb Behav 16:519–533

Moscovitch M (2008) The hippocampus as a "stupid", domain-specific module: implications for theories of recent and remote memory, and of imagination. Can J Exp Psychol 62:62–79

Moscovitch M, Cabeza R, Winocur G, Nadel L (2016) Episodic memory and beyond: the hippocampus and neocortex in transformation. Annu Rev Psychol 67:105–134

Muzzio IA, Kentros C, Kandel E (2009a) What is remembered? Role of attention on the encoding and retrieval of hippocampal representations. J Physiol 12:2837–2854

Muzzio IA, Levita L, Kulkarni J, Monaco J, Kentros C, Stead M, Abbott LF, Kandel ER (2009b) Attention enhances the retrieval and stability of visuospatial and olfactory representations in the dorsal hippocampus. PLoS Biol 7(e1000140):1–20

Newman EL, Gupta K, Climer JR, Monaghan CK, Hasselmo ME (2012) Cholinergic modulation of cognitive processing: insights drawn from computational models. Front Behav Neurosci 6:1–19 , Article 24

O'Keefe J, Dostrovsky J (1971) The hippocampus as a spatial map. Preliminary evidence from unit activity in the freely-moving rat. Brain Res 34:171–175

Olsen RK, Moses SN, Riggs L, Ryan JD (2012) The hippocampus supports multiple cognitive processes through relational binding and comparison. Front Hum Neurosci 6:1–13 , Article 146

Otten LJ, Henson RNA, Rugg MD (2001) Depth of processing effects on neural correlates of memory encoding: relationship between findings from across- and within-task comparisons. Brain 124:399–412

Parent MB, Baxter MG (2004) Septohippocampal acetylcholine: involved in but not necessary for learning and memory? Learn Mem 11:9–20

Park H, Quinlan J, Thornton E, Reder LM (2004) The effect of midazolam on visual search: implications for understanding amnesia. Proc Natl Acad Sci 101:17879–17883

Posner MI, Rothbart MK (2014) Attention to learning of school subjects. Trends Neurosci Educ 3:14–17

Preston AR, Gabrieli JDE (2008) Dissociation between explicit memory and configural memory in the human medial temporal lobe. Cereb Cortex 18:2192–2207

Ranganath C (2010) A unified framework for the functional organization of the medial temporal lobes and the phenomenology of episodic memory. Hippocampus 20:1263–1290

Ranganath C, Blumenfeld RS (2005) Doubts about double dissociations between short- and long-term memory. Trends Cogn Sci 9:374–380

Ranganath C, Ritchey M (2012) Two cortical systems for memory-guided behavior. Nat Rev Neurosci 13:713–726

Rensink RA, O'Regan JK, Clark JJ (1997) To see or not to see: the need for attention to perceive changes in scenes. Psychol Sci 8:368–373

Rosen ML, Stern CE, Somers DC (2014) Long-term memory guidance of visuospatial attention in a change-detection paradigm. Front Psychol 5 , Article 266:1–8. doi:10.3389/fpsyg.00266

Rosen ML, Stern CE, Michalka SW, Devaney KJ, Somers DC (2015) Cognitive control network contributions to memory-guided visual attention. Cereb Cortex. doi:10.1093/cercor/bhv028

Rowland DC, Kentros CG (2008) Potential anatomical basis for attentional modulation of hippocampal neurons. Ann N Y Acad Sci 1129:213–224

Ryals AJ, Wang JX, Polnaszek KL, Voss JL (2015) Hippocampal contribution to implicit configuration memory expressed via eye movements during scene exploration. Hippocampus. doi:10.1002/hipo.22425

Ryan JD, Althoff RR, Whitlow S, Cohen NJ (2000) Amnesia is a deficit in relational memory. Psychol Sci 11:454–461

Schapiro AC, Turk-Browne NB (2015) Statistical learning. In: Toga AW (ed) Brain mapping: an encyclopedic reference. Academic Press: Elsevier, New York, NY, pp 501–506

Schapiro AC, Kustner LV, Turk-Browne NB (2012) Shaping of object representations in the human medial temporal lobe based on temporal regularities. Curr Biol 22:1622–1627

Schapiro AC, Gregory E, Landau B, McCloskey M, Turk-Browne NB (2014) The necessity of the medial temporal lobe for statistical learning. J Cogn Neurosci 26:1736–1747

Schott BH, Wustenberg T, Wimber M, Fenker DB, Zierhut KC, Seidenbecher CI, Heinze HJ, Walter H, Düzel E, Richardson-Klavehn A (2013) The relationship between level of processing and hippocampal-cortical functional connectivity during episodic memory formation in humans. Hum Brain Mapp 34:407–424

Seidl-Rathkopf K, Turk-Browne NB, Kastner S (2015) Automatic guidance of attention during real-world visual search. Atten Percept Psychophys 77:1881–1895

Sheldon S, Moscovitch M (2012) The nature and time-course of medial temporal lobe contributions to semantic retrieval: an fMRI study on verbal fluency. Hippocampus 22:1451–1466

Soto D, Humphreys GW, Rotshtein P (2007) Dissociating the neural mechanisms of memory-based guidance of visual selection. Proc Natl Acad Sci 104:17186–17191

Soto D, Hodsoll J, Rotshtein P, Humphreys GW (2008) Automatic guidance of attention from working memory. Trends Cogn Sci 12:342–348

Soto D, Greene CM, Kiyonaga A, Rosenthal CR, Egner T (2012) A parieto-medial temporal pathway for the strategic control over working memory biases in human visual attention. J Neurosci 32:17563–17571

Sprague TC, Saproo S, Serences JT (2015) Visual attention mitigates information loss in small- and large-scale neural codes. Trends Cogn Sci 19:215–226

Stokes MG, Atherton K, Patai EZ, Nobre AC (2012) Long-term memory prepares neural activity for perception. Proc Natl Acad Sci 109:E360–E367

Strange BA, Dolan RJ (2001) Adaptive anterior hippocampal responses to oddball stimuli. Hippocampus 11:690–698

Sulzer J, Haller S, Scharnowski F, Weiskopf N, Birbaumer N, Blefari ML, Bruehl AB, Cohen LG, deCharms RC, Gassert R et al (2013) Real-time fMRI neurofeedback: progress and challenges. Neuroimage 76:386–399

Summerfield JJ, Lepsien J, Gitelman DR, Mesulam MM, Nobre AC (2006) Orienting attention based on long-term memory experience. Neuron 49:905–916

Suthana NA, Ekstrom A, Moshirvaziri S, Knowlton B, Bookheimer S (2011) Dissociations within human hippocampal subregions during encoding and retrieval of spatial information. Hippocampus 21:694–701

Suthana NA, Donix M, Wozny DR, Bazih A, Jones M, Heidemann RM, Trampel R, Ekstrom AD, Scharf M, Knowlton B, Turner R, Bookheimer SY (2015) High-resolution 7-tesla fMRI of human hippocampal subfields during associative learning. J Cogn Neurosci 27:1194–1206

Torralba A, Oliva A, Castelhano MS, Henderson JM (2006) Contextual guidance of eye movements and attention in real-world scenes: the role of global features in object search. Psychol Rev 113:766–786

Turk-Browne NB, Jungé JA, Scholl BJ (2005) The automaticity of visual statistical learning. J Exp Psychol Gen 134:552–564

Turk-Browne NB, Scholl BJ, Chun MM, Johnson MK (2009) Neural evidence of statistical learning: efficient detection of visual regularities without awareness. J Cogn Neurosci 21:1934–1945

Turk-Browne NB, Scholl BJ, Johnson MK, Chun MM (2010) Implicit perceptual anticipation triggered by statistical learning. J Neurosci 30:11177–11187

Turk-Browne NB, Golomb JD, Chun MM (2013) Complementary attentional components of successful memory encoding. Neuroimage 66:553–562

Uncapher MR, Rugg MD (2005) Effects of divided attention on fMRI correlates of memory encoding. J Cogn Neurosci 17:1923–1935

Uncapher MR, Rugg MD (2006) Episodic encoding is more than the sum of its parts: an fMRI investigation of multifeatural contextual encoding. Neuron 52:547–556

Uncapher MR, Rugg MD (2008) Fractionation of the component processes underlying successful episodic encoding: a combined fMRI and divided-attention study. J Cogn Neurosci 20:240–254

Uncapher MR, Rugg MD (2009) Selecting for memory? The influence of selective attention on the mnemonic binding of contextual information. J Neurosci 29:8270–8279

Uncapher MR, Hutchinson JB, Wagner AD (2011) Dissociable effects of top-down and bottom-up attention during episodic encoding. J Neurosci 31:12613–12628

Vilberg KL, Rugg MD (2007) Dissociation of the neural correlates of recognition memory according to familiarity, recollection, and amount of recollected information. Neuropsychologia 45:2216–2225

Vilberg KL, Rugg MD (2012) The neural correlates of recollection: transient versus sustained fMRI effects. J Neurosci 32:15679–15687

Vilberg KL, Rugg MD (2014) Temporal dissociations within the core recollection network. Cogn Neurosci 5:77–84

Vincent JL, Kahn I, Snyder AZ, Raichle ME, Buckner RL (2008) Evidence for a frontoparietal control system revealed by intrinsic functional connectivity. J Neurophysiol 100:3328–3342

Voss JL, Gonsalves BD, Federmeier KD, Tranel D, Cohen NJ (2010) Hippocampal brain-network coordination during volitional exploratory behavior enhances learning. Nat Neurosci 14:115–120

Wagner AD, Schacter DL, Rotte M, Koutstaal W, Maril A, Dale AM, Rosen BR, Buckner RL (1998) Building memories: remembering and forgetting of verbal experiences as predicted by brain activity. Science 281:1188–1191

Wais PE, Gazzaley A (2011) The impact of auditory distraction on retrieval of visual memories. Psychon Bull Rev 18:1090–1097

Wais PE, Rubens MT, Boccanfuso J, Gazzaley A (2010) Neural mechanisms underlying the impact of visual distraction on retrieval of long-term memory. J Neurosci 29:8541–8550

West Channon V, Hopfinger JB (2008) Memory's grip on attention: the influence of item memory on the allocation of attention. Vis Cogn 16:325–340

Wimber M, Alink A, Charest I, Kriegeskorte N, Anderson MC (2015) Retrieval induces adaptive forgetting of competing memories via cortical pattern suppression. Nat Neurosci 18:582–589

Wolfe JM (2012) Saved by a log: how do humans perform hybrid visual and memory search? Psychol Sci 23:698–703

Wolosin SM, Zeithamova D, Preston AR (2013) Distributed hippocampal patterns that discriminate reward context are associated with enhanced associative binding. J Exp Psychol Gen 142:1264–1276

Yamaguchi S, Hale LA, D'Esposito M, Knight RT (2004) Rapid prefrontal-hippocampal habituation to novel events. J Neurosci 24:5356–5363

Yi DJ, Chun MM (2005) Attentional modulation of learning-related repetition attenuation effects in human parahippocampal cortex. J Neurosci 25:3593–3600

Yonelinas AP (2002) The nature of recollection and familiarity: a review of 30 years of research. J Mem Lang 46:441–517

Yonelinas AP (2013) The hippocampus supports high-resolution binding the service of perception, working memory and long-term memory. Behav Brain Res 252:34–44

Yonelinas AP, Aly M, Wang WC, Koen JD (2010) Recollection and familiarity: examining controversial assumptions and new directions. Hippocampus 20:1178–1194

Yu RQ, Zhao J (2015) The persistence of the attentional bias to regularities in a changing environment. Atten Percept Psychophys. doi:10.3758/s13414-015-0930-5

Zeineh MM, Engel SA, Thompson PM, Bookehimer SY (2003) Dynamics of the hippocampus during the encoding and retrieval of face-name pairs. Science 299:577–580

Zhao J, Al-Aidroos N, Turk-Browne NB (2013) Attention is spontaneously biased toward regularities. Psychol Sci 24:667–677

The Hippocampus and Memory Integration: Building Knowledge to Navigate Future Decisions

Margaret L. Schlichting and Alison R. Preston

Abstract Everyday behaviors require a high degree of flexibility, in which prior knowledge is applied to inform behavior in new situations. Such flexibility is thought to be supported in part by memory integration, a process whereby related memories become interconnected in the brain through recruitment of overlapping neuronal populations. Mechanistically, integration is thought to occur through specialized hippocampal encoding processes that integrate related events during learning. By recalling past events during new experiences, connections can be created between newly formed and existing memories. The resulting integrated memory traces would extend beyond direct experience in anticipation of future judgments that require consideration of multiple learned events. Recent advances in cognitive and behavioral neuroscience have provided empirical evidence for the existence of such a mechanism, with hippocampal encoding mechanisms—in coordination with medial prefrontal cortex—supporting memory integration. Emerging research suggests that abstracted representations in medial prefrontal cortex guide reactivation of related memories during new encoding events, thus promoting hippocampal integration of related experiences. Moreover, recent work indicates that integrated memories can impact a host of behaviors, from promoting spatial navigation and imagination to resulting in memory distortion and deletion.

M.L. Schlichting
Department of Psychology, University of Toronto, 100 St. George St., Sidney Smith Hall, Toronto, ON M5S 3G3, Canada
e-mail: mschlichting@utexas.edu

A.R. Preston (✉)
Center for Learning and Memory, The University of Texas at Austin, 1 University Station C7000, Austin, TX 78712, USA

Department of Psychology, The University of Texas at Austin, 1 University Station C7000, Austin, TX 78712, USA

Department of Neuroscience, The University of Texas at Austin, 1 University Station C7000, Austin, TX 78712, USA
e-mail: apreston@utexas.edu

© Springer International Publishing AG 2017
D.E. Hannula, M.C. Duff (eds.), *The Hippocampus from Cells to Systems*,
DOI 10.1007/978-3-319-50406-3_13

Introduction

Decades' worth of research documents the involvement of the hippocampus in rapidly encoding new episodes, which are then transferred (i.e., *consolidated*) to neocortex over time. However, memory is a dynamic phenomenon. The once widely accepted view that such consolidated memories are immune to modification has long since been refuted. Consolidated memories may be reactivated during new experiences, at which point they are susceptible to distortion, deletion, or updating (Nadel and Hardt 2011; McKenzie and Eichenbaum 2011; Nadel et al. 2012). Conversely, reactivated memories may also influence how new content is encoded (Zeithamova et al. 2012a; Gershman et al. 2013). Here, we review the recent work in cognitive and behavioral neuroscience that investigates the complex ways in which memories influence one another and change over time. One way by which such mutual influence may occur is through *memory integration*.

Memory integration refers to the idea that memories for related experiences are stored as overlapping representations in the brain, forming memory networks that span events and support the flexible extraction of novel information. For example, imagine you see a woman walking her dog in the park near your house (Fig. 1).

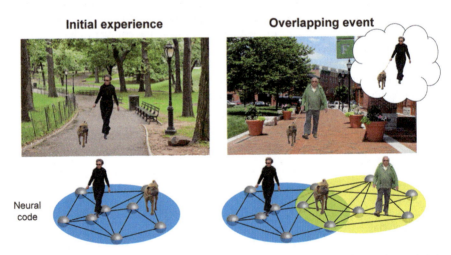

Fig. 1 Schematic depiction of related events that might lead to memory integration and their associated neural codes. One day while walking in the park, you encounter a woman and her dog (initial experience, *left*). Connections are formed among a group of simultaneously activated neurons, coding the woman–dog association (*blue network*). A few days later, you encounter the same dog in town, this time with a man (overlapping event, *right*). The dog (overlapping element) triggers reactivation of your initial experience in the park (woman–dog association). Such reactivation enables connections to be formed among neural representations of the woman, dog, and man, linking the related events across time (overlapping *blue* and *yellow networks*). The resulting integrated memories are hypothesized to support novel judgments that require consideration of both events; here, for instance, you may infer a relationship between the woman and the man despite never having seen them together. Figure adapted with permission from Schlichting and Preston (2015)

During this experience, you form a memory for the event that represents the relationship among the woman, the dog, and the park. The next day, you see the same dog out for a walk in town with a man. The familiar element (the dog) during this second experience may serve as a cue for hippocampal pattern completion, triggering the reactivation of your prior experience with the woman and dog. The new event (the man walking the dog in town) is then encoded in the presence of the reactivated information about your first experience with the dog. In this way, a link between the woman, the man, and the dog can be formed during encoding, despite the fact that you have never seen the woman with the man.

The notion that new encoding and prior knowledge interact with one another is by no means new (Bartlett 1932; Tolman 1948; Cohen and Eichenbaum 1993); yet, the neural mechanisms and behavioral implications of memory integration have only recently become the subject of empirical investigation. The field's growing interest in understanding these complex, real-world aspects of episodic memory has been realized thanks to the advent of elegant behavioral paradigms and advanced analysis methods for neural data. We first review evidence for the neural mechanisms that underlie memory integration. We then turn to a discussion of the range of behaviors that might be supported by integration, from flexible navigation to imagination and creativity. Finally, we set forth questions and considerations for future research.

Neural Mechanisms

Memory integration has been studied in both rodents and humans using highly controlled experimental paradigms in which subjects make decisions that span multiple learned experiences. In one example task, the associative inference paradigm (Preston et al. 2004), participants encode a series of overlapping events: AB followed later by BC, where 'AB' denotes a studied arbitrary association between items A and B. Participants are later tested on their memory for directly experienced information (AB, BC) as well as on their ability to make novel inferences (AC) that require consideration of two events. In this task, performance on the AC inference test serves as the critical behavioral index of memory integration. By recalling past (AB) events during new (BC) experiences, knowledge structures are formed that integrate the newly learned information into prior memories (Fig. 1). The resulting integrated memories would allow for direct extraction of novel inferences that cross event boundaries, thereby promoting performance on the AC test. In addition to a behavioral index of integration, such experimental designs allow researchers to index the neural processes specific to encoding of the second (BC) overlapping event, during which there is a unique opportunity to integrate across related memories. A number of similar paradigms have been used in the literature (Eichenbaum et al. 1996; Shohamy and Wagner 2008), all of which require participants to make novel decisions spanning learned pieces of information. For simplicity, in this section we refer to behaviors thought to index memory integration as *integration behaviors*; see section "Implications for Behavior" for a detailed discussion of the diverse set of behaviors potentially impacted by memory integration. In section "Mechanism Overview", we provide an overview of the neural mechanisms underlying memory integration. We then describe examples of empirical evidence for these processes in sections "Evidence for a Hippocampal Role in Integration" and "Hippocampal-Medial Prefrontal Interactions".

Mechanism Overview

Human and animal lesion work highlights the critical role of the hippocampus and an interconnected structure, the medial prefrontal cortex (Iordanova et al. 2007; DeVito et al. 2010b; Koscik and Tranel 2012; Ghosh et al. 2014; Warren et al. 2014), in memory integration (Fig. 2). Damage to either of these structures impairs the ability to combine information acquired during different episodes despite intact memory for the previously learned individual events (Bunsey and Eichenbaum 1996; DeVito et al. 2010b; Koscik and Tranel 2012). Work in rodents also demonstrates dynamic interactions between these structures during memory updating, perhaps reflecting the flow of information from hippocampus to MPFC (Tse et al.

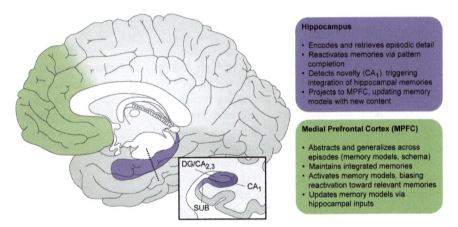

Fig. 2 Proposed roles of human hippocampus and MPFC in memory integration. Locations and hypothesized functions of regions critical for memory integration in the human brain. *Purple*, hippocampus; *green*, medial prefrontal cortex. Here, we intentionally provide a broad definition of MPFC due to high variability in the precise location of effects reported across studies. For instance, we include anterior cingulate cortex, which has been implicated in memory integration (Wang et al. 2012) and the formation of memory models (Roy et al. 2012). Inset, cross section through the hippocampus (*purple*) highlighting area CA_1 (*dark purple portion*). Approximate hippocampal subfield boundaries are indicated with thin *dashed lines*. Location of cross section along hippocampal axis is indicated with a thick *dashed line*. *MPFC* medial prefrontal cortex, CA_1 Cornu ammonis field 1, $DG/CA_{2,3}$ dentate gyrus and Cornu ammonis fields 2 and 3, *SUB* subiculum. Figure adapted with permission from Schlichting and Preston (2015)

2011). However, while these data underscore the importance of hippocampus and MPFC in memory integration, the precise mechanisms by which these regions contribute have only recently started to become clear.

One period during which memory integration may take place is when new learning experiences share content (e.g., a person, place, or thing) with existing memory traces (Fig. 1). Through this process, termed *integrative encoding*, memories are formed that integrate information across distinct experiences (Nadel and Hardt 2011; McKenzie and Eichenbaum 2011; Nadel et al. 2012) in anticipation of future use. This constructive, or prospective, nature of memory (Klein et al. 2002b; Buckner 2010; Addis and Schacter 2012) dates back to Tolman's concept of a "cognitive map" (Tolman 1948) and is reflected in modern memory theories including relational memory theory (Eichenbaum et al. 1999), multiple trace theory (Nadel and Moscovitch 1997), and schema theory (Bartlett 1932; van Kesteren et al. 2012). Memory integration has been proposed as a key mechanism underlying a host of flexible behaviors, including inferring novel relationships (O'Reilly and Rudy 2001; Shohamy and Wagner 2008; Zeithamova and Preston 2010; Zeithamova et al. 2012a), determining new routes through the environment (Gupta et al. 2010), and making adaptive decisions (Wimmer and Shohamy 2012). These ideas are also highly related to the influential temporal context model (Kahana 1996; Howard et al. 2005), in which items are bound to the learning context in which they occur. In this case, learning context may include related content that has been reactivated.

When new event relates to prior experience, pattern completion mechanisms supported by the hippocampus reactivate the previously stored, overlapping memory (Zeithamova et al. 2012b; Preston and Eichenbaum 2013). Empirical support for reactivation of prior memories during overlapping learning experiences has recently been garnered using neural decoding of fMRI data (Fig. 3) (Kuhl et al. 2012; Zeithamova et al. 2012a; Gershman et al. 2013). With the related content reinstated in the brain, hippocampal area CA_1 (Fig. 2) is thought to compare prior memories with incoming information from the environment (Hasselmo and Schnell 1994). CA_1 may signal the presence of associative novelty (i.e., when new experiences violate memory-based predictions) and facilitate new encoding (Hasselmo et al. 1996; Larkin et al. 2014). Models of hippocampal subfield function have suggested that CA_1 novelty signals may influence neural dynamics via feedback connections to the medial septum, modulating acetylcholine levels and setting appropriate dynamics for learning (i.e., encoding rather than retrieval; Hasselmo and Schnell 1994; Hasselmo et al. 1996). The resulting integrated memories are highly structured, with shared elements coded similarly across experiences (McKenzie et al. 2013, 2014). One recent study (McKenzie et al. 2014) has shown that hippocampal CA field firing patterns for overlapping events reflect a hierarchy of features coded according to their behavioral relevance. This organization scheme could then be exploited to extract commonalities across episodes and support a host of behaviors, as discussed below.

Hippocampal mechanisms may be additionally influenced by operations in MPFC. While its specific role in memory is only starting to be uncovered, at least

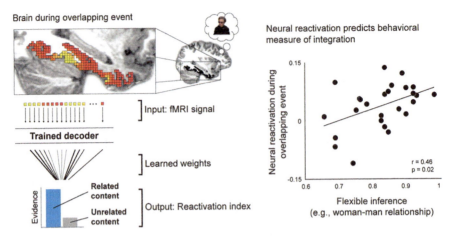

Fig. 3 Example use of neural decoding to quantify memory reactivation. *Left panel*, depiction of a neural decoding approach quantifying the degree of memory reactivation during learning. The neural pattern evoked during the overlapping event (man-dog from Fig. 1) is hypothesized to reflect reinstatement of the related—but not presently viewed—element (the woman). The fMRI signal is extracted for each voxel in a region of interest (here, ventral temporal cortex is used as an example). This information is then input into a neural decoder trained to recognize activation patterns associated with different kinds of stimuli (e.g., faces). Based on the weights for each voxel learned during training, the decoder outputs a value reflecting the degree to which the neural pattern reflects reactivation of the related versus unrelated content. These evidence scores can then be used as an index of reactivation. *Right panel*, reactivation during encoding of overlapping events predicts later flexible inference (woman-man association), a behavioral index of memory integration. Data are adapted with permission from Zeithamova et al. (2012a). Figure adapted with permission from Schlichting and Preston (2015)

two functions have been proposed for MPFC that are of relevance to the present discussion. First, MPFC is thought to represent mental models that guide behavior across a number of domains (Roy et al. 2012; Wilson et al. 2014). With regards to memory, some suggest that MPFC encodes interconnected information to form mental models based on mnemonic content (i.e., memory models) (Schacter et al. 2012; St. Jacques et al. 2013), which may include features such as behavioral relevance and appropriate response associated with a particular context (Miller and Cohen 2001; Euston et al. 2012; Kroes and Fernández 2012). This functionality may explain the involvement of MPFC in reinforcement learning, which has been hypothesized to reflect its coding of action-outcome associations. Anatomical features of MPFC may make it especially well suited to form such complex representations of goals or task rules, as it receives a broad range of input from sensory and limbic regions (Price and Drevets 2009).

A second possible function of MPFC is in biasing learning-phase retrieval toward the most behaviorally relevant memories, thereby influencing what will be integrated (van Kesteren et al. 2012; Kroes and Fernández 2012; Preston and Eichenbaum 2013). This may be conceptualized as the deployment of memory models to resolve conflict among related experiences. Memory models are thought to be activated when incoming information relates to existing knowledge. MPFC may then select specific task-relevant memories for reactivation (van Kesteren et al. 2012; Kroes and Fernández 2012; Wilson et al. 2014), perhaps via white matter projections to the medial temporal lobe (MTL) cortical structures that provide the major input to hippocampus (Cavada et al. 2000). Hippocampus may then bind reactivated content to current experience, resulting in an integrated trace. Following integration in hippocampus, memory models may be updated with new content as needed through direct hippocampal inputs to MPFC (van Kesteren et al. 2012). Through this process, MPFC may come to represent integrated memories that have been abstracted away from individual episodes (i.e., schema) over time (van Kesteren et al. 2012; Richards et al. 2014).

Of course, the possibilities we describe here are neither exhaustive nor mutually exclusive. Future research will be needed to fully understand the role of MPFC in memory integration, and assess whether its functionality might differ across subregions.

Evidence for a Hippocampal Role in Integration

Electrophysiological studies in rodents have shown hippocampal-mediated replay of prior event sequences in new spatial contexts (Karlsson and Frank 2009) and never-experienced spatial trajectories that represent a shortcut through a well-learned environment (Gupta et al. 2010; although note that novel routes represented a very small proportion of all replay events), consistent with the idea that memories extend beyond direct experience. Furthermore, in environments with overlapping elements, individual hippocampal neurons demonstrate experience-dependent generalized firing patterns that respond in multiple similar locations (Singer et al. 2010; McKenzie et al. 2013) or to the overlapping features themselves (Wood et al. 1999). Such generalized firing patterns suggest that hippocampal neurons develop representations that code the similarities between events. By representing features common to multiple events similarly, hippocampal codes can capture regularities shared across different experiences and, in doing so, may act as "nodes" that link distinct behavioral episodes (Fig. 1) (Eichenbaum et al. 1999).

Behavioral work in humans suggests that reactivating related memories immediately prior to a new learning experience increases the likelihood that new content will be integrated into existing memories (Hupbach et al. 2007). Using neuroimaging, researchers have also related the degree of reactivation of prior experience during encoding of new overlapping events to evidence for integration (Kuhl et al. 2010; Zeithamova et al. 2012a). In one study, the evidence for hippocampus-mediated reactivation of prior memories was associated with greater retention of the reactivated information (Kuhl et al. 2010), demonstrating that reactivating memories during new learning helps reduce forgetting of past events. In an associative inference paradigm, another study demonstrated that reactivation of existing knowledge during new learning of overlapping associations predicted superior integration behavior, suggesting that combining related memories during learning might underlie successful inferential reasoning (Fig. 3; Zeithamova et al. 2012a; see also Richter et al. 2015).

With relevant prior experience reactivated in the brain, hippocampus is then thought to bind or integrate current and prior experience (Shohamy and Wagner 2008; Zeithamova and Preston 2010; Zeithamova et al. 2012a). In one study (Shohamy and Wagner 2008), increases in hippocampal activation across the learning phase were associated with individual differences in integration behavior, even when accounting for performance differences on trained associations (Fig. 4a). Changes in hippocampal activation over learning in the associative inference task were also related to integration behavior across participants, even when accounting for differences in memory for single events (Zeithamova et al. 2012a). Moreover, interrogation of trial-by-trial neural engagement revealed that hippocampal activation during encoding of overlapping associations (BC), but not initially acquired associations (AB), differentiated between subsequently correct and incorrect inference judgments (AC; Fig. 4b-i) (Zeithamova et al. 2012a). Collectively, these findings highlight the importance of a hippocampal encoding mechanism whereby overlapping experiences are integrated into a network of related memories as they are learned.

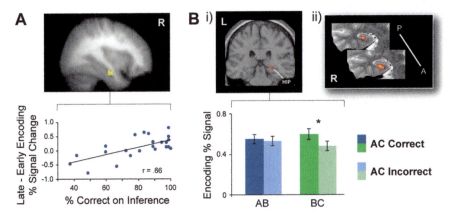

Fig. 4 Learning-phase integration signatures in hippocampus. (**a**) Activation in both left (not shown) and right hippocampus during encoding of overlapping associations was correlated with individual differences in inference performance. Specifically, increases in hippocampal activation from the early to late portion of the training phase were associated with superior performance on inferential probe trials. (**b**-i) In an associative inference task, right hippocampal activation during encoding of overlapping associations (BC) was greater for trials in which the corresponding inference judgment (AC) was later correct relative to trials on which the inference judgment was later incorrect. Hippocampal activation during initially acquired associations (AB) was not related to subsequent inferential performance. (ii) Using high-resolution fMRI, this signature was isolated to the CA_1 subfield of the hippocampus. Data are adapted with permission from: Shohamy and Wagner (2008) (panel **a**), Zeithamova and Preston (2010) (panel **b**-i), and Schlichting et al. (2014) (panel **b**-ii)

Integration is thought to be triggered by hippocampal comparator processes, with hippocampal area CA_1 signaling deviations between current events and reactivated content (i.e., associative novelty signaling; Lisman and Grace 2005). Empirical work in humans has supported the notion that area CA_1 signals deviations from prior experience, with engagement of this region increasing as the number of changes to a studied stimulus increase (i.e., with an increasing degree of mismatch; Duncan et al. 2012a). Activation in human CA_1 during the encoding of events that overlap with prior experiences has been shown to relate to a behavioral measure of memory integration (Fig. 4b-ii; Schlichting et al. 2014), consistent with the notion that novelty signals triggers the formation of links between new content and prior memories.

Hippocampal-Medial Prefrontal Interactions

Recent evidence suggests that hippocampus interacts with MPFC to support memory integration in many circumstances (Fig. 2). One possible explanation for this region's involvement in encoding-phase memory updating lies in its pattern of anatomical connectivity: MPFC is directly connected to the hippocampus, receiving inputs primarily from the anterior portion of CA_1 (Barbas and Blatt 1995; Cavada et al. 2000). MPFC also has extensive connections with a diverse set of sensory, limbic, and subcortical structures (Cavada et al. 2000), suggesting that it might be important for combining across episodic memories, represented in the brain across distributed cortical and subcortical networks. Consistent with this idea, recent studies have observed encoding-phase engagement (Zeithamova et al. 2012a) and evidence for reactivation of prior memories in MPFC (Richter et al. 2015), demonstrating the importance of this region for memory integration during encoding. Moreover, enhanced functional coupling of hippocampus and MPFC has been shown when new learning can be integrated into prior knowledge (Schlichting and Preston 2016), consistent with the notion that MPFC interacts with hippocampus to promote integration. Integration behavior has also been linked to individual differences in the intrinsic functional connectivity (Gerraty et al. 2014) and structural connectivity (Schlichting and Preston 2016) of hippocampus and MPFC, highlighting that even static neural characteristics might render some individuals better suited for combining across related events.

Learning Factors Promoting Integration

A number of studies have investigated the learning factors that influence integration. For instance, while there is evidence that integration can occur in the absence of conscious awareness (Shohamy and Wagner 2008; Wimmer and Shohamy 2012; Henke et al. 2013; Munnelly and Dymond 2014), studies have shown that integration may be facilitated when subjects become aware of the task structure (either via instructional manipulations or spontaneously) (Kumaran and Melo 2013; Richter et al. 2015). In fact, one experiment (Kumaran and Melo 2013) demonstrated that such knowledge specifically benefitted judgments that spanned episodes with no effect on memory for the individual episodes themselves, suggesting that integration does not necessarily emerge with learning of the underlying experiences. One possibility is that awareness constrains MPFC control processes, which in turn biases hippocampal reactivation during learning toward task-relevant memories, allowing for integration across events.

It has been hypothesized that being reminded of related memories prior to a new learning experience also increases the likelihood of integration, as the reactivated memories become labile and readily updated. Consistent with this idea, behavioral work in humans (Hupbach et al. 2007) found more intrusions from a second learned list (List 2) when recalling the initial list (List 1) if participants had been reminded of List 1 before encoding List 2. This finding was recently replicated in rodents using "lists" of ordered feeder locations (Jones et al. 2012), with animals who learned two lists in the same relative to different spatial contexts producing more intrusions. Another study manipulated the degree of retrieval on a trial-by-trial basis within participants (Duncan et al. 2012b). That study similarly found superior integration performance for learning experiences that followed an old item (i.e., when retrieval was possible) versus those that followed a new item (when retrieval was not possible). These findings are consistent with the proposal that integration occurs via reactivation of prior memories; this work further highlights that reminding the learner of the prior related memory may encourage integration.

The strength of existing memories may be an additional factor mediating integration. In particular, stronger memories might be more readily reactivated during learning, thereby allowing for integration across memories. One neuroimaging study showed that offline processing of initial memories was associated with more evidence for reactivation and superior integration behavior during a subsequent learning experience, suggesting that memory strengthening during rest facilitates integration (Schlichting and Preston 2014). Integration signatures have also been preferentially observed when initial memories are well-learned at the time of the first overlapping event, as is the case in blocked learning (i.e., multiple AB learning opportunities occurring before any BC learning; Schlichting et al. 2015). These results suggest that strong prior memories may promote reactivation during learning, thereby allowing for integration across memories. This work underscores that integration may be especially likely when initial memories are well established prior to new learning.

Other factors hypothesized to impact integration include (1) the nature of the underlying memory representations, with more distributed as opposed to localized representations proposed to promote integration (Schiller and Phelps 2011); and (2) the degree of competition between new content and prior memories (i.e., whether or not the two memories can coexist), with integration preferentially occurring in cases when competition is minimal (Hupbach 2011).

Offline Processes Promoting Integration

Numerous empirical studies (Tambini et al. 2010; Jadhav et al. 2012; Deuker et al. 2013; Staresina et al. 2013) and theoretical accounts (Marr 1970; McClelland et al. 1995) highlight the importance of offline processes—such as reinstatement of recent experience and enhanced interregional communication—for episodic memory. It has been proposed that through hippocampal-neocortical interactions (McClelland et al. 1995; Nadel et al. 2000), memories are reactivated during periods of sleep and awake rest. Such reactivation (or *replay*) is thought to support the strengthening and transfer of memory traces from the hippocampus to neocortical regions for long-term storage (i.e., consolidation).

These mechanisms may also support the integration of memories across experiences (Kumaran and McClelland 2012). Recent theories suggest that hippocampus-mediated replay of event sequences during sleep (Hoffman and McNaughton 2002; Ji and Wilson 2007) provides a potential mechanism for constructing networks of related memories that anticipate future decisions and actions (Sara 2010; Diekelmann and Born 2010; Lewis and Durrant 2011)—a process referred to as *prospective consolidation* (Buckner 2010). Such theories propose that by reactivating memories during sleep, representations are recombined and recoded, resulting in rich networks of related memories that extend beyond initially encoded events (Kumaran and McClelland 2012). This process is thought to promote both the integration of new information into existing memories and abstraction across episodes in neocortical regions, particularly MPFC (Lewis and Durrant 2011). According to this view, stored memories are not veridical representations of events, but rather derived representations formed in anticipation of future use. Sleep-based replay of hippocampal memory traces, therefore, could enhance integration behaviors that tap knowledge about the relationships among events experienced at different times (Ellenbogen et al. 2007; Werchan and Gómez 2013; Coutanche et al. 2013). Consistent with this notion, one study (Ellenbogen et al. 2007) demonstrated that participants who slept following learning showed better integration behavior relative to a comparison group who remained awake.

In addition to sleep-based mechanisms that might promote integration, offline processes occurring during periods of awake rest have also been suggested to be important for memory. The mnemonic consequences of reactivation of recent experience has been demonstrated during awake rest using neurophysiological techniques in rodents (Jadhav et al. 2012) and, more recently, in humans using pattern information analysis of fMRI data (Deuker et al. 2013; Staresina et al. 2013). For instance, more delay period reactivation was observed for stimuli that were remembered relative to those that were forgotten in a subsequent test (Staresina et al. 2013). Moreover, studies have shown that the degree of hippocampal-neocortical functional coupling during rest periods following learning relates to later memory for the learned content (Tambini et al. 2010).

Recent evidence suggests that similar rest-phase mechanisms may promote the integration of memories that span related events (van Kesteren et al. 2010; Craig et al. 2015; Schlichting and Preston 2016), with integration-related neural signatures persisting into offline periods following encoding. For example, one study showed increased hippocampal-MPFC functional coupling during encoding conditions that necessitate schema reorganization and updating; interestingly, this pattern persisted during the post-encoding rest period (van Kesteren et al. 2010). These findings are consistent with the idea that neural patterns evoked during encoding are reactivated during offline rest periods, potentially reflecting early-phase consolidation mechanisms. Similar neural signatures have been reported following memory updating in an associative inference task, with the degree of hippocampal-MPFC connectivity enhancements during awake rest following an integration opportunity predicting individual differences in behavior (Schlichting and Preston 2016). While the precise effect of these rest-phase processes for memory integration is yet to be determined, it may be the case that memory reactivation and increased interregional coupling may strengthen connections among related memories, thereby further promoting the formation of integrated memory representations.

Neural Representations

Initial research suggests that one way in which the hippocampus supports behavioral flexibility is by integrating information across multiple experiences to establish links between related events, either during new learning or offline through replay of related experiences during sleep and rest. However, questions remain regarding the precise nature of the underlying hippocampal representations. Several theoretical and computational frameworks have proposed alternate accounts of the properties of memory representations that can support inference, which we describe here.

One hypothesized representational structure supporting integration is one in which new events are incorporated into existing memory traces to be parsimoniously represented in a single, composite memory representation (Fig. 5a). For instance, consider the simple example of two events that share a common element (AB, BC) as in the associative inference paradigm. When a new event occurs that contains an element overlapping with a previous event (e.g., BC after encoding AB), the overlapping element (B) can trigger pattern completion of the previously encoded memory (AB). According to this hypothesized representational structure, elements from the new, overlapping event (in this case, C) would be encoded into the existing, reactivated memory (AB) to form a single integrated representation that combines the two experiences (ABC). Because these integrated representations directly code the novel relationship between A and C along with the original experiences, this representational format provides a basis for the inferential use of memory, but has a notable cost in that details of the individual experiences may not be preserved (e.g., the knowledge that A and C were presented in two different temporal contexts).

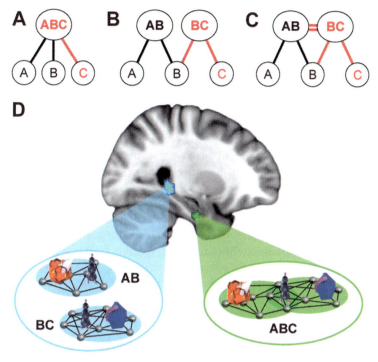

Fig. 5 Schematic depiction of alternative accounts of hippocampal representation in memory integration. Representations of overlapping events (here AB, BC in the associative inference paradigm) are shown using a simplified two-layer architecture. The *bottom* layer contains units for each event element; the *top* layer contains hypothesized patterns of hippocampal representation. (**a**) Single integrated representation for overlapping events. According to this hypothesized structure, new, overlapping event elements (C) are encoded into an existing, reactivated memory (AB) to form a single composite representation for the two related associations. (**b**) Pattern separated representations of individual events. In this view, a new event (BC) with partial overlap to a previous memory (AB) would recruit a distinct hippocampal representation that preserves the details of each individual experience. Links between the common element (B) and each of the individual experiences could be used to mediate inference at encoding or retrieval. (**c**) Relational representation of overlapping events. In this framework, separate representations are maintained for overlapping events (AB, BC) and direct links between those events (at the level of the hippocampus) code their relationship to one another. (**d**) Evidence for dissociable coding schemes for indirectly related (A, C) items in an associative inference task across the anterior-posterior axis of hippocampus. While posterior hippocampus showed that A and C items became more dissimilar following overlapping encoding (*blue cluster*), anterior hippocampus coded A and C items more similarly, particularly when memories were strong (*green cluster*). These findings suggest simultaneous separated and integrated representation of overlapping memories in posterior and anterior hippocampus, respectively. Panels (**a–c**) are as originally published in Zeithamova D, Schlichting ML, Preston AR (2012) The hippocampus and inferential reasoning: building memories to navigate future decisions. *Frontiers in Human Neuroscience* 6:70. doi:10.3389/fnhum.2012.00070. Panel (**d**) is adapted from Schlichting et al. (2015)

The influential cognitive map theory (Tolman 1948; O'Keefe and Nadel 1978)—which first sparked interest in the flexible functions of the hippocampus—implicitly assumes such integrated representations. In the context of this theoretical framework, memory traces for newly learned individual events (i.e., recently traveled routes) are combined with memories of previously traveled routes to allow for the creation of an integrated map of the environment, including information about paths not traveled. As a cognitive map of an environment becomes established, it can be reactivated when an animal enters the same environment at a later point and updated with new experiences in that environment. When familiar routes to a goal are blocked, the cognitive map will enable navigation to the goal via an alternate route because information about this novel (i.e., never before traveled) route is included in a single representational structure of the environment.

In the context of non-spatial integration tasks, there is some evidence to support this hypothesized ABC representational structure. For example, one study showed that successful participants perform as quickly on integration judgments as on explicitly trained associations (Shohamy and Wagner 2008), suggesting similar representations for both directly learned and inferential associations. Moreover, informal assessment suggested that the majority of participants in this study failed to recognize the inferential probes as novel combinations of items, perhaps indicating that some contextual details of original experiences were lost. Returning to the dog-walking example, you may remember that the woman and man are a couple with a dog, but may not remember specific details about how you first encountered them. Future studies may provide a more detailed account of the circumstances under which memory for original experience may become degraded.

The loss of experiential detail is a significant downside to the single, composite representational structure linking elements of discrete events. Other computational perspectives propose a different representational structure for hippocampus, with pattern separation processes preserving distinct individual experiences and recurrent connections between the element and event representations allowing inference across experiences (Fig. 5b; Kumaran and McClelland 2012; McClelland et al. 1995). In our example, this representational structure would predict that a new event partially overlapping with a previous event (i.e., BC) would recruit a different hippocampal representation to make it distinct from the originally experienced event (AB). The two events would be linked through their individual connections to the shared event element (B). Because of the recurrent connections between individual element and event representations (ascribed to entorhinal cortex and hippocampus, respectively), such a hypothesized structure allows for preservation of event details while also supporting inferential judgments about the relationship between experiences. For example, when presented with a novel inferential probe (AC), each individual element (A and C) may serve as a partial cue leading to the reactivation of the originally experienced events (AB and BC). Activation of the common item (B) in both cases would lead to successful inference.

Results showing unique hippocampal responses during integration behavior itself (Preston et al. 2004; Zalesak and Heckers 2009; Zeithamova and Preston 2010) might reflect the use of such pattern separated inputs to support performance.

This representational structure can also explain recruitment of the hippocampus during encoding of overlapping events (Shohamy and Wagner 2008; Zeithamova and Preston 2010), which potentially reflects changes in the weights linking common elements to the individually experienced events. It is important to note that even such pattern-separated representations would be expected to change over time and become more generalized as follows (Kumaran and McClelland 2012). Reactivation of these memory representations during the consolidation process or during offline replay would result in more frequent reactivation of common elements and strengthening of their connections to event representations. In contrast, idiosyncratic elements unique to individual events would be reactivated less frequently and gradually lose their connections to event representations (Lewis and Durrant 2011). This process would lead to the gradual loss of episodic details in favor of abstracted representations that capture regularities across experiences (McClelland et al. 1995).

An alternate view that combines elements of both of these frameworks stems from relational memory theory (Cohen and Eichenbaum 1993). Relational memory theory proposes that the hippocampus maintains representations of individual events while also directly encoding relationships between separate experiences (Eichenbaum et al. 1999). In our symbolic representation of this theory, different hippocampal units are recruited to represent individual events, but a lateral connection exists at the second level, linking the representations of overlapping events together (Fig. 5c). Both pattern separation and pattern completion at the level of the hippocampus would contribute to the formation of such networks of related memories. For example, a new overlapping event (BC) would recruit a hippocampal representation distinct from the originally experienced event (AB). Simultaneously, the overlapping element (B) serves as a partial cue that reactivates the prior event (AB). Based on a Hebbian learning rule, the connection between the two hippocampal memory traces would be strengthened and an explicit link between the overlapping events would be formed. Like the representational structure above, such relational networks would support mnemonic inference while simultaneously preserving memory for individual experiences.

Different coding strategies may be preferred across subregions of the hippocampus. Prior work has implicated anterior hippocampus in processing relational information (Schacter and Wagner 1999; Kirwan and Stark 2004; Chua et al. 2007) and combining information across episodes (Preston et al. 2004; Addis et al. 2007; Barron et al. 2013), typically on the basis of activation enhancements during tasks that require consideration of multiple episodes. Mechanistically, anterior hippocampus might form generalized representations promoting behavioral flexibility using its broad place fields (Poppenk et al. 2013; Preston and Eichenbaum 2013; Strange et al. 2014). In contrast, posterior hippocampus, with its more finely tuned place fields, is thought to code event specifics. Consistent with this notion, rodent work has shown that while anterior hippocampal neurons respond similarly across related episodes, posterior hippocampal firing patterns are event-specific (Komorowski et al. 2013). Moreover, the ability to retrieve details has been differentially related to hippocampal volumes across the long axis, with smaller anterior and larger posterior regions being associated with superior recollection across individuals (Poppenk et al. 2013). These findings and others (Demaster et al. 2013) suggest dissociable functions along the hippocampal anterior-posterior axis in humans, with anterior generalizing across events and posterior representing event details (Poppenk et al. 2013). Anterior hippocampus also shares the strongest anatomical connections with MPFC (Barbas and Blatt 1995), making it a good candidate region for integrating across related experiences.

Despite the prominence of these theories, empirical evidence as to how elements of overlapping events are coded in human hippocampus has been demonstrated only recently (Collin et al. 2015; Schlichting et al. 2015). One study (Fig. 5d; Schlichting et al. 2015) scanned participants during viewing of individual items both before and after encoding of overlapping (AB, BC) associations to quantify how the representations of individual memory elements shift as a function of learning. In anterior hippocampus, indirectly related (A and C) items became more similar to one another following learning, consistent with integration across the related AB and BC events. In contrast, indirectly related items became more *dissimilar* in posterior hippocampus, suggesting separation of the overlapping events in this region. Similar findings were reported in another fMRI study (Collin et al. 2015) using a paradigm involving related events that could be combined to form narratives. Results revealed a gradient in the granularity of memory representations across the anterior-posterior axis of hippocampus, with individual events (small scale network) coded in posterior hippocampus and indirect relationships among related events (large scale network) represented only in anterior hippocampus. Neural codes also related to behavior, with only participants showing behavioral evidence of integration demonstrating a gradient in memory representation granularity. The results of both studies demonstrate that there are important regional differences in neural codes that allow for the simultaneous representation of integrated and separated memories within the hippocampus.

Implications for Behavior

Forming memories that integrate across related episodes is thought to confer a degree of mnemonic flexibility. For instance, by coding the relationships that span events, memories may be formed in anticipation of future decisions. In this section, we discuss the behavioral implications of memory integration across a number of cognitive domains. In section "Behavioral Benefits", we focus on the various benefits conferred by integrated memories on behavior. However, memory integration may also yield undesirable mnemonic consequences, which we describe in section "Behavioral Consequences".

Behavioral Benefits

Inferring Relationships

Inference is typically conceptualized as a logical, effortful process by which multiple memories are recombined to make a novel decision. In line with this intuition, initial studies of inference focused on hippocampal contributions to successful inference at the time of retrieval (Heckers et al. 2004; Preston et al. 2004; Zalesak and Heckers 2009; DeVito et al. 2010a). More recently, however, research attention has turned to the specialized hippocampal encoding mechanisms supporting the formation of integrated memories well suited to later decisions. Integrated memories may facilitate a host of novel judgments that require knowledge of the relationships among events, such as in associative inference, transitive inference, and acquired equivalence paradigms (Fig. 6; Zeithamova et al. 2012b; c.f. Kumaran 2012). These judgments tap memory flexibility, requiring participants to make novel inferences on the basis of trained associations; for simplicity, we group these behaviors under the term "inference." Because integrated memories code for the relationships among learned associations (Fig. 1), they may be reinstated and the new information directly extracted during an inference judgment itself (Shohamy and Wagner 2008).

Empirical evidence using neural decoding of cognitive states has demonstrated that an integration state can be differentiated from both pure retrieval and pure encoding states (Richter et al. 2015), suggesting that integration is neurally distinct from its underlying components. Evidence for an integration state in that study also predicted performance on the subsequent inference test both within and across participants, demonstrating that fluctuations in learning-phase integration impact subsequent behavioral flexibility. Recent work has also directly linked the degree of neural evidence for learning-phase reactivation of related memories to subsequent behavior (Kuhl et al. 2010; Zeithamova et al. 2012a; Richter et al. 2015). For instance, the degree to which previously encoded content is reactivated during new events has been shown to predict both subsequent memory for the reactivated content itself (Kuhl et al. 2010) and later inference (Fig. 3; Zeithamova et al. 2012a; Richter et al. 2015), consistent with the notion that reactivation supports memory strengthening and flexibility via integration. One study (Zeithamova et al. 2012a) also demonstrated that engagement of hippocampus and ventral MPFC related to later inference performance. Moreover, that study observed functional connectivity enhancements across learning repetitions, suggesting that memories bound in hippocampus may come to depend on MPFC as they are integrated and strengthened (Zeithamova et al. 2012a). Within the hippocampus, CA_1 engagement during overlapping events has been shown to predict subsequent inference (Schlichting et al. 2014). The degree to which learning-phase CA_1 patterns are reinstated during inference has also been shown to relate to speed and accuracy, consistent with ideas regarding this region's role in integration (Schlichting et al. 2014).

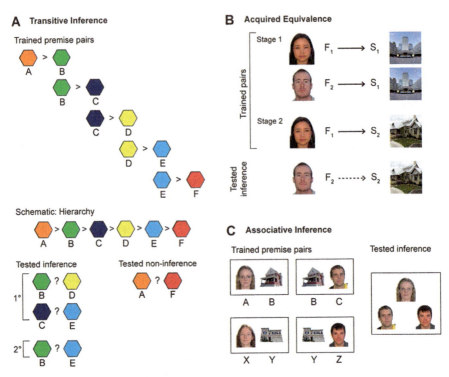

Fig. 6 Inference tasks. (**a**) Transitive inference task with six elements. A set of overlapping training pairs forms an ordered hierarchy of relationships. Participants learn each individual training pair via feedback (e.g., A > B) and are then tested on novel inference and novel non-inference judgments. Items in inferential probe trials may be separated by one element in the hierarchy (e.g., B ? D, indicated as 1°) or two elements (e.g., B ? E, indicated as 2°). Novel non-inferential probes test knowledge of the relationship between the end items of the hierarchy (A ? F). (**b**) Acquired equivalence task. In stage one of training, participants are trained via feedback to associate two faces (F_1 and F_2) with a particular scene (S_1). In stage two, participants learn to select a second scene (S_2) when cued with one of the faces (F_1). Inference is then measured as the proportion of trials on which participants choose S_2 when cued with F_2. The schematic depicts trained stimulus–response relationships (*solid black arrows*) and inferential relationships (*dashed black arrows*). (**c**) Associative inference task. Participants learn an overlapping set of associations (here, face–house associations), in which two stimuli (a man and a woman) are associated with a common third item (a house). Novel inference trials evaluate knowledge for the indirect relationship between items (who lives together in the same house). Figure as originally published in Zeithamova D, Schlichting ML, Preston AR (2012) The hippocampus and inferential reasoning: building memories to navigate future decisions. *Frontiers in Human Neuroscience* 6:70. doi:10.3389/fnhum.2012.00070

Recent work has also shown that inference is impaired in patients with lesions to ventral MPFC (Koscik and Tranel 2012). Furthermore, novel inference judgments are selectively facilitated following sleep (Ellenbogen et al. 2007; Werchan and Gómez 2013), emphasizing the importance of offline processes in integration.

Spatial Navigation

Perhaps the most familiar and widely studied form of memory integration stems from Tolman's seminal work on cognitive maps (Tolman 1948). Tolman proposed that navigation relies on the coherent representation of spatial layouts, which can flexibly give rise to new inferences about the relative locations of landmarks in the environment (Tolman 1948). One mechanism by which cognitive maps may be formed is by representing both recent past and future experience in the hippocampus at the same time. One rodent study demonstrated such simultaneous coding of retrospective and prospective paths leading up to a choice point in a continuous T-maze (Catanese et al. 2014), consistent with the notion that the hippocampus forms an ongoing representation of space including both past and future routes (see also e.g., Johnson and Redish 2007). Another recent report (Wu and Foster 2014) suggests that rather than separately representing the temporal structure of multiple traversed paths, the hippocampus integrates across overlapping routes to accurately codes the overall spatial topology of the environment. This type of representational scheme might support the ability to generate novel paths when, for instance, there is an obstacle blocking a learned route.

Recent work in humans has demonstrated a relationship between hippocampal volumes and the ability to infer novel spatial relationships among a set of trained landmarks (Schinazi et al. 2013), consistent with the idea that the hippocampus constructs integrated spatial maps. Behavioral studies have further found sleep- (Coutanche et al. 2013) and rest- (Craig et al. 2015) related increases in spatial relational inference performance. For instance, participants who passively rested for 10 min following route learning through a virtual environment had better memory for the spatial layout relative to participants who engaged in a 10-min distractor task (Craig et al. 2015). Importantly, the memory test tapped the formation of a cognitive map by assessing knowledge of routes that had never been directly experienced. Similar behavioral benefits have been reported in a group of participants who slept relative to a group who remained awake (Coutanche et al. 2013), indicating that early phase consolidation processes engaged during offline periods may facilitate the construction of cognitive maps.

Work in rodents demonstrates that the firing patterns of hippocampal CA_1 neurons predict future routes (Pfeiffer and Foster 2013). In one study, hippocampal trajectory events predicted rats' immediate future behavior as they navigated to a previously learned goal location in a familiar open arena. Trajectory events were more consistent with future than with previously traveled routes (Pfeiffer and Foster 2013), suggesting a role for hippocampal processing in planning future navigation through a familiar environment. Interestingly, trajectories can represent even novel future paths (although this is rare; Gupta et al. 2010; Pfeiffer and Foster 2013), suggesting that the hippocampus—perhaps guided by MPFC (de Bruin et al. 1994)—may support flexible navigation by simulating and evaluating possible trajectories in the context of current goals.

It is of note that uncertainty remains in the literature about precisely how the hippocampus encodes cognitive maps. For instance, it has been proposed that CA cells code the transition among locations ("transition cells"). Thus, the relationship between the memory integration mechanisms described here and the emergence of a cognitive map remain unclear at this point, and will be an important avenue of future investigations.

Mapping Social "Space"

The role of the hippocampus in integration is thought to be domain-general, with recent work extending this idea into studying social relationships. For example, one study (Kumaran et al. 2012) taught participants both social and non-social hierarchies in a transitive inference paradigm. Results showed that while fMRI activation and volume of the amygdala was specifically related to performance on the social hierarchy, the hippocampus represented the hierarchical structure for both social and non-social scenarios. In another recent experiment (Tavares et al. 2015), participants performed a role-playing task comprising a series of interactions with fictional characters. Over the course of the experiment, characters moved across social space due to changes in their power over and affiliation with the participant. Hippocampal engagement was modulated by the position of the character in social space, suggesting that the hippocampus codes for characters' relative positions as a function of their social attributes (i.e., power and affiliation). Hippocampal activation was also correlated with behavioral measures of social skills across participants, consistent with the notion that hippocampal representations of social space may explain some variability in real-world social behaviors. Taken together, these studies suggest that hippocampal integration mechanisms may aid us in forming a cognitive map of social space.

Decision Making

Integrated memories may also influence non-mnemonic decision making. For example, one recent fMRI study (Wimmer and Shohamy 2012) suggests that the hippocampus supports the transfer of monetary value across related experiences through additional recruitment of reward regions. Participants first learned a series of arbitrary $S_1 S_2$ associations. They then learned that half of the S_2 stimuli predicted a monetary reward (S_2+). During the critical decision phase, participants chose between two S_1 stimuli, only one of which was indirectly associated with a monetary reward (S_1+) through its association with a rewarded S_2. Value transfer was operationalized as the tendency to choose S_1+ over S_1-; importantly, though neither S_1 stimulus had been directly associated with a reward, one was indirectly predictive of monetary gain via S_2+. The researchers showed greater reactivation of prior related knowledge during encoding of new reward information for stimuli that showed more evidence of subsequent preference shifts toward S_1+.

Hippocampal-striatal functional coupling was also associated with value-related preference changes (Wimmer and Shohamy 2012), suggesting that hippocampus may interact with domain-specific regions (e.g., striatum in value learning tasks) in service of integration.

Consistent with a domain-general role for hippocampus in memory integration, rodent work (Blanquat et al. 2013) found that the hippocampus is necessary for updating a known goal location with new value information. These updated memories may then be transferred to neocortex, as MPFC was necessary for retaining the updated knowledge to support performance on the next day (Blanquat et al. 2013). Thus, integrated memories incorporating value information may be maintained as memory models in MPFC that will later bias behavior. We note that this role for MPFC is likely also domain-general given its documented involvement in a number of tasks lacking an explicit value component.

Schemas

Schemas are knowledge frameworks that capture regular patterns in the environment by abstracting information across experiences (Bartlett 1932) and represent features common to multiple different events while discarding idiosyncratic details. For example, a "restaurant schema" may contain commonly experienced elements such as sitting at a table, ordering from a menu, and paying the bill, but not one-time elements such as the waiter spilling water on you. We suggest that while the specific paradigms typically used to study memory schema are quite different from the associative learning tasks that are the focus of this chapter, these bodies of work share important features and the behaviors may be supported by a common neural mechanism. Like memory integration, building upon an existing knowledge structure (schema) to incorporate new information in particular has been shown to involve both hippocampus and MPFC.

Schemas guide behavior by providing a set of expectations for a given experience. Like integrated memory representations, schemas also contain information derived from multiple events that may support inferential decisions. Specifically, schemas represent relationships among elements commonly associated with certain types of situations, despite the fact that these elements have not necessarily been experienced together. Moreover, encoding new events in the context of a reactivated schema may provide an additional mechanism for inferential reasoning. For example, a person may come to your table at the end of your meal and inquire about the quality of the food and service. In the absence of an introduction, you may infer that this person is the owner or manager of the restaurant because your restaurant schema contains information about who is likely to ask for feedback about your dining experience.

Recent attention has focused on the behavioral benefits conferred by memory schema. For instance, research in rodents has demonstrated that reactivation of an existing task schema (in this case, a well-learned spatial layout) allowed for rapid acquisition of new flavor–place associations in a single trial (Tse et al. 2007, 2011).

Without an existing schema, such associative learning required repeated training across multiple weeks. Importantly, rats with hippocampal lesions failed to show facilitated learning of new information in the presence of reactivated schemas, highlighting a critical role for this region in the rapid incorporation of new information into existing knowledge frameworks. Echoing these results, a number of human studies have reported behavioral benefits in learning and memory when new information can be incorporated into an existing schema (Kumaran 2013; van Kesteren et al. 2013, 2014).

Rodent (Tse et al. 2011) and human (van Kesteren et al. 2010, 2013, 2014) work suggests that both MPFC and hippocampus are engaged during learning of schema-related information (i.e., schema updating). Recent empirical data indicate that one factor that may influence the relative engagement of MTL and MPFC is the degree of consistency between new information and existing schema. Specifically, one study (van Kesteren et al. 2013) demonstrated that MPFC engagement was more predictive of subsequent memory for information congruent with existing schema, perhaps reflecting direct encoding of new content into prior knowledge. Note that this idea contrasts with standard views of consolidation, which propose that hippocampal memories are transferred to neocortex after long time periods; however, recent work suggests the possibility of neocortical encoding of new information independent of the hippocampus (Sharon et al. 2011; see however Smith et al. 2014; Warren and Duff 2014). Conversely, MTL engagement was more predictive of successful encoding of incongruent information. Application of a schema to a new scenario has also been shown to primarily recruit hippocampus (Kumaran et al. 2009; de Hoz and Martin 2014). For example, one fMRI study (Kumaran et al. 2009) found that while engagement and connectivity of hippocampus and ventral MPFC was enhanced during generation of a task schema, the application of schema to guide behavior in a novel but similarly structured task selectively recruited hippocampus.

One theory (van Kesteren et al. 2012) of schema-dependent learning suggests that with increasing congruency, MPFC becomes increasingly able to bias reactivation toward related memories. Increasing congruency would also be associated with decreasing novelty, which may result in diminished reliance on hippocampal integration triggered by area CA_1. In such cases, MPFC memory models may guide reactivation and be updated directly, thus bypassing hippocampal involvement. In contrast, when an existing memory model is weak or nonexistent, MPFC would play no role in guiding memory retrieval. In this case, new content would be encoded by hippocampus. Across multiple related experiences (i.e., when forming a new schema), MPFC may come online (Zeithamova et al. 2012a), reflecting the emergence of guided reactivation and the abstraction across experiences. However, in many cases, new events are likely to be neither entirely novel nor identical replications of prior experience. These events will instead share a moderate level of congruency with existing memory models, and would thus be expected to involve both MPFC and hippocampus.

While one important characteristic typically ascribed to schemas is the loss of idiosyncratic details that code the differences among events, it remains unknown whether the same is true of integrated memory representations. Anecdotal evidence from the acquired equivalence paradigm suggests that some event details may also be lost during integration, as participants failed to recognize inferential probe trials as novel pairings of stimuli (Shohamy and Wagner 2008). This finding suggests that details about directly experienced events may sometimes be lost in favor of an abstracted, generalized framework that codes consistencies among distinct stimulus-response relationships. However, whether a similar loss of detailed event information is typical in other inference paradigms, especially those that utilize rapid acquisition procedures (e.g., single-trial learning), is not known. More research is needed to understand how the processes supporting inference are related to those implicated in the formation and use of schemas. Consideration of how task dynamics influence the type of representational structure formed may provide important insights into how the hippocampus codes overlapping event information and interacts with MPFC to support mnemonic flexibility. Moreover, it is noteworthy that the operational definition of "schema" varies across species (e.g., spatial layouts in rodents versus movie knowledge in humans) and across studies within a species (e.g., movie knowledge versus semantic knowledge in humans). Future work should seek to bridge the gap between animal and human work to better specify the conditions and mechanisms that support the building, updating, and use of memory schemas.

Learning and Associative Facilitation

Recent work suggests that new learning can be promoted by integrating new information into existing knowledge structures. This phenomenon is highly related to findings in the schema literature showing a behavioral benefit to encoding schema-congruent information (described in section "Schemas"). However, here we make no assumptions about the level of detail retained in the existing knowledge structure; prior memories need not be generalized.

The observation that prior knowledge can boost learning is by no means new; classic studies have shown that prior knowledge is beneficial to new learning under some circumstances (Bransford and Johnson 1972). For example, one such classic study showed a memory advantage for new responses paired with well-learned old stimuli (i.e., stimuli previously learned with a different response), a phenomenon known as associative facilitation (Underwood 1949). These observations appear robust across species, with existing knowledge of a spatial layout shown to facilitate acquisition of new related associations in rodents (Tse et al. 2007), for example. Such facilitation may also extend to novel judgments that require the simultaneous consideration of multiple memories (e.g., inferences).

Behaviorally, memory integration has been shown to have a protective effect on memory; instructing participants to integrate is associated with better memory for both the initial and newly encoded content (Anderson and McCulloch 1999; Forcato et al. 2010; see however Richter et al. 2015). Neuroimaging studies using the associative inference paradigm have shown that memory integration mechanisms may underlie associative facilitation (Schlichting and Preston 2014, 2016). Participants first formed strong memories for (AB) face-object pairs across four study-test iterations in a pre-training phase. They then encoded new object-object associations in a single exposure, half of which overlapped with (BC) and half of which did not overlap with (XY) prior knowledge. Importantly, overlapping and non-overlapping pairs were matched in terms of content type (two objects) and number of exposures (one per pair); thus, any differences in neural or behavioral signatures are attributable to the presence or absence of prior related knowledge. Results showed that the degree of evidence for memory reactivation during a rest period following AB pre-training predicted individual differences in the ability to later encode the new overlapping associations. Moreover, neural signatures during rest predicted engagement of face-sensitive regions at task, suggesting that offline memory processing promotes reactivation during the new learning phase (Schlichting and Preston 2014). Successful overlapping pair encoding was also associated with engagement of the hippocampal-MPFC circuit (Schlichting and Preston 2016). These findings suggest that the same memory integration mechanisms that support the ability to make novel inferences spanning events may also facilitate the encoding of new, related information.

Creativity and Imagination

Memory integration may also underlie the ability to recombine prior memories to construct new ideas and imagine future scenarios (Schacter et al. 2012). Consistent with this notion, recent work (Duff et al. 2013) has demonstrated that hippocampal damage results in impaired performance on creativity tasks in which participants generate novel responses on the basis of existing knowledge. MPFC may also support performance in such tasks; one fMRI study (Takeuchi et al. 2012) showed that individual differences in resting state functional connectivity of MPFC with posterior cingulate cortex predicted creativity.

Hippocampus and MPFC are also engaged during imagination (Martin et al. 2011; Barron et al. 2013), particularly when imagined scenarios are rich in episodic detail. One human fMRI study showed enhanced connectivity between hippocampus and MPFC during imagination of future scenarios that were later remembered (Martin et al. 2011), consistent with the notion that these regions are important for creating and maintaining integrated memories—even those representing imagined events. Another study (Barron et al. 2013) required participants to construct mental representations of novel foods from two familiar ingredients. Using an fMRI adaptation paradigm, researchers found that imagining novel foods engaged the same neuronal populations as did the ingredients in both hippocampus and MPFC, reflecting retrieval and recombination of prior memories during mental construction. The ingredient items themselves also came to recruit overlapping neuronal populations, perhaps reflecting integration of the simultaneously reactivated memories (Fig. 5). Interestingly, the degree of representational overlap of the ingredients in hippocampus and MPFC tracked across participants with subjective value of the imagined foods, suggesting that integration may be enhanced according to behavioral relevance (here, for high value items).

Behavioral Consequences

While we focus primarily on the positive outcomes associated with memory integration, a few noteworthy studies have highlighted its negative behavioral consequences. For example, integration may lead to the formation of false memories (i.e., through overgeneralization) (Cabeza et al. 2001; Warren et al. 2014), memory misattributions (Hupbach et al. 2007; Jones et al. 2012; Gershman et al. 2013; St. Jacques et al. 2013), and interference (Chan and LaPaglia 2013).

Both MTL and MPFC have been implicated in the formation of false memories. Neuroimaging studies have reported similar MTL engagement during recognition of both studied items ("true" memories) and unstudied lures ("false" memories) (Cabeza et al. 2001; Slotnick and Schacter 2004; Abe et al. 2008), suggesting that integrated hippocampal representations might underlie the tendency to incorrectly identify conceptually similar items as having been studied. Interestingly, these effects appear somewhat specific to anterior aspects of both hippocampus and MTL cortex (Cabeza et al. 2001; Abe et al. 2008; McTighe et al. 2010), while more posterior MTL regions (e.g., parahippocampal cortex) typically differentiate true from false memories based on activation (Cabeza et al. 2001; Okado and Stark 2003; Kim and Cabeza 2007a, b). These results are broadly consistent with the notion that anterior hippocampus in particular is well suited to integrate across related memories, perhaps at the cost of memory specificity. Ventral MPFC has also been implicated in constructing generalized memory representations; patients with ventral MPFC lesions show reduced false memories relative to healthy control participants for words that were never seen but are thematically related to a studied word list (Warren et al. 2014).

Integration may also explain the phenomenon of memory misattribution, in which an episodic experience is incorrectly attributed to a different encoding context than the one in which it occurred (e.g., as measured by intrusions). Misattributions may result when prior knowledge is reactivated and updated with the current experience to the detriment of memory accuracy. One fMRI study (Gershman et al. 2013) used neural decoding to quantify the reinstatement of the context associated with prior memories (List 1) during new learning (List 2). Results showed that greater evidence for reactivation of the List 1 context was associated with more misattributions of List 2 words to List 1. Another study (St. Jacques et al. 2013) showed that when participants reactivated a prior experience during new encoding, engagement of both hippocampus and ventral MPFC was associated with later memory misattributions, consistent with a role for these regions in linking experiences across time.

Memory integration mechanisms may also lead to interference or forgetting. When a memory is retrieved during a new learning experience, that memory becomes malleable and susceptible to change as a function of the current experience. One possible outcome of learning-phase reactivation is integration—that is, prior memories are updated to incorporate the new information. However, learning-phase reactivation can also lead to forgetting of the initial memory under some conditions (Walker et al. 2003; Forcato et al. 2007; Chan and LaPaglia 2013). For example, one behavioral study had participants watch a movie of a crime, which served as the initial memory. Later, participants listened to a narrative describing the crime that included misinformation: the crime was committed with a different weapon than the one depicted in the movie. Critically, reactivation of the initial memory prior to hearing the narrative resulted in forgetting of the initial, "true" memory of the crime (Chan and LaPaglia 2013). By design, the newly learned information in that study directly competed with or replaced the prior knowledge. Thus, whether memories for the original events are "overwritten" or simply updated to incorporate the new information may depend largely on the degree to which the two memories are compatible (Hupbach 2011).

It is notable that in the hippocampus, forgetting has typically been attributed to passive decay rather than interference due to the strong hippocampal tendency to pattern separate (Hardt et al. 2013). However, recent work suggesting that hippocampus—particularly its anterior portion—can form integrated codes that span related memories (Collin et al. 2015; Schlichting et al. 2015) calls this view into question. That is, memory integration predicts that even hippocampal memories may be forgotten when related content is incorporated into existing memory traces (i.e., through interference).

Conclusions

In summary, extensive evidence indicates that the hippocampus and its interactions with MPFC promote memory integration processes that support flexible cognition. Hippocampus does so by building memory representations that code not only associations within individual events, but also relationships spanning multiple episodes. In this way, the function of the hippocampus is not merely to enable the retrospective use of memory; rather, hippocampal function is "intrinsically prospective" (Klein et al. 2002a), aimed at constructing representations that can be used to successfully negotiate future judgments and actions. Integration tasks thus provide a powerful tool for studying the adaptive nature of memory and how the computational properties of the hippocampus allow memories to be reconstructed into prospectively useful formats.

The findings described here collectively suggest the importance of hippocampal encoding processes in linking related experiences. Integrated memories may support a host of flexible behaviors, from navigating our environment to imagining our future. Importantly, hippocampus does not work in isolation; rather, it communicates with other cortical regions to facilitate reactivation of memories, encoding of new memories, and updating of existing representations to incorporate new information. In doing so, it plays a key role in the extraction of knowledge across learning events.

Acknowledgments The authors were supported by the National Institute of Mental Health of the National Institutes of Health under award number R01MH100121 and by the National Science Foundation under CAREER award number 1056019 to Alison R. Preston during the writing of this chapter.

References

Abe N, Okuda J, Suzuki M et al (2008) Neural correlates of true memory, false memory, and deception. Cereb Cortex 18:2811–2819. doi:10.1093/cercor/bhn037

Addis DR, Schacter DL (2012) The hippocampus and imagining the future: where do we stand? Front Hum Neurosci 5:173. doi:10.3389/fnhum.2011.00173

Addis DR, Wong AT, Schacter DL (2007) Remembering the past and imagining the future: common and distinct neural substrates during event construction and elaboration. Neuropsychologia 45:1363–1377. doi:10.1016/j.neuropsychologia.2006.10.016

Anderson MC, McCulloch KC (1999) Integration as a general boundary condition on retrieval-induced forgetting. J Exp Psychol Learn Mem Cogn 25:608–629. doi:10.1037/0278-7393.25.3.608

Barbas H, Blatt GJ (1995) Topographically specific hippocampal projections target functionally distinct prefrontal areas in the rhesus monkey. Hippocampus 5:511–533

Barron HC, Dolan RJ, Behrens TEJ (2013) Online evaluation of novel choices by simultaneous representation of multiple memories. Nat Neurosci 16:1492–1498. doi:10.1038/nn.3515

Bartlett F (1932) Remembering: a study in experimental and social psychology. Cambridge University Press, Cambridge

Blanquat PDS, Hok V, Save E et al (2013) Differential role of the dorsal hippocampus, ventro-intermediate hippocampus, and medial prefrontal cortex in updating the value of a spatial goal. Hippocampus 23:342–351. doi:10.1002/hipo.22094

Bransford JD, Johnson MK (1972) Contextual prerequisites for understanding: some investigations of comprehension and recall. J Verbal Learn Verbal Behav 11:717–726

Buckner RL (2010) The role of the hippocampus in prediction and imagination. Annu Rev Psychol 61:27–48. doi:10.1146/annurev.psych.60.110707.163508

Bunsey M, Eichenbaum H (1996) Conservation of hippocampal memory function in rats and humans. Nature 379:255–257. doi:10.1038/379255a0

Cabeza R, Rao SM, Wagner AD et al (2001) Can medial temporal lobe regions distinguish true from false? An event-related functional MRI study of veridical and illusory recognition memory. Proc Natl Acad Sci U S A 98:4805–4810. doi:10.1073/pnas.081082698

Catanese J, Viggiano A, Cerasti E et al (2014) Retrospectively and prospectively modulated hippocampal place responses are differentially distributed along a common path in a continuous T-maze. J Neurosci 34:13163–13169. doi:10.1523/JNEUROSCI.0819-14.2014

Cavada C, Compañy T, Tejedor J et al (2000) The anatomical connections of the macaque monkey orbitofrontal cortex. A review. Cereb Cortex 10:220–242

Chan JCK, LaPaglia JA (2013) Impairing existing declarative memory in humans by disrupting reconsolidation. Proc Natl Acad Sci U S A 110:9309–9313. doi:10.1073/pnas.1218472110

Chua EF, Schacter DL, Rand-Giovannetti E, Sperling RA (2007) Evidence for a specific role of the anterior hippocampal region in successful associative encoding. Hippocampus 17:1071–1080. doi:10.1002/hipo.20340

Cohen NJ, Eichenbaum H (1993) Memory, amnesia, and the hippocampal system. The MIT Press, Cambridge, MA

Collin SHP, Milivojevic B, Doeller CF (2015) Memory hierarchies map onto the hippocampal long axis in humans. Nat Neurosci 18:1–5. doi:10.1038/nn.4138

Coutanche MN, Gianessi CA, Chanales AJH et al (2013) The role of sleep in forming a memory representation of a two-dimensional space. Hippocampus 23:1189–1197. doi:10.1002/hipo.22157

Craig M, Dewar M, Harris MA, Della Sala S (2015) Wakeful rest promotes the integration of spatial memories into accurate cognitive maps. Hippocampus. doi:10.1002/hipo.22502

de Bruin JPC, Sànchez-Santed F, Heinsbroek RPW et al (1994) A behavioural analysis of rats with damage to the medial prefrontal cortex using the morris water maze: evidence for behavioural flexibility, but not for impaired spatial navigation. Brain Res 652:323–333

de Hoz L, Martin SJ (2014) Double dissociation between the contributions of the septal and temporal hippocampus to spatial learning: the role of prior experience. Hippocampus 24:990–1005. doi:10.1002/hipo.22285

Demaster DM, Pathman T, Lee JK, Ghetti S (2013) Structural development of the hippocampus and episodic memory: developmental differences along the anterior/posterior axis. Cereb Cortex 24:3036–3045

Deuker L, Olligs J, Fell J et al (2013) Memory consolidation by replay of stimulus-specific neural activity. J Neurosci 33:19373–19383. doi:10.1523/JNEUROSCI.0414-13.2013

DeVito LM, Kanter BR, Eichenbaum H (2010a) The hippocampus contributes to memory expression during transitive inference in mice. Hippocampus 20:208–217. doi:10.1002/hipo.20610

DeVito LM, Lykken C, Kanter BR, Eichenbaum H (2010b) Prefrontal cortex: role in acquisition of overlapping associations and transitive inference. Learn Mem 17:161–167

Diekelmann S, Born J (2010) The memory function of sleep. Nat Rev Neurosci 11:114–126. doi:10.1038/nrn2762

Duff MC, Kurczek J, Rubin R et al (2013) Hippocampal amnesia disrupts creative thinking. Hippocampus 23:1143–1149. doi:10.1002/hipo.22208

Duncan K, Ketz N, Inati SJ, Davachi L (2012a) Evidence for area CA1 as a match/mismatch detector: a high-resolution fMRI study of the human hippocampus. Hippocampus 22:389–398. doi:10.1002/hipo.20933

Duncan K, Sadanand A, Davachi L (2012b) Memory's penumbra: episodic memory decisions induce lingering mnemonic biases. Science 337:485–487. doi:10.1126/science.1221936

Eichenbaum H, Schoenbaum G, Young B, Bunsey M (1996) Functional organization of the hippocampal memory system. Proc Natl Acad Sci U S A 93:13500–13507

Eichenbaum H, Dudchenko PA, Wood E et al (1999) The hippocampus, memory, and place cells: is it spatial memory or a memory space? Neuron 23:209–226

Ellenbogen JM, Hu PT, Payne JD et al (2007) Human relational memory requires time and sleep. Proc Natl Acad Sci U S A 104:7723–7728. doi:10.1073/pnas.0700094104

Euston DR, Gruber AJ, McNaughton BL (2012) The role of medial prefrontal cortex in memory and decision making. Neuron 76:1057–1070. doi:S0896-6273(12)01108-7

Forcato C, Burgos VL, Argibay PF et al (2007) Reconsolidation of declarative memory in humans. Learn Mem 14:295–303. doi:10.1101/lm.486107

Forcato C, Rodríguez MLC, Pedreira ME, Maldonado H (2010) Reconsolidation in humans opens up declarative memory to the entrance of new information. Neurobiol Learn Mem 93:77–84. doi:10.1016/j.nlm.2009.08.006

Gerraty RT, Davidow JY, Wimmer GE et al (2014) Transfer of learning relates to intrinsic connectivity between hippocampus, ventromedial prefrontal cortex, and large-scale networks. J Neurosci 34:11297–11303. doi:10.1523/JNEUROSCI.0185-14.2014

Gershman SJ, Schapiro AC, Hupbach A, Norman KA (2013) Neural context reinstatement predicts memory misattribution. J Neurosci 33:8590–8595. doi:10.1523/JNEUROSCI.0096-13.2013

Ghosh VE, Moscovitch M, Melo Colella B, Gilboa A (2014) Schema representation in patients with ventromedial PFC lesions. J Neurosci 34:12057–12070. doi:10.1523/JNEUROSCI.0740-14.2014

Gupta AS, van der Meer MAA, Touretzky DS, Redish AD (2010) Hippocampal replay is not a simple function of experience. Neuron 65:695–705. doi:10.1016/j.neuron.2010.01.034

Hardt O, Nader K, Nadel L (2013) Decay happens: the role of active forgetting in memory. Trends Cogn Sci 17:109–118. doi:10.1016/j.tics.2013.01.001

Hasselmo ME, Schnell E (1994) Laminar selectivity of the cholinergic suppression of synaptic transmission in rat hippocampal region CA1: computational modeling and brain slice physiology. J Neurosci 14:3898–3914

Hasselmo ME, Wyble BP, Wallenstein GV (1996) Encoding and retrieval of episodic memories: role of cholinergic and GABAergic modulation in the hippocampus. Hippocampus 6:693–708. doi:10.1002/(SICI)1098-1063(1996)6:6<693::AID-HIPO12>3.0.CO;2-W

Heckers S, Zalesak M, Weiss AP et al (2004) Hippocampal activation during transitive inference in humans. Hippocampus 14:153–162. doi:10.1002/hipo.10189

Henke K, Reber TP, Duss SB (2013) Integrating events across levels of consciousness. Front Behav Neurosci 7:68. doi:10.3389/fnbeh.2013.00068

Hoffman KL, Mcnaughton BL (2002) Sleep on it: cortical reorganization after-the fact. Trends Neurosci 25:1–2

Howard MW, Fotedar MS, Datey AV, Hasselmo ME (2005) The temporal context model in spatial navigation and relational learning: toward a common explanation of medial temporal lobe function across domains. Psychol Rev 112:75–116. doi:10.1037/0033-295X.112.1.75

Hupbach A (2011) The specific outcomes of reactivation-induced memory changes depend on the degree of competition between old and new information. Front Behav Neurosci 5:33. doi:10.3389/fnbeh.2011.00033

Hupbach A, Gomez R, Hardt O, Nadel L (2007) Reconsolidation of episodic memories: a subtle reminder triggers integration of new information. Learn Mem 14:47–53. doi:10.1101/lm.365707

Iordanova MD, Killcross AS, Honey RC (2007) Role of the medial prefrontal cortex in acquired distinctiveness and equivalence of cues. Behav Neurosci 121:1431–1436. doi:10.1037/0735-7044.121.6.1431

Jadhav SP, Kemere C, German PW, Frank LM (2012) Awake hippocampal sharp-wave ripples support spatial memory. Science 336:1454–1457. doi:10.1126/science.1217230

Ji D, Wilson MA (2007) Coordinated memory replay in the visual cortex and hippocampus during sleep. Nat Neurosci 10:100–107

Johnson A, Redish AD (2007) Neural ensembles in CA3 transiently encode paths forward of the animal at a decision point. J Neurosci 27:12176–12189. doi:10.1523/JNEUROSCI.3761-07.2007

Jones B, Bukoski E, Nadel L, Fellous J-M (2012) Remaking memories: reconsolidation updates positively motivated spatial memory in rats. Learn Mem 19:91–98. doi:10.1101/lm.023408.111

Kahana M (1996) Associative retrieval processes in free recall. Mem Cognit 24:103–109. doi:10.3758/BF03197276

Karlsson MP, Frank LM (2009) Awake replay of remote experiences in the hippocampus. Nat Neurosci 12:913–918. doi:10.1038/nn.2344

Kim H, Cabeza R (2007a) Differential contributions of prefrontal, medial temporal, and sensory-perceptual regions to true and false memory formation. Cereb Cortex 17:2143–2150. doi:10.1093/cercor/bhl122

Kim H, Cabeza R (2007b) Trusting our memories: dissociating the neural correlates of confidence in veridical versus illusory memories. J Neurosci 27:12190–12197. doi:10.1523/JNEUROSCI.3408-07.2007

Kirwan CB, Stark CEL (2004) Medial temporal lobe activation during encoding and retrieval of novel face-name pairs. Hippocampus 14:919–930. doi:10.1002/hipo.20014

Klein SB, Cosmides L, Tooby J, Chance S (2002a) Decisions and the evolution of memory: multiple systems, multiple functions. Psychol Rev 109:306–329. doi:10.1037//0033-295X.109.2.306

Klein SB, Loftus J, Kihlstrom JF (2002b) Memory and temporal experience: the effects of episodic memory loss on an amnesic patient's ability to remember the past and imagine the future. Soc Cogn 20:353–379

Komorowski RW, Garcia CG, Wilson A et al (2013) Ventral hippocampal neurons are shaped by experience to represent behaviorally relevant contexts. J Neurosci 33:8079–8087. doi:10.1523/JNEUROSCI.5458-12.2013

Koscik TR, Tranel D (2012) The human ventromedial prefrontal cortex is critical for transitive inference. J Cogn Neurosci 24:1191–1204. doi:10.1162/jocn_a_00203

Kroes MCW, Fernández G (2012) Dynamic neural systems enable adaptive, flexible memories. Neurosci Biobehav Rev 36:1646–1666. doi:10.1016/j.neubiorev.2012.02.014

Kuhl BA, Shah AT, DuBrow S, Wagner AD (2010) Resistance to forgetting associated with hippocampus-mediated reactivation during new learning. Nat Neurosci 13:501–506. doi:10.1038/nn.2498

Kuhl BA, Bainbridge WA, Chun MM (2012) Neural reactivation reveals mechanisms for updating memory. J Neurosci 32:3453–3461. doi:10.1523/JNEUROSCI.5846-11.2012

Kumaran D (2012) What representations and computations underpin the contribution of the hippocampus to generalization and inference? Front Hum Neurosci 6:157. doi:10.3389/fnhum.2012.00157

Kumaran D (2013) Schema-driven facilitation of new hierarchy learning in the transitive inference paradigm. Learn Mem 20:388–394. doi:10.1101/lm.030296.113

Kumaran D, McClelland JL (2012) Generalization through the recurrent interaction of episodic memories: a model of the hippocampal system. Psychol Rev 119:573–616. doi:10.1037/a0028681

Kumaran D, Melo HL (2013) Transitivity performance, relational hierarchy knowledge and awareness: results of an instructional framing manipulation. Hippocampus 23:1259–1268. doi:10.1002/hipo.22163

Kumaran D, Summerfield JJ, Hassabis D, Maguire EA (2009) Tracking the emergence of conceptual knowledge during human decision making. Neuron 63:889–901. doi:10.1016/j.neuron.2009.07.030

Kumaran D, Melo H, Duzel E (2012) The emergence and representation of knowledge about social and nonsocial hierarchies. Neuron 76:653–666. doi:10.1016/j.neuron.2012.09.035

Larkin MC, Lykken C, Tye LD et al (2014) Hippocampal output area CA1 broadcasts a generalized novelty signal during an object-place recognition task. Hippocampus 24:773–783. doi:10.1002/hipo.22268

Lewis PA, Durrant SJ (2011) Overlapping memory replay during sleep builds cognitive schemata. Trends Cogn Sci 15:343–351. doi:10.1016/j.tics.2011.06.004

Lisman JE, Grace AA (2005) The hippocampal-VTA loop: controlling the entry of information into long-term memory. Neuron 46:703–713. doi:10.1016/j.neuron.2005.05.002

Marr D (1970) A theory for cerebral neocortex. Proc R Soc Lond B Biol Sci 176:161–234

Martin VC, Schacter DL, Corballis MC, Addis DR (2011) A role for the hippocampus in encoding simulations of future events. Proc Natl Acad Sci U S A 108:13858–13863. doi:10.1073/pnas.1105816108

McClelland JL, McNaughton BL, O'Reilly RC (1995) Why there are complementary learning systems in the hippocampus and neocortex: insights from the successes and failures of connectionist models of learning and memory. Psychol Rev 102:419–457. doi:10.1037//0033-295X.102.3.419

McKenzie S, Eichenbaum H (2011) Consolidation and reconsolidation: two lives of memories? Neuron 71:224–233. doi:10.1016/j.neuron.2011.06.037

McKenzie S, Robinson NTM, Herrera L et al (2013) Learning causes reorganization of neuronal firing patterns to represent related experiences within a hippocampal schema. J Neurosci 33:10243–10256. doi:10.1523/JNEUROSCI.0879-13.2013

McKenzie S, Frank AJ, Kinsky NR et al (2014) Hippocampal representation of related and opposing memories develop within distinct, hierarchically organized neural schemas. Neuron 83:202–215. doi:10.1016/j.neuron.2014.05.019

McTighe SM, Cowell RA, Winters BD et al (2010) Paradoxical false memory for objects after brain damage. Science 330:1408–1410. doi:10.1126/science.1194780

Miller EK, Cohen JD (2001) An integrative theory of prefrontal cortex function. Annu Rev Neurosci 24:167–202. doi:10.1146/annurev.neuro.24.1.167

Munnelly A, Dymond S (2014) Relational memory generalization and integration in a transitive inference task with and without instructed awareness. Neurobiol Learn Mem 109:169–177. doi:10.1016/j.nlm.2014.01.004

Nadel L, Hardt O (2011) Update on memory systems and processes. Neuropsychopharmacology 36:251–273. doi:10.1038/npp.2010.169

Nadel L, Moscovitch M (1997) Memory consolidation, retrograde amnesia, and the hippocampal complex. Curr Opin Neurobiol 7:217–227

Nadel L, Samsonovich A, Ryan L, Moscovitch M (2000) Multiple trace theory of human memory: computational, neuroimaging, and neuropsychological results. Hippocampus 10:352–368

Nadel L, Hupbach A, Gomez R, Newman-Smith K (2012) Memory formation, consolidation and transformation. Neurosci Biobehav Rev 36:1640–1645. doi:10.1016/j.neubiorev.2012.03.001

O'Keefe J, Nadel L (1978) The hippocampus as a cognitive map. Clarendon, London

O'Reilly RC, Rudy JW (2001) Conjunctive representations in learning and memory: principles of cortical and hippocampal function. Psychol Rev 108:311–345. doi:10.1037//0033-295X.108.2.311

Okado Y, Stark C (2003) Neural processing associated with true and false memory retrieval. Cogn Affect Behav Neurosci 3:323–334. doi:10.3758/CABN.3.4.323

Pfeiffer BE, Foster DJ (2013) Hippocampal place-cell sequences depict future paths to remembered goals. Nature 497:74–81. doi:10.1038/nature12112

Poppenk J, Evensmoen HR, Moscovitch M, Nadel L (2013) Long-axis specialization of the human hippocampus. Trends Cogn Sci 17:230–240. doi:10.1016/j.tics.2013.03.005

Preston AR, Eichenbaum H (2013) Interplay of hippocampus and prefrontal cortex in memory. Curr Biol 23:R764–R773. doi:10.1016/j.cub.2013.05.041

Preston AR, Shrager Y, Dudukovic N, Gabrieli JDE (2004) Hippocampal contribution to the novel use of relational information in declarative memory. Hippocampus 14:148–152. doi:10.1002/hipo.20009

Price JL, Drevets WC (2009) Neurocircuitry of mood disorders. Neuropsychopharmacology 35:192–216. doi:10.1038/npp.2009.104

Richards BA, Xia F, Santoro A et al (2014) Patterns across multiple memories are identified over time. Nat Neurosci 17:981–986. doi:10.1038/nn.3736

Richter FR, Chanales AJH, Kuhl BA (2015) Predicting the integration of overlapping memories by decoding mnemonic processing states during learning. Neuroimage. doi:10.1016/j.neuroimage.2015.08.051

Roy M, Shohamy D, Wager TD (2012) Ventromedial prefrontal-subcortical systems and the generation of affective meaning. Trends Cogn Sci 16:147–156. doi:10.1016/j.tics.2012.01.005

Sara SJ (2010) Reactivation, retrieval, replay and reconsolidation in and out of sleep: connecting the dots. Front Behav Neurosci 4:185. doi:10.3389/fnbeh.2010.00185

Schacter DL, Wagner AD (1999) Medial temporal lobe activations in fMRI and PET studies of episodic encoding and retrieval. Hippocampus 9:7–24

Schacter DL, Addis DR, Hassabis D et al (2012) The future of memory: remembering, imagining, and the brain. Neuron 76:677–694

Schiller D, Phelps EA (2011) Does reconsolidation occur in humans? Front Behav Neurosci 5:24. doi:10.3389/fnbeh.2011.00024

Schinazi VR, Nardi D, Newcombe NS et al (2013) Hippocampal size predicts rapid learning of a cognitive map in humans. Hippocampus 23:515–528. doi:10.1002/hipo.22111

Schlichting ML, Preston AR (2014) Memory reactivation during rest supports upcoming learning of related content. Proc Natl Acad Sci U S A 111:15845–15850. doi:10.1073/pnas.1404396111

Schlichting ML, Preston AR (2015) Memory integration: neural mechanisms and implications for behavior. Curr Opin Behav Sci 1:1–8. doi:10.1016/j.cobeha.2014.07.005

Schlichting ML, Preston AR (2016) Hippocampal-medial prefrontal circuit supports memory updating during learning and post-encoding rest. Neurobiol Learn Mem. doi:10.1016/j.nlm. 2015.11.005

Schlichting ML, Zeithamova D, Preston AR (2014) CA1 subfield contributions to memory integration and inference. Hippocampus 24:1248–1260. doi:10.1002/hipo.22310

Schlichting ML, Mumford JA, Preston AR (2015) Learning-related representational changes reveal dissociable integration and separation signatures in the hippocampus and prefrontal cortex. Nat Commun 6:8151. doi:10.1038/ncomms9151

Sharon T, Moscovitch M, Gilboa A (2011) Rapid neocortical acquisition of long-term arbitrary associations independent of the hippocampus. Proc Natl Acad Sci U S A 108:1146–1151. doi:10.1073/pnas.1005238108/-/DCSupplemental.   www.pnas.org/cgi/doi/10.1073/pnas. 1005238108

Shohamy D, Wagner AD (2008) Integrating memories in the human brain: hippocampal-midbrain encoding of overlapping events. Neuron 60:378–389. doi:10.1016/j.neuron.2008.09.023

Singer AC, Karlsson MP, Nathe AR et al (2010) Experience-dependent development of coordinated hippocampal spatial activity representing the similarity of related locations. J Neurosci 30:11586–11604. doi:10.1523/JNEUROSCI.0926-10.2010

Slotnick SD, Schacter DL (2004) A sensory signature that distinguishes true from false memories. Nat Neurosci 7:664–672. doi:10.1038/nn1252

Smith CN, Urgolites ZJ, Hopkins RO, Squire LR (2014) Comparison of explicit and incidental learning strategies in memory-impaired patients. Proc Natl Acad Sci U S A 111:475–479. doi:10.1073/pnas.1322263111

St. Jacques PL, Olm C, Schacter DL (2013) Neural mechanisms of reactivation-induced updating that enhance and distort memory. Proc Natl Acad Sci U S A 110:19671–19678. doi:10.1073/pnas.1319630110

Staresina BP, Alink A, Kriegeskorte N, Henson RN (2013) Awake reactivation predicts memory in humans. Proc Natl Acad Sci U S A 110:21159–21164

Strange BA, Witter MP, Lein ES, Moser EI (2014) Functional organization of the hippocampal longitudinal axis. Nat Rev Neurosci 15:655–669. doi:10.1038/nrn3785

Takeuchi H, Taki Y, Hashizume H et al (2012) The association between resting functional connectivity and creativity. Cereb Cortex 22:2921–2929. doi:10.1093/cercor/bhr371

Tambini A, Ketz N, Davachi L (2010) Enhanced brain correlations during rest are related to memory for recent experiences. Neuron 65:280–290. doi:10.1016/j.neuron.2010.01.001

Tavares RM, Mendelsohn A, Grossman Y et al (2015) A map for social navigation in the human brain. Neuron 87:231–243. doi:10.1016/j.neuron.2015.06.011

Tolman EC (1948) Cognitive maps in rats and men. Psychol Rev 55:189–208. doi:10.1037/h0061626

Tse D, Langston RF, Kakeyama M et al (2007) Schemas and memory consolidation. Science 316:76–82. doi:10.1126/science.1135935

Tse D, Takeuchi T, Kakeyama M et al (2011) Schema-dependent gene activation and memory encoding in neocortex. Science 333:891–895. doi:10.1126/science.1205274

Underwood BJ (1949) Proactive inhibition as a function of time and degree of prior learning. J Exp Psychol 39:24–34

van Kesteren MTR, Fernández G, Norris DG, Hermans EJ (2010) Persistent schema-dependent hippocampal-neocortical connectivity during memory encoding and postencoding rest in humans. Proc Natl Acad Sci U S A 107:7550–7555. doi:10.1073/pnas.0914892107

van Kesteren MTR, Ruiter DJ, Fernández G, Henson RN (2012) How schema and novelty augment memory formation. Trends Neurosci 35:211–219. doi:10.1016/j.tins.2012.02.001

van Kesteren MTR, Beul SF, Takashima A et al (2013) Differential roles for medial prefrontal and medial temporal cortices in schema-dependent encoding: from congruent to incongruent. Neuropsychologia 51:2352–2359. doi:10.1016/j.neuropsychologia.2013.05.027

van Kesteren MTR, Rijpkema M, Ruiter DJ et al (2014) Building on prior knowledge: schema-dependent encoding processes relate to academic performance. J Cogn Neurosci 26:2250–2261. doi:10.1162/jocn

Walker MP, Brakefield T, Hobson JA (2003) Dissociable stages of human memory consolidation and reconsolidation. Nature 425:616–620. doi:10.1038/nature01951.1

Wang S-H, Tse D, Morris RGM (2012) Anterior cingulate cortex in schema assimilation and expression. Learn Mem 19:315–318. doi:10.1101/lm.026336.112

Warren DE, Duff MC (2014) Not so fast: hippocampal amnesia slows word learning despite successful fast mapping. Hippocampus 24:920–933. doi:10.1002/hipo.22279

Warren DE, Jones SH, Duff MC, Tranel D (2014) False recall is reduced by damage to the ventromedial prefrontal cortex: implications for understanding the neural correlates of schematic memory. J Neurosci 34:7677–7682. doi:10.1523/JNEUROSCI.0119-14.2014

Werchan DM, Gómez RL (2013) Generalizing memories over time: sleep and reinforcement facilitate transitive inference. Neurobiol Learn Mem 100:70–76. doi:10.1016/j.nlm.2012.12.006

Wilson RC, Takahashi YK, Schoenbaum G, Niv Y (2014) Orbitofrontal cortex as a cognitive map of task space. Neuron 81:267–279. doi:10.1016/j.neuron.2013.11.005

Wimmer GE, Shohamy D (2012) Preference by association: how memory mechanisms in the hippocampus bias decisions. Science 338:270–273. doi:10.1126/science.1223252

Wood ER, Dudchenko PA, Eichenbaum H (1999) The global record of memory in hippocampal neuronal activity. Nature 397:613–616. doi:10.1038/17605

Wu X, Foster DJ (2014) Hippocampal replay captures the unique topological structure of a novel environment. J Neurosci 34:6459–6469. doi:10.1523/JNEUROSCI.3414-13.2014

Zalesak M, Heckers S (2009) The role of the hippocampus in transitive inference. Psychiatry Res Neuroimaging 172:24–30. doi:10.1016/j.pscychresns.2008.09.008

Zeithamova D, Preston AR (2010) Flexible memories: differential roles for medial temporal lobe and prefrontal cortex in cross-episode binding. J Neurosci 30:14676–14684. doi:10.1523/JNEUROSCI.3250-10.2010

Zeithamova D, Dominick AL, Preston AR (2012a) Hippocampal and ventral medial prefrontal activation during retrieval-mediated learning supports novel inference. Neuron 75:168–179. doi:10.1016/j.neuron.2012.05.010

Zeithamova D, Schlichting ML, Preston AR (2012b) The hippocampus and inferential reasoning: building memories to navigate future decisions. Front Hum Neurosci 6:70. doi:10.3389/fnhum.2012.00070

Escaping the Past: Contributions of the Hippocampus to Future Thinking and Imagination

Daniel L. Schacter, Donna Rose Addis, and Karl K. Szpunar

Abstract The hippocampus has long been of interest to memory researchers, but recent studies have also implicated the hippocampus in various aspects of future thinking and imagination. Here we provide an overview of relevant studies and ideas that have attempted to characterize the contributions of the hippocampus to future thinking and imagination, focusing mainly on neuroimaging studies conducted in our laboratories that have been concerned with *episodic simulation* or the construction of a detailed mental representation of a possible experience. We briefly describe a multi-component model of hippocampal contributions to episodic simulation, and also consider the hippocampal contributions in the context of a recent taxonomy that distinguishes several forms of future thinking.

Introduction

It is difficult to think of a topic in cognitive neuroscience that has been investigated more extensively than the role of the hippocampus in memory. The range of questions posed about the hippocampus and memory is vast, covering just about all key aspects of memory research: What role does the hippocampus play in the consolidation of memories over time? Is the hippocampus critical for recall of only relatively recent memories, or is it also critical for recalling remote memories? What contribution does the hippocampus make to the initial encoding of memories? Is the hippocampus important for item memories or just for relational/associative

D.L. Schacter (✉)
Department of Psychology and Center for Brain Science, Harvard University, Cambridge, MA 02138, USA
e-mail: dls@wjh.harvard.edu

D.R. Addis
School of Psychology and Centre for Brain Research, The University of Auckland, Auckland 1142, New Zealand

K.K. Szpunar
Department of Psychology, University of Illinois-Chicago, Chicago, IL, USA

© Springer International Publishing AG 2017
D.E. Hannula, M.C. Duff (eds.), *The Hippocampus from Cells to Systems*,
DOI 10.1007/978-3-319-50406-3_14

memories, and is it more important for recollection than familiarity? Is hippocampal involvement restricted to the domain of long-term memory or is it also involved in short-term, working memory? Is the hippocampus critical only for conscious, explicit or declarative memories or does its influence extend to non-conscious, implicit, or non-declarative memories? Does the hippocampus play a special role in spatial memory and knowledge? The list could go on and on.

During the past decade or so, however, the range of questions about the hippocampus has expanded into new domains focusing on future thinking and imagination—topics that had hardly been considered in mainstream hippocampus research in previous decades. This expansion was fueled in large part by a convergence of findings from studies using different approaches and methods that revealed striking similarities between the cognitive and neural processes that support remembering past experiences and imagining possible future experiences. Thus, for example, behavioral studies revealed that remembered past events and imagined future events share phenomenological features, as exemplified by the finding that temporally close events in either the past or future include more episodic, sensory, and contextual details than more temporally distant events (e.g., Addis et al. 2008; D'Argembeau and Van der Linden 2004). Several different populations that show reduced retrieval of episodic details when remembering past experiences exhibit comparable reductions in episodic details when imagining future experiences, including older compared with younger adults (e.g., Addis et al. 2008) as well as patients with Alzheimer's disease (e.g., Addis et al. 2009b), mild cognitive impairment (Gamboz et al. 2010), depression (e.g., Williams et al. 1996), schizophrenia (e.g., D'Argembeau et al. 2008), bipolar disorder (King et al. 2011), Parkinson's disease (de Vito et al. 2012), and post-traumatic stress disorder (e.g., Brown et al. 2014). Linking more directly to the hippocampus, a number of studies have reported that amnesic patients with hippocampal damage also exhibit deficits when imagining future experiences and novel scenes (e.g., Andelman et al. 2010; Hassabis et al. 2007b; Kurzcek et al. 2015; Race et al. 2011; but for evidence of intact future imagining in amnesics, see Squire et al. 2010). Similarly, some evidence from developmental amnesics with hippocampal damage points toward impaired future imagining (Kwan et al. 2010) whereas other studies suggest spared capacities for imagining novel scenes and future scenarios in such patients (Cooper et al. 2011; Hurley et al. 2011). Although the exact reasons for the contrasting findings in hippocampal patients are still being debated (for discussion, see Addis and Schacter 2012; Maguire and Hassabis 2011; Schacter et al. 2012; Squire et al. 2011), numerous neuroimaging studies have shown that when healthy individuals are asked to remember past experiences and imagine future experiences, a common core network of regions is recruited that includes the hippocampus and medial temporal lobes (for review and discussion, see Benoit and Schacter 2015; Buckner and Carroll 2007; Mullally and Maguire 2013; Schacter et al. 2007a; Schacter et al. 2012). These kinds of observations have led to a dramatic increase in cognitive neuroscience research aimed at future thinking and imagination, with much of it directed at attempting to understand what role is played by the hippocampus in these processes, and how it is related to the more traditional role ascribed to the hippocampus in explicit or declarative memory.

We have previously written several reviews that have provided relatively comprehensive coverage of research from many laboratories that has examined hippocampal contributions to imagination and future thinking (Addis and Schacter 2012; Schacter and Addis 2009; Schacter et al. 2012; see also, Buckner 2010; Mullally and Maguire 2013). In the current chapter, we do not attempt to replicate this broad coverage of the entire field. Instead, we will focus mainly on reviewing studies of imagination and future thinking conducted in our own laboratories that have provided evidence relevant to conceptualizing the nature of hippocampal contributions to these processes. In so doing we will attempt to highlight key questions and issues that we have attempted to address, take stock of our findings to-date, and consider critical open questions that we think need to be pursued in future research. Before discussing our experimental observations concerning the role of the hippocampus in future thinking and imagination, however, we will first consider some general conceptual issues that are relevant to our research.

Imagination, Prospection, and Varieties of Future Thinking

As we have noted, the recent uptick in research concerning the role of the hippocampus in imagination and future thinking is attributable in part to the demonstration of striking similarities between remembering the past and imagining the future in neuroimaging studies, including common activation of the hippocampus. However, as we have discussed elsewhere (Addis et al. 2009a; Schacter et al. 2012), the distinction between "past events" and "future events" in many neuroimaging (and cognitive) studies is confounded with the distinction between "remembering" and "imagining". Remembered events must, of course, refer to past experiences. However, neural activity or cognitive properties that are associated with "future events" could be associated with "imagined events", regardless of whether the imagined events refer to the future, the past, or the present (see also, Hassabis and Maguire 2009). In Schacter et al. (2012), we argued that in light of these considerations, it is important to ask whether experiments that examine the relation between remembering the past and imagining the future inform our understanding of the relation between past and future, or whether they inform our understanding of the relation between memory and imagination, regardless of the temporal properties of imagined events. We reviewed relevant evidence and concluded that while there is some evidence of a role for temporal factors—that is, there is evidence that "imagining the future" differs in some respects from "atemporal imagining"—many of the documented similarities between remembering the past and imagining the future reflect commonalities between memory and imagination, independent of temporal factors (Schacter et al. 2012). We will return to this issue later in the chapter in relation to observations of hippocampal activations in neuroimaging studies.

A second general conceptual issue has to do with what we mean when we talk about "imagining the future" or "future thinking". Thinking about the future—often referred to by the term "prospection" (Gilbert and Wilson 2007; Seligman et al.

2013)—can take different forms. We (Szpunar et al. 2014a) have recently proposed a taxonomy of prospection that distinguishes among four basic modes of future thinking: *simulation* or the construction of a detailed mental representation of the future; *prediction* or the estimation of the likelihood of and/or one's reaction to a particular future outcome; *intention* or the mental act of setting a goal; and *planning* or the identification and organization of steps toward achieving a goal state. We further proposed that each of these four basic modes of prospection varies in the extent to which they are based on *episodic* or *semantic* information (Tulving 1983, 2002). In the context of our taxonomy, *episodic* refers to simulations, predictions, intentions, or plans concerning specific autobiographical events that might occur in the future (e.g., an upcoming vacation that will take place next month), whereas *semantic* refers to simulations, predictions, intentions, and plans that relate to more general or abstract states of the world that might arise in the future (e.g., thinking about what the world economy will be like 10 years from now). We conceived of this episodic-semantic dimension as continuous (vs. categorical) in order to allow for what we called "hybrid" forms of knowledge that combine episodic and semantic elements, such as personal semantics (Grilli and Verfaellie 2014; Renoult et al. 2012), which involves general but personal bits of knowledge (e.g., "I am a good golfer") that people can think about prospectively ("e.g., Someday I want to play golf on the PGA tour").

With respect to the present chapter, it is important to note that most research on the hippocampus and future thinking in our laboratories, as well as in the field more generally, has focused on episodic simulation (Schacter et al. 2008), that is, the construction of a detailed representation of a specific future personal experience. Thus, our discussion will necessarily focus primarily on the role of the hippocampus in episodic simulation. However, towards the end of the chapter we will also briefly discuss research that has provided evidence concerning the involvement of the hippocampus in prediction, intention, and planning. Note also that there is some evidence relevant to our understanding of possible contributions of the hippocampus to semantic simulation. In an early study, Klein et al. (2002) found that an amnesic patient who exhibited impaired episodic simulation of personal future events was nonetheless able to produce semantic simulations regarding problems that might face the world in the future, such as global warming. More recently, Race et al. (2013) showed that amnesic patients with medial temporal lobe damage (including hippocampal damage), and who were characterized by significant deficits in episodic simulation, could generate semantic simulations regarding issues that the world might face in the future. However, Race and colleagues found that these patients were impaired in their ability to elaborate on those issues. Thus, amnesic patients with episodic simulation deficits may also possess fine-grained deficits in semantic simulation, but the exact relation of these deficits to hippocampal function remains unclear, as Klein et al. (2002) did not report any neuroanatomical findings concerning their patient, and only one of the eight patients studied by Race et al. (2013) had damage restricted to the hippocampus (for detailed discussion of issues related to amnesic patients and future thinking, see Addis and Schacter 2012).

The Constructive Episodic Simulation Hypothesis

Our theoretical approach to conceptualizing hippocampal activations during imagination and future thinking has been defined by an idea that we have referred to as the *constructive episodic simulation hypothesis* (Schacter and Addis 2007a, b). This view emphasizes the key role played by episodic memory in supporting simulations of future experiences, although as acknowledged in our recent taxonomy (Szpunar et al. 2014a), it is clear that semantic memory also contributes critically to future thinking (see also Irish et al. 2012; Klein 2013). The constructive episodic simulation hypothesis holds that past and future events typically draw on similar information stored in episodic memory and rely on many of the same underlying constructive processes. Thus episodic memory is thought to support the construction of future events by extracting and recombining stored information into a simulation of a novel event. We have argued that this arrangement is adaptive because it enables past experiences to be used flexibly in simulating alternative future scenarios without engaging in actual behavior. Importantly, there is considerable evidence pointing toward adaptive functions of episodic simulation (for review and discussion, see Schacter 2012).

However, one potential cost of such a flexible system is that it is vulnerable to memory errors that result from miscombining elements of past experiences, such as misattribution and false recognition. Thus, Schacter and Addis (2007a, b) claimed that the constructive, error-prone nature of episodic memory is at least partly attributable to the key role of the episodic system in allowing people to construct simulations of their personal futures by drawing flexibly on elements of past experiences (for related ideas, see Dudai and Carruthers 2005; Suddendorf and Busby 2003; Suddendorf and Corballis 1997). Indeed, recent experimental evidence has shown that when people recombine elements of actual memories into novel simulations of possible experiences, they are sometimes prone to autobiographical memory conjunction errors, where a simulated experience is mistaken for an actual past experience (Devitt et al. 2015). Moreover, experiments by Carpenter and Schacter (2016) have provided evidence linking flexible recombination processes that support an adaptive cognitive function—associative inferences about relations between separate episodes that share a common element (e.g., Zeithamova and Preston 2010)—to source memory errors that result from mixing up elements of these episodes.

The emphasis placed by the constructive episodic simulation hypothesis on flexibly retrieving and recombining information from past episodes into future simulations provides a theoretical link to a conceptualization of hippocampal functions that naturally allows for its contributions to episodic simulation. Specifically, Eichenbaum and Cohen (2001, 2014) have proposed and provided evidence for the idea that the hippocampal region supports relational memory processes that link together disparate bits of information. Schacter and Addis (2007a) argued that these relational binding processes could support the function of recombining elements of information from episodic memory into simulations of events that might occur in the future, thereby suggesting at least one way in which the hippocampus might contribute to future event simulation. As we will see in subsequent sections, however, there are other ways in which the hippocampus may also contribute to imagining and future thinking.

Hippocampal Activity and Imagining the Future: Initial Observations

Our research on the relationship between remembering the past and imagining the future began with fMRI studies published in 2007, one that provided striking evidence of hippocampal activation during future imagining (Addis et al. 2007) and another that did not (Szpunar et al. 2007). Differences in the experimental designs used in the two studies, however, are the likely source of the different patterns of results regarding the hippocampus.

Our studies had been preceded by a positron emission tomography (PET) study from Okuda et al. (2003) that examined brain activity when people were asked to talk about past or future experiences that were either temporally close (i.e., last or next few days) or distant (i.e., last or next few years). Numerous brain regions showed common activation during these tasks compared with a control task that required semantic retrieval, including the hippocampus and other regions within the medial temporal lobe (MTL). These observations were important in suggesting a hippocampal contribution to future thinking, but the requirement to use a blocked design did not allow analysis of brain activity in relation to specific events. The relatively unconstrained nature of the task also made it difficult to discern whether participants were recalling and imagining specific experiences or providing more generic or semantic information about their pasts and futures. Thus these results could only provide limited evidence for the contribution of the hippocampus to imagining the future.

In an attempt to gain more experimental control over the nature of participants' memories and future imaginings, Addis et al. (2007) used event-related fMRI, which allowed separation of the past and future tasks into two phases: (1) an initial construction phase during which participants were instructed to remember a past event or imagine a future event in response to a cue word (e.g., "dress") and make a button-press when they had an event in mind; and (2) an elaboration phase during which participants mentally generated as much detail as they could about the event. We compared activity during the past and future tasks with two control tasks that required semantic and/or imagery processing. The main result of the experiment was a striking overlap during both construction and elaboration phases (more so during the elaboration phase) in a core network of regions that was similarly active when participants remembered the past and imagined the future, including medial prefrontal, medial temporal, and posterior parietal cortices (for discussion of this core network, see Benoit and Schacter 2015; Schacter et al. 2007a). Most important for the present purposes, the left hippocampus was robustly engaged during both the construction and elaboration phases in both the past and future tasks. Perhaps even more striking, the right hippocampus was selectively engaged during the construction phase of the future imagining task.

Addis et al. (2007) proposed that the common engagement of the left hippocampus during past and future tasks could reflect the retrieval of episodic details that are required both to remember a past event and imagine a future event. This finding and interpretation is consistent with the traditional characterization of the hippocampus as primarily a "memory region". However, the selective right hippocampal activation observed for future event construction fits well with the idea from the constructive episodic simulation hypothesis that the hippocampus may support a process of recombining details into a novel event, which is critical when imagining the future but not recruited to the same extent when remembering the past. In the next section, we will consider a series of subsequent studies that have explored alternative explanations and attempted to provide a more stringent direct test of the idea that the hippocampus contributes to recombination processes that are critical to future event simulation.

It is also useful to consider the previously mentioned study on future event simulation by Szpunar et al. (2007) in light of the preceding ideas. Participants were instructed to remember personal past events, imagine personal future events, or imagine events involving a familiar individual (Bill Clinton) in response to event cues (e.g., past birthday, retirement party). Consistent with the results of Okuda et al. (2003) and Addis et al. (2007), there was clear overlap in activity associated with remembering past events and imagining personal events in many core network regions. Importantly, these regions were not recruited to the same extent when participants imagined events involving Bill Clinton, thus providing evidence that the activated core network regions were specifically linked to the construction of events in their *personal* pasts or futures. However, there was no evidence in the experiment by Szpunar et al. (2007) for greater hippocampal activity for personal past or future events than for "Bill Clinton" events. Although we must be cautious about interpreting a negative finding, it is plausible that the "Bill Clinton" control task required the kinds of relational processing and recombining of event details that are associated with hippocampal activation. If so, significant hippocampal activations during the personal event task would not be evident in a comparison with the Bill Clinton control task.

Simulation or Prospection? Further Characterizations of Hippocampal Activity

These early observations established that hippocampal activity can be observed when people imagine future events, but left open many questions concerning how to interpret such activity. In particular, the idea that hippocampal activity during future imagining reflects, at least in part, recombination processes that are central to episodic simulation and play a relatively more important role in simulation than in remembering, is consistent with the initial results reported by Addis et al. (2007) indicating selective right hippocampal recruitment during construction of imagined future events. Several subsequent studies addressed the issue more directly, and also examined whether such activity is specific to prospection or can be observed when episodic simulations are not focused on the future.

A study by Addis and Schacter (2008) analyzed further hippocampal activity during the elaboration phase of the past and future event tasks that had been reported initially by Addis et al. (2007), focusing in particular on hippocampal responses associated with increasing amounts of rated detail for past and future events. Addis and Schacter (2008) suggested that when participants remember past events, details are primarily *reintegrated* (i.e., details that have been retrieved together previously are further integrated during retrieval), whereas when they imagine future events, additional processes are recruited that involve *recombining* details into a coherent event. Thus, hippocampal responses to increasing detail in past and future events should be distinguishable. A parametric modulation analysis showed that, on the one hand, the left posterior hippocampus was responsive to the amount of detail for *both* past and future events, probably reflecting the retrieval of details from episodic memory that are important for both tasks. On the other hand, a distinct region in the left anterior hippocampus responded more strongly to the amount of detail comprising future events, which we hypothesized reflects the recombination of details into a novel future event. An additional parametric modulation analysis focused on hippocampal responses associated with the temporal distance of events (i.e., recent or remote) in the past and future. Whereas increasing recency of past events was associated with activity in the right parahippocampal gyrus, increasing remoteness of future events was associated with activity in bilateral hippocampus. Addis and Schacter (2008) suggested that the stronger hippocampal response to more distant than closer future events reflects the increasing disparateness of details that participants included in remote future events, which in turn required more intensive relational and recombination processing to integrate these disparate details into a coherent future simulation.

In an attempt to link hippocampal activity and recombination processing even more closely, Addis et al. (2009a) developed a new *experimental recombination paradigm* that more clearly and directly elicits recombination processes than do standard paradigms that only require participants to imagine a future event. While it is typically assumed that participants engage in recombination processing in these standard paradigms, it is also possible that participants simply remember an entire past event and recast it as a possible future event. To address this issue, the experimental recombination paradigm requires participants to create a novel event from three details that they are recombining for the first time in the experiment. The procedure involves multiple stages. First, prior to scanning participants provide a long list of episodic memories comprised of a key *person, object,* and *place.* Second, the experimenter randomly recombines details across different memories into novel person-object-place arrangements. Third, during scanning participants imagine novel future events that include the recombined person-object-place details. A key finding from the Addis et al. (2009a) study was that of robust hippocampal activation when participants recombined event details on imagination trials, effectively ruling out the possibility that prior observations of hippocampal activity during future imagining reflects only recasting of entire actual past events into the future. Moreover, Addis et al. (2009a) also provided evidence that the activity in the right hippocampus was preferentially associated with imagining recombined events versus remembering actual events, in line with earlier observations from Addis et al. (2007).

This study also investigated another key question concerning the characteristics of hippocampal activation during episodic simulation: Is such activity specific to imagining *future* events, or is it more broadly associated with imagination irrespective of temporal considerations? To address the question, Addis et al. (2009a) included conditions in which participants were instructed to use person-object-place cues to imagine events that might occur in the future or might have occurred in the past (but had not). The result was clear-cut: the hippocampus was recruited to a similar extent when participants imagined both future and past events, suggesting that these regions are used for event simulation regardless of the temporal location of the event. These results dovetail nicely with findings from studies by Hassabis and Maguire and their colleagues showing that the hippocampus is strongly engaged when people are asked to imagine atemporal scenes that are not specifically linked to the past or future, suggesting that the hippocampus contributes importantly to a process of *scene construction* that is central to both remembering and imagining (e.g., Hassabis et al. 2007a; for review and discussion, see Hassabis and Maguire 2009; Mullally and Maguire 2013). Note that this scene construction hypothesis is quite similar to the constructive episodic simulation hypothesis, in that both ideas emphasize the contribution of the hippocampus to the construction of mental events. The scene construction idea places greater emphasis on the role of spatial information in constructed events, whereas the constructive episodic simulation hypothesis places greater emphasis on the contribution of the hippocampus to flexible recombination of various kinds of episodic details (e.g., people, objects, actions, places), with less focus on spatial details in particular.

Additional data indicating that the hippocampus serves a role in episodic simulation that is not exclusively prospective comes from studies that used fMRI to probe the neural correlates of *episodic counterfactual thinking* (De Brigard and Giovanello 2012): when people simulate an alternative outcome to a specific event that occurred in their personal pasts. De Brigard et al. (2013a) used a variant of the experimental recombination procedure in which participants initially provided detailed episodic memories of specific past experiences that had a particular outcome (e.g., "Last summer I went horseback riding with my sister in Virginia and I fell off my horse."). The experimenter then decomposed each memory into three components: a *context* (e.g., Last summer, Virginia), *action* ("Horse riding"), and *outcome* ("Fell off horse"). In the scanner, participants either recalled the memory in response to these three components, or constructed a counterfactual version of the memory with a different outcome (provided by the experimenter) that could involve changing a negative outcome to a positive one, changing a positive outcome to a negative one, or changing a peripheral detail of the memory that did not affect the outcome. De Brigard et al. (2013a) found that the right hippocampus (as well as many other regions in the core network noted earlier) was recruited during the construction of episodic counterfactual simulations where the outcome of the memory changed (from either positive to negative or vice versa). In a closely related study, Van Hoeck et al. (2013) directly compared brain activity when participants remembered past events, imagined possible future events, or constructed counterfactual simulations in which they mentally changed the outcome of a past event. They found that the left hippocampus was robustly engaged during the past, future, and counterfactual trials compared with a semantic control condition. In addition, Van Hoeck et al. reported that left hippocampus was more strongly engaged during the future than the past condition, thereby extending similar earlier observations from Addis et al. (2007), but did not find evidence for greater engagement of the hippocampus in the counterfactual than in the past condition. They suggested that because counterfactual simulations are more constrained by what actually happened in the past than are future simulations, there might be lesser recombination demand during counterfactual than future simulations.

In a subsequent study, De Brigard et al. (2015) examined counterfactual simulations involving self and others using an experimental paradigm that draws on autobiographical memories of events about which participants felt regret because of the outcome of a choice they made. For example, if a participant reported a memory where they missed an important appointment because they decided to take a bus instead of the subway, in the self condition they would be asked to construct a counterfactual simulation with a different outcome, i.e., "If only I had taken the subway instead of the bus." There were also several "other" conditions where participants constructed counterfactual simulations about people they knew well, or unfamiliar fictitious individuals. Compared with a control condition in which participants imagined changes to objects, there was evidence for robust recruitment of the right hippocampus and other core network regions for counterfactual simulation involving self and others, thereby extending the earlier results of De Brigard et al. (2013a). In addition, right hippocampus showed increased recruitment for counterfactual simulations about the self, compared with counterfactual simulations about others. Overall, then, the findings from the studies by De Brigard et al. (2013a, 2015), and Van Hoeck et al. (2013) provide further support for the idea that the hippocampus contributes broadly to the construction of episodic simulations of personal events regardless of whether those simulations entail novel future events or altered past events, although there may be important differences between future and counterfactual simulations (see also De Brigard et al. 2013b, for relevant behavioral evidence).

Additional evidence indicating that the hippocampus is not recruited to the same extent for all types of imagined events comes from a study by Addis et al. (2011a) that contrasted imagining (and remembering) specific events, as in the aforementioned studies, with remembering or imagining general or routine events that occurred frequently in the past or might occur frequently in the future (e.g., reading the newspaper each morning). Given prior evidence that the hippocampus is responsive to the amount of recombined detail in an imagined future event, we hypothesized that the hippocampus would show heightened activity for imagined specific events compared with routine events. If, by contrast, the hippocampus is mainly responsive to the prospective nature of future events, then it should be more engaged during the construction of both specific and general future events compared with past events.

Addis et al. (2011a) replicated the previously discussed finding from Addis et al. (2007) of increased right hippocampal activity for future versus past event construction. Critically, this increased right hippocampal activity was evident only for specific future events; there was no evidence for right hippocampal activity during construction of generic future events. Thus, consistent with results from Addis et al. (2009a) and De Brigard et al. (2013a), these data provide evidence against the idea that right hippocampal activation for specific future events reveals a uniquely prospective function for this region. Instead, it appears to respond to the amount of specific detail contained in an imagined event.

Encoding Processes and Memory for the Future

The studies reviewed in the previous section point toward a close link between hippocampal activity and episodic simulation that includes, but is not restricted to, imagined future events. The evidence is also consistent with the idea suggested by the constructive episodic simulation hypothesis that the hippocampus is linked to flexible recombination of event details. However, another possibility more closely linked with traditional views of hippocampal function is that activation of the hippocampus during episodic simulation reflects successful encoding of a novel simulated event into memory. Several decades ago, the Swedish neuroscientist David Ingvar recognized that in order for a future event simulation to be useful, it is important to encode the simulation into memory so that the information contained in the simulation could be retrieved at a later time when the simulated behavior is actually carried out. Ingvar (1985) termed this process "memory of the future" (for further discussion, see Szpunar et al. 2013). Given extensive evidence that the hippocampus contributes to successful encoding, especially of relational information (for review, see Davachi 2006), it is possible that some or all of the hippocampal activity observed in episodic simulation studies could be attributed to successful encoding.

To address this issue, Martin et al. (2011) used the experimental recombination paradigm described earlier together with a subsequent memory approach, where brain activity at the time of encoding is analyzed according to whether a particular item is subsequently remembered or forgotten on a memory test (e.g., Wagner et al. 1998). One desirable feature of the experimental recombination paradigm is that it provides a means to assess retention of the details that comprise an episodic simulation: specific details from the simulation can be provided as retrieval cues for other details. In the study by Martin et al. (2011), participants were scanned while they imagined future events involving person-object-place details that the experimenter had recombined from autobiographical memories that participants provided prior to scanning. Ten minutes after scanning, participants were given a cued recall test that included two details from each simulated event, and they were asked to recall the missing third detail (each type of detail served equally often as a cue as a memory target). When participants provided the correct detail, a simulation was scored as "remembered". When participants did not come up with a detail, or generated an incorrect detail, a simulation was scored as "forgotten". Of course, failing to generate the missing detail need not mean that the participant completely forgot all aspects of that simulation, but it seems reasonable to assume that participants retained more information from "remembered" than "forgotten" simulations, which is crucial to the logic of the experiment.

Martin et al. (2011) replicated previous findings of hippocampal activation during episodic simulation compared with a control condition. Critically, simulations that were successfully remembered were associated with greater activity at the time of encoding in the right anterior and posterior hippocampus than simulations that were later forgotten. An additional functional connectivity analysis showed that during successful encoding of a simulation, both anterior and posterior hippocampus exhibited connectivity with each other and with other core network regions. By contrast, when encoding was not successful this pattern of connectivity was no longer observed in the posterior hippocampus, whereas the anterior region still exhibited connectivity with the broader core network. Martin et al. (2011, see also Addis and Schacter 2012) suggested that the connectivity of the anterior hippocampus with the broader core network even during unsuccessful encoding might reflect the attempt to construct episodic simulation, even if it is encoded only to a level that is not sufficient to support subsequent recall.

Martin et al. (2011) also reported that successfully remembered episodic simulations were rated by participants during encoding as more detailed than subsequently forgotten ones, and that activity in both anterior and posterior hippocampal clusters was modulated by the level of detail (though the effect was significant only in the anterior hippocampus). Thus the contributions of the hippocampus to encoding success in this context might be related to construction of a detailed simulation of a future event.

Hippocampus, Event Novelty, and Repetition Suppression

Although it has been well established that the hippocampus is involved in future event simulation, not all studies reveal greater neural activity in this region when contrasting future with past events. For instance, Botzung et al. (2008) asked participants to provide detailed descriptions and summaries (e.g., museum-exposition) of 20 past and 20 future events one day prior to scanning. The summary cues were meant to (re-)evoke past and future events and were subsequently re-presented to participants in the scanner. In contrast to studies discussed earlier (e.g., Addis et al. 2007), there was no indication of greater activity in the hippo-campus for the future relative to past events. One possibility for the lack of a future > past pattern in hippocampus was that the novelty of future events as compared to past events had been eliminated by the provision for participants to generate future simulations outside the scanner. As a result, participants in this study may have been simulating memories of actual events and *memories* of simulated events. One implication of this pattern of data is that the hippocampus is involved in constructing *novel* future events.

To test this idea, van Mulukom et al. (2013) had participants simulate novel future events multiple times. Specifically, participants provided details about familiar people, places, and objects from 100 personal memories that were later used to generate 60 person-location-object simulation cues. One week later, these simulation cues were used to evoke 60 novel simulations of future events in the scanner. Critically, each simulation cue was presented three times in order to assess the extent to which hippocampal contributions to simulation were modulated by event repetition. The results of this study showed that, indeed, increases in simu-lation frequency were associated with decreases in hippocampal response (see Fig. 1), thus showing that hippocampus is especially responsive to initial as compared with repeated simulations of future events. These data suggest that future investigations of the role of the hippocampus in future event simulation should take care to ensure that the simulated events under consideration are sufficiently novel.

In addition to their findings associated with the hippocampus, van Mulukom et al. (2013) found a similar reduction in neural activity across the entire core network of regions generally associated with future event simulation (Benoit and Schacter 2015; Schacter et al. 2007a). This finding makes sense in light of extant work on the concept of repetition suppression, which states that regions or sets of regions responsible for representing particular stimuli demonstrate reduced neural responding with repeated presentations to those stimuli (Grill-Spector et al. 2006;

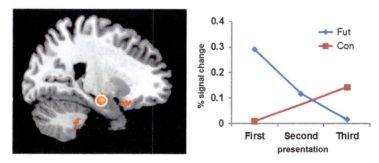

Fig. 1 Reduction in anterior right hippocampal activity across repeated simulations of future simulations (Fut) versus repeated presentations of the control task (Con). Adapted from van Mulukom et al. (2013)

Schacter et al. 2007b). Although prior work on repetition suppression had been mainly conducted using basic stimuli such as pictures of scenes (e.g., Epstein 2008), the results of van Mulukom et al. (2013) clearly demonstrated that a network of regions responsible for representing simulations of complex events abides by a similar principle.

Further evidence that bears on the interpretation of the hippocampal novelty effects reported by van Mulukom et al. (2013) comes from a study by Szpunar et al. (2014b), who assessed whether repetition suppression could be used to isolate the contributions of specific core network regions to future event simulation. Simulations of future events often involve details about people, places, and scenarios that tie those details together. The premise of the study by Szpunar et al. (2014b) was to manipulate the frequency with which specific elements of a complex event were simulated in order to assess which aspects of the core network would show repetition suppression in response to those particular elements. Among other findings, the results of this study neatly demonstrated that regions of the core network commonly associated with representing information about people, such as medial prefrontal cortex (e.g., Raposo et al. 2011), showed repetition suppression when people were repeated but not when locations or scenarios tying people and locations together were repeated. Moreover, regions commonly associated with representing information about places, such as retrosplenial, parahippocampal, and lateral parietal cortices (e.g., Epstein 2008), showed repetition suppression when places were repeated but not when people or scenarios tying people and locations together were repeated. Lastly, regions commonly associated with representing information about social scenarios, such as medial prefrontal, posterior cingulate, temporoparietal and lateral temporal cortices (e.g., Van Overwalle 2009), showed repetition suppression when particular scenarios were repeated but not when people or locations in isolation were repeated. Importantly, as was the case with the results of van Mulukom et al. (2013), Szpunar et al. (2014b) found that the hippocampus was particularly responsive to entirely novel events for which *all* elements (i.e., person, place, and scenario) had not been previously simulated, underscoring once again the link between hippocampal activity and event novelty.

A Multicomponent Account of Hippocampal Contributions to Episodic Simulation and Some Implications

The evidence that we have considered so far indicates that the hippocampus may contribute to episodic simulation in multiple ways. Addis and Schacter (2012) suggested that different regions within the hippocampus might support distinct component processes that play a role in imagining and remembering, including retrieving episodic details, recombining those details into coherent scenarios, and encoding novel scenarios into episodic memory. An important implication of this perspective is that hippocampal activations in neuroimaging studies could potentially reflect the contributions of some or all of these component processes, depending on the extent to which experimental conditions recruit each component. A further implication is that attempts to isolate the contribution of any one particular component need to control for the potential contributions of the others.

A study by Gaesser et al. (2013) illustrates these points. Gaesser et al. attempted to isolate the contributions of the hippocampus to the process of detail recombination during construction of an episodic simulation by using three procedures: the experimental recombination paradigm and a subsequent memory approach, as in the previously discussed study by Martin et al. (2011), together with a task switching procedure that was used in an attempt to control for novelty processing. As in the Martin et al. (2011) study, participants imagined novel future events based on person, object, and place details taken from autobiographical memories that participants had previously provided. Participants imagined some of these events for the first time in the scanner, and re-imagined other events that they previously imagined the day before. Gaesser et al. (2013) reasoned that events imagined for the first time should require greater recombination processing than re-imagined events because they require the initial integration of disparate details into a coherent simulation, whereas this process has already been carried out once for re-imagined events. However, as discussed in the previous section, simulations that are imagined for the first time are also more novel than re-imagined simulations, thereby making it difficult to determine whether any increased hippocampal activity for newly imagined compared with re-imagined simulations reflects differences in recombination demand or differences in event novelty (Szpunar et al. 2014b; van Mulukom et al. 2013).

Gaesser et al. (2013) attempted to control for such novelty differences by presenting recombined person, object, and place detail sets in a pre-exposure session the day before scanning (participants had provided person-object-place autobiographical memories a week earlier). During this pre-scan session, participants imagined future events for some detail sets and performed a different task— judging the relative pleasantness of the details—for other detail sets. The central idea here is that the novelty of the event details could be held constant across these two conditions by equating pre-exposure to the detail sets. Critically, however, the details were integrated into a coherent episodic future simulation during the imagine pre-exposure condition only; they were not integrated into a coherent episodic simulation in the pleasantness pre-exposure condition. When participants entered the scanner the following day, critical trials involved either switching tasks using the same detail sets as the previous day, or repeating the imagining task. Thus, differences between the imagine condition (i.e., imagine an event for the first time) and re-imagine condition during scanning could be attributed to recombination demand rather than to the novelty of event details in the imagine condition, because event details in this condition had been judged for pleasantness in the pre-scan session.

Ten minutes after the conclusion of scanning, participants received a cued recall test identical to the one used by Martin et al. (2011), in which two event details served as retrieval cues for the third event detail. Gaesser et al. attempted to hold constant encoding success in the neuroimaging analyses by focusing only on simulations that were successfully remembered. The key analysis thus focused on regions that showed increased activity for the imagine condition compared with the re-imagine condition, examining only successfully remembered items. This analysis revealed an effect in left posterior hippocampus, which Gaesser et al. (2013) attributed to a constructive process of recombining event details into a coherent episodic simulation of a future event. However, Gaesser et al. (2013) also pointed out that the precise localization of this activity differs from observations linking anterior hippocampus, and in some cases right anterior hippocampus, with flexible recombination of event details during episodic simulation (e.g., Addis et al. 2007, 2009a; Addis and Schacter 2008). One possible approach to reconciling the contrasting observations is that the anterior and primarily right-sided hippocampal activity in earlier studies reflects primarily successful encoding of novel episodes, consistent with the aforementioned results of Martin et al. (2011) and other evidence linking the anterior hippocampus with successful associative encoding (e.g., Chua et al. 2007; Jackson and Schacter 2004; Kirwan and Stark 2004; for review and discussion, see Davachi 2006; Poppenk et al. 2013; Schacter and Wagner 1999).

We are hesitant to attempt to draw any firm conclusions concerning the precise localization within the hippocampus (i.e., anterior-posterior, right-left) of component processes that support episodic simulation based on current neuroimaging evidence. We believe that progress on this issue should be facilitated by the use of high-resolution imaging protocols that allow more fine-grained distinctions among hippocampal subfields than are possible with the standard resolution techniques discussed so far (see also, Addis and Schacter 2012). Preliminary evidence along these lines is provided by a study from our laboratory by Stein et al. (2014) using the experimental recombination/subsequent memory paradigm from Martin et al. (2011) together with high-resolution imaging of the hippocampus. Consistent with previous results, Stein et al. found that the CA_1 hippocampal subfield, part of the anterior hippocampus, was associated with successful encoding of simulations into episodic memories. By contrast, activity in the dentate gyrus and $CA_{2/3}$ subfields, which occupy more posterior regions of the hippocampus, was linked with increasingly detailed episodic simulations, which Stein et al. (2014) hypothesized could reflect the operation of retrieval and recombination processes. Thus, although much more evidence is needed before firm conclusions can be drawn regarding intra-hippocampal localization of components processes that support episodic simulation, a preliminary sketch is beginning to emerge. Further research and theorizing on this issue will do well to consider research on episodic simulation in the broader context of studies examining possible function distinctions between anterior and posterior regions of the hippocampus and their connectivity with other brain regions (Poppenk et al. 2013).

Hippocampal Activity During Episodic Simulation in Aging and Depression

In another line of research, we have examined future simulation in populations that exhibit deficits in remembering past events, including older adults and individuals with depression. Consistent with the constructive episodic simulation hypothesis, our general hypothesis for these studies has been that if access to memory is impaired, then a parallel deficit for future simulation should also be evident. Moreover, given hippocampal dysfunction evident in these individuals, the ability to recombine any details accessed from episodic memory should also be compromised. We initially examined these questions across two behavioral studies with healthy older adults, where we used the Autobiographical Interview (AI; Levine et al. 2002) to distinguish between the "internal" or episodic details and "external" or semantic details that comprise autobiographical memories and simulations. In the first study (Addis et al. 2008) we examined the episodic content comprising past and future events generated by younger and older adults in response to single cue words (akin to the design of our first fMRI study described earlier—Addis et al. 2007). Our findings confirmed that older adults showed parallel deficits for past and future events, showing a significant reduction in the number of internal details generated for both past and future events. Surprisingly, however, older adults also exhibited an increase in external content (including semantic details and generic events) for both past and future events, which we suggested might occur to offset the decrease in episodic details or alternatively may reflect a change in communicative goals. Whatever the mechanism underlying the increase in non-episodic content, we had confirmed that the reduction in the episodic content of past events extended to future events.

In a follow-up study, we sought to investigate whether the ability to recombine details into a coherent simulation was also affected in healthy aging (Addis et al. 2010). To this end, we utilized the aforementioned experimental recombination paradigm, using recombined sets of person-object-place details as cues for past and future event trials. We replicated the overall finding of reduced internal and increased external details in descriptions of both past and future events. Importantly, however, the presentation of three simulation details meant we could also determine, for each trial, how many of these critical details were actually integrated into a single simulated event (i.e., one specific event occurring in a specific spatiotemporal context). The key finding here was that older adults integrated significantly fewer details into a single future event than did younger adults, suggesting impaired recombinatory processes likely due to reduced hippocampal function with advancing age.

To directly test the idea that age-related hippocampal dysfunction plays a role in these changes in episodic simulation, we (Addis et al. 2011b) conducted an fMRI study based on our original fMRI task (Addis et al. 2007). Overall, when remembering past and imagining future events older adults engaged many core network regions to a similar extent as young adults. Critically, however, older adults exhibited reduced activity in medial temporal regions, including the bilateral hippocampus, supporting the notion that reduced hippocampal activation is associated with reduced episodic content of past and future simulations. Indeed, ratings for the amount of detail comprising past and future events was only correlated with hippocampal activity in younger adults. In older adults, detail ratings were associated with increased activity in anterolateral temporal cortex (BA 20), likely reflecting increased non-episodic detail.

Individuals with depression also exhibit parallel changes in past and future events, such that the ability to generate *specific* events (i.e., events that are temporally and spatially specific) is reduced. Instead, depressed individuals typically generate "overgeneral" past and future events (e.g., "I am always late" vs. "I was late to work last Monday due to a traffic jam on the Northern Motorway"). Initially described in suicidally-depressed patients for past events (Williams and Broadbent 1986), this phenomenon has since been observed across the spectrum of depression, in individuals who are subclinically depressed (e.g., Dagleish et al. 2007), dysphoric (e.g., Dickson and Bates 2006), at risk of depression (e.g., Young et al. 2013) or currently in remission (e.g., Brittlebank et al. 1993; Mackinger et al. 2000). Moreover, this overgenerality extends to future events (Dickson and Bates 2006; Williams et al. 1996), consistent with our findings for older adults and the notion that remembering and imagining are closely related.

While much of the literature on overgeneral past and future events has attributed this impairment to the effects of rumination, functional avoidance and executive dysfunction (i.e., the CaRFaX model; Williams et al. 2007), few studies have considered the impact of hippocampal atrophy and dysfunction which is often evident in depression (Campbell and MacQueen 2004; Fairhall et al. 2010). Existing fMRI studies of past events in depression had not controlled for event specificity (Whalley et al. 2012; Young et al. 2012, 2013) and no imaging study had examined the neural correlates of future events in depression. Thus, we conducted a study in which individuals with and without a history of depression retrieved past events and imagined future events (Hach et al. 2014). Importantly, non-specific events were removed from the analysis to ensure specificity was matched, enabling us to compare group differences in the neural correlates of event construction rather than specificity per se. We found that the depression group not only exhibited reduced activity in the right hippocampus, but that right hippocampal connectivity with other core network regions was reduced relative to the control group. However, the depression group did show increased recruitment of lateral and medial frontal regions during the past and future tasks, as well as unique hippocampal connectivity with the dorsal attention network during the future task. It is possible that the additional neural resources recruited by the depression group, particularly during the future condition, may reflect greater effort given that the behavioral results from this fMRI study indicated that the deficit for specific events was significantly greater in the future than the past condition (Hach et al. 2014). That is, while the depression group generated significantly fewer specific future events than controls, this group difference was not significant for past events. Preliminary findings from a follow-up study we have conducted suggest that non-hippocampal factors such as strategic retrieval abilities may also contribute to this differential deficit of future simulation in depression (Hach et al. 2013).

However, for both depression and healthy aging, fMRI studies that decompose the component processes of future simulation (i.e., access to episodic details, recombination, novelty and encoding) are yet to be conducted. Such studies would provide a fuller and more nuanced picture of the changes in hippocampal function across these different groups and different types of future simulation deficits.

Future Directions and Concluding Comments

The findings and ideas discussed in this chapter indicate clearly that much has been learned about the contributions of the hippocampus to episodic simulation. Although it is equally clear that much remains to be learned, given that research on this topic only began in earnest within the past decade, the rapid recent increase in relevant data and theorizing is impressive and suggests that interest and activity will only continue to increase during the coming years. We conclude by briefly considering a couple of possible directions for future studies.

A recurring theme running through this chapter centers on the importance of distinguishing among component processes that support episodic simulation. Thus we have focused processes such as relational encoding, novelty processing, detail retrieval and recombination, and also referred to related concepts such as scene construction, all of which are thought to rely on the hippocampus. But it is important to note that experimental paradigms used to assess episodic simulation may also be influenced by other factors that have not been linked specifically to hippocampal function. The point is well illustrated by behavioral studies from our laboratory focused on aging and episodic simulation. As noted earlier, in studies using the AI (Levine et al. 2002) we found that older adults reported fewer internal (episodic) details and more external (semantic) details than younger adults both when they remembered past experiences and imagined future experiences (Addis et al. 2008, 2010). We initially interpreted these findings as support for the constructive episodic simulation hypothesis—i.e., that age-related changes in episodic memory are responsible for reduced internal details in older adults during both remembering and imagining. However, a subsequent study from our laboratory (Gaesser et al. 2011) showed that when older adults were asked to describe a picture of a complex scene—a task that we assumed would not recruit episodic memory mechanisms—they also produced fewer internal details (i.e., details present in the picture) and more external details (i.e., commentary and inferences about the picture) than did younger adults. These findings suggest that changes in such non-episodic processes as narrative style or communicative goals that occur with aging (see, for example, Adams et al. 1997; Labouvie-Vief and Blanchard-Fields 1982) impact both memory and simulation tasks, and thus contribute to the observed similarities between memory and simulation as a function of aging.

They also raise the possibility that even in studies that are not focused on aging, similarities between remembering the past and imagining the future might reflect primarily the influence of general, non-episodic processes, such as communicative goals or narrative style. If this is the case, it could have implications for interpreting hippocampal activations during episodic simulation which, contrary to theoretical approaches such as the constructive episodic simulation hypothesis, might be related to these non-episodic processes.

To begin to address the issue, we have carried out a series of recent studies in our laboratory that have allowed us to distinguish the impact of general, non-episodic processes such as narrative style or communicative goals from processes more closely related to episodic retrieval. We have done so by using what we refer to as an *episodic specificity induction*: brief training in recollecting details of a recent experience (Madore et al. 2014; Madore and Schacter 2016; for review and discussion, see Schacter and Madore 2016). In these studies, participants receive either an episodic specificity induction, where they are guided to focus on retrieving specific details from a recently viewed video (i.e., details of people, objects, and actions), or a control induction, where they are guided to provide general impressions of a video (i.e., how much they liked it, how well made they thought it was). The critical finding from these studies is that after receiving the specificity induction, participants later provide more internal or episodic details, but not external or semantic details, on subsequent tasks that involve remembering the past or imagining the future than after receiving the control induction. By contrast, the specificity induction has had no impact on a picture description task (Madore et al. 2014) or another semantic task that requires providing definitions of words (Madore and Schacter 2016).

These findings indicate that a specificity induction can dissociate the contributions of episodic retrieval and closely related processes (e.g., event or scene construction; see Schacter and Madore 2016) on the one hand from more general narrative or semantic processes on the other. Linking back to the hippocampus, in light of these behavioral results we have hypothesized that after receiving a specificity induction, hippocampal activity should increase when participants are scanned as they perform an episodic simulation task. We have recently reported an fMRI study that indeed provides evidence for increased activity in the hippocampus and other core network regions during an episodic future simulation task after a specificity induction versus after a control induction (Madore et al. 2016).

These preliminary findings suggest that specificity inductions could prove to be useful tools in helping to pinpoint the processes supported by the hippocampus during episodic simulation. We have also shown that the specificity induction can impact related tasks, such as means-end problem solving (Madore and Schacter 2014) and divergent creative thinking (Madore et al. 2015), for which there is also evidence of hippocampal or medial temporal lobe involvement (e.g., for means-end problem solving see Sheldon et al. 2011; for divergent creative thinking, see Benedek et al. 2014; Duff et al. 2013).

Finally, we began by noting at the outset of the chapter that according to a recent taxonomy of prospection, four basic modes of future thinking can be distinguished that vary along an episodic-semantic gradient: simulation, prediction, intention, and planning (Szpunar et al. 2014b). We have discussed only studies of episodic simulation in the current chapter because that has been the major focus on research related to the hippocampus in our lab and in other labs. But an intriguing question for future research concerns the extent and nature of hippocampal involvement in other forms of prospection. For example, there is an extensive literature from cognitive and social psychology concerning what is termed *episodic prediction* in our taxonomy, that is, estimating the likelihood of an outcome to a particular future autobiographical event or one's subjective response to that outcome. Studies of affective forecasting have shown that when making predictions about how they would feel in upcoming situations, people often overestimate or underestimate their future happiness (Gilbert and Wilson 2007). Gilbert and Wilson (2007) have linked these mistaken predictions to limitations on the kinds of episodic simulations that people construct regarding future scenarios, e.g., they have suggested that simulations sometimes capture the most salient but not the most likely elements of an experience, and at other times omit nonessential details that can impact future happiness. We are not aware of any evidence linking hippocampal activity to these kinds of episodic predictions, but given hippocampal involvement in episodic simulations, we expect that the hippocampus would also be involved in episodic predictions of future affective states.

Other modes of future thinking have received somewhat more attention in cognitive neuroscience research. For instance, studies of prospective memory have demonstrated a clear role for the hippocampus in encoding and retrieving delayed intentions for specific autobiographical events, or what we refer to as *episodic intentions* in our taxonomy (e.g., Cohen and O'Reilly 1996; Kliegel et al. 2008; Poppenk et al. 2010). Nonetheless, next to nothing is currently known about whether the hippocampus plays a similar role in processing intentions about non-specific autobiographical goals (e.g., forming an intention to become a better student; *hybrid intentions* in the taxonomy) or specific but non-autobiographical goals (e.g., setting a fiscal goal for a sales team; *semantic intentions*). Moreover, whether the hippocampus plays a role in processing delayed intentions beyond ensuring that those intentions are successfully encoded and retrieved remains to be elucidated in the literature.

Recent evidence also suggests a role for the hippocampus in episodic or autobiographical planning, which involves the organization of steps that need to be executed in order to attain a specific autobiographical future event or outcome. A series of studies conducted in our laboratory by Spreng and colleagues have used an autobiographical planning task in which participants are scanned while they mentally formulate plans to achieve specified goals. For example, a participant might be asked to formulate a plan to achieve the goal of academic success, and to integrate into the plan designated steps (attend class, study) and obstacles to be overcome (taking tests). Spreng et al. (2010) found that such autobiographical planning recruited all of the key regions within the core network discussed earlier including the hippocampus, and that activity within the core network during planning was coupled with activity in executive regions of the frontoparietal control network (see also, Spreng and Schacter 2012; Spreng et al. 2013). More recent analyses indicate that hippocampal activity during autobiographical planning is associated with more detailed and specific autobiographical plans (Spreng et al. 2015).

It is interesting to note in relation to the foregoing studies that there have been numerous studies of maze learning and spatial navigation in rats that suggest that activity in hippocampal neurons can serve predictive and planning functions via a neural "preplay" of upcoming events that allow the rat to use past experiences to plan future actions (for review and discussion, see Buckner 2010; Wikenheiser and Redish 2015). Although the relation between these studies and research on human future thinking and imagination is not fully understood, the two lines of research converge in that they point toward an important prospective function for the hippocampus. We expect that during the coming years, studies of both humans and non-human animals will continue to provide novel insights into the contribution of the hippocampus to imagination, future thinking, and related forms of cognition.

Acknowledgements Preparation of this chapter was supported by National Institute of Mental Health Grant MH060941 and National Institute on Aging Grant AG08441 to DLS and Rutherford Discovery Fellowship (RDF-10-UOA-024) to DRA.

References

Adams C, Smith MC, Nyquist L, Perlmutter M (1997) Adult age-group differences in recall for the literal and interpretive meanings of narrative texts. J Gerontol Psychol Sci 52B:187–195

Addis DR, Schacter DL (2008) Constructive episodic simulation: temporal distance and detail of past and future events modulate hippocampal engagement. Hippocampus 18:227–237

Addis DR, Schacter DL (2012) The hippocampus and imagining the future: where do we stand? Front Hum Neurosci 5:173

Addis DR, Wong AT, Schacter DL (2007) Remembering the past and imagining the future: common and distinct neural substrates during event construction and elaboration. Neuropsychologia 45:1363–1377

Addis DR, Wong AT, Schacter DL (2008) Age-related changes in the episodic simulation of future events. Psychol Sci 19:33–41

Addis DR, Pan L, Vu MA, Laiser N, Schacter DL (2009a) Constructive episodic simulation of the future and the past: distinct subsystems of a core brain network mediate imagining and remembering. Neuropsychologia 47:2222–2238

Addis DR, Sacchetti DC, Ally BA, Budson AE, Schacter DL (2009b) Episodic simulation of future events is impaired in mild Alzheimer's disease. Neuropsychologia 47:9–2671

Addis DR, Musicaro R, Pan L, Schacter DL (2010) Episodic simulation of past and future events in older adults: evidence from an experimental recombination task. Psychol Aging 25:369–376

Addis DR, Cheng T, Roberts R, Schacter DL (2011a) Hippocampal contributions to the episodic simulation of specific and general future events. Hippocampus 21:1045–1052

Addis DR, Roberts RP, Schacter DL (2011b) Age-related neural changes in remembering and imagining. Neuropsychologia 49:3656–3669

Andelman F, Hoofien D, Goldberg I, Aizenstein O, Neufeld MY (2010) Bilateral hippocampal lesion and a selective impairment of the ability for mental time travel. Neurocase 16:426–435

Benedek M, Jauk E, Fink A, Koschutnig K, Reishofer G, Ebner F, Neubauer AC (2014) To create or recall? Neural mechanisms underlying the generation of creative new ideas. Neuroimage 88:125–133

Benoit RG, Schacter DL (2015) Specifying the core network supporting episodic simulation and episodic memory by activation likelihood estimation. Neuropsychologia 75:450–457

Botzung A, Denkova E, Manning L (2008) Experiencing past and future events: functional neuroimaging evidence on the neural bases of mental time travel. Brain Cogn 66:202–212

Brittlebank AD, Scott J, Williams JMG, Ferrier IN (1993) Autobiographical memory in depression: state or trait marker? Br J Psychiatry 162:118–121

Brown AD, Addis DR, Romano TA, Marmar CR, Bryant RA, Hirst W, Schacter DL (2014) Episodic and semantic components of autobiographical memories and imagined future events in posttraumatic stress disorder. Memory 22:594–604

Buckner RL (2010) The role of the hippocampus in prediction and imagination. Annu Rev Psychol 61:27–48

Buckner RL, Carroll DC (2007) Self-projection and the brain. Trends Cogn Sci 11:49–57

Campbell S, MacQueen GM (2004) The role of the hippocampus in the pathophysiology of major depression. J Psychiatry Neurosci 29:417–426

Carpenter, AC, Schacter, DL (2016) Flexible retrieval: when true inferences produce false memories. J Exp Psychol Learn Mem Cogn. doi:10.1037/xlm0000340

Chua EF, Schacter DL, Rand-Giovannetti E, Sperling RA (2007) Evidence for a specific role of the anterior hippocampal region in successful associative encoding. Hippocampus 17:1071–1080

Cohen JD, O'Reilly RC (1996) A preliminary theory of the interactions between prefrontal cortex and hippocampus that conrtibute to planning and prospective memory. In: Brandimonte M, Einstein GO, McDaniel MA (eds) Prospective memory: theory and applications. Erlbaum, Hillsdale, NJ

Cooper JM, Vargha-Khadem F, Gadian DG, Maguire EA (2011) The effect of hippocampal damage in children on recalling the past and imagining new experiences. Neuropsychologia 49:1843–1850

D'Argembeau A, Van der Linden M (2004) Phenomenal characteristics associated with projecting oneself back into the past and forward into the future: influence of valence and temporal distance. Conscious Cogn 13:844–858

D'Argembeau A, Raffard S, Van der Linden M (2008) Remembering the past and imagining the future in schizophrenia. J Abnorm Psychol 117:247–251

Dagleish T, Golden AJ, Barrett LF, Yueng CA, Murphy V, Tchanturia K, Watkins E (2007) Reduced specificity of autobiographical memory and depression: the role of executive control. J Exp Psychol Gen 136:23–42

Davachi L (2006) Item, context and relational episodic encoding in humans. Curr Opin Neurobiol 16:693–700

De Brigard F, Giovanello KS (2012) Influence of outcome valence in the subjective experience of episodic past, future and counterfactual thinking. Conscious Cogn 21:1085–1096

De Brigard F, Addis DR, Ford JH, Schacter DL, Giovanello KS (2013a) Remembering what could have happened: neural correlates of episodic counterfactual thinking. Neuropsychologia 51:2401–2414

De Brigard F, Szpunar KK, Schacter DL (2013b) Coming to grips with the past: effects of repeated simulation on the perceived plausibility of episodic counterfactual thoughts. Psychol Sci 24:1329–1334

De Brigard F, Spreng RN, Mitchell JP, Schacter DL (2015) Neural activity associated with self, other, and object-based counterfactual thinking. Neuroimage 109:12–26

de Vito S, Gamboz N, Brandimonte MA, Barone P, Amboni M, Della Sala S (2012) Future thinking in Parkinson's disease: an executive function? Neuropsychologia 50:1494–1501

Devitt AL, Monk-Fromont E, Schacter DL, Addis DR (2015) Factors that influence the generation of autobiographical memory conjunction errors. Memory:1–19

Dickson JM, Bates GW (2006) Autobiographical memories and views of the future: in relation to dysphoria. Int J Psychol 41:107–116

Dudai Y, Carruthers M (2005) The Janus face of mnemosyne. Nature 434:567

Duff MC, Kurzcek J, Rubin R, Cohen NJ, Tranel D (2013) Hippocampal amnesia disrupts creative thinking. Hippocampus 23(12):1143–1149

Eichenbaum HE, Cohen NJ (2001) From conditioning to conscious recollection: memory systems of the brain. Oxford University Press, New York, NY

Eichenbaum HE, Cohen NJ (2014) Can we reconcile the declarative memory and spatial navigation views on hippocampal function. Neuron 83:764–770

Epstein RA (2008) Parahippocampal and retrosplenial contributions to human spatial navigation. Trends Cogn Sci 12:388–396

Fairhall SL, Sharma S, Magnusson J, Murphy B (2010) Memory related dysregulation of hippocampal function in major depressive disorder. Biol Psychol 85:499–503

Gaesser B, Sacchetti DC, Addis DR, Schacter DL (2011) Characterizing age-related changes in remembering the past and imagining the future. Psychol Aging 26:80–84

Gaesser B, Spreng RN, McLelland VC, Addis DR, Schacter DL (2013) Imagining the future: evidence for a hippocampal contribution to constructive processing. Hippocampus 12:1150–1161

Gamboz N, De Vito S, Brandimonte MA, Pappalardo S, Galeone F, Iavarone A, Della Sala S (2010) Episodic future thinking in amnesic mild cognitive impairment. Neuropsychologia 48:2091–2097

Gilbert DT, Wilson T (2007) Prospection: experiencing the future. Science 317:1351–1354

Grilli MD, Verfaellie M (2014) Personal semantic memory: insights from neuropsychological research on amnesia. Neuropsychologia 61:56–64

Grill-Spector K, Henson R, Martin A (2006) Repetition and the brain: neural models of stimulus-specific effects. Trends Cogn Sci 10:14–23

Hach S, Tippett LJ, Smith J, Bird-Ritchie T, Addis DR (2013) Staging the future: contributions of executive dysfunction to low specificity of future thinking in depression. Poster presented at the Annual Meeting of the Cognitive Neuroscience Society, San Francisco, USA

Hach S, Tippett LJ, Addis DR (2014) Neural changes associated with the generation of specific past and future events in depression. Neuropsychologia 65:41–55

Hassabis D, Maguire EA (2009) The construction system of the brain. Philos Trans R Soc (B) 364:1263–1271

Hassabis D, Kumaran D, Maguire EA (2007a) Using imagination to understand the neural basis of episodic memory. J Neurosci 27:14365–14374

Hassabis D, Kumaran D, Vann SD, Maguire EA (2007b) Patients with hippocampal amnesia cannot imagine new experiences. Proc Natl Acad Sci U S A 104(5):1726–1731

Hurley NC, Maguire EA, Vargha-Khadem F (2011) Patient HC with developmental amnesia can construct future scenarios. Neuropsychologia 49:3620–3628

Ingvar DH (1985) "Memory of the future": an essay on the temporal organization of conscious awareness. Hum Neurobiol 4:127–136

Irish M, Addis DR, Hodges JR, Piguet O (2012) Considering the role of semantic memory in episodic future thinking: evidence from semantic dementia. Brain 135:2178–2191

Jackson O, Schacter DL (2004) Encoding activity in anterior medial temporal lobe supports associative recognition. Neuroimage 21:456–464

King MJ, Williams LA, MacDougall AG, Ferris S, Smith JRV, Ziolkowski N, McKinnon MC (2011) Patients with bipolar disorder show a selective deficit in the episodic simulation of future events. Conscious Cogn 20:1801–1807

Kirwan CB, Stark CEL (2004) Medial temporal lobe activation during encoding and retrieval of novel face-name pairs. Hippocampus 14:919–930

Klein SB (2013) The complex act of projecting oneself into the future. Wiley Interdiscip Rev Cogn Sci 4:63–79

Klein SB, Loftus J, Kihlstrom JF (2002) Memory and temporal experience: the effects of episodic memory loss on an amnesic patient's ability to remember the past and imagine the future. Social Cogn 20:353–379

Kliegel M, McDaniel MA, Einstein G (eds) (2008) Prospective memory: an overview and synthesis of an emerging field. Erlbaum, Mahwah, NJ

Kurzcek J, Wechsler E, Ahuja S, Jensen U, Cohen NJ, Tranel D, Duff MC (2015) Differential contributions of the hippocampus and medial prefrontal cortex to self-projection and self-referential processing. Neuropsychologia 73:116–126

Kwan D, Carson N, Addis DR, Rosenbaum RS (2010) Deficits in past remembering extend to future imagining in a case of developmental amnesia. Neuropsychologia 48:3179–3186

Labouvie-Vief G, Blanchard-Fields F (1982) Cognitive aging and psychological growth. Aging Soc 2:183–209

Levine B, Svoboda E, Hay JF, Winocur G, Moscovitch M (2002) Aging and autobiographical memory: dissociating episodic from semantic retrieval. Psychol Aging 17:677–689

Mackinger HF, Pachinger MM, Leibetseder MM, Fartacek RR (2000) Autobiographical memories in women remitted from major depression. J Abnorm Psychol 109:331–334

Madore KP, Schacter DL (2014) An episodic specificity induction enhances means-end problem solving in young and older adults. Psychol Aging 29:913–924

Madore KP, Schacter DL (2016) Remembering the past and imagining the future: selective effects of an episodic specificity induction on detail generation. Q J Exp Psychol 69:285–298

Madore KP, Gaesser B, Schacter DL (2014) Constructive episodic simulation dissociable effects of a specificity induction on remembering, imagining, and describing in young and older adults. J Exp Psychol Learn Mem Cogn 40:609–622

Madore KP, Addis DR, Schacter DL (2015) Creativity and memory: effects of an episodic specificity induction on divergent thinking. Psychol Sci 26:1461–1468

Madore KP, Szpunar K, Addis DR, Schacter DL (2016) Episodic specificity induction impacts activity in a core brain network during construction of imagined future experiences. Proc Natl Acad Sci USA 113:10696–10701

Maguire, E.A., and Hassabis, D. (2011). Role of the hippocampus in imagination and future thinking. Proc Natl Acad Sci U S A, 108, E39.

Martin VC, Schacter DL, Corballis M, Addis DR (2011) A role for the hippocampus in encoding simulations of future events. Proc Natl Acad Sci U S A 108:13858–13863

Mullally SL, Maguire EA (2013) Memory, imagination, and predicting the future: a common brain mechanism? Neuroscientist 20:220–234

Okuda J, Fujii T, Ohtake H, Tsukiura T, Tanji K, Suzuki K, Kawashima R, Fukuda H, Itoh M, Yamadori A (2003) Thinking of the future and past: the roles of the frontal pole and the medial temporal lobes. Neuroimage 19:1369–1380

Poppenk J, Moscovitch M, McIntosh AR, Ozcelik E, Craik FIM (2010) Encoding the future: successful processing of intentions engages predictive brain networks. Neuroimage 49:905–913

Poppenk J, Evensmoen HR, Moscovitch M, Nadel L (2013) Long-axis specialization of the human hippocampus. Trends Cogn Sci 5:230–240

Race E, Keane MM, Verfaellie M (2011) Medial temporal lobe damage causes deficits in episodic memory and episodic future thinking not attributable to deficits in narrative construction. J Neurosci 31:10262–10269

Race E, Keane MM, Verfaellie M (2013) Losing sight of the future: impaired semantic prospection following medial temporal lobe lesions. Hippocampus 23:268–277

Raposo A, Vicens L, Clithero JA, Dobbins IG, Huettel SA (2011) Contributions of frontopolar cortex to judgments about self, others, and relations. Soc Cogn Affect Neurosci 6:260–269

Renoult L, Davidson PSR, Palombo DJ, Moscovitch M, Levine B (2012) Personal semantics: at the crossroads of sematic and episodic memory. Trends Cogn Sci 16:550–558

Schacter DL (2012) Adaptive constructive processes and the future of memory. Am Psychol 67:603–613

Schacter DL, Addis DR (2007a) The cognitive neuroscience of constructive memory: remember-
   ing the past and imagining the future. Philos Trans R Soc (B) 362:773–786
Schacter DL, Addis DR (2007b) The ghosts of past and future. Nature 445:27
Schacter DL, Addis DR (2009) On the nature of medial temporal lobe contributions to the
   constructive simulation of future events. Philos Trans R Soc (B) 364:1245–1253
Schacter DL, Madore KP (2016) Remembering the past and imagining the future: identifying and
   enhancing the contribution of episodic memory. Mem Stud 9:245–255
Schacter DL, Wagner AD (1999) Medial temporal lobe activations in fMRI and PET studies of
   episodic encoding and retrieval. Hippocampus 9:7–24
Schacter DL, Addis DR, Buckner RL (2007a) Remembering the past to imagine the future: the
   prospective brain. Nat Rev Neurosci 8:657–661
Schacter DL, Wig GS, Stevens WD (2007b) Reductions in cortical activity during priming. Curr
   Opin Neurobiol 17:171–176
Schacter DL, Addis DR, Buckner RL (2008) Episodic simulation of future events: concepts, data,
   and applications. Year Cogn Neurosci, Ann N Y Acad Sci 1124:39–60
Schacter DL, Addis DR, Hassabis D, Martin VC, Spreng RN, Szpunar KK (2012) The future of
   memory: remembering, imagining, and the brain. Neuron 76:677–694
Seligman MEP, Railton P, Baumeister RF, Sripada C (2013) Navigating into the future or driven
   by the past. Perspect Psychol Sci 8:119–141
Sheldon S, McAndrews MP, Moscovitch M (2011) Episodic memory processes mediated by the
   medial temporal lobes contribute to open-ended problem solving. Neuropsychologia
   49:2439–2447
Spreng RN, Schacter DL (2012) Default network modulation and large-scale network interactivity
   in healthy young and old adults. Cereb Cortex 22:2610–2621
Spreng RN, Stevens WD, Chamberlain JP, Gilmore AW, Schacter DL (2010) Default network
   activity, coupled with the frontoparietal control network, supports goal-directed cognition.
   Neuroimage 31:303–317
Spreng RN, Sepulcre J, Turner GR, Stevens WD, Schacter DL (2013) Intrinsic architecture
   underlying the relations among the default, dorsal attention, and frontoparietal control networks of the human brain. J Cogn Neurosci 25(2):74–86
Spreng RN, Gerlach KD, Turner GR, Schacter DL (2015) Autobiographical planning and the
   brain: activation and its modulation by qualitative features. J Cogn Neurosci. doi:10.1162/
   jocn_a_00846
Squire LR, van der Horst AS, McDuff SGR, Frascino JC, Hopkins RO, Mauldin KN (2010) Role
   of the hippocampus in remembering the past and imagining the future. Proc Natl Acad Sci U S
   A 107:19044–19048
Squire, L.R., McDuff, S.G., and Frascino, J.C. (2011). Reply to Maguire and Hassabis: autobio-
   graphical memory and future imagining. Proc Natl Acad Sci U S A, 108, E40.
Stein EM, McLelland VC, Devitt A, Schacter DL, Preston AR, Addis DR (2014) Dissociable roles
   of hippocampal subfields in episodic simulation. Presented at the Annual Meeting of the
   Society for Neuroscience, Washington, DC
Suddendorf T, Busby J (2003) Mental time travel in animals. Trends Cogn Sci 9:391–396
Suddendorf T, Corballis MC (1997) Mental time travel and the evolution of the human mind.
   Genet Soc Gen Psychol Monogr 123:133–167
Szpunar KK, Watson JM, McDermott KB (2007) Neural substrates of envisioning the future. Proc
   Natl Acad Sci U S A 104:642–647
Szpunar KK, Addis DR, McLelland VC, Schacter DL (2013) Memories of the future: new insights
   into the adaptive value of episodic memory. Front Behav Neurosci 7:47. doi:10.3389/fnbeh.
   2013.00047
Szpunar KK, Spreng RN, Schacter DL (2014a) A taxonomy of prospection: introducing an
   organizational framework for future-oriented cognition. Proc Natl Acad Sci U S A
   111:18414–18421

Szpunar KK, St. Jacques PL, Robbins CA, Wig GS, Schacter DL (2014b) Repetition-related reductions in neural activity reveal component processes of mental simulation. Soc Cogn Affect Neurosci 9:712–722

Tulving E (1983) Elements of episodic memory. Oxford University Press, New York, NY

Tulving E (2002) Episodic memory: from mind to brain. Annu Rev Psychol 53:1–25

Van Hoeck N, Ma N, Ampe L, Baetens K, Vanderkerchove M, Van Overwalle F (2013) Counterfactual thinking: an fMRI study of changing the past for a better future. Soc Cogn Affect Neurosci 8:556–564

van Mulukom V, Schacter DL, Corballis MC, Addis DR (2013) Re-imagining the future: repetition decreases hippocampal involvement in future simulation. PLoS One 8:e69596

Van Overwalle F (2009) Social cognition and the brain: a meta-analysis. Hum Brain Mapp 30:829–858

Wagner AD, Schacter DL, Rotte M, Koutstaal W, Maril A, Dale AM, Rosen BR, Buckner RL (1998) Building memories: remembering and forgetting of verbal experiences as predicted by brain activity. Science 281:1188–1190

Whalley MG, Rugg MD, Brewin CR (2012) Autobiographical memory in depression: an fMRI study. Psychiatry Res 201:98–106

Wikenheiser AM, Redish AD (2015) Decoding the cognitive map: ensemble hippocampal sequences and decision making. Curr Opin Neurobiol 32:8–15

Williams JMG, Broadbent K (1986) Autobiographical memory in suicide attempters. J Abnorm Psychol 95:144–149

Williams JMG, Ellis NC, Tyers C, Healy H, Rose G, MacLeod AK (1996) The specificity of autobiographical memory and imageability of the future. Mem Cognit 24:116–125

Williams JM, Barnhofer T, Crane C, Herman D, Raes F, Watkins E, Dalgleish T (2007) Autobiographical memory specificity and emotional disorder. Psychol Bull 133:122–148

Young KD, Erickson K, Nugent AC, Fromm SJ, Mallinger AG, Furey ML, Drevets WC (2012) Functional anatomy of autobiographical memory recall in depression. Psychol Med 42:345–357

Young KD, Bellgowan SF, Bodurka J, Drevets WC (2013) Behavioral and neurophysiological correlates of autobiographical memory deficits in patients with depression and individuals at high risk for depression. J Am Med Assoc Psychiatry 70:698–708

Zeithamova D, Preston AR (2010) Flexible memories: differential roles for medial temporal lobe and prefrontal cortex in cross-episode binding. J Neurosci 30:14676–14684

Distinct Medial Temporal Lobe Network States as Neural Contexts for Motivated Memory Formation

Vishnu P. Murty and R. Alison Adcock

Abstract In this chapter we examine how motivation creates a neural context for learning by dynamically engaging medial temporal lobe (MTL) systems. We review findings demonstrating that distinct modulatory networks, centered on the ventral tegmental area (VTA) and amygdala, are coherently recruited during specific motivational states and shunt encoding to hippocampal versus cortical MTL systems during learning. We posit that these shifts in encoding substrate serve to tailor both the content and form of memory representations, and speculate that these different representations support current and future adaptive behavior.

Memories are not veridical, but rather selective representations of the environment. Understanding memory selectivity is a fundamental aim of memory research, and a rich literature accumulated over a century has documented properties of external events that are likely to change the brain to create lasting memories. Events that are distinctive, salient, or emotional, are better remembered. The intrinsic properties of events explain many features of memory selectivity, but the brain is a dynamic system. In order to understand how experience become memories, we must begin to characterize not only the properties of external events, but also the state of the brain during encoding.

Motivation as an Adaptive Neural Context for Encoding

One potentially powerful taxonomy of brain states defines them in relationship to motivated behaviors. Animals actively acquire information from the environment—both intentionally and incidentally—as they strive to achieve their goals.

V.P. Murty
Department of Psychiatry, Center for the Neural Basis of Cognition, University of Pittsburgh, Pittsburgh, PA 15123, USA

R. Alison Adcock (✉)
Departments of Psychiatry and Behavioral Sciences, Department of Neurobiology, Department of Psychology, Center of Cognitive Neuroscience, Duke University, Durham, NC, USA
e-mail: alison.adcock@duke.edu

© Springer International Publishing AG 2017   467
D.E. Hannula, M.C. Duff (eds.), *The Hippocampus from Cells to Systems*,
DOI 10.1007/978-3-319-50406-3_15

Neural systems underlying motivation translate goals into actions that in turn support adaptive behavior. A long history of learning theory has examined how motivation supports learning (Daw and Doya 2006; Wise 2004; Schultz 2016). This literature has mainly focused on simple Pavlovian (*stimulus-stimulus*) and instrumental (*stimulus-response-outcome*) learning. Although it has long been known that motivational states modulate neurophysiology throughout the brain, their influence on memory formation remained relatively unexamined for decades.

Emerging research has now demonstrated a critical role for motivation in guiding forms of memory that rely on engagement of the hippocampus and the surrounding medial temporal lobe (MTL) cortex. These studies have investigated how motivation, mostly in the reward domain, influences declarative memory, relational learning, and generalization (Shohamy and Adcock 2010; Miendlarzewska et al. 2016). This literature broadly suggests that systems underlying motivation interact with the MTL to support episodic memory, and generally enhance, episodic memory encoding. We propose that ultimately it is the state of an individual's neuromodulatory circuits in response to motivational incentives, and hence the MTL memory systems engaged during encoding, that precisely determine the content and form of memory.

Below we review the recent evidence indicating that motivation supports memory formation, and that distinct motivational states may correspond to distinct neural contexts for memory formation. We propose a model in which encoding during motivated behavior reflects the specific interactions of neuromodulatory and MTL memory systems engaged by motivational states, to create adaptive memory.

Motivational Goals Are Complex and Encompass Both Action and Learning

An important assumption of our proposed model is that motivation is not a unitary construct, but rather encompasses multiple dimensions. These dimensions include not only characteristics of actions, which include energization (*vigor*) and orientation (*approach, avoidance*), but also the incentives posed by accumulating evidence and by information itself. We refer to these latter information-based motivational states as *interrogative* and *imperative* motivational states. Within our framework, interrogative states reflect information processing that not only supports an individual's immediate goals but also supports resolving goal conflicts and future goal attainment. Imperative states reflect information processing that is predominantly focused on supporting an individual's immediate, unconflicted goals. As examples, when an individual encounters a threatening snake on a hike, she may have an imperative goal of avoiding the present, immediate threat. However, if a fellow hiker mentions the great view of the sunset from the trail, she may have an interrogative goal of learning all of the best locations to capture this view.

It has been extensively argued that objective descriptors of external incentives are insufficient characterizations of motivational states, in that many incentive structures may be framed as either approaching good or avoiding bad outcomes (Strauman and Wilson 2010; Higgins 1998). Assessing individuals' incentives for information processing may be similarly complex. A reward may be likely to evoke an interrogative state, but in the face of high stakes opportunities or social evaluation, the same reward incentive could evoke an imperative goal (Mobbs et al. 2009a; Yu 2015; Ariely et al. 2009). For example, if the hiker found out there was only five minutes before sunset, her information processing may reflect an imperative goal state. Similarly, prior knowledge that snakes could appear somewhere on a trail may evoke an interrogative goal state for avoiding threats. Despite these complexities, however, substantial evidence supports a predisposition for reward incentives to evoke interrogative goal states and punishment incentives to evoke imperative goal states, as we review below.

An Investigative Approach to Motivated Memory

Compared to motivation for action, operationalizing the outcomes of motivation to learn presents additional challenges. Motivation for action can be directly manipulated by the nature of the incentive (i.e., punishment versus reward) and assessed by measuring behavior (i.e., reaction time or number of button presses as measures of effort or vigor). Motivation for learning, in contrast, does not have an established behavioral signature. To address this challenge, our approach emphasizes the activation of discrete neuromodulatory brain systems as the most direct indicator of distinct motivational states for learning, with the form and content of memory serving as their behavioral read-out.

We focus our review and our recent experimental work on two discrete neuromodulatory systems, centered on the ventral tegmental area (VTA) and the amygdala. The VTA has been associated with relatively interrogative goal-oriented behaviors, such as exploration. The amygdala has been associated with relatively imperative goal-oriented behaviors, such as freezing. (See Fig. 1). Literature in the present review is organized, in part, based on the use of different incentive conditions, namely reward and punishment. These incentives have been shown to reliably, albeit not uniquely, engage distinct brain centers for motivation: the VTA and amygdala, respectively (described below). Thus, characterizing discrete influences of reward and punishment allows us to examine the neural architecture of encoding and the declarative memories formed as participants engage in similar encoding tasks while under varying modulatory influences that reflect interrogative and imperative goal states (See Fig. 4 for an example of our own approach.) Note that when the evidence is available, however, we focus on the neuromodulatory system that is engaged, rather than the valence of the incentive.

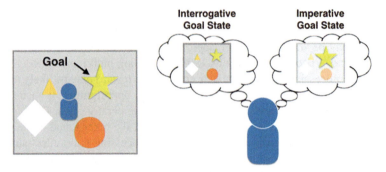

Fig. 1 Characterizing motivational states related to information-seeking. Here we elaborate on a conventional view of motivation as a valenced state of approach or avoidance. We posit an additional characterization of motivational states as they relate to information-seeking in a complex environment. We assume first that multiple goals compete and second that the strength of evidence for actions varies. *Interrogative goal states* emerge when actions to achieve goals are diffuse, conflicting, or under-determined, requiring resolution and active information seeking. *Imperative states* emerge when goals are salient, unconflicted, or urgent and additional information is of limited utility

Operationalizing Memory as an Outcome of Motivational State

To understand the memory outcomes of specific motivational states requires characterizing memory beyond binary construct of remembering versus forgetting to examine the form and content of memories. We predict that interrogative motivational states will support rich, relational memories that support later wayfinding (or disambiguation), whereas imperative motivational states will support sparse, feature-focused, item-based memories that support decisive action. While the entirety of the MTL is typically engaged during episodic memory encoding, a large body of research has shown that MTL subdivisions are specialized to support distinct forms of memory representations, with important distinctions between the MTL cortex and the hippocampus proper(Davachi 2006; Eichenbaum et al. 2007; Ranganath 2010; Eichenbaum et al. 1994). The behavioral patterns we describe are closely aligned with these MTL specializations and make clear anatomical predictions about motivated encoding, as follows.

The cortical MTL, including perirhinal and parahippocampal cortices, supports the encoding of unitized, isolated representations of items or contexts, respectively. This encoding supports familiarity-based memory judgments, item-specific details, and priming. Conversely, the hippocampus proper supports encoding of the relationships between unitized constructs. Encoding in the hippocampus supports more relational forms of memory, such as recollection-based memory judgments, memory for the relationships amongst items, and the binding of items in their broader contexts. For example, patients with non-specific MTL resections show deficits in memory for items and details about the context in which they were encoded, whereas patients with specific hippocampal resections have intact memory for items but show deficits in identifying the broader contexts associated with individual items (Giovanello et al. 2003; Quamme et al. 2007; Hannula et al. 2015). This organizational schema allows us to test mechanistic hypotheses about the precise impact of motivation on memory by investigating how motivation influences encoding supported by the hippocampus versus MTL cortex.

Specifically, we predict that to the degree that interrogative motivational goals engage the hippocampus proper during encoding, memories will include detailed, relational representations of the environment. These representations will include items associated with reward incentives and features of the environment in which they were encountered. Conversely, we predict that to the degree that imperative motivational goals engage the cortical MTL during encoding, memories will include sparse feature-based memories associated with threats, absent the details of the surrounding environment. We further predict that the engagement of the hippocampus and cortical MTL during these motivational states will depend on engagement of discrete neuromodulatory systems. Specifically, we predict that VTA neuromodulation will facilitate hippocampal encoding during interrogative goal states and amygdala neuromodulation will facilitate cortical MTL encoding during imperative goal states. While there will be strong tendencies for reward and punishment incentives to elicit interrogative and imperative goal states, respectively, we predict that the final determinant of memory specialization will be the neuromodulatory systems engaged during encoding. For example, rewarding contexts that engage amygdala and cortical MTL would still result in feature-based mnemonic representations, while punishing contexts that engage VTA and hippocampus would still result in relational mnemonic representations.

Considerable empirical evidence from animal and human literatures supports these general predictions. First, we will review literatures demonstrating that VTA engagement, which is reliably associated with reward-based motivation, biases information processing toward hippocampal-dependent memory encoding. Next, we will review literature demonstrating that amygdala engagement, which is reliably associated with punishments, biases information processing toward cortical MTL-dependent encoding. Finally, we will then expand our arguments to discuss how large scale networks beyond the MTL, VTA, and amygdala, in conjunction with activation of neuromodulatory neurotransmitter systems, support motivated memory encoding.

The Role of the VTA in Motivated Behavior

This section reviews literatures implicating the role of the VTA in motivated behavior associated with interrogative goal states. In our model, interrogative goal states reflect when individuals' motivational drive is oriented not only towards obtaining an immediate goal but also gathering information about the environment to support future adaptive behavior. Thus, this state reflects a balance between initiating motivated behaviors and an active state of environmental exploration. Given the extant literature, we focus this review on reward-motivated behaviors, as the majority of studies investigating the VTA have characterized this neuromodulatory system using reward incentives (i.e., food, drugs, monetary rewards). There are instances, however, when punishment incentives can engage the VTA and evoke interrogative goal states (Salamone 1994; Bromberg-Martin et al. 2010).

A long history of animal research has implicated the VTA (and other midbrain dopamine nuclei), in motivated behavior. Observations that animals will forgo ecologically important rewards, such as sex or food, to self-stimulate the VTA or its targets in the ventral striatum provided early evidence that the mesolimbic dopamine system is intimately involved with motivation (Olds and Milner 1954). Since these early observations, a large literature has associated VTA activity with the initiation and propagation of motivated behaviors. Dopamine release in the rodent ventral striatum, including dopamine release in response to optogenetic stimulation of the VTA, initiates a variety of motivated behaviors (Ikemoto and Panksepp 1999; Adamantidis et al. 2011; Fink and Smith 1980) consistent with interrogative goal states. Along with immediate goal pursuit, VTA activation or dopaminergic stimulation of efferent projection targets results in behavioral activation and a variety of exploratory behaviors, including orientation to novel stimuli and environments (Ikemoto and Panksepp 1999; Düzel et al. 2010; Kakade and Dayan 2002). Theoretical models have suggested that these exploratory behaviors emerge in motivationally-relevant environments in order to support future goal attainment (Kakade and Dayan 2002).

A large body of research has also characterized a role for the VTA in motivated learning, particularly associative learning. VTA firing tracks learning about associations between neutral cues and their rewarding outcomes in both Pavlovian conditioning (stimulus-outcome), and instrumental learning (stimulus-response-outcome; Schultz 2016). Specifically, VTA neurons increase their firing in response to cues that predict rewards and track prediction errors between anticipated rewards and actual reward outcomes. Activation of dopamine neurons in the VTA via optogenetic stimulation results in patterns of behavior similar to those evoked by VTA prediction errors (Steinberg et al. 2013; Adamantidis et al. 2011; Kim et al. 2012), implying that VTA activation is sufficient for motivated learning. These learning-related patterns are not specific to rewards, as VTA neurons have also been shown to respond to novel, salient, and punishing stimuli and to code for information that contributes to future goal attainment, including motivational value, environmental orientation, and increasing the precision of predictions (Bromberg-Martin et al. 2010). For example, VTA neurons have been shown to respond to cues indicating the nature of upcoming rewards (Bromberg-Martin and Hikosaka 2009). Together, these studies demonstrate that VTA neurons not only track the "value" of cues for reward outcomes, but also encode general properties of the environment that relate to obtaining future rewards.

In parallel with this animal literature, human neuroimaging has supported a role for the VTA in both motivated behaviors and motivated learning. Functional magnetic resonance imaging (fMRI) and electrophysiological studies have reliably documented VTA activation during reward motivated behaviors (D'Ardenne et al. 2008; Murray et al. 2008; Zaghloul et al. 2009; Carter et al. 2009; Krebs et al. 2011). While the majority of these studies were performed in the domain of reward motivation, research has also shown VTA activation in humans in response to other salient events, such as novelty, surprise, and loss avoidance (Krebs et al. 2011; Krebs et al. 2009; Bunzeck and Düzel 2006; Boll et al. 2013; Carter et al. 2009). In addition to reward-motivated behaviors, both the VTA and its targets in the striatum have been associated with motivated learning in Pavlovian and instrumental conditioning paradigms (D'Ardenne et al. 2008; Murray et al. 2008; McClure et al. 2003; Pagnoni et al. 2002). Further, pharmacological challenges have shown that both behavioral and neural associative learning signals are amplified when participants are given dopamine agonists versus dopamine antagonists (Pessiglione et al. 2006).

Together, these findings show that the VTA not only supports the initiation of motivated behaviors, but also supports learning about salient features of the environment that predict rewards, or more generally, information relevant to future goal attainment. Further, these properties of the VTA seem to be homologous across rodents, non-human primates, and humans.

VTA Activation Facilitates Hippocampus-Dependent Encoding

We next review literatures implicating a role for the VTA in hippocampal-dependent motivated memory encoding. As detailed above, reward motivation has been shown to reliably engage the VTA. Thus in this section, we include studies that directly characterize VTA activation as it relates to memory encoding, as well as studies that characterize how reward motivation influences memory encoding. Together, the reviewed research demonstrates that VTA activity, which is reliably engaged during reward (or interrogative) motivation, facilitates hippocampus-dependent encoding and results in rich relational memories of motivationally-relevant environments. Notably, prior research has also demonstrated contexts in which reward incentives can actually engage more imperative goal states, (i.e., reward-induced anxiety, "choking"), which we detail in a subsequent section (see section "Dissecting the Relationship Between Valence and Incentivized Information Processing").

Structural connectivity between the VTA and hippocampus has been shown across species. Neuroanatomical studies in rodent and non-human primate have documented monosynaptic, afferent projections from dopamine neurons in the VTA to the hippocampus (Amaral and Cowan 1980; Samson et al. 1990). Dopamine receptors, predominantly D1/D5 like receptors, have been identified throughout the hippocampus in both rodents and non-human primates (Ciliax et al. 2000;

Khan et al. 2000; Bergson et al. 1995). More recently, human neuroimaging has characterized indirect markers of structural connectivity between the VTA and hippocampus. FMRI studies have documented significant connectivity between these regions at rest, which are thought to reflect the intrinsic connectivity of the human brain (Murty et al. 2014; Kahn and Shohamy 2013). Further, diffusion tensor imaging (DTI) studies have documented white matter tracts originating in the dopaminergic midbrain and terminating in the hippocampus (Kwon and Jang 2014). Finally, post-mortem studies and PET studies have provided evidence for expression of dopamine receptors in the human hippocampus (Mukherjee et al. 2002; Khan et al. 2000; Little et al. 1995; Camps et al. 1990). Together, these studies show that the neuroanatomical architecture of the brain supports structural connectivity between the VTA and hippocampus.

In addition to neuroanatomy, VTA projections to the hippocampus are known to dynamically modulate hippocampal neurophysiology in a manner which could facilitate memory encoding. Dopamine agonists that mimic VTA activation have been shown to lower hippocampal firing thresholds (Hammad and Wagner 2006). Further, dopamine has been shown to stabilize hippocampal place fields, ensembles of neurons that represent the environment (Tran et al. 2008; Martig and Mizumori 2011). Both of these neuromodulatory processes should directly influence the initial encoding of memory traces within the hippocampus.

Beyond modulating encoding, dopamine and VTA activation have each been shown to facilitate long-term plasticity in the hippocampus (Lisman et al. 2011; Wang and Morris 2010). For example, dopaminergic stimulation results in LTP-like enhancements in the firing of hippocampal neurons, and the blockade of dopamine receptors in the hippocampus abolishes the effect of standard LTP-induction procedures (Huang and Kandel 1995). More recent evidence suggests that the VTA could also support systems-level memory consolidation, a process by which memory representations are reactivated and stabilized throughout the brain after encoding. Specifically, rodent studies have shown a preferential 'replay' of rewarding events after encoding, such that the sequence of neurons firing in the VTA and hippocampus during motivated encoding is repeated offline (Valdes et al. 2011; Singer and Frank 2009; Gomperts et al. 2015). In conjunction with enhancing encoding, these plasticity-related processes could stabilize long-term representations of motivationally-relevant environments.

Finally, dopaminergic neuromodulation and VTA activation have been shown to directly affect hippocampal-dependent memories. Rodent studies have shown engagement of the mesolimbic dopamine system, including the VTA, during successful spatial memory encoding (DeCoteau et al. 2007; Martig et al. 2009; Khan et al. 2000; Rossato et al. 2009). Further, dopamine release prior to and during encoding strengthens hippocampus-dependent memory representations, and dopamine antagonism at the time of encoding can disrupt long-term memory (Wang and Morris 2010; Salvetti et al. 2014; O'Carroll et al. 2006). For example, exposure to novel environments, which is known to engage the VTA, enhances performance on a hippocampal-dependent spatial learning task; further, this effect is abolished by dopamine antagonists (Li et al. 2003). More recently, it has been shown that optogenetic stimulation of VTA afferents to the hippocampus stabilized neural place fields and increased performance on a hippocampal-dependent spatial navigation task (McNamara et al. 2014). Thus, the rodent literature demonstrates that both VTA activation and dopaminergic neuromodulation support hippocampal-based memory encoding.

Similar to these animal studies, an emerging literature in humans has supported a role for VTA engagement in supporting hippocampal-dependent memories during motivated learning. We focus our review on neurobehavioral studies looking at the influence of reward motivation, a putative proxy for VTA engagement, on declarative memory encoding. Critically, these human studies have provided a more detailed characterization of how VTA activation and/or reward motivation influences the form and content of the memories. Initial studies investigating reward's influence on episodic memory tested memory for information that was explicitly incentivized. In an early study in this literature, participants were presented with either high ($5) or low ($0.10) reward cues, which indicated how much participants could earn if they could successfully remember target images that followed each cue (Adcock et al. 2006). At a 24-h memory test, participants had significantly better memory for items associated with high versus low rewards, demonstrating that reward motivation enhanced episodic memory. Interestingly, the benefits of reward motivation enhanced recollection-based memory judgments, which are thought to contain information about the item being tested and details of the broader encoding context. Similarly, studies have shown participants to have better memory for pictures that were predictive of high versus low rewards, and for the temporal context in which they were encoded (Wittmann et al. 2005). These initial studies suggest that reward motivation not only enhances memory for rewarding items, but also supports relational memory between items and their broader context.

Since these initial studies, research has provided additional evidence that reward motivation enhances relational memory. For example, a recent study demonstrated that incentivizing encoding of pairs of object images resulted in better memory for the relationship between those images (Wolosin et al. 2012). Specifically, participants were able to discriminate pairs of items that appeared together versus pairs of items that were rearranged (i.e., presented during encoding but not together). In our own work, we found that reward motivation improved memory for spatial locations and broader environmental contexts during a spatial navigation task (Murty et al. 2011). In this study, participants completed a virtual Morris Water Task, in which hidden platforms had to be identified by successfully encoding relationships between discrete environmental cues. Within this task, participants had better spatial navigation when incentivized with monetary rewards compared to a non-rewarded control condition (and compared to punishment incentives). Together these findings show that reward motivation facilitates relational memory for items that are explicitly rewarded as well as rewarded items within their broader spatial context.

Further work in human fMRI supports the assertion that reward motivation facilitates memory encoding via interactions between the VTA and hippocampus, via mechanisms which are convergent with the extant animal research. Specifically, fMRI studies have demonstrated that activation of the VTA and hippocampus predict declarative memory both for trial-unique cues that predict reward (Bunzeck et al. 2012; Wittmann et al. 2005), as well as for information that is explicitly incentivized during encoding (Adcock et al. 2006; Cohen et al. 2014; Callan and Schweighofer 2008; Wolosin et al. 2012). In the reward-motivated memory encoding paradigm described above (Adcock et al. 2006), successful memory encoding (i.e. subsequently remembered versus subsequently forgotten items) in the high reward condition was associated with greater activation in VTA and hippocampus, as well as increased connectivity between these regions. In contrast, there was no increase in activation or connectivity between VTA and hippocampus in the low reward condition. Subsequent research has bolstered this conclusion that interactions between the VTA and hippocampus predict the successful encoding of information incentivized with monetary rewards (Adcock et al. 2006; Cohen et al. 2014; Callan and Schweighofer 2008; Wolosin et al. 2012). These studies suggest that reward-motivated enhancements in relational memory are supported by VTA neuromodulation over the hippocampus.

More recently, research has begun to characterize how rewarding contexts not only enhance memory for reward-associated items but also memory for neutral information presented in rewarding contexts, i.e. information that was not explicitly incentivized or predictive of upcoming rewards. For example, in a recent study we placed individuals in either high or low rewarding context, by having them anticipate making instrumental responses to earn either $2 (high reward) or $0.10 (low rewards; Murty and Adcock 2014). During these states of either high or low reward anticipation, participants were incidentally presented with novel, salient images. On a surprise memory test, we found that individuals had better memory for this neutral information when they were anticipating high versus low reward. Similarly, a recent study had participants incidentally encode neutral images, unrelated to reward outcomes, that were embedded in either high or low reward predicting scenes (Gruber et al. 2016). During a surprise memory test, participants had better memory for the neutral images embedded in the high versus low reward context. Together, these findings extend the domains in which reward motivation can support memory. Where prior studies demonstrated better relational memory for reward-related items, these studies show that reward can also enhance memory for neutral information embedded in rewarding contexts. This latter observation is consistent with the proposed role of reward motivation, and its putative activation of the VTA, in supporting interrogative goal pursuit. Neutral information that is embedded in motivationally-relevant contexts theoretically could act as reward-predicting cues or provide information of the acquisition of future goals (Fu and Anderson 2008).

Human neuroimaging studies have related enhancements in memory for neutral information presented in rewarding contexts to interactions between the VTA and hippocampus. In our work, we found that in rewarding contexts, VTA activity predicts subsequent increases in hippocampal sensitivity to surprising neutral information (Murty and Adcock 2014). Similarly, a recent study showed enhanced memory for neutral images embedded in reward-predicting contexts

(Loh et al. 2015). Authors found that these increases in hippocampal-dependent memory were only evident when the reward-predicting cues engaged the VTA; this suggests that VTA activation rather than intrinsic properties of the incentives determined enhanced memory formation. Together, these studies demonstrate that the same circuitry guiding reward-motivated memory enhancements also supports enhanced memory for neutral information embedded in rewarding contexts.

In sum, evidence across human, non-human primate, and rodent studies suggest that VTA activation, often elicited by reward, promotes relational memory via engagement of hippocampal-dependent encoding. Animal studies have shown that dopamine promotes better encoding in the hippocampus, and direct dopamine modulation enhances the stabilization of rewarding items in long-term memory. Human research has further demonstrated that reward motivation, as well as VTA-hippocampal interactions, specifically support rich mnemonic representations of motivationally-relevant environments, that include (1) reward-associated items, (2) relationships amongst items in rewarding environments, (3) relationships of rewarded items in their broader environmental context, and (4) neutral items encountered during reward pursuit.

The Role of the Amygdala in Motivated Behavior

This section reviews literatures implicating a role for the amygdala in motivated behavior associated with imperative goal states. Here, we operationalize imperative goal states as motivational drive oriented towards obtaining an immediate, compulsory goal, and not the surrounding motivationally-relevant environment. Given the extant literature, we focus this review on punishment-motivated and threat-related behaviors, which have been shown to both induce imperative goal states and engage the amygdala. We note that many of the behaviors described involve coordinated interactions between the VTA and amygdala (Salamone 1994). In this section, however, we focus on the amygdala, as its engagement is necessary and sufficient to engage a variety of behaviors associated with imperative goal states.

A long history of animal research has implicated the amygdala in behaviors associated with imperative goal states, starting with early observations that animals with amygdala lesions fail to exhibit stereotypical responses to imminent threats. Within rodent and non-human primate literatures, amygdala activity and its functional afferents have been implicated in the generation of freezing and startle behaviors (Blanchard and Blanchard 1969; Fendt 2001; Davis 1992), sympathetic arousal in response to threats and punishment (Korte 2001), and active avoidance (Reilly and Bornovalova 2005). Rodents with amygdala lesions fail to show typical avoidance of open fields and elevated arms of mazes (Davis 1992), environments where they may be more vulnerable to threat. Similarly, rodents will typically avoid spatial locations that were previously associated with punishment; however, following amygdala lesions the animals re-approach these areas (Xue et al. 2012). Together, these findings suggest that the amygdala supports reflexive behaviors that contribute to the animal's immediate goal (in these studies, typically avoiding threats); this in turn reduces exploration of goal-relevant environments.

Interestingly, the role of the amygdala in punishment-motivated behavior may also depend on dopaminergic projections from the VTA. Dopamine depletion can disrupt and dopamine administration can facilitate a variety of punishment-motivated behaviors (Salamone 1994). Interestingly, when punishment avoidance and reward pursuit coincide, amygdala lesions can actually result in increased reward-motivated behaviors (Xue et al. 2012). Thus, the amygdala may play a role in determining the orientation of dopamine's role in motivation. These findings suggest that amygdala neuromodulation may asymmetrically promote the instantiation of punishment motivation at the expense of reward/approach-related behaviors; how this balance relates to interrogative versus imperative goal states remains to be investigated.

Like the VTA, the amygdala is also centrally implicated in motivated learning. Early rodent and non-human primate studies show that lesions to the amygdala result in deficits in fear conditioning (LeDoux 1992; LaBar and Cabeza 2006), the learned association of a reflexive response to intrinsically threatening stimuli with a neutral stimulus that predicts punishment (Choi et al. 2010; Holahan and White 2004; Rorick-Kehn and Steinmetz 2005). Animals with amygdala lesions fail to show the typical freezing or startle response elicited by such cues that predict threats (Blanchard and Blanchard 1969; Hitchcock and Davis 1986; Campeau and Davis 1995; Kim and Davis 1993; Phillips and LeDoux 1992; Kim et al. 1993).

Further analysis of amygdala subregions during fear conditioning demonstrate that the behaviors associated with imperative goal states may map most closely onto central regions of the amygdala, implicated in noradrenergic responses and arousal. Interestingly, these central regions promote freezing/behavioral inhibition and may actually inhibit active avoidance (Choi et al. 2010; Davis 1992). Active avoidance of threats may depend instead on basolateral portions of the amygdala. Thus, the basolateral portions of the amygdala, which have also been shown to track associations between neutral cues and reward (Murray 2007; Baxter and Murray 2002), may thus contribute to a subset of avoidance behaviors that are more interrogative in nature.

Research in humans has bolstered support for the role of the amygdala in imperative goal states. Patients with amygdala lesions have deficits in perceiving and reflexively responding to imminent environmental threats (Broks et al. 1998; Adolphs et al. 2005; Scott et al. 1997), such as eliciting startle responses to neutral cues predicting threat. Functional neuroimaging studies have further demonstrated amygdala activation with the anticipation (Hahn et al. 2010) and active avoidance of punishments (Mobbs et al. 2007, 2009b; Schlund and Cataldo 2010). Similarly, the human amygdala has been associated with punishment-motivated reinforcement learning, as both lesion and human neuroimaging studies have implicated the amygdala during Pavlovian fear conditioning and instrumental avoidance

(LaBar et al. 1995, 1998; Büchel et al. 1998; Prevost et al. 2011, 2012), particularly regions within the central amygdala. However, the spatial resolution of many of these lesion and neuroimaging studies have made it difficult to discern the contributions of central and basolateral portions of the amygdala.

Together, these findings show that the amygdala supports a variety of motivated behaviors associated with imperative goal states, mainly in the domain of threat and punishment avoidance. Fast stereotyped responses provide a means to fulfill an individual's immediate goals to avoid a threat at the expense of gaining information about the surrounding environment. Further, these properties of the amygdala appear to be homologous across rodents, non-human primates, and humans.

Amygdala Activation Facilitates Cortical-MTL Dependent Encoding

We next review literatures on the role of the amygdala in cortical MTL-dependent memory encoding. As detailed above, punishment motivation and threats have been shown to reliably engage the amygdala, thus we include studies that either modulate amygdala activation or investigate memory encoding in these contexts. The reviewed research demonstrates that amygdala activation, which is reliably engaged during threat processing and punishment motivation, facilitates cortical MTL-dependent encoding. Further, engagement of cortical MTL results in sparse, de-contextualized, item-based representations of potential threats. Notably, prior research has demonstrated contexts in which reward incentives can elicit imperative goal states, (i.e., reward-induced anxiety, "choking"), and result in cortical MTL-dependent encoding (reviewed in section "Dissecting the Relationship Between Valence and Incentivized Information Processing").

Structurally, the amygdala has direct efferent projections throughout both the hippocampus and surrounding cortical MTL (McGaugh 2004), and stimulation of the amygdala can increase long-term potentiation in both of these regions (Ikegaya et al. 1995; Akirav and Richter-Levin 1999; Frey et al. 2001). Further, rodent studies have demonstrated that stimulation of the amygdala during and after encoding can facilitate memory encoding across the MTL, including both cortical MTL and hippocampus (McGaugh 2004). For example, pharmacological activation of the amygdala enhanced memory for safety platforms in a MTL-dependent spatial navigation task (Roozendaal and McGaugh 1997; Roozendaal et al. 1999) and these enhancements in memory were blocked by amygdala lesions (Roozendaal et al. 1996; Roozendaal and McGaugh 1997). Similarly, amygdala modulation over the MTL, including both the hippocampus and cortical MTL, has been shown to support contextual conditioning, in which threatening stimuli become associated with the surrounding environment (Rudy 2009; Fanselow 2000). These early rodent studies detailing the functional and structural connectivity of the amygdala reveal an organization that cannot discriminate between cortical MTL versus hippocampus encoding.

In spite of this evidence, there is accumulating evidence that amygdala neuromodulation may bias encoding towards cortical MTL structures. In rodents, activation of the amygdala results in increased coupling amongst cortical MTL regions, which was subsequently related to memory enhancements (Paz et al. 2006). Further, lesions to the rodent amygdala have been shown to selectively impair memory processes depending on cortical MTL-dependent encoding while sparing hippocampal-dependent encoding (Farovik et al. 2011). Through a series of behavioral modelling techniques, the authors demonstrated that amygdala lesions following encoding resulted in a failure to retrieve memories putatively stored in cortical MTL regions. Complementing these findings, research has demonstrated that amygdala engagement can interfere with the use of hippocampal-dependent memories during motivated behaviors. In these studies, rodents performed spatial navigation tasks that depended on using hippocampal-dependent memories. Critically, lesions of the amygdala increased the use of hippocampal-dependent memories during motivated behaviors. Conversely, stimulation of the amygdala decreased the use of hippocampal dependent memories.(Kim et al. 2001; McDonald and White 1993; Roozendaal et al. 2003). Together these studies suggest that amygdala activation promotes cortical MTL dependent encoding over hippocampal-dependent encoding.

Research from humans has similarly indicated that amygdala activation may shunt encoding towards cortical-MTL, resulting in item-based, sparse representations of the environment that are focused on the immediate goals of the individual. The majority of the human evidence for amygdala involvement in memory encoding emerges from studies testing memory for intrinsically aversive items such as trial-unique pictures of snakes and spiders (LaBar and Cabeza 2006). We first review these studies of memory for intrinsically threatening items, before reviewing the emerging literature explicitly investigating the influence of punishment motivation on encoding.

Human studies reliably show a memory advantage for intrinsically aversive memoranda or aversive environments compared to neutral memoranda (LaBar and Cabeza 2006; Bennion et al. 2013). However, these studies typically only probe item-based memory, which could be supported by either cortical MTL- or hippocampus-dependent encoding. A growing body of literature, however, has shown that threat-related stimuli actually disrupt relational memory processes when it is explicitly probed. For example, behavioral research has shown intrinsically threatening items result in worse source memory for encoding contexts (Dougal et al. 2007; Rimmele et al. 2011, 2012) and worse recognition memory for contexts presented simultaneous to threats (Steinmetz and Kensinger 2013; Kensinger et al. 2007a). Similarly, individuals are impaired at relational binding of threatening items with each other in memory (Onoda et al. 2009), as well as relational updating of memories that once contained a threatening item (Sakaki et al. 2014). Together, this research suggests that environmental threat, which is strongly associated with amygdala activation, can support item-based memories but actually disrupts relational memory processes.

In parallel to these behavioral findings, neuroimaging research has begun to characterize MTL engagement during memory encoding for intrinsically threatening items. We recently conducted a meta-analysis of successful memory encoding for intrinsically threatening items. We found that successful memory for threatening versus neutral items was associated with amygdala, but not VTA, activation. Further, we also revealed reliable engagement of both the cortical MTL and hippocampus during emotional memory encoding (Murty et al. 2010). Thus, this meta-analysis suggested that the amygdala neuromodulation could support both cortical MTL and hippocampus activation during encoding. There are important caveats to consider when addressing the role of amygdala neuromodulation on MTL encoding from this meta-analysis. Firstly, the spatial resolution of a meta-analysis is somewhat poor, as spatial information becomes blurred by combining data across multiple studies (Nee et al. 2007). Thus it may be difficult to delineate hippocampal engagement from amygdala and cortical MTL given their spatial proximity. Further, the meta-analysis does not avail the opportunity to investigate dynamic relationships between the amygdala and MTL during encoding (i.e. amygdala neuromodulation of the MTL). Reliable activation of the amygdala, cortical MTL, and hippocampus during memory encoding could be evoked by different subsets of trials within the same study. Finally, this meta-analysis was not able to dissect how different memory tasks and/or arousal levels influenced engagement of different MTL regions. Thus, this early meta-analysis provides evidence for amygdala, cortical MTL, and hippocampal engagement during memory encoding, but cannot speak to the relationship amongst these regions.

When studies have directly examined the relationship amongst these structures in detail, they have found that amygdala engagement selectively increases cortical MTL-dependent encoding but not hippocampus-dependent encoding. Amygdala activation during encoding was found to predict memory for threatening items, a cortical MTL-dependent process; but did not predict relational memory for items and their surrounding contexts, a hippocampal-dependent measure (Dougal et al. 2007; Kensinger and Schacter 2006). Similarly, studies directly investigating interactions of the amygdala and MTL have demonstrated that successful encoding of emotional memories were associated with amygdala-cortical MTL functional interactions (Dolcos et al. 2004; Ritchey et al. 2008), but not amygdala-hippocampus interactions. These neuroimaging results are corroborated by the human lesion literature, which shows that patients with hippocampal lesions that spare amygdala and cortical MTL, still show a memory advantage for intrinsically threatening items (Hamann et al. 1997a, b). Together, these findings suggest that amygdala engagement facilitates cortical-MTL supported, item-based representations of the environment, devoid of relationships between items and their surrounding environment.

The studies reviewed above focus on memory for intrinsically emotional information. Recently, our laboratory and others have begun to investigate the specific role of punishment motivation on MTL-dependent memory encoding. Dovetailing well with the emotional memory literature, we find that punishment motivation results in item-based representations of threatening stimuli devoid of information about the surrounding environment. In an early study, we directly compared the influence of reward and punishment motivation on allocentric spatial navigation during a virtual reality water task paradigm (Murty et al. 2011). In this study, we found that, compared to reward and neutral (no incentive) motivation, punishment motivation impaired encoding of the environment in which threatening items existed. However, this first study was purely behavioral and thus could not relate these behavioral patterns to amygdala-cortical MTL interactions.

To characterize the neural architecture underlying encoding, we next turned to a punishment-motivated encoding paradigm (Murty et al. 2012), by modifying the design of Adcock et al. (2006) described above. In this paradigm, before each item to be memorized, participants saw cues that indicated whether or not forgetting the image would be punished by a shock; thus, the shock could be avoided by successful encoding. We found that punishment motivation enhanced memory for items directly associated with threat. However, neuroimaging revealed circuitry distinct from those identified in reward-motivated memory encoding. Whereas reward motivation was associated with VTA-hippocampal interactions, we found that successful punishment-motivated memory was associated with amygdala-cortical MTL interactions. Although punishment motivation was still associated with enhanced recognition memory for motivationally relevant items, successful encoding was predicted by amygdala interactions with cortical MTL.

We observed a similar neuromodulation of cortical MTL in a study comparing incidental encoding in rewarding versus punishing contexts (Murty et al. 2016a). Specifically, we adapted the paradigm utilized by Murty and Adcock (2014), in which neutral surprising items were embedded in states of high or low reward motivation contexts, to test the impact of punishment incentives (See Fig. 2). Using a configural memory task that specifically indexed hippocampal representations, we saw that while reward incentives enhanced memory for neutral items relative to no rewards, we found no motivation benefit on memory when participants were avoiding punishments. Directly comparing the encoding-related fMRI activations under the two incentive conditions revealed discrete states of encoding under reward versus punishment motivation: a double dissociation of MTL-dependent encoding, such that reward facilitated hippocampus activation without any modulation of cortical MTL, and punishment motivation facilitated cortical MTL activation without any modulation of the hippocampus. Thus, across multiple studies, we have found that punishment motivation facilitates learning via mechanisms distinct from reward motivation, enhancing memory for items and not the surrounding contextual details.

Similar behavioral and neural profiles have also been demonstrated in studies investigating punishment's influence on memory. In line with our own findings, emerging research shows that memory encoding of neutral items associated with threat engages both the amygdala and cortical MTL-dependent encoding, and in some contexts actually impairs hippocampal-dependent encoding (Schwarze et al. 2012; Qin et al. 2012). For example, a recent study had participants encode neutral

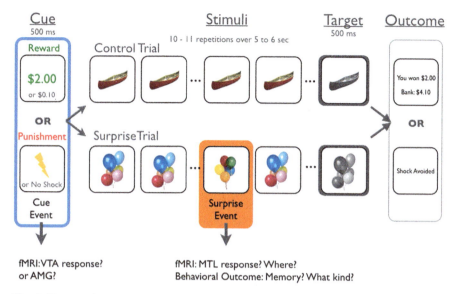

Fig. 2 Example investigative approach to motivated memory. Memory formation for neutral stimuli is examined in the same behavioral task under contrasting incentive structures. In the illustrated paradigm, incidental encoding of unexpected events is examined during a reaction time task incentivized in two different motivational contexts: reward incentives or punishment incentives (Each context includes a low-motivation control condition). Both groups are working to press a button when the repeating stimulus turns to a grayscale image. On some trials a surprise occurs, irrelevant to performance. In these contrasting motivational contexts, we examine anticipatory engagement of brain networks implicated in motivation (indexed by fMRI activation in the VTA or amygdala (AMG above). We then relate the state of these modulatory networks to MTL responses elicited by surprising events and to memory outcomes in each context

information that was associated with varying levels of threat of unavoidable shock (Bauch et al. 2014). The authors found that increasing threat of shock enhanced item-based memory responses and associated encoding-related activity in cortical MTL. Further, the authors found that increased threat of shock impaired associative-based memory (i.e., relational memories) and was associated with decreased encoding-related activity in the hippocampus.

In sum, evidence across human, primate, and rodent studies suggests that motivational contexts that engage the amygdala result in item-based memory encoding via neuromoduation of cortical MTL-dependent encoding. The amygdala has strong projections to cortical MTL, and, has been shown to facilitate encoding by cortical MTL. These lines of research suggest that punishment motivation, as well as amygdala cortical-MTL interactions, specifically supports representations in line with imperative goal orientations. Specifically, memory is enhanced for the targets of goal orientation (i.e., threatening items or items directly associated with punishments), but impaired memory for contextual information and other aspects of relational memory. Thus, the literature offers mounting evidence that, in general, amygdala activation results in a state of learning distinct from that resulting from VTA activation.

Dissecting the Relationship Between Valence and Incentivized Information Processing

Thus far, we have reviewed literatures characterizing how the VTA and amygdala, respectively, support memory encoding. Most studies have characterized VTA engagement using reward incentives and amygdala engagement using punishment incentives and/or threat. Notable exceptions exist in the literature, however, in which reward incentives can engage the amygdala and punishment incentives can engage the VTA. Within our theoretical perspective, it is not the valence of the incentive which dictates targets in the MTL during encoding; rather it is the neuromodulatory system which is engaged by the goal state of the motivational context. Below, we provide examples of reward eliciting amygdala engagement/imperative goal states and punishment eliciting VTA engagement/interrogative goal states. Further, where available, we discuss the downstream consequences on memory encoding within these contexts.

While reward incentives may generally foster interrogative goal states, high reward salience may elicit an imperative goal states. One domain in which this has been well studied is addiction. Specifically, while VTA reward systems (and BL amygdala) are implicated in initiation of drug use, in addicted individuals, well-learned drug cues result in central amygdala activation. These highly salient drug cues, which elicit amygdala activation, result in devaluation of other motivationally-relevant goals to solely orient animals towards the drug reward (Lesscher and Vanderschuren 2012). Similarly, exogenous stimulation of the amygdala during reward conditioning results in compulsive, reflexive reward seeking (Robinson et al. 2014). Critically, contexts and neurobehavioral states (Wingard and Packard 2008; Packard 2009) associated with highly salient drugs of abuse putatively impair hippocampal-dependent encoding in favor of striatal learning. Motivated learning in these contexts has been shown to result in rigid representations between drug cues and actions to obtain them, which are insensitive to information about the surrounding contextual environment (i.e., habits) (Yin and Knowlton 2006). Thus, in the context of addiction, a reward incentive can actually engage an imperative goal state and disrupt hippocampus-dependent encoding.

Similarly, research has shown that increasing the salience of incentives during reward-motivated behavior can actually induce states of perceived threat, that is, "choking" (Mobbs et al. 2009a; Yu 2015; Ariely et al. 2009)—implying imperative goals. For example, one study demonstrated that offering people rewards in a high-stakes situation resulted in greater errors on a variety of both motor and cognitive tasks (Ariely et al. 2009). These deficits were interpreted to result from individuals perceiving rewards as stressful, yielding states of high physiological arousal; this threat-like state may be associated with engagement of the amygdala, particularly the central amygdala (as reviewed above).

This concept of "choking" has also been demonstrated in the domain of reward-motivated memory encoding. For example, research has shown that incentives to encode information for monetary rewards can induce anxiety in some participants (Callan and Schweighofer 2008). Critically, participants that reported greater states of anxiety showed reduced engagement of the VTA and hippocampus, and had worse memory performance. Similarly, when we used a complementary approach to investigate this phenomenon by measuring individuals' physiological arousal during our motivated spatial navigation paradigm (Murty et al. 2011), we found that hippocampal-dependent memory encoding was worse when individuals had high arousal responses. This pattern held both within and across participants, and even in a reward context: The sub-group of participants who reliably showed hippocampal-memory deficits in reward contexts showed physiological responses indistinguishable from the group of participants who performed learning in a punishment context. Together, these findings show that when reward incentives induce states of anxiety or high physiological arousal (which are both associated with amygdala activation), their memory profiles resemble those associated with imperative goal states.

While punishment incentives reliably engage the amygdala and imperative goal states, there are also contexts in which punishments can evoke interrogative goal states. One context in which this emerges is when individuals have warning about a distal punishment, and do not have any imminent potentials of harm (Mobbs et al. 2015). In this context, the punishment incentive may be less salient, and thus individuals' goal orientation can be divided both between the threat and other features of the environment, i.e. an interrogative goal state. In line with this interpretation, human neuroimaging has demonstrated that when a threat is distal and avoidable, and individuals are not susceptible to immediate harm, there is robust engagement of the hippocampus; but, as the threat approaches hippocampal engagement diminishes (Mobbs et al. 2009b).

This prior study suggests that when punishment incentives do not induce a state of immediate threat in an individual, they may engage interrogative goal states. In line with this interpretation, research using less salient punishment incentives, such as monetary loss, have shown engagement of VTA instead of amygdala neuromodulation (Carter et al. 2009; Delgado et al. 2011). For example, amygdala activation has been shown to track the avoidance of electrical shock punishments which may be more salient and elicit imperative goal states, but not monetary loss punishments which are less salient and may elicit interrogative goals despite being negatively valenced incentives (Delgado et al. 2011). Interestingly, a recent study investigating the influence of punishment motivation on memory encoding, showed that the threat of monetary losses facilitated VTA and hippocampal engagement and further resulted in better relational memory. Thus, even a punishment incentive can result in better relational memory if it engages the VTA and hippocampus.

These findings highlight that the relationship between an incentive's valence and downstream neuromodulatory engagement is not direct. Reward incentives can reliably elicit interrogative goal states and engage the VTA. However, in contexts when a reward incentive becomes highly salient or is interpreted as threatening, the incentive can elicit an imperative goal state and facilitate amygdala neuromodulation. Similarly, punishment incentives reliably elicit imperative goal states and engage the amygdala. However, in contexts when a punishment incentive is less salient and may not directly threaten an individual, the incentive can elicit an interrogative goal state and facilitate VTA neuromodulation. These findings support a key facet of our model: the form and content of memory will be determined by the neuromodulatory systems engaged during encoding rather than the valence of the incentive states.

Proposed Model: Current Motivation Organizes MTL Networks to Shape the Content and Form of Memory

Together the reviewed findings support a nuanced model of motivated memory in which the motivational state of an individual during encoding shapes the neural substrates supporting memory (Fig. 3). This, in turn leads to qualitatively different mnemonic representations of the environment. Specifically, this model proposes that memory encoding under interrogative motivation is supported by VTA and dopaminergic neuromodulation, and is more common under reward incentives. In contrast, memory encoding under imperative motivation is supported by amygdala neuromodulation, and is more common under punishment incentives.

Engagement of these distinct neuromodulatory systems shunts memory processing towards different MTL encoding substrates. Interrogative motivation facilitates hippocampal-dependent encoding processes whereas imperative motivation facilitates cortical MTL-encoding processes. Finally, the model predicts that differential engagement of these MTL substrates results in the storage of quantitatively and qualitatively different representations of the environment in long-term memory. Specifically, under interrogative motivation, environmental representations are relational, such that relationships between individual items and their surrounding contexts are maintained. Conversely, under imperative motivation, environmental representations are reduced, such that features directly associated with goals are extracted and stored without relational context.

The majority of research supporting this model comes from studies investigating reward incentives that engage states of interrogative motivation, and punishment incentives that engage states of imperative motivation. However, high salience rewards and low salience punishments may engage imperative and interrogative goal states, respectively. One open question is under what conditions VTA and amygdala-based networks might both be engaged during encoding. Active avoidance is one candidate context: during active avoidance of a discretely localized threat, both interrogative (way-finding) and imperative (flight) motives and behaviors are appropriate with co-activation of VTA and amygdala. We propose that final

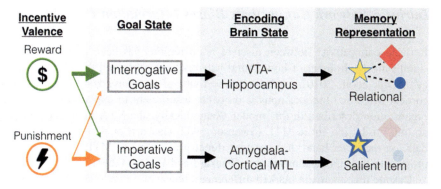

Fig. 3 A neurobehavioral model of motivation's influence on MTL-dependent memory. In the current model, we propose that the valence of motivational incentives (Incentive Valence above) can drive discrete states of MTL-dependent memory encoding. We delineate motivational states (Goal States) that are centered on 'interrogative' goals, which are associated with exploratory behaviors that support disambiguation of goal conflict and future goal attainment, versus 'imperative' states in which resources are captured by a highly salient and proximal immediate goal. We propose that interrogative and imperative goals are strongly, but not exclusively, associated with states of reward and punishment motivation, respectively. Critically, we believe that these motivational states result in neuromodulation over discrete MTL targets (Encoding Brain State). Namely, interrogative goals facilitate VTA-hippocampal interactions, whereas imperative goals facilitate 'amygdala cortical-MTL' interactions. In turn, engagement of these learning systems result in distinct representations of the environment: with hippocampal engagement supporting relational representations of multiple aspects of the environment, but cortical MTL engagement supporting unitized extraction of salient features directly relevant to an individual's goals (Memory Representations)

determinant of memory encoding will be the neuromodulatory systems engaged; thus, an important open question is how such joint activation would influence memory.

Mechanisms of Motivated Memory Specialization

The research reviewed above provides evidence that incentive contexts engage distinct coherent network states, including distinct regions in the MTL, during encoding. However, these prior studies do not offer a mechanistic account of how or why VTA versus amygdala activation would bias the MTL toward hippocampal versus cortical encoding, respectively. Accumulating evidence from psychology and neuroscience literatures provides several potential mechanisms for this functional organization. Below, we detail three possibilities. They are not mutually exclusive and are potentially synergistic. These mechanisms are as follows: intrinsic organization of functional neuroanatomy, differences in neurochemical engagement, and activation of distinct behaviors.

Intrinsic Network Connectivity Biases Information Flow

Intrinsic connectivity between discrete brain regions has been proposed to be a significant determinant of how neural networks are organized to guide cognition (Van Dijk et al. 2010). Intrinsic connectivity of the amygdala and VTA with cortical MTL and hippocampus is probably insufficient to explain the functional organization described in this model. Anatomically, the VTA innervates, though not uniformly, the entire MTL (Swanson 1982; Gasbarri et al. 1994). The amygdala, on the other hand, has stronger direct projections to MTL cortex, but also projects to the hippocampus proper (Packard and Wingard 2004).

Despite this, there are marked differences, however, in the relative connectivity of VTA and amygdala with broader cortical regions. These cortical regions may act as intermediaries in evoking preferential engagement of hippocampus or cortical MTL (Fig. 4). For example, activation of lateral prefrontal cortex (PFC) has been implicated in hippocampal-dependent encoding and relational memory (Blumenfeld and Ranganath 2007), whereas item-related memory can occur in its absence (Blumenfeld et al. 2011). Dopaminergic inputs from the VTA are thought to modulate PFC activity as it relates to a variety of executive functions (Bergson et al. 1995; Williams and Goldman-Rakic 1995; Sawaguchi and Goldman-Rakic 1991; Durstewitz et al. 2000). On the contrary, amygdala activation has been demonstrated to impede PFC activation during working and episodic memory (Dolcos and McCarthy 2006) and patients with amygdala lesions show enhanced PFC-dependent working memory performance (Morgan et al. 2012), suggesting that amygdala engagement during encoding could inhibit PFC function.

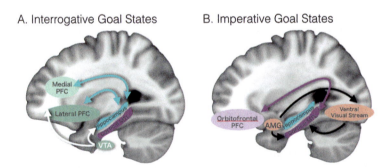

Fig. 4 Intrinsic network connectivity may delineate medial temporal lobe targets during motivated learning. Beyond direct modulation of the MTL, the amygdala and VTA could support differential MTL-dependent encoding by modulating discrete cortical targets. Here, we propose that the VTA may indirectly support hippocampal-dependent encoding by supporting working memory functions in the dorsomedial, dorsolateral and ventrolateral prefrontal cortex (white), which have been shown to support relational memory processes. We further propose that the amygdala supports cortical MTL dependent encoding by facilitating perceptual processing throughout the ventral visual stream, which has strong connectivity with the cortical MTL. Note that the arrows in this schematic are intended to indicate functional relationships shaped via neuromodulation, not monosynaptic projections; see text for details

The VTA has rich anatomical connectivity with dorsal and ventrolateral prefrontal cortex (Durstewitz et al. 2000; Bergson et al. 1995; Williams and Goldman-Rakic 1993, 1998), whereas the amygdala has only sparse anatomical connectivity with these regions (Amaral and Price 1984). This anatomical bias suggests that facilitation of these prefrontal regions by the VTA, versus the sparse projections from the amygdala, could specifically support relational memory encoding in the hippocampus.

Conversely, the amygdala has significant connectivity with the ventral visual stream (Amaral and Cowan 1980; Amaral and Price 1984), but VTA projections to these regions are sparse (Swanson 1982). Amygdala projections to targets in the ventral visual stream have been implicated in supporting enhanced detection and perception of environmental threats (Pessoa and Adolphs 2010). Anatomically, the cortical MTL is thought to be the most anterior portion of the ventral visual stream and has rich connectivity (Suzuki and Amaral 1994), while the hippocampus does not receive any direct inputs from posterior regions of the ventral visual stream. Selective enhancement of ventral visual-stream processing via amygdala connectivity could facilitate cortical MTL-dependent encoding over hippocampal-dependent encoding.

These differential connectivity patterns of the VTA and amygdala with PFC and ventral visual stream are well-positioned to shunt encoding activity toward specific MTL subdivisions during motivated encoding. In our own study of MTL network responses to surprising events under reward versus punishment incentives, we observed these predicted dissociations in prefrontal and visual ventral stream connectivity (Murty et al. 2016a). Specifically, under reward incentives, we observed hippocampal responses to surprising events and functional connectivity between hippocampus and dorsomedial PFC; whereas under punishment incentives, we observed parahippocampal cortex responses to surprises and connectivity with orbitofrontal cortex. Similarly, research has shown simultaneous engagement of the VTA and lateral PFC during reward-motivated encoding (Cohen et al. 2014, 2016), as well as co-engagement of the amygdala and ventral visual stream during the encoding of intrinsically threatening items (Kensinger et al. 2007b; Mickley Steinmetz et al. 2010). Together, these findings provide preliminary evidence of regions outside of the VTA, amygdala, and MTL supporting motivated memory encoding, but future research needs to detail their exact role.

Broadcast Actions of Neuromodulatory Transmitters Reconfigure Networks

In the imaging data discussed above, we have used fMRI activation of VTA and amygdala as indices of neuromodulation during motivational states; these robust fMRI signals can be related to both interrogative and imperative motivational states and to memory outcomes. Excitation of these nuclei is closely associated with release of dopamine and norepinephrine, although via different mechanisms. VTA activation may indirectly reflect dopamine release from VTA terminals: it has been shown to correlate with displacement (presumably by endogenous dopamine) of radioligand from dopamine receptors in striatum (Schott et al. 2008), but activity in non-dopaminergic neurons within VTA would also contribute to this signal. Amygdala activation as detected by conventional fMRI is likely to reflect the larger central nucleus, which is closely associated with arousal and increased noradrenergic activity. Both dopamine and norepinephrine act as global neuromodulators capable of rapidly reconfiguring neural networks.

Direct evidence of these network effects of neuromodulators has come primarily from invertebrate models (see Marder 2012 for review), but a few studies in humans have used analyses that characterize topology within and between brain networks to quantitate configural shifts associated with changes in motivational context (Kinnison et al. 2012) and used pharmacological challenges in fMRI to demonstrate their neurochemical origins. For example, dopaminergic enhancement and antagonism have opposing effects on resting state networks centered on the midbrain (Cole et al. 2013a, b). One compelling pharmacological fMRI study has demonstrated rapid reconfiguration of network connectivity in response to acute stressors; these increases were diminished by beta-adrenergic blockade (Hermans et al. 2011).

It should be noted that neuromodulatory transmitters alter brain function at multiple levels of functioning. Dopamine, for example, impacts cellular-level physiology, modulating synaptic learning signals (Calabresi et al. 2007; Lisman et al., 2011; Reynolds and Wickens 2002), altering neuronal excitability (Henze et al. 2000; Nicola et al. 2000), enhancing the signal-to-noise ratio (Durstewitz and Seamans 2008; Thurley et al. 2008), impacting the temporal patterning of neural activity (Walters et al. 2000), and sharpening cortical tuning (Hains and Arnsten 2008). These cellular changes necessarily translate to changes at the circuit and network levels and may be synergistic.

In summary, because neuromodulators such as dopamine and norepinephrine can rapidly reconfigure brain networks in response to the organism's current environment, they are well suited to establishing large-scale dynamic neural contexts during interrogative and imperative states. These neuromodulators act at multiple levels, an important open question is disentangling the actions of anatomical nuclei detectable with fMRI (or specific subdivisions of these like the central nucleus) from the broadcast actions of neuromodulators they are associated with. One early effort on this front (de Voogd et al. 2016) suggests that for memory formation, it is not arousal or noradrenergic tone, but amygdala activation *per se,* that is key.

Distinct Behavioral Responses to Incentives Could Separately Influence Memory

Early evidence from behavioral neuroscience has shown that motivational contexts change how organisms interact with their environments. Specifically, reward motivation has been associated with increased exploratory and novelty-seeking behaviors (Ikemoto and Panksepp 1999), whereas punishment motivation has been associated with increased freezing and escape behaviors (Davis 1992). Similarly, social and cognitive psychology literatures have demonstrated that individuals change their orientation and interactions with the environment under states of approaching rewards and avoiding punishments. Validated models have shown that positive affect and reward motivation promote exploration and active engagement with the environment, whereas negative affect and punishment motivation draw a response specifically to environmental threats (Elliot 2008; Elliot and Thrash 2002). Changing individuals' interactions with their environment changes the information available for encoding into long-term memory, and could, as a result, modulate the locus of MTL-dependent encoding. Changes in how individuals interact with their environment could thus potentially guide the organization of memory systems.

In line with these models, behavioral studies in humans have demonstrated attentional broadening during reward-focused states versus attentional narrowing during punishment-focused states (Fredrickson 2004). Specifically, during states of broadened attention and exploration elicited by reward motivation, individuals have increased capacity to attend to multiple features of the environment. This type of attentional state to multiple features provides the opportunity for the hippocampus to construct a more elaborated, integrative representation of the environment. This proposal suggests that manipulations that taxed attentional systems would in turn result in deficits in reward-motivated memory enhancements. Conversely, during avoidance of punishment motivation, individuals may narrowly attend to environmental threats or avenues for escape. Given narrow attention, only itemized constructs are available to encode into long-term memory, a process specialized in the cortical MTL. This interpretation converges with literatures showing a prioritization of attention towards threatening stimuli (Pessoa et al. 2010). This proposal suggests that manipulations that tax attentional systems may not affect punishment-motivated memory enhancements, as potentially threatening stimuli would remain prioritized (Dolcos et al. 2011).

A framework that considers information-seeking, for example interrogative and imperative motivational states, allows for more complex predictions about relationships between incentive valence, attention, and behavior. As noted above, in addiction or other compulsive (imperative) reward-seeking, we would expect amygdala activation and thus narrowed attention. On the other hand, during active avoidance of a discretely localized threat, both interrogative (way-finding) and imperative (flight) motives and behaviors are appropriate.

Open Questions and Future Directions

The model presented here offers first, a theoretical framework for understanding motivation to learn as it relates to complex incentives; and second, a systems-level characterization of motivational states as specific neural contexts for memory formation. We describe the impact of distinct motivational states on medial temporal lobe function, and we further propose that the specifics of the neural contexts will serve to encode memories structured to support similar future behavior. Our model implies differential behavioral impacts for memories formed under interrogative versus imperative goal orientation incentives, based on broad correlations between incentive structures and these states, but holds that the neural responses to incentives are the ultimate determinants of memory modulation.

Extensions of the work into more ecologically valid domains may help isolate the environmental determinants of these states, but a key constraint on these efforts is the lability and state-dependence of motivation itself. Advances in methods for decoding motivational states from the brain are needed both to better predict responses to extrinsic motivators, including money and primary rewards, and to understand intrinsic motivational drives, like curiosity.

Our proposed model implies distinct effects of motivational context at memory encoding on future behavior. Extant research has focused on how motivation influences memory, and here we specify how different motivational states influence its content and form. Emerging research has begun to investigate how memories encoded in motivationally-relevant contexts support adaptive behavior and decision-making (Murty et al. 2016b; Wimmer and Shohamy 2012; Gluth et al. 2015). In line with our model predictions, emerging research shows that reward-motivated, hippocampal-dependent (versus cortical MTL-dependent) memories preferentially support adaptive decision-making to obtain rewards (Murty et al. 2016b; Wimmer and Shohamy 2012; Gluth et al. 2015). Future research will need to test the converse: whether cortical MTL-dependent (versus hippocampally-encoded) memories preferentially support future behaviors specifically consistent with imperative motivation.

This evolving picture of how motivational states impact the medial temporal lobe system complements the long-established body of research on the role of motivation in associative and skill learning. With a precise and nuanced understanding of the antecedents, neural mechanisms and behavioral impact of motivation on memory formation, researchers are positioned to help develop tools to optimize learning for a wide range of contexts, from education to learning-based psychotherapies for mental disorders.

References

Adamantidis AR, Tsai H-C, Boutrel B, Zhang F, Stuber GD, Budygin EA, Touriño C, Bonci A, Deisseroth K, de Lecea L (2011) Optogenetic interrogation of dopaminergic modulation of the multiple phases of reward-seeking behavior. J Neurosci 31(30):10829–10835

Adcock RA, Thangavel A, Whitfield-Gabrieli S, Knutson B, Gabrieli JDE (2006) Reward-motivated learning: mesolimbic activation precedes memory formation. Neuron 50 (3):507–517

Adolphs R, Gosselin F, Buchanan TW, Tranel D, Schyns P, Damasio AR (2005) A mechanism for impaired fear recognition after amygdala damage. Nature 433(7021):68–72

Akirav I, Richter-Levin G (1999) Priming stimulation in the basolateral amygdala modulates synaptic plasticity in the rat dentate gyrus. Neurosci Lett 270(2):83–86

Amaral DG, Cowan WM (1980) Subcortical afferents to the hippocampal formation in the monkey. J Comp Neurol 189(4):573–591

Amaral DG, Price JL (1984) Amygdalo-cortical projections in the monkey (*Macaca Fascicularis*). J Comp Neurol 230(4):465–496

Ariely D, Gneezy U, Loewenstein G, Mazar N (2009) Large stakes and big mistakes. Rev Econ Stud 76(2):451–469

Bauch EM, Rausch VH, Bunzeck N (2014) Pain anticipation recruits the mesolimbic system and differentially modulates subsequent recognition memory. Hum Brain Mapp 35(9):4594–4606

Baxter MG, Murray EA (2002) The amygdala and reward. Nat Rev Neurosci 3(7):563–573

Bennion KA, Ford JH, Murray BD, Kensinger EA (2013) Oversimplification in the study of emotional memory. J Int Neuropsychol Soc 19(9):953–961

Bergson C, Mrzljak L, Smiley JF, Pappy M, Levenson R, Goldman-Rakic PS (1995) Regional, cellular, and subcellular variations in the distribution of D1 and D5 dopamine receptors in primate brain. J Neurosci 15(12):7821–7836

Blanchard RJ, Blanchard DC (1969) Passive and active reactions to fear-eliciting stimuli. J Comp Physiol Psychol 68(1):129–135

Blumenfeld RS, Ranganath C (2007) Prefrontal cortex and long-term memory encoding: an integrative review of findings from neuropsychology and neuroimaging. Neuroscientist 13 (3):280–291

Blumenfeld RS, Parks CM, Yonelinas AP, Ranganath C (2011) Putting the pieces together: the role of dorsolateral prefrontal cortex in relational memory encoding. J Cogn Neurosci 23 (1):257–265

Boll S, Gamer M, Gluth S, Finsterbusch J, Büchel C (2013) Separate amygdala subregions signal surprise and predictiveness during associative fear learning in humans. Eur J Neurosci 37 (5):758–767

Broks P, Young AW, Maratos EJ, Coffey PJ, Calder AJ, Isaac CL, Mayes AR et al (1998) Face processing impairments after encephalitis: amygdala damage and recognition of fear. Neuropsychologia 36(1):59–70

Bromberg-Martin ES, Hikosaka O (2009) Midbrain dopamine neurons signal preference for advance information about upcoming rewards. Neuron 63(1):119–126

Bromberg-Martin ES, Matsumoto M, Hikosaka O (2010) Dopamine in motivational control: rewarding, aversive, and alerting. Neuron 68(5):815–834

Büchel C, Morris J, Dolan RJ, Friston KJ (1998) Brain systems mediating aversive conditioning: an event-related fmri study. Neuron 20(5):947–957

Bunzeck N, Düzel E (2006) Absolute coding of stimulus novelty in the human substantia nigra/VTA. Neuron 51(3):369–379

Bunzeck N, Doeller CF, Dolan RJ, Duzel E (2012) Contextual interaction between novelty and reward processing within the mesolimbic system. Hum Brain Mapp 33(6):1309–1324

Calabresi P, Picconi B, Tozzi A, Di Filippo M (2007) Dopamine-mediated regulation of corticostriatal synaptic plasticity. Trends Neurosci 30(5):211–219

Callan DE, Schweighofer N (2008) Positive and negative modulation of word learning by reward anticipation. Hum Brain Mapp 29(2):237–249

Campeau S, Davis M (1995) Involvement of the central nucleus and basolateral complex of the amygdala in fear conditioning measured with fear-potentiated startle in rats trained concurrently with auditory and visual conditioned stimuli. J Neurosci 15(3 Pt 2):2301–2311

Camps M, Kelly PH, Palacios JM (1990) Autoradiographic localization of dopamine D1 and D2 receptors in the brain of several mammalian species. J Neural Transm 80(2):105–127

Carter RM, MacInnes JJ, Huettel S, Alison Adcock R (2009) Activation in the VTA and nucleus accumbens increases in anticipation of both gains and losses. Neuroeconomics 3:21

Choi J-S, Cain CK, LeDoux JE (2010) The role of amygdala nuclei in the expression of auditory signaled two-way active avoidance in rats. Learn Mem 17(3):139–147

Ciliax BJ, Nash N, Heilman C, Sunahara R, Hartney A, Tiberi M, Rye DB, Caron MG, Niznik HB, Levey AI (2000) Dopamine D(5) receptor immunolocalization in rat and monkey brain. Synapse 37(2):125–145

Cohen MS, Rissman J, Suthana NA, Castel AD, Knowlton BJ (2014) Value-based modulation of memory encoding involves strategic engagement of fronto-temporal semantic processing regions. Cogn Affect Behav Neurosci 14(2):578–592

Cohen MS, Rissman J, Suthana NA, Castel AD, Knowlton BJ (2016) Effects of aging on value-directed modulation of semantic network activity during verbal learning. NeuroImage 125 (January):1046–1062

Cole DM, Beckmann CF, Oei NYL, Both S, van Gerven JMA, Rombouts SARB (2013a) Differential and distributed effects of dopamine neuromodulations on resting-state network connectivity. NeuroImage 78(September):59–67

Cole DM, Oei NYL, Soeter RP, Both S, van Gerven JMA, Rombouts SARB, Beckmann CF (2013b) Dopamine-dependent architecture of cortico-subcortical network connectivity. Cereb Cortex 23(7):1509–1516

D'Ardenne K, McClure SM, Nystrom LE, Cohen JD (2008) BOLD responses reflecting dopaminergic signals in the human ventral tegmental area. Science 319(5867):1264–1267

Davachi L (2006) Item, context and relational episodic encoding in humans. Curr Opin Neurobiol 16(6):693–700

Davis M (1992) The role of the amygdala in fear and anxiety. Annu Rev Neurosci 15(1):353–375

Daw ND, Doya K (2006) The computational neurobiology of learning and reward. Curr Opin Neurobiol 16(2):199–204

de Voogd LD, Fernández G, Hermans EJ (2016) Disentangling the roles of arousal and amygdala activation in emotional declarative memory. Soc Cogn Affect Neurosci. doi:10.1093/scan/nsw055

DeCoteau WE, Thorn C, Gibson DJ, Courtemanche R, Mitra P, Kubota Y, Graybiel AM (2007) Learning-related coordination of striatal and hippocampal theta rhythms during acquisition of a procedural maze task. Proc Natl Acad Sci U S A 104(13):5644–5649

Delgado MR, Jou RL, Phelps EA (2011) Neural systems underlying aversive conditioning in humans with primary and secondary reinforcers. Front Neurosci 5(May):71

Dolcos F, McCarthy G (2006) Brain systems mediating cognitive interference by emotional distraction. J Neurosci 26(7):2072–2079

Dolcos F, LaBar KS, Cabeza R (2004) Interaction between the amygdala and the medial temporal lobe memory system predicts better memory for emotional events. Neuron 42(5):855–863

Dolcos F, Iordan AD, Dolcos S (2011) Neural correlates of emotion–cognition interactions: a review of evidence from brain imaging investigations. J Cogn Psychol 23(6):669–694

Dougal S, Phelps EA, Davachi L (2007) The role of medial temporal lobe in item recognition and source recollection of emotional stimuli. Cogn Affect Behav Neurosci 7(3):233–242

Durstewitz D, Seamans JK (2008) The dual-state theory of prefrontal cortex dopamine function with relevance to catechol-o-methyltransferase genotypes and schizophrenia. Biol Psychiatry 64(9):739–749

Durstewitz D, Seamans JK, Sejnowski TJ (2000) Neurocomputational models of working memory. Nat Neurosci 3(November):1184–1191

Düzel E, Bunzeck N, Guitart-Masip M, Düzel S (2010) NOvelty-related motivation of anticipation and exploration by dopamine (NOMAD): implications for healthy aging. Neurosci Biobehav Rev 34(5):660–669

Eichenbaum H, Otto T, Cohen NJ (1994) Two functional components of the hippocampal memory system. Behav Brain Sci 17(03):449–472

Eichenbaum H, Yonelinas AP, Ranganath C (2007) The medial temporal lobe and recognition memory. Annu Rev Neurosci 30:123–152

Elliot AJ (2008) Handbook of approach and avoidance motivation. Taylor & Francis, New York

Elliot AJ, Thrash TM (2002) Approach-avoidance motivation in personality: approach and avoidance temperaments and goals. J Pers Soc Psychol 82(5):804–818

Fanselow MS (2000) Contextual fear, gestalt memories, and the hippocampus. Behav Brain Res 110(1–2):73–81

Farovik A, Place RJ, Miller DR, Eichenbaum H (2011) Amygdala lesions selectively impair familiarity in recognition memory. Nat Neurosci 14(11):1416–1417

Fendt M (2001) Injections of the NMDA receptor antagonist aminophosphonopentanoic acid into the lateral nucleus of the amygdala block the expression of fear-potentiated startle and freezing. J Neurosci 21(11):4111–4115

Fink JS, Smith GP (1980) Mesolimbicocortical dopamine terminal fields are necessary for normal locomotor and investigatory exploration in rats. Brain Res 199(2):359–384

Fredrickson BL (2004) The broaden-and-build theory of positive emotions. Philos Trans R Soc Lond Ser B Biol Sci 359(1449):1367–1378

Frey S, Bergado-Rosado J, Seidenbecher T, Pape H-C, Uwe Frey J (2001) Reinforcement of early long-term potentiation (early-LTP) in dentate gyrus by stimulation of the basolateral amygdala: heterosynaptic induction mechanisms of late-LTP. J Neurosci 21(10):3697–3703

Fu W-T, Anderson JR (2008) Solving the credit assignment problem: explicit and implicit learning of action sequences with probabilistic outcomes. Psychol Res 72(3):321–330

Gasbarri A, Verney C, Innocenzi R, Campana E, Pacitti C (1994) Mesolimbic dopaminergic neurons innervating the hippocampal formation in the rat: a combined retrograde tracing and immunohistochemical study. Brain Res 668(1–2):71–79

Giovanello KS, Verfaellie M, Keane MM (2003) Disproportionate deficit in associative recognition relative to item recognition in global amnesia. Cogn Affect Behav Neurosci 3(3):186–194

Gluth S, Sommer T, Rieskamp J, Büchel C (2015) Effective connectivity between hippocampus and ventromedial prefrontal cortex controls preferential choices from memory. Neuron 86 (4):1078–1090

Gomperts SN, Kloosterman F, Wilson MA (2015) VTA neurons coordinate with the hippocampal reactivation of spatial experience. eLife 4(October). doi:10.7554/eLife.05360

Gruber MJ, Ritchey M, Wang S-F, Doss MK, Ranganath C (2016) Post-learning hippocampal dynamics promote preferential retention of rewarding events. Neuron. doi:10.1016/j.neuron. 2016.01.017

Hahn T, Dresler T, Plichta MM, Ehlis A-C, Ernst LH, Markulin F, Polak T et al (2010) Functional amygdala-hippocampus connectivity during anticipation of aversive events is associated with gray's trait 'sensitivity to punishment. Biol Psychiatry 68(5):459–464

Hains AB, Arnsten AF (2008) Molecular mechanisms of stress-induced prefrontal cortical impairment: implications for mental illness. Learn Mem 15(8):551–564

Hamann SB, Cahill L, McGaugh JL, Squire LR (1997a) Intact enhancement of declarative memory for emotional material in amnesia. Learn Mem 4(3):301–309

Hamann SB, Cahill L, Squire LR (1997b) Emotional perception and memory in amnesia. Neuropsychology 11(1):104–113

Hammad H, Wagner JJ (2006) Dopamine-mediated disinhibition in the CA1 region of rat hippocampus via D3 receptor activation. J Pharmacol Exp Ther 316(1):113–120

Hannula DE, Tranel D, Allen JS, Kirchhoff BA, Nickel AE, Cohen NJ (2015) Memory for items and relationships among items embedded in realistic scenes: disproportionate relational memory impairments in amnesia. Neuropsychology 29(1):126–138

Henze DA, González-Burgos GR, Urban NN, Lewis DA, Barrionuevo G (2000) Dopamine increases excitability of pyramidal neurons in primate prefrontal cortex. J Neurophys 84(6): 2799–2809

Hermans EJ, van Marle HJF, Ossewaarde L, Henckens MJAG, Qin S, van Kesteren MTR, Schoots VC et al (2011) Stress-related noradrenergic activity prompts large-scale neural network reconfiguration. Science 334(6059):1151–1153

Higgins ET (1998) Promotion and prevention: regulatory focus as a motivational principle. In: Zanna MP (ed) Advances in experimental social psychology, vol 30. Academic Press, New York, pp 1–46

Hitchcock J, Davis M (1986) Lesions of the amygdala, but not of the cerebellum or red nucleus, block conditioned fear as measured with the potentiated startle paradigm. Behav Neurosci 100 (1):11–22

Holahan MR, White NM (2004) Amygdala inactivation blocks expression of conditioned memory modulation and the promotion of avoidance and freezing. Behav Neurosci 118(1):24–35

Huang YY, Kandel ER (1995) D1/D5 receptor agonists induce a protein synthesis-dependent late potentiation in the CA1 region of the hippocampus. Proc Natl Acad Sci U S A 92 (7):2446–2450

Ikegaya Y, Abe K, Saito H, Nishiyama N (1995) Medial amygdala enhances synaptic transmission and synaptic plasticity in the dentate gyrus of rats in vivo. J Neurophysiol 74(5):2201–2203

Ikemoto S, Panksepp J (1999) The role of nucleus accumbens dopamine in motivated behavior: a unifying interpretation with special reference to reward-seeking. Brain Res Brain Res Rev 31 (1):6–41

Kahn I, Shohamy D (2013) Intrinsic connectivity between the hippocampus, nucleus accumbens, and ventral tegmental area in humans. Hippocampus 23(3):187–192

Kakade S, Dayan P (2002) Dopamine: generalization and bonuses. Neural Netw 15(4–6):549–559

Kensinger EA, Schacter DL (2006) Amygdala activity is associated with the successful encoding of item, but not source, information for positive and negative stimuli. J Neurosci 26 (9):2564–2570

Kensinger EA, Garoff-Eaton RJ, Schacter DL (2007a) Effects of emotion on memory specificity: memory trade-offs elicited by negative visually arousing stimuli. J Mem Lang 56(4):575–591

Kensinger EA, Garoff-Eaton RJ, Schacter DL (2007b) How negative emotion enhances the visual specificity of a memory. J Cogn Neurosci 19(11):1872–1887

Khan ZU, Gutiérrez A, Martín R, Peñafiel A, Rivera A, de la Calle A (2000) Dopamine D5 receptors of rat and human brain. Neuroscience 100(4):689–699

Kim M, Davis M (1993) Electrolytic lesions of the amygdala block acquisition and expression of fear-potentiated startle even with extensive training but do not prevent reacquisition. Behav Neurosci 107(4):580–595

Kim JJ, Rison RA, Fanselow MS (1993) Effects of amygdala, hippocampus, and periaqueductal gray lesions on short- and long-term contextual fear. Behav Neurosci 107(6):1093–1098

Kim JJ, Lee HJ, Han JS, Packard MG (2001) Amygdala is critical for stress-induced modulation of hippocampal long-term potentiation and learning. J Neurosci 21(14):5222–5228

Kim KM, Baratta MV, Yang A, Lee D, Boyden ES, Fiorillo CD (2012) Optogenetic mimicry of the transient activation of dopamine neurons by natural reward is sufficient for operant reinforcement. PLoS One 7(4):e33612

Kinnison J, Padmala S, Choi JM, Pessoa L (2012) Network analysis reveals increased integration during emotional and motivational processing. J Neurosci 32(24):8361–8372

Korte SM (2001) Corticosteroids in relation to fear, anxiety and psychopathology. Neurosci Biobehav Rev 25(2):117–142

Krebs RM, Schott BH, Düzel E (2009) Personality traits are differentially associated with patterns of reward and novelty processing in the human substantia nigra/ventral tegmental area. Biol Psychiatry 65(2):103–110

Krebs RM, Heipertz D, Schuetze H, Duzel E (2011) Novelty increases the mesolimbic functional connectivity of the substantia nigra/ventral tegmental area (SN/VTA) during reward anticipation: evidence from high-resolution fMRI. NeuroImage 58(2):647–655

Kwon HG, Jang SH (2014) Differences in neural connectivity between the substantia nigra and ventral tegmental area in the human brain. Front Hum Neurosci 8(February):41

LaBar KS, Cabeza R (2006) Cognitive neuroscience of emotional memory. Nat Rev Neurosci 7 (1):54–64

LaBar KS, LeDoux JE, Spencer DD, Phelps EA (1995) Impaired fear conditioning following unilateral temporal lobectomy in humans. J Neurosci 15(10):6846–6855

LaBar KS, Gatenby JC, Gore JC, LeDoux JE, Phelps EA (1998) Human amygdala activation during conditioned fear acquisition and extinction: a mixed-trial fMRI study. Neuron 20 (5):937–945

LeDoux JE (1992) Brain mechanisms of emotion and emotional learning. Curr Opin Neurobiol 2 (2):191–197

Lesscher HMB, Vanderschuren LJMJ (2012) Compulsive drug use and its neural substrates. Rev Neurosci 23(5–6):731–745

Li S, Cullen WK, Anwyl R, Rowan MJ (2003) Dopamine-dependent facilitation of LTP induction in hippocampal CA1 by exposure to spatial novelty. Nat Neurosci 6(5):526–531

Lisman J, Grace AA, Duzel E (2011) A neoHebbian framework for episodic memory; role of dopamine-dependent late LTP. Trends Neurosci 34(10):536–547

Little KY, Carroll FI, Cassin BJ (1995) Characterization and localization of [125I]RTI-121 binding sites in human striatum and medial temporal lobe. J Pharmacol Exp Ther 274 (3):1473–1483

Loh E, Kumaran D, Koster R, Berron D, Dolan R, Duzel E (2015) Context-specific activation of hippocampus and SN/VTA by reward is related to enhanced long-term memory for embedded objects. Neurobiol Learn Mem. doi:10.1016/j.nlm.2015.11.018

Marder E (2012) Neuromodulation of neuronal circuits: back to the future. Neuron 76(1):1–11

Martig AK, Mizumori SJY (2011) Ventral tegmental area disruption selectively affects CA1/CA2 but not CA3 place fields during a differential reward working memory task. Hippocampus 21 (2):172–184

Martig AK, Jones GL, Smith KE, Mizumori SJY (2009) Context dependent effects of ventral tegmental area inactivation on spatial working memory. Behav Brain Res 203(2):316–320

McClure SM, Berns GS, Read Montague P (2003) Temporal prediction errors in a passive learning task activate human striatum. Neuron 38(2):339–346

McDonald RJ, White NM (1993) A triple dissociation of memory systems: hippocampus, amygdala, and dorsal striatum. Behav Neurosci 107(1):3–22

McGaugh JL (2004) The amygdala modulates the consolidation of memories of emotionally arousing experiences. Annu Rev Neurosci 27:1–28

McNamara CG, Tejero-Cantero Á, Trouche S, Campo-Urriza N, Dupret D (2014) Dopaminergic neurons promote hippocampal reactivation and spatial memory persistence. Nat Neurosci 17 (12):1658–1660

Mickley Steinmetz KR, Addis DR, Kensinger EA (2010) The effect of arousal on the emotional memory network depends on valence. NeuroImage 53(1):318–324

Miendlarzewska EA, Bavelier D, Schwartz S (2016) Influence of reward motivation on human declarative memory. Neurosci Biobehav Rev 61(February):156–176

Mobbs D, Petrovic P, Marchant JL, Hassabis D, Weiskopf N, Seymour B, Dolan RJ, Frith CD (2007) When fear is near: threat imminence elicits prefrontal-periaqueductal gray shifts in humans. Science 317(5841):1079–1083

Mobbs D, Hassabis D, Seymour B, Marchant JL, Weiskopf N, Dolan RJ, Frith CD (2009a) Choking on the money reward-based performance decrements are associated with midbrain activity. Psychol Sci 20(8):955–962

Mobbs D, Marchant JL, Hassabis D, Seymour B, Tan G, Gray M, Petrovic P, Dolan RJ, Frith CD (2009b) From threat to fear: the neural organization of defensive fear systems in humans. J Neurosci 29(39):12236–12243

Mobbs D, Hagan CC, Dalgleish T, Silston B, Prévost C (2015) The ecology of human fear: survival optimization and the nervous system. Front Neurosci 9(March):55

Morgan B, David T, Thornton HB, Stein DJ, van Honk J (2012) Paradoxical facilitation of working memory after basolateral amygdala damage. PLoS One 7(6):e38116

Mukherjee J, Christian BT, Dunigan KA, Shi B, Narayanan TK, Satter M, Mantil J (2002) Brain imaging of 18F-fallypride in normal volunteers: blood analysis, distribution, test-retest studies, and preliminary assessment of sensitivity to aging effects on dopamine D-2/D-3 receptors. Synapse 46(3):170–188

Murray EA (2007) The amygdala, reward and emotion. Trends Cogn Sci 11(11):489–497

Murray GK, Corlett PR, Clark L, Pessiglione M, Blackwell AD, Honey G, Jones PB, Bullmore ET, Robbins TW, Fletcher PC (2008) Substantia nigra/ventral tegmental reward prediction error disruption in psychosis. Mol Psychiatry 13(3) 239, 267–276

Murty VP, Adcock RA (2014) Enriched encoding: reward motivation organizes cortical networks for hippocampal detection of unexpected events. Cereb Cortex 24(8):2160–2168

Murty VP, Ritchey M, Alison Adcock R, LaBar KS (2010) fMRI studies of successful emotional memory encoding: a quantitative meta-analysis. Neuropsychologia 48(12):3459–3469

Murty VP, LaBar KS, Hamilton DA, Alison Adcock R (2011) Is all motivation good for learning? dissociable influences of approach and avoidance motivation in declarative memory. Learn Mem 18(11):712–717

Murty VP, Labar KS, Alison Adcock R (2012) Threat of punishment motivates memory encoding via amygdala, not midbrain, interactions with the medial temporal lobe. J Neurosci 32 (26):8969–8976

Murty VP, Shermohammed M, Smith DV, Mckell Carter R, Huettel SA, Alison Adcock R (2014) Resting state networks distinguish human ventral tegmental area from substantia nigra. NeuroImage 100(October):580–589

Murty VP, LaBar KS, Alison Adcock R (2016a) Distinct medial temporal networks encode surprise during motivation by reward versus punishment. Neurobiol Learn Mem. doi:10.1016/j.nlm.2016.01.018

Murty VP, Oriel FH, Hunter LE, Phelps EA, Davachi L (2016b) Episodic memories predict adaptive value-based decision-making. J Exp Psychol Gen 145:548–558

Nee DE, Wager TD, Jonides J (2007) Interference resolution: insights from a meta-analysis of neuroimaging tasks. Cogn Affect Behav Neurosci 7(1):1–17

Nicola SM, Surmeier DJ, Malenka RC (2000) Dopaminergic modulation of neuronal excitability in the striatum and nucleus accumbens. Annu Rev Neurosci 23(1):185–215

O'Carroll CM, Martin SJ, Sandin J, Frenguelli B, Morris RGM (2006) Dopaminergic modulation of the persistence of one-trial hippocampus-dependent memory. Learn Mem 13(6):760–769

Olds J, Milner P (1954) Positive reinforcement produced by electrical stimulation of septal area and other regions of rat brain. J Comp Physiol Psychol 47(6):419–427

Onoda K, Okamoto Y, Yamawaki S (2009) Neural correlates of associative memory: the effects of negative emotion. Neurosci Res 64(1):50–55

Packard MG (2009) Anxiety, cognition, and habit: a multiple memory systems perspective. Brain Res 1293(October):121–128

Packard MG, Wingard JC (2004) Amygdala and 'emotional' modulation of the relative use of multiple memory systems. Neurobiol Learn Mem 82(3):243–252

Pagnoni G, Zink CF, Read Montague P, Berns GS (2002) Activity in human ventral striatum locked to errors of reward prediction. Nat Neurosci 5(2):97–98

Paz R, Pelletier JG, Bauer EP, Paré D (2006) Emotional enhancement of memory via amygdala-driven facilitation of rhinal interactions. Nat Neurosci 9(10):1321–1329

Pessiglione M, Seymour B, Flandin G, Dolan RJ, Frith CD (2006) Dopamine-dependent prediction errors underpin reward-seeking behaviour in humans. Nature 442(7106):1042–1045

Pessoa L, Adolphs R (2010) Emotion processing and the amygdala: from a 'low road' to 'many roads' of evaluating biological significance. Nat Rev Neurosci 11(11):773–783

Pessoa L, Pereira MG, Oliveira L (2010) Attention and emotion. Scholarpedia J 5(2):6314

Phillips RG, LeDoux JE (1992) Differential contribution of amygdala and hippocampus to cued and contextual fear conditioning. Behav Neurosci 106(2):274–285

Prevost C, McCabe JA, Jessup RK, Bossaerts P, O'Doherty JP (2011) Differentiable contributions of human amygdalar subregions in the computations underlying reward and avoidance learning. Eur J Neurosci 34(1):134–145

Prévost C, Liljeholm M, Tyszka JM, O'Doherty JP (2012) Neural correlates of specific and general pavlovian-to-instrumental transfer within human amygdalar subregions: a high-resolution fMRI study. J Neurosci 32(24):8383–8390

Qin S, Hermans EJ, van Marle HJF, Fernández G (2012) Understanding low reliability of memories for neutral information encoded under stress: alterations in memory-related activation in the hippocampus and midbrain. J Neurosci 32(12):4032–4041

Quamme JR, Yonelinas AP, Norman KA (2007) Effect of unitization on associative recognition in amnesia. Hippocampus 17(3):192–200

Ranganath C (2010) A unified framework for the functional organization of the medial temporal lobes and the phenomenology of episodic memory. Hippocampus 20(11):1263–1290

Reilly S, Bornovalova MA (2005) Conditioned taste aversion and amygdala lesions in the rat: a critical review. Neurosci Biobehav Rev 29(7):1067–1088

Reynolds JN, Wickens JR (2002) Dopamine-dependent plasticity of corticostriatal synapses. Neural Netw 15(4):507–521

Rimmele U, Davachi L, Petrov R, Dougal S, Phelps EA (2011) Emotion enhances the subjective feeling of remembering, despite lower accuracy for contextual details. Emotion 11(3):553–562

Rimmele U, Davachi L, Phelps EA (2012) Memory for time and place contributes to enhanced confidence in memories for emotional events. Emotion 12(4):834–846

Ritchey M, Dolcos F, Cabeza R (2008) Role of amygdala connectivity in the persistence of emotional memories over time: an event-related fMRI investigation. Cereb Cortex 18 (11):2494–2504

Robinson MJF, Warlow SM, Berridge KC (2014) Optogenetic excitation of central amygdala amplifies and narrows incentive motivation to pursue one reward above another. J Neurosci 34 (50):16567–16580

Roozendaal B, McGaugh JL (1997) Basolateral amygdala lesions block the memory-enhancing effect of glucocorticoid administration in the dorsal hippocampus of rats. Eur J Neurosci 9 (1):76–83

Roozendaal B, Portillo-Marquez G, McGaugh JL (1996) Basolateral amygdala lesions block glucocorticoid-induced modulation of memory for spatial learning. Behav Neurosci 110 (5):1074–1083

Roozendaal B, Nguyen BT, Power AE, McGaugh JL (1999) Basolateral amygdala noradrenergic influence enables enhancement of memory consolidation induced by hippocampal glucocorticoid receptor activation. Proc Natl Acad Sci U S A 96(20):11642–11647

Roozendaal B, Griffith QK, Jason B, Dominique J-F, McGaugh JL (2003) The hippocampus mediates glucocorticoid-induced impairment of spatial memory retrieval: dependence on the basolateral amygdala. Proc Natl Acad Sci U S A 100(3):1328–1333

Rorick-Kehn LM, Steinmetz JE (2005) Amygdalar unit activity during three learning tasks: eyeblink classical conditioning, pavlovian fear conditioning, and signaled avoidance conditioning. Behav Neurosci 119(5):1254–1276

Rossato JI, Bevilaqua LRM, Izquierdo I, Medina JH, Cammarota M (2009) Dopamine controls persistence of long-term memory storage. Science 325(5943):1017–1020

Rudy JW (2009) Context representations, context functions, and the parahippocampal-hippocampal system. Learn Mem 16(10):573–585

Sakaki M, Ycaza-Herrera AE, Mather M (2014) Association learning for emotional harbinger cues: when do previous emotional associations impair and when do they facilitate subsequent learning of new associations? Emotion 14(1):115–129

Salamone JD (1994) The Involvement of nucleus accumbens dopamine in appetitive and aversive motivation. Behav Brain Res 61(2):117–133

Salvetti B, Morris RGM, Wang S-H (2014) The role of rewarding and novel events in facilitating memory persistence in a separate spatial memory task. Learn Mem 21(2):61–72

Samson Y, Wu JJ, Friedman AH, Davis JN (1990) Catecholaminergic innervation of the hippocampus in the cynomolgus monkey. J Comp Neurol 298(2):250–263

Sawaguchi T, Goldman-Rakic PS (1991) D1 dopamine receptors in prefrontal cortex: involvement in working memory. Science 251(4996):947–950

Schlund MW, Cataldo MF (2010) Amygdala involvement in human avoidance, escape and approach behavior. NeuroImage 53(2):769–776

Schott BH, Minuzzi L, Krebs RM, Elmenhorst D, Lang M, Winz OH, Seidenbecher CI et al (2008) Mesolimbic functional magnetic resonance imaging activations during reward anticipation correlate with reward-related ventral striatal dopamine release. J Neurosci 28 (52):14311–14319

Schultz W (2016) Dopamine reward prediction-error signalling: a two-component response. Nat Rev Neurosci 17(3):183–195

Schwarze U, Bingel U, Sommer T (2012) Event-related nociceptive arousal enhances memory consolidation for neutral scenes. J Neurosci 32(4):1481–1487

Scott SK, Young AW, Calder AJ, Hellawell DJ, Aggleton JP, Johnson M (1997) Impaired auditory recognition of fear and anger following bilateral amygdala lesions. Nature 385(6613):254–257

Shohamy D, Adcock RA (2010) Dopamine and adaptive memory. Trends Cogn Sci 14 (10):464–472

Singer AC, Frank LM (2009) Rewarded outcomes enhance reactivation of experience in the hippocampus. Neuron 64(6):910–921

Steinberg EE, Keiflin R, Boivin JR, Witten IB, Deisseroth K, Janak PH (2013) A causal link between prediction errors, dopamine neurons and learning. Nat Neurosci 16(7):966–973

Steinmetz KRM, Kensinger EA (2013) The emotion-induced memory trade-off: more than an effect of overt attention? Mem Cogn 41(1):69–81

Strauman TJ, Wilson WA (2010) Behavioral activation/inhibition and regulatory focus as distinct levels of analysis. In: Hoyle RH (ed) Handbook of personality and self-regulation. Blackwell, Malden, MA, pp 447–473

Suzuki WA, Amaral DG (1994) Perirhinal and parahippocampal cortices of the macaque monkey: cortical afferents. J Comp Neurol 350(4):497–533

Swanson LW (1982) The projections of the ventral tegmental area and adjacent regions: a combined fluorescent retrograde tracer and immunofluorescence study in the rat. Brain Res Bull 9(1–6):321–353

Thurley K, Senn W, Lüscher HR (2008) Dopamine increases the gain of the input-output response of rat prefrontal pyramidal neurons. J Neurophys 99(6):2985–2997

Tran AH, Uwano T, Kimura T, Hori E, Katsuki M, Nishijo H, Ono T (2008) Dopamine D1 receptor modulates hippocampal representation plasticity to spatial novelty. J Neurosci 28 (50):13390–13400

Valdes J, McNaughton B, Fellous J-M (2011) Experience-dependent reactivations of ventral tegmental area neurons in the rat. BMC Neurosci. doi:10.1186/1471-2202-12-S1-P107

Van Dijk KRA, Hedden T, Venkataraman A, Evans KC, Lazar SW, Buckner RL (2010) Intrinsic functional connectivity as a tool for human connectomics: theory, properties, and optimization. J Neurophysiol 103(1):297–321

Walters JR, Ruskin DN, Allers KA, Bergstrom DA (2000) Pre-and postsynaptic aspects of dopamine-mediated transmission. Trends Neurosci 23:S41–S47

Wang S-H, Morris RGM (2010) Hippocampal-neocortical interactions in memory formation, consolidation, and reconsolidation. Annu Rev Psychol 61(49–79):C1–C4

Williams SM, Goldman-Rakic PS (1993) Characterization of the dopaminergic innervation of the primate frontal cortex using a dopamine-specific antibody. Cereb Cortex 3(3):199–222

Williams GV, Goldman-Rakic PS (1995) Modulation of memory fields by dopamine D1 receptors in prefrontal cortex. Nature 376(6541):572–575

Williams SM, Goldman-Rakic PS (1998) Widespread origin of the primate mesofrontal dopamine system. Cereb Cortex 8(4):321–345

Wimmer GE, Shohamy D (2012) Preference by association: how memory mechanisms in the hippocampus bias decisions. Science 338(6104):270–273

Wingard JC, Packard MG (2008) The amygdala and emotional modulation of competition between cognitive and habit memory. Behav Brain Res 193(1):126–131

Wise RA (2004) Dopamine, learning and motivation. Nat Rev Neurosci 5(6):483–494

Wittmann BC, Schott BH, Guderian S, Frey JU, Heinze H-J, Düzel E (2005) Reward-related fMRI activation of dopaminergic midbrain is associated with enhanced hippocampus-dependent long-term memory formation. Neuron 45(3):459–467

Wolosin SM, Zeithamova D, Preston AR (2012) Reward modulation of hippocampal subfield activation during successful associative encoding and retrieval. J Cogn Neurosci 24 (7):1532–1547

Xue Y, Steketee JD, Sun W (2012) Inactivation of the central nucleus of the amygdala reduces the effect of punishment on cocaine self-administration in rats. Eur J Neurosci 35(5):775–783

Yin HH, Knowlton BJ (2006) The role of the basal ganglia in habit formation. Nat Rev Neurosci 7 (6):464–476

Yu R (2015) Choking under pressure: the neuropsychological mechanisms of incentive-induced performance decrements. Front Behav Neurosci 9(February):19

Zaghloul KA, Blanco JA, Weidemann CT, McGill K, Jaggi JL, Baltuch GH, Kahana MJ (2009) Human substantia nigra neurons encode unexpected financial rewards. Science 323 (5920):1496–1499

Hippocampal Contributions to Language Use and Processing

Melissa C. Duff and Sarah Brown-Schmidt

Abstract Recent advances in understanding the functionality of the human hippo-campus has led to a number of proposals for how hippocampus may support a range of cognitive abilities beyond memory. Building on these advances, we offered a new account of the memory-language interface [Duff and Brown-Schmidt (Front Hum Neurosci 6:69, 2012)]. We proposed that the same processes by which the hippocampal declarative memory system creates and flexibly integrates represen-tations across diverse sources in the formation of new memories, and maintains representations on-line to be evaluated and used in service of behavioral perfor-mance, are the same processes necessary for the flexible use and on-line processing of language. This proposal leads to a set of testable predictions and hypotheses about how language and memory work together and argues that efforts to examine the relationship between memory and language are best served by broad-scope approaches that include the study of a range of communicative activities, including those that are characteristic of everyday language use. In this chapter we review the evidence for hippocampal contributions to language use across communicative phenomena (e.g., semantic representation, gesture, perspective-taking) and a range of language related processes (e.g., on-line processing, statistical learning). The present represents a time of tremendous potential for discovery and progress in the study of memory and language and for more representative, biologically plausible, and ecologically valid investigations of memory-and-language-in-use in every-day life.

Memory and language are two quintessential human abilities that enrich our daily experience. Memory, like language, allows us to mentally represent objects and events from other times and places, to think about the relationship between the past and the present—*I haven't seen a tree like this in years!*—and to express thoughts

M.C. Duff (✉)
Department of Hearing and Speech Sciences, Vanderbilt University Medical Center, 10328 Medical Center East, South Tower, 1215 21st Avenue South, Nashville, TN 37232, USA
e-mail: melissa.c.duff@vanderbilt.edu

S. Brown-Schmidt
Department of Psychology and Human Development, Vanderbilt University, Nashville, TN, USA

© Springer International Publishing AG 2017
D.E. Hannula, M.C. Duff (eds.), *The Hippocampus from Cells to Systems*,
DOI 10.1007/978-3-319-50406-3_16

503

that draw on previous records of personal experience, and project into the future—
"If we get some lumber this weekend we could build another treehouse." While
basic aspects of everyday language use necessarily draw on memorial representations (of language, of events and experiences), and functional memory use often
draws on language (reminder notes, verbal encoding and rehearsal, autobiographical narrative to represent life events), more often than not, memory and language
are studied in isolation. To be sure, the interface and interactions of memory and
language are not uncharted territory. Instead, we argue that there is a huge potential
for future discovery and forward progress in each field, by building on existing
approaches to the study of memory and language, while moving towards more
representative, biologically plausible, and ecologically valid investigations of
memory-and-language-in-use in every-day life.

In the domain of language, a thriving literature studies the interface of language
with the construct of *working memory* and related executive constructs. These
approaches can be broadly construed as exploring the influence of limited storage
and/or executive resources on the way in which humans process language. Influential empirical findings and theories explore contributions of working memory to
processing syntactically complex sentences (e.g., Gibson 1998; Just and Carpenter
1992; Lewis et al. 2006; Huettig and Janse 2016), as well as the roles of executive
constructs such as cognitive control in syntactic ambiguity resolution (e.g., Novick
et al. 2005). An alternative approach has re-characterized the demonstrated relationship between working memory and sentence processing as a relationship
between *language experience* and sentence processing (MacDonald and
Christiansen 2002). Emerging research is now exploring the role of variability in
language experience in how language is processed in real-time (e.g., Mishra
et al. 2012).

Particularly exciting are ideas about the role of memory retrieval phenomena in
shaping what people say and how they say it (e.g., Oppenheim et al. 2010). For
example, MacDonald (2013) proposes that a speaker's relative ease of memory
retrieval for her to-be-uttered concepts shapes utterance form, in turn influencing
expectations that listeners have when comprehending language. In a different
approach, McElree (2000; McElree et al. 2003) borrow a technique from the
memory literature that captures both the accuracy and timing of retrieval from
memory to develop a theory of how sentential elements are accessed from memory
during sentence comprehension. In yet another approach, Chang et al. (2006)
develop an error-based learning account of language production that accounts for
key ideas in language acquisition. Each of these approaches posits a clear role for
learning and memory processes in the way language is processed.

Classic studies of memory for discourse explored what is remembered from
conversation, using measures such as free recall (e.g., Stafford and Daly 1984;
Isaacs 1990; Ross and Sicoly 1979), as well as recognition memory measures that
can control for response biases (Brown et al. 1995; Fischer et al. 2015). While this
work provides important insights into what is and is not remembered, a limitation is
that the way in which memory for conversation reveals itself when a person is
tasked with explicit memory retrieval may be quite different than how memory

guides everyday behaviors such as future conversation (Hyman 2000). A potentially fruitful way forward would be through the combination of language measures that tap memory, along with memory measures that tap language (e.g., Yoon et al. 2016).

A different approach to the study of language-memory interactions is Ullman's declarative-procedural model. Ullman et al. (1997)'s influential theory characterizes language as composed of words (the lexicon) and rules (the grammar). Studies of multiple patient populations in the production of regular (*to look*) and irregular (*to dig*) verbs provide evidence of dissociations between deficits in production of irregular past-tense forms (*dug*) and production of rule-governed forms (*looked*). Ullman's dual-system theory assumes that irregular forms are stored "words" in the declarative memory system, and are impaired amnesia, Wernicke's aphasia, and Alzheimer's disease. By contrast, regular past tense forms are generated using rules that draw on a frontal/ basal ganglia procedural system, a process which is impaired in Broca's aphasia and Parkinson's disease. Ullman, et al. (also Ullman 2001, 2004) conclude that the patterns of deficits in these populations support theories that grammar and lexicon are separate components of language, and contrasts this view with theories of language in which words and rules are part of a single system (sometimes described with connectionist architectures). In the domain of verb production, the necessity of a dual route to explaining patterns in patients' production of regular and irregular verb forms is contested, and the evidence supporting the dual route model has been critiqued (Joanisse and Seidenberg 1999; also see Gordon and Dell 2003; and Seidenberg and Plaut 2014).

This narrow focus on the lexicon and grammar is useful in that it allows the theory to make specific predictions about production deficits in different patient populations. This narrow focus also afforded the development of multiple computational accounts of verb production which provides testable, falsifiable theories. Yet, a limitation of this approach is that it abstracts away from the natural uses of language in communicative settings. Here we argue that this limited focus not only limits the generalizability of the theory but in addition ignores the breadth of communicative phenomena.

Communicative language use is a multisensory, interactive process that draws on multiple linguistic and non-linguistic cognitive abilities and behaviors. Speakers use disfluency (e.g., *"thee uh, the large martini glass"*) in specific and predictable locations in speaking (Fox Tree and Clark 1997; Fraundorf and Watson 2014; Brown-Schmidt and Tanenhaus 2006), and listeners make use of these patterns to guide understanding (Fox Tree 2001; Arnold et al. 2004) and shape memory for discourses (Fraundorf and Watson 2011). The prosodic form of utterances is shaped by speakers' communicative intent (Snedeker and Trueswell 2003; Kraljic and Brennan 2005) and similarly shapes both real-time processing of language (Dahan et al. 2002; Kurumada et al. 2014), as well as memory for what has been said (Fraundorf et al. 2015). This acoustic variability with which words and utterances are produced does not so neatly fit into the circumscribed domains of lexicon and grammar, yet disfluency and prosody affect language use, language processing, and memory for language.

Beyond the acoustic form of spoken language is the physical context in which it is used. A large body of work now shows that a wide range of linguistic processes are affected by the physical context of language use. These findings include evidence that real-time linguistic ambiguity resolution is guided by information gleaned from the visual world, such as the number of potential referents in a visual scene (Tanenhaus et al. 1995; Novick et al. 2008), as well as the physical location of characters in described events (Greene et al. 1994; Nieuwland et al. 2007). Moreover, the physical realm provides a means by which spoken language users can express and understand meaning through visual information such as gesture, pointing, and placing of objects (Cook and Tanenhaus 2009; Clark and Krych 2004; Wu and Coulson 2007). Along with linguistically shared information, jointly experienced physical events, such as spotting a friend at a party, e.g., *"You see the man with the martini--isn't that John?"*, provide a basis for interlocutors to establish common ground (Horton and Gerrig 2005; Clark and Marshall 1978). Physical and spoken communication can be used in concert to express rich meanings that integrate the two domains. For example, Clark (2016) develops a theory of how people use *depictions* as a means to illustrate or act out a person or event from another place or time. Clark describes how speakers may interrupt an on-going sentence to assume a different tone of voice or interpose with a gesture to depict a meaning; the lead example is a sentence that could have continued ... *"shot those falcons"*, but instead ends with an acting-out with the fingers and a mouth-click, the action of shooting at the birds. While the physical world, both as a source of context and as means through which to communicate, clearly shapes language use, neither seems to be captured by the lexicon-grammar distinction.

The methods we use to communicate, including the use of words, gesture, emotional expression, et cetera can be fruitfully viewed through the lens of the goals of the communication. Speech Act theory (Austin 1962; Searle 1969) emphasized the communicative intentions of the speaker in understanding what was said and what was communicated. We propose that understanding basic aspects of communicative language use must consider the communicative event more broadly, including the person's goals or preferences (e.g., Ferguson and Breheny 2011, 2012; Creel 2014), their communicative style (e.g., literal or ironic, Regel et al. 2010) and identity (e.g., Van Berkum et al. 2008). Putting the reason for the communication in the foreground might explain basic aspects of communication. For example, Yoon et al. (2012) argue that failures to take into consideration the listener's perspective when speaking can, in some cases, be explained by that speaker's communicative goals. When precise communication was important for the speaker to achieve her goals, Yoon, et al. found significantly better use of perspective-taking abilities. This result suggests that conclusions regarding, e.g., whether perspective-taking is a routine part of language processing (cf. Keysar et al. 2003) must take into consideration whether doing so was relevant to the person's behavioral goals at the time. Ross and Sicoly (1979) similarly found an influence of personal motivations in *memory* for conversation: When recalling who-said-what in a past conversation, conversational partners were more likely to self-attribute when to do so would be flattering (e.g., reporting that it was you and not your partner who first proposed the solution to the problem). Thus the way in which we communicate, and our memory for past conversations may be similarly shaped by our goals and desires.

In sum, understanding how language and memory work together to support language use would benefit from a broad-scope approach that includes the study of a range of communicative activities, including those that are characteristic of everyday language use, coupled with the most recent understandings of the biological and functional components of memory. In what follows we present our first steps towards such an approach.

A Proposal About Hippocampal Contributions to Language Use and Processing

As a starting point, our proposal recognizes that the acquisition and use of mental representations for any complex and multidimensional behavioral phenomenon will not be the purview of a single memory system. In broad strokes, the declarative and non-declarative memory systems are traditionally thought to constitute the long-term-memory system. These systems are now known to be specialized in the type of information they process, in the time course over which they encode information, and in the flexibility with which they deploy this information (Eichenbaum and Cohen 2001; Henke 2010; Reber et al. 1996). Yet, these two systems operate in parallel: Declarative and non-declarative processes and knowledge are simultaneously activated in real-world behaviors (Voss et al. 2012) and interact in their support of complex behavior (Poldrack et al. 2011). We argue that both memory systems contribute to multiple aspects of language (i.e., declarative and non-declarative memory systems make unique contribution to multiple aspects of language) rather than classic descriptions of one-to-one mappings of between memory systems and domains of language. Building on recent theoretical advances in the understanding of the hippocampal dependent declarative memory system, our work has closely addressed how the processing capacity of the human hippocampus is particularly well suited for meeting the demands of a number of complex linguistic and communicative phenomenon.

Historical Considerations

As briefly reviewed above, there have been many attempts to link forms of memory with aspects of language. Yet, with a few limited exceptions, the hippocampal declarative memory system has not received serious consideration as a key neural/cognitive system involved in language use and processing. Furthermore, for those frameworks that do posit a role for hippocampus in language, that role is often conceptualized as being circumscribed to a single domain of language (the lexicon) and only for a limited period of time (the acquisition of new words but not their long-term maintenance) (e.g., Ullman et al. 1997).

It should be noted that there are conceptual and historical reasons that likely kept researchers from considering a more significant role for the hippocampal declarative memory system in language use and processing. Methodologically, isolating the unique contributions of a single memory system, or adjudicating alternative accounts in healthy participants is particularly challenging given that all forms of memory, as well as other cognitive abilities are largely intact. In neurological patients with profound memory impairment, the magnitude of the observed memory deficit outweighs any disruption one might observe in language. That is, patients with hippocampal amnesia do not have aphasia and consistently perform well on standardized neuropsychological tests of language (although the majority of these standardized tests were designed to assess for aphasic language disorders; i.e., they lack the sensitivity and specificity to non-aphasic language disorder).

Theoretically, hippocampus would seem to be an attractive candidate for meeting the demands of language processing. The hippocampus has long been associated with the formation of new enduring (long-term) memories and their subsequent retrieval. Two hallmark features of the hippocampal declarative memory system that are of relevance to language include its role in the creation and integration of *relational representations* and the *flexible expression* of those representations. Relational representations are created and supported through the binding of the arbitrary co-occurrences of people, places, and things of a scene or event. These binding operations link the spatial, temporal and interactional relations the components of an event, thus establishing the larger record of one's experience over time (e.g., object and its location; a word and its meaning; a pronoun and its referent) (Davachi 2006; Eichenbaum and Cohen 2001; Konkel et al. 2008; Ranganath 2010). These hippocampal dependent relational representations are *uniquely flexible, permitting integration with other types of representations.* Through the interaction of the hippocampal system with various neocortical storage sites that are also involved in the initial encoding of stimuli, representations are accessible to other processing systems (as when a rich autobiographical memory is evoked by the sound of a familiar song), and are readily extended to use in novel contexts (Cohen 1984; Dusek and Eichenbaum 1997; Eichenbaum and Cohen 2001; O'Keefe and Nadel 1978; Squire 1992).

Yet, the traditional view has been that these processing features support relational representations exclusively for long-term memory and that hippocampus does not support information processing in the moment, or on the time scale of working memory. Given the spoken language is processed as it unfolds over time, it makes good sense that researchers would <u>not</u> look to the hippocampus, a neural system associated with long-term memory, as a key substrate involved in on-line language processing. Thus, most accounts of a role of hippocampus in cognitive processes have been limited to its role in relational binding during the acquisition of new semantic information.

New Discoveries About Hippocampal Functionality: Implications for the Neurobiology of Language

Recent discoveries have expanded the breadth of hippocampal functionality beyond just its historic role in long-term memory and have significant implications for theories of language processing. A growing body of evidence has challenged the view of hippocampal declarative memory as contributing exclusively to long-term memory. Empirical work with hippocampal amnesic patients reveals deficits in declarative memory when there are minimal delays, and even when all the necessary information is immediately available (Barense et al. 2007; Hannula et al. 2006; Olsen et al. 2006; Shrager et al. 2008; Warren et al. 2011). Converging evidence from fMRI shows hippocampal activation for relational learning over similarly short delays (e.g., Hannula and Ranganath 2008; Ranganath and D'Esposito 2001). These findings suggest that new hippocampal-dependent representations are available rapidly enough to influence ongoing processing when: new information is perceived; old information is retrieved; and representations are held on-line to be evaluated, manipulated, integrated, and used in service of behavioral performance. Although the work in this area comes largely from studies using visual or visuo-spatial stimuli, the strong implication is that the hallmark flexibility and integration of hippocampal-dependent representations will be deployed and are rapidly available when information is processed in an ongoing fashion. These new findings also suggest that the performance of patients with hippocampal lesions and declarative memory deficits should consequentially suffer under such conditions. These provocative new findings regarding the time course of hippocampal contributions to on-line cognitive processes have profound implications for neurobiological theories of language use and open the possibility of hippocampal involvement in real-time language processing.

Another line of work expanding the breadth of hippocampal functionality reveals hippocampal participation in a neural network that supports the flexible and creative (re)construction and use of mental representation for remembering the past, imagining the future, simulating the thoughts of others and creativity more broadly (Addis et al. 2007; Buckner and Carrol 2007; Duff et al. 2013; Hassabis et al. 2007; Kurczek et al. 2015; Madore and Schacter 2014; Schacter and Addis 2007). Researchers have argued that the processes that enable a person to combine and recombine individual elements of real or imagined experience may also serve flexible and creative cognition including performance in decision making, navigation, social cognition, and language (for review see Rubin et al. 2014). Notably, language use involves the flexible and creative use of words and thoughts to recreate, repurpose, and recontextualize language for the specific situation and communicative partner. If hippocampus contributes to flexible and creative (re) construction and use of mental representations, then such contributions may also extend to supporting creative flexibility in discourse and interactional language use.

The Hippocampal-Dependent Declarative Memory System and Language Processing: A Proposal

These recent advances in our understanding of hippocampal functioning, along with a line of work demonstrating a range of deficits in language use and communication in patients with hippocampal damage and severe declarative memory deficits (Duff et al. 2006, 2007, 2008, 2009, 2011; Kurczek and Duff 2011), lead us to propose that the hippocampus is a key contributor to language use and processing (Duff and Brown-Schmidt 2012). Across a number of studies, we have shown that patients with hippocampal amnesia have difficulty using language when the demands on establishing, recovering, maintaining and using declarative memory representations are high in both off-line tasks (Klooster and Duff 2015) and in real-time language processing (Kurczek et al. 2013; Rubin et al. 2011).

At the heart of our proposal is the notion that the same processes by which the hippocampal declarative memory system creates and flexibly integrates representations across diverse sources in the formation of new memories, and maintains representations on-line to be evaluated and used in service of behavioral performance, are the same processes necessary for the flexible use and on-line processing of language. Our proposal, and our methodological approach, goes beyond previous models and accounts of the memory-language interface in several substantive ways.

First, rather than isolating the contribution of the hippocampal declarative memory system to a single domain of language (e.g., words or grammar), our proposal argues that any aspect of language use and processing that places high demands on the processing features afforded by the hippocampal declarative memory (e.g., relational binding, representational (re)construction, flexibility and integration, and on-line maintenance of relational representations) should recruit hippocampus in service of meeting those demands. In this way, our proposal allows for the generation and testing of hypotheses about hippocampal contributions to a breadth of communicative phenomena (e.g., gesture, reference, prosody, perspective-taking) and to a wide range of language related processes (e.g., on-line processing, statistical learning).

Second, the bulk of our work has been conducted in neurological patients with hippocampal damage and severe declarative memory impairment. This approach allows for a direct test of the contribution of a given memory system to language use and offers a biologically and psychological plausible account of the contribution of hippocampus to language and communication. From this perspective, we can test the necessity of hippocampus. Yet, those occasions where patients with amnesia are successful in using language are as informative as their failures in mapping the role of multiple memory systems to language use. Patterns of successes and failures in these patients help flesh out a theory of the contribution and coordination of multiple cognitive and neural systems to the larger language network. Finally, in recognition that language and memory are often studied in isolation, our proposal bridges the theoretical literatures and methods of both disciplines. Given significant advances in our understanding of the functional and biological mechanisms of memory, the present is a unique and timely opportunity to offer a fresh look at, and offer new proposals for, the memory and language interface.

In the sections that follow, we review work on hippocampal contributions to language use and processing across a wide range of linguistic and communicative domains. We first consider the contributions of hippocampus to lexical semantics, then move to evaluate how hippocampal-dependent processes might contribute to the integration of words in discourses, including face-to-face conversation where adjustments to one's partner become necessary. We then discuss the potential contributions of hippocampus to on-line language processing, statistical learning, and multimodal language use.

Hippocampal Contributions to Word Learning and Semantic Representation

The role of hippocampus in language has historically been confined to word learning and interpreted in the context of semantic memory. Acquisition of new vocabulary requires the binding of arbitrarily related information such as the relation between word form (phonological and orthography) and the associated object or concept. There is an extensive literature demonstrating deficits in word learning in patients with hippocampal damage and declarative memory impairment (e.g., Bayley and Squire 2002, 2005; Gabrieli et al. 1988; Manns et al. 2003; O'Kane et al. 2004; Schmolck et al. 2002; Verfaellie et al. 2000; Warren and Duff 2014) (although see Sharon et al. 2011).

While there is strong evidence that hippocampus supports word learning, the traditional view has been that this support is limited and that over time, through neocortical consolidation, semantic memory becomes independent of hippocampus (McClelland et al. 1995; O'Reily and Rudy 2000). Support for this view comes from findings that individuals with hippocampal amnesia, who are impaired in the acquisition of new vocabulary, do not exhibit deficits in their knowledge of remote vocabulary (i.e., semantic knowledge acquired long before the onset of their amnesia). More recently, however, Klooster and Duff (2015) revisited the notion that remote semantic knowledge is intact in amnesia. They argue that recent developments in our understanding of the breadth of hippocampal function support the prediction that deficits in remote semantic knowledge in amnesia may be revealed with more sensitive measures.

In unpacking this prediction, we first consider how remote semantic memory is assessed. The amnesia literature has focused primarily on vocabulary breadth and surface-level pairings of vocabulary or lexical information. Participants might be shown a picture of apple and asked to name it or asked to match the label 'apple' to a short definition. Patients with hippocampal amnesia do not differ from healthy participants on these types of tasks. Yet, what it means to know a word can extend beyond surface-level pairings to include how much information is associated with a word or concept and how words are used and processed across contexts. Under the umbrella term of semantic richness (defined as the amount of semantic information associated with a word or concept; Pexman et al. 2002), measures of depth of knowledge include the number of features of a word or concept (e.g., grows on trees, eaten in pies, has seeds) (Pexman et al. 2002; Yap et al. 2011) and the number of different senses a word can take (e.g., fruit, tree, body type, tech company) (Taler et al. 2013). Little is known about how hippocampus may contribute to the depth of semantic knowledge of individual words over the lifespan.

The time course over which a word is learned is another consideration. In amnesia word learning studies, participants either fail or succeed to reach criterion for surface-level pairings within several trials or sessions, where the learning events seldom stretch more than a day or two (e.g., Gabrieli et al. 1988; Duff et al. 2006). Yet, word learning is considered a protracted process, spanning days, weeks, and even years (Carey 2010; McMurray et al. 2012). The idea is that over time, and with extensive experience with a word or concept, people associate more and more information with each concept (McGregor et al. 2002). If word learning is a protracted process, one that is possibly never fully complete (Ryskin et al. 2016; also see Chang et al. 2006; Goldinger 1998), then hippocampus may support previously acquired knowledge by strengthening and creating new connections among and between words and adding new features or senses to existing representations. This view would fit with work demonstrating hippocampal contribution to the updating and maintenance of relational information in the moment (Hannula et al. 2006; Warren et al. 2011), to the updating and strengthening of previously acquired information through reconsolidation (McKenzie and Eichenbaum 2011; Lee 2008), and to the integration of relational representations across time (Zeithamova and Preston 2010).

Klooster and Duff (2015) examined vocabulary depth and semantic richness in patients with bilateral hippocampal damage and severe declarative memory impairment, as well as brain-damaged comparison (BDC) participants (those with bilateral damage to the ventromedial prefrontal cortex (vmPFC)), and healthy comparison participants. In the "features" task, participants were presented with a word and asked to verbally list as many features of the word as possible. For the "senses" task, participants were given a word and instructed to provide as many senses of the word as possible (e.g., the word 'bank' can mean a financial intuition or the bank of a river). Across both tasks, the BDC and healthy participant groups produced significantly more responses than participants with hippocampal amnesia (Fig. 1). For the features task, the comparison groups produced more than twice as many features, on average, (BDC M = 20.20 features, NC M = 22.25) than the HC group (M = 9.92). When asked to rate the familiarity of the words, there were—critically—no significant differences in familiarity ratings between groups. All groups had high average familiarity ratings for the words in each task, indicating that the words were subjectively (and equally) familiar to the participants.

These findings documenting impoverished remote word knowledge (i.e., knowledge of words learned early in life and well before hippocampal insult) suggests

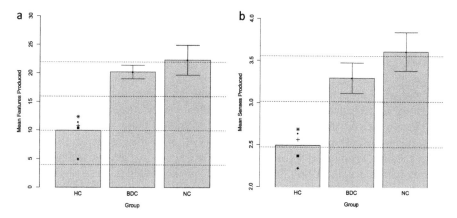

Fig. 1 Impoverished semantic knowledge in hippocampal amnesia. (**a**) Features Task. Mean number of features per target word. (**b**) Senses task. Senses per target word produced. HC; *black symbols* for each HC participant and comparison groups (with 95 % CI error bars). The dotted lines represent the standard deviation for the combined comparison group (BDC and NC). Figure adapted from Klooster and Duff 2015, Neuropsychologia

that hippocampal contributions to word learning and semantic representation are not limited to new information. Rather, hippocampus may play a protracted and sustained role in the life-long development of semantic knowledge by strengthening and enriching existing semantic representations, and through the addition and integration of new information as words are retrieved and used in novel contexts (Klooster and Duff 2015). While such an interpretation is at odds with the traditional view that word knowledge becomes independent of the hippocampus over time (e.g., McClelland et al. 1995; O'Reily and Rudy 2000), it does fit with more recent conceptualizations of hippocampal functionality. Consistent with our proposal regarding hippocampal contributions to language more broadly (Duff and Brown-Schmidt 2012), the hallmark processing features of the hippocampus, including its capacity for relational binding (Cohen and Eichenbaum 1993; Eichenbaum and Cohen 2001), reconsolidation (McKenzie and Eichenbaum 2011; Lee 2008) and flexible integration of information (Zeithamova and Preston 2010), situate it well to meet the demands of maintaining, updating, and using semantic memory across language experiences, contexts, and over time.

These findings point to a number of avenues worthy of further investigation.

First, an open question that is ripe for future study is whether previously acquired semantic representations become impoverished in amnesia due to a failure to update and enrich knowledge (little or no new learning), a failure to maintain and strengthen existing knowledge (some degree of decay), or some combination of processes.

Second, given previous work pointing to hippocampal involvement in the retrieval of semantic information (e.g., Sheldon and Moscovitch 2012) and the creative and flexible (re)construction, and integration of relational representations, hippocampus may also contribute to online semantic processing. The mechanism of this contribution may be through the processing of relations between incoming words, and the generation of a coherent semantic understanding across phrases, sentences, and discourse histories. Such a finding would further extend the role of hippocampus in semantic representation and processing.

Finally, documenting disruptions in the remote semantic knowledge of individuals with adult onset hippocampal pathology and amnesia raise questions about the intact status of semantic memory of individuals with *developmental* amnesia. Individuals with developmental amnesia, who sustained hippocampal damage and subsequent episodic memory impairment very early in life, are reported to have intact semantic memory and vocabulary (i.e., average to low average performance; Vargha-Khadem et al. 1997). To the best of our knowledge, the assessments of word knowledge in developmental amnesia have been limited to performance on standardized language and intelligence tests and have not included assessment of semantic depth or richness. If developmental hippocampal pathology affects long-term vocabulary acquisition (in size, depth, semantic processing, or learning rate) it would have significant theoretical implications for debates regarding the neural substrates of semantic and episodic memory and educational ramifications for such children, thus warranting investigation.

Hippocampal Contributions to Creative and Flexible Discourse

All language use, to some degree, involves creativity and flexibility. We see this in the way speakers rhetorically, or even poetically, select and craft particular meanings and details to represent for a specific listener on a particular occasion (e.g., Norrick 1998, Tannen 1989; Clark and Murphy 1982). Creativity requires the rapid combination and recombination of existing mental representations to create new ways of thinking and to generate novel ideas (Bristol and Viskontas 2006; Damasio 2001). Hippocampal declarative memory, in its capacity for creating, updating, and in the juxtaposition of mental representations as well as their flexible and novel use, has been linked to creativity and creative thinking (Duff et al. 2013; Madore and Schacter 2014; Madore et al. 2015; Warren et al. 2016). According to our proposal, the hippocampal declarative memory system directly supports discursive creativity and flexibility across communication partners and discourse contexts (Duff and Brown-Schmidt 2012).

In our studies of creative language use we find support for this hypothesis: Quite simply, patients with hippocampal amnesia exhibit disruptions in the creative and flexible use of language. In our first study examining the creative and flexible use of discourse we examined the conversational use of reported speech. Reported speech is a discourse practice in which speakers represent, or reenact, words or thoughts from other times and/or places (e.g., *Eleanor said, if I see Otto this week I'm going to tell him, 'You have to get us tickets to see Neko Case at the Ryman next weekend'*). We found that in the conversational interactions of the patients with hippocampal amnesia, there were only half as many reported speech episodes (RSEs) ($M = 30.3$; $SD = 16.9$) as there were in conversations with healthy comparison participants ($M = 61.5$; $SD = 30.1$) (Fig. 2; Duff et al. 2007). Interestingly, the reduced use of reported speech was <u>not</u> limited to events since the onset of their amnesia. Rather, even when producing detailed and vivid memories from their remote past, the individuals with hippocampal amnesia were less likely to use reported speech when representing these memories in communicative interactions. We interpreted this finding as evidence for a role of the hippocampus in the flexible and creative expression of declarative memory in novel situations (Eichenbaum and Cohen 2001; see Duff et al. 2007). Reported speech seems to place high demands on the hippocampal declarative memory system as its use requires flexible access to our autobiographical experiences as well as the ability to flexibly and creatively generate unique combinations of the reconstructed elements (what details to represent, what details to omit to meet the specific interactional goals of this telling, on this occasion, with this communication partner). We propose that any discursive practices that place such high demands on the processing features of the hippocampus would similarly be disrupted. It should be

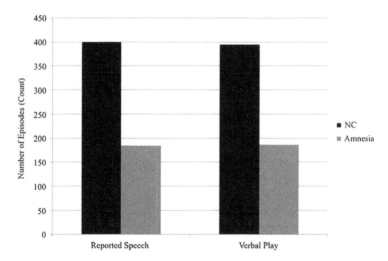

Fig. 2 Reported speech and verbal play. In conversational interactions with a clinician, patients with hippocampal amnesia produce significantly fewer episodes of reported speech (185) than do normal comparisons (400). In the interactions with a familiar communication partner while completing trials of a collaborative referencing game, patients with hippocampal amnesia produced significantly fewer episodes of verbal play (187) than do normal comparisons (395). Data presented are group totals for patients with hippocampal amnesia (Amnesia) and demographically matched healthy normal comparison (NC) participants. Data of interactional partners (clinician; familiar communication partner) are not presented. Figure from Rubin et al. (2014), Frontiers in Human Neuroscience

noted that there appears to be some specificity to these findings. Damage to vmPFC, a neural system shown repeatedly to be involved in social and emotional processing and future thinking, does not impair reported speech use (Duff et al. 2015).

In a second study of creative language use we examined verbal play; the playful use of sounds and meanings of words through the use of puns, voices and sound effects, teasing, and telling funny stories. Verbal play is ubiquitous in everyday language use and requires a range of creative and interpretive functions (Crystal 1998; Sherzer 2002). We found that in face to face interactions, individuals with amnesia produced significantly fewer verbal play episodes (Total episodes = 187) than healthy comparison participants (Total = 395). Of note, the quality of verbal play episodes produced by the participants with amnesia also differed (Fig. 2; Duff et al. 2009). In contrast to comparison participants, verbal play episodes produced by the patients were rote and repetitive. A common occurrence was for an amnesia patient to reproduce the same joke, nearly verbatim, multiple times in the same conversation. In addition, these episodes were less richly and skillfully deployed (i.e., fewer productions combining verbal, prosodic, and gestural resources). In the amnesic sessions, collaborative playful themes, where each participant contributes to the verbal play, were not sustained across stretches of interaction or returned to in subsequent interactions. These findings suggest that the ability to creatively and flexibility deploy the communicative and cognitive resources necessary to meet the moment-to-moment demands of interactional discourse is impaired following hippocampal damage (see Rubin et al. 2014, for discussion). We also note that, in terms of specificity, verbal play is apparently not impaired by damage to the vmPFC (Gupta et al. 2012).

Language use places high demands on creativity and flexibility. We propose that the hippocampal declarative memory system supports the creative and flexible use of language across discourse histories, contexts, and communication partners and that any aspect of language use that places high demands on creative and flexible uses of language should recruit hippocampus (Duff and Brown-Schmidt 2012). One particularly rich arena of language use to explore such demands is narrative construction. Often taking the form of autobiographical narrative, conversational narrative involves the use of language to represent life events, providing a temporal order and coherence across past, present, and future experiences. In addition, conversational narrative extends to the description of scenes and pictures, retelling stories and retelling stories for different audiences across conversational settings (i.e., the creative selection of details for conveying information to an adult peer vs a child). Systematic study of narrative, and its many forms and contexts of use, would provide a robust test of our proposal and address questions in the literature regarding the (in)dependence of narrative construction and episodic memory (Race et al. 2011).

Hippocampal Contributions to Common Ground Processes

Classic theories of conversational language use posit that we form representations of the perspectives of our communicative partners and then use these representations to guide both language production and comprehension (Clark 1996; Clark and Murphy 1982; Wilkes-Gibbs and Clark 1992; Hanna et al. 2003). The fact that conversational language use is guided by knowledge of what information is shared by the conversational partners—their *common ground*—as well as knowledge of what information is not shared—their *privileged ground points* to a high degree of flexibility in conversational language use. It is this flexibility that leads to the prediction that the flexible use of common and privileged ground in communication is likely to be impaired in amnesia (Duff and Brown-Schmidt 2012).

The first study in this line of work investigated the ability of individuals with hippocampal amnesia to acquire and use referential labels for a set of abstract figures while interacting with a familiar communication partner (Duff et al. 2006). This study was modeled after Clark and Wilkes-Gibbs (1986) who gave pairs of healthy young adults a referential communication task in which they had to repeatedly described a series of abstract "tangram" images to each other. Clark and Wilkes-Gibbs (also Wilkes-Gibbs and Clark 1992) found that through the process of conversation, the pairs developed brief referential labels for abstract images (e.g., "the ice skater"). Duff et al. (2006) asked whether the same acquisition of novel labels could be achieved in the absence of declarative memory. In a modified version of the referential communication task, patients with hippocampal amnesia played the role of Director, and were seated across from familiar partner (friend, spouse) who played the role of Matcher in the task. Both the Director and Matcher each had a board with 12 numbered spaces and a set of 12 cards displaying the abstract tangram images. On the table between them was a low barrier preventing a view of the others' cards. The amnesic patients always played the role of Director, communicating to the Matcher how to arrange their tangram images so that at the end of the trial the two boards looked alike. The pairs communicated freely across 24 trials, each of which involved sorting the same 12 tangram images into a different random order.

Despite severe declarative memory impairments, the patients with hippocampal amnesia demonstrated robust learning, arriving at increasingly concise and stable labels (e.g., "siesta man", "angel") that facilitated rapid and efficient communication with their communication partner (Duff et al. 2006). In fact, the rate of learning exhibited by amnesic participants (measured by the reduction in time and words necessary to complete each trial), did not differ from that of healthy participants. On one hand, this learning would seem to constitute an example of intact formation of common ground. If so, this would suggest that at least some forms of common ground can be supported outside of explicit recollection and declarative memory, opening the possibility that non-declarative mechanisms (which are intact in this participant population) play some role in acquiring and representing information in these communicative interactions. On the other hand, these findings do not indicate whether or not the amnesic patients represented these labels as knowledge that is jointly shared with their partners. If the patients do not represent, on some level, that this information is part of their joint knowledge, then common ground would not have been formed. Instead, the amnesic patients would have developed egocentric representations of these image labels.

A second study examined the referential form of the labels used by the patients with hippocampal amnesia to determine if they used definite reference to signal to their partner that the referent was part of shared knowledge (Duff et al. 2011). Healthy speakers tend to signal when information is shared, using a definite reference, such as *The* Viking ship, and indefinite reference, *A* Viking ship, when it is not (see Issacs and Clark 1987). Whereas the healthy comparison pairs marked shared terms with definite reference (90 %) amnesia patients used definite reference significantly less often (56 %). That is, even though the participants with amnesia had described the tangrams multiple times and were using concise labels, they did not consistently mark these labels as part of the common ground through definite reference. Even during the final round of matching—the 24th time they had described the images—the patients still used indefinite references, as if they were encountering the tangrams and deriving the descriptors for the very first time. By contrast, comparison directors nearly always used definite references to refer to the tangrams after the first trial.

This result argues strongly that the ability to signal to the listener that a referent is mutually known through use of a definite reference depends on declarative memory. An open question, however, is whether the patients with hippocampal amnesia are able to form common ground representations with specific conversational partners. That is, can these patients know that the information is shared with a specific person in spite of not being able to mark this shared knowledge through the use of definite reference? The failure to use definite reference could be taken as evidence that the ability to form partner-specific representations is encoded by in declarative memory, a form of memory not available to these patients. Yet, it is still possible that partner-specific common ground representations are formed via the non-declarative memory system. Indeed, patients with hippocampal amnesia have been shown to acquire a range of person-specific information (e.g., accent, political preference, personality attributes, Coronel et al. 2012; Johnson et al. 1985; Trude et al. 2014; Tranel and Damasio 1993). Whether such person-specific common ground information can be encoded and used to drive behavior warrants further study.

Hippocampal Contributions to Online Language Processing

Language processing is highly incremental. Spoken language unfolds over time, at a rate of about 150–200 words per minute (Tauroza and Allison 1990; Levelt 1989) and many words and phrases are, at least temporarily, ambiguous. Because the meaning of many words is unclear until later in the sentence, multiple sources of information must be generated, integrated, and maintained in real-time to create meaning. How this is accomplished in the brain is not fully understood, although links to cognitive control and working memory, putatively associated with prefrontal cortex mechanisms, are long standing (e.g., Novick et al. 2005; Gibson 1998; MacDonald et al. 1992; Lewis et al. 2006). Recent behavioral and neuroimaging findings demonstrating hippocampal relational memory recruitment and processing over very short delays, and even no delays, have significant implications for theories of language processing. Specifically, we argue that new hippocampal-dependent representations are available rapidly enough to influence ongoing language processing making it a key contributor to meeting the demands of incremental and online language processing (Duff and Brown-Schmidt 2012).

In the first test of hippocampal contributions to online language processing, we assessed the role of hippocampal declarative memory in the use of common ground during on-line referential ambiguity resolution (Rubin et al. 2011). In addition to being relevant to the way in which language is produced in conversation (Wilkes-Gibbs and Clark 1992), representations of common ground guide on-line ambiguity resolution processes in language comprehension as well (Hanna et al. 2003). Thus in this project we assessed the role of hippocampus in the use of common ground during on-line language processing. Individuals with amnesia sat across a table from an experimenter, each in front of a computer that displayed a 3-D rendering of nine cubbyholes (Fig. 3). While each cubbyhole contained a picture of an object, not all of the pictures were visible to both partners. Of the nine objects, *three* objects were visible to both the experimenter and the participant and were thus in common ground. Three further objects were visible to the participant only and were therefore in their *privileged ground*. The final three objects were visible only to the experimenter. The task was a conversational variant of the visual-world paradigm (Tanenhaus et al. 1995), thus we monitored the participants' eye-movements as they engaged in the task. In particular, we were interested in the eye fixations that the patients made to the objects in the display as they interpreted the experimenter's sentences. On critical trials, displays contained two identical objects (e.g., two ducks). In one condition, common ground was cued visually, such that one duck was in an open cubbyhole (common ground), and the other duck was in the participant's privileged ground (i.e., designed to appear as if the partner seated opposite could not view it). In another condition, both ducks were in the participants' privileged ground, and one duck was brought into common ground verbally: The experimenter asked about it, *What's in your bottom left cubby?*, so that the participant's response ("a duck") would bring that object into common ground. Then, immediately, or after a brief delay, the experimenter gave the critical instruction, *Look at the duck,* during which the participant's eye movements were monitored. Linguistic ambiguity (which duck was being referenced) can be eliminated by taking into account common ground information (Hanna et al. 2003).

We found that, like healthy comparison participants, the patients with amnesia were able to rapidly use common ground when common ground was cued visually. The patients could also successfully use verbal common ground when there was no delay (Fig. 4a, c). However, when there was a short (~40 s) delay between

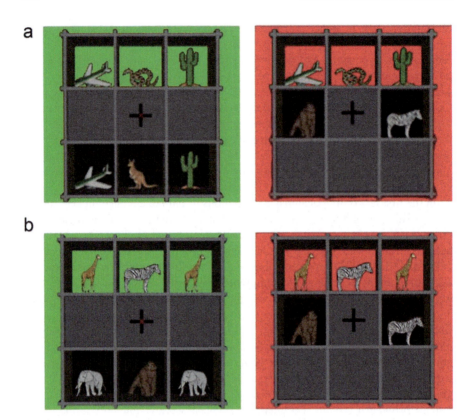

Fig. 3 Examples of the displays viewed by participants (*left column*) and the experimenter (*right column*) in the visual common-ground conditions. Each display consisted of a 3 × 3 grid of cubbyholes with a fixation cross in the center. Objects shared between the participant and the experimenter were presented in open cubbyholes in the top row. Objects that could be seen by only the participant or only the experimenter (i.e., privileged objects) were hidden by closed cubbyholes in the display's bottom row and middle row, respectively. In visually unambiguous displays (**a**), a target object (in this example, the cactus) was presented once among the shared objects and once among the participant's privileged objects. In visually ambiguous displays (**b**), a target object (in this example, the giraffe) was presented twice among the shared objects. The same displays used in the visually ambiguous condition were also used in the two linguistic common-ground conditions. Figure from Rubin et al. (2011); Psychological Science

establishing common ground, and the ambiguous reference (e.g., "*Look at the duck*), individuals with amnesia were equally likely to look at the two ducks (Fig. 4b, d). There was no evidence they knew that one of the ducks had just been brought into common ground less than a minute earlier. We also observed that even on trials when the target was successfully fixated, amnesic patients were more likely to glance at the competitor object than comparison participants. This lingering competition raised questions about the distinctness of the representations in the amnesic patients, and suggests that hippocampus may influence the on-line resolution of competition during language processing.

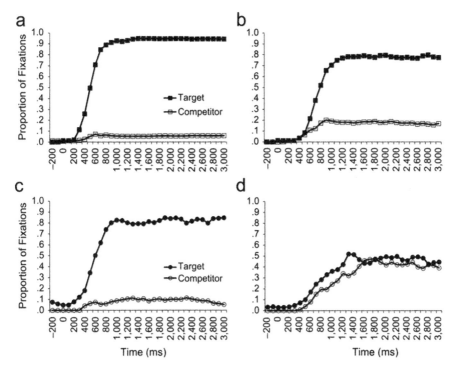

Fig. 4 Results for the linguistic common-ground conditions: proportion of fixations to targets and to competitor objects as a function of time. Time 0 is the onset of the critical word. Results are shown separately for healthy comparison participants in (**a**) the no-delay condition and (**b**) the filled-delay condition and for amnesic participants in (**c**) the no-delay condition and (**d**) the filled-delay condition. The filled delay was <40 s. Figure from Rubin et al. (2011); Psychological Science

In another line of work, we examined hippocampal contributions to the ability to understand a personal pronoun such as "he" or "she" when its referent appeared in a previous sentence. Personal pronouns can be highly ambiguous, and successful interpretation of the pronoun typically draws on representations of the discourse history, built up using contextual information as well as past utterances. In Example (1), the pronoun "he" in the second sentence is typically interpreted as referring to the subject of the first sentence:

(1)
    Forrest is playing cello for Charles as the sun is shining overhead.
    <u>He</u> is wearing a purple shirt, and it looks like it will be a nice day.

While the pronoun itself is highly ambiguous, the subject of the prior sentence, "Forrest", who was also mentioned first in that sentence, is likely to be the intended referent. Indeed, in a variant of the visual world paradigm, Arnold et al. (2000) (also see Kaiser et al. 2009) find that in situations like this, healthy young adults show a

Table 1 Narrative Design

		Order of mention	
		First (1)	Second (2)
Gender	Same (S)	S1: Melissa is playing the violin for Sarah as the sun is shining overhead. She is wearing a yellow bracelet and it looks like the song is being played well.	S2: Melissa is playing the violin for Sarah as the sun is shining overhead. She is wearing a yellow bracelet and it looks like the song is being played well.
	Different (D)	D1: Melissa is playing the violin for James as the sun is shining overhead. She is wearing a yellow bracelet and it looks like the song is being played well.	D2: Melissa is playing the violin for James as the sun is shining overhead. He is wearing a yellow bracelet and it looks like the song is being played well.

Table modified from Kurczek et al. (2013); Journal of Experimental Psychology: General

preference to fixate the subject of the first sentence, e.g., "Forrest", when interpreting the ambiguous pronoun. Using a similar paradigm (Table 1), Kurczek et al. (2013) asked whether this ability to draw on information about the relative prominence of two candidate referents introduced in a previous sentence, and who are depicted in a co-present scene, is impaired in hippocampal amnesia (Fig. 5). Despite the fact that these two-sentence stories were only about 10 s long, individuals with amnesia showed significant impairment relative to healthy comparison participants. When interpreting the critical pronoun, e.g., "he", the amnesic patients only showed a slight tendency to fixate the first-mentioned character (e.g., "Forrest") over the second-mentioned character (e.g., "Charles"). When gender provided an additional cue to the intended referent (e.g., "Forrest is playing cello for Eleanor... he..."), the patients were more successful, but not nearly at the levels of healthy comparison participants. A group of individuals with bilateral vmPFC damage served as an additional comparison group; these individuals performed similarly to healthy comparisons.

Together, this pair of studies points to key contributions of the hippocampal-dependent declarative memory system to on-line ambiguity resolution processes in spoken language comprehension. In particular, impairments in amnesia seem to surface when information must be integrated across sentences or across temporal boundaries between discourse segments. By contrast, the availability of information in the immediate environment promotes some success, but not nearly at the level of healthy individuals. These findings provide initial insights into how hippocampus might contribute to on-line sentence processing more generally. Promising lines of inquiry include the investigation of how information from distinct contexts (temporal or physical) are integrated on-line, and how features of candidate referents are bound to that referent.

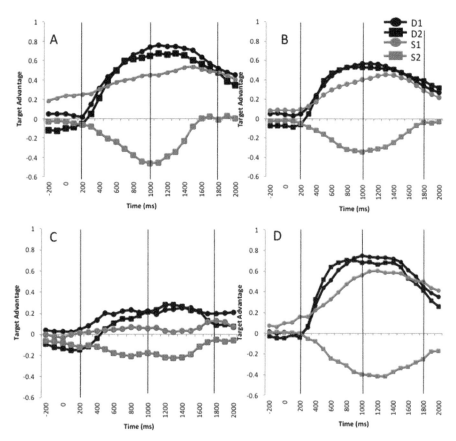

Fig. 5 Time-course of fixation preferences plotted as the difference between target and competitor fixations (proportion target minus proportion competitor), separately by condition and group. Positive values indicate target preference. 0 ms = pronoun onset. Panel A = undergraduate participants; Panel B = healthy comparison participants matched to amnesia patients; Panel C = amnesic patients. Panel D = brain-damaged comparison participants; D1 = Different Gender, First Mention; D2 = Different Gender, Second Mention; S1 = Same Gender, First Mention; S2 = Same Gender, Second Mention. Vertical lines denote analysis time-windows. Figure from Kurczek et al. (2013); Journal of Experimental Psychology: General

Hippocampal Contributions to Grammatical Processing and Statistical Learning

As we move through the world, we are confronted with a barrage of stimuli in multiple modalities. To make sense of the environment, our brains look for underlying patterns and regularities that allow us to generate categories and rules that guide our perception, learning, and behavior. Language learning offers a particularly salient example. As we are bombarded with the sounds of our language, cognitive and neural systems track the statistical frequencies with which these stimuli occur, as well as the frequency of co-occurrence between stimuli. From this information, we are able to uncover words from streams of sounds and acquire the grammatical rules of our language. This capacity of learners to track the statistical regularities of input to uncover underlying patterns in a set of stimuli is referred to as statistical learning. It is a robust phenomenon that has been shown to occur in humans from infancy through adulthood (Saffran et al. 1996; Baldwin et al. 2008; Conway and Christiansen 2005). Although statistical learning has been well characterized behaviorally, little work has examined its neural correlates.

Given the incremental and implicit nature of the learning process, the prominent view has been that statistical learning is most akin to the known processing capacities of non-declarative memory or procedural memory and basal ganglia (Evans et al. 2009; Perruchet and Pacton 2006; Kim et al. 2009). Neuroimaging studies have provided support for a link to the basal ganglia (e.g., Karuza et al. 2013; Turk-Browne et al. 2009) as have behavioral studies with brain-damaged individuals. In fact, it is the abundance of behavioral data from various patient populations linking grammatical deficits to the basal ganglia, and not the hippocampus, that is thought to provide some of the most compelling support for Ullman's (2004) declarative-procedural model of language (c.f., Joanisse and Seidenberg 1999). For example, in striking contrast to patients with basal ganglia pathology (most often in Parkinson's patients), individuals with adult onset hippocampal amnesia perform well on artificial grammar learning tasks (Knowlton et al. 1992; Knowlton and Squire 1994, 1996) and patient H.M. did not differ from healthy comparison participants on various tests of grammatical use and processing for previously acquired constructs (Kensington et al. 2001; although see MacKay et al. 1998a, b for description of grammatical errors and processing deficits in H. M.). Over the past 20 years, there has been general consensus that hippocampus does not appear to be involved in statistical learning or grammatical processing.

More recent evidence suggests that it may be time to revisit the role that hippocampus may play in such aspects of language. There is a growing body of work suggesting that hippocampus plays a critical role in the rapid extraction of regularities from the environment (e.g., Schapiro et al. 2016; Turk-Browne et al. 2009). Building on prior work demonstrating hippocampal activation during statistical learning tasks in healthy participants, Schapiro and colleagues reported on a patient (LSJ) with medial temporal lobe damage, including damage to hippocampus, who failed to show statistical learning across stimulus modalities (shapes, syllables, scenes, tones) (Schapiro et al. 2014; Fig. 6). Potentially related to the tracking of co-occurrence frequencies, Klooster and Duff (2015) reported that patients with hippocampal amnesia performed significantly worse than healthy comparison participants at identifying acceptable collocates (i.e., words that often follow the target in a phrase or sentence; "sudden noise" is permissible in English while "sudden doctor" is not). Hippocampal pathology in Alzheimer's disease has been linked to disruptions in past tense verb use in narrative production. Finally, in an individual differences study, Lee et al. (2013a, b) compared hippocampal volumes and language scores of a group of individuals with developmental language impairment (DLI) and a healthy comparison group. DLI is a developmental

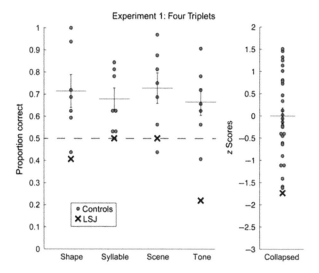

Fig. 6 Test performance for control participants and LSJ in Experiment 1. For each of the shape, syllable, scene, and tone tasks, LSJ's score (*black cross*) and the individual control scores (*gray dots*) are plotted against chance (*dashed line*), along with the mean and standard error of the controls. To visualize LSJ's performance with respect to the entire sample of controls, the data in each task were converted to z scores and then collapsed across task (LSJ's four z scores were averaged). She fell 1.74 SDs below the mean on average. Figure from Schapiro et al. (2014); Journal of Cognitive Neuroscience

disorder characterized by difficulty learning and using language with particularly deficits in the morphosyntactic components of language thought to be acquired via statistical learning (Evans et al. 2009). Prior work with individuals with DLI has demonstrated increased hippocampal volume relative to comparison participants, suggesting that hippocampal enlargement in these individuals might be a form of neural compensation for an impaired corticostriatal system (e.g., Ullman and Pierpont 2005). However, Lee, et al. found that the volume of the hippocampus was significantly enlarged in the DLI group than in the healthy comparison group, and it was strongly correlated with individual differences in language, with the larger hippocampal volumes in DLI group associated with poor language scores. Lee and colleagues interpret these findings as evidence against a compensatory account and suggest that disruptions outside the corticostriatal system (e.g., hippocampus) may also contribute to language deficits in DLI. Evidence of failures in statistical learning in patients with hippocampal damage taken together with findings of enlarged hippocampal volumes in individuals with DLI suggest a link between aberrant hippocampal structure (whether significantly reduced or enlarged) and poor statistical learning and language outcome.

Neuroimaging findings of hippocampal activation during statistical learning tasks and data from populations with hippocampal pathology suggest that hippocampus plays a critical role in at least some aspects of grammatical processing and statistical learning. These results challenge proposed divisions of labor for hippocampus (vocabulary) and basal ganglia (grammar) in language processing (e.g., Ullman 2004). Yet, when considering the demands of statistical learning and the functionality of the hippocampus, the potential link makes good sense. Statistical learning depends on the ability to encode and track the relations between individual stimuli. The hallmark processing feature of the hippocampus is its ability to support the rapid binding of arbitrarily related elements that make up a scene or event and their temporal, spatial, and interactional relations. A role for hippocampus in statistical learning and grammatical processing, abilities previously linked to the incremental and implicit/unconscious capacities of non-declarative memory, is also a more palatable fit given a growing body of work demonstrating hippocampal contributions to unconscious processing of relational binding in certain contexts (for review see Hannula and Greene 2012).

These provocative findings reopen an area of investigation regarding an aspect of the memory-language interface for which there had been long standing consensus. While many previous studies of adult onset hippocampal amnesia have documented intact grammatical knowledge, an open question is if such individuals retain the ability to acquire *new* types of grammatical information. For example, do individuals with amnesia show the same malleability in verb bias through language experience as do healthy individuals (Ryskin et al. 2016)? Another possibility is that in patients with hippocampal amnesia, adjacent dependences can be acquired through non-declarative mechanisms of unitization (e.g., Ryan et al. 2013) whereas non-adjacent dependences require more hippocampal processing. Given recent links between hippocampal abnormalities and language outcome in DLI (Lee et al. 2013a) and the well-established deficit in morphosyntactic aspects of language learning in DLI, investigating statistical learning and grammatical processing in individuals with developmental amnesia is warranted and would be particularly informative. Finally, documenting a role for hippocampus in statistical learning does not exclude or negate the contribution of the basal ganglia (i.e., Ullman 2004). Indeed, these structures may make joint or complementary contributions to the neural mechanisms of statistical learning. Understanding how multiple neural systems work in concert to support statistical learning and grammatical processing is critical in moving towards a more integrated model of the neurobiology of language.

Hippocampal Contributions to Hand Gestures

Language use is multi-modal, encompassing multiple parallel input and output modalities and mechanisms, including verbal (or signed) language, manual gestures, facial expressions, body movements and interactions with the physical world. The integration and use of multimodal information is integral to language use and processing. We propose that co-speech gesture may provide a particularly unique window into the role that hippocampus plays in the integration of multimodal relational representations. Hand gesture is ubiquitous with spoken language.

Despite being related both temporally and semantically to language (McNeill 1992), gesture communicates in a different manner than the symbolic and sequential speech stream. Gesture communicates information imagistically and iconically, translating mental representations into hand movements that reflect properties of these representations (Kendon 1997). For example, when giving directions people tend to list in words the turns that should be made, mentioning landmarks and cardinal directions. Concurrently, speakers produce gestures that *show* the listener the way, reflecting on their hands the representation of the route that they have in their mind.

Although it is clear that gestures reflect mental representations, the functional links between the nature of the underlying mental representation and gesture production are less clear. We propose that the hippocampal declarative memory system may, in part, support co-speech gesture. In the first study testing this hypothesis we examined the gesture production of patients with hippocampal amnesia as they produced narratives in response to four prompts. Two were autobiographical experience prompts from before their injury (JFK's assassination, most frightening experience) and two were procedural discourse prompts (how to go grocery shopping, how to make a sandwich). Consistent with previous studies (e.g., Kurczek et al. 2015; Race et al. 2011), patients with amnesia provided fewer episodic details in their spoken productions compared to healthy comparison participants. The critical question, though, was whether these disruptions extended to their gesture production. The answer is yes. Despite comparable amounts of spoken language across narratives, the patients with hippocampal amnesia gestured at a lower rate (# gestures/word) than healthy comparison participants (Hilverman et al. 2016). Moreover, for healthy comparisons, the number of episodic features in a narrative was a positive predictor of gesture rate; participants gestured more as they provided more episodic detail. This relationship between episodic details and gesture production was not present in the patients with amnesia.

The hippocampus is thought to play a critical role in relational binding and flexibility for the reconstruction and recreation of richly detailed, multimodal, mental representations of experience (Eichenbaum and Cohen 2001). In this context, studies of gesture in amnesia show that relational memory representations can contain information conveyed in gesture and that disruptions or deficits in relational memory representations extend to gesture production. These exciting findings open the door to a number of future lines of work. For example, studies directly relating the imagistic properties of these representations to physical properties of gesture— and demonstrating how impoverishment affects gesture properties—will be paramount to uncovering precisely how hippocampal representations support gesture production. It is also possible, given the motoric components of gesture, that non-declarative or procedural memory may contribute to gesture production under certain conditions. More broadly, documenting a deficit in gesture production in patients with hippocampal amnesia suggests that the disruption in language following hippocampal pathology may extend to other non-verbal means of communication.

Conclusions and Future Directions

The memory-language interface and its instantiation in the brain are core themes in neuroscience, psychology, and linguistics. Indeed, there are a number of proposals regarding the relationship between aspects of language and distinct forms of memory. Building on new and emerging theoretical and empirical evidence expanding the functionality of the hippocampus, we have proposed that the hippocampus plays a critical role in meeting the demands of language use across a broad range of communicative phenomena (Duff and Brown-Schmidt 2012). Specifically, we argue that the same processes by which the hippocampal declarative memory system creates and flexibly integrates representations across diverse sources in the formation of new memories, and maintains representations on-line to be evaluated and used in service of behavioral performance, are the same processes necessary for the flexible use and on-line processing of language. Support for hippocampal contributions to language comes from work by our group and others (e.g., MacKay et al. 1998a, b; MacKay and James 2001; Schapiro et al. 2014) demonstrating that individuals with hippocampal damage and declarative memory impairment show disruptions and deficits across a range of language domains (e.g., semantic representation, gesture, perspective-taking) and to a wide range of language related processes (e.g., on-line processing, statistical learning).

A critical component of our proposal is the notion that a single component of language will not be the purview of a single memory system. Rather, we argue that multiple memory systems contribute to multiple aspects of language. That is, declarative and non-declarative memory systems make unique contributions to multiple aspects of language (e.g., hippocampus contributes to word and grammatical learning) and complementary contributions to the same aspects of language (e.g., declarative and non-declarative memory both contribute to common ground representation). There are aspects of language use and processing that appear to be independent of the declarative memory system and intact in certain neurological patients [e.g., patients with amnesia are intact at acquiring person-specific accent information (Trude et al. 2014) and achieving conversational synchrony]. Given our proposal that complex behavioral phenomena are supported by multiple memory systems, we predict that patients with only one intact memory system do not use language normally in communicative settings. For example, even when patients with hippocampal amnesia are successful at acquiring partner-specific accent information, we would predict that their hippocampal damage and declarative memory impairment would prevent them from using such information flexibility and across communicative contexts. Moving forward, documenting and characterizing the nature and time course of these contributions and the interactions among memory systems will be an important future direction in advancing our understanding of the memory-language interface and of memory-and-language-in-use in every-day communicative settings.

Use of converging methods to support and extend the work from lesion patients on hippocampal contributions to language is also warranted. One important direction is the continued and expanded use of neuroimaging methods to examine hippocampal involvement in language use and processing. Some notable examples to date include fMRI studies revealing hippocampal activation during statistical learning tasks (Turk-Browne et al. 2009) and direct recordings from hippocampus in implanted epilepsy patients demonstrating increased hippocampal theta oscillations during on-line language processing (Piai et al. 2016). Neuroimaging studies linking the structure and function of hippocampus to language use and processing provide critical evidence for proposals about the memory-language interface. Neuroimaging studies, and fMRI in particular, can also begin to document how, and over what time course, hippocampus interacts with the rest of the canonical language network, which is an important step for understanding the role of the hippocampus in the neurobiology of language more broadly.

In summary, the recent advances in understanding the functionality of the human hippocampus has led to a number of proposals for how hippocampus may support a range of cognitive abilities beyond memory (for review see Rubin et al. 2014). Building on these advances, we have offered a new account of the memory-language interface (Duff and Brown-Schmidt 2012). This proposal leads to a set of testable predictions and hypotheses about how language and memory work together and argues that efforts to examine the relationship between memory and language are best served by broad-scope approaches that include the study of a range of communicative activities, including those that are characteristic of every-day language use. Indeed, the present represents a time of tremendous potential for discovery and progress in the study of memory and language and for more representative, biologically plausible, and ecologically valid investigations of memory-and-language-in-use in every-day life.

Acknowledgements Support by NIDCD R01 DC011755.

References

Addis DR, Wong AT, Schacter DL (2007) Remembering the past and imagining the future: common and distinct neural substrates during event construction and elaboration. Neuropsychologia 45:1363–1377

Arnold J, Eisenband J, Brown-Schmidt S, Trueswell J (2000) The rapid use of gender information: evidence of the time course of pronoun resolution from eyetracking. Cognition 76:813–826

Arnold JE, Tanenhaus MK, Altmann RJ, Fagnano M (2004) The old and thee, uh, new— Disfluency and reference resolution. Psychol Sci 15:578–582

Austin JL (1962) How to do things with words. Clarendon, Oxford

Baldwin D, Andersson A, Saffran J, Meyer M (2008) Segmenting dynamic human action via statistical structure. Cognition 106:1382–1407

Barense MD, Gaffan D, Graham KS (2007) The human medial temporal lobe processes online representations of complex objects. Neuropsychologia 45:2963–2297

Bayley P, Squire L (2002) Medial temporal lobe amnesia: gradual acquisition of factual information by nondeclarative memory. J Neurosci 22(13):5741–5748

Bayley PJ, Squire LR (2005) Failure to acquire new semantic knowledge in patients with large medial temporal lobe lesions. Hippocampus 15:273–280

Bristol, AS, Viskontas, IV (2006) Dynamic processes within associative memory stores: piecing together the neural basis of creative cognition. In: Kaufman JC, Baer J (eds) Creativity, knowledge and reason. Cambridge University Press, pp 60–80

Brown AS, Jones EM, Davis TL (1995) Age differences in conversational source monitoring. Psychol Aging 10:111–122

Brown-Schmidt S, Tanenhaus MK (2006) Watching the eyes when talking about size: an investigation of message formulation and utterance planning. J Mem Lang 54:592–609

Buckner RL, Carroll DC (2007) Self-projection and the brain. Trends Cogn Sci 11(2):49–57

Carey S (2010) Beyond fast mapping. Lang Learn Dev 6(3):184–205

Chang F, Dell GS, Bock K (2006) Becoming syntactic. Psychol Rev 113(2):234–272

Clark HH (2016) Depicting as a method of communication. Psychol Rev 123(3):324–347

Clark HH (1996) Using language. Cambridge University Press, Cambridge

Clark HH, Krych MA (2004) Speaking while monitoring addressees for understanding. J Mem Lang 50:62–81

Clark HH, Marshall CR (1978) Reference diaries. In: Waltz DL (ed) Theoretical issues in natural language processing, vol 2. Association for Computing Machinery, New York, pp 57–63

Clark HH, Murphy GL (1982) Audience design in meaning and reference. In: Ny J-FL, Kintsch W (eds) Language and comprehension. North Holland, New York

Clark H, Wilkes-Gibbs D (1986) Referring as a collaborative process. Cognition 22:1–39

Cohen NJ (1984) Preserved learning capacity in amnesia: evidence for multiple memory systems. In: Butters N, Squire L (eds) The neuropsychology of memory. Guilford Press, New York, pp 83–103

Cohen NJ, Eichenbaum H (1993) Memory, amnesia and the hippocampalsystem. MIT Press, Cambridge, MA

Conway C, Christiansen MH (2005) Modality constrained statistical learning of tactile, visual, and auditory sequences. J Exp Psychol Learn Mem Cogn 31:24–39

Cook S, Tanenhaus MK (2009) Embodied communication: speakers' gestures affect listeners' actions. Cognition 113:98–104

Coronel JC, Duff MC, Warren DE, Gonsalves BD, Tranel D, Cohen NJ (2012) Remembering and voting: theory and evidence from amnesic patients. Am J Polit Sci 56(4):837–848

Creel SC (2014) Preschoolers' flexible use of talker information during word learning. J Mem Lang 73:81–98

Crystal D (1998) Language play. University of Chicago Press, Chicago

Dahan D, Tanenhaus MK, Chambers CG (2002) Accent and reference resolution in spokenlanguage comprehension. J Mem Lang 47:292–314

Damasio AR (2001) Some notes on brain, imagination, and creativity. In: Pfenninger KH, Shubik VR (eds) The origins of creativity. Oxford University Press, Oxford

Davachi L (2006) Item, context, and relational episodic encoding in humans. Curr Opin Neurobiol 16:693–700

Duff MC, Brown-Schmidt S (2012) The hippocampus and the flexible use and processing of language. Front Hum Neurosci 6:69. doi:10.3389/fnhum.2012.00069

Duff MC, Hengst J, Tranel D, Cohen NJ (2006) Development of shared information in communication despite hippocampal amnesia. Nat Neurosci 9(1):140–146

Duff MC, Hengst J, Tranel D, Cohen NJ (2007) Talking across time: using reported speech as a communicative resource in amnesia. Aphasiology 21(6–8):1–14

Duff MC, Hengst J, Tranel D, Cohen NJ (2008) Collaborative discourse facilitates efficient communication and new learning in amnesia. Brain Lang 106:41–54

Duff MC, Hengst J, Tranel D, Cohen NJ (2009) Hippocampal amnesia disrupts verbal play and the creative use of language in social interaction. Aphasiology 23(7):926–939

Duff MC, Gupta R, Hengst J, Tranel D, Cohen NJ (2011) The use of definite references signals declarative memory: evidence from hippocampal amnesia. Psychol Sci 22(5):666–673

Duff MC, Kurczek J, Rubin R, Cohen NJ, Tranel D (2013) Hippocampal amnesia impairs creative thinking. Hippocampus 23(12):1143–1149

Duff MC, Kurczek J, Miller M (2015) Use of reported speech in the communicative interactions of individuals with ventromedial prefrontal cortex damage. J Interact Res Commun Disord 6 (1):97–114

Dusek J, Eichenbaum H (1997) The hippocampus and memory for orderly stimulus relations. Proc Natl Acad Sci U S A 13:7109–7114

Eichenbaum H, Cohen NJ (2001) From conditioning to conscious recollection: memory systems of the brain. Oxford University Press, New York

Evans JL, Saffran JR, Robe-Torres K (2009) Statistical learning in children with specific language impairment. J Speech, Lang, Hear Res 52(April):321–336

Ferguson HJ, Breheny R (2011) Eye movements reveal the time-course of anticipating behaviour based on complex, conflicting desires. Cognition 119(2):179–196

Ferguson HJ, Breheny R (2012) Listeners' eyes reveal spontaneous sensitivity to others' perspectives. J Exp Soc Psychol 48:257–263

Fischer NM, Schult JC, Steffens MC (2015) Source and destination memory in face-to-face interaction: a multinomial modeling approach. J Exp Psychol Appl 21(2):195

Fox Tree JE (2001) Listeners' uses of um and uh in speech comprehension. Mem Cogn 29:320–326

Fox Tree JE, Clark HH (1997) Pronouncing "the" as "thee" to signal problems in speaking. Cognition 62:151–167

Fraundorf SH, Watson DG (2011) The disfluent discourse: effects of filled pauses on recall. J Mem Lang 65:161–175

Fraundorf SH, Watson DG (2014) Alice's adventures in um-derland: psycholinguistic sources of variation in disfluency production. Lang Cogn Neurosci 29:1083–1096

Fraundorf SH, Watson DG, Benjamin A (2015) Reduction in prosodic prominence predicts speakers' recall: implications for theories of prosody. Lang Cogn Neurosci 20(5):606–619

Gabrieli JD, Cohen NJ, Corkin S (1988) The impaired learning of semantic knowledge following bilateral medial temporal-lobe resection. Special issue: single-case studies in Amnesia: theoretical advances. Brain Cogn 7:157–177

Gibson E (1998) Linguistic complexity: locality of syntactic dependencies. Cognition 68:1–76

Goldinger SD (1998) Echoes of echoes? An episodic theory of lexical access. Psychol Rev 105:251–279

Gordon JK, Dell GS (2003) Learning to divide the labor: an account of deficits in light and heavy verb production. Cogn Sci 27(1):1–40

Greene SB, Gerrig RJ, McKoon G, Ratcliff R (1994) Unheralded pronouns and management by common ground. J Mem Lang 33:511–526

Gupta R, Tranel D, Duff MC (2012) Ventromedial prefrontal cortex damage does not impair the development and use of common ground in social interaction: implications for cognitive theory of mind. Neuropsychologia 25(2):137–146

Hanna JE, Tanenhaus MK, Trueswell JC (2003) The effects of common ground and perspective on domains of referential interpretation. J Mem Lang 49:43–61

Hannula DE, Greene A (2012) The hippocampus reevaluated in unconscious learning and memory: at a tipping point? Front Hum Neurosci 6:80. doi:10.3389/fnhum.2012.00080

Hannula D, Raganath C (2008) Medial temporal lobe activity predicts successful relational memory binding. J Neurosci 28:116–124

Hannula D, Tranel D, Cohen NJ (2006) The long and the short of it: relational memory impairments in amnesia, even at short lags. J Neurosci 26(32):8352–8259

Hassabis D, Kumaran D, Vann S, Maguire EA (2007) Patients with hippocampal amnesia cannot imagine new experiences. Proc Natl Acad Sci U S A 104(5):1726–1731

Henke K (2010) A model for memory systems based on processing modes rather than consciousness. Nat Rev Neurosci 11(7):523–532

Hilverman C, Cook SW, Duff MC (2016) Hippocampal declarative memory supports gesture production: evidence from amnesia. Cortex 85:25–36

Horton WS, Gerrig RJ (2005) Conversational common ground and memory processes in language production. Discourse Processes 40:1–35

Huettig F, Janse E (2016) Individual differences in working memory and processing speed predict anticipatory spoken language processing in the visual world. Lang Cogn Neurosci 31(1):80–93

Hyman IE Jr (2000) Conversational remembering. In: Neisser U, Hyman Jr IE (eds) Memory observed: remembering in natural contexts, 2nd edn. Worth, New York

Isaacs, E. A. (1990). Mutual memory for conversation. Unpublished doctoral dissertation, Stanford University.

Issacs EA, Clark HH (1987) References in conversation between experts and novices. J Exp Psychol Gen 116:26–37

Joanisse MF, Seidenberg MS (1999) Impairments in verb morphology after brain injury: a connectionist model. Proc Natl Acad Sci U S A 96:7592–7597

Johnson MK, Kim JK, Risse G (1985) Do alcoholic Korsakoff's syndrome patients acquire affective reactions? J Exp Psychol Learn Mem Cogn 11:22–36

Just MA, Carpenter PA (1992) A capacity theory of comprehension: individual differences in working memory. Psychol Rev 99:122–149

Kaiser E, Runner JT, Sussman RS, Tanenhaus MK (2009) Structural and semantic constraints on the resolution of pronouns and reflexives. Cognition 112:55–80

Karuza E, Newport EL, Aslin RN, Starling SJ, Tivarus ME, Bavelier D (2013) The neural correlates of statistical learning in a word segmentation task: an fMRI study. Brain Lang 127 (1):46–54

Kendon A (1997) Gesture. Annual Rev Anthropol 26:109–128

Kensington EA, Ullman MT, Corking S (2001) Bilateral medial temporal lobe damage does not affect lexical or grammatical processing: evidence from the amnesic patient H.M. Hippocampus 11(4):347–360

Keysar B, Lin S, Barr DJ (2003) Limits on theory of mind use in adults. Cognition 89:25–41

Kim R, Seitz A, Feenstra H, Shams L (2009) Testing assumptions of statistical learning: is it long-term and implicit? Neurosci Lett 461(2):145–149

Klooster N, Duff MC (2015) Remote semantic memory is impoverished in hippocampal amnesia. Neuropsychologia 79(Part A):42–52

Knowlton BJ, Squire LR (1994) The information acquired during artificial grammar learning. J Exp Psychol Learn Mem Cogn 20:79–91

Knowlton BJ, Squire LR (1996) Artificial grammar learning depends on implicit acquisition of both abstract and exemplar-specific information. J Exp Psychol Learn Mem Cogn 22(1):169

Knowlton B, Ramus S, Squire L (1992) Intact artificial grammar learning in amnesia: dissociation of classification learning and explicit memory for specific instances. Psychol Sci 3(3):172–179

Konkel A, Warren DE, Duff MC, Tranel D, Cohen NJ (2008) Hippocampal amnesia impairs all manner of relational memory. Front Hum Neurosci 2:15

Kraljic T, Brennan SE (2005) Prosodic disambiguation of syntactic structure: for the speaker or for the addressee? Cogn Psychol 50:194–231

Kurczek J, Duff MC (2011) Cohesion, coherence, and declarative memory: discourse patterns of patients with hippocampal amnesia. Aphasiology 25(6–7):700–712

Kurczek J, Brown-Schmidt S, Duff MC (2013) Hippocampal contributions to language: evidence of referential processing deficits in amnesia. J Exp Psychol Gen 142(4):1346–1354

Kurczek J, Wechsler E, Ahuja S, Jensen U, Cohen N, Tranel D, Duff MC (2015) Differential contributions of hippocampus and medial prefrontal cortex to self-projection and self-referential processing. Neuropsychologia 73:116–126

Kurumada C, Brown M, Bibyk S, Pontillo DF, Tanenhaus MK (2014) Is it or isn't it: listeners make rapid use of prosody to infer speaker meanings. Cognition 133(2):335–342

Lee JL (2008) Memory reconsolidation mediates the strengthening of memories by additional learning. Nat Neurosci 11(11):1264–1266

Lee J, Nopoulous P, Tomblin B (2013a) Abnormal subcortical components of the corticostriatal system in young adults with DLI: a combined structural MRI and DTI study. Neuropsychologia 51(11):2154–2161

Lee EK, Brown-Schmidt S, Watson DG (2013b) Ways of looking ahead: hierarchical planning in language production. Cognition 129:544–562

Levelt WJM (1989) Speaking: from intention to articulation. MIT Press, Cambridge, MA

Lewis RL, Vasishth S, Van Dyke J (2006) Computational principles of working memory in sentence comprehension. Trends Cogn Sci 10:447–454

MacDonald MC (2013) How language production shapes language form and comprehension. Front Psychol 4(226):1–16

MacDonald MC, Just MA, Carpenter PA (1992) Working memory constraints on the processing of syntactic ambiguity. Cogn Psychol 24:56–98

MacDonald MC, Christiansen MH (2002) Reassessing working memory: comment on Just and Carpenter (1992) and Waters and Caplan 2002. Psychol Rev 109:35–54

MacKay DG, James LE (2001) The binding problem for syntax, semantics, and prosody: H.M.'s selective sentence-reading deficits under the theoretical-syndrome approach. Lang Cogn Process 16(4):419–460

MacKay DG, Burke DM, Stewart R (1998a) H.M's language production deficits: implications for relations between memory, semantic binding, and the hippocampal system. J Mem Lang 38:28–69

MacKay DG, Stewart R, Burke DM (1998b) H.M. revisited: relations between language, comprehension, memory and the hippocampal system. J Cogn Neurosci 10(3):377–394

Madore KP, Schacter DL (2014) An episodic specificity induction enhances means-end problem solving in young and older adults. Psychol Aging 29:913–924

Madore KP, Addis DR, Schacter DL (2015) Creativity and memory: effects of an episodic-specificity induction on divergent thinking. Psychol Sci:1461–1468

Manns JR, Hopkins RO, Squire LR (2003) Semantic memory and the human hippocampus. Neuron 38(1):127–133

McClelland JL, McNaughton BL, O'Reilly RC (1995) Why there are complementary learning systems in the hippocampus and neocortex: insights from the successes and failures of connectionist models of learning and memory. Psychol Rev 102:419–457

McElree B (2000) Sentence comprehension is mediated by content-addressable memory structures. J Psycholinguist Res 29:111–123

McElree B, Foraker S, Dyer L (2003) Memory structures that subserve sentence comprehension. J Mem Lang 48:67–91

McGregor KK, Newman RM, Reilly RM, Capone NC (2002) Semantic representation and naming in children with specific language impairment. J Speech Lang Hear Res 45(5):998–1014

McKenzie S, Eichenbaum H (2011) Consolidation and reconsolidation: two lives of memories? Neuron 71(2):224–233

McMurray B, Horst JS, Samuelson LK (2012) Word learning emerges from the interaction of online referent selection and slow associative learning. Psychol Rev 119:831–877

McNeill D (1992) Hand and mind: what gestures reveal about thought. The University of Chicago Press, Chicago

Mishra RK, Singh N, Pandey A, Huettig F (2012) Spoken language-mediated anticipatory eye movements are modulated by reading ability: evidence from Indian low and high literates. J Eye Mov Res 5(1):1–10

Nieuwland MS, Otten M, Van Berkum JJA (2007) Who are you talking about? Tracking discourse-level referential processing with event-related brain potentials. J Cogn Neurosci 19:228–236

Norrick N (1998) Retelling stories in spontaneous conversation. Discourse Process 25:75–97

Novick JM, Trueswell JC, Thompson-Schill SL (2005) Cognitive control and parsing: reexamining the role of Broca's area in sentence comprehension. Cogn Affect Behav Neurosci 5:263–281

Novick JM, Thompson-Schill SL, Trueswell JC (2008) Putting lexical constraints in context into the visual-world paradigm. Cognition 107:850–903

O'Kane G, Kensinger EA, Corkin S (2004) Evidence for semantic learning in profound amnesia: an investigation with patient H.M. Hippocampus 14:417–425

O'Keefe J, Nadel L (1978) The hippocampus as a cognitive map. Oxford University Press, Oxford

O'Reilly RC, Rudy JW (2000) Computational principles of learning in the neocortex and hippocampus. Hippocampus 10:389–397

Olsen IR, Page K, Sledge Moore K, Chatterjee A, Verfaellie M (2006) Working memory for conjunctions relies on the medial temporal lobe. J Neurosci 26:4596–4601

Oppenheim GM, Dell GS, Schwartz MF (2010) The dark side of incremental learning: a model of cumulative semantic interference during lexical access in speech production. Cognition 114:227–252

Perruchet P, Pacton S (2006) Implicit learning and statistical learning: one phenomenon, two approaches. Trends Cogn Sci 10(5):233–238

Pexman PM, Lupker SJ, Hino Y (2002) The impact of feedback semantics in visual word recognition: number-of-features effects in lexical decision and naming tasks. Psychon Bull Rev 9(3):542–549

Piai V, Anderson KL, Lin JJ, Dewar C, Parvizi J, Dronkers NF, Knight RT (2016) Direct brain recordings reveal hippocampal rhythm underpinnings of language processing. Proc Natl Acad Sci 113:11366–11371

Poldrack RA, Kittur A, Kalar D, Miller E, Seppa C, Gil Y, Parker DS, Sabb FW, Bilder RM (2011) The cognitive atlas: toward a knowledge foundation for cognitive neuroscience. Front Neuroinform 5:17

Race E, Keane M, Verfaellie M (2011) Medial temporal lobe damage causes deficits in episodic memory and episodic future thinking not attributable to deficits in narrative construction. J Neurosci 31(28):10262–10269

Ranganath C (2010) A unified framework for the functional organization of the medial temporal lobes and the phenomenology of episodic memory. Hippocampus 20:1263–1290

Ranganath C, D'Esposito M (2001) Medial temporal lobe activity associated with active maintenance of novel information. Neuron 31:865–873

Reber PJ, Knowlton BJ, Squire LR (1996) Dissociable properties of memory systems: differences in the flexibility of declarative and nondeclarative knowledge. Behav Neurosci 110:859–869

Regel S, Coulson S, Gunter TC (2010) The communicative style of a speaker can affect language comprehension? ERP evidence from the comprehension of irony. Brain Res 1311:121–135

Ross M, Sicoly F (1979) Egocentric biases in availability and attribution. J Pers Soc Psychol 37:322–336

Rubin RD, Brown-Schmidt S, Duff MC, Tranel D, Cohen NJ (2011) How do I remember that I know you know that I know? Psychol Sci 22:1574–1582

Rubin R, Watson P, Duff MC, Cohen NJ (2014) The role of the hippocampus in flexible cognition and social behavior. Front Hum Neurosci 8. doi:10.3389/fnhum.2014.00742

Ryan JD, Moses SN, Barense M, Rosenbaum RS (2013) Intact learning of new relations in amnesia as achieved through unitization. J Neurosci 33:9601–9613

Ryskin RA, Qi Z, Duff MC, Brown-Schmidt S (2016) Verb biases are shaped through lifelong learning. J Exp Psychol Learn Mem Cogn, Epub ahead of print. doi: 10.1037/xlm0000341

Saffran JR, Aslin RN, Newport EL (1996) Statistical learning by 8-month-old infants. Science 274 (5294):1926–1928

Schacter DL, Addis DR (2007) The cognitive neuroscience of constructive memory: remembering the past and imagining the future. Philos Trans R Soc B 362:773–786

Schapiro AC, Gregory E, Landau B, McCloskey M, Turk-Browne NB (2014) The necessity of the medial temporal lobe for statistical learning. J Cogn Neurosci 26(8):1736–1747

Schapiro AC, Turk-Browne NB, Norman KA, Botvinick MM (2016) Statistical learning of temporal community structure in the hippocampus. Hippocampus 26(1):3–8

Schmolck H, Kensinger E, Corkin S, Squire L (2002) Semantic knowledge in patient H.M. and other patients with bilateral medial and lateral temporal lobe lesions. Hippocampus 12:520–533

Searle J (1969) Speech acts. Cambridge University Press, Cambridge

Sharon T, Moscovitch M, Gilboa A (2011) Rapid neocortical acquisition of long-term arbitrary associations independent of the hippocampus. Proc Natl Acad Sci U S A 108:1146–1151

Sheldon S, Moscovitch M (2012) The nature and time-course of medial temporal lobe contributions to semantic retrieval: an fMRI study on verbal fluency. Hippocampus 22(6):1451–1466

Seidenberg MS, Plaut DC (2014) Quasiregularity and its discontents: the legacy of the past tense debate. Cogn Sci 38(6):1190–1228

Sherzer J (2002) Speech play and verbal art. University of Texas Press, Austin, TX

Shrager Y, Levy DA, Hopkins RO, Squire LR (2008) Working memory and the organization of brain systems. J Neurosci 28:4818–4822

Snedeker J, Trueswell J (2003) Using prosody to avoid ambiguity: effects of speaker awareness and referential context. J Mem Lang 48:103–130

Squire LR (1992) Memory and the hippocampus: a synthesis from findings with rats, monkeys, and humans. Psychol Rev 99:195–231

Stafford L, Daly JA (1984) Conversational memory: the effects of recall mode and memory expectancies on remembrances of natural conversations. Hum Commun Res 10:379–402

Taler V, Kousaie S, Lopez Zunini R (2013) ERP measures of semantic richness: the case of multiple senses. Front Hum Neurosci 7:5

Tanenhaus MK, Spivey-Knowlton MJ, Eberhard KM, Sedivy JC (1995) Integration of visual and linguistic information in spoken language comprehension. Science 268:1632–1634

Tannen D (1989) Talking voices: repetition, dialogue, and imagery in conversational discourse. Harvard University Press, Cambridge, MA

Tauroza S, Allison D (1990) Speech rates in British english. Appl Linguist 11:90–105

Tranel D, Damasio AR (1993) The covert learning of affective valence does not require structures in hippocampal system or amygdala. J Cogn Neurosci 5:79–88

Trude A, Duff MC, Brown-Schmidt S (2014) Talker-specific learning in amnesia: insight into mechanisms of adaptive speech perception. Cortex 54:117–123

Turk-Browne NB, Scholl BJ, Chun MM, Johnson MK (2009) Neural evidence of statistical learning: efficient detection of visual regularities without awareness. J Cogn Neurosci 21:1934–1945

Ullman MT (2001) The declarative/procedural model of lexicon and grammar. J Psycholinguist Res 30(1):37–69

Ullman MT (2004) Contributions of memory circuits to language: the declarative/procedural model. Cognition 92:231–270

Ullman MT, Pierpont E (2005) Specific language impairment is not specific to language: the procedural deficit hypothesis. Cortex 41:399–433

Ullman MT, Corkin S, Coppola M, Hickok G, Growdon JH, Koroshetz WJ, Pinker S (1997) A neural dissocation within language: evidence that the mental dictionary is part of declarative memory and that grammatical rules are processed by the procedural system. J Cogn Neurosci 9 (2):266–276

Van Berkum JJA, Van den Brink D, Tesink CMJY, Kos M, Hagoort P (2008) The neural integration of speaker and message. J Cogn Neurosci 20:580–591

Vargha-Khadem F, Gadian DG, Watkins KE, Connely A, Van Paesschen W, Mishkin M (1997) Differential effects of early hippocampal pathology on episodic and semantic memory. Science 277:376–380

Verfaellie M, Koseff P, Alexander MP (2000) Acquisition of novel semantic information in amnesia: effects of lesion location. Neuropsychologia 38:484–492

Voss J, Lucas H, Paller K (2012) More than a feeling: pervasive influences of memory without awareness of retrieval. Cogn Neurosci 3(3–4):196–207

Warren D, Duff MC (2014) Not so fast: hippocampal amnesia slows word learning despite successful fast mapping. Hippocampus 24(8):920–933

Warren D, Duff MC, Tranel D, Cohen NJ (2011) Observing degradation of visual representations over short intervals when medial temporal lobe is damaged. J Cogn Neurosci 23 (12):3862–3873 PMCID: PMC3521516

Warren D, Kurczek J, Duff MC (2016) What relates newspaper, definite, and clothing? An article describing deficits in convergent problem solving and creativity following hippocampal damage. Hippocampus 26:835–840

Wilkes-Gibbs D, Clark HH (1992) Coordinating beliefs in conversation. J Mem Lang 31:183–194

Wu YC, Coulson S (2007) How iconic gestures enhance communication: an ERP study. Brain Lang 101:234–245

Yap MJ, Tan SE, Pexman PM, Hargreaves IS (2011) Is more always better? Effects of semantic richness on lexical decision, speeded pronunciation, and semantic classification. Psychon Bull Rev 18(4):742–750

Yoon SO, Koh S, Brown-Schmidt S (2012) Influence of perspective and goals on reference production in conversation. Psychon Bull Rev 19:699–707

Yoon SO, Benjamin AS, Brown-Schmidt S (2016) The historical context in conversation: lexical differentiation and memory for the discourse history. Cognition 154:102–117

Zeithamova D, Preston AR (2010) Flexible memories: differential roles for medial temporal lobe and prefrontal cortex in cross-episode binding. J Neurosci 30(44):14676–14684

The Hippocampus and Social Cognition

Anne C. Laurita and R. Nathan Spreng

Abstract The function of memory is not simply to recall the past but also to form and update models of our experiences that help us navigate the complexities of the social world. In the present chapter, we review behavioral, neuroimaging, and neuropsychological evidence that suggest an important role for memory—and the hippocampus—in social cognition.

Introduction

> "Lastly, she pictured to herself how this same little sister of hers would, in the after-time, be herself a grown woman; and how she would keep, through all her riper years, the simple and loving heart of her childhood: and how she would gather about her other little children, and make their eyes bright and eager with many a strange tale, perhaps even with the dream of Wonderland of long ago: and how she would feel with all their simple sorrows, and find a pleasure in all their simple joys, remembering her own child-life, and the happy summer days."
>
> —Lewis Carroll, Alice's Adventures in Wonderland

So ends Lewis Carroll's classic novel, with a nostalgic introspection by Alice's older sister. This passage illustrates how social cognition and memory intertwine to create a rich, social connectedness, infusing one's personal experience into reflections about the thoughts, aspirations and motivations of others. Just as Alice's sister draws from childhood memories to create her wistful vision of Alice's future, we too utilize our personal experiences to envision the inner worlds of those around us. Memory thus provides an essential footing from which we are able to reach out to and engage with our social environment.

Memory is not simply a static representation of the past. It is a surprisingly flexible account of our accumulated experience and knowledge—a record of our past measured in space, time and context. Functionally, we access, reconfigure and

A.C. Laurita • R. Nathan Spreng (✉)
Department of Human Development, Cornell University, Martha Van Rensselaer Hall, Ithaca, NY 14950, USA
e-mail: nathan.spreng@gmail.com

© Springer International Publishing AG 2017
D.E. Hannula, M.C. Duff (eds.), *The Hippocampus from Cells to Systems*,
DOI 10.1007/978-3-319-50406-3_17

537

re-encode these representations as we rely on our memory to guide our present thoughts and actions and plan for our future. As memory plays a critical role in constructing our personal past, present and future, it is perhaps not surprising that memory also plays a crucial role in how we construct, interact with and predict the thoughts and actions of others. In this chapter, we explore this relationship between memory and social cognition. Specifically we discuss how hippocampal-mediated memory processes influence social functioning. We examine evidence from behavioral, neuroimaging, and neuropsychological studies that point to a strong bond between our ability to recollect and reconstruct our personal past and our capacity to imagine, infer and ultimately interact with the intentions of others. The hippocampal memory system is highly attuned to this kind of social information, for which relational binding is critical. Although there is insufficient evidence thus far to draw the conclusion that the hippocampus carries a special function for social cognition above and beyond what it does for memory, we examine recent evidence regarding how the hippocampus is recruited in social contexts.

We begin the chapter with a broad discussion of social cognition and memory. We next review the importance of the hippocampus in contextual and relational processing and explore its critical role in navigating physical and temporal contexts. Although consideration of a role for the hippocampus in social cognition is relatively novel, a growing body of evidence suggests that social cognition depends on the binding of discrete elements of social interaction. As binding is considered to be a central function of the hippocampus, and the medial temporal lobe memory system more broadly, this raises the intriguing possibility that social cognition may depend on the functioning of the hippocampal memory system and associated brain regions. We next suggest that this functional role of the hippocampal memory system be expanded to include navigating social contexts. We discuss experimental lesion, neuropsychological research and functional neuroimaging investigations that are providing increasingly convergent evidence pointing to an important role for the hippocampal memory system in social cognition. Increasingly, the neural basis of social cognition is associated with functionally connected brain networks. These networks are defined by correlated oscillations of activity in spatially-distributed brain regions observed during task or at rest. The default network, a set of brain regions implicated in mnemonic and associative processes, has been specifically implicated in the processing of socially-relevant information. As the hippocampus is a core node within this network, in the final section of the chapter we extend our review to studies of the default network and social cognition.

Memory and Social Cognition in Everyday Function

Social cognition, broadly defined, describes the way in which people understand themselves and other people. It encompasses the cognitive processes used to decode and encode the social world (Beer and Oschner 2006). These include perception of self and others, the incorporation of social information into existing knowledge structures, and the selection of actions based on social cues. Here we focus on how social cognition involves the binding of basic social percepts and their integration into stored memory representations to guide future thoughts and behavior. In this respect we do not review early sensory processing or action planning, but rather focus on the constructive and binding aspects of social cognition in which hippocampal structures are likely to play a more direct role. As much of the research literature focuses on hippocampal functions, we restrict our discussion here to explicit encoding and retrieval processes.

In this first section we lay the conceptual foundation for the remainder of the chapter by suggesting four points of intersection between social cognition and memory in everyday life: perceiving interpersonal cues, constructing complex social representations, navigating social relations, and forming close personal bonds. Put another way, how does memory influence how we perceive, construct, interact and, ultimately connect with, our social world?

Perceiving Interpersonal Cues

At a fundamental level, social thinking requires the ability to perceive, disambiguate and ultimately categorize social stimuli. Some of these basic perceptual categories include living versus non-living, human versus non-human, friend versus enemy, and same versus other. The hippocampal memory system is critical for encoding and retrieval of these social percepts (Rubin et al. 2014). These memory processes are necessary to identify an acquaintance's face in a crowd or to differentiate your friend's from your sister's voice when answering a phone call. Accurately perceiving and recognizing social stimuli requires forming and accessing person cues, and developing a store of person-specific knowledge—both stable (such as personality traits) and transient (such as affective states). Through repeated exposure across multiple contexts, these cues form patterns and provide the basis of a more abstract sense of person identity (see Carlston and Smith 1996, for review). Explicit encoding and retrieval of person-specific knowledge and the formation of context-independent, person-schema depend on the hippocampal memory system (Eichenbaum and Cohen 2014; Ochsner et al. 2005). Other social perceptions, such as impression formation, are formed rapidly, often in a single exposure. These implicit associations occur outside of conscious awareness and are likely not dependent on the hippocampus (e.g. Freeman et al. 2014).

Constructing Complex Mental Representations

Across time, perceptions of personal identity are imbued with learned positive, negative, and neutral associations, linking stimuli and situational contexts with specific social actors (Carlston and Smith 1996). Through repeated interactions, complex person-specific schema are formed including judgments of self-similarity and attributions of intent. Forming social relationships involves the development of complex mental representations, also known as "internal working models", of relationship partners (Bowlby 1969; Carlston 2010; Collins and Read 1994; Pietromonaco and Feldman Barrett 2000). Hippocampal memory systems play a critical role in forming these representations, weaving together past experiences and extracting stable patterns across time. Such mental representations consist of extensive interpersonal memories of figures in our social world that are integrated with affective associations (e.g., Zayas and Shoda 2005). These patterns of social expectation and behavior facilitate the development of long-lasting dyadic social relationships. In such close relationships, the utilization of an innate bonding system is also based upon the development of complex cognitive representations (Zayas and Shoda 2005; Zayas et al. 2002). These representations facilitate forward-modeling of behavior enabling individuals to predict the actions of others and guide their own actions in dynamic social contexts (Holmes 2002).

Navigating Social Relationships

As we discuss in more detail in the following section, the hippocampal memory system is critically involved in spatial processing (see Eichenbaum and Cohen 2014 for a review). This role has recently been extended to navigating social distances and social hierarchies (Tavares et al. 2015), insofar as the hippocampus binds various components of social information. From this perspective social cognition is considered analogous to navigating a social landscape with distance measured along two dimensions: power and social affiliation (Tavares et al. 2015). A parent, who possesses both high power status and high social-affiliation, would be close to their child in terms of social distance. In contrast, a friend who may be high in social affiliation but equivalent with respect to social power hierarchy would be considered more socially distant. As with mapping physical space, the hippocampus is important for charting and navigating the myriad social distances and hierarchies that make up our social milieu (Tavares et al. 2015; see also Kumaran et al. 2016). More specifically, hippocampal involvement in mapping social space hinges on representing others in multi-dimensional social spaces. Previously, others examining the neural basis of social distance found little evidence of hippocampal recruitment, when considering the tracking of only one dimension (Muscatell et al. 2012; Parkinson et al. 2014; Tamir and Mitchell 2011). Tavares et al. (2015) assert that the role of the hippocampus involves mapping the combination of social-dimensions rather than individual social-dimensions.

Social Bonding

Highly salient social memories comprise the mental representations we form of close others—children, parents, romantic partners (Pietromonaco and Feldman Barrett 2000). Romantic partner mental representations in particular promote the formation of stable, mutually-beneficial bonds with relationship partners. Within the context of pair-bonds, romantic partner mental representations have been further conceptualized as cognitive expansions of the self (Aron and Aron 1986). These partner representations have been demonstrated to play a role in subconscious pursuit of partner-specific interpersonal goals (Fitzsimons and Bargh 2003) and to inherently intertwine the cognitive and emotional contexts of both relationship partners (Zayas et al. 2002). Long-term declarative and relational memory, supported by the hippocampus, is crucial for forming and maintaining and accessing these 'other' representations, which form the basis of complex interpersonal bonds (Rubin et al. 2014). Moreover, these close-other representations can influence our perceptions, judgments and responses to others in our social world—a process known as social-cognitive transference (Anderson and Cole 1990; Günaydin et al. 2012).

Hippocampal Function and Social Cognition

The influence of hippocampally mediated memory processes on social cognition is an emerging area of inquiry. However, two well established accounts of hippocampal memory function: relational integration and constructive memory have provided a theoretical bridge between memory and social cognition. We review each of these theories in turn and discuss how they have been used to characterize this relationship. In the following sections, we draw from several different theoretical perspectives of hippocampal function in order to provide a comprehensive account of the potential role of the hippocampus in social cognition. However, for the purposes of this review, we remain agnostic with respect to the merit of these individual perspectives as theories of hippocampal functioning per se.

Relational Integration and Social Cognition: The Role of Spatial and Social Navigation

The role of the hippocampus in relational processing was first posited by O'Keefe and Nadel (1978), who proposed the cognitive mapping hypothesis of hippocampal functioning. This theory suggested that the hippocampal system forms mnemonic representations by linking stimuli to specific locations through a process of allocentric mapping of distance and direction of an object within its spatial environment. Building from this earlier work, Eichenbaum et al. (1996) argued that the hippocampal memory system, conceptualized as a functional grouping of the hippocampus, medial temporal lobe, and cortical regions, was critically involved in relational processing as well as mapping stimuli to specific spatial contexts. The authors suggested that these representations were flexible and could be dynamically reconfigured to reflect changing contexts. Relational memory theory proposes that the hippocampus is responsible for computing an associative scaffold, linking items and events in "memory space" (Eichenbaum 2004; Eichenbaum et al. 1999). Others have hypothesized that memory space can include relational information beyond spatial location such as temporal, emotional or configural associations, implicating the hippocampal memory system in a range of complex cognitive processes that depend upon binding of relational information (e.g. Zeithamova et al. 2012). Next we review evidence that relational processing theory extends to the binding of social relationships—drawing a direct association between hippocampal memory and social cognition.

Social cognition is just one of several domains in which humans demonstrate active engagement with their environments, through dynamic representation, manipulation, and flexible updating to match action and context (Rubin et al. 2014). Hippocampal memory can be understood as a map, constructed from past experience, that guides our personal actions, as well as our interactions with the social world in the present and future (Rubin et al. 2014; Eichenbaum and Cohen 2014; Wang et al. 2015). In this respect, the hippocampal system performs a crucial role in constructing and navigating a much more complex memory space, one that includes an expansive map of personal and interpersonal experience (Eichenbaum and Cohen 2014). Indeed it has been suggested that the hippocampus builds a "currency" of spatially—or otherwise connected—scenes (Maguire and Mullally 2013). Furthermore, this conceptualization of hippocampal function overlaps significantly with the theory of constructive memory, discussed below.

More recently, research has suggested that social relations may occupy a significant portion of the human 'memory space', positioning the hippocampus as a hub for social navigation. In the social domain the hippocampal memory system would bind and dynamically reconfigure various elements of social relationships such as social distance and hierarchies, social bonds and transgressions. These relational scaffolds or schema are then accessed to guide behavior in social contexts (Zayas et al. 2002). Support for this idea was recently demonstrated in an fMRI investigation of hippocampal functioning and social relatedness (Tavares et al. 2015). In this study, social distance was manipulated along two primary dimensions: power (including competence, dominance, and hierarchy) and affiliation (including warmth, intimacy, trustworthiness, and love). Participants were presented with fictional characters in a virtual role-playing game. Hippocampal activity predicted changes in the interaction of self-reported affiliation and power between the participants and the fictional characters. Results were characterized in terms of vectors through social space along the two social-relationship dimensions (power-ranking and affiliation), with hippocampus activity associated with vector angles, and posterior cingulate cortex (PCC) associated with vector length—i.e. social distance. This geometric representation of social distance is consistent with the relational theory of hippocampal function and represents an extension to the realm of social cognition (Eichenbaum and Cohen 2014; Tavares et al. 2015). Hippocampal activation during social navigation also correlated with individual differences in social skills. Greater activation was associated with reduced avoidance and neuroticism and increased conscientiousness, providing a further link between hippocampal function, social navigation and social capacity in the real world (Tavares et al. 2015). The authors also suggest that deficits in social cognition may be a direct consequence of hippocampal dysfunction. We discuss this further in the section on hippocampal amnesia below.

Constructive Memory and Social Cognition

Constructive memory theory (Schacter 2012) suggests that memories are not veridical presentations of the past but rather reconfigurations of related mnemonic features that are continuously re-shaped by retrieval and re-encoding processes. These same processes are posited to support the constructive nature of imagination, in which features of disparate prior experiences are re-integrated in novel ways such that new, imagined "experiences" can be creatively processed (Schacter 2012). Imagination, however, is not limited to the process of musing on personal pasts, presents, and futures; imagination also shares its inventive, additive nature with how we envision the experiences of other individuals in our social world. In this respect the constructive nature of memory supports social cognition, enabling us to predict social interactions and prepare adaptive responses. The concept of memory construal raises the possibility that how we represent our personal past will influence our actions and thoughts about others. In the next sections we briefly review how constructive memory shapes social cognition by influencing our self-perception, empathy towards others and group social behavior.

Self-Perception

Representations of the self have been conceptualized as a "cognitive filter", through which we see and understand others in our social world (Beer and Ochsner 2006). Individuals draw from remembered experiences and introspective thoughts to infer motivations and affective states of others (e.g. Meltzoff and Brooks 2001; Nickerson 1999). Constructed representations of the self can serve as reference points for characterizing and framing others in terms of similar personality traits or shared preferences. These notions of self serve to anchor perceptions of others' feelings and experiences (Epley et al. 2004). Moreover the influence of self-representation and memory on social cognition is likely reciprocal. There is a deep history of cultural and developmental psychological theory arguing that self-concept is defined and enacted through social settings (e.g. Bem 1972; Sampson 1977; Vygotsky 1978; Markus and Cross 1990).

Empathy

The process of imagining the experiences of social others can facilitate empathy and prosocial behavior (Gaesser 2012). Vivid imagining has been associated with increased prosocial motivation and this relationship appears to be mediated by the hippocampal memory system. Memories of helping others that are recalled in greater detail and more coherently increase prosocial motivation (Gaesser and Schacter 2014). Individual differences in the capacity to vividly recollect past experiences have also been shown to modulate empathic responding (Ciaramelli et al. 2013).

Group Social Behavior

The hippocampal memory system also plays a role in group dynamics. Collective identity can be achieved through the merging of personal memory content (Brown et al. 2012, for review). Collective identity suggests that social group members can form shared memories through their social interactions (Bartlett 1932). This notion of a shared personal past may emerge from common childhood experiences, daily activities or major life experiences. These commonalities promote the construction of shared in-group schemata, leading to collective representations of a personal past. These shared schemata shape how group members remember their personal and group pasts, although group status (in/out) appears to moderate this effect (Lindner et al. 2012).

Thus far in the chapter we have provided a theoretical framework relating hippocampal memory system functioning to social cognition. Relational integration theory suggests that hippocampal-mediated memory processes are necessary to navigate social distance and complex social hierarchies. Constructive memory theory argues that how we retrieve, reconfigure and re-encode our past experiences can influence our imagined social future, influencing our sense of social proximity to the 'other', our capacity for prosocial behavior and our collective memory. Together these theories point to a critical role for memory in imagining our social future and successfully navigating our way there.

Mnemonic Contributions to Social Cognition

In this next section we review the experimental evidence linking hippocampal memory function and social cognition. We will begin with an evolutionary perspective, examining comparative psychological evidence. Next we will review animal and human lesion studies. Experimental neuropsychology in animal models allows for direct, experimental manipulation of the neural regions involved in cognitive processes; however, this methodology is limited in its applicability to human models. Human neuropsychology—particularly, lesion studies—provides evidence for the cognitive and behavioral results of neural abnormalities in humans, but it is more difficult to ascribe cognitive processes to specific brain regions due to poorly-defined lesion boundaries. Finally, we review functional neuroimaging evidence. While these studies provide only correlational data, they enable more precise topographical mapping of cognitive processes in vivo. Further, by simultaneously recording data across the whole brain, functional neuroimaging enables network-level analysis, describing cognitive functions as emergent properties of spatially-distributed, yet functionally connected, brain regions. Here we review insights from each of these methodologies to characterize the role of the hippocampus and functionally-connected brain regions in social cognition.

Comparative Psychology

The capacity to successfully maneuver through our social world is fundamental to human survival. Basic social competency is thought to be fundamental for defining one's sense of self, surviving to mate and raise young and bolstering physical and mental health throughout the lifespan (e.g. Cohen 2004; House et al. 1988; Kiecolt-Glaser and Newton 2001; Vygotsky 1978). This social capacity to represent, reflect upon and anticipate the intentions of others differentiates the human species from our primate relatives (Tomasello 1999). Evolutionary theorists have proposed the "social brain hypothesis" to describe why humans have become comparatively more reliant upon social cognition for survival. This hypothesis suggests that throughout human evolution, an increasing number of social relationships and complexity of social hierarchies was associated with a rapid increase in brain size (Dunbar 1998; Humphrey 1976).

As humans gathered in tribal groupings, social capacity was needed to differentiate oneself from others and to represent, reflect upon and anticipate the intentions of those 'others' to optimize survival. This required tracking complex social dynamics including group sizes, inter-connectedness of members, and dominance hierarchies. Thus each incremental increase in group membership imposed exponentially greater mnemonic demands to encode these relationships and cognitive flexibility to update shifting relationships. According to social brain theorists, this rapid increase in cognitive load was an important factor in human cortical expansion. Evidence for such an association has been observed in non-human primates where affiliation with larger social groups is positively correlated with cortical volume (Dunbar 1998).

Experimental Neuropsychology

Although the enhanced capacity for social cognition in humans may reflect the "social brain hypothesis", other animal models can provide us with information about the specific recruitment of the hippocampus in social processing. In humans and non-human animals alike, the ability to remember different social individuals is essential for the formation of social relationships and groups. For example, social recognition in mice involves the capability to identify and recognize conspecifics. Social recognition in mice appears to be organized in a manner similar to that of other hippocampus-dependent memory capabilities; in one study, hippocampal lesions in mice disrupted social cognition after a 30-min delay (Kogan et al. 2000). Social processing has also been investigated using animal models of Alzheimer's Disease (AD). In one recent study, increasing the social demands by co-housing AD model mice with non-AD animals reversed memory deficits in the AD cohort (Hsiao et al. 2014). These memory gains were attributed to increases in brain-derived neurotrophic factor (BDNF) as well as hippocampal neurogenesis.

Hitti and Siegelbaum (2014) isolated the important role of the CA2 subfield of the hippocampus in social recognition. Using a novel transgenic adult mouse line, the authors reported that selective genetic inactivation of CA2 neurons resulted in the loss of social recognition; however, there was no observed change in other hippocampally-mediated behaviors. Other studies have demonstrated that lesions within CA2 selectively impair social recognition (Leser and Wagner 2015; Stevenson and Caldwell 2014), giving rise to the idea that this hippocampal subfield may be a 'social cognition' area.

Human Neuropsychology: Hippocampal Amnesia and Social Cognition

Human neuropsychological studies have examined the role of memory in social cognition and the underlying hippocampal mechanisms for social cognitive processes. Much of the existing neuropsychological research focuses on how the loss of detailed memories and experiences from one's personal past—due to hippocampal lesions or damage—is correlated with social impairments. As we have discussed earlier hippocampal damage would be expected to disrupt relational processing or memory construal, leading to deficits in social cognition (Eichenbaum and Cohen 2014; Schacter 2012). In one recent study, patients with hippocampal damage showed impairments in episodic recall but preserved ability to generate coherent concepts of self (Kurczek et al. 2015). In contrast, those with medial prefrontal cortex (mPFC) damage were able to reconstruct detailed episodic events but were unable to integrate self concepts into the recollections (Kurczek et al. 2015). In a similar study, hippocampal and bilateral ventromedial PFC (vmPFC) patients were asked to make moral judgments about unfamiliar individuals before and after learning about the individuals' behavior in previous moral situations. Interestingly, the hippocampal damage group showed the most amount of change (shift from good to bad or vice versa) in character judgments from pre-test to post-test. The vmPFC damage group demonstrated the least amount of change in moral character judgments. These findings suggest that the vmPFC is important for encoding affective context, whereas the hippocampus encodes situational context (Croft et al. 2010). Together, these studies suggest that the hippocampus is involved in social functioning but may interact with other brain regions to mediate social cognition in more real-world contexts. We will explore the role of functional brain networks further in the final section of the chapter.

As discussed in the constructive memory section above, the ability to use flexible cognition processes to imagine is closely linked with the social cognitive skills of theory of mind and mentalizing about the thoughts of others. Some research suggests that individuals need to be able to construct and process imagined scenes in order to represent the perspective of an 'other'. Several studies have documented deficits in the construction of imagined events following bilateral hippocampal damage (Andelman et al. 2010; Hassabis et al. 2007; Lee et al. 2012; Mullally et al. 2012; Race et al. 2011; Rosenbaum et al. 2009; Tulving 1985). Hassabis et al. (2007) tested a group of patients with hippocampal damage on an imagination-related cognitive task. These patients showed marked impairments in their abilities to create imagined, novel stories based on short verbal prompts. Furthermore, the imagined experiences lacked coherency with respect to the spatial or environmental setting, compared with healthy control subjects. The authors posited that this spatial fragmentation of imagined components was attributable to missing input from the hippocampal memory system. Within this group of patients, their lack of hippocampal function was most evident in the functional losses of the ability to bind together disparate aspects of experiences (real or imagined), a critical capacity for social cognition (see section "Constructing Complex Mental Representations"). Individuals may need to be able to construct and process imagined scenes in order to represent the perspective of an 'other', although this has not been tested directly. Further, it appears this relationship may be moderated by personal familiarity. While episodic scene reconstruction may be necessary to take the perspective of a familiar other, this does not appear to be the case for unknown others (Rabin and Rosenbaum 2012).

Hippocampal patients are also more likely to have poor social functioning in real-world contexts, with few strong social bonds and smaller social network size (Davidson et al. 2012; Warren et al. 2012). Patients reported making very few close friends and being less involved with neighborhood, religious, and community groups. Deficits in the ability to use hippocampal memory representations in the processes of encoding, updating or retrieving models of social others is a significant contributor in these patients' struggles to develop and maintain close social connections (Davidson et al. 2012).

Davidson et al. (2012) examined the close relationships of three amnesic individuals. The patients in this study showed less involvement in community groups than their demographically-matched control subjects. Two patients with adult-onset hippocampal amnesia had made very few new friends since their injuries. In contrast, the third patient, with developmental amnesia, had fostered several close relationships over the time of the study. The authors concluded that social network size and social bonding is impaired in acquired hippocampal amnesia. However, Warren et al. (2012), demonstrated positive social outcomes in a case of severe hippocampal amnesia. These were attributed to the strength of the existing social networks (husband, extended family), which relieved the patient of many functional responsibilities, enabling her to focus on maintaining or expanding her social relationships. Additionally, Duff et al. (2008) reported on an amnesic patient who was successful in forming new close social relationships, despite her memory impairment. This stands as a second counterexample to the finding that those with hippocampal amnesia generally have great difficulty with everyday social tasks such as learning new names, consciously remembering sharing experiences with others, and updating mental representations of existing social relationships (for review, see Rubin et al. 2014). Clearly, more work is necessary to examine the social ramifications of amnesia. In healthy adults, the range of individual differences in memory ability predicts social network size (Stiller and Dunbar 2007), suggesting an augmenting function.

Hippocampal amnesia patients also demonstrate deficits in trait and state empathy (Beadle et al. 2013): to imagine the life events of unfamiliar others (Rabin and Rosenbaum 2012) and to make complex social judgments (Staniloiu et al. 2013). This last study provides an interesting perspective on social cognitive deficits in amnesia. They reported that their developmental amnesia patient was impaired on complex social judgment task but not on empathy or theory of mind tasks. The authors suggested that their findings implicated the hippocampus in more complex relational integration processes.

Duff and colleagues (2013b) studied several female patients with early stage AD and their interactions with familiar conversation partners. Somewhat surprisingly, the patients displayed significant learning on a cognitive task when paired with communication partners. The authors argued that the social interactions likely recruited neural resources outside of the medial temporal regions to support non-hippocampally-mediated learning (Duff et al. 2013b). These findings demonstrate how differences in social cognitive task demands relate to the recruitment of differential functional networks in the brain. The hippocampal memory system is not the only region involved in social processing, and this study raises important questions about the integration of various memory processes.

Functional Neuroimaging

The field of social cognitive neuroscience has undergone almost exponential growth over the last decade. In this section we limit our review to studies that specifically investigate the role of the hippocampal memory system in social cognition. Specifically we review two common paradigms: (1) perception of self versus others and (2) recognition of social cues.

Perception of Self vs. Other

The hippocampus has been implicated in many processes which support differentiation of self versus other perceptions: recalling and reconstructing personal past events and imagining potential personal futures (e.g., Addis et al. 2007; Gaesser et al. 2013; Hassabis et al. 2007; Okuda et al. 2003; Race et al. 2011; Szpunar et al. 2007), scene construction (e.g., Hassabis and Maguire 2009; Mullally and Maguire 2014), creative thinking and imagination (Duff et al. 2013a). Other kinds of relational processes help to further differentiate amongst others without necessarily referencing the self: for example, combining information about the relationships between objects across time (e.g., Davachi 2006; Duff et al. 2007; Konkel and Cohen 2009; Ranganath 2010).

Brain regions recruited in self- and other- perception are distinctive (for review, see Beer and Ochsner 2006). Notions of the self become more semantic over time as experiences are relatedly accessed and re-encoded. However, judgments about non-close others are more dependent on episodic recollection (e.g. Klein et al. 1999; Klein et al. 1996), suggesting that the hippocampal memory system is implicated in 'other' more than 'self' perception. In a similar finding, Ochsner et al. (2005) looked at patterns of functional activation associated with reflecting on a close other individual's opinions about oneself or reflecting on one's own opinions of oneself. They reported that reflections on close others' judgments, but not those of the self, were associated with activation of the hippocampal memory system, suggesting that this system is engaged both by perception and judgment of the social 'other'. Yet, the existing literature on self and other referencing contains some inconsistencies; as mentioned earlier in this chapter, Kurczek et al. (2015) reported that hippocampal amnesia patients did not show a significant difference from healthy controls on a measure of self-referential processing. This raises the possibility that social-perception and social referencing may be discrete processes. While this question is beyond the scope of this review, it remains an important question for future research.

Rabin and Rosenbaum (2012) reported that the hippocampus is involved in theory of mind for familiar but not unfamiliar others. Perry et al. (2011) also demonstrated a differentiated role of the hippocampus based on the nature of other-oriented thought. In this study, subjects selected individuals who were similar versus dissimilar from themselves from a pool of varied protagonists. During scanning subjects were asked to imagine how themselves or their selected protagonists would feel in certain situations (for instance, losing a wallet). After scanning, subjects were led through interviews about their autobiographical memories. Specifically, the interviewers asked subjects if they remembered whether or not each event that had been presented in the scanning session had ever personally happened to them before. They further divided the subsequent data into "remembered events" and "not remembered events". Results indicated that magnitude of the hippocampal activity was highest in the self condition, second highest in the similar other condition, and lowest in the dissimilar other condition. Additionally, there was a significant correlation between the self ratings and the similar-other ratings for "remembered events" but not for "not remembered events". This result demonstrates that personal episodic memory was recruited when subjects judged the protagonists' emotional states. The authors concluded that the hippocampus was involved in subjects' emotional judgments about the self and similar others.

Recognition of Social Cues

The ability to adapt our behavior based on dynamic social cues, such as changing facial expressions, relies heavily on working memory (Gobbini and Haxby 2007). We need to be able to distinguish between cognitive representations of different individuals' faces and those of different expressions from the same individual. Ross et al. (2013) examined neural activity in regions contributing to the processes of encoding, maintaining, and retrieving overlapping facial expression—here, two different affective expressions by the same or another individual. They utilized a match-to-sample task, contrasting conditions of overlap (two faces from the same individual, with different expressions) and non-overlap (two faces from different individuals, with different expressions). The authors found that, whereas lateral orbitofrontal cortex contributes to encoding and maintaining mental representations of overlapping stimuli, the hippocampus was engaged during retrieval. This suggests that retrieval of overlapping social percepts, as is likely required to differentiate facial expressions, is hippocampally dependent.

Taken together, the studies reviewed here provide converging evidence of mnemonic contributions to social cognition. From an evolutionary perspective, with increasing hippocampal volume linked to increasing social network size, to animal and human neuropsychological studies providing more causal evidence linking memory and social cognition and finally to functional neuroimaging studies providing a more precise topographical mapping of social cognition and hippocampal activation. These lines of evidence suggest a critical role for memory in navigating our complex and constantly shifting social milieu. In the final section of the chapter we will examine the contribution of memory to social cognition from a network neuroscience perspective. Specifically we will examine the role of the default network—a collection of brain regions functionally connected to the hippocampus that have been implicated in social cognition.

Social Cogntion and the Default Network

The recent discovery of a common functional anatomy for autobiographical memory (recalling personally experienced events) and theory of mind (inferring the mental states of others) suggested that memory and social cognitive processes share a common functional architecture that extends beyond the hippocampal memory system (Buckner and Carroll 2007; Spreng and Grady 2010; Rabin et al. 2010). This common functional architecture overlapped almost completely with a collection of functionally connected brain regions referred to as the default network (Spreng et al. 2009). Core brain areas within the default network include the medial temporal lobes, mPFC, lateral prefrontal cortex, lateral temporal cortices, PCC, and lateral parietal cortices (Addis et al. 2007; Buckner and Carroll 2007; Spreng et al. 2009). In this final section of the chapter we first describe how core nodes of the default network have been directly implicated in social processing. Then we review how default network and hippocampal memory systems interact with other brain regions and other functional networks during social cognition.

Default Network Brain Regions and Social Cognition

The default network has been implicated in processes including recollection and future thinking (Schacter 2012), autobiographical planning (Spreng et al. 2015) and mind-wandering (Fox et al. 2015). More recently the role of the default network has been associated with several aspects of social cognition. It has been implicated in the integration of personal and interpersonal information. Personal experiences are used to generate social conceptual knowledge, which in turn, leads to the development and implementation of strategic social behavior (Spreng 2013, for review; see also Spreng and Mar 2012). The integrity of vmPFC, a core node of the default network, predicts ability to retrieve impressions of others (Cassidy and Gutchess 2012). Attributional decisions and judgments of others' emotional states recruited areas of the default network, such as vmPFC, in a recent study (Haas et al. 2015). The default network also enables us to imagine the experiences of others. Hassabis et al. (2013) taught participants the personalities (based on the two dimensions of agreeableness and extraversion) of four characters. They then imagined their behavior across different situations. Results showed that activity in the mPFC reliably predicted which characters participants were imagining.

Furthermore, other core regions and subsystems of the default network have been specifically linked with social cognitive processes (for review, see Spreng and Andrews-Hanna 2015). Saxe (2010) found that activity in the right temporoparietal junction is associated with reflecting upon other individuals' beliefs. Inferior frontal and lateral temporal regions also show activation during social tasks, and have been specifically implicated in the semantic aspects of mentalizing (e.g., Binder and Desai 2011). Others still show that the PCC is active across a wide variety of self-

and other- related cognitive processes including self-referential processing, familiarity representation and theory of mind (Binder et al. 2009; Brewer et al. 2013; Qin and Northoff 2011; Spreng et al. 2009).

Default Network Functional Connectivity and Social Cognition

Throughout the chapter we have implicated the hippocampal memory system, a component of the default network, in social cognition. In the previous section we reviewed how specific nodes of the default network were implicated in various aspects of social cognition. Here we review how functional connectivity within the default network, and specifically between hippocampal memory systems and other default brain regions, supports social cognitive processing. Increased functional connectivity between hippocampal regions of interest and default network nodes indicates correlated neural activation that is associated with social-cognitive processes.

In a recent meta-analysis of functional neuroimaging studies of perspective-taking (Bzdok et al. 2013), ventral mPFC was robustly functionally connected with the hippocampal component of the default network. Functional connectivity between these regions was associated with reward-associations and evaluation-related processes. In contrast, dorsal mPFC showed greater functional connectivity to inferior frontal gyrus, temporal-parietal junction, and middle temporal gyrus regions. Functional connectivity within this aspect of the default network was associated with perspective-taking and episodic memory retrieval. These findings were convergent with a recent review which suggested that functional connectivity between the hippocampus and mPFC was important for future thinking and imagination (Buckner 2010). A study by Perry et al. (2011) also observed that the mPFC and the PCC are crucial for processing self-relevant information. The PCC plays an additional role in encoding information about others, while the hippocampus is engaged by internal mentation about oneself and differentiating self- from other-focused experiences. Specifically the hippocampus—in conjunction with the broader default network—served to mediate judgments of self versus others with respect to events in memory (Perry et al. 2011). These results highlight the importance of hippocampal interactions with other default network regions in mediating social cognition.

Conclusion

In this chapter we provided a broad overview of the research literature ascribing a role for the hippocampal memory system in social cognition. The hippocampus is highly attuned to social information, which is inherently composed of distinct components requiring relational binding. We suggest that social cognition is a form of mental processing that places high demands on the type of processing supported by the hippocampus. Currently however, there is insufficient evidence to suggest that the hippocampus plays a unique role for social cognition. We began the review by describing four areas in which memory and social cognition overlap in everyday functioning. Next we examined two theories of hippocampal memory, relational integration and constructive memory, and described how these memory theories readily extend to the domain of social cognition. In the third section we briefly surveyed the research literature investigating the association between memory and social cognition, reviewing results from evolutionary psychology, experimental and human neuropsychology, and functional neuroimaging. Finally we examined how the hippocampal memory system, working in concert with default network brain regions, was involved in social cognition.

Memory and social cognition are complex cognitive functions, each encompassing different processes and engaging numerous brain regions. That these complex functions interact or overlap at the level of the brain is perhaps not surprising. What is surprising is the extent of the overlap. Do they share a common psychological and neural architecture? Is social cognition simply a projection of personal memory and experience onto an external 'other'? Or might the social content engage different cognitive processes and brain regions? These remain active questions of research. Our ability to step outside ourselves, to appreciate, understand, predict and adapt to the thoughts, intentions and actions of others makes us truly human. Understanding how our store of experience and memory influences our perceptions of and engagement with the 'other' will only become more important in our increasingly interconnected world.

References

Addis DR, Wong AT, Schacter DL (2007) Remembering the past and imagining the future: common and distinct neural substrates during event construction and elaboration. Neuropsychologia 45(7):1363–1377

Andelman F, Hoofien D, Goldberg I, Aizenstein O, Neufeld MY (2010) Bilateral hippocampal lesion and a selective impairment of the ability for mental time travel. Neurocase 16 (5):426–435

Andersen SM, Cole SW (1990) "Do I know you?": The role of significant others in general social perception. J Pers Soc Psychol 59:384–399. doi:10.1037/0022-3514.59.3.384

Aron A, Aron EN (1986) Love and the expansion of self: understanding attraction and satisfaction. Hemisphere/Harper & Row, New York

Bartlett F (1932) Remembering: a study in experimental and social psychology. Cambridge University Press, New York

Beadle JN, Tranel D, Cohen NJ, Duff MC (2013) Empathy in hippocampal amnesia. Front Psychol 4:69. doi:10.3389/fpsyg.2013.00069

Beer JS, Ochsner KN (2006) Social cognition: a multi level analysis. Brain Res 1079(1):98–105

Bem DJ (1972) Constructing cross-situational consistencies in behavior: some thoughts on Alker's critique of Mischel. J Pers 40(1):17–26

Binder JR, Desai RH (2011) The neurobiology of semantic memory. Trends Cogn Sci 15:527–536

Binder JR, Desai RH, Graves WW, Conant LL (2009) Where is the semantic system? A critical review and meta-analysis of 120 functional neuroimaging studies. Cereb Cortex 19:2767–2796

Bowlby J (1969) Attachment and loss: attachment (vol. 1). Basic Books, New York

Brewer JA, Garrison KA, Whitfield-Gabrieli S (2013) What about the "self" is processed in the posterior cingulate cortex? Front Hum Neurosci 7:647

Brown AD, Kouri N, Hirst W (2012) Memory's malleability: its role in shaping collective memory and social identity. Front Psychol 3:257. doi:10.3389/fpsyg.2012.00257

Buckner RL (2010) The role of the hippocampus in prediction and imagination. Annu Rev Psychol 61:27–48

Buckner RL, Carroll DC (2007) Self-projection and the brain. Trends Cogn Sci 11:49–57. doi:10. 1016/j.tics.2006.11.004

Bzdok D, Langner R, Schilbach L, Engemann DA, Laird AR, Fox PT, Eickhoff SB (2013) Segregation of the human medial prefrontal cortex in social cognition. Front Hum Neurosci 7:232

Carlston D (2010) Models of implicit and explicit mental representation. In: Gawronski B, Payne KB (eds) Handbook of implicit social cognition: measurement, theory, and applications. Guilford Press, New York, pp 38–61

Carlston DE, Smith ER (1996) Principles of mental representation. In: Higgins ET, Kruglanski AW (eds) Social psychology: handbook of basic principles. Guilford, New York, pp 184–210

Cassidy BS, Gutchess AH (2012) Structural variation within the amygdala and ventromedial prefrontal cortex predicts memory for impressions in older adults. Front Psychol 3:319. doi:10.3389/fpsyg.2012.00319

Ciaramelli E, Bernardi F, Moscovitch M (2013) Individualized theory of mind (iToM): when memory modulates empathy. Front Psychol 4:4. doi:10.3389/fpsyg.2013.00004

Cohen S (2004) Social relationships and health. Am Psychol 59(8):676

Collins NL, Read SJ (1994) Cognitive representations of adult attachment: the structure and function of working models. In: Bartholomew K, Perlman D (eds) Advances in personal relationships: vol. 5. Attachment processes in adulthood. Jessica Kingsley, London, pp 53–90

Croft KE, Duff MC, Kovach CK, Anderson SW, Adolphs R, Tranel D (2010) Detestable or marvelous? Neuroanatomical correlates of character judgments. Neuropsychologia 48 (6):1789–1801

Davachi L (2006) Item, context and relational episodic encoding in humans. Curr Opin Neurobiol 16(6):693–700

Davidson PS, Drouin H, Kwan D, Moscovitch M, Rosenbaum RS (2012) Memory as social glue: close interpersonal relationships in amnesic patients. Front Psychol 3:531. doi:10.3389/fpsyg. 2012.00531

Duff MC, Hengst JA, Tranel D, Cohen NJ (2007) Talking across time: using reported speech as a communicative resource in amnesia. Aphasiology 21(6–8):702–716

Duff MC, Wszalek TW, Tranel D, Cohen NJ (2008) Successful life outcome and management of real-world memory demands despite profound anterograde amnesia. J Clin Exp Neuropsychol 30:931–945

Duff MC, Kurczek J, Rubin R, Cohen NJ, Tranel D (2013a) Hippocampal amnesia disrupts creative thinking. Hippocampus 23(12):1143–1149

Duff MC, Gallegos D, Cohen NJ, Tranel D (2013b) Learning in Alzheimer's disease is facilitated by social interaction and common ground. J Comp Neurol 521(18):4356–4369

Dunbar R (1998) The social brain hypothesis. Evol Anthropol 6:178–190

Eichenbaum H (2004) Hippocampus: cognitive processes and neural representations that underlie declarative memory. Neuron 44(1):109–120

Eichenbaum H, Cohen NJ (2014) Can we reconcile the declarative memory and spatial navigation views on hippocampal function? Neuron 83(4):764–770

Eichenbaum H, Schoenbaum G, Young B, Bunsey M (1996) Functional organization of the hippocampal memory system. Proc Natl Acad Sci U S A 93(24):13500–13507

Eichenbaum H, Dudchenko P, Wood E, Shapiro M, Tanila H (1999) The hippocampus, memory, and place cells: is it spatial memory or a memory space? Neuron 23(2):209–226

Epley N, Keysar B, Van Boven L, Gilovich T (2004) Perspective taking as egocentric anchoring and adjustment. J Pers Soc Psychol 87(3):327

Fitzsimons GM, Bargh JA (2003) Thinking of you: nonconscious pursuit of interpersonal goals associated with relationship partners. J Pers Soc Psychol 84:148–164

Fox KCR, Spreng RN, Ellamil M, Andrews-Hanna JR, Christoff K (2015) The wandering brain: meta-analysis of functional neuroimaging studies of mind-wandering and related spontaneous thought processes. NeuroImage 111:611–621

Freeman JB, Stolier RM, Ingbretsen ZA, Hehman EA (2014) Amygdala responsivity to high-level social information from unseen faces. J Neurosci 34(32):10573–10581

Gaesser B (2012) Constructing memory, imagination, and empathy: a cognitive neuroscience perspective. Front Psychol 3:576. doi:10.3389/fpsyg.2012.00576

Gaesser B, Schacter DL (2014) Episodic simulation and episodic memory can increase intentions to help others. Proc Natl Acad Sci U S A 111(12):4415–4420

Gaesser B, Spreng RN, McLelland VC, Addis DR, Schacter DL (2013) Imagining the future: evidence for a hippocampal contribution to constructive processing. Hippocampus 23(12):1150–1161

Gobbini MI, Haxby JV (2007) Neural systems for recognition of familiar faces. Neuropsychologia 45(1):32–41

Günaydin G, Zayas V, Selcuk E, Hazan C (2012) I like you but I don't know why: objective facial resemblance to significant others influences snap judgments. J Exp Soc Psychol 48:350–353

Haas BW, Anderson IW, Filkowski MM (2015) Interpersonal reactivity and the attribution of emotional reactions. Emotion 15(3):390

Hassabis D, Maguire EA (2009) The construction system of the brain. Philos Trans R Soc Lond B Biol Sci 364(1521):1263–1271

Hassabis D, Kumaran D, Maguire EA (2007) Using imagination to understand the neural basis of episodic memory. J Neurosci 27(52):14365–14374

Hassabis D, Spreng RN, Rusu AA, Robbins CA, Mar RA, Schacter DL (2013) Imagine all the people: how the brain creates and uses personality models to predict behavior. Cereb Cortex. doi:10.1093/cercor/bht042

Hitti FL, Siegelbaum SA (2014) The hippocampal CA2 region is essential for social memory. Nature 508(7494):88–92

Holmes JG (2002) Interpersonal expectations as the building blocks of social cognition: an interdependence theory perspective. Pers Relat 9:1–26

House JS, Landis KR, Umberson D (1988) Social relationships and health. Science 241:540–545. doi:10.1126/science.3399889

Hsiao YH, Hung HC, Chen SH, Gean PW (2014) Social interaction rescues memory deficit in an animal model of Alzheimer's disease by increasing BDNF-dependent hippocampal neurogenesis. J Neurosci 34(49):16207–16219

Humphrey NK (1976) The social function of intellect. Cambridge University Press, Cambridge

Kiecolt-Glaser JK, Newton TL (2001) Marriage and health: his and hers. Psychol Bull 127(4):472

Klein SB, Sherman JW, Loftus J (1996) The role of episodic and semantic memory in the development of trait self-knowledge. Soc Cogn 14(4):277–291

Klein SB, Chan RL, Loftus J (1999) Independence of episodic and semantic self-knowledge: the case from autism. Soc Cogn 17(4):413–436

Kogan JH, Frankland PW, Silva AJ (2000) Long-term memory underlying hippocampus-dependent social recognition in mice. Hippocampus 10(1):47–56

Konkel A, Cohen NJ (2009) Relational memory and the hippocampus: representations and methods. Front Neurosci 3(2):166

Kumaran D, Banino A, Blundell C, Hassabis D, Dayan P (2016) Computations underlying social hierarchy learning: distinct neural mechanisms for updating and representing self-relevant information. Neuron 92:1135–1147

Kurczek J, Wechsler E, Ahuja S, Jensen U, Cohen NJ, Tranel D, Duff M (2015) Differential contributions of hippocampus and medial prefrontal cortex to self-projection and self-referential processing. Neuropsychologia 73:116–126

Lee AC, Yeung LK, Barense MD (2012) The hippocampus and visual perception. Front Hum Neurosci 6:91

Leser N, Wagner S (2015) The effects of acute social isolation on long-term social recognition memory. Neurobiol Learn Mem 124:97–103

Lindner I, Schain C, Kopietz R, Echterhoff G (2012) When do we confuse self and other in action memory? Reduced false memories of self-performance after observing actions by an out-group vs. ingroup actor. Front Psychol 3:467. doi:10.3389/fpsyg.2012.00467

Maguire EA, Mullally SL (2013) The hippocampus: a manifesto for change. J Exp Psychol Gen 142(4):1180

Markus H, Cross S (1990) The interpersonal self. In: Pervin LA (ed) Handbook of personality: theory and research. Guilford Press, New York, pp 576–608

Meltzoff AN, Brooks R (2001) "Like me" as a building block for understanding other minds: bodily acts, attention, and intention. In: Malle BF, Moses LJ, Baldwin DA (eds) Intentions and intentionality: foundations of social cognition. MIT Press, Cambridge, MA, pp 171–191

Mullally SL, Maguire EA (2014) Learning to remember: the early ontogeny of episodic memory. Dev Cogn Neurosci 9:12–29

Mullally SL, Intraub H, Maguire EA (2012) Attenuated boundary extension produces a paradoxical memory advantage in amnesic patients. Curr Biol 22(4):261–268

Muscatell KA, Morelli SA, Falk EB, Way BM, Pfeifer JH, Galinsky AD et al (2012) Social status modulates neural activity in the mentalizing network. NeuroImage 60:1771–1777

Nickerson RS (1999) How we know—and sometimes misjudge—what others know: imputing one's own knowledge to others. Psychol Bull 125:737

O'keefe J, Nadel L (1978) The hippocampus as a cognitive map, vol 3. Clarendon Press, Oxford, pp. 483–484

Ochsner KN, Beer JS, Robertson ER, Cooper JC, Gabrieli JD, Kihsltrom JF, D'Esposito M (2005) The neural correlates of direct and reflected self-knowledge. NeuroImage 28(4):797–814

Okuda J, Fujii T, Ohtake H, Tsukiura T, Tanji K, Suzuki K, Yamadori A (2003) Thinking of the future and past: the roles of the frontal pole and the medial temporal lobes. NeuroImage 19 (4):1369–1380

Parkinson C, Liu S, Wheatley T (2014) A common cortical metric for spatial, temporal, and social distance. J Neurosci 34:1979–1987

Perry D, Hendler T, Shamay-Tsoory SG (2011) Projecting memories: the role of the hippocampus in emotional mentalizing. NeuroImage 54:1669–1676. doi:10.1016/j.neuroimage.2010.08.057

Pietromonaco PR, Feldman Barrett L (2000) The internal working models concept: what do we really know about the self in relation to others? Rev Gen Psychol 4:155–175. doi:10.1037/1089-2680.4.2.155

Qin P, Northoff G (2011) How is our self related to midline regions and the default-mode network? NeuroImage 57:1221–1233

Rabin JS, Rosenbaum RS (2012) Familiarity modulates the functional relationship between theory of mind and autobiographical memory. NeuroImage 62:520–529. doi:10.1016/j.neuroimage.2012.05.002

Rabin JS, Gilboa A, Stuss DT, Mar RA, Rosenbaum RS (2010) Common and unique neural correlates of autobiographical memory and theory of mind. J Cogn Neurosci 22:1095–1111. doi:10.1162/jocn.2009.21344

Race E, Keane MM, Verfaellie M (2011) Medial temporal lobe damage causes deficits in episodic memory and episodic future thinking not attributable to deficits in narrative construction. J Neurosci 31(28):10262–10269

Ranganath C (2010) A unified framework for the functional organization of the medial temporal lobes and the phenomenology of episodic memory. Hippocampus 20(11):1263–1290

Rosenbaum RS, Gilboa A, Levine B, Winocur G, Moscovitch M (2009) Amnesia as an impairment of detail generation and binding: evidence from personal, fictional, and semantic narratives in KC. Neuropsychologia 47(11):2181–2187

Ross RS, LoPresti ML, Schon K, Stern CE (2013) Role of the hippocampus and orbitofrontal cortex during the disambiguation of social cues in working memory. Cogn Affect Behav Neurosci 13(4):900–915

Rubin RD, Watson PD, Duff MC, Cohen NJ (2014) The role of the hippocampus in flexible cognition and social behavior. Front Hum Neurosci 8

Sampson EE (1977) Psychology and the American ideal. J Pers Soc Psychol 35(11):767

Saxe R (2010) The right temporo-parietal junction: a specific brain region for thinking about thoughts. In: Leslie A, German T (eds) Handbook of theory of mind. Erlbaum, Hillsdale, NJ, pp 1–35

Schacter DL (2012) Adaptive constructive processes and the future of memory. Am Psychol 67:603–613. doi:10.1037/a0029869

Spreng RN (2013) Examining the role of memory in social cognition. Front Psychol 4:437

Spreng RN, Andrews-Hanna JR (2015) The default network and social cognition. In: Toga AW (ed) Brain mapping: an encyclopedic reference. Academic Press, Elsevier, pp 165–169

Spreng RN, Grady C (2010) Patterns of brain activity supporting autobiographical memory, prospection and theory-of-mind and their relationship to the default mode network. J Cogn Neurosci 22:1112–1123. doi:10.1162/jocn.2009.21282

Spreng RN, Mar RA (2012) I remember you: a role for memory in social cognition and the functional neuroanatomy of their interaction. Brain Res 1428:43–50. doi:10.1016/j.brainres. 2010.12.024

Spreng RN, Mar RA, Kim AS (2009) The common neural basis of autobiographical memory, prospection, navigation, theory of mind, and the default mode: a quantitative meta-analysis. J Cogn Neurosci 21:489–510. doi:10.1162/jocn.2008.21029

Spreng RN, Gerlach KD, Turner GR, Schacter DL (2015) Autobiographical planning and the brain: activation and its modulation by qualitative features. J Cogn Neurosci. doi:10.1162/jocn_a_00846

Staniloiu A, Woermann F, Borsutzky S, Markowitsch HJ (2013) Social cognition in a case of amnesia with neurodevelopmental mechanisms. Front Psychol 4:342. doi:10.3389/fpsyg.2013. 00342

Stevenson EL, Caldwell HK (2014) Lesions to the CA2 region of the hippocampus impair social memory in mice. Eur J Neurosci 40(9):3294–3301

Stiller J, Dunbar RI (2007) Perspective-taking and memory capacity predict social network size. Soc Networks 29:93–104

Szpunar KK, Watson JM, McDermott KB (2007) Neural substrates of envisioning the future. Proc Natl Acad Sci U S A 104(2):642–647

Tamir DI, Mitchell JP (2011) The default network distinguishes construals of proximal versus distal events. J Cogn Neurosci 23:2945–2955

Tavares RM, Mendelsohn A, Grossman Y, Williams CH, Shapiro M, Trope Y, Schiller D (2015) A map for social navigation in the human brain. Neuron 87(1):231–243

Tomasello M (1999) The human adaptation for culture. Annu Rev Anthropol 28:509–529

Tulving E (1985) Elements of episodic memory. Oxford University Press, London

Vygotsky L (1978) Interaction between learning and development. Read Dev Child 23(3):34–41

Wang JX, Cohen NJ, Voss JL (2015) Covert rapid action-memory simulation (CRAMS): a hypothesis of hippocampal–prefrontal interactions for adaptive behavior. Neurobiol Learn Mem 117:22–33

Warren DE, Duff MC, Magnotta V, Capizzano AA, Cassell MD, Tranel D (2012) Long-term neuropsychological, neuroanatomical, and life outcome in hippocampal amnesia. Clin Neuropsychol 26(2):335–369

Zayas V, Shoda Y (2005) Do automatic reactions elicited by thoughts of romantic partner, mother, and self relate to adult romantic attachment? Personal Soc Psychol Bull 31:1011–1025

Zayas V, Shoda Y, Ayduk ON (2002) Personality in context: an interpersonal systems perspective. J Pers 70:851–898

Zeithamova D, Dominick AL, Preston AR (2012) Hippocampal and ventral medial prefrontal activation during retrieval-mediated learning supports novel inference. Neuron 75(1):168–179

Dynamic Cortico-hippocampal Networks Underlying Memory and Cognition: The PMAT Framework

Marika C. Inhoff and Charan Ranganath

Abstract Models seeking to explain the neural basis of memory have long focused on the individual roles of the hippocampus and medial temporal lobe (MTL) cortex. Many such models argue that MTL areas form a memory system that is anatomically and functionally separate from the surrounding neocortex. In this review, we critically assess this idea in light of empirical evidence from neuroanatomy, neurophysiology, neuropsychology, and neuroimaging. In each case, the evidence suggests that neocortical regions in the MTL can be more accurately depicted as subcomponents of two functionally and anatomically distinct, large-scale networks—a posterior medial (PM) and an anterior-temporal (AT) system—that extend beyond the MTL. According to our "PMAT" framework, the PM and AT networks are not sensory processing streams, nor are they dedicated memory systems. Instead, they are networks that contribute to a wide range of cognitive tasks, including episodic and semantic memory, perception, language, navigation, and reasoning. We argue that the PMAT framework is an important advance over memory systems theories in that it can explain a larger breadth of phenomena and provides a larger number of predictions and testable hypotheses.

Introduction

> *"In the cortex, as in the rest of the brain, there are no 'systems of memory,' but there is the memory of systems. All cortical systems have their own memory, which is inextricable from the operations they perform. The substrate for process is inseparable from the substrate for representation."* (Fuster 2009, p. 2063).

M.C. Inhoff (✉)
Department of Psychology, University of California at Davis, Davis, CA 95616, USA
e-mail: inhoff@ucdavis.edu

C. Ranganath (✉)
Department of Psychology, University of California at Davis, Davis, CA 95616, USA

Center for Neuroscience, University of California at Davis, 1544 Newton Court, Davis, CA 95618, USA
e-mail: cranganath@ucdavis.edu

© Springer International Publishing AG 2017   559
D.E. Hannula, M.C. Duff (eds.), *The Hippocampus from Cells to Systems*,
DOI 10.1007/978-3-319-50406-3_18

In the 1980s, neuropsychologists and neurophysiologists generally embraced the idea that the brain has multiple "memory systems", and that the "Medial Temporal Lobe Memory System" (MTLMS) is dedicated to the formation and consolidation of declarative and/or episodic memory. At present, the MTLMS framework has transitioned from a concept to a basic principle that is taught in contemporary psychology and neuroscience textbooks. Although the MTLMS and related frameworks have been of tremendous value, it is reasonable to revisit these ideas and assess how well they align with recent discoveries about neuroanatomy, functional brain organization, and human memory processes.

In this chapter, we will briefly review the development of memory systems frameworks to explain MTL involvement in memory. As we describe in this chapter, models that propose a distinct MTLMS have limited explanatory power and they do not capture what is known about the anatomical and functional organization of the brain. Instead, a significant body of empirical data can be accounted for by a framework in which the hippocampus interfaces with, and coordinates between, a posterior-medial (PM), and an anterior temporal (AT) - cortico-hippocampal network. Experimental evidence from neuroanatomy, neuropsychology and functional imaging demonstrates that these networks are separable in terms of connectivity, function, and involvement in neurological disorders. We conclude by presenting new ways PMAT can guide our understanding of important topics in memory research, predictions about cortico-hippocampal interactions in memory, and open questions that need to be addressed in future studies.

A Brief History of Hippocampo-centric Memory Systems

Scoville and Milner's (1957) seminal paper on patient H.M. (and other patients) established the idea that bilateral damage to the hippocampus can cause severe memory impairments. In that paper, however, Milner remained agnostic about the relationship between H.M.'s amnesia and the underlying anatomy. Milner's caution was justified by later work showing that H.M. primarily had damage to the head, but the not tail, of the hippocampus (Corkin et al. 1997). Outside of the hippocampus, H.M. had bilateral lesions to the amygdala and anterior temporal neocortex, along with massive white matter damage in the uncinate fasciculus (Annese et al. 2014; Corkin et al. 1997). To account for the possibility that memory could be supported by brain areas extending beyond the hippocampus, early theories described the hippocampus as part of a broader "memory system". Frameworks that predated work with H.M., such as the Papez Circuit of Emotion (1937) and the Limbic System (MacLean 1955), also proposed that the hippocampus was part of a distributed neural circuit. However, the Papez circuit and limbic system were distributed networks that were associated with emotional processing, unlike newer frameworks that assigned a central role for the hippocampus in memory.

Initially, memory systems frameworks served as simple heuristics for differentiating between functions that were dependent on the hippocampus and functions that were not dependent on the hippocampus. For instance, O'Keefe and Nadel (1979) proposed that the hippocampus was the center of a "locale" system that represented maps of physical and semantic space, unlike the hippocampally-independent "taxon" system, which supported slow learning of simple behavioral responses. Other frameworks suggested that the hippocampus was the center of a system that supported contextual memory (Hirsh 1974), working memory (Olton et al. 1979), configural learning (Sutherland and Rudy 1989), or declarative memory (Cohen and Squire 1980).

Results from lesion studies enabled researchers to develop more anatomically-specific memory systems frameworks. Mishkin (1982), drawing from his seminal research on visual recognition memory in monkeys, proposed a hierarchical model that characterized the transition from visual perception to recognition memory. According to this model, high level cortical areas, including area TE, the hippocampus, and amygdala, are privileged regions for memory that exist at the top of a hierarchy of earlier visual areas. Building on this idea of a hierarchically-organized memory system, Squire and Zola-Morgan (1991) proposed a framework in which the hippocampus was situated at the apex of a system that integrates information from neocortical MTL regions, including perirhinal cortex (PRC) and parahippocampal cortex (PHC). The MTLMS framework assumes a privileged role for both the hippocampus and neocortical MTL areas in memory, although the specific functions of MTL cortical regions are not specified (Wixted and Squire 2011).

The MTLMS framework differed from Mishkin's model in several important ways. First, the MTLMS incorporated the entorhinal cortex (EC), PRC, and PHC and it excluded the amygdala and inferior temporal cortex. These aspects of the MTLMS were based on lesion studies in monkeys, which revealed severe recognition memory deficits only when lesions included the PRC and/or PHC. Squire and Zola Morgan (1991) also emphasized that, "the ability to acquire new memories is a distinct cerebral function, separable from other perceptual and cognitive abilities" (p. 1380). Accordingly, regions of the MTLMS were proposed to specifically support the formation and temporary storage of memories for facts and events (i.e. "declarative memory"). MTL regions were also separated from "unimodal and polymodal association areas" that were considered to support specific aspects of cognition that are separate from declarative memory. Although the framework has been largely unchanged over the past 25 years, it has been refined by clarifying the cortical inputs to PRC and PHC (e.g., Mishkin et al. 1997) and by differentiating between medial and lateral EC (van Strien et al. 2009; Witter et al. 2000b).

Although Squire and Zola-Morgan's (1991) framework depicted the MTLMS as an anatomically-interconnected region with a common purpose, subsequent research revealed evidence suggesting that unique memory impairments can be observed following damage to different regions within the proposed MTL memory system (Eacott et al. 1994). Building on these and similar results, newer models proposed important functional distinctions between the hippocampus and neocortical areas of the MTLMS. Mishkin et al. (1997) adapted the ideas of Tulving and proposed that PRC is essential for learning and retaining facts and general knowledge ("semantic memory"), whereas the hippocampus is selectively important for remembering specific events ("episodic memory"). Eichenbaum and Cohen (2004) proposed that neocortical areas (including PRC and PHC) represent information about specific items, whereas the hippocampus primarily supports information about the relationships between items that are encountered during an event. Finally, other frameworks proposed that recollection and familiarity or spatial and object processing depend on the hippocampus and PRC, respectively (Aggleton and Brown 1999).

More recent frameworks proposed that the hippocampus, PHC, and PRC contribute to memory in different ways (Bird and Burgess 2008; Davachi 2006; Diana et al. 2007; Eacott and Gaffan 2005; Ellenbogen et al. 2007; Knierim et al. 2006; Montaldi and Mayes 2010; Ranganath 2010). One influential account is the "what"/ "where" model, which holds that the PRC represents information about objects ("what"), the PHC represents information about spatial contexts ("where"), and the hippocampus specifically integrates information from the two streams (Knierim et al. 2006). Extending this idea to human episodic memory, frameworks such as the Binding of Items and Context (BIC) model propose that the PHC and PRC are involved in the encoding and retrieval of contextual and item information, respectively, while the hippocampus is involved in binding item and context information (Diana et al. 2007; Eichenbaum et al. 2007; see also Davachi 2006). According to this view, the hippocampus and PHC would be essential for recollection of contextual information associated with a particular item, but the PRC would be sufficient to support recognition of a specific item.

Proponents of the original MTLMS framework acknowledge the possibility of functional heterogeneity between MTL subregions, but they argue that such functional differences are subtle and unlikely to be captured by a simple model (Squire et al. 2007). Although there is still no consensus regarding this issue, it is notable that all of the major frameworks discussed above adopt the hierarchically organized anatomical framework of the MTLMS, and they focus primarily on delineating roles for specific MTL subregions in memory. As such, these models incorporate the hierarchical organization of information flow in the brain noted by Mishkin (1982), but they do not specify the involvement of extra-MTL regions in memory processes.

Some researchers, however, have eschewed the anatomical organization of the MTLMS by proposing frameworks that do not treat the MTL separately from other neocortical networks. For instance, Mishkin and colleagues (Kravitz et al. 2011, but see also Bar 2007; Nadel and Peterson 2013; Ranganath and Ritchey 2012; Ritchey et al. 2015) proposed a "parieto-medial temporal pathway", which included the hippocampus, PHC, retrosplenial cortex (RSC), posterior cingulate, angular gyrus (AnG) and precuneus. Aggleton (2012) also proposed that MTL subregions are situated within multiple, functionally separable, memory systems. Gaffan (2002), on the other hand, argued for a stronger focus on the role of regions beyond the MTL, suggesting that that the entire concept of a memory system is misguided because, "memory traces are stored in widespread cortical areas rather than in a specialized memory system restricted to the temporal lobe" (p. 1111).

In short, there are several views about the hippocampus and memory systems, and there are fundamental disagreements about whether the MTL should be treated as a distinct network or whether it should be incorporated in a broader anatomical framework. We believe that it is possible to adjudicate between these perspectives, but progress cannot be made if one starts with a preconceived notion of an anatomically and functionally encapsulated memory system. Instead, it is necessary to step back and consider the overall organization and function of brain networks and determine how these networks might contribute to memory processes. In the next section, we will consider the organization that emerges when the anatomical properties of memory processes are considered in the context of connectivity both within and outside the MTL.

Common Patterns of Cortical and Neocortical Organization Revealed by Neuroanatomy and Functional Connectivity

Although traditional approaches have generally considered the PRC and PHC in the context of their relationship to the hippocampus, an alternative approach is to start by considering the broader functional and anatomical connections of these regions. Most theories focus on differences between visual inputs to the PRC and PHC, however, both PRC and PHC receive extensive inputs from overlapping ventral visual stream areas. Both PHC and PRC are interconnected with higher-level ventral temporal areas (e.g., areas TE and TEO), though PHC is more extensively interconnected with lower-level visual areas, including V3 and V4 (Lavenex et al. 2004; Suzuki and Amaral 1994). Some have argued that the connectivity profile of the PRC positions it at the apex of the ventral visual processing stream, such that it represents highly specific visual information about objects (Bussey and Saksida 2007; Lavenex et al. 2002; Mishkin 1982; Suzuki and Amaral 1994). The anatomical evidence, however, suggests that PRC also functions as a site of integration for visual, auditory, olfactory, and gustatory information (Suzuki and Amaral 1994).

The most striking differences between PHC and PRC are evident in connectivity with cortical association areas and subcortical regions. For instance, anatomical studies in non-human primates have suggested that PRC is primarily interconnected with lateral orbital prefrontal cortex (PFC), whereas PHC is primarily interconnected with medial prefrontal regions (Carmichael and Price 1995, 1996; Kondo et al. 2005; Ongur and Price 2000) and the posterior parietal lobe (Suzuki and Amaral 1994). In addition to differential connectivity with frontal regions, Kondo et al. (2005) also noted that PHC and PRC display dissociable connectivity patterns with regions that are themselves anatomically interconnected with larger orbital and medial prefrontal networks. On the basis of this evidence, Kondo et al. (2005) proposed that PRC and PHC are key regions in two distinct neocortical circuits.

FMRI-based analyses of intrinsic functional connectivity in humans have also repeatedly demonstrated that human PHC and PRC are situated in different networks, consistent with the framework proposed by Kondo et al. (2005). PHC exhibits strong intrinsic connectivity with the RSC, posterior cingulate, AnG, precuneus, and ventromedial PFC, whereas PRC exhibits stronger connectivity with anterior fusiform cortex, posterior lateral and inferior temporal cortices, temporoparietal cortex, orbitofrontal cortex (OFC), and prefrontal regions (Kahn et al. 2008; Libby et al. 2012; Ritchey et al. 2014). FMRI-based parcellations of the MTL have arrived at similar conclusions. In two independent investigations, researchers found evidence for preferential PHC connectivity with posterior medial regions, while PRC is strongly connected with more anterior-temporal areas (see Fig. 1a; Wang et al. 2016; Zhuo et al. 2016). Additional human evidence for dissociable networks can also be seen in investigations of the 'default mode network' (DMN). Although this network is typically linked with resting state activity, the regions that appear in this network have a strong overlap with regions that demonstrate significant connectivity with PHC (Greicius et al. 2003; Raichle et al. 2001).[1]

The convergence between intrinsic functional connectivity relationships in humans (e.g., Kahn et al. 2008) and the neocortical connections revealed by tract-tracing in non-human primates (Kondo et al. 2005) likely reflects the fact that many of the connections of PHC and PRC correspond to the targets of the cingulum bundle and the uncinate fasciculus—two major long-range white matter pathways common to both the monkey and human brain (Schmahmann et al. 2007). The cingulum bundle connects PHC with medial parietal cortex, precuneus, ventrolateral PFC, and medial PFC (Mufson and Pandya 1984). The uncinate fasciculus, in turn, links PRC with temporopolar cortex, lateral orbital PFC, and the amygdala (Schmahmann et al. 2007; Von Der Heide et al. 2013).

Given the significant differences in cortico-cortical connectivity between PRC and PHC, it is not surprising that these areas show different connectivity patterns within the medial temporal lobes (see Fig. 2, Lavenex and Amaral 2000; Muñoz and Insausti 2005; Witter et al. 2000a). Anatomical work in rodents has revealed that EC, a major input structure into the hippocampus, can be divided into two parallel information streams, medial EC (MEC) and lateral EC (LEC) (Knierim 2006; van Strien et al. 2009; Witter et al. 2013). Whereas MEC is preferentially interconnected with PRC, LEC is preferentially interconnected with PHC (Burwell and Amaral 1998). Using ultra-high resolution fMRI at 7 T, intrinsic connectivity studies have revealed an analogous distinction in humans. Analyses across

---

[1] It should be noted that while regions with significant PHc connectivity are typically included in the DMN, the inclusion of the specific MTL regions, including PHC, in the DMN is still unclear (Greicius et al. 2003; Raichle et al. 2001; Schulman et al. 1997).

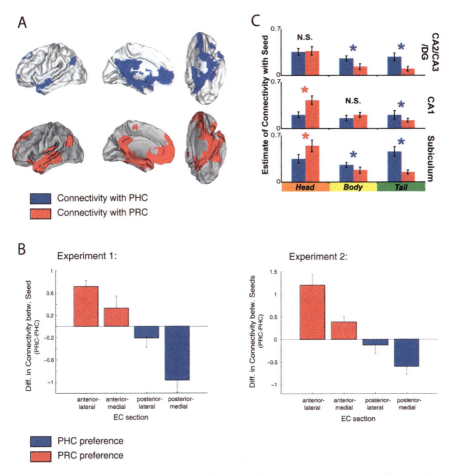

Fig. 1 Evidence for extended and separable cortico-hippocampal networks (**a**) Slice-by-slice functional connectivity-based parcellation of the MTL reveals that strikingly different large-scale networks are connected with PHC and PRC in humans. Depicted voxels displayed significant connectivity with either the PHC (*blue*) or posterior PRC (*red*). Adapted from Wang et al. (2016). (**b**) Comparable differences in connectivity have also been observed in EC. PRC is preferentially connected with anterior-lateral EC, while PHC is preferentially connected with posterior-medial EC. These connectivity differences were replicated across two independent datasets. Adapted from Maass et al. (2015). (**c**) Connectivity differences are also visible at the level of the hippocampus and its subfields. PRC-hippocampal connectivity (*red*) is more prominent in the anterior hippocampus (*head*) across CA1 and subiculum, while increased PHC-hippocampal connectivity (*blue*) is visible across CA1, subiculum, and CA2/3/DG in the posterior hippocampus (*tail*). Figure adapted from Libby et al. (2012)

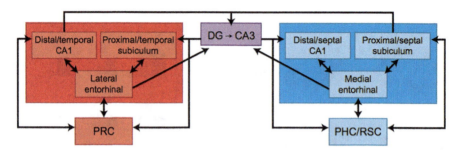

Fig. 2 Summary of anatomical investigations revealing segregated inputs into the hippocampus. PRC relays information to LEC, while inputs from the PHC and RSC generally converge on MEC. Connectivity patterns also suggest that subfields CA3 and DG in the hippocampus may be sites of integration. Figure reproduced from Ritchey et al. (2015)

independent data sets have confirmed preferential connectivity between PRC and anterior-lateral EC, and preferential connectivity between PHC and posterior-medial EC (see Fig. 1b; Maass et al. 2015). Another study replicated this finding, additionally reporting that anterior-lateral and posterior-medial EC exhibit differential connectivity patterns with neocortical regions outside of the MTL (Navarro Schröder et al. 2015). Mirroring the cortical connectivity patterns of PRC and PHC, anterior-lateral EC was preferentially connected to medial prefrontal and orbitofrontal cortex, while posterior-medial EC displayed stronger connectivity with occipital and posterior-parietal areas.

In line with their observed patterns of neocortical connectivity, LEC and MEC also exhibit different topographies of connections to hippocampal subfields (see Fig. 2). Specifically, anatomical tracer studies in rodents have revealed that inputs from LEC and MEC to CA1 exhibit a topographic gradient along the transverse (distal-proximal) axis, with distal CA1 and proximal subiculum receiving input from LEC, and proximal CA1 and distal subiculum receiving MEC inputs (Witter et al. 2000a, b). Additional segregation can be seen in the dentate gyrus (DG), where LEC projects to the outer third of the superficial layer, while MEC projects to the middle third (Witter et al. 2000a). Patterns of connectivity between EC and hippocampus have not been extensively studied in humans, but at least one study demonstrated that anterior-lateral and posterior-medial EC exhibit preferential connectivity with proximal and distal subiculum, respectively (Maass et al. 2015).

Consistent with the distinction between MEC and LEC, direct connections between PRC and PHC and hippocampus in rodents also exhibit different topographies (Aggleton 2012; Furtak et al. 2007; Witter et al. 2000a). Intrinsic connectivity analyses in humans indicate that these differences primarily fall along the longitudinal axis of the hippocampus, such that PRC exhibits preferential connectivity with the anterior hippocampus, and PHC exhibits preferential connectivity with the posterior hippocampus (see Fig. 1c; Kahn et al. 2008; Libby et al. 2012). At the level of hippocampal subfields, Libby et al. (2012) found that these connectivity differences are most pronounced in human CA1 and subiculum, which might be expected given that these are the primary sites of cortical connectivity within the hippocampus proper.

Despite the relative segregation of cortico-hippocampal pathways,[2] the hippocampus is nonetheless in a position to support integration across networks. Work in rodents suggests that DG and CA3 may be particularly well situated to act as convergence zones for information from LEC and MEC (Knierim et al. 2006). This idea is also supported by evidence in humans suggesting that the hippocampus has comparable functional connectivity relationships with PRC and PHC (Libby et al. 2012).

Reconceptualizing Memory Systems in Light of Large-Scale Cortico-hippocampal Networks: The PMAT Framework

The patterns of cortico-hippocampal connectivity described above do not suggest the existence of a single hippocampal memory system. Instead, the data support the existence of at least two distinct cortico-hippocampal networks: a posterior medial (PM) network that includes PHC, RSC, posterior cingulate, precuneus, AnG, and ventromedial PFC, and an anterior temporal (AT) network that includes PRC, ventral temporopolar cortex, lateral OFC and amygdala. We have proposed that the distinction between the PM and AT networks—the "PMAT" framework (Ritchey et al. 2015)—could be an alternative to the memory systems accounts described above (Fig. 3).

PMAT is based on findings showing that within-network connectivity is strong, both amongst regions within the PM network and amongst regions within the AT network. Connections between the two networks, however, are relatively sparse. These patterns of high-within network connectivity and sparse between-network connectivity correspond with what has been described as a "small-world" network architecture (Hilgetag et al. 2000; Sporns et al. 2004). It is widely assumed that functional specialization in the neocortex emerges from the unique pattern of connectivity exhibited by a particular region (Passingham et al. 2002). Consequently, regions in small-world networks—such as the AT and PM networks— should exhibit similar functional properties, whereas regions in distinct networks should exhibit different functional properties.

The PMAT and MTLMS frameworks differ in some important ways. Although MTLMS frameworks do not exclude the potential involvement of other brain regions in memory, they do not specify any role for extra-MTL regions in memory. Indeed, the standard model is that MTL regions play a privileged role in supporting

---

[2]These differences are "relative," as quantified in the studies described above. That said, LEC and MEC are extensively interconnected in rodents (Witter et al. 2000a, b), and are PRC and PHC are interconnected in rodents and primates (with PHC sending feedforward connections to PRC; Lavenex et al. 2004).

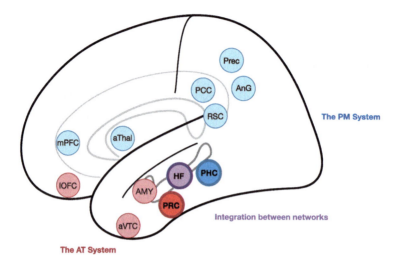

Fig. 3 The PMAT Framework. The AT network includes lateral orbitofrontal cortex (lOFC), amygdala (AMY), anterior ventral temporal cortex (aVTC), and PRC. The PM network includes medial prefrontal cortex (mPFC), anterior thalamus and mammillary bodies (aThal), posterior cingulate cortex (PCC), retrosplenial cortex (RSC), angular gyrus (AnG), precuneus (Prec) and PHC. PMAT proposes that the hippocampal formation (HF) allows for integration and coordination across networks, while the PRC and PHC serve as "connector hubs" that integrate information within their respective networks. Adapted from Ritchey et al. (2015)

memory for facts and events, whereas regions outside the MTL "store distinct components of information [sic]" (Wixted and Squire 2011). Thus, according to MTLMS theories, integrated declarative memory representations are distributed within the MTL and distinct from other neocortical areas that only represent discrete aspects of experience. The PMAT framework, on the other hand, suggests that declarative memory functions are distributed across the PM and AT networks, and that the functions of an MTL region will depend on its preferential network affiliation. Thus, PMAT differs from the MTLMS in that it assumes that integrated mnemonic representations are distributed across neocortical networks that span both MTL and extra-MTL areas (see section "What Do We Gain by Using the PMAT Framework?" for further discussion).

Functional Properties of the PM and AT Networks

Consistent with the anatomical and functional connectivity differences between the PM and AT networks, there is ample evidence to suggest that they can be differentiated functionally as well. Many studies have shown that PRC and PHC are differentially involved in the processing of specific types of information. Overall, these investigations have focused largely on differences in representations of object and spatial context information. For instance, lesion studies in monkeys have yielded evidence that PHC is involved in representing spatial information, while PRC supports item-based information. Evidence for this dissociation comes from lesion work with non-human primates, where animals were run through a delayed visual paired comparison procedure. After assessing looking times, damage to PHC was found to be associated with deficits in detecting location changes of a single familiar stimulus as well as changes in relative locations amongst familiar objects in a group, whereas damage to PRC was associated only with deficits in detecting changes in the relative locations amongst familiar objects in a group. These results have been interpreted as evidence that PRC is only involved in spatial location tasks that involve stimuli that share overlapping features, while PHC is more broadly involved in representing spatial information (Bachevalier and Nemanic 2008). Complimentary dissociations have also been reported in rodent work, where PRC lesions have been associated with impairments in object recognition, and postrhinal lesions, thought to be the rodent homologue of PHC, are associated with severe deficits for object-in-context associations in a task where rodents were exposed to familiar and novel configurations of objects in different contexts (Norman and Eacott 2005).

While the firing characteristics of PHC have not been studied as extensively as those of PRC, evidence suggests that place cells in PHC code for environmental cues. fMRI studies in humans have largely echoed rodent and non-human primate work, supporting the idea that PRC is involved in supporting subsequent memory for items, while PHC is involved in supporting subsequent memory for spatial or contextual details (Awipi and Davachi 2008; Pihlajamaki et al. 2004; Staresina et al. 2011).

Neuropsychological investigations in individuals with lesions offer additional support for dissociations between PRC and PHC. Investigations in a patient with damage restricted to the left PRC, amygdala, temporopolar cortex and EC, but sparing the hippocampus and PHC, demonstrated impairments when making familiarity-based recognition memory judgments on items, but was intact on measures of recollection (Bowles et al. 2007). These results have been interpreted as support for the idea that PRC is involved in supporting item-based familiarity, but PHC and the hippocampus are involved in representing the contextual processes that support recollection (Brown and Aggleton 2001; Eichenbaum et al. 2007; Montaldi and Mayes 2010; Ranganath 2010). Recent work has also shown that differences between PRC and PHC extend far beyond object and spatial differences, and that the properties of these structures closely correspond to the properties of other regions in the AT and PM networks, respectively.

Tasks That Engage the PM Network

Research has largely implicated the PHC in representing contextual information, and this function seems to underlie many of the tasks that activate regions in the PM network. Specifically, regions in the PM network are particularly engaged during episodic memory tasks and successful retrieval of contextual details, including memory for spatial and temporal information. Consistent with a role in representing context, the PM network, including the PHC, RSC and AnG, have been hypothesized to comprise a 'core recollection network,' based on observations that these regions are active when participants successfully recollect detailed information associated with a study episode (Johnson and Rugg 2007; Vilberg and Rugg 2008). In line with these findings, regions in the PM network have also been implicated in retrieval of autobiographical memories (St. Jacques et al. 2013), which tend to be particularly rich in contextual detail. Specifically, in a task where participants took a real-life museum tour, activity in a network of regions, including posterior parietal cortex, RSC and posterior PHC showed significantly higher activation when participants reported a high degree of reliving on a subsequent memory test, which may be associated with higher levels of contextual detail. Additionally, connectivity between the MTL and occipital and parietal cortices in the PM network is associated with better memory for detailed autobiographical events (Sheldon et al. 2016).

The core recollection network has also been implicated in representing information about different types of context, including spatial and temporal context. Specifically, an investigation by Hsieh and Ranganath (2015) recently examined multi-voxel patterns of activity while participants retrieved learned and random sequences of objects. Results suggested that regions in the core recollection network represent information about position in a temporal sequence, but not features of individual objects that comprised each sequence. In line with these results, a study recently published by Peer and colleagues (Peer et al. 2015) probed the neural basis of orientation along environmental dimensions of space, time, and person. While these processes have largely been studied independently, all three of these domains can be conceptualized as interacting cognitive contexts in the environment. Consistent with the idea that the core recollection network is involved in representing contextual information, Peer and colleagues, found a common neural basis for representations of these different domains in the core recollection network.

Evidence also suggests that the PM network plays an important role in episodic simulation, or the use of prior experiences to imagine or predict possible future events (Szpunar et al. 2007). It has been noted that individuals with Alzheimer's disease (AD), a disease primarily affecting the PM network in advanced stages, display impairments in tasks that involve episodic future thinking and scene construction (Irish et al. 2015) (see section "What Do We Gain by Using the PMAT Framework?" for additional information on AD and the PM network). Interestingly, activation in PHC and RSC is stronger for remembered events compared to imagined future events, possibly reflecting the increased contextual richness that can accompany the recall of a past episode (Gilmore et al. 2014).

An emerging literature on event segmentation, or the parsing of continuous perceptual and cognitive processes into meaningful event representations, also suggests a key role for the PM network. In one investigation, participants viewed a series of short clips of everyday actions and were asked to indicate the start of a new event sequence, or event boundary (Zacks et al. 2001). Neuroimaging results revealed that activity in the PM network, including posterior occipital temporal regions, parietal cortex, and lateral frontal cortex, were activated at event boundary locations. Similar regions were found to be active in tasks where participants read sentences or narratives that could be segmented into a number of events (Ezzyat and Davachi 2010; Speer et al. 2007). In this same vein, regions in the PM network, including the AnG, RSCf, precuneus, and posterior cingulate cortex have been implicated in integrating information across large temporal windows, suggesting that these areas are involved in representing meaningful, continuous contextual information (Hasson et al. 2015).

When considering the complete pattern of tasks that engage regions in the PM network, an overarching theme is the construction and use of 'situation models. A situation model is an on-line representation of an event that is constructed by retrieving knowledge about classes of events (i.e., an event schema or a script) and representations of specific past events (i.e., episodic memories) (Ranganath and Ritchey 2012).[3] For example, a situation model of a regular morning train commute might encompass information about the relevant spatial information, including the city or town where the train was boarded, the path of the train to the final destination, and the final destination itself, as well as temporal information relevant to the commute, including the temporal sequence of events (board train, scan ticket, open laptop, respond to email, close laptop, disembark), and temporal context (weekday mornings). Any social context, which might include conversing with the train's conductor or another commuter, would also be included in this situation model. As such, the situation model provides a framework for spatio-temporal and causal relationships in the environment.

Tasks That Engage the AT Network

The functions of the AT network have not been extensively studied in humans, in part because fMRI scanning protocols often have inadequate coverage of the anterior temporal and orbitofrontal cortex. That said, available evidence from animal models, fMRI and electrophysiological recording studies in humans suggest that the AT network is heavily recruited during a number of tasks and cognitive processes with common features.

Although PRC has historically been seen as solely involved in memory, accumulating evidence suggests that it is also necessary for perceptual processing of objects. Evidence supporting this view comes from work in humans and animals,

---

[3]Situation models are related to cognitive constructs such as mental models (Johnston-Laird 1983) schemas (Bartlett 1932), and scripts (Schank and Abelson 1977). In the literature, these terms are often defined differently, and they are sometimes used interchangeably. See Radvansky and Zacks (2011) for a fuller discussion of situation models and related concepts in cognitive science.

where damage to PRC has been linked with impairments in perceptual oddity discriminations (Barense et al. 2007; Bussey et al. 2005; see Bussey et al. 2002 for a review). Deficits are usually most pronounced in situations where a large amount of featural overlap exists between presented items or when discriminations cannot be made on the basis of simple features such as size or color.

In addition to visual features, AT regions have been implicated in representing conceptual information about objects (Patterson et al. 2007). Patients with damage to the anterior temporal lobe display deficits in making fine semantic discriminations and are also impaired when differentiation between visually similar objects requires the use of semantic knowledge (Hennies et al. 2016; Lambon Ralph et al. 2012; Moss 2004). Findings from studies of conceptual priming, or facilitated semantic processing after prior exposure to an item, also suggest a role for AT regions in semantic processing. For instance, a number of fMRI studies have shown that PRC activity is modulated by repeated conceptual processing of an item (O'Kane et al. 2005), even across different modalities (Heusser et al. 2013). Another recent study found a direct relationship between PRC deactivation during conceptual priming and deactivation during familiarity-based recognition (Wang et al. 2014). Some of the strongest evidence for PRC involvement in conceptual priming has been demonstrated in a parallel fMRI and neuropsychological study by Wang et al. (2010). Using two different paradigms, Wang and colleagues found that individuals with left PRC damage exhibited severe impairments on conceptual priming tasks (Wang et al. 2010). The findings were unlikely to reflect hippocampal damage, because patients with likely bilateral hippocampal damage showed intact conceptual priming. Consistent with the neuropsychological results, a parallel fMRI study in healthy adults revealed that left PRC activation during encoding predicted conceptual priming, and the activated region coincided with the site of maximal lesion overlap in the patient group. Additional evidence for conceptual representation in the AT network comes from intracranial EEG studies that have identified a field potential, the AMTL N400, that indexes meaningful word and item-based semantic processing (Nobre and Mccarthy 1995). Converging evidence has also come from fMRI studies that have reliably implicated the PRC and temporopolar cortex in semantic processing (Bruffaerts et al. 2013; Clarke and Tyler 2014; Kivisaari et al. 2012).

In addition to processing conceptual and semantic information, the AT network, and PRC in particular, has been implicated in memory for objects or other items. One form of item memory that has been strongly linked to PRC is familiarity-based item recognition. Numerous fMRI studies have shown that PRC activity during encoding or retrieval is predictive of item familiarity and familiarity-based learning of item-item associations (Daselaar 2006; Davachi et al. 2003; Haskins et al. 2008; Kirwan et al. 2008; Ranganath et al. 2004; see also Montaldi and Mayes 2010; see Diana et al. 2007 for a review). In line with fMRI work in humans, lesion studies in rats and monkeys corroborate PRC involvement in item familiarity (Brown and Aggleton 2001; Eichenbaum et al. 2007). Other regions in the AT network, including the amygdala, have also been implicated in representing familiarity. Neural firing patterns in the amygdala reflect object novelty or familiarity (Wilson and Rolls 1993), and amygdala damage has also been shown to impair familiarity in rodents (Farovik et al. 2011). In addition to familiarity, the AT network has been implicated in representing other types of mnemonic information. In monkeys, single unit recordings suggest that PRC is particularly involved in supporting memory tasks involving objects, including working memory maintenance and the formation of item-item associations (Miyashita 1988; Nakamura and Kubota 1995).

Regions in the AT network have also been directly implicated in representing information about the affective or motivational significance of objects or entities. Early evidence for this idea came from studies showing that extensive anterior temporal lobe lesions caused a syndrome of altered perceptual, exploratory, dietary, and sexual behavior indicative of impaired representation of the significance of objects (Klüver and Bucy 1937). More recent investigations have implicated PRC and EC in the ability to learn associations between individual visual cues and information about the amount of work needed to obtain a reward (Liu et al. 2000). This effect may have been driven by input about affective significance from the amygdala (see Phelps and LeDoux 2005 for a review) and by dopaminergic input from the ventral tegmental area (VTA). In support of the latter idea, Liu et al. (2000) found that performance on this task was impaired by blocking activity in dopamine D2 receptor proteins in PRC and EC. These findings are consistent with recent work by Eradath et al. (2015), where monkeys learned to associate visually similar cue objects with either a reward or no reward. Erdath et al. found that neurons in visual area TE were principally involved in signaling perceptual similarity, and neurons in PRC signaled the reward status of objects[4] (but see also Inhoff and Ranganath 2015). Additional regions in the AT network, specifically the amygdala and OFC, have also been implicated in representing motivational significance (see Dolan 2007 for a review). In line with this idea, amygdala deactivation in rodents has been shown to decrease reward-seeking behaviors in response to a cue (Ishikawa et al. 2008) and the activity of neurons in the amygdala to a cue stimulus has been linked to motivated, reward-seeking behaviors (Tye and Janak 2007). Similarly, the OFC has also been implicated in signaling the reward value of choice options (Rushworth et al. 2011).

Taken together, the AT network seems to be principally involved in representing item information, including semantic and perceptual information, as well as salience and reward value. Returning to the morning train commute example, the PMAT framework posits that the AT network is responsible for representing conceptual information such as, "train," "ticket," "laptop," "conductor" as well as the salience and recognition of each of these objects. It also possible that the AT network may be involved in representing negative values like punishments or fear, however additional research is required on this topic.

---

[4]Like the Nucleus Basalis and the Locus Coeruleus, the VTA provides modulatory inputs to a broad range of cortical areas. Accordingly, the VTA may modulate processes in the AT network.

Testing the PMAT Framework

In practice, the MTLMS framework is used to summarize evidence from selected lesion studies (e.g. Squire et al. 2007; Wixted and Squire 2011), rather than to generate new hypotheses. One clear prediction of the MTLMS framework, however, is that MTL regions are distinct from regions outside of the MTL, in terms of both anatomical organization and function. This idea is made clear both in visual depictions of the MTLMS, and in papers arguing that the MTLMS and extra-MTL regions differ in their relative involvement in memory and perception (Squire et al. 2007). The PMAT framework, in contrast, argues that MTL neocortical regions are situated within distinct networks that each include MTL regions and neocortical areas that are outside of the MTL. Moreover the PMAT framework suggests that there are key differences between the PM and AT networks, but, within each network, MTL and extra-MTL regions should exhibit similar properties.

To test these competing predictions, Ritchey et al. (2014) conducted an fMRI study in order to relate separable memory processes to cortical systems identified on the basis of functional connectivity relationships. In this investigation, subjects first participated in a baseline rest scan, which was used to provide a measure of intrinsic functional connectivity in absence of any task demands. Data from this scan was submitted to a data-driven, "community detection" algorithm and used to identify separate modules of regions that exhibit high within-network connectivity and low connectivity with regions in other modules. Consistent with the PMAT framework, two of the three distinct modules detected by the algorithm corresponded closely with the PM and AT networks. The algorithm did not identify a module corresponding to the MTLMS, and instead MTL regions were situated either in the PM or AT modules.

In order to examine whether the PM and AT modules identified by the community detection algorithm might also support different memory processes, as predicted by the PMAT framework, the authors examined activity in these modules during a memory encoding task. Participants were scanned while encoding sentences that included a concrete noun referring to an object and either a fact about its appearance (turquoise, purple, soft or bumpy), situational context (contest prize, birthday present, new purchase, or rental), or spatial context (yogurt shop, pizzeria, science lab, lecture hall). Within each region in the three modules, Ritchey et al. (2014) estimated activation during each encoding condition, and quantified the similarity of the corresponding task activation profiles across all of the regions that were entered into the community detection algorithm. This analysis revealed that, for every participant, task activation similarity was significantly higher for regions that had been identified by the algorithm as within the same module than across different modules. As would be predicted by the PMAT framework, the PM module was preferentially active during trials where correct spatial context was subsequently remembered relative to appearance details, whereas the AT module demonstrated the opposite effect. Ritchey and colleagues also estimated multivariate pattern information on a trial-by-trial basis, and identified that, for every participant, voxel pattern information was more similar between regions in the same module than between regions in different modules (see Fig. 4). Given that the

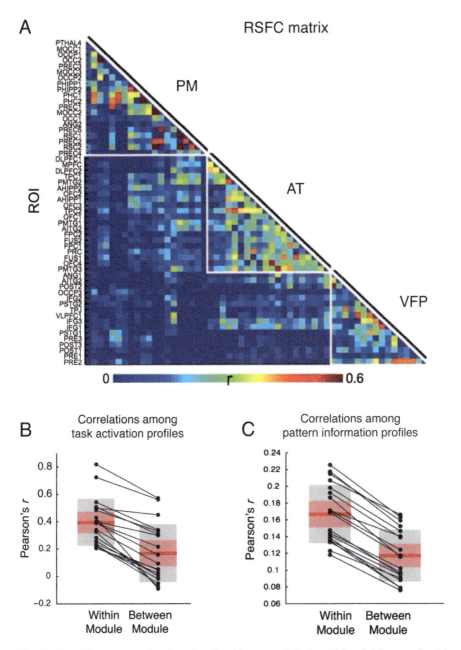

Fig. 4 Data-driven community detection algorithms reveal distinct PM and AT networks. (**a**) Resting state functional connectivity analyses have revealed three separate modules, two corresponding to the PM and AT networks, and an additional module including ventral frontal and parietal (VFP) regions. (**b**) Task activation and (**c**) multivariate pattern profiles measured during an encoding task are significantly more similar within module compared to between module. Adapted from Ritchey et al. (2014)

modules were identified in an independent resting state scan, the results provide strong support for the PMAT framework and are the first demonstration, to our knowledge, that regional specialization of function is directly related to network affiliations.

Although the analyses described above demonstrate that the PMAT model is able to capture both network affiliations and functional properties of regions within and outside of the MTL, the authors also considered whether the MTLMS framework could better explain the results. To test this idea, MTL subregions were removed from the three modules defined from resting state functional connectivity, and assigned to their own "MTL memory system" module. Results from this analysis revealed that adding a separate MTL module to the PM and AT networks did not provide a better explanation of the data, as might be expected if MTL regions constitute a specialized memory network. To ensure that results from these analyses were not driven solely by extra-MTL areas, the authors conducted the same analyses on regions within the MTL. Consistent with the predictions of the PMAT framework, even regions within the MTL showed higher within-module than across module similarity, in terms of activity profiles and pattern information during memory encoding.[5] Taken together, results of Ritchey et al. (2014) strongly support the ideas put forth in the PMAT framework.

What Do We Gain by Using the PMAT Framework?

A core difference between the PMAT framework and MTLMS theories is a focus on situating individual regions in terms of their common connectivity profiles and positions in broader cortico-hippocampal networks. This approach gives us a unique perspective on many central findings in memory research. We explore two of these topics here, and use PMAT to both explain observed patterns in the literature and to generate novel predictions.

Memory Deficits and Disorders

Traditionally, memory deficits in amnesia have been used to motivate the idea that integrated declarative or episodic memory representations are localized to regions

---

[5]The MTLMS framework is compatible with findings of functional distinctions within the MTL, but as noted earlier, the MTLMS framework depicts MTL subregions as being more related to one another than to extra-MTL regions, both in terms of connectivity and memory functions. This prediction was not consistent with the results of Ritchey et al. (2014), which repeatedly demonstrated strong relationships between MTL and extra-MTL regions within the AT and PM modules, and weak differentiation between MTL and extra-MTL regions. The results fail to support the idea that there is any added value in distinguishing between MTL and extra-MTL regions.

in the MTL. For instance, studies of densely amnestic patients like H.M. and E.P. led many to believe that MTL regions play a privileged role in memory by associating individual elements of experience. This highly modular, localizationist view of brain organization can be contrasted with the PMAT framework, which situates MTL subregions within extended networks and *does not assign a special functional role for MTL regions in memory*. To better understand how PMAT explains amnesia, it is necessary to consider how focal lesions can affect network-level interactions.

In the field of neurology, it has long been known that a focal lesion can have effects that expand far from the lesion site (Mesulam 1990), a phenomenon known as "diaschisis". Empirical research has confirmed this concept, but it has also shown that the extent of disruption varies according to the region is damaged. Specifically, the severity of behavioral deficits following brain damage can be significantly predicted by the extent of damage to "connector hubs"— areas that have both strong within-network connections and strong connections with nodes in one or more other networks (Gratton et al. 2012; Warren et al. 2014). Recent work has demonstrated that MTL damage is associated with diaschisis (Henson et al. 2016), suggesting that the behavioral effects of a focal lesion may be secondary to network-level disruption.

Following this logic, we propose that MTL amnesia results in a disconnection syndrome (cf. Warrington and Weiskrantz 1982), in which the PM and AT networks are isolated from critical inputs, resulting in significant alteration of activity patterns within and between these networks. Thus, MTL neocortical areas are special, not by virtue of the information they represent, but rather by their roles in directing information flow to other nodes in the PM and AT networks. We predict that PRC and PHC, along with RSC (and possibly other areas), serve as connector hubs, in that they are strongly interconnected with other PM and AT network nodes, and they have strong connections with networks that process sensory information during the experience of an event. Thus, damage to these areas would be expected to have a disproportionate effect on cognition by compromising the ability of the AT and PM networks to encode critical elements of experience.[6] Conversely, memory and other cognitive functions could be more robust in the face of damage to other AT and PM network regions if the connections of those regions are relatively redundant with spared regions in each network.

If mnemonic representations are indeed distributed across the AT and PM networks, then one would expect that widespread damage to extra-MTL nodes of these networks should severely impair memory performance. Moreover, PMAT suggests that different effects should be seen following disruption of the AT and PM networks. These predictions are very difficult to test, but studies of Semantic Dementia (SD) and Alzheimer's disease (AD) strongly support the idea that

---

[6]This effect is analogous to what can be observed in airline traffic. For instance, weather-related flight cancellations in Chicago O'Hare airport elicit a chain reaction of delays in smaller regional airports where weather conditions are normal.

Greymatter atrophy: AD vs. healthy controls

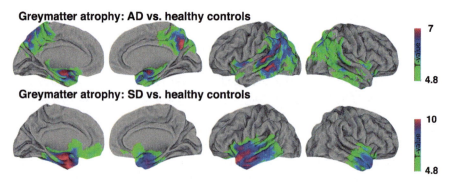

Greymatter atrophy: SD vs. healthy controls

Fig. 5 Neurodegenerative disease profiles in AD and SD follow the PM and AT network architecture, respectively. Adapted from La Joie et al. (2014)

disruptions of the AT and PM networks cause severe and distinct patterns of memory deficits. Both AD and SD are associated with atrophy in the hippocampus and MTL neocortex,[7] but they are associated with different patterns of atrophy in the AT and PM networks (see Fig. 5). Individuals with preclinical AD (i.e., asymptomatic individuals identified before progression to amnestic mild cognitive impairment [MCI] or full AD status) show atrophy in CA1, LEC, and PRC, as well as in the PHC and precuneus (Khan et al. 2013). At the MCI stage and beyond, atrophy extends throughout the PM network, including RSC, posterior cingulate, medial parietal cortex, and ventrolateral parietal cortex (Grothe and Teipel 2016; Seeley et al. 2009). Like AD patients, SD patients also have prominent atrophy in CA1, EC, and PRC, but atrophy is also evident in the ventral temporopolar cortex, OFC, and amygdala (La Joie et al. 2014). In summary, MTL damage is present even in the early stages of AD and SD, but these disorders differentially affect the PM and AT networks.

Although semantic and episodic memory are affected in both disorders, episodic memory impairments are more prominent in AD (along with MCI and preclinical AD) and semantic memory impairments are more prominent in SD (Nestor et al. 2006). These differences are difficult to reconcile with the MTLMS framework, which predicts a common amnestic syndrome following MTL damage. The observed pattern of behavioral deficits in AD and SD is exactly what would be predicted based on the PMAT framework, however, because PMAT attributes mnemonic functions to both MTL and non-MTL components of the AT and PM networks.

It should be noted that AD and SD are not the only disorders that disproportionately target one of the two cortico-hippocampal networks (see Ranganath and Ritchey 2012, for a review). For instance, herpes simplex encephalitis ravages the hippocampus and anterior MTL, but damage extends throughout the AT network, including the amygdala, ventral temporopolar cortex, and lateral OFC. Korsakoff's

---

[7]MTL Atrophy tends to be more left-lateralized in SD than in AD.

syndrome, in contrast, is primarily associated with damage or dysfunction to PM system regions, including RSC and posterior cingulate[8] (Bernasconi 2003; Bonilha et al. 2003; Noppeney et al. 2006). Additionally, patients who have reported episodes of transient global amnesia (TGA) display hyperperfusion in the precuneus, suggesting that irregularities in the PM system could contribute to TGA (Jang et al. 2015).

In addition to providing a parsimonious explanation for memory disorders, the PMAT framework also offers predictions about the network-level effects of different kinds of lesions. If PHC and PRC indeed serve as "connector hubs" within their respective networks, damage to either structure should lead to mnemonic deficits that are commensurate with disrupted connectivity within remaining nodes of their corresponding networks. More broadly, one might expect alterations of within-network and between-network connectivity in the PM network following damage to PHC, and in the AT network following damage to PRC. Damage to the hippocampus, in turn, is predicted to lead to increased separation of the PM and AT networks.

Systems Consolidation

According to the standard model of systems consolidation, the hippocampus plays a temporary role in memory by associating aspects of an event that are distributed across the entire cortex (see Frankland and Bontempi 2005, for review). Over time, it is expected that either cortical representations become directly associated with one another, or alternatively, that the ventromedial PFC plays a more permanent role in associating these representations. The transformation model, in contrast, argues that the hippocampus always retains a representation that is necessary for episodic memory and other context-specific expressions of memory (Winocur et al. 2010). Through extended cortico-hippocampal interactions, the transformation model predicts that neocortical regions can develop generalized representations of past events, but these representations will not support retrieval of context-specific details.

Although the standard and transformation models have explanatory power, they are both limited by the fact that they do not say much about the neocortical areas (either in or outside the MTL) that are proposed to support lasting memory traces. PMAT, however, provides a framework to better understand the effects of systems consolidation. For instance, the standard model assumes that a memory can become independent of the hippocampus if the distributed cell assemblies that store event features become linked via Hebbian plasticity. This model requires one to assume that any neocortical area can form strong and lasting connections with any other neocortical area—this assumption is highly implausible given the semi-modular

---

[8]Contrary to popular belief, evidence is inconclusive about hippocampal damage or dysfunction in Korsakoff's syndrome.

network organization of the human brain[9] (Sporns et al. 2004). The PMAT framework, in contrast, does not depend on this assumption, because it is based on the known organization of cortico-hippocampal pathways in the primate brain.

According to PMAT model, the PM and AT networks bind features of memories distributed across different cortical pathways, both for pre-existing and novel representations. In this framework, the role of the hippocampus is to interface between these networks and to accelerate incorporation of new information via "systems consolidation". Models of systems consolidation propose that hippocampal representations are reactivated during sleep or waking rest, leading to offline "replay" of recent memories (Walker and Stickgold 2004). Computational models suggest that offline replay can strengthen new memories while minimizing interference with other established memories (McClelland et al. 1995). Other work suggests that systems consolidation can establish links between interrelated memories (Ellenbogen et al. 2007). The PMAT model is fully compatible with the idea that offline replay helps to reinforce representations of recent experiences in the PM and AT networks, and it suggests that replay could have different effects on representations of situations and entities.

Conclusions and Future Directions

Taken together, the PMAT framework accounts for evidence from studies of intrinsic connectivity, univariate, and multivariate neuroimaging analyses, single-unit recordings in rodents and non-human primates, and cognitive deficits observed in patients with memory disorders. The broad scope of the PMAT framework facilitates integration of concepts and findings from different methods and species. In doing so, PMAT offers a unified and parsimonious framework for explaining cognition in the healthy brain and in patient populations. The PMAT framework can be seen as a logical progression from memory systems models, in that it accounts for the same lesion data that inspired memory systems models, but it also incorporates data showing involvement of MTL regions in perception, semantic cognition, and many forms of high-level cognition. Moreover, unlike MTLMS and related models, PMAT does not rely on the assumption (McClelland et al. 1995) that neocortical MTL areas have computational characteristics, or follow plasticity rules that are fundamentally different from those of extra-MTL areas.

PMAT generates several testable hypotheses, and future research is likely to motivate refinement or modifications to the framework. In particular, it is unclear whether the PM network is involved in representing aspects of situation models that extend beyond space and time (e.g., goal contexts or motivational states). Additionally, the role of prediction errors in updating and refining situation models

---

[9]The transformation model is not burdened by this assumption, but only because it does not address what is represented in extra-MTL regions.

represented by the PM network has yet to be fully addressed (Kurby and Zacks 2008). More work is also necessary to characterize the AT network, because it is not clear whether the network supports cognition beyond perceptual and semantic characteristics of entities. Additional technical challenges exist in elucidating the precise functions of the AT network; regions such as the ventral temporopolar cortex are particularly affected by MRI signal distortion and dropout because of their anatomical location. Solutions to these challenges will enable a broader range of studies to characterize functional characteristics of the AT network.

Although our review has highlighted the existence of two cortico-hippocampal networks, we expect that other important cortico-hippocampal networks will be identified. Support for this idea comes from a recent data-driven slice-by-slice parcellation of the MTL based on intrinsic functional connectivity patterns (Wang et al. 2016). This study identified regions corresponding to PHC and PRC, as well as a more anterior PRC region that showed stronger correlations with regions in the ventrolateral prefrontal cortex, insula and auditory cortex (see also Lavenex et al. 2004). We expect that targeted analyses with sophisticated imaging methods will reveal more information about anterior PRC and its corresponding network.

Future research might also reveal that, as potential "connector hubs", regions like PRC, PHC, and RSC could change their network affiliations under certain circumstances. For instance, during tasks that necessitate attention solely to sensory information, these areas might affiliate primarily with sensory networks, as suggested by theories that emphasize MTL interactions with ventral stream visual areas (Mishkin 1982; Bussey and Saksida 2007). For instance, PHC and RSC reliably co-activate with medial ventral stream scene processing regions during tasks that solely involve attention to visual features of scenes (Epstein 2008), but activations extend throughout the PM network during tasks that require comprehension of the context depicted in a scene (Hassabis and Maguire 2007). Likewise, PRC co-activates with lateral ventral stream areas during tasks that solely involve attention to visual features of a face or object (Gomez et al. 2015), but other areas in the AT network are additionally recruited during tasks that require analysis of the properties of a person or object (Olson et al. 2007). Further work is needed to determine whether the structure of the PM and AT networks is altered during perceptually-driven tasks, as compared with tasks that also involve comprehension of sensory stimuli.

Motivational states might also alter network-level organization, particularly for the hippocampus (Aggleton 2012). For instance, studies of fear learning and anxiety have identified interactions between the hippocampus, medial prefrontal cortex and amygdala (Maren et al. 2013), and studies of reward-motivated learning have identified hippocampal interactions with the ventral striatum and VTA (Axmacher et al. 2010; Gruber et al. 2016; Lisman and Grace 2005). Given that these network interactions are evident during tasks that involve motivationally-salient events, it is possible that neuromodulation by norepinephrine or dopamine could transiently alter the organization of hippocampal networks.

Further research is also needed to better characterize whether or how the hippocampus contributes to different forms of cognition through interactions with different networks. There is growing acceptance of the idea that the reach of the hippocampus extends beyond traditional memory paradigms (Cohen 2015), and it is possible that hippocampal contributions to different cognitive domains will be driven by different kinds of cortico-hippocampal interactions. For instance, in some cases, the hippocampus might act to modulate activity patterns within the PM or AT networks, and in others it might act to facilitate interactions between these networks. If hippocampal network affiliations are flexible and context-dependent, then it will be essential to determine how these affiliations are determined. Several findings suggest that the prefrontal cortex could play an important role in modulating hippocampal network interactions (Shin and Jadhav 2016).

In conclusion, the PMAT framework offers a compelling alternative to traditional memory models that focus on the hippocampus as the apex of a dedicated, hierarchically-organized memory system. PMAT abandons distinctions between memory and other forms of cognition, and instead attempts to carve nature at its joints by drawing on contemporary knowledge about the organization and functions of different cortico-hippocampal networks. Critically, the PMAT framework provides an integrated account of memory and cognition in terms of representations that are distributed across large-scale networks of cortical association areas (Fuster 2009). Consequently, it accounts for a large amount of data across human, animal, and patient work, and it provides critical insights into cognitive deficits seen in a wide range of disorders that affect memory.

Acknowledgements We thank Dr. Debbie Hannula and an anonymous reviewer for helpful suggestions on a previous draft of this manuscript. Work on this manuscript was supported by a Vannevar Bush Faculty Fellowship to C.R. (Office of Naval Research Grant N00014-15-1-0033) and an NSF Graduate Research Fellowship to M.C.I. (#1148897). Any opinions, findings, and conclusions or recommendations expressed in this material are those of the author(s) and do not necessarily reflect the views of the Office of Naval Research or the U.S. Department of Defense.

References

Aggleton JP (2012) Multiple anatomical systems embedded within the primate medial temporal lobe: implications for hippocampal function. Neurosci Biobehav Rev 36(7):1579–1596. doi:10.1016/j.neubiorev.2011.09.005

Aggleton JP, Brown MW (1999) Episodic memory, amnesia, and the hippocampal–anterior thalamic axis. Behav Brain Sci 22(03):425–444

Annese J, Schenker-Ahmed NM, Bartsch H, Maechler P, Sheh C, Thomas N et al (2014) Postmortem examination of patient H.M.'s brain based on histological sectioning and digital 3D reconstruction. Nat Commun 5. doi:10.1038/ncomms4122

Awipi T, Davachi L (2008) Content-specific source encoding in the human medial temporal lobe. J Exp Psychol Learn Mem Cogn 34(4):769–779. doi:10.1037/0278-7393.34.4.769

Axmacher N, Cohen MX, Fell J, Haupt S, Dümpelmann M, Elger CE et al (2010) Intracranial EEG correlates of expectancy and memory formation in the human hippocampus and nucleus accumbens. Neuron 65(4):541–549. doi:10.1016/j.neuron.2010.02.006

Bachevalier J, Nemanic S (2008) Memory for spatial location and object-place associations are differently processed by the hippocampal formation, parahippocampal areas TH/TF and perirhinal cortex. Hippocampus 18(1):64–80. doi:10.1002/hipo.20369

Bar M (2007) The proactive brain: using analogies and associations to generate predictions. Trends Cogn Sci 11(7):280–289. doi:10.1016/j.tics.2007.05.005

Barense MD, Gaffan D, Graham KS (2007) The human medial temporal lobe processes online representations of complex objects. Neuropsychologia 45(13):2963–2974. doi:10.1016/j.neuropsychologia.2007.05.023

Bartlett FC (1932) Remembering: a study in experimental and social psychology. University Press, Cambridge

Bernasconi N (2003) Mesial temporal damage in temporal lobe epilepsy: a volumetric MRI study of the hippocampus, amygdala and parahippocampal region. Brain 126(2):462–469. doi:10.1093/brain/awg034

Bird CM, Burgess N (2008) The hippocampus and memory: insights from spatial processing. Nat Rev Neurosci 9(3):182–194. doi:10.1038/nrn2335

Bonilha L, Kobayashi E, Rorden C, Cendes F, Li LM (2003) Medial temporal lobe atrophy in patients with refractory temporal lobe epilepsy. J Neurol Neurosurg Psychiatry 74 (12):1627–1630

Bowles B, Crupi C, Mirsattari SM, Pigott SE, Parrent AG, Pruessner JC et al (2007) Impaired familiarity with preserved recollection after anterior temporal-lobe resection that spares the hippocampus. Proc Natl Acad Sci U S A 104(41):16382–16387

Brown MW, Aggleton JP (2001) Recognition memory: what are the roles of the perirhinal cortex and hippocampus? Nat Rev Neurosci 2(1):51–61

Bruffaerts R, Dupont P, Peeters R, De Deyne S, Storms G, Vandenberghe R (2013) Similarity of fMRI activity patterns in left perirhinal cortex reflects semantic similarity between words. J Neurosci 33(47):18597–18607. doi:10.1523/JNEUROSCI.1548-13.2013

Burwell RD, Amaral DG (1998) Perirhinal and postrhinal cortices of the rat: interconnectivity and connections with the entorhinal cortex. J Comp Neurol 391(3):293–321

Bussey TJ, Saksida LM (2007) Memory, perception, and the ventral visual-perirhinal-hippocampal stream: thinking outside of the boxes. Hippocampus 17(9):898–908. doi:10.1002/hipo.20320

Bussey TJ, Saksida LM, Murray EA (2002) Perirhinal cortex resolves feature ambiguity in complex visual discriminations. Eur J Neurosci 15(2):365–374

Bussey TJ, Saksida LM, Murray EA (2005) The perceptual-mnemonic/feature conjunction model of perirhinal cortex function. Q J Exp Psychol B 58(3–4):269–282

Carmichael ST, Price JL (1995) Sensory and premotor connections of the orbital and medial prefrontal cortex of macaque monkeys. J Comp Neurol 363(4):642–664

Carmichael ST, Price JJ (1996) Connectional networks within the orbital and medial prefrontal cortex of macaque monkeys. J Comp Neurol 371:179–207

Clarke A, Tyler LK (2014) Object-specific semantic coding in human perirhinal cortex. J Neurosci 34(14):4766–4775. doi:10.1523/JNEUROSCI.2828-13.2014

Cohen NJ (2015) Navigating life: navigating life. Hippocampus 25(6):704–708. doi:10.1002/hipo.22443

Cohen NJ, Squire LR (1980) Preserved learning and retention of pattern-analyzing skill in amnesia: dissociation of knowing how and knowing that. Science 210(4466):207–210

Corkin S, Amaral DG, González RG, Johnson KA, Hyman BT (1997) HM's medial temporal lobe lesion: findings from magnetic resonance imaging. J Neurosci 17(10):3964–3979

Daselaar SM (2006) Triple dissociation in the medial temporal lobes: recollection, familiarity, and novelty. J Neurophysiol 96(4):1902–1911. doi:10.1152/jn.01029.2005

Davachi L (2006) Item, context and relational episodic encoding in humans. Curr Opin Neurobiol 16(6):693–700. doi:10.1016/j.conb.2006.10.012

Davachi L, Mitchell JP, Wagner AD (2003) Multiple routes to memory: distinct medial temporal lobe processes build item and source memories. Proc Natl Acad Sci U S A 100(4):2157

Diana RA, Yonelinas AP, Ranganath C (2007) Imaging recollection and familiarity in the medial temporal lobe: a three-component model. Trends Cogn Sci 11(9):379–386

Dolan R (2007) The human amygdala and orbital prefrontal cortex in behavioural regulation. Philos Trans R Soc Lond B Biol Sci 362(1481):787–799. doi:10.1098/rstb.2007.2088

Eacott MJ, Gaffan EA (2005) The roles of perirhinal cortex, postrhinal cortex, and the fornix in memory for objects, contexts, and events in the rat. Q J Exp Psychol B 58(3–4):202–217. doi:10.1080/02724990444000203

Eacott MJ, Gaffan D, Murray EA (1994) Preserved recognition memory for small sets, and impaired stimulus identification for large sets, following rhinal cortex ablations in monkeys. Eur J Neurosci 6(9):1466–1478

Eichenbaum H, Cohen NJ (2004) From conditioning to conscious recollection: memory systems of the brain. Oxford University Press, New York

Eichenbaum H, Yonelinas AR, Ranganath C (2007) The medial temporal lobe and recognition memory. Annu Rev Neurosci 30:123

Ellenbogen JM, Hu PT, Payne JD, Titone D, Walker MP (2007) Human relational memory requires time and sleep. Proc Natl Acad Sci U S A 104(18):7723–7728

Epstein RA (2008) Parahippocampal and retrosplenial contributions to human spatial navigation. Trends Cogn Sci 12(10):388–396. doi:10.1016/j.tics.2008.07.004

Eradath MK, Mogami T, Wang G, Tanaka K (2015) Time context of cue-outcome associations represented by neurons in perirhinal cortex. J Neurosci 35(10):4350–4365. doi:10.1523/JNEUROSCI.4730-14.2015

Ezzyat Y, Davachi L (2010) What constitutes an episode in episodic memory? Psychol Sci 22 (2):243–252. doi:10.1177/0956797610393742

Farovik A, Place RJ, Miller DR, Eichenbaum H (2011) Amygdala lesions selectively impair familiarity in recognition memory. Nat Neurosci 14(11):1416–1417. doi:10.1038/nn.2919

Frankland PW, Bontempi B (2005) The organization of recent and remote memories. Nat Rev Neurosci 6(2):119–130. doi:10.1038/nrn1607

Furtak SC, Wei S-M, Agster KL, Burwell RD (2007) Functional neuroanatomy of the parahippocampal region in the rat: the perirhinal and postrhinal cortices. Hippocampus 17 (9):709–722. doi:10.1002/hipo.20314

Fuster JM (2009) Cortex and memory: emergence of a new paradigm. J Cogn Neurosci 21 (11):2047–2072

Gaffan D (2002) Against memory systems. Philos Trans R Soc Lond B Biol Sci 357 (1424):1111–1121. doi:10.1098/rstb.2002.1110

Gilmore AW, Nelson SM, McDermott KB (2014) The contextual association network activates more for remembered than for imagined events. Cereb Cortex 26:611–617. doi:10.1093/cercor/bhu223

Gomez J, Pestilli F, Witthoft N, Golarai G, Liberman A, Poltoratski S et al (2015) Functionally defined white matter reveals segregated pathways in human ventral temporal cortex associated with category-specific processing. Neuron 85(1):216–227. doi:10.1016/j.neuron.2014.12.027

Gratton C, Nomura EM, Pérez F, D'Esposito M (2012) Focal brain lesions to critical locations cause widespread disruption of the modular organization of the brain. J Cogn Neurosci 24 (6):1275–1285. doi:10.1162/jocn_a_00222

Greicius MD, Krasnow B, Reiss AL, Menon V (2003) Functional connectivity in the resting brain: a network analysis of the default mode hypothesis. Proc Natl Acad Sci U S A 100(1):253–258

Grothe MJ, Teipel SJ (2016) Spatial patterns of atrophy, hypometabolism, and amyloid deposition in Alzheimer's disease correspond to dissociable functional brain networks: network-specificity of AD pathology. Hum Brain Mapp 37(1):35–53. doi:10.1002/hbm.23018

Gruber MJ, Ritchey M, Wang S-F, Doss MK, Ranganath C (2016) Post-learning hippocampal dynamics promote preferential retention of rewarding events. Neuron 89(5):1110–1120. doi:10.1016/j.neuron.2016.01.017

Haskins AL, Yonelinas AP, Quamme JR, Ranganath C (2008) Perirhinal cortex supports encoding and familiarity-based recognition of novel associations. Neuron 59(4):554–560. doi:10.1016/j.neuron.2008.07.035

Hassabis D, Maguire EA (2007) Deconstructing episodic memory with construction. Trends Cogn Sci 11(7):299–306. doi:10.1016/j.tics.2007.05.001

Hasson U, Chen J, Honey CJ (2015) Hierarchical process memory: memory as an integral component of information processing. Trends Cogn Sci 19(6):304–313. doi:10.1016/j.tics. 2015.04.006

Hennies N, Lambon Ralph MA, Kempkes M, Cousins JN, Lewis PA (2016) Sleep spindle density predicts the effect of prior knowledge on memory consolidation. J Neurosci 36 (13):3799–3810. doi:10.1523/JNEUROSCI.3162-15.2016

Henson RN, Greve A, Cooper E, Gregori M, Simons JS, Geerligs L et al (2016) The effects of hippocampal lesions on MRI measures of structural and functional connectivity: connectivity change after hippocampal lesion. Hippocampus 26:1447–1463. doi:10.1002/hipo.22621

Heusser AC, Awipi T, Davachi L (2013) The ups and downs of repetition: modulation of the perirhinal cortex by conceptual repetition predicts priming and long-term memory. Neuropsychologia 51(12):2333–2343. doi:10.1016/j.neuropsychologia.2013.04.018

Hilgetag C, Burns GAPC, O'Neill MA, Scannell JW, Young MP (2000) Anatomical connectivity defines the organization of clusters of cortical areas in the macaque and the cat. Philos Trans R Soc Lond B Biol Sci 355(1393):91–110. doi:10.1098/rstb.2000.0551

Hirsh R (1974) The hippocampus and contextual retrieval of information from memory: a theory. Behav Biol 12:421–444

Hsieh L-T, Ranganath C (2015) Cortical and subcortical contributions to sequence retrieval: schematic coding of temporal context in the neocortical recollection network. NeuroImage 121:78–90. doi:10.1016/j.neuroimage.2015.07.040

Inhoff MC, Ranganath C (2015) Significance of objects in the perirhinal cortex. Trends Cogn Sci 19(6):302–303. doi:10.1016/j.tics.2015.04.008

Irish M, Halena S, Kamminga J, Tu S, Hornberger M, Hodges JR (2015) Scene construction impairments in Alzheimer's disease – a unique role for the posterior cingulate cortex. Cortex 73:10–23. doi:10.1016/j.cortex.2015.08.004

Ishikawa A, Ambroggi F, Nicola SM, Fields HL (2008) Contributions of the amygdala and medial prefrontal cortex to incentive cue responding. Neuroscience 155(3):573–584. doi:10.1016/j. neuroscience.2008.06.037

Jang J-W, Park YH, Park SY, Wang MJ, Lim J-S, Kim S-H et al (2015) Longitudinal cerebral perfusion change in transient global amnesia related to left posterior medial network disruption. PLoS One 10(12):e0145658

Johnson JD, Rugg MD (2007) Recollection and the reinstatement of encoding-related cortical activity. Cereb Cortex 17(11):2507–2515. doi:10.1093/cercor/bhl156

Johnston-Laird PN (1983) Mental models: toward a cognitive science of language, inference, and consciousness. Harvard University Press, Cambridge, MA

Kahn I, Andrews-Hanna JR, Vincent JL, Snyder AZ, Buckner RL (2008) Distinct cortical anatomy linked to subregions of the medial temporal lobe revealed by intrinsic functional connectivity. J Neurophysiol 100(1):129–139. doi:10.1152/jn.00077.2008

Khan UA, Liu L, Provenzano FA, Berman DE, Profaci CP, Sloan R et al (2013) Molecular drivers and cortical spread of lateral entorhinal cortex dysfunction in preclinical Alzheimer's disease. Nat Neurosci 17(2):304–311. doi:10.1038/nn.3606

Kirwan CB, Wixted JT, Squire LR (2008) Activity in the medial temporal lobe predicts memory strength, whereas activity in the prefrontal cortex predicts recollection. J Neurosci 28 (42):10541–10548. doi:10.1523/JNEUROSCI.3456-08.2008

Kivisaari SL, Tyler LK, Monsch AU, Taylor KI (2012) Medial perirhinal cortex disambiguates confusable objects. Brain 135(12):3757–3769. doi:10.1093/brain/aws277

Klüver H, Bucy PC (1937) "Psychic blindness" and other symptoms following bilateral temporal lobectomy in Rhesus monkeys. Am J Physiol 119:352–353

Knierim JJ (2006) Neural representations of location outside the hippocampus. Learn Mem 13 (4):405–415. doi:10.1101/lm.224606

Knierim JJ, Lee I, Hargreaves EL (2006) Hippocampal place cells: parallel input streams, subregional processing, and implications for episodic memory. Hippocampus 16(9):755–764. doi:10.1002/hipo.20203

Kondo H, Saleem KS, Price JL (2005) Differential connections of the perirhinal and parahippocampal cortex with the orbital and medial prefrontal networks in macaque monkeys. J Comp Neurol 493(4):479–509. doi:10.1002/cne.20796

Kravitz DJ, Saleem KS, Baker CI, Mishkin M (2011) A new neural framework for visuospatial processing. Nat Rev Neurosci 12(4):217–230. doi:10.1038/nrn3008

Kurby CA, Zacks JM (2008) Segmentation in the perception and memory of events. Trends Cogn Sci 12(2):72–79. doi:10.1016/j.tics.2007.11.004

La Joie R, Landeau B, Perrotin A, Bejanin A, Egret S, Pélerin A et al (2014) Intrinsic connectivity identifies the hippocampus as a main crossroad between Alzheimer's and semantic dementia-targeted networks. Neuron 81(6):1417–1428. doi:10.1016/j.neuron.2014.01.026

Lambon Ralph MA, Ehsan S, Baker GA, Rogers TT (2012) Semantic memory is impaired in patients with unilateral anterior temporal lobe resection for temporal lobe epilepsy. Brain 135 (1):242–258. doi:10.1093/brain/awr325

Lavenex P, Amaral DG (2000) Hippocampal-neocortical interaction: a hierarchy of associativity. Hippocampus 10(4):420–430. doi:10.1002/1098-1063(2000)10:4<420::AID-HIPO8>3.0. CO;2-5

Lavenex P, Suzuki WA, Amaral DG (2002) Perirhinal and parahippocampal cortices of the macaque monkey: projections to the neocortex. J Comp Neurol 447(4):394–420. doi:10. 1002/cne.10243

Lavenex P, Suzuki WA, Amaral DG (2004) Perirhinal and parahippocampal cortices of the macaque monkey: intrinsic projections and interconnections. J Comp Neurol 472 (3):371–394. doi:10.1002/cne.20079

Libby LA, Ekstrom AD, Ragland JD, Ranganath C (2012) Differential connectivity of perirhinal and parahippocampal cortices within human hippocampal subregions revealed by high-resolution functional imaging. J Neurosci 32(19):6550–6560. doi:10.1523/JNEUROSCI. 3711-11.2012

Lisman JE, Grace AA (2005) The hippocampal-VTA loop: controlling the entry of information into long-term memory. Neuron 46(5):703–713. doi:10.1016/j.neuron.2005.05.002

Liu Z, Murray EA, Richmond BJ (2000) Learning motivational significance of visual cues for reward schedules requires rhinal cortex. Nat Neurosci 3(12):1307–1315

Maass A, Berron D, Libby LA, Ranganath C, Düzel E (2015) Functional subregions of the human entorhinal cortex. ELife 4:e06426

MacLean PD (1955) The limbic system ('visceral brain') and emotional behavior. Arch Neurol Psychiatr 73(2):130

Maren S, Phan KL, Liberzon I (2013) The contextual brain: implications for fear conditioning, extinction and psychopathology. Nat Rev Neurosci 14(6):417–428. doi:10.1038/nrn3492

McClelland JL, McNaughton BL, O'Reilly RC (1995) Why there are complementary learning systems in the hippocampus and neocortex: insights from the successes and failures of connectionist models of learning and memory. Psychol Rev 103(2):419–457

Mesulam M (1990) Large-scale neurocognitive networks and distributed processing for attention, language, and memory. Ann Neurol 28(5):597–613

Mishkin M (1982) A memory system in the monkey. Philos Trans R Soc Lond B Biol Sci 298 (1089):85–95

Mishkin M, Suzuki WA, Gadian DG, Vargha–Khadem F (1997) Hierarchical organization of cognitive memory. Philos Trans R Soc Lond B Biol Sci 352(1360):1461–1467

Miyashita Y (1988) Neuronal correlate of visual associative long-term memory in the primate temporal cortex. Nature 335:817–820

Montaldi D, Mayes AR (2010) The role of recollection and familiarity in the functional differentiation of the medial temporal lobes. Hippocampus 20(11):1291–1314. doi:10.1002/hipo. 20853

Moss HE (2004) Anteromedial temporal cortex supports fine-grained differentiation among objects. Cereb Cortex 15(5):616–627. doi:10.1093/cercor/bhh163

Mufson EJ, Pandya DN (1984) Some observations on the course and composition of the cingulum bundle in the rhesus monkey. J Comp Neurol 225(1):31–43

Muñoz M, Insausti R (2005) Cortical efferents of the entorhinal cortex and the adjacent parahippocampal region in the monkey (Macaca fascicularis): entorhinal cortex output to neocortex in monkeys. Eur J Neurosci 22(6):1368–1388. doi:10.1111/j.1460-9568.2005.04299.x

Nadel L, Peterson MA (2013) The hippocampus: part of an interactive posterior representational system spanning perceptual and memory systems. J Exp Psychol Gen 142(4):1242–1254

Nakamura K, Kubota K (1995) Mnemonic firing of neurons in the monkey temporal pole during a visual recognition memory task. J Neurophysiol 74(1):162–178

Navarro Schröder T, Haak KV, Jimenez NIZ, Beckmann CF, Doeller CF (2015) Functional topography of the human entorhinal cortex. Elife 4:e06738

Nestor PJ, Fryer TD, Hodges JR (2006) Declarative memory impairments in Alzheimer's disease and semantic dementia. NeuroImage 30(3):1010–1020. doi:10.1016/j.neuroimage.2005.10.008

Nobre AC, Mccarthy G (1995) Language-related field potentials in the anterior-medial temporal lobe: II. Effects of word type and semantic priming. J Neurosci 15(2):1090–1098

Noppeney U, Patterson K, Tyler LK, Moss H, Stamatakis EA, Bright P et al (2006) Temporal lobe lesions and semantic impairment: a comparison of herpes simplex virus encephalitis and semantic dementia. Brain 130(4):1138–1147. doi:10.1093/brain/awl344

Norman G, Eacott MJ (2005) Dissociable effects of lesions to the perirhinal cortex and the postrhinal cortex on memory for context and objects in rats. Behav Neurosci 119(2):557–566. doi:10.1037/0735-7044.119.2.557

O'Kane G, Insler RZ, Wagner AD (2005) Conceptual and perceptual novelty effects in human medial temporal cortex. Hippocampus 15(3):326–332. doi:10.1002/hipo.20053

O'Keefe J, Nadel L (1979) The hippocampus as a cogitive nap. Clarendon Press, Oxford

Olson IR, Plotzker A, Ezzyat Y (2007) The enigmatic temporal pole: a review of findings on social and emotional processing. Brain 130(7):1718–1731. doi:10.1093/brain/awm052

Olton DS, Becker JT, Handelmann GE (1979) Hippocampus, space and memory. Behav Brain Sci 2:313–322. doi:10.1017/S0140525X00062713

Ongur D, Price JL (2000) The organization of networks within the orbital and medial prefrontal cortex of rats, monkeys and humans. Cereb Cortex 10:206–219

Papez JW (1937) A proposed mechanism of emotion. Arch Neurol Psychiatr 38(4):725

Passingham RE, Stephan KE, Kötter R (2002) The anatomical basis of functional localization in the cortex. Nat Rev Neurosci 3(8):606–616. doi:10.1038/nrn893

Patterson K, Nestor PJ, Rogers TT (2007) Where do you know what you know? The representation of semantic knowledge in the human brain. Nat Rev Neurosci 8(12):976–987. doi:10.1038/nrn2277

Peer M, Salomon R, Goldberg I, Blanke O, Arzy S (2015) Brain system for mental orientation in space, time, and person. Proc Natl Acad Sci U S A 112(35):11072–11077. doi:10.1073/pnas.1504242112

Phelps EA, LeDoux JE (2005) Contributions of the amygdala to emotion processing: from animal models to human behavior. Neuron 48(2):175–187. doi:10.1016/j.neuron.2005.09.025

Pihlajamaki M, Tanila H, Kononen M, Hanninen T, Hamalainen A, Soininen H, Aronen HJ (2004) Visual presentation of novel objects and new spatial arrangements of objects differentially activates the medial temporal lobe subareas in humans. Eur J Neurosci 19(7):1939–1949. doi:10.1111/j.1460-9568.2004.03282.x

Radvansky GA, Zacks JM (2011) Event perception. Wiley Interdiscip Rev Cogn Sci 2(6):608–620. doi:10.1002/wcs.133

Raichle ME, MacLeod AM, Snyder AZ, Powers WJ, Gusnard DA, Shulman GL (2001) A default mode of brain function. Proc Natl Acad Sci U S A 98(2):676–682

Ranganath C (2010) A unified framework for the functional organization of the medial temporal lobes and the phenomenology of episodic memory. Hippocampus 20(11):1263–1290. doi:10.1002/hipo.20852

Ranganath C, Ritchey M (2012) Two cortical systems for memory-guided behaviour. Nat Rev Neurosci 13(10):713–726. doi:10.1038/nrn3338

Ranganath C, Yonelinas AP, Cohen MX, Dy CJ, Tom SM, D'Esposito M (2004) Dissociable correlates of recollection and familiarity within the medial temporal lobes. Neuropsychologia 42(1):2–13. doi:10.1016/j.neuropsychologia.2003.07.006

Ritchey M, Yonelinas AP, Ranganath C (2014) Functional connectivity relationships predict similarities in task activation and pattern information during associative memory encoding. J Cogn Neurosci 26(5):1085–1099. doi:10.1162/jocn_a_00533

Ritchey M, Libby LA, Ranganath C (2015) Cortico-hippocampal systems involved in memory and cognition: the PMAT framework. In Progress in brain research (Vol. 219, pp 45–64). Elsevier. Retrieved from http://linkinghub.elsevier.com/retrieve/pii/S0079612315000588

Rushworth MFS, Noonan MP, Boorman ED, Walton ME, Behrens TE (2011) Frontal cortex and reward-guided learning and decision-making. Neuron 70(6):1054–1069. doi:10.1016/j.neuron.2011.05.014

Schank RC, Abelson RP (1977) Scripts, plans, goals and understanding: an inquiry into human knowledge structures. Lawrence Erlbaum Associates, Mahway, NJ

Schmahmann JD, Pandya DN, Wang R, Dai G, D'Arceuil HE, de Crespigny AJ, Wedeen VJ (2007) Association fibre pathways of the brain: parallel observations from diffusion spectrum imaging and autoradiography. Brain 130(3):630–653. doi:10.1093/brain/awl359

Schulman GL, Fiez JA, Corbetta M, Buckner RL, Miezin FM, Raichle ME, Petersen SE (1997) Common blood flow changes across visual tasks: II. Decreases in cerebral cortex. J Cogn Neurosci 9(5):648–663

Scoville WB, Milner B (1957) Loss of recent memory after bilateral hippocampal lesions. J Neurol Neurosurg Psychiatry 20:11–21

Seeley WW, Crawford RK, Zhou J, Miller BL, Greicius MD (2009) Neurodegenerative diseases target large-scale human brain networks. Neuron 62(1):42–52. doi:10.1016/j.neuron.2009.03.024

Sheldon S, Farb N, Palombo DJ, Levine B (2016) Intrinsic medial temporal lobe connectivity relates to individual differences in episodic autobiographical remembering. Cortex 74:206–216. doi:10.1016/j.cortex.2015.11.005

Shin JD, Jadhav SP (2016) Multiple modes of hippocampal–prefrontal interactions in memory-guided behavior. Curr Opin Neurobiol 40:161–169. doi:10.1016/j.conb.2016.07.015

Speer NK, Zacks JM, Reynolds JR (2007) Human brain activity time-locked to narrative event boundaries. Psychol Sci 18(5):449–455

Sporns O, Chialvo D, Kaiser M, Hilgetag C (2004) Organization, development and function of complex brain networks. Trends Cogn Sci 8(9):418–425. doi:10.1016/j.tics.2004.07.008

Squire LR, Zola-Morgan S (1991) The medial temporal lobe memory system. Science 253 (5026):1380–1386

Squire LR, Wixted JT, Clark RE (2007) Recognition memory and the medial temporal lobe: a new perspective. Nat Rev Neurosci 8(11):872–883. doi:10.1038/nrn2154

St. Jacques PL, Olm C, Schacter DL (2013) Neural mechanisms of reactivation-induced updating that enhance and distort memory. Proc Natl Acad Sci U S A 110(49):19671–19678

Staresina BP, Duncan KD, Davachi L (2011) Perirhinal and parahippocampal cortices differentially contribute to later recollection of object- and scene-related event details. J Neurosci 31 (24):8739–8747. doi:10.1523/JNEUROSCI.4978-10.2011

Sutherland RJ, Rudy JW (1989) Configural association theory: the role of the hippocampal formation in learning, memory, and amnesia. Psychobiology 17(2):129–144

Suzuki WA, Amaral DG (1994) Perirhinal and parahippocampal cortices of the macaque monkey: cortical afferents. J Comp Neurol 350:497–533

Szpunar KK, Watson JM, McDermott KB (2007) Neural substrates of envisioning the future. Proc Natl Acad Sci U S A 104(2):642–647

Tye KM, Janak PH (2007) Amygdala neurons differentially encode motivation and reinforcement. J Neurosci 27(15):3937–3945. doi:10.1523/JNEUROSCI.5281-06.2007

van Strien NM, Cappaert NLM, Witter MP (2009) The anatomy of memory: an interactive overview of the parahippocampal–hippocampal network. Nat Rev Neurosci 10(4):272–282. doi:10.1038/nrn2614

Vilberg KL, Rugg MD (2008) Memory retrieval and the parietal cortex: a review of evidence from a dual-process perspective. Neuropsychologia 46(7):1787–1799. doi:10.1016/j.neuropsychologia.2008.01.004

Von Der Heide RJ, Skipper LM, Klobusicky E, Olson IR (2013) Dissecting the uncinate fasciculus: disorders, controversies and a hypothesis. Brain 136(6):1692–1707. doi:10.1093/brain/awt094

Walker MP, Stickgold R (2004) Sleep-dependent learning and memory consolidation. Neuron 44 (1):121–133

Wang W-C, Lazzara MM, Ranganath C, Knight RT, Yonelinas AP (2010) The medial temporal lobe supports conceptual implicit memory. Neuron 68(5):835–842. doi:10.1016/j.neuron.2010.11.009

Wang W-C, Ranganath C, Yonelinas AP (2014) Activity reductions in perirhinal cortex predict conceptual priming and familiarity-based recognition. Neuropsychologia 52:19–26. doi:10.1016/j.neuropsychologia.2013.10.006

Wang S-F, Ritchey M, Libby LA, Ranganath C (2016) Functional connectivity based parcellation of the human medial temporal lobe. Neurobiol Learn Mem. doi:10.1016/j.nlm.2016.01.005

Warren DE, Power JD, Bruss J, Denburg NL, Waldron EJ, Sun H et al (2014) Network measures predict neuropsychological outcome after brain injury. Proc Natl Acad Sci U S A 111 (39):14247–14252. doi:10.1073/pnas.1322173111

Warrington EK, Weiskrantz L (1982) Amnesia: a disconnection syndrome? Neuropsychologia 30 (3):233–248

Wilson FAW, Rolls ET (1993) The effects of stimulus novelty and familiarity on neuronal activity in the amygdala of monkeys performing recognition memory tasks. Exp Brain Res 93 (3):367–382

Winocur G, Moscovitch M, Bontempi B (2010) Memory formation and long-term retention in humans and animals: convergence towards a transformation account of hippocampal–neocortical interactions. Neuropsychologia 48(8):2339–2356. doi:10.1016/j.neuropsychologia.2010.04.016

Witter MP, Naber PA, van Haeften T, Machielsen WC, Rombouts SA, Barkhof F et al (2000a) Cortico-hippocampal communication by way of parallel parahippocampal-subicular pathways. Hippocampus 10(4):398–410

Witter M, Wouterlood FG, Naber PA, Van Haeften T (2000b) Anatomical organization of the parahippocampal-hippocampal network. Ann NY Acad Sci 911(1):1–24

Witter MP, Canto CB, Couey JJ, Koganezawa N, O'Reilly KC (2013) Architecture of spatial circuits in the hippocampal region. Philos Trans R Soc Lond B Biol Sci 369 (1635):20120515–20120515. doi:10.1098/rstb.2012.0515

Wixted JT, Squire LR (2011) The medial temporal lobe and the attributes of memory. Trends Cogn Sci 15(5):210–217. doi:10.1016/j.tics.2011.03.005

Zacks JM, Braver TS, Sheridan MA, Donaldson DI, Snyder AZ, Ollinger JM et al (2001) Human brain activity time-locked to perceptual event boundaries. Nat Neurosci 4(6):651–655

Zhuo J, Fan L, Liu Y, Zhang Y, Yu C, Jiang T (2016) Connectivity profiles reveal a transition subarea in the parahippocampal region that integrates the anterior temporal-posterior medial systems. J Neurosci 36(9):2782–2795. doi:10.1523/JNEUROSCI.1975-15.2016

Printed by Printforce, the Netherlands